"The 7th edition of *Environmental Hazards* is like having coffee with an old friend - the coffee is just better. This timely update of this classic text brings to the fold the latest in hazard, risk, and disaster risk reduction. I believe this is, and remains, an essential text for students, academics, and policy makers alike."

Dewald van Niekerk, African Centre for Disaster Studies, North-West University, South Africa.

"This book provides a clear explanation of hazard and resilience typology in environmental science so that readers can completely understand the terminology and its usage. The updates of the current pandemic and "new" types of disasters are also covered, providing insights and connecting the events with the existing theory on disaster management and environmental issues. Highly recommended for undergraduates and graduates alike to have an excellent understanding on the subject on hazard, resiliency and environmental issues."

Elisabeth Rianawati, Director, Resilience Development Initiative (RDI).

"This book provides one of the most comprehensive compilations on the important topic of Environmental Hazards with cross cutting relations with disaster risk reduction. The book eloquently describes hazards, risks and its assessment and management. I find this book a great reference for the students, researchers and practitioners in the related field."

Professor Rajib Shaw, Graduate School of Media and Governance, Keio University, Japan.

"This timely new edition of a now classic text forefronts the latest perspectives on hazards and disasters without losing sight of the salient history of ideas and interventions. It is a tremendous learning resource, and an invaluable reference for anyone interested in, or involved with, disaster risk reduction."

Clive Oppenheimer, Cambridge University, UK.

Environmental Hazards

The seventh edition of *Environmental Hazards* provides a much expanded and fully up-to-date overview of all the extreme environmental events that threaten people and what they value in the 21st century globally. It integrates cutting-edge materials to provide an interdisciplinary approach to environmental hazards and their management, illustrating how natural and human systems interact to place communities of all sizes, and at all stages of economic development, at risk. Part 1 defines basic concepts of hazard, risk, vulnerability and disaster and explores the evolution of hazards theory. Part 2 employs a consistent chapter structure to demonstrate how individual hazards occur, their impacts and how the risks can be assessed and managed. Part 3 brings the material together to understand and improve the present and the future of environmental hazards.

This extensively revised edition includes:

- Fresh perspectives on the reliability of disaster data, disaster risk reduction, risk and disaster perception and communication, and new technologies available to assist with environmental hazard management
- The addition of several new environmental hazards, including landslides and avalanches, cryospheric hazards, subsidence hazards, and hazards of the Anthropocene
- More boxed sections with a focus on both generic issues and the lessons from a carefully selected range of environmental hazards
- Lists of further reading and relevant websites
- More colour diagrams and photographs, and more than 1,000 references to some of the most significant and recent published material

This carefully structured and balanced textbook captures the complexity and dynamism of environmental hazards and is essential reading for students across many disciplines, including geography, environmental science, environmental studies and natural resources.

Keith Smith was Emeritus Professor at the University of Stirling, UK.

Carina J. Fearnley is Professor in Warnings and Science Communication at University College London, UK in the Department of Science and Technology Studies.

Deborah Dixon is Professor of Geography in the School of Geographical & Earth Sciences at the University of Glasgow, UK.

Deanne K. Bird is Adjunct Research Fellow at the Monash University Disaster Resilience Initiative, Australia, and a Research Specialist with the Nordic Centre of Excellence on Resilience and Societal Security (NORDRESS), University of Iceland.

Ilan Kelman is Professor of Disasters and Health in the Institute for Risk and Disaster Reduction and Institute for Global Health at University College London, UK, and Professor II at the University of Agder, Norway.

Environmental Hazards

Assessing Risk and Reducing Disaster

Seventh edition

Keith Smith, Carina J. Fearnley, Deborah Dixon, Deanne K. Bird and Ilan Kelman

LONDON AND NEW YORK

Cover image: © Getty Images/Sander Meertins

Seventh edition published 2024
by Routledge
4 Park Square, Milton Park, Abingdon, Oxon, OX14 4RN

and by Routledge
605 Third Avenue, New York, NY 10158

Routledge is an imprint of the Taylor & Francis Group, an informa business

First edition published by Routledge 1991
Sixth edition published by Routledge 2013

British Library Cataloguing-in-Publication Data
A catalogue record for this book is available from the British Library

Library of Congress Cataloging-in-Publication Data
Names: Smith, Keith, 1938- author. | Fearnley, Carina J., author. | Kelman, Ilan, author. | Dixon, Deborah P., author. | Bird, Deanne K., author.
Title: Environmental hazards : assessing risk and reducing disaster / Keith Smith, Carina J. Fearnley, Ilan Kelman, Deborah Dixon and Deanne K. Bird.
Description: Seventh edition. | Abingdon, Oxon ; New York, NY : Routledge, 2023. | Includes bibliographical references and index.
Identifiers: LCCN 2023018317 (print) | LCCN 2023018318 (ebook) | ISBN 9780815365402 (hardback) | ISBN 9780815365419 (paperback) | ISBN 9781351261647 (ebook)
Subjects: LCSH: Natural disasters. | Environmental disasters. | Environmental risk assessment.
Classification: LCC GB5014 .S6 2023 (print) | LCC GB5014 (ebook) | DDC 363.34—dc23/eng/20230419
LC record available at https://lccn.loc.gov/2023018317
LC ebook record available at https://lccn.loc.gov/2023018318

ISBN: 978-0-815-36540-2 (hbk)
ISBN: 978-0-815-36541-9 (pbk)
ISBN: 978-1-351-26164-7 (ebk)

DOI: 10.4324/9781351261647

Typeset in Minion
by Apex CoVantage, LLC

Printed in Great Britain by Bell and Bain Ltd, Glasgow

Contents

Figures

Tables

About the Authors

Keith Smith was Emeritus Professor at the University of Stirling, UK. His research was primarily in the fields of hydrology, climatology and environmental hazards.

Carina J. Fearnley is Professor in Warnings and Science Communication at the University College London in the Department of Science and Technology Studies, and Director of the University College London Warning Research Centre. She is an active interdisciplinary researcher in two areas: first, early warning systems (particularly volcanoes), focusing specifically on alert level systems; and second, in art/science projects that address some of the complex issues surrounding our understanding of environmental hazards.

Deborah Dixon is Professor of Geography in the School of Geographical & Earth Sciences at the University of Glasgow, UK. An internationally recognised scholar in feminist geopolitics, she has been key to the emergence of 'geohumanities' as an interdisciplinary field of research and practice. She is the co-founder and editor of the new interdisciplinary (American Association of Geography) journal *GeoHumanities*, which publishes analytic and practice-based research, as well as accounts of arts performances and outputs, and hosts online art exhibitions.

Deanne K. Bird is Adjunct Research Fellow at the Monash University Disaster Resilience Initiative, Monash University, Australia. She is also an active member of the Nordic Centre of Excellence on Resilience and Societal Security (NORDRESS), a specialised project under the Social Security Programme of NordForsk, based at the University of Iceland. Deanne's research interests focus on human-centred, place-based initiatives aimed at enhancing community preparedness, response and recovery across all hazard environments. Much of this work has focused on regional and rural communities along with the tourism sector.

Ilan Kelman (www.ilankelman.org and Twitter/Instagram @ILANKELMAN) is Professor of Disasters and Health at University College London, England, and a Professor II at the University of Agder, Norway. His overall research interest is linking disasters and health, integrating climate change into both. Three main areas are (i) disaster diplomacy and health diplomacy (www.disasterdiplomacy.org); (ii) island sustainability involving safe and healthy communities in isolated locations (www.islandvulnerability.org); and (iii) risk education for health and disasters (www.riskred.org).

Preface to the Seventh Edition of *Environmental Hazards*

The power of the Earth is both extraordinary and overwhelming. From the violent explosions of volcanoes to the giant waves of a tsunami and the scouring blast of tornadoes the planet inspires and creates awe. For thousands of years people have studied environmental hazards and adapted to the forces and energies of nature to try and live safely, and harmoniously, whilst reaping the benefits that environmental hazards can also bring. Some people have given their lives to the endeavour of enhancing knowledge of hazards.

In 2022 the movie *Fire of Love* (Dosa, 2022) was released, documenting the love of Maurice and Katia Krafft for each other, and for volcanoes. They spent their professional lives dedicated to understanding the science of volcanoes to help protect people. A legacy of the Kraffts remains their educational video on volcanoes that was used during the 1991 eruption of Mt Pinatubo, which helped save countless lives in what was the second largest eruption of the last century. Sadly, they never witnessed their achievement as only weeks beforehand they were killed at an eruption of Mt Unzen in Japan. There are always new things to observe and learn, and new emerging hazards and risks that challenge our ability to live safely.

This seventh edition of a remarkable textbook that has helped shape an entire generation of scholars working in the hazard, risk and disaster world aims to capture some of these lessons on the need to use science wisely, but also to understand that scientific knowledge is only part of the issue. People – through their risk perceptions and decision-making, all while embedded in complex socio-economic and political structures – drive disasters and new emerging risks. This is why we do not use the phrase 'natural disaster' in this book: disasters are made by societies, not nature.

This edition addresses efforts to decolonise our understandings and explanations of hazards and disasters; foster inclusivity in managing environmental hazards and disaster risk reduction; and integrate our different capacities and awareness of constraints into these actions. Certainly the scale and diverse scope of environmental hazards is ever more apparent in what some call an 'Anthropocene' era, and the range of exposures and vulnerabilities of groups and individuals to these needs to be urgently addressed. There is no need for despair, however, given previous achievements in tackling environmental hazards and mitigating – and even avoiding – disasters. These challenges raise questions about what our future looks like, but they also bring into focus how that future is shaped by actions taken now. If there is one key take-home message from this edition, it is that while environmental hazards wax and wane, surge and subside, disasters are not inevitable.

For the first time since this book was first published in 1991, Prof Keith Smith, who devised and wrote all six editions of *Environmental Hazards*, was unable to contribute. Sadly Keith passed away, handing the baton on to us to continue the original aim, which was to 'provide an introductory text to environmental hazards of university and college students' of related disciplines (first edition foreword). We have come a long way in the last 32 years, and we know this book has grown in scope and will be of value to a wide range of actors within environmental hazards and inspire many to help devise more effective approaches to reducing risks, vulnerabilities and uncertainties to help prevent disasters in the future. It is only right that this seventh edition is dedicated to Keith and his desire to provide a useful road map through the challenging complexities of today. Thank you for taking this journey with us.

Carina J. Fearnley London, UK
Deanne K. Bird Melbourne, Australia
Deborah Dixon Glasgow, UK
Ilan Kelman London, UK
November 2022

REFERENCE

Dosa, S. (Director). (2022). *Fire of love* [Film]. Sandbox Films (New York); Intuitive Pictures (Montreal); Cottage M (Los Angeles and New York).

Foreword for the Seventh Edition of *Environmental Hazards*

As the world struggles to deal with the impacts of the climate emergency, conflict and the socio-economic fallout of the COVID-19 pandemic, it is becoming increasingly clear that we are living in an era where risk is complex and interconnected.

External shocks interact with existing drivers of risk, such as environmental degradation and poor urban planning, to create recipes for disasters large and small.

The compounded impacts of these disasters are cascading to a point where lives and livelihoods are undermined, poverty is deepening and countries are derailed from achieving the Sustainable Development Goals.

To stop the spiral of disasters, countries must accelerate the implementation of the globally agreed Sendai Framework for Disaster Risk Reduction 2015–2030, which now is entering its midpoint.

With seven years remaining to achieve the framework's targets and goals, countries must seek out opportunities to accelerate progress.

A key recommendation in the 2022 Report of the Secretary-General on the Implementation of the Sendai Framework is around the need for countries to expand their approaches to risk assessment and modelling to include the full scope of hazards and risks outlined in the Sendai Framework.

Attaining this full understanding of risk is the first priority of action in the framework because it underpins all risk mitigation and resilience-building actions. Risk reduction strategies, development and investment decisions, and climate adaptation plans should all be based on a solid understanding of risk, including hazard characteristics and the environment.

To that end, we welcome the publication of the seventh edition of *Environmental Hazards* as a valuable contribution to enriching the understanding and management of disaster risks.

We especially appreciate the textbook's effort to address the complex interaction of human and environmental hazards.

This is in line with the key message from the Global Assessment Report on Disaster Risk Reduction 2019 that countries ought to consider how risks interact in a complex environment, rather than follow the traditional 'hazard-by-hazard' approach to risk management.

We hope students, practitioners, and, most importantly, policy-makers consider and apply the knowledge contained in this textbook to leapfrog progress towards the implementation of the Sendai Framework.

It is the only way countries can move from a state of constantly responding to growing disasters to a state of resilience built around prevention and proactive risk mitigation. Our future depends on it.

水鳥真美

Mami Mizutori

Special Representative of the UN Secretary-General for Disaster Risk Reduction and Head of UNDRR

Part One

THE NATURE OF HAZARD

Amina J. Mohammed, Deputy Secretary-General of the United Nations

Source: © Getty Images/Monica Schipper/Stringer

Part 1:

We must now focus on prevention and break the cycle of disaster, response, and recovery.

– Amina J. Mohammed, United Nations Deputy Secretary-General

DOI: 10.4324/9781351261647-1

CHAPTER ONE

Hazards in the Environment

1

OVERVIEW

This chapter provides an overview of environmental hazards, exploring what hazards are, how they are formed and how they are interconnected. Contemporary socio-environmental challenges, alongside a rethinking of the nature of academia and the role of disaster-focused organisations, have meant revising and expanding on many concepts such as hazard, disaster, risk, vulnerability and resilience. Charting the chronology of key perspectives and approaches adopted over hundreds of years – from observational to engineering, behavioural, developmental, feminist, decolonial and complexity frameworks – provides valuable insight into how we understand, research, respond to and manage environmental hazards. The chapter concludes by noting how in recent decades international organisations (private sector and non-profit as well as governmental and inter-governmental) have increased their scope. A key example is the United Nations (UN) and its associated programmes such as the Sustainable Development Goals, the Sendai Framework and the Paris Agreement, which are increasingly shaping top-down approaches and policies to understand, manage and reduce hazards and disasters.

A. INTRODUCTION: A NEW ENVIRONMENTAL HAZARDS AGENDA

As we move further into the 21st century the established academic fields that focus on environmental hazards (such as the geosciences, environmental sciences, civil engineering, geography, public health, science and technology studies, and the arts/humanities) and their management (disaster studies and management, civil protection, disaster risk reduction, risk studies and humanitarian studies) are transforming. New (or revived) connections with disciplines across and beyond academia are being made, and previously marginalised knowledges are increasingly valued – though much work still needs to

DOI: 10.4324/9781351261647-2

be done to diversify the environmental hazards agenda.

At one level, this transformation has been prompted by a recognition that the specialist training provided by academia does not always provide the breadth of expertise needed to deal with the complex character of hazards, including their emergence over time and space and their many social and physical impacts. In response, researchers have considered how best to combine their expertise, methods, techniques and technologies so that the relationships between different aspects of an environmental challenge can be addressed.

This *interdisciplinarity* has been facilitated to an extent by funding programmes, new academic outlets such as journals and books, and popular science outlets for disseminating scientific findings, from podcasts and vlogs to microblogging and art exhibitions. Additionally, a proliferation of undergraduate and graduate programmes aim to make students aware of the breadth of topics relevant to environmental hazards while providing in-depth training. Nongovernmental organisations (NGOs), government organisations and global institutions increasingly work across disciplines and stakeholders to listen to and act on more of the voices of those living with and impacted by environmental hazards – which means everyone.

At a broader level, this move towards interdisciplinarity has itself been prompted by two key factors. First, although the environmental consequences of colonialism, industrialisation and urbanisation (alongside the profound reworking of agriculture) have emerged over centuries, the claim that we are now witnessing *global-scale* hazardous circumstances, the so-called Anthropocene (Chapter 16), has taken hold across some academic and policy circles. While the debates on whether the Anthropocene constitutes a new geological epoch are ongoing – as are the debates on whether 'Anthropocene' is a useful term – academics in all fields are more united in turning their attention to explaining the scope and impact of observed planet-wide conditions. In particular, we work to characterise and explain the apparent reconfiguration not only of the atmosphere, hydrosphere, lithosphere and biosphere of the Earth, but also of bodies, genomes and human ways of living. This complex condition influences and is influenced by the hazards that we have long focused on, such as flooding, earthquakes, fires and disease outbreaks, while pushing us to expand our understanding of what is hazardous and why.

One of the main consequences of this focus on planetary conditions is that the traditional terms used to classify hazards have come under scrutiny. For a long time a distinction has been made between 'natural hazards' and 'human hazards'. 'Natural hazards' referred to 'extreme events' ostensibly from physical processes, with typical examples being earthquakes, volcanic eruptions, floods, landslides and storms. 'Human hazards' or 'technological hazards' were so called because they were associated with industrial, agricultural and domestic landscapes. Some of these 'human hazards' are slow-paced or 'creeping', such as toxic build-up in the environment, while others are produced by the often rapid failure of technologies and infrastructures, possibly under the influence of 'natural hazards'. Yet, the hazards that impact so many people – such as the complex processes and events comprising contemporary climate change (Chapter 18) – do not fit easily into either 'natural' or 'human' categories. These include sea-level rise; ocean acidification; changes to the global oceanic circulation; biodiversity, ecodiversity and geodiversity loss; rapid disease spread; and widespread pollution, including the accumulation of plastics, nanoparticles throughout our bodies and the biomagnification of persistent organic pollutants.

Second, those involved in researching environmental hazards have been confronted with calls to *decolonise* their work. Decolonisation refers to much more than the formal end of colonial rule. It includes the recognition that many of the assumptions colonisers made about the people and lands of colonised countries were supported by academics trained and teaching within the 'home' countries (see Figure 1.1). Many were sent to investigate the social and environmental conditions of the colonies with a view to how these could be better exploited. Environmental racism, for example – wherein the environment was understood to shape not only the physiognomies but also the temperaments and intellects of whole populations – was a prevailing concept at the turn of the 20th century in Geography and helped justify claims that indolence and poverty were ingrained in particular parts of the world. Decolonisation also refers to the need to understand the continuing impacts of such theories in the form of (1) the systematic belittling of indigenous knowledges of the environment, as well as other knowledge forms such as local, traditional and vernacular. Certainly, the manner in which many explanations of why some people suffer more than others – including explanations of why some are more exposed or vulnerable to hazards than

Figure 1.1 Distribution of famine relief in the Madras Presidency, India, from the *Illustrated London News* on 26 May 1877.

Source: Wikimedia Commons

Note: These publicised hazard management efforts obscure the ways in which British colonial governance (1) re-engineered a flood-dependent agrarian environment in order to usher in a taxable commercial agriculture, and (2) used famine to argue for a 'scientific irrigation' that would be carried out by British forces (Tozzi et al., 2022).

others – has often excluded how patterns of inequality across the globe have been shaped by colonialism, and (2) the continued justification of ownership of lands accrued as part of colonialism.

One of the main consequences of this call to decolonise is that many of the categories used to discuss the impact of environmental hazards have come under scrutiny. A profound gap exists, for example, between 'the economy', as measured by a country's gross domestic product (GDP) and which focuses on the goods and services a population consumes, and 'economic security', which is an individual or community's livelihood capacity to withstand hazards and, at a broader level, to live well. While both terms use the word 'economy', they refer to very different scales and objects and to very different understandings of what is at stake when a hazard appears. Certainly, 'the economy' is impacted by disasters in the sense that more personal and corporate wealth is placed at risk, and regular hazards without preparedness can stall production regionally, nationally and even internationally. Services that rely on travel or face-to-face interactions (such as some in leisure and sport, tourism and retail) can be especially hard hit. Economic security, however, focuses on the ability of people – individually or as part of families, households and communities – to cover their essential needs sustainably. This capacity is compromised when livelihood resources are destroyed and when the commodity chains that link production, distribution and consumption across the globe are broken. Job losses erode economic security and are especially devastating for those living paycheck to paycheck and/or in debt.

This closer examination of how the economic security of individuals is impacted by environmental hazards allows a consideration of how inequalities, inequities, marginalisations and exclusions are produced and maintained to shape an individual's or community's vulnerability. It also allows us to focus on how individuals and communities are able to plan for and respond to specific hazards by drawing on particular resources and networks that are accessible to them. And, we can address how specific vulnerabilities are produced by the intersection of racialised, sexed, aged, classed, abled and other social relations that together erode the capability of individuals and groups to deal with environmental hazards.

The COVID-19 pandemic, starting in 2020, demonstrates the globalised nature of our lives that allows new diseases to become epidemics and then pandemics, as well as our dependency on workers who may be on the other side of the planet for providing goods or enabling the transport of cargo or medical supplies (see Box 1.1). While economic metrics such as GDP may be rallied by central banks stepping in to ensure the flow of credit, an economic *insecurity* proliferates as unemployment levels rise. Similarly, we might blame rising levels of consumption and the heavy ecological burden this produces on 'the economy'. Earth's ecologies are overexploited and/or subject to degradation and pollution, producing a suite of hazardous conditions such as biodiversity loss and increased toxicity. Economic security foregrounds the fact that these ecological impacts are felt more keenly by people dependent on subsistence livelihoods, many of whom continue to live with the political and economic legacies of colonialism and postcolonialism. Many people are already experiencing economic insecurity because of austerity policies that have stripped back social service and public health infrastructures. Rapid-onset events such as earthquakes and flash floods then

Box 1.1 COVID-19 – A Global Threat

The COVID-19 pandemic was caused by the transmission of the virus called severe acute respiratory syndrome coronavirus 2 (SARS-CoV-2) and its impacts on the human body as a disease called COVID-19. SARS-CoV-2 is part of a large family of viruses and is one of several coronaviruses that are 'zoonotic', meaning that they have jumped from animal populations into the human population. SARS, MERS and COVID-19 all jumped into the human population post-2000. Driving this growing emergence of zoonotic diseases is:

- increased urbanisation/livestock farming via deforestation, which fragments habitats and forces animals to adapt to urban conditions;
- the rise of intensive animal husbandry, which involves keeping a large number of animals confined to a small space in close physical contact with people;
- changes in food consumption patterns, which increase the chances of exposing consumers to food-borne pathogens; and
- the rapid proliferation of low-cost intercontinental flights, as well as the global transport of animals.

In December 2019, cases of a contagious disease were diagnosed in the city of Wuhan, the capital of Hubei province in China. From respiratory samples of a Wuhan patient, scientists identified the new coronavirus as belonging to the coronavirus family. After sequencing the SARS-CoV-2 genome, findings were made available to the World Health Organization (WHO) on 12 January2020. Given the spread of the virus and its impacts, the WHO declared COVID-19 a Public Health Emergency of International Concern on 30 January 2020. On 11 March 2020, COVID-19 was declared a 'pandemic' by the WHO, meaning a disease outbreak that occurs over a very wide region or even worldwide, crossing international boundaries and affecting a large number of people.

At the start of May 2022, 6.25 million deaths had been recorded from COVID-19. This death rate averages to 739 per million across the globe; other deaths will have had COVID-related complications, and many will be unreported (Our World in Data, 2022). Concerning those infected – with 65,701 per million confirmed cases at the start of May 2022 – many will continue to be impacted by 'long COVID' symptoms. The death toll from COVID-19 makes it the most deadly pandemic since the 1918–1920 flu pandemic that killed 50–100 million people out of a population of around 2 billion (see Table 2.1 in Chapter 2). The WHO estimates that the full death toll associated directly or indirectly with the COVID-19 pandemic between 1 January 2020 and 31 December 2021 was approximately 14.9 million (see Figure 1.2). The majority of the excess deaths (84%) are concentrated in Southeast Asia, Europe and the Americas.

The pandemic severely limited travel and social interaction, with many people working from home, children missing in-classroom education for significant periods of time, drops in GDP and stock exchanges, and a fundamental shift and evaluation of modern-day life. COVID-19 has made clear our interconnectedness with our environment but has also highlighted the increased density of our social relations. The pandemic has crossed distance and borders rapidly, reduced manufacturing and the

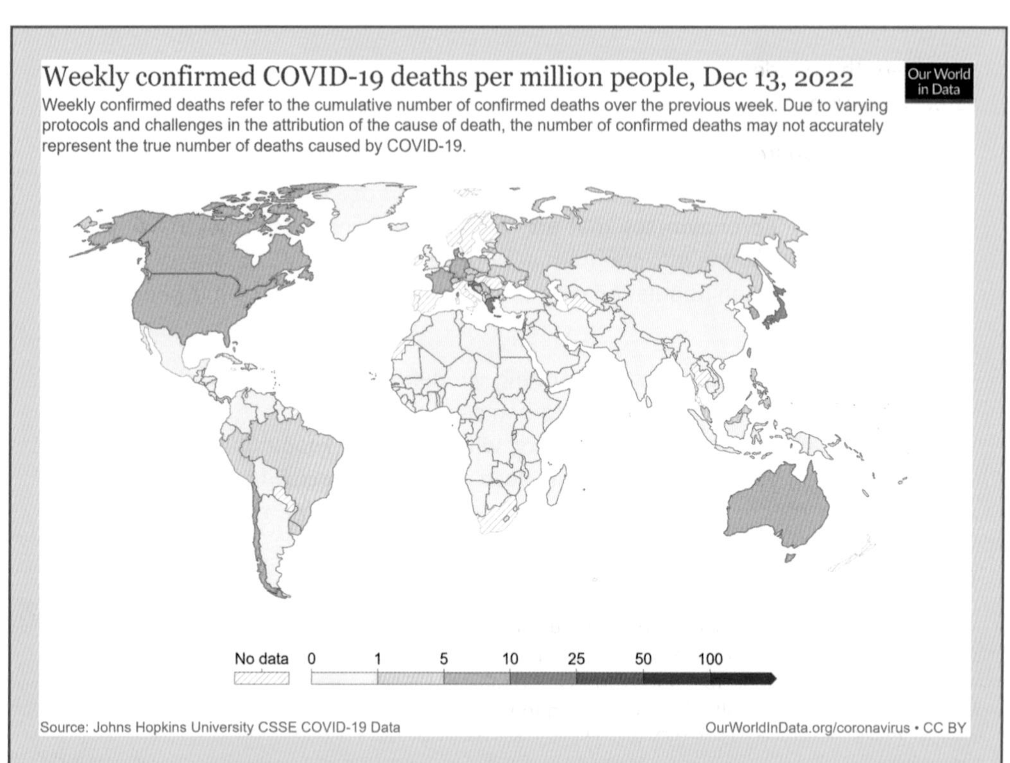

Figure 1.2 The cumulative number of confirmed deaths per million people.

Source: Our World in Data (2022)

export of goods due to factory closures and reduced staffing, and has had a severe impact on the service and tourism sectors (including aviation and shipping).

In terms of economic security, data on the impact of COVID-19 is fragmentary. According to UN Women (2021), women have faced a disproportionate impact of COVID-19, with job loss rates around 1.8 times higher than that of men globally (5.7% versus 3.1% respectively) (Madgavkar et al., 2020). The reduction of the service and tourism sector has fallen disproportionality on female employment. More non-paid caring hours and duties have fallen disproportionately on women and girls. Women and girls have experienced higher levels of domestic violence, reduced access to clean and safe health facilities, and severed reductions in the number studying at school.

Due to the internet enhancing the ability to communicate and advances in modern science that enable testing and the development of vaccines, people in some countries and with a certain pre-pandemic level of affluence were able to adjust while awaiting a comparatively quick vaccination sequence. Nevertheless, there is a profound disparity in access to modern communications and to vaccinations. While 65.5% of the world population has received at least one dose of a COVID-19 vaccine as of the beginning of May 2022, only 15.9% of people in low-income countries have received at least one dose. There have been profound problems in communicating pandemic information.

Indeed, the pandemic has highlighted weaknesses in disaster risk reduction measures, including:

i. A lack of *preparedness* for pandemics that resulted in untested, non-comprehensive plans that were not coordinated between key stakeholders.
ii. A lack of *planning* for zoonotic disease by *investing* in food safety policy, animal control, public health, hygiene and welfare initiatives and medical science research, in favour of *responding* to crisis by curtailing movement via border controls and quarantines, alongside animal culling.
iii. Poor *warning* systems to forewarn, track and trace the spread of the virus as it spread both within and between nations.
iv. Scientific *uncertainty* around COVID-19 leading to either too much dependence on uncertain science to guide policy or the exclusion of scientific data.
v. A lack of integration of *expertise* in the policy development and management of the crisis, most notably of disaster management professionals.

REFERENCES

The Our World in Data project team provides, and is a collection center for, data on the various impacts of COVID-19 at diverse scales. For updated figures go here: https://ourworldindata.org/coronavirus. For more on the One World One Health approach see S. Hinchliffe 'More than one world, more than one health: Re-configuring interspecies health'. *Social Science & Medicine* 129 (2015): 28–35. For more on COVID-19, risk and resilience please see M. K. Goyal, and A. K. Gupta, eds. *Integrated Risk of Pandemic: Covid-19 Impacts, Resilience and Recommendations*. Springer, 2020. For more on pandemic warning systems please see Carina J. Fearnley and Deborah Dixon. 'Early warning systems for pandemics: Lessons learned from natural hazards'. *International Journal of Disaster Risk Reduction* 49 (2020): 101674.

prove hazardous and ultimately devastating to those already vulnerable.

The power of modern communications, including non-stop news coverage and the rise of social media and connectedness via smartphones that allow for a variety of applications, permit the rapid dissemination of information – and misinformation and disinformation – about the latest disaster, often in graphic detail. Despite – or perhaps because of – this constant flow of information, it is difficult to place individual disasters in context and make broader assessments of the scope and impact of hazards. Is the world really becoming more hazardous? If so, what kinds of hazards are proliferating and what are the causes? Who is at risk and why? What are trends and changes in these profiles? Why do some disasters create much greater losses than the physical scale of the hazard suggests? What are the best means of planning for hazards, warning those at most risk and reducing the impacts of hazards and disasters in the future?

We all regularly face some degree of risk, whether it is to life and limb in a vehicle crash, to our possessions from theft or to our well-being from pollution. In some cases, we might even seek risks, such as participating in dangerous sports and leisure activities. Some risks, like smoking tobacco or driving a car, seem at first glance to be an individual 'lifestyle' choice. Yet our environments are designed to encourage particular ways of acting over and against others, such as cities supporting driving in private vehicles over bicycling and walking. Also, considerable effort has been made to advertise and sell products that carry a risk to others, such as second-hand smoke from tobacco or unrecyclable plastics. Indeed, it is because these risks have become so familiar and dispersed through a population that their cumulative toll on health and well-being are difficult to convey. By contrast, large-scale disasters such as the 2011 Fukushima Nuclear Power Plant breakdown (Box 1.2) are perceived to be disasters because they are so concentrated in time and space while rapidly disrupting the everyday lives of vast numbers of people.

Such industrial disasters are not isolated events. They occur within the context of a national planning process predicated on growing the economy. This planning process directs finance, design and engineering to produce particular technologies and infrastructures that will undergird industry, such as the power plants and energy grids that transmit electricity to buildings and transport systems. Nuclear power – as with coal, gas, wind, water, solar and others – allows for domestic consumption as well as manufacturing and services. All of these energy sources support particular kinds of technologies – from cars and aeroplanes to smartphones and supercomputers – in operating, also underpinning those heavy-duty technologies required to substantially use our environments, from tunnelling and land clearance to building and bridging. Any of these energy sources and other technologies can 'fail' in particular ways, from breakdown to loss of economic value, with varying physical and social (including economic) impacts that can stretch across space and time. These failures present risks that range from the mildly annoying to the catastrophic.

B. WHAT ARE ENVIRONMENTAL HAZARDS?

While physical events and human activities can produce hazardous conditions, it is only when people, infrastructure and livelihoods are impacted on a substantial scale that we refer to a disaster. It has been discussed from the beginning of formal disaster studies whether humans must be affected for a disaster to occur or whether disasters can occur from environmental hazards affecting only non-human entities, such as animals or rivers. The United Nations International Strategy for Disaster Reduction (UNDRR, 2017) specifically implies human impacts in their definition of a hazard:

> A process, phenomenon or human activity that may cause loss of life, injury or other health impacts, property damage, social and economic disruption or environmental degradation.

This book addresses a wide range of environmental hazards. Predominantly, the field has focused on the more extreme, often rapid-onset hazardous events that emerge quickly or ostensibly unexpectedly to directly threaten human life, livelihoods, physical and mental well-being, property and other assets by means of acute physical or chemical trauma on a relatively large scale, such as an earthquake, flood, vegetation fire or chemical explosion. These events comprise the release of energy or materials or the involvement of forces, in concentrations greatly in excess of typical background levels.

Slow-onset hazards remain important in terms of their impact on people. Environmental hazards such as drought, desertification, famine and epidemic disease typically develop over months or years, and therefore may be harder to track despite their heavy toll. While the gradually evolving nature of slow-onset hazards provides plenty of lead-in time to prepare and mitigate, there may be a lack of political will to recognise a set of conditions as hazardous. There may also be a political or geopolitical situation (such as armed conflict or a trans-border dispute) that stalls action, while complacency can prevail due to a lack of urgency. People and organisations might know exactly what they must do and how yet lack the power, resources and opportunities to do so. All hazards, irrespective of onset speed and rate, are complex phenomena, and it is important to address that complexity when understanding how hazardous conditions emerge and become disastrous to people.

Environmental hazards may be single, sequential or combined in their origin and effects. Each hazard can be characterised by its location, intensity, magnitude, frequency and probability among other parameters (UNDRR, 2017). Environmental hazards can occur over a small scale, affecting only nearby locations that nonetheless require assistance beyond the affected community (such as a hillside landslide), or they can be large scale, requiring national or international assistance (such as an ocean-wide tsunami). They can present a pattern over time (such as seasonal hurricanes) or none (such as pandemics). Impacts of frequent hazards can be cumulative or become chronic for individuals and communities as well as countries and continents.

Typically, hazards are assumed to divide into two not necessarily distinct categories: so-called natural hazards and technological hazards, the latter of which are often conflated with 'human hazards' (Table 1.1).

Table 1.1 Commonly used (but limited) major categories of environmental hazards with examples

Natural hazards
Geological: earthquakes, volcanic eruptions, tsunamis, landslides, karsts, subsidence, heave, land degradation
Atmospheric: cyclones, storm surge, tornadoes, storms, extreme temperatures, wildfires, lightning, precipitation
Hydrological: drought, flood, glacial, periglacial, wildfires, ice and snow avalanches
Biological: outbreaks, epidemics, pandemics, infectious diseases, dangerous plants and animals
Technological hazards
Transport: aircraft, car, bus, train and other vehicle crashes, shipwrecks
Industrial: Explosions and fires, release of toxic or radioactive materials, hazardous materials (storage, transport or misuse of)
Unsafe buildings and facilities: structural collapse, fire
Specific human activities: pollution, mining, land contamination, soil degradation, fracking, oil spills

As noted in Section A, the field of environmental hazards has become complicated with increasing recognition of human influences on nature. The ever-increasing capacity of people to impact physical processes through the cumulative impacts of technology and infrastructure, and the events and landforms associated with these, means that hazards such as storms, droughts and bushfires have become altered by anthropogenic climate change; hazards such as landslides and subsidence can be by-products of mass earth-moving activities associated with construction; and epidemiological hazards, from crop pests to antibiotic pollution, have been shaped by a global-scale organisation of species and organisms. Engineering rivers and coasts affects flood hazards; urban expansion creates microclimates; and the design of a building and the areas immediately around it affect the shaking it experiences during an earthquake. Furthermore, many of the technological hazards that were previously associated with local or regional scales, such as the release of toxic or radioactive materials, are now recognised as permeating ecologies that stretch horizontally across the globe, and vertically, from the high altitudes to the depths of the oceans.

The debates regarding the Anthropocene are examined in Chapter 16. Many of the hazards associated with the contemporary world are located within what has been called the Critical Zone, where chemical, biological, physical and geological processes operate to support life at or near the Earth's surface. The use of the term 'critical' indicates not only that this zone supports human life, but also that it is under severe threat from pollution and degradation.

This anthropogenic forcing has foregrounded the need to examine classificatory names for particular environmental hazards, along with theoretical approaches that emphasise complexity, non-linearity and contextuality. Nevertheless, many organisations and literatures retain the use of 'natural' and 'technological' as classificatory names for the reasons outlined later.

1. Natural Hazards

A key group of hazards has traditionally been classed as 'natural' because it is 'predominantly associated with natural processes and phenomena' (UNDRR, 2017). This type of description is well rooted in science, policy and practice but fails to provide a sense of the complexity of factors that makes an environmental hazard be involved in a disaster. It is most suitable for hazards like earthquakes and volcanic eruptions, where the fundamental processes remain largely unaffected by human actions, even while human actions can trigger earthquakes, change the shaking a building experiences and modify specific volcanic hazards such as lahar and lava flow direction and intensity.

The complexity of many of these hazards lies in the fact that the Earth's surface, atmosphere and subsurface are increasingly accepted as being subject to anthropogenic change. This suggests that, although all 'natural hazards' are assumed to be triggered by physical processes, particular events and their outcomes are influenced by human actions, whether deliberate or unintended. In other words, some types of natural hazards become quasi-natural hazards. For example, the disaster impact of a river flood may be increased by deforestation or a dam in the catchment. Where an increase in the frequency and/or severity of hazardous physical events can be attributed to degraded land or overexploited resources, the term socionatural hazard is sometimes used.

Furthermore, the physical events that trigger a hazard are by no means localised; they are linked to wider-scale processes. The slope failure that produces a landslide, or the rainstorm that produces a river flood, may originate respectively through tectonic and ocean-atmosphere mechanisms operating far beyond the mountain range or the river valley where the impact occurs. Many physical processes in the Earth's crust and its atmosphere are driven by forces that operate on hemispheric or even planetary scales, deploying vast amounts of energy and materials. Seemingly isolated events occur as part of a deeply complex, interlinked processes that have been termed *Earth systems*.

The physical processes that operate around the Earth exhibit some degree of variability, but most socio-economic activities are geared to an expectation of 'average', 'stable' and 'normal' environmental conditions. These are poor assumptions, but their basis is worth exploring and understanding, since they often form a foundation for policy and practice.

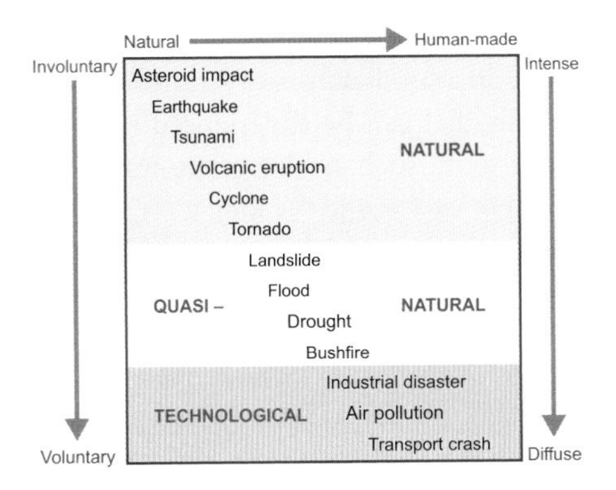

Figure 1.3 Presumed sensitivity to environmental hazards expressed as a function of annual rainfall and societal tolerance. Within the yellow band of tolerance, variations are perceived as resources. Beyond the damage thresholds, they are perceived as hazards that have the potential to create disasters.

Source: Adapted from K. Hewitt and I. Burton, The Hazardousness of a Place: A Regional Ecology of Damaging Events (University of Toronto Press, 1971). Reprinted with permission of the publisher.

In Figure 1.3, the central yellow zone is said to represent an acceptable, or tolerable, range of fluctuation for any 'element' vital for human survival or well-being, such as rainfall. Within this zone, the element is perceived as a beneficial resource. When it fluctuates over a critical (that word again) threshold beyond the 'normal' band of tolerance, the element becomes a hazard. Thus, very high or very low rainfall will be deemed to create a flood or a drought respectively. The hazard magnitude can be determined by the peak deviation beyond the threshold on the vertical scale, and the hazard duration is determined from the length of time the threshold is exceeded on the horizontal scale. The potential timescale of environmental hazard duration ranges over at least seven orders of magnitude, from the fractions of a second of lightning to drought conditions persisting for decades.

2. Technological Hazards

Technological hazards are defined by the UNDRR (2021) as threats that:

> originate from technological or industrial conditions, dangerous procedures, infrastructure failures or specific human activities.

Most of the hazards in this category arise from the risk-laden, technological trajectories that have been planned for and engineered as a means of pursuing economic progress. Built structures, industrial-scale processes, extensive transformation of ecosystems and waged labour – all facilitated by banking, taxation and credit systems, some form of state welfare and education system, and the increasing use of algorithms in decision-making – have produced a diverse range of hazardous conditions. Crashes, wrecks, substance releases, collapses and explosions are some obvious rapid-onset events that are produced through these technological trajectories. All of these hazards are seemingly one-off incidents, yet they emerge within a built environment that has been designed to function, and thus to fail to function, in particular ways.

More difficult to identify and track are the slow-onset technological hazards that take a toll on life and well-being. These include, but are not limited to, the long-term exposure of individuals to pollutants associated with industry, manufacturing and transport, such as through air, water, food and everyday products; truncated lives from a combination of unsanitary conditions, poverty and lack of access to health care; the effects of sedentary work practices and manual labour practices; and mental and physical health issues relating to economic insecurity.

3. Socionatural Hazards

Socionatural hazards are said to be induced by human activities that have major impacts on local, regional or even planetary environments. These can include specific, rapid-onset events, such as landslides that are triggered by road building, and collapses caused by tunnelling. Socionatural hazards can also work slowly over time and over larger spatial scales. Deforestation, for example, can set in train numerous hazardous conditions ranging from the closer proximity of pathogen-bearing animals to human populations through to soil erosion and flooding far from the ruined forest.

Radically transformed environments, with examples being urban (Figure 1.4) and agricultural (Figure 1.5) landscapes, usher in a proliferation of hazardous conditions that can be experienced over the short term, such as changes to water and nutrient cycles, and the long term, such as the gradual accumulation of toxic pollutants and the release of significant amounts of carbon dioxide, methane and nitrous oxide. The anthropogenic forcing of physical processes can produce socionatural hazards on a planetary scale, such as warming, acidifying and rising oceans along with wider climate change. The aggregation of these socionatural hazards, and their proliferation over time, has led many scholars to talk about a new Anthropocene epoch for the planet marked by the generation of anthropogenic hazards, defined as human-induced hazards that are induced entirely or predominantly by human activities and choices (UNDRR, 2017).

Figure 1.4 Landscapes transformed into a city: Macau.

Source: Photo: Ilan Kelman

Another phrase used occasionally is natural-technological ('na-tech') hazards to describe the impact of physical processes on infrastructure, industrial structures and other land use systems. As these hazards occur sequentially, they may also be referred to as cascading hazards; that is, where one hazard triggers another. Although triggered by physical processes, the main threat from na-tech hazards often comes from pollution due to inadvertent releases of dangerous substances (Showalter & Myers, 1994). As such, and especially with technology being a product of society, they are no different from socionatural hazards and do not require a different category.

Further examples include the following:

- In August 1975 the Banqiao Dam and 61 other dams in Henan, China, collapsed under the influence of Typhoon Nina. It affected a total population of

Figure 1.5 Landscapes transformed into agricultural land: the Netherlands.

Source: Photo: Ilan Kelman

10.15 million and inundated around 30 cities and counties of 12,000 km^2 with an estimated death toll ranging from 26,000 to 240,000. Typhoons, floods and earthquakes destroyed river dams, causing the uncontrolled release of stored water to create damage downstream.

- In September 2003, Italy suddenly lost over one-quarter of its electricity supply when storm-force winds in Switzerland damaged the transmission line importing power into the country.
- Contaminated drinking water was a major issue following Typhoon Haiyan in the Philippines in November 2013. In addition to pollution infiltrating the drinking water supply system due to infrastructure damage, surface water and groundwater resources were contaminated by pollution from industrial and domestic waste and septic systems, further exacerbating the potential spread of waterborne diseases.
- In February 2015, Tropical Cyclone Marcia dumped intense rainfall over a short period in the Callide Dam catchment in Australia, causing the gates of the dam to automatically open. The resultant flood caused considerable damage to homes, properties and businesses and placed human life at risk.
- In late August 2017, Hurricane Harvey caused significant flooding in Houston, USA, causing large explosions and fires at the Arkema chemical plant. The explosions were caused by the loss of refrigeration to a warehouse storing highly volatile and extremely flammable chemicals, which then spewed toxic chemicals into the air and water. The chemical fires lasted nearly a week,

resulting in respiratory illnesses and headaches for local populations and emergency responders. A full investigation found that the plant did not fully consider flood risk, despite the plant's site on a floodplain and previous storms hitting the area.

- The radioactive pollution, evacuation and associated social impacts that followed the nuclear power plant disaster in Fukushima, Japan. This disaster was triggered by the inundation of water from the tsunami caused by the earthquake of 11 March 2011. More detail on the complex nature of this hazard is noted in Box 1.2.

4. The Interrelation of Environmental Hazards

As the prior examples help to demonstrate, there are challenges in classifying specific hazardous events and conditions. Hazards and disasters are not one-off, site-specific, unique, isolated situations capable of regulation and mitigation by local responses alone. While explicit causal linkages may be difficult to identify, it is possible to draw out the complex physical and social conditions within which hazards are produced and which contribute to the making of disasters that

Box 1.2 Japan's 2011 Earthquake, Tsunami and Nuclear Power Plant

The residents of Fukushima, Japan, experienced a cascading sequence of hazards, some of which were rapid onset, and others that continue over a longer timeframe.

Event One: On 11 March 2011, the island of Honshu, northeast Japan, was rocked by an earthquake of moment magnitude Mw = 9.0. At the time, this was the largest instrumentally recorded earthquake ever to hit Japan. The offshore epicentre was roughly 70 km east of the Oshika Peninsula of Tōhoku, on the Sanriku coast, with a hypocentre 30 km below sea level. The earthquake resulted from thrust faulting at the plate boundary between the Pacific and Indo-Australian–Fiji plates (see Figure 1.6). At this point, the Pacific plate is moving west at a rate of 83 mm per year and descends beneath the Japan landmass in a subduction zone.

Event Two: The earthquake triggered tsunami waves of almost 40 m high that struck the Japanese coast several minutes later, overrunning sea walls. The Sanriku coast has many deep coastal bays that constrain and amplify the height of approaching tsunami waves, sending some up to 10 km inland. Water inundated about 560 sq km, and resulted in an estimated death toll of around 20,000 and extensive damage to coastal ports and towns, with over a million buildings destroyed or partly collapsed. The Japanese Red Cross sent 230 response teams, and over 2,000 evacuation centres were set up in northeast Japan. Total economic damage was estimated at US$366 billion, with insured property losses of US$20–$30 billion. About US$4,000 billion was wiped off the Nikkei 225 stock market index, which initially fell by over 6%.

Event Three: Three reactors at the Fukushima Daiichi nuclear power plant – owned by the Tokyo Electric Power Company (TEPCO) – lost power when the entire site was

Figure 1.6 A tsunami wave generated by the Mw = 9.0 magnitude Great Tōhoku earthquake on 11 March 2011 crashes over a sea wall, carrying small boats, vehicles and other debris to devastate Miyako City, Iwate Prefecture, in northeast Japan.

Source: Photo: © Getty Images/JIJI PRESS/Contributor

flooded by a 15-metre tsunami wave. As a result of generator failure, four reactors began to overheat and three ultimately suffered meltdown. Explosions caused by a build-up of hydrogen gas in the outer containment buildings released radioactive material. Levels of 400 millisieverts (mSv) were recorded at No 4 reactor. This compares with the 350 mSv criterion adopted for evacuation at Chernobyl in 1986. Initially, the incident was rated 5 on the 7-point International Nuclear and Radiological Event Scale (INES), but later it was reassessed at the highest level. This was the first Category 7 nuclear disaster since the Chernobyl disaster, indicating substantial risks to human health and environmental contamination from leakage of cooling water and contamination of coastal waters.

Event Four: Immediately after the disaster, the government ordered residents in a 3 km radius around the Fukushima Daiichi Nuclear Power plant to evacuate. The next day this was expanded to a 20 km radius; later this would become a 'restricted area'. In addition, residents at sites where the annual cumulative dose of radiation was expected to reach 20 mSv/year were asked to evacuate. In all, around 80,000 residents from this area were forced to live in emergency shelters, triggering a series of social crises. According to Hasegawa et al. (2016, pp. 237–238):

> Evacuation of the inpatients and elderly residents of nursing care facilities was hurriedly carried out by buses shortly after the accident. No medical personnel accompanied the evacuees who were laid down on the seats of the jam-packed buses with full protective suits on. No medical care, even food or water, was provided for many hours during the evacuation. As a result, scores of patients died in an evacuation that was supposedly intended to minimise radiation exposure. . . . the severe health risk associated with the rapid evacuation of elderly residents from nursing care facilities after the Fukushima accident was 30 times higher than the radiation risk of the reference levels for evacuation that are recommended by the International Committee for Radiological Protection.

Many households were separated after the disaster and individuals had to move several times, while evacuees have suffered from stigma, social isolation and discrimination. In 2017, evacuation orders were lifted for several towns, ending subsidies and housing support and exacerbating fears around delayed and inadequate decontamination procedures. The entire Tohoku region in northern Japan, of which Fukushima is a part, saw a population decline of about 5% after the disaster.

Full decommissioning of the plant would include decontaminating an area extending to 2,000 sq km, the disposal of an estimated 90,000 tonnes of contaminated seawater and the removal of millions of cubic metres of topsoil. This process could cost up to US$50 billion and take 40 years to accomplish.

REFERENCES

For more detail on the events leading up to the nuclear plant shutdown, see N. Akiyama et al. 'The Fukushima nuclear accident and crisis management-lessons for Japan-US alliance cooperation'. *The Sasakawa Peace Foundation* (2012). On the social impacts of the evacuation, see A. Hasegawa, et al. 'Emergency responses and health consequences after the Fukushima accident; evacuation and relocation'. *Clinical Oncology* 28.4 (2016): 237–244 as well as Sudeepa Abeysinghe, Claire Leppold, Akihiko Ozaki, Alison Lloyd Williams (eds.), 2022, *Health, Wellbeing and Community Recovery in Fukushima*, Routledge, Abingdon, U.K.

affect some more than others. In many instances, the wider links and consequences are fairly clear. It is often useful, then, to think about how hazards cascade from one to the next, as with the events outlined in Box 1.2.

There may be a simple sequence as a multi-hazard event, indicating primary, secondary, tertiary and more hazards. The hazard event, for example an earthquake (usually known as a primary hazard), generates a secondary hazard such as a landslide, which can generate a tertiary hazard of a tsunami. There may also be a multi-hazard event with concurrent hazards, when more than one hazard happens at the same time and impacts the same locations. Mt Pinatubo in the Philippines erupted in 1991 at the same time that Typhoon Yunya passed over Luzon Island, creating more challenging conditions than expected, notably in rain- and ash-slicked roads as well as ash absorbing rainwater increasing the load on roofs (read more in *Fire and Mud: Eruptions and Lahars of Mt Pinatubo* by Newhall & Punyongbayan, 1996).

These cascading hazards then impact lives, livelihoods, health and economic security of individuals and communities multiple times, simultaneously and/or sequentially. Events further down the cascading chain can cause disastrous conditions that are at least as serious as the first one, and may well unfold over a much longer timeframe. Cascading events often include the disruption of key infrastructure, such as energy, water, food, transport and health. A prime example of this complex scenario is the 2010 earthquake that struck Haiti (see Box 1.3).

The examples of Haiti 2010 and Japan 2011 indicate that, while the initial trigger of a hazard cascade may be located in an extreme physical event, the disaster really comes from pre-existing social structures creating the conditions for disaster. How has the colonial history of Haiti and the policies of France, the USA and the UN entrenched poverty there? How has the decision to rely on nuclear power for Japan's economic growth, and the siting of coastal nuclear power plants by a then privately owned company, placed nearby residents – and ultimately the country – at risk?

Chronic hazards or creeping hazards: this term is rarely used in the academic

Box 1.3 Cascading Hazards in Haiti

A magnitude 7.0 Mw earthquake struck Haiti on the afternoon of 12 January2010. Occurring 9.9 km below the surface – which is a shallow depth – the energy released from the earthquake caused widespread destruction that killed over 250,000 people and displaced over 1.5 million into makeshift shelters. Building and infrastructure collapses not only led to loss of life, but also severely disrupted energy and water networks, which were in poor condition to begin with, along with the everyday operation of government, health, education and business sectors. As such, transportation and communication networks that humanitarian organisations and personnel could use to organise and distribute aid were in poor condition. Individual countries such as Canada, Cuba, the Dominican Republic, Iceland and Israel dispatched rescuers, medical teams and engineers, and many pledged funds for relief and rebuilding. In the earthquake's aftermath, the UN extended its presence by bringing in more peacekeeping troops, who had first arrived in 2004 as a means of 'stabilising' the country.

On 20 October2010, an outbreak of cholera was confirmed in Haiti for the first time in more than a century, ten months after the catastrophic earthquake.This cholera outbreak led to over 665,000 cases and over 10,000 deaths. How did this environmental hazard emerge?The bacterium was brought in by UN peacekeepers at their camp in Mirebalais in the heart of the country who then put untreated sewage waste from the camp into a nearby river. Soon after, in a nearby village, doctors of the Cuban Medical Brigade recorded the first cholera cases.

On 1 December 2016, after immense pressure, United Nations Secretary-General Ban Ki-moon finally apologised for the cholera as his term was nearing its end and indicated the possibility of some form of recompense, with many commitments remaining unfulfilled a decade after cholera appeared. According to the US CDC (Centers for Disease Control and Prevention), 'Experience from past cholera outbreaks around the world suggests that Haiti may have ongoing cholera transmission for years to come. Improving Haiti's water and sanitation infrastructure is critical to achieving large health gains and reducing the opportunity for cholera to spread' (see Figure 1.7).

The events triggering and subsequent to the 2010 situation can be considered cascading hazards, insofar as the epidemic that emerged further down the cascading chain has caused disastrous conditions that are at least as serious in impact as the original earthquake. The impacts from them all will continue to unfold over generations. Yet, while the 2010 earthquake has been noted as the initial triggering hazard, the vulnerabilities of people to disaster were set in train long before the earthquake struck.

A 1791–1803 rebellion freed the Haitians from French colonialism, with independence declared on 1 January 1804. France and the US punished this effort at

Figure 1.7 Haiti during the cholera outbreak.

Source: © Alamy Images/ZUMA Press, Inc.

self-governance, with France taking reparations until 1947 and the US government interfering in Haitian politics, in effect, until the UN took control in 2004. These actions, and more afterwards, led to the creation and perpetuation of disaster risk in Haiti to the advantage of the external powers and to the disadvantage of Haitians. This long-term construction of disaster was seen in the 2010 and 2021 earthquakes – as well as multiple hurricanes that have hit the country.

REFERENCES

For more on the seismological background see Bilham, Roger. 'Lessons from the Haiti earthquake'. *Nature* 463.7283 (2010): 878–879. The UN's own assessment of the organisation's role in introducing cholera can be found in the *Report of the Special Rapporteur on extreme poverty and human rights* (August 26, 2016) accessible here: https://reliefweb.int/sites/reliefweb.int/files/resources/N1627119.pdf. For more on Haiti and cascading hazards see Mika, Kasia. (2019) *Disasters, Vulnerability, and Narratives Writing Haiti's Futures.* Routledge, Abingdon.

risk literature but is an important issue for communities exposed to long-term hazards such as volcanic gases and water contamination. A chronic or creeping hazard has the potential for cumulative, long-term impacts. Often the effects of the hazard are hidden until they cause serious health complications, as seen by respiratory problems in those living near active volcanoes or digestive issues in the case of water contamination. Chronic hazards are typically seen as slow-onset hazards, but they remain present during rapid-onset hazards, exacerbating their impacts. Chronic hazards can also be the result of cascades. The impact of tropical cyclones on permanent displacement, significant personal trauma (e.g. the loss of a loved one), and reduced or no access to social services can be seen as a 'chronic disaster syndrome' whereby affected people are in a constant state of disaster (Adams et al., 2009).

In summary, the following are the chief features of environmental hazards:

- The cause(s) and the conditions from which hazards emerge can be traced back to physical processes and/or human activities, producing known threats to human life or well-being.
- Exposure to hazards is largely involuntary, normally due to the location of people in a hazardous area without power or resources to change their situation for the better.
- The impacts of hazards are exacerbated by inequality and inequity, including poverty, marginalisation and oppression.
- The disaster can be used to justify an emergency response, sometimes on an international humanitarian scale.

C. UNDERSTANDING DISASTER AND RISK

The preceding section looked at various types of hazards and emphasised that these become disasters when individuals and communities are impacted by them in the form of loss of life, reduced well-being, disruption and damage. Environmental hazards create the following threats:

- To people: death, injury, disease, mental stress and other negative impacts on

mental health and well-being, and livelihood disruption.
- To goods: property and infrastructure damage, economic loss and breakdown of supply chains.
- To the environment: loss of flora and fauna, pollution, toxicity and loss of amenity.

Given these impacts, a disaster can be defined as 'serious disruption of the functioning of a community or a society at any scale due to hazardous events interacting with conditions of exposure, vulnerability and capacity, leading to one or more of the following: human, material, economic and environmental losses and impacts' (UNDRR, 2017), although many books have been published examining the definition of 'disaster'.

Despite the diversity of impacts and threats, it is typically those phenomena that lend themselves to quantitative measurement that have been used to assess the magnitude and scope of disasters. Traditionally, disasters are measured in terms of lives lost (mortality rates), the physical harm caused to individuals (morbidity rates) and the losses estimated by insurance companies (as discussed further in Chapter 2). Quantification allows for different events to be compared – ostensibly objectively – over space and across time. They can give some insight into how the interconnectedness of economies and the mobilities of people and goods are impacted by specific disasters. As an example, the 2010 Eyjafjallajökull volcanic eruption in Iceland closed much of European airspace to commercial flights for five days, resulting in significant economic loss, livelihood interruption and disruption worldwide. The flower, fruit and vegetable industries in Kenya suffered significant losses as perishable goods were unable to be air freighted to European markets.

The danger of relying on quantifiable measures to assess the scope of a disaster is threefold. First, when phenomena such as deaths are counted, it removes the deeply personal and emotional context within which these occurred. Statistical measures do not adequately capture the nature and experience of the phenomenon that they tabulate. Second, quantification relies on recorded observations, which then become sorted and made searchable under particular protocols and depend on resources available to make and keep these observations. At each stage – recording, sorting and searching – choices are made as to how to count and manage data. Few of these choices are made transparent to those using the aggregated statistics. Then, when assessing the relative severity of disasters using measured losses, choices can produce datasets that are not comparable. Third, an emphasis on quantifiable measures tends to devalue those phenomena that do not lend themselves to being counted, such as loss of nature, cultural heritage, social connections and feelings of security. It is certainly difficult to calculate the value of environmental and cultural amenities and people's connection to nature, history and the sense of connection to place they have which is disrupted.

Assessing and responding to the threat a particular hazard poses to society can be undertaken using a range of approaches:

- Understanding the hazard: this involves developing knowledge around hazards, monitoring (data collection and analysis), forecasting and prediction and communicating this information and interpretations.
- Managing the hazard: many tools can be adopted, such as engineering, working

with people to determine what they seek for their own safety, and warning systems. These actions commonly come under the banner of disaster risk reduction (DRR) and involve preparedness, planning, prevention and mitigation measures while being ready for response, recovery and reconstruction (see Chapter 5).

- Investing in the hazard: cost-benefit analysis is one approach for estimating the financial strengths and weaknesses of actions to invest in achieving benefits whilst reducing costs over multiple timeframes. In addition, insurance and money-based risk management strategies support businesses, organisations, families and individuals against certain events, although many natural hazards are bizarrely classed as 'acts of God'.

In sum, the preceding points relate to managing threats that a hazard might pose. Risk is sometimes taken as synonymous with hazard, but risk has a different meaning. In the academic fields dealing with hazard and risk, the latter has frequently been defined as *the combination of the probability of a hazardous event occurring and its negative consequences* calculated in the form of losses. This understanding of risk has driven quantitative approaches to researching and communicating the level of exposure that people have to hazards.

The difference between hazard and risk can be illustrated by two people crossing an ocean, one in a large ship and the other in a rowing boat (Okrent, 1980). Hazards of deep water and large waves are the same in both cases, but risks of capsizing and drowning are much greater for the person in the rowing boat. This analogy shows that, whilst some shaking from earthquakes, for example, is similar in many places around the world, the dangers and threats differ, especially since marginalised and poor communities and individuals are often more vulnerable and at greater risk than others. When large numbers of people are killed, injured or otherwise adversely affected, the differing vulnerabilities are exposed and it becomes a disaster.

Risk is nonetheless inherently difficult to define, because it is understood by different disciplines in different ways, as explored further in Chapter 4. Key terms, as defined by the UN, are in Table 1.2, but all have extensive scientific work and legislative and regulatory approaches, with many debates and often irreconcilable disagreements. Here, it is important to note that in any one situation, the different components comprising risk could be specific to that situation or context.

While Table 1.2 draws on definitions of key terms provided by UNDRR (2019), they must be used with caution. Many of these terms have different meanings in different fields – academic, official, government, the private sector and the non-profit sector. As a result, there have been intense debates around their application in discussions on whether and how they actually do reduce the risk of disaster (Lewis & Kelman, 2010). To ensure comprehension and usability, there is a need to clearly define these terms when they are used.

Who is most at risk from environmental hazards? In what places could we expect to find that a hazardous condition becomes a disaster?

Exposure (sometimes separated from vulnerability and sometimes included within it) can be understood as the location, attributes and value of assets that are important to communities (e.g. people, infrastructure, factories, businesses, public resources and land) and that could be

Table 1.2 Often-debated definitions from the UNDRR (2019)

Risk	Defined as 'disaster risk' being 'The potential loss of life, injury, or destroyed or damaged assets which could occur to a system, society or a community in a specific period of time, determined probabilistically as a function of hazard, exposure, vulnerability and capacity'.
Hazard	'A process, phenomenon or human activity that may cause loss of life, injury or other health impacts, property damage, social and economic disruption or environmental degradation.'
Exposure	'The situation of people, infrastructure, housing, production capacities and other tangible human assets located in hazard-prone areas.'
Vulnerability	'The conditions determined by physical, social, economic and environmental factors or processes which increase the susceptibility of an individual, a community, assets or systems to the impacts of hazards.'
Resilience	'The ability of a system, community or society exposed to hazards to resist, absorb, accommodate, adapt to, transform and recover from the effects of a hazard in a timely and efficient manner, including through the preservation and restoration of its essential basic structures and functions through risk management.'
Capacity	'The combination of all the strengths, attributes and resources available within an organization, community or society to manage and reduce disaster risks and strengthen resilience.'
Disaster	'A serious disruption of the functioning of a community or a society at any scale due to hazardous events interacting with conditions of exposure, vulnerability and capacity, leading to one or more of the following: human, material, economic and environmental losses and impacts.'

affected by a hazard. The physical environment is a key factor in producing a high level of exposure to risk, as polluted and degraded ecosystems will have deleterious impacts. These exposures may take place at the local scale (from the release of chemicals in fields and waterways), regional scale (such as widespread air pollution) or planetary scale (a key example being climate change and its impacts on weather and the oceans). Some individuals and communities – because of poverty, marginalisation, oppression, inequity, inequality and exclusion – have little choice other than to occupy or work in such environments and will thus be more exposed to particular hazards. As such, exposure overlaps so much with vulnerability that it can be difficult to differentiate them. Debates continue regarding whether separate concepts of exposure and vulnerability are needed.

Vulnerability can be understood as the susceptibility of people and their assets to being harmed by particular hazards. More so, vulnerability is the long-term societal processes creating and perpetuating this susceptibility. A high vulnerability to hazards generally depends on poverty and other long-term social factors. Traditionally measured in terms of a national GDP and through the classification of 'low-income' or 'developing' countries, poverty is more usefully understood as the livelihood insecurity of individuals, as per the earlier economic security discussion. Poverty levels depend on how international, national, regional and local economies have been and continue to be organised, who has access to particular livelihoods and resources, in what forms such access occurs and how money and other forms of wealth are redistributed through services such as public health, education, infrastructure maintenance and social welfare. Vulnerable cohorts can include the economically disadvantaged, minorities, the uninsured, the elderly, children, the homeless and those with chronic health conditions.

An example of vulnerability can be drawn from the COVID-19 pandemic, introduced in Box 1.1. Despite the global spread of the virus, some are more vulnerable than others to infection and adverse effects due to a lack of access to vaccinations; a lack of access to health care including protection measures and treatments; the targeting of misinformation and disinformation campaigns; underlying health conditions; poverty levels associated with a structural racism that undermines the health of those racialised; and a lack of investment in public health.

Resilience is a highly contentious term, with one class of definitions referring to the ability of individuals, communities and organisations to adapt to and recover from hazards without compromising long-term prospects for sustainable living. Resilience requires better anticipation of disasters and planning to reduce disaster losses. Implementing disaster resilience is complex and politically challenging but can be very effective and inexpensive as an investment – if the understanding of 'resilience' is agreed and is relevant. Building resilience requires an understanding that hazards can be cascading in different ways while vulnerabilities interweave and overlay on each other. Resilience therefore must be integrated with wider concepts such as vulnerability, development and sustainability – especially noting that vulnerability and resilience are not opposites or in opposition – alongside developing effective disaster risk reduction and management. Resilience is related to 'capacity'.

Capacity is also highly contentious as a term. It is often related to the strengths, attributes and resources within a community, organisation, group or society which, in this context, would be applied for managing and reducing disaster risks – and for strengthening resilience while reducing vulnerability. Rather than just focusing on vulnerabilities, it addresses people's abilities to anticipate, cope with, resist and recover from disasters. Creating and sustaining capacity is key to successful disaster risk reduction. Often vulnerability and capacity are seen as two sides of a coin, with other models presenting different relationships and interconnections. Discussions continue on how to differentiate capacity, capability and ability – or the need for doing so.

D. PERSPECTIVES AND APPROACHES

Our understanding of hazards – as well as the risks that are present and the disasters that can ensue – depends on the broader views we hold as to how the physical world works and interacts with society to produce difficulties and ways of addressing those difficulties, often expressed as exposures, vulnerabilities, resiliences and capacities. This section focuses on relatively recent perspectives and approaches associated with the emergence of an environmental hazards-focused academia summarised in Table 1.3. There are, however, many examples of hazard management from various societies at various time, space and governance scales. Indeed, societies have a strong and complex relationship with hazards from rapid physical events to slow-onset hazardous conditions. Empires were built around sites of great hazards and threats, such as earthquakes across the Middle East,

Table 1.3 A broad overview of the emergence of environmental hazard paradigms and approaches and their core questions and concerns

Period	Paradigm name	Main questions	Main responses
Pre-1950	Engineering	What are the physical causes for the magnitude and frequency of natural hazards at certain sites and how can protection be provided against them?	Scientific weather forecasting and large structures designed and built to defend against natural hazards, especially those of hydrometeorological origin.
1950–onwards	Behavioural	Why do natural hazards produce deaths and economic damage and how can changes in human behaviour minimise risk?	Improved short-term warning and better longer-term land planning so that humans can adapt and avoid sites prone to natural hazards.
1970–onwards	Underdevelopment	Why do people suffer so severely in disasters and what are the historical and current socio-economic causes of this situation?	Greater understanding of how exploitative economic and political relations produce vulnerability to disaster and how poverty constrains preparedness and resilience.
1990–onwards	Feminist and intersectional Settler colonialism, postcolonialism and decolonialism Complexity	How do sexism, racism, homophobia and other forms of discrimination produce systematic inequalities that are experienced by individuals in their everyday lives, such that some are more vulnerable than others are to hazard and risk? How has colonialism and postcolonialism produced exposure to hazards and diverse geographies of vulnerability; how can marginalised understandings of people, environment and hazard be centred in academia and policy; and how can exploitative social relations be transformed as a means of reducing exposure, risk and vulnerability? How can disaster impacts be reduced in a sustainable way in the future, especially for the poorest people in an unequal and rapidly changing world?	Greater understanding of the diverse experiences people have of disaster and the need to look beyond preconstructed datasets to investigate this diversity. Greater appreciation of how academic knowledge is produced, and with what effect on other forms of knowledge; and a commitment to academic research that helps transform the societies we are a part of, and the ways in which those societies engage with and impact the environment, for the well-being and health of all rather than the few. Emphasis on the complicated interactions between natural and human systems, leading to improvement in the long-term management of hazards according to local needs.

as they provided opportunities for water resources and access to minerals that provided resources to trade. In turn, societies have responses to disasters in many ways, including accepting them as divine acts, building back bigger, better and stronger, or abandoning sites. Numerous disasters were witnessed at Pompeii in today's Italy, with various eruptions of Mt Vesuvius volcano during the start, middle and end of the Roman Empire, each affecting the populations in different ways (Grattan & Torrence, 2016). We have much to learn through archaeological and anthropological studies of how societies interacted with hazards.

1. Observing and Measuring

While there are many different understandings of how hazards are produced and appropriately managed and responded to – often involving systematic observation and data collection as well as scientific experimentation and other approaches to observing and interpreting the environment – the emergence of Environmental Hazards as a modern-day field of enquiry is generally associated with an Anglo-European Enlightenment that came to fruition in the 18th century. The systematisation of knowledge into university-based disciplines, alongside the organisation of Learned Societies for sharing information amongst colleagues, allowed for particular knowledges about hazards to become valued as 'expert', and other knowledges, associated with colonised peoples, women and the working classes, to be dismissed as 'inexpert'.

The environmental hazard research that emerged within universities and societies was quite different in approach and scope to that which dominates our classrooms, journals and textbooks today. This is because the study of hazards was not limited to one specific area but was found throughout what we now think of as the arts, humanities and social sciences as well as other disciplines such as engineering and earth sciences. Furthermore, the techniques used to undertake research were a mix of activities that many scholars would now consider as niche; that is, to be found only within either the arts or the sciences.

Contributions to these scientific societies, for example, came from wealthy men holding power who were undertaking the 'Grand Tour' or who worked abroad in a diplomatic capacity, and who provided accounts of events and specimens to audiences back home. What the Enlightenment philosopher and politician Edmund Burke was to refer to as sublime landscapes – that is, capable of producing awe and even terror in the observer – were understood not only as visual displays, but also as complex scenes that could be described and even explained by attending closely to their sweep and detail. Sketches, paintings and letters sought to capture as faithful a rendition as possible of what had been observed in the rural margins of Europe, as well as the newly colonised lands of America and Australasia. These were sent to organisations such as the Royal Society (or, 'The President, Council, and Fellows of the Royal Society of London for Improving Natural Knowledge'), established in 1660.

One of the key observers of volcanic hazards was Sir William Hamilton (1730–1803), the British Envoy to the Spanish Court at Naples, 1764–1800, and elected a Fellow of the Royal Society in 1766. With a house in the foothills of Vesuvius, as well as a residence in Naples, Hamilton explored this volcano's craters and associated features while also collecting gemstones and geological samples. Letters recounting his experiences were read at weekly meetings in London, where the audience could peruse the paintings, sketches, pieces of lava and soil samples that Hamilton had sent over. Hamilton's production in 1776 of the richly illustrated folio collection of letters, *Campi Phlegraei: Observations on the Volcanos of the Two Sicilies*, is testament to his desire to convey to readers what it was like to be immersed in volcanic landscapes. Despite the terrors that such hazards may evoke, Hamilton understood volcanoes to be part of a pastoral relationship between environment and society. 'There is no doubt, but that the neighbourhood of an active Volcano, must suffer from time to time the most dire calamities, the natural attendants of earthquakes, and eruptions', he wrote, 'But to consider such misfortunes, on the great scale of nature, it was no more than the chance or ill fate of those cities to have stood in the line of its operations; intended perhaps for some wise purpose, and the benefit of future generations' (pp. 3–4) including, he noted, rich soils for agriculture.

The importance of witnessing phenomena first-hand, and of accurately describing them, was to remain a key objective of scientific practice into the 19th century. The aim in doing so was to shift, in that these observations were to be sorted and synthesised so as to produce an understanding of the physical processes at work in producing events. Textbooks, such as Mary Somerville's (1854) *Physical Geography*, sought to present an overarching explanation of how the world worked in order to situate specific events. In contrast to the efforts of Hamilton and the Royal Society, such knowledge, Somerville argued, should be 'widely diffused amongst all ranks of society' (1854, p. 395 (Box 1.4)).

The late 19th and early 20th centuries saw a flourishing of experimental observational techniques aimed at recording

Box 1.4 Mary Somerville, Observer and Synthesiser

Mary Somerville was one of the few women to achieve international recognition for her research in the 19th century (see Figure 1.8). She was elected honorary member of Société de Physique et d'Histoire Naturelle de Genève and the Royal Irish Academy in 1834, and member of the Royal Astronomical Society in 1835. She was also elected to the American Geographical and Statistical Society in 1857, and the Italian Geographical Society in 1870, receiving the Victoria Gold Medal of the Royal Geographical Society that same year. She moved to Rome for her husband's health in 1838, where she began work on the two-volume *Physical Geography*, published in 1854. She travelled throughout Italy, maintaining a travel diary as well as correspondence with scientists across Europe. Somerville's *Personal Recollections* (1873) include a record of her descent into the crater of Vesuvius in 1818 and her witnessing of the April 1872 eruption. 'On Sunday, 28th', she wrote, 'I was surprised at the extreme darkness . . . the fall was a little less dense during the day, but at night it was worse than ever . . . certainly the constant loud roaring of Vesuvius was appalling enough amidst the darkness and gloom of the falling ashes' (Somerville, Mary. Personal Recollections, from Early Life to Old Age, of Mary Somerville . Biblio Bazaar, LLC, Originally published 1873, reprinted in 2009).

Figure 1.8 Painting of Mary Somerville by Thomas Phillips (1834).

Source: Wikimedia Commons

REFERENCES

Somerville, Mary. *Physical Geography*. Blanchard and Lea, 1854.

Somerville, Mary. *Personal Recollections, from Early Life to Old Age, of Mary Somerville*. BiblioBazaar, LLC, 2009.

and monitoring the signs of physical activity that heralded hazards and hence potential disasters. The study of volcanoes over centuries has included theological and philosophical debates about the nature and significance of fire, as well as meticulous observation and documentation of volcanic events. One early, known event was that of the AD 79 eruption of Mt Vesuvius by Pliny the Elder and Pliny the Younger that resulted in the publication of one of the largest Roman publications to survive – *Naturalis Historia*.

An effort to create a global-scale observation of volcanoes emerged at the beginning of the 20th century. The earliest observatories for such tasks were established in Asama, Japan, and Hawaii, USA. In 1912, this led to the establishment of the Hawaii Volcano Observatory (HVO), dedicated to studying volcanism at Kilauea volcano by developing and modernising equipment to systematically monitor phenomena related to the volcanic eruption and to use this to develop knowledge about volcanic processes and eruptions. HVO remains a pioneer in instrumentation for monitoring hazards, and volcanologists around the world have trained there. With over 80 observatories operating globally now, data from ground equipment and satellite data is shared. Where progress still needs to be made, however, is in listening to and valuing as expert diverse past and present indigenous, local, vernacular and traditional knowledges at observatory sites and within this network.

Mechanisms to record the size and travelling of earthquake waves can be traced back to at least AD 132 in China. Within European science, the work of John Milne systematised seismology into a global-scale field of inquiry. In 1876, Milne was appointed as the Professor of Geology and Mining at the Imperial College of Engineering in Tokyo, and his work on geologic faults developed into an interest in the volcanic and earthquake hazards associated with these. On his return to England in 1895 he established an observatory at Shide on the Isle of Wight, which became the centre of a global earthquake recording network. His bulletin, the Shide Circulars, became the forerunner for the bulletins of the International Seismological Summary and eventually the International Seismological Centre. Today the United States Geological Survey (USGS), in partnership with Incorporated Research Institutions for Seismology (IRIS), has established the Global Seismic Network. This provides a near-standardised monitoring of the Earth, with over 150 seismic stations globally, and acts as a multi-use scientific facility and societal resource for monitoring, research and education on seismicity.

2. Engineering and Its Critics

Efforts to monitor for signs of impending hazards, and to record them as they happened, have been matched by efforts to mitigate their impacts. The development of what has been termed the engineering paradigm has been traced back to the river dams constructed in the Middle East over 4,000 years ago, although it is recognised that this itself has precedents. Attempts to fortify buildings against earthquakes date back at least 2,000 years. This approach is based on 'hardening' built structures to withstand most hazard stresses and evacuating people from harm by emergency action. The growth of the earth sciences and civil engineering practices over the following centuries, in concert with governmental concerns to enhance resource use and reduce the number and impact of disasters, has led to increasingly large-scale structural responses. These range from the building of extensive sea walls and levee systems to the relocation of substantial portions of cities underground to avoid winter snowstorms.

Engineering does not simply refer to fortification. The word is derived from the Latin meaning to devise or create. As such, engineering also refers to working with, rather than against, the environment to enhance safety and security, meaning adapting to the physical processes at work in a landscape for producing solutions that can be low tech and low cost with minimal impact on ecologies and environments (Box 1.5).

An influential critique of structural engineering approaches came from the work of American geographer Gilbert White (1936, 1942/1945). Developing what became known as the behavioural paradigm, White argued that natural hazards are not purely geophysical phenomena outside of society, but are also linked to societal decisions to settle and develop hazard-prone land, often for economic motives (Box 1.6). He was critical of the undue reliance placed on wall-type engineered structures to attempt to control floods and

Box 1.5 Example of Low Tech for Living With Earthquakes

Buildings can be designed to withstand earthquakes. Engineering technology has been successfully employed to build high-rise buildings that use base isolators (effectively a shock-absorbing and moving foundation to a building) and also structural elements to keep the building together, as commonly seen across Japan and California. Building codes determine levels of earthquake-resistant construction to prevent building collapse and help save the lives of those working or living within the building – and of those outside and nearby. Whilst this process incurs expenses, it is generally cost effective as soon as an earthquake hits.

How does this relate to earthquake-resistant homes for poorer people and structures with typically low levels of formal engineering? Low-tech solutions can be applied. When a 7.8 Mw earthquake hit Nepal in 2015 killing 8,790 people and affecting over 8 million people (nearly a third of the country's population), many family homes were destroyed. Organisations such as the Earthquake Housing Reconstruction Project (EHRP) worked with village organisations to foreground carpentry and earthquake-resistant techniques to reconstruct housing. Many small furniture enterprises were supported by funds to enable local communities to rebuild their homes and livelihoods.

The Kashmir earthquake of 2005 killed over 70,000 people and left over 3 million people homeless. The NGO Article 25 worked with Muslim Aid to develop earthquake-resistant buildings. By hosting workshops to enable local people to take part in the rebuilding process, many were able to rebuild their own homes. New designs were made using simple seismic mitigation techniques and traditional construction techniques, materials and skills.

Figure 1.9 Photo from Article 25: low-tech engineering to rebuilding housing in North Pakistan.

Source: www.article-25.org/pakistan-earthquake-housing

Through these low-tech options for seismic-resistant buildings, many globally are now in safer housing for future earthquakes, although there is a long way to go to make everyone safe. Low tech can provide the means for many to benefit from local skills, resources and funding (see Figure 1.9).

Box 1.6 Gilbert White and Planning for Disasters

Following the 1936 Flood Control Act, the US Army Corps of Engineers constructed major flood-related engineering works (dams and levees) throughout the country. This strategy appeared rational during the 1930s and 1940s, due to growing confidence in the relevant scientific fields of meteorology and hydrology, political demands for greater development of natural resources and land, and the availability of capital for public works.

Gilbert White was at first a lone voice in arguing that this approach to engineering flood control works should be integrated with governance methods like land use planning, to produce more comprehensive floodplain management. Urban development of flood-prone land was attributed to 'behavioural' faults, including a misperception – by developers and homeowners alike – of the risk/reward balance that exists when hazardous land is occupied for economic gain. Within what he thought of as 'developing countries', other forms of behaviour, such as deforestation or the overgrazing of land, were

Figure 1.10 Gilbert F. White.

Source: Ken Abbott/UCB

considered irrational as they contributed to disaster. The universal purpose of addressing hazards and reducing disaster risk was to prevent temporary disruptions to 'normal' life.

'It has become common in scientific as well as popular literature to consider floods as great natural adversaries which man seeks persistently to overpower. . . . This simple and prevailing view neglects in large measure the possible feasibility of other forms of adjustment.' Taken from Human Adjustment to Floods, 1945. University of Chicago Department of Geography Research Paper No. 29. Chicago: University of Chicago Department of Geography.

Although White's ideas gained some attention, 'technical fix' solutions dominated in many places. It was believed that, in the fullness of time, the transfer of technology to the 'developing' world, as part of an overall 'modernisation' process, would solve its problems too. Many centralised organisations were created because only government-backed bodies had the financial resources and expertise needed to apply science and engineering on the required scale. The UN, in particular, sprouted a number of agencies responsible for international disaster mitigation at this time.

other hazards in the USA and so he highlighted the social perspective of human ecology. This interpretation stems from earlier work in the 1920s, notably by Harlan H. Barrows, who applied concepts from ecology – such as interconnectivity, spatial organisation and system behaviour – to the functioning of human communities (Barrows, 1923). The basic idea was that the interactive nature of human–environment relations defines the well-being of both. In other words, human ecology links the physical and social sciences to provide a more balanced approach to resolving the conflicts that arise between human needs and the sustainability of the environment.

Building on White's work and ideas, Hewitt (1983) suggested that the behavioural paradigm had three main thrusts:

- Despite some acknowledgement of the role of human behaviour in the occupation of hazard-prone land, the prime aim was to contain nature through engineering works, such as flood embankments and earthquake-'proofed' buildings, allied with land use controls.
- Other measures included field monitoring and the scientific explanation and statistical assessment of geophysical processes. Modelling and prediction of damaging events followed the introduction of advanced technical tools, for example remote sensing and telemetry.
- Priority was given to strengthening bureaucracy for disaster planning and emergency responses, mostly operated by the armed forces. The notion that only a military-style organisation could function in a disaster area was attractive to governments because it emphasised the authority of the state when re-imposing order.

The paradigm makes a number of assumptions about how individuals behave that have implications for assigning responsibility for disasters and for communicating the potential for disasters to occur. In focusing on people as individuals who make decisions based on correct or incorrect information, the paradigm assumes that decision-making is an individualised effort unswayed by ideologies that seek to disseminate a particular worldview or by advertising that seeks to sell a particular lifestyle. It further assumes that people have the opportunities to obtain and retain comprehensive knowledge and to act on that knowledge. This means that simply disseminating the correct likelihood of a particular risk – what has been termed bridging the credibility gap – is not a solution. Nor does blaming incorrect decision-making help us to identify why disasters impact some and not others.

A second critique of structural engineering approaches has emerged with the development of a *wicked problems approach*. This approach has its origins in the social protests of the 1960s in the USA. Here, the money spent on the most ambitious and costly scientific challenges to date – placing people on the moon – was compared with the state's inability to deal with tumultuous social problems such as structural racism, widespread poverty and urban deprivation. As part of an effort to mitigate this criticism, the National Aeronautics and Space Administration (NASA) funded a series of activities that aimed to transfer expertise as well as technology from the space programme into urban design, planning and management.

At a weekly seminar held in 1967 at the University of California-Berkeley, mathematician Horst Rittel, who taught in the College of Environmental Design on the intersection of Planning, Architecture, Engineering and Policy, presented on the differences between social and what he called scientific, or technical, problems. Here, and in a later paper co-authored with Melvin Webber (Rittel & Webber, 1973), Rittel argued that while the natural sciences and engineering can use hypothesis testing to work towards solving what are called 'tame problems' as though they were 'puzzles', the social sciences operate very differently. This is because they deal with complex and power-laden social issues and with complicated individuals, so that there are competing explanations as to what the puzzle itself is. In the planning field there is no such thing as a 'solution' that can be found for problems; there is instead an effort to resolve what Rittel and Webber call 'wicked problems'. These have multiple and conflicting criteria for defining what a solution would be, have solutions that create further problems and should be approached not as if they can be solved by the finding of a truth, but as situations within which the consequences of decisions made impact the lives of people. By this definition, and the characteristics in Box 1.7, all societal problems are, in effect, 'wicked problems', which raises the issue regarding why they should be termed 'wicked problems' rather than just 'problems'.

Box 1.7 Ten Characteristics of a Wicked Problem, From Rittel and Webber (1973)

1. There is no definite formulation of a wicked problem.
2. Wicked problems have no stopping rule.
3. Solutions to wicked problems are not true-or-false, but good-or-bad.
4. There is no immediate and no ultimate test of a solution to a wicked problem.
5. Every solution to a wicked problem is a 'one-shot operation'; because there is no opportunity to learn by trial-and-error, every attempt counts significantly.
6. Wicked problems do not have an enumerable (or an exhaustively describable) set of potential solutions, nor is there a well-described set of permissible operations that may be incorporated into the plan.
7. Every wicked problem is essentially unique.
8. Every wicked problem can be considered to be a symptom of another problem.
9. The existence of a discrepancy representing a wicked problem can be explained in numerous ways. The choice of explanation determines the nature of the problem's resolution.
10. The planner has no right to be wrong.

Source: Rittel, Horst WJ, and Melvin M. Webber. 'Dilemmas in a general theory of planning' *Policy Sciences* 4.2 (1973): 155–169.

While Rittel and Webber (1973) noted only in passing conflicts over environmental management, the concept of wicked problems has been taken up extensively across disciplines. Air and water pollution, hazardous waste sites and watershed management have all been researched as complex constellations of interlinked problems that involve diverse cohorts and communities with widely varying understandings, as well as differing goals, values and life experiences, and which thus have no single solution.

Indeed, some environmental problems, such as human-caused climate change and land degradation, are so large in scope and impact so many people that the concept of *super wicked problems* (Levin et al., 2012) has been proposed to up the ante in rhetoric. In addition to the characteristics noted previously, we are running out of time to adequately address these problems, since significant aspects might not be reversible in the near term or medium term; sometimes there is no central authority to manage the definition of what these problems are, or those authorities achieve little; the political infrastructures that we need are geared to the short term and thus push the work of dealing with long-term problems into the future; and those seeking to solve the problem are also causing it, at individual and collective levels, especially through misplaced assumptions around, for instance, economies of production and consumption.

3. Underdevelopment and Vulnerability

Both the behavioural paradigm and the wicked problem approach outlined earlier deal with social issues at a superficial level. That is, they take for granted that when a crisis occurs and is seen to impact particular individuals and groups, then an explanation of what has occurred must lie within the actions, and inactions, of those groups. What is not present in either approach is a broad understanding of how individuals and groups are situated within complex, dynamic societies that place constraints on individual actions and capacity to act. A broader understanding of society, including its relationship with nature, is the focus of *social theory*.

This is a highly diverse field built from (a) abstract theorisations and grounded observations that describe society and advance explanations for social phenomena, events and problems, and (b) political activism that focuses on how societies *should be*, either through a defence of existing order or a call to transformation. The latter is often referred to as *critical social theory* and includes various forms of, and responses to, Marxist, feminist and postcolonial theories on society and society-nature relations. Key themes within social theory are power, inequality, exploitation and suffering, but also solidarity, care and ethics.

As with the field of environmental hazards, what academia takes as social theory emerged from the Anglo-European Enlightenment and came to fruition in the 19th century within and outside of universities. Though very much concerned with the environmental and health impacts of urbanisation, agricultural land use and industrialisation, social theory was not taken up by scholars working on hazards until the end of the 20th century. This lack of attention might be explained by the increasing separation between the earth sciences, which remained focused on physical processes and events, and the social sciences, which tended to treat nature as a backdrop to social events and relations. Additionally, the engineering solutions so often proposed to tackle environmental hazards were regarded as 'quick fixes' with no need for a 'politicised' debate.

In the 1970s, however, the field of environmental hazards was radically recast. The end of World War II and reconstruction, the collapse of colonial empires and the onset of a Cold War, protests for civil rights and the end of race, class and gender-based inequalities, as well as a burgeoning environmental movement all propelled critical social theory on the making of disasters into hazards research and teaching. Key to this approach is explaining disasters as arising principally from the combined workings of colonialism and capitalism over centuries, and the marginalisation of disadvantaged people in environments that have been profoundly altered by resource overextraction and the urban and transport infrastructures that facilitate this, the introduction of commodity crops through land clearances, the making of people into landless labourers and the pollution of environments. All of these allowed for an 'economic progress' to emerge in the colonising countries. Yet, development in these countries was very much at the expense of *underdevelopment* in the colonies.

The unequal trading conditions established under colonialism remain largely unchanged. Large trading companies with access to global finance still control the imports and exports of ex-colonies. Ongoing underdevelopment, then, and the economic insecurities it proliferates, cannot be solved by increased investment, a greater output per capita or higher GDP. Nor can low-income locations be examined as though they existed in isolation or were somehow on their way to becoming as 'developed' as high-income places. Wealth and poverty are inextricably linked.

Environmental events and processes may 'trigger' rapid-onset hazards, but these become disasters amid these deeply rooted and long-standing problems, especially poverty. Contrary to the behavioural paradigm and wicked problem thinking, the limits to individual actions imposed by powerful financial and political interests are acknowledged as part of the problem.

In consequence, human vulnerability – a feature of the poorest and the most disadvantaged people in the world – became an important concept for understanding disaster impacts (Blaikie et al., 2014; Wisner et al., 2004). The poorest sections of society are forced to overuse the land and other resources, so that this behaviour cannot be regarded as 'irrational' nor can they be blamed. Indeed, farmers and those working with herds have a substantial repertoire of adaptations to dynamic environmental conditions – what Wisner et al. (2004) called a 'people's science' – that is gradually undermined by colonialism and capitalism. Specifically, rural landlessness and migration to unplanned, hazard-prone cities are said to be the inevitable outcomes of these destructive systems, which are the root cause of disaster.

The underdevelopment model was summarised in the work of Wisner et al. (2004), who envisage disasters as the outcome of a direct clash between the socio-economic processes that produce human vulnerability and the physical processes that cause hazards. There are several key points:

- Disasters are caused largely by human exploitation rather than by physical or technological processes. Macro-scale root causes of vulnerability lie in the

economic and political systems that exercise power and influence, both nationally and globally, and result in marginalising poor people.
- Ongoing pressures, such as chronic malnutrition, disease, armed conflict and other forms of violence channel the most vulnerable people into unsafe environments, such as flimsy housing and steep-sloped and flood-prone areas, either as a rural proletariat (dispossessed of land) or as an urban proletariat (forced into informal settlements). Effective local responses to hazards are limited by a lack of resources at all levels.
- 'Normality' as stability is an illusion. Frequent disaster strikes are the 'normality'. They are characteristic, rather than unusual, and reinforce socio-economic inequalities.
- Disaster risk reduction in poor places depends on fundamental changes and a redistribution of wealth and power. Modernisation relying on imported technology and 'quick fix' measures must backfire. Instead, economic autonomy and the protection of local knowledges and adaptations are seen as more successful.

Yet underdevelopment and other processes leading to vulnerability and hence disasters are not unique to colonial or capitalist societies (see Figure 1.11). Disasters occurred prior to colonisation and after independence, within and irrespective of postcolonialism. Marxist, socialist and communist

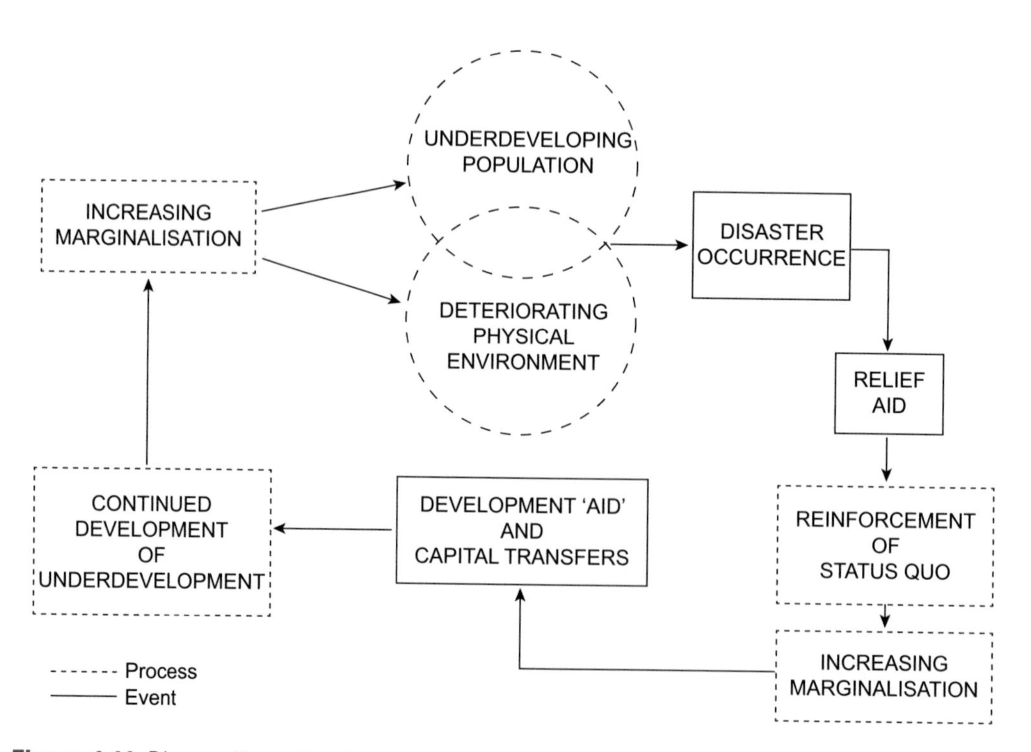

Figure 1.11 Diagram illustrating the process of marginalisation and the relationship to disaster.

Source: Susman, O'Keefe and Wisner, 1983, p. 279. Susman, Paul, Phil O'Keefe and Ben Wisner. 'Global disasters, a radical interpretation'. *Interpretations of Calamity: from the viewpoint of human ecology* 4 (1983): 263.

societies experience disasters as do places with capitalism, neoliberalism and social democracy. Monarchies, republics, democracies and dictatorships, and their combinations such as constitutional monarchies, are far from immune to vulnerability, disaster risk creation and disasters.

4. Feminist and Intersectional Approaches

Feminist approaches have no single set way of looking at the world or research techniques for investigating it, but rather are collectively concerned with researching and designing solutions to inequalities and inequities. The disparities are often gender-based, and feminist approaches have certainly picked up on ways in which the contributions and potential of women have been largely ignored not only in the hazards and risk field, but also across academia, policy and practice more broadly. Feminist scholars have produced research that foregrounds the often overlooked everyday experiences of those identified as women and girls before, during and after disasters, as well as recognising the importance of including those experiences in dealing with disasters. The term 'double disaster', for example, refers to what happens when the losses associated with a hazard (such as destruction of homes and livelihoods) produce secondary impacts on those identified as women and girls who already are marginalised and might well bear the brunt of managing how a family responds to it (Bradshaw & Fordham, 2015). Secondary impacts can include increased economic insecurity, restricted mobility, sexual exploitation and abuse and gender-based violence. These concerns do not only apply to those identified as women and girls, but also can adversely affect those identified as men and boys.

Feminist approaches to environmental hazards can also address how much of the reporting on hazards and disasters, the relief efforts of organisations and descriptions in some academic subfields is gender biased. Women are portrayed in a reductionist and pejorative way as 'Third World Women' while men can be seen as not needing help. Lumped into one group, those identified as women are caricatured as the impoverished victims of patriarchy and/or capitalism while men are lumped into one group as leaders, controllers and perpetrators. Feminist approaches deconstruct these assumptions of homogeneity.

Gender-based inequalities and inequities rarely emerge independently of other forms of discrimination and bias. *Intersectionality* is an approach developed by Crenshaw (1989) and is often used by feminist scholars to understand how the unequal and inequitable treatment of people because of sexism, racism, ageism and ableism, alongside other forms of discrimination that target sexual orientation, religion and ethnicity among other characteristics, intensifies the vulnerability of individuals and their capacity to plan for risk and respond to disasters. These systematic inequalities and inequities are closely tied to economic security insofar as they shape the conditions within which people are able to pursue livelihoods, borrow money, pay a fair rate for goods and services and take their own actions to reduce vulnerabilities of themselves and those around them (Arora et al., 2020; Jacobs & Hens, 2018; Ryder, 2017).

5. Settler Colonialism, Postcolonialism and Decolonialism

Settler colonialism refers to the displacement of existing peoples of a colonised region and their replacement by settlers who transform environments to establish a permanent society and connect up governance and trade routes between the 'home country' and colonised region. The land set aside for the displaced peoples is often regarded as marginal and even unusable, producing vulnerabilities, including exposures to environmental hazards that are passed on from generation to generation. Industrial hazards can accumulate in these places, as polluting sectors are often located next to the sites where peoples have been displaced to (Box 1.8).

Tracing its roots to *anti-colonial* revolutions, protests and commentaries that came from those suffering colonialisation, the 'post' in postcolonialism implies leaving behind the formal apparatus of colonialism. Postcolonialism is also a coming-to-terms with what happened under the formal colonialism apparatus, and its many political, economic, cultural and environmental legacies and ramifications. Postcolonialism is thus much more than an approach to viewing and researching the world. As a form of critical social theory – what can be called 'postcolonial

Box 1.8 Canada's Chemical Valley

Forty per cent of Canada's chemical industry is located adjacent to the Aamjiwnaang First Nation's reserve, a space subject to treaty negotiation by the British Crown in 1827. Over 60 refineries and chemical plants stretch 30 km along the St Clair River and produce the worst air pollution in the country. Oil was located here in 1858, and that discovery was quickly followed by the construction of a large concentration of oil wells. Since the 1940s, petrochemical, chemical and polymer industries have released substantial toxic waste into the soil and water, including the contaminants mercury, nickel, arsenic, cadmium, cyanide, PCBs (polychlorinated biphenyls), polynuclear aromatic hydrocarbons and organochlorine hydrocarbons.

In September 2008, the Ontario Ministry of the Environment and Climate Change (MOECC) established an air quality monitoring station in the area in partnership with the Aamjiwnaang First Nation. The community here has fought to mitigate air pollution through the courts, including a (discontinued) lawsuit against the Province of Ontario's regulation of industrial emissions, which looks only at the contaminants produced by individual facilities rather than the cumulative effect on air quality of a large number of polluters. Noting the long history of resistance to the colonial governance of Canada by indigenous populations and the difficulties faced by them in dealing with a legal system that systematically marginalises colonised people, Bagelman and Wiebe write, 'the toxic lifestyle that is endemic to the environment inside the reserve is largely rendered invisible, or at the least pushed out of public sight, and removed from the forefront of Canadian consciousness outside the reserve space' (2017, p. 80).

REFERENCES

The Aamjiwnaang News Centre can be accessed directly here: www.aamjiwnaang.ca/. For more on exposure to environmental hazards in Chemical Valley see Booth, Annie. 'Everyday Exposure:

Indigenous Mobilization and Environmental Justice in Canada's Chemical Valley'. *The Canadian Journal of Native Studies* 37.2 (2017): 220–222. For more on the politics of Chemical Valley see Bagelman, Jen and Sarah Marie Wiebe. 'Intimacies of global toxins: Exposure & resistance in "Chemical Valley"'. *Political Geography* 60 (2017): 76–85. For more on community resistance to cumulative environmental hazards see Wiebe, Sarah Marie. *Everyday exposure: Indigenous mobilization and environmental justice in Canada's chemical valley*. UBC Press, 2016.

theory' – it embodies the imperative to actively improve the world. What is more, ideas and practices labelled 'postcolonial' extend far beyond academia, and indeed there may be little desire to connect these with academic research.

This is a wide-ranging effort, and it has had profound impacts on how the history and present of academia itself is understood. Certainly, a growing number of studies on environmental hazards draws on postcolonial ideas and practices, or is at least aware of them. Key challenges that postcolonialism has presented to those working on environmental hazards are the following:

- Recognising that the theories, ideas and techniques taught in this field have emerged from an Anglo-European context that includes colonialism and male domination. Thus the material has served the interests of particular groups and is neither unbiased nor universally applicable.
- Accepting the urgent need to diversify knowledges of hazards by including voices and experiences of those experiencing the worst disaster impacts in tandem with practices from other regions of the world, including indigenous peoples.

For some, this can become a *decolonial* approach, as it connects with the proliferation of networks of mutual support to challenge and overcome the forms of social organisation creating and maintaining vulnerabilities (including exposures) to environmental hazards.

6. Complexity

By the late 1990s, there was growing acceptance of the truism that interactions between natural environments and human perceptions, actions and organisations are part of a genuinely 'complex' system (Mileti, 1999), where lines of causality were no longer clear to distinguish. In an environmental hazard, and any subsequent disaster, the whole system of both environment/environments and society/societies is interconnected and inseparable and each element influences the other– exactly as various systems theories had expressed for decades and exactly as many indigenous societies had lived for millennia. The idea of a systems theory is that the components of a system yield more than the sum of its parts. While a system, by definition, is complex, complex systems theories emerged focusing on interactions among the system elements and the effects of these interactions. Complexity can be seen to occur during a disaster and as exemplified by the Bam earthquake example (see Box 1.9).

A complexity-based perspective integrating society and the environment is hardly a revelation, being rather a basis

Box 1.9 Complexity in the Real World: The Bam Earthquake

On 26 December 2003, an earthquake struck southern Iran at 5:26 a.m. local time near the city of Bam (Figure 1.12). The Mw 6.6 earthquake magnitude was not particularly large since such earthquakes are recorded almost every week worldwide, but the disaster impact here ended up being huge. Most of the 140,000 residents were asleep at the time, with little chance of escape as their dwellings crumbled, and an estimated 26,000 people died (Bouchon et al., 2006). There was almost total destruction of the ancient citadel of Bam (see Figure 1.13), the world's largest adobe building complex, parts of which were 2,400 years old. About 70% of all buildings collapsed completely, together with the three main hospitals and the fire station. It was estimated that 90% of the building stock of the city was damaged by up to 60%–100%, whilst the remaining buildings were damaged by 40%–60% (EERI, 2005). What were the reasons for this exceptional loss?

There were a series of uncertainties and feedback loops that made this disaster complex, as can be identified for any disaster. The earthquake was fairly shallow, with a hypocentre about 7 km below the ground surface that created a rupture along a 15 km stretch of the Bam Fault. The city was subjected to just 15 seconds of shaking, but a strong-motion seismometer located within the city recorded very high peak ground accelerations (Ahmadizadeh & Shakib, 2004). The earthquake damage zone was limited to an area of 16 km^2, demonstrating intense localised shaking, although it is not fully clear why. Peyret et al. (2007) suggested that the rupture occurred not on the previously known Bam Fault but about 5 km to the west, at a location where no surface

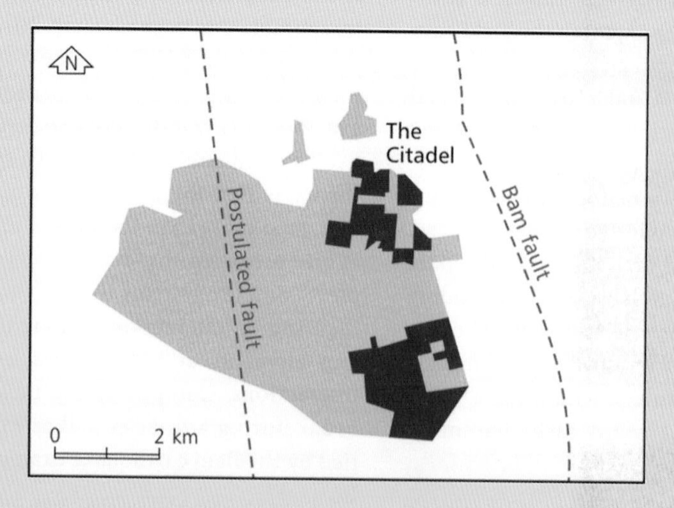

Figure 1.12 Location map of the city of Bam, Iran. The city was largely destroyed in the December 2003 earthquake.

Source: After Petley (2009)

Figure 1.13a The Arg-e-Bam before the earthquake.

Source: © Getty Images/Dea/W. Buss

Figure 1.13b The Arg-e-Bam after the earthquake.

Source: © Getty Images/Atta Kenare

evidence of faulting existed. This would mean that the seismic waves occurred almost directly under the city. Conversely, Bouchon et al. (2006) argued that the rupture initiated on a fault to the south of the city and propagated seismic waves straight at the urban area. Whichever process prevailed, the net result was an unusually high intensity of ground shaking.

In light of the uncertainty and unknown intensity of shaking that could be generated for an earthquake in the region, the poor quality of the building stock, including the widespread presence of adobe-constructed buildings in the Bam citadel (EERI, 2005) without adequate seismic resistance measures, meant severe structural failures, resulting in the significant death toll. Langenbach (2005) noted that, although the majority of these older buildings did collapse, only three people were trapped in the rubble within the citadel itself. Overall, most of the fatalities were recorded in buildings that were fewer than 30 years old, and the traditionally built and unrepaired buildings fared much better than those either constructed or repaired in the late 20th century. Renovation of some traditional adobe buildings was found to be deficient, due to the use of cement containing too much sand (Kiyono & Kalantari, 2004). Adobe buildings have heavy roofs and, in the absence of any reinforcement of the walls, the roofs fall in if a single wall collapses. Many of these structures had been further weakened by termite activity, a situation reflecting more failures in the implementation and enforcement of the Iranian seismic building code. In turn, this neglect may have been because the area had not suffered a large earthquake in historical times and was not considered to be at high seismic risk. In other words, a combination of code enforcement failures interacted with weaknesses in the buildings and earthquake mechanics to create a major disaster.

The emergency services were inadequately prepared for an effective response and suffered their own losses. Initial search-and-rescue operations were hindered by the destruction of key facilities, including the main fire station, which collapsed, crushing the fire engines and killing some firefighters. The three local hospitals, all the urban health centres and 95% of the rural health centres were destroyed (Akbari et al., 2004). This demonstrates the value of land use planning and the need to invest in key infrastructure. One-fifth of health professionals in the area were killed. Many of the remainder were incapable of providing full support due to injuries and post-traumatic stress. Many casualties with severe injuries failed to receive urgently needed medical attention and died. Weather data are missing from Bam on 26–27 December due to the earthquake, but the day before, the minimum nighttime temperature was just above freezing, while the Kerman weather station 190 km away recorded below freezing temperatures for 22–27 December. Some of the trapped people who died might have succumbed to hypothermia while awaiting rescue during the first night (Moszynski, 2004).

The first rescue teams did not arrive until nightfall on the first day, 12 hours after the earthquake, and were unable to work properly until daylight the next morning. As often happens, most of the trapped people were rescued by other survivors working through the rubble with their bare hands. The subsequent arrival of international rescue teams was of limited practical value. A total of 34 international teams eventually arrived in Bam but, in total, these teams saved just 22 people. In contrast, it is estimated that local people recovered over 2,000 people from damaged buildings in the first few hours afterward.

Further complications arose due to poor cooperation between the international aid agencies and the Iranian army. Under legislation passed in 2003, the Iranian Red Crescent Society was mandated to play the lead role in disaster response, but this led to tensions with the Iranian army, especially over the use of aircraft. Interviews with primary care nurses suggested that medical care was hindered by a lack of prior training and an inability of health workers to collaborate effectively, especially when overseas health teams were involved (Nasrabadi et al., 2007). These many facets of the disaster, especially those after the shaking, demonstrate the complexities, uncertainties, feedback loops and interdependencies of aspects that may not seem apparent prior to a disaster.

REFERENCES

Kiyono, J., & Kalantari, A. (2004). Collapse Mechanism of Adobe and Masonry Structures During the 2003 Iran Bam Earthquake. *Bulletin of the Earthquake Research Institute*, 79, 157–161.

Akbari, M.E., Farshad, A.A., & Asadi-Lari, M. (2004). The devastation of Bam: an overview of health issues 1 month after the earthquake. *Public Health, 118*, 403–408.

for human living. It suggests that, for such a disaster to develop, a series of cumulative events and processes must occur. One way of representing this is by the Swiss cheese model of disaster (Box 1.10 and Figure 1.14). In the case of Bam, some events in the chain, like the timing of the earthquake, could not have been modified by human action, but factors such as the termite infestation could have been. If certain factors had operated in different ways – if the earthquake had occurred at another time of day, if the rupture had propagated from the north, if the buildings had offered greater resistance to seismic stress, if nighttime temperatures had been warmer, if the coordination of the response had been stronger and quicker or if long-term preparedness and seismic resistance had been enacted – then the outcome would have been quite different.

Complexity is difficult to define yet is ever present. Whether it is the way that insect colonies function building large intricate structures, how the brain works and its neural plasticity or how the internet and national economies operate, there are a number of common properties that complex systems have. They all exhibit nontrivial emergent and self-organising behaviours (Mitchell, 2009, p. 13) including the following:

- Collective action of vast numbers of components leading to changing and evolving patterns of behaviour.
- Signalling and information processes for and from both their internal and external environments.
- Adaptation and flexibility, changing behaviour to improve survival chances and success through learning or evolutionary processes.

A complexity science approach effectively means accepting and adopting the daily life experience of being part of a co-evolving

Box 1.10 The Swiss Cheese Model

The Swiss cheese model of disaster was proposed by Reason (1990) following the Bhopal, India, chemical leak in 1984, the Challenger shuttle explosion in Florida in 1986 and the Chernobyl, USSR, nuclear power plant disaster in 1986. It aims to explain causation and to contribute to risk analysis and risk management. Reason (1990) examined the strategies put in place by organisations to prevent such disasters and concluded that they could be thought of as slices of Swiss cheese lined up one behind the other (Figure 1.14). Holes in the individual pieces of cheese were considered weak points in each 'line of defence', and it was argued that a disaster occurs when the holes in all the slices align. If even one hole is out of line, then the defence works and a developing disaster is blocked. That is, if one defence lapses then other defences exist to help prevent a single point of failure. In addition, the model represents the errors that can accumulate and lead to adverse events. Consequently, the model is also called the 'cumulative act effect'.

This model of disaster causation has been applied in the aviation industry, which is highly safety conscious (Reason et al., 2006). Many barriers are erected to prevent disasters, on the principle that no single component should cause the system to fail. Measures include conservative aircraft design, careful selection and training of pilots, well-established emergency procedures and a culture of learning to improve. This

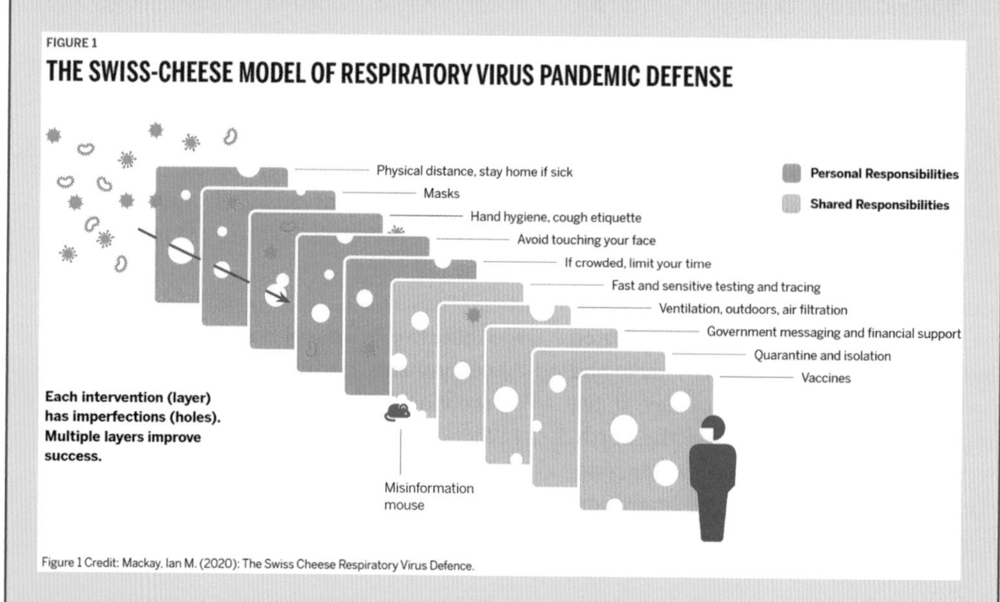

Figure 1.14 The Swiss cheese model of pandemic defence. A disaster can occur only when several vulnerable circumstances arise simultaneously. The vulnerabilities are represented by holes in the cheese; disaster occurs when holes line up.

Source: Siobhan Roberts, 5 December 2020, www.nytimes.com/2020/12/05/health/coronavirus-swiss-cheese-infection-mackay.html

approach has shown that, even when a disaster can be attributed to a single mistake, there is a surrounding framework that should prevent a disaster.

The model can also be applied to disasters such as the COVID-19 pandemic (see Figure 1.14) to highlight that no single intervention is sufficient to prevent the spread of the virus. Each intervention has holes, and both personal and shared responsibilities via actions are needed to prevent the spread of the virus. By adding protective layers, such as emerging infection surveillance, vaccinations, face coverings, lockdowns, tracking and tracing and financial support – none of which can be perfect and all of which have adverse consequences – an impenetrable barrier will eventually form, but it requires all of these actions, not just one. This shifts the thinking that there is a magic solution to the problem. It is argued that there is a 'misinformation mouse' busy eating new holes for the virus to pass through, including non-expert or unofficial information; for example, anti-vaccination groups.

system that self-organises, learns and adapts to its dynamically changing situations, inextricably linking environments and societies. Given the many uncertainties related to environmental hazards and disasters, complexities depend on locations, timings, scales, societal characteristics, environmental conditions and the political, cultural and natural contexts of the phenomena and entities involved (see Figure 1.15).

Consequently, hallmarks of dealing with environmental hazards are adaptability, flexibility and acceptance of an ever-changing situation. Many aspects of predictability should not be expected, but the end goals of fairness and equitability are part of the overall approach – perhaps moving from the buzzword of 'complex adaptive systems (CAS)' to 'complex equitable systems' or similar. The concept of systemic risk is based on the notion that the risk of an adverse outcome of a policy, action or hazard event can depend on how the elements of the affected systems interact with each other. This can either aggravate or reduce the overall effect of the constituent parts.

Given this basis, when dealing with environmental hazards, feedback is required to monitor the progress of actions. This feedback should be within the system to enable the system to self-correct or modify behaviour, learning from experience and improving. Complexity theories and models can provide a framing or tool for practitioners, policy-makers, managers and researchers to reflect collectively on how they are trying to solve problems, improve awareness and seek to always do better.

E. THE ORGANISATIONAL CONTEXT

Current efforts to mobilise a global governance of environmental hazards and deal with disasters have their origins in the 19th century, and in particular the work of colonial powers and philanthropic organisations to maintain the movement of troops and protect and grow trade by controlling infectious disease. The International Sanitary Conferences held between 1851 and 1938, for example, promised an organised response to infectious diseases

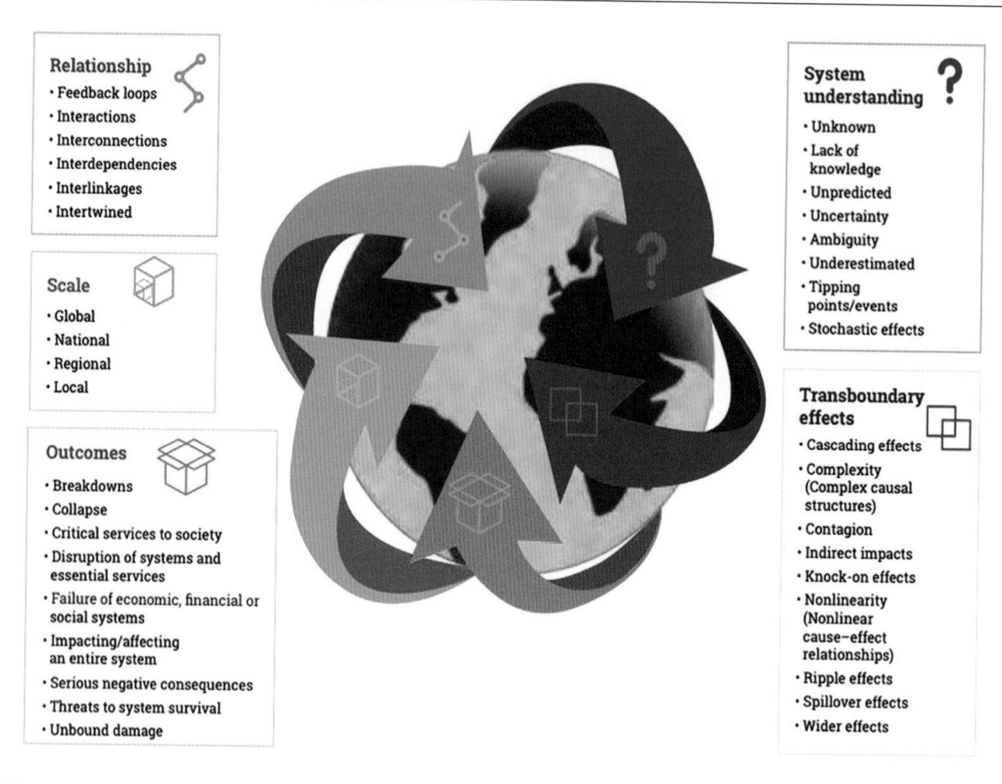

Figure 1.15 Terminology for key attributes of systemic risk.

Source: Based on Sillmann et al. (2022)

including cholera, yellow fever and the bubonic plague. The League of Nations' Health Organisation (1919–1946) drew heavily on this precedent, as did philanthropic initiatives carried out by organisations such as the Rockefeller Foundation.

The League of Nations (1920–1946) was superseded by the United Nations (1945 onwards) as the world's major intergovernmental organisation. The personnel and data collecting process and files of the League of Nation's Health Organisation were subsequently incorporated into the World Health Organization (WHO), established in 1948 as a specialised agency of the United Nations responsible for international public health. In line with the development of colonial medical practice, the WHO determined a policy of disease containment, with an emphasis on venereal disease as well as malaria and tuberculosis, and the support of maternal and child health, nutrition and environmental hygiene. The WHO relies on contributions from member states (both assessed and voluntary) and private donors such as the Gates Foundation. To combat infectious disease, the WHO works with other international organisations such as the World Organisation for Animal Health (OIE) and the Food and Agricultural Organisation (FAO).

The United Nations (UN), within which the WHO sits, has taken on the responsibility for creating an international framework for disaster reduction (see Figure 1.16).

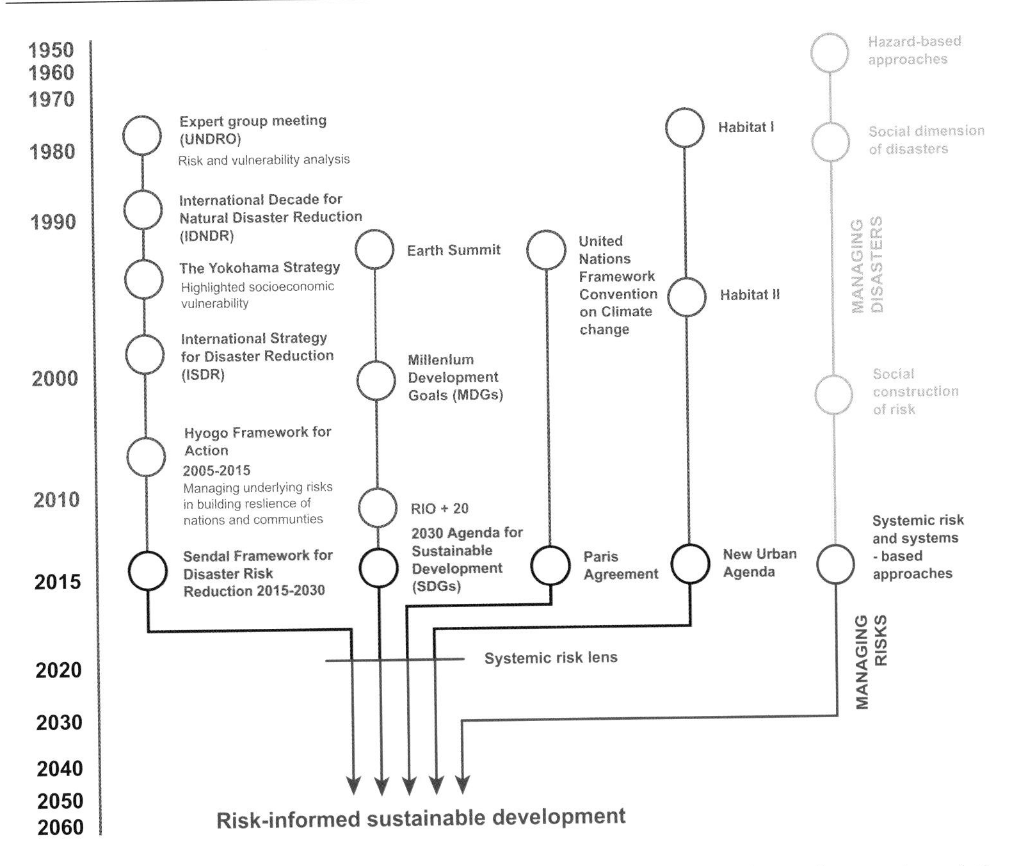

Figure 1.16 Background to UN endeavours on DRR – disaster risk reduction – a journey through time and space.

Source: UN Global Assessment Report on Disaster Risk Reduction, 2019, https://gar.undrr.org

The most recent formal process began in 1990 with the International Decade for Natural Disaster Reduction (IDNDR), a programme driven by concerns that disaster losses threatened especially what was then termed the 'developing countries'. At the 1994 midterm World Conference on Natural Disaster Reduction, held in Yokohama, Japan, several policy failings were highlighted, including an excessive emphasis on scientific solutions, reliance on the transfer of hazard mitigation technologies to 'developing' countries and a relative neglect of the social (including economic and political) aspects of disaster.

In December 1999, the Secretariat for the United Nations International Strategy for Disaster Reduction (UNISDR) was created to act as the focal point within the UN system for coordinating disaster reduction and to ensure that disaster reduction would become integral to all sustainable development, environmental protection and humanitarian policies. UNISDR consisted of a wide array of partnerships comprising governments, intergovernmental and nongovernmental organisations, financial institutions, scientific and technical bodies and the private and non-profit sectors. In May 2019, UNISDR changed its acronym to

UNDRR to represent its latest name the United Nations Office for Disaster Risk Reduction.

In the meantime, further developments took place at the Hyogo World Conference on Disaster Reduction held in Kobe, Japan, in 2005. This meeting produced the Hyogo Framework for Action (HFA), adopted by 168 countries for actions during 2005–2015 aiming to:

- ensure that disaster risk reduction is a national and local priority supported by strong institutions;
- identify, assess and monitor disaster risks and enhance the provision of early warning;
- increase capacity, knowledge and innovation to build a culture of safety and hazard resilience at all levels;
- integrate all disaster reduction measures – preparedness, mitigation and programmes to lower vulnerability – into sustainable development policies; and
- add risk reduction into the design and implementation of disaster and emergency response, recovery and reconstruction programmes.

The HFA encourages a collaborative strategy that includes national governments, regional bodies and local communities. It was voluntary with no legal status.

In addition to reviewing the implementation of the HFA, the Third UN World Conference on Disaster Risk Reduction was convened in 2015 to establish and adopt a post-2015 framework. The Sendai Framework for Disaster Risk Reduction 2015–2030 (Sendai Framework), named after the conference's host city, Sendai, Japan, was endorsed by the UN General Assembly as the successor to the HFA in 2015. The Sendai Framework identifies four priorities for action and seven targets (see Figure 1.17):

- understanding disaster risk;
- strengthening disaster risk governance to manage disaster risk;
- investing in disaster risk reduction for resilience; and
- enhancing disaster preparedness for effective response and to 'Build Back Better' in recovery, rehabilitation and reconstruction.

Parallel efforts to mobilise action at the international level have reflected concerns about hazard- and disaster-related topics. The United Nations Framework Convention on Climate Change (UNFCCC) is an international treaty, established in 1992,

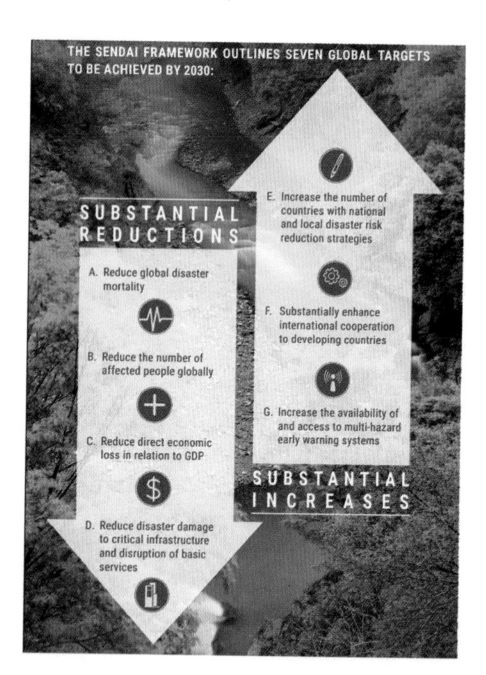

Figure 1.17 The Sendai Framework for Disaster Risk Reduction 2015–2030. Seven Global Targets to be achieved by 2030 (UNDRR).

Source: www.undrr.org/implementing-sendai-framework/what-sendai-framework

supporting the global response to the threat of human-caused climate change. Attempts at effective international laws include the Kyoto Protocol (1997), the Paris Agreement (2005) and the Glasgow Climate Pact (2021). The Millennium Declaration of September 2000 was an ambitious agenda that formalised eight Millennium Development Goals (MDGs) set out by world leaders for reducing poverty and improving lives. When they ended in 2015, the 17 Sustainable Development Goals (SDGs) began, many of which included aspects of addressing environmental hazards and disaster risk reduction.

Overall, the current UN agreements form the 2030 Agenda, including the Sendai Framework, the Paris Agreement, the Glasgow Climate Pact, the Addis Ababa Action Agenda on Financing for Development, the New Urban Agenda and the SDGs. A key focus within the Sendai Framework is putting people's health at the centre of global policy and action to reduce disaster risks. It recognises that the state has the primary role to reduce disaster risk but that responsibility should be shared with others, including local government and the private sector. Progress of the Sendai Framework goals are measured by the Global Assessment Report on Disaster Risk Reduction (GAR), which is published every few years by UNDRR. The GAR adopts a complexity approach to disaster risk as it considers the pluralistic nature of risk in multiple dimensions, at multiple scales and with multiple impacts.

It would be tempting to conclude that disaster risks are different from, and are treated less seriously than, the adverse effects of climate change. This would be inaccurate. Previously, responses to disaster have often been viewed as short-term emergency actions, whilst climate change

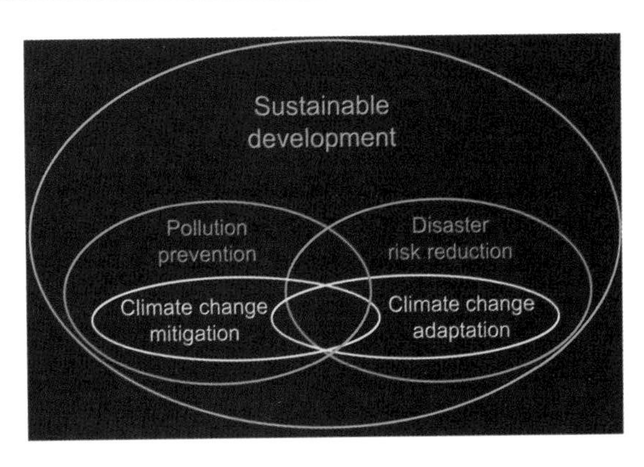

Figure 1.18 The interlinkages among global topics.

Source: Kelman, I. 2017. 'Linking disaster risk reduction, climate change, and the sustainable development goals'. *Disaster Prevention and Management*, vol. 26, no. 3, pp. 254–258.

has been presented as a slow-onset, multi-generational problem. In truth, they are interrelated. Climate change adaptation (CCA) sits as a subset of disaster risk reduction which sits as a subset of sustainable development (Figure 1.18), highlighting the complex and interdependent relationships between the environment and society that can never be separated or seen in an isolated manner. It is important that this complexity is recognised so that synergies can be harnessed to reduce environmental risks from all sources in a more comprehensive and sustainable manner for a better future.

KEY TAKE AWAY POINTS

- The hazard and risk subfield is undergoing significant change in response to complex, socio-environmental conditions but also as academia itself strives to become more inclusive and to value diversity.
- As the hazard and risk subfield changes, the definitions of key terms – including those which are pervasive yet recent arrivals, such as vulnerability, resilience and adaptation – are debated.

- Definitions of key terms hinge on the conceptual approach used by those working on hazard and risk. There is no one dominant conceptual approach, although some are more taken for granted than others when used in a policy and humanitarian setting.
- The ever-increasing capacity of people to impact physical processes means that hazards such as storms, droughts and bushfires that have traditionally been considered 'natural' no longer sit easily in this category.
- While physical processes can produce a range of hazardous conditions, it is only when people are exposed to and vulnerable to these that a disaster can occur.
- Hazards can emerge quickly and with devastating consequences but can also emerge slowly over time, impacting several generations. Understanding the temporal and spatial scales across which hazards unfold and impact helps for tackling vulnerabilities, including through disaster preparation and mitigation.

BIBLIOGRAPHY

Adams, V., Van Hattum, T., & English, D. (2009). Chronic disaster syndrome: Displacement, disaster capitalism, and the eviction of the poor from New Orleans. *American Ethnologist, 36*(4), 615–636.

Ahmadizadeh, M., & Shakib, H. (2004). On the December 26, 2003, Southeastern Iran earthquake in Bam region. *Engineering Structures, 26*(8), 1055–1070.

Arora, A., Belenzon, S., Patacconi, A., & Suh, J. (2020). The changing structure of American innovation: Some cautionary remarks for economic growth. *Innovation Policy and the Economy, 20*(1), 39–93.

Bagelman, J., & Wiebe, S. M. (2017). Intimacies of global toxins: Exposure & resistance in 'Chemical Valley'. *Political Geography, 60*, 76–85.

Barrows, H. H. (1923). Geography as human ecology. *Annals of the Association of American Geographers, 13*(1), 1–14. DOI: 10.1080/00045602309356882

Blaikie, P., Cannon, T., Davis, I., & Wisner, B. (2014). *At risk: Natural hazards, people's vulnerability and disasters*. Routledge.

Bouchon, M., Hatzfeld, D., Jackson, J. A., & Haghshenas, E. (2006). Some insight on why Bam (Iran) was destroyed by an earthquake of relatively moderate size. *Geophysical Research Letters, 33*(9).

Bradshaw, S., & Fordham, M. (2015). *Double disaster: Disaster through a gender lens. In Hazards, risks, and disasters in society* (pp. 233-251). Academic Press.

CDC. (2021). Cholera in Haiti. www.cdc.gov/cholera/haiti/index.html. Accessed 11 May 2022.

Crenshaw, K. (1989). Demarginalizing the intersection of race and sex: A black feminist critique of antidiscrimination doctrine, feminist theory and antiracist politics. *University of Chicago Legal Forum, 1*, 139–167.

Earthquake Engineering Research Institute (EERI). (2005). *Special Earthquake Report —May 2005.*

Grattan, J., & Torrence, R. (Eds.). (2016). *Living under the shadow: Cultural impacts of volcanic eruptions* (Vol. 53). Abingdon, UK: Routledge.

Hasegawa, A., Ohira, T., Maeda, M., Yasumura, S., & Tanigawa, K. (2016). Emergency responses and health consequences after the Fukushima accident: Evacuation and relocation. *Clinical Oncology, 28*(4), 237–244.

Hewitt, K. (Ed.). (1983). *Interpretations of calamity: From the viewpoint of human ecology*. Routledge.

Jacobs, D., & Hens, K. (2018). Love, neuroparenting and autism: From individual to collective responsibility towards parents and children. *Analize Journal of Gender and Feminist Studies, 2018*(11), 102–124.

Langenbach, R. (2005). Performance of the Earthen Arg-e-Bam (Bam Citadel) during the 2003 Bam, Iran, Earthquake. *Earthquake Spectra, 21*(1_suppl), 345–374. https://doi.org/10.1193/1.2113167

Levin, K., Cashore, B., Bernstein, S., & Auld, G. (2012). Overcoming the tragedy of super wicked problems: Constraining our future selves to ameliorate global climate change. *Policy Sciences, 45*(2), 123–152.

Lewis, J., & Kelman, I. (2010). Places, people and perpetuity: Community capacities in ecologies of catastrophe. *ACME: An International E-Journal for Critical Geographies, 9*, 191–220.

Madgavkar, A., White, O., Krishnan, M., Mahajan, D., & Azcue, X. (2020). COVID-19 and gender equality: Countering the regressive effects. *McKinsey Global Institute, 15*. www.mckinsey.com/featured-insights/future-of-work/covid-19-and-gender-equality-countering-the-regressive-effects. Accessed 11 May 2022.

Mileti, D. (1999). *Disasters by design: A reassessment of natural hazards in the United States*. Joseph Henry Press.

Mitchell, M. (2009). *Complexity: A guided tour*. Oxford University Press.

Moszynski, P. (2004). Cold is the main health threat after the Bam earthquake. *BMJ: British Medical Journal, 328*(7431), 66.

Nasrabadi, A. N., Naji, H., Mirzabeigi, G., & Dadbakhs, M. (2007). Earthquake relief: Iranian nurses' responses in Bam, 2003, and lessons learned. *International Nursing Review, 54*(1), 13–18.

Newhall, C. G., & Punyongbayan, R. S. (1996). *Fire and mud: Eruptions and lahars of Mt Pinatubo*. Philippines and London: University of Washington Press.

Okrent, D. (1980). Comment on societal risk. *Science, 208*(4442), 372–375.

Peyret, M., Chéry, J., Djamour, Y., Avallone, A., Sarti, F., Briole, P., & Sarpoulaki, M. (2007). The source motion of 2003 Bam (Iran) earthquake constrained by satellite and ground-based geodetic data. *Geophysical Journal International, 169*(3), 849–865.

Phillips, T. (1834). Mary Somerville. https://artsandculture.google.com/asset/mary-somerville-thomas-phillips/LAH6LDlkQO8NMg

Reason, J. (1990). *Swiss cheese model. Human error*. Cambridge: Cambridge University Press.

Reason, J., Hollnagel, E., & Paries, J. (2006). Revisiting the Swiss cheese model of accidents. *Journal of Clinical Engineering, 27*(4), 110–115.

Rittel, H. W., & Webber, M. M. (1973). Dilemmas in a general theory of planning. *Policy Sciences, 4*(2), 155–169.

Ryder, E. (2017). Political solidarity in the twenty-first century US feminist movement: How far have we come?. *de genere-Rivista di studi letterari, postcoloniali e di genere*, (3).

Showalter, P. S., & Myers, M. F. (1994). Natural disasters in the United States as Release agents of oil, chemicals, or radiological materials between 1980–1989: Analysis and recommendations. *Risk Analysis, 14*(2), 169–182.

Somerville, M. (1854). *Physical geography*. Blanchard and Lea.

Somerville, M. (1873). *Personal recollections, from early life to old age, of Mary Somerville with selections from her correspondence by her daughter, Martha Somerville*. J. Murray.

Tozzi, A., Bouzarovski, S., & Henry, C. (2022). Colonizing the rains: Disentangling more-than-human technopolitics of drought protection in the archive. *Geoforum, 135*, 12–24.

UN Women. (2021). https://interactive.unwomen.org/multimedia/explainer/covid19/en/index.html?gclid=Cj0KCQjw-NaJBhDsARIsAAja6dPUpQypQAIlnps3v61EtssJb2djH7oy-mTnt0gHiwZIV0eWQmLYqilaAnsOEALw_wcB. Accessed 11 May 2022.

UNDRR. (2017). *Sendai framework terminology on disaster risk reduction*. https://www.undrr.org/terminology. Accessed 11 May 2022.

UNDRR. (2019). *Sendai framework terminology on disaster risk reduction*. https://www.preventionweb.net/terminology/hazard. Accessed 11 May 2022.

White, G. F. (1936). The limit of economic justification for flood protection. *Journal of Land & Public Utility Economics, 12*, 133.

White, G. F. (1942/1945). Human adjustment to floods: a geographical approach to the flood problem in the United States. Doctoral Dissertation, The University of Chicago (1942) and republished as Research Paper No. 29 (1945), University of Chicago, IL, Department of Geography.

FURTHER READING

Bobrowsky, P. T. (Ed.). (2013). *Encyclopedia of natural hazards*. London: Springer. A good overview of natural hazards in the context of disasters and DRR.

Carrigan, A. (2015). Towards a post-colonial disaster studies. In E. Deloughrey, J. Didur, & A. Carrigan (Eds.), *Global ecologies and the environmental humanities: Postcolonial approaches* (pp. 117–138). Abingdon, UK: Routledge.

Lewis, J. (1999). *Development in disaster-prone places: Studies of vulnerability*. London: Intermediate Technology Publications. A solid foundation of connecting disaster-related work to development from a vulnerability perspective.

UNDRR. (2022). The early engagement of the United Nations in disaster risk reduction (1970–2000): A brief history. United Nations, Geneva. Describes the beginnings and evolution of the UN in DRR endeavours.

UN/ISDR. (2004). Living with risk: A global review of disaster reduction initiatives. United Nations, Geneva. Sets out the general direction of international action for disaster reduction.

Wisner, B., Blaikie, P., Cannon, T., & Davis, I. (2004). *At risk: Natural hazards, people's vulnerability and disasters*. London and New York: Routledge. An overview of hazards and disasters with a focus firmly on human vulnerability.

WEB LINKS

Global Assessment Report on Disaster Risk Reduction. https://gar.undrr.org

Global Network of Civil Society Organisations for Disaster Reduction: Views from the Frontline. www.gndr.org/project/views-from-the-frontline

Our World in Data. https://ourworldindata.org/covid-vaccinations

UCL Institute for Risk and Disaster Reduction. www.ucl.ac.uk/risk-disaster-reduction

UN Agenda 2030. https://sdgs.un.org/2030agenda

UN Office for Disaster Risk Reduction. www.undrr.org

CHAPTER TWO

Dimensions of Disaster

2

OVERVIEW

This chapter explores how to understand what a disaster is by examining its different dimensions. Defining a disaster has typically focused on quantifying its impact in terms of lives lost, injuries and people affected, as well as different mechanisms of costing. These approaches to understanding a disaster leave out multiple incidences happening at the same time, along with harder-to-identify consequences such as job losses, livelihood degradation, secondary health impacts, and intangible losses such as social connections, cultural heritage, and people's photographs and heirlooms. Data on disaster impacts depends on the quality of its collection and the classification scheme used by the data collectors, and whether data are comparable across disasters. Moreover, data on the scope and impact of disasters is no longer a matter of key statistics provided by insurance companies and international institutions. Increasingly, data is being provided by individuals via social media and crowdsourced data platforms (often run by universities), and often in 'real time' as a disaster unfolds. With this increase in data from multiple sources comes the increased circulation of misinformation/disinformation, a hazard that has its own damaging impacts. Our understanding of the dimensions of disasters is constantly evolving as technologies transform the production, collection and distribution of data.

A. INTRODUCTION

In the years between 1900 and 2020 more than 15,552 natural hazard events were catalogued in the Emergency Events Database (EM-DAT) produced by the Centre for Research on the Epidemiology of Disasters (CRED). A non-profit organisation established in Belgium in 1973, CRED became a World Health Organization (WHO) Collaborating Centre in 1980, supporting the WHO Global Programme for Emergency Preparedness and Response. The category called 'natural hazards' in the database has

DOI: 10.4324/9781351261647-3

changed over the years with efforts continuing to list and classify hazards (UNDRR, 2020). Broadly speaking, hazards have tended – with many revisions, disputes and exceptions – to comprise the following categories: geophysical (originating from the solid earth), meteorological (weather and atmospheric conditions), hydrological (surface and subsurface water), climatological (atmospheric processes that are intra-seasonal and multi-decadal), biological/ecological (from exposure to living organisms and their toxic substances) and extraterrestrial (namely space objects and radiation, with the possibility of space-living entities, especially microbes). Together, EM-DAT reports that these hazards resulted in over 32 million deaths and over 10 million injured. This data indicate that over 5 billion people have collectively been affected by these hazards, costing over US$3.5 trillion and leaving nearly 177 million homeless.

These are profound losses, but all disaster data should be interpreted with care. As noted in Chapter 1, the 'environmental hazards' or 'natural hazards' category itself is not clear cut. As well, data generally overrepresent rapid-onset, fairly large-scale hazard events where a physical trigger can be clearly identified. Comprehensive records are difficult to assemble, and hazard and disaster impacts, trends and patterns are complex and often controversial. Nevertheless, many disaster losses are due to a small number of very high-magnitude events, even as research indicates that the cumulative effect of continual 'smaller' disasters can overtake the impact of the less frequent 'larger' ones, as shown by another disaster database called DesInventar (Marulanda et al., 2010).

As another example of the limitations of disaster data, Table 2.1 lists an arbitrary snapshot from human history covering the millennium from 1020 to 2020 of recorded hazards and associated disasters that have claimed at least 100,000 lives – also an arbitrary number that does not account for changes in human population numbers. Together these have been responsible for over 22 million deaths.

What is most important in Table 2.1 is what is missing. Stopping at 2020 misses the COVID-19 pandemic with its over 6 million deaths, effectively 100% of the world's population affected, adding nearly 8 billion people to the toll of 'disaster-affected individuals', and economic damage perhaps an order of magnitude greater than the total economic damage listed for all disasters in EM-DAT (except COVID-19). Furthermore, COVID-19 is far from the worst pandemic in human history. Spanish flu from 1918 to 1920 is estimated to have killed 50–100 million people and would have had similar global impacts to COVID-19 regarding people affected and economic damage. Same with plague outbreaks, notably the Black Death in Europe in the 14th century – not to mention the genocide by Europeans of indigenous peoples around the world, with disease often killing more people than the violent conflicts did. Malaria today is estimated at over 200 million cases per year and over 500,000 deaths per year with many other diseases being prevalent over the last millennium. Sadly more 'banal', the table lists the 1985 drought in Mozambique, but not across the rest of Africa with the death toll in Ethiopia alone potentially being over 1 million. Disaster databases such as EM-DAT and the reinsurance companies are notoriously arbitrary, imprecise and inaccurate, sidelining the real human impacts and experiences of disaster.

For disaster data, it is as important to always consider what is not provided as much as what is actually provided. In the meantime, work continues towards more

Table 2.1 A comprehensive, but not exhaustive list of disasters by death toll that have killed over 100,000 people. The table illustrates how disaster data can be arbitrarily rounded up, and often the data do not include precise figures (particularly for epidemics and famines that are difficult to account for). For some hazards such as volcanic eruptions, the death tolls from the consequential failure of crops are not included as they are unknown. Therefore, this table can only represent the data collected.

Year	Country of origin	Details	Type of disaster	Fatalities
1346–1353	Europe, Asia and North Africa	Black Death	Pandemic	75–200 million
1918–1920	Worldwide	Spanish flu	Pandemic	50 million
1981– present	Worldwide	HIV/AIDS	Pandemic	40.1 million
541–542	Europe and West Asia	Plague of Justinian	Pandemic	30–50 million
1906–1907	China	Chinese	Famine	25 million
1959–1961	China	Great Chinese	Famine	15–85 million
1855–1960	Worldwide	Third plague	Pandemic	12–15 million
1789–1793	India	Doji bara or Skull	Famine	11 million
1783–1784	India	Chalisa	Famine	11 million
1769–1773	British India	Great Bengal	Famine	10 million
1876–1879	China	Northern Chinese	Famine	9–13 million
1315–1317	Europe	Great European	Famine	7.5 million
1630–1632	Mughal Empire (now India)	Deccan	Famine	7.4 million
2019– present	Worldwide	COVID-19	Pandemic	7–25 million
1545–1548	Mexico	Cocoliztli epidemic	Pandemic	5–15 million
165–180 approx	Roman Empire	Antonine plague	Pandemic	5–10 million
1519–1520	Mexico	Mexico smallpox epidemic	Pandemic	5–8 million
1932–1933	Soviet Union	–	Famine	5–8 million
1876–1878	British India	Indian Great	Famine	5.5 million
1928	China	Northwest China famine	Drought	3 million
1931	China	Yangtze-Huai River	Flood	3.7–4 million
1942	Bengal (now Bangladesh and India)	Bengal	Drought	1.5–3.8 million
1918–1922	Russia	Russia typhus epidemic	Pandemic	2.5 million
1921	Soviet Union	Povolzhye	Drought	1.2 million
1887	China	Yellow River	Flood	900,000–2 million
1556	China	Shaanxi	Earthquake	830,000
1918	Bangladesh	Spanish flu	Epidemic	393,000
2010	Haiti	–	Earthquake	316,000

(*Continued*)

Table 2.1 (Continued)

Year	Country of origin	Details	Type of disaster	Fatalities
1737	India	Calcutta	Tropical cyclone	300,000
1850	China	Xichang	Earthquake	300,000
1881	Vietnam	Haiphong	Tropical cyclone	300,000
1970	Bangladesh	Bhola	Tropical cyclone	300,000–500,000
1983–84	Ethiopia	–	Drought	300,000
526	Byzantine Empire (now Turkey)	Antioch	Earthquake	300,000
1839	India	Coringa	Tropical cyclone	300,000
1976	China	Tangshan	Earthquake	290,000–655,000
115	Roman Empire (now Turkey)	Antioch	Earthquake	260,000
1920	China	Haiyuan	Earthquake	235,000–273,400
2004	Indian Ocean	Indian Ocean	Tsunami	230,210
1138	Zengid dynasty (now Syria)	Aleppo	Earthquake	230,000
1138	Azerbaijan and Georgia	Ganja	Earthquake	230,000
1975	China	Banqiao	Dam failure	230,000
1920	China	Nina	Tropical cyclone	229,000
1876	British Raj (now Bangladesh)	Backerganj	Tropical cyclone	215,000
1303	Mongol Empire (now China)	Hongdong	Earthquake	200,000
1901	Uganda	Sleeping sickness	Epidemic	200,000
856	Abbasid Caliphate (now Iran)	Damghan	Earthquake	200,000
1780	Iran	Tabiz	Earthquake	200,000
1622	China	North Guyuan	Earthquake	150,000
1983–1985	Sudan	–	Drought	150,000
1935	China	Yangtze	Flood	145,000
1923	Japan	Kant	Earthquake	142,800
1991	Bangladesh	Bangladesh	Tropical cyclone	139,000
2008	Myanmar	Nargis	Tropical cyclone	138,000
1948	Soviet Union	Ashgabat	Earthquake	110,000
1920	China	Haiyuan	Mass movement	100,000
1290	China	Chihli	Earthquake	100,000
1786	China	Dadu River	Landslide	100,000

Year	Country of origin	Details	Type of disaster	Fatalities
1362	Germany	Saint Marcellus's	Flood	100,000
1421	Netherlands	St Elizabeth's	Flood	100,000
1731	China	Mount Tai	Earthquake	100,000
1852	China	Yellow River	Flood	100,000
1882	India	Bombay	Tropical cyclone	100,000
1922	China	Swatow	Tropical cyclone	100,000
1923	Niger		Epidemic	100,000
1981–1985	Mozambique	–	Drought	100,000
1971	Vietnam	Hanoi and Red River Delta	Flood	100,000
1911	China	Yangtze	Flood	100,000
1530	Holy Roman Empire	St Felix's	Flood	100,000
1780	Philippines	–	Tropical cyclone	100,000

reliable, accurate, precise and comprehensive disaster data in order to identify time trends and spatial patterns that should inform disaster risk reduction. The lack of standardised methods for data collection, and inconsistencies in defining and assessing key impacts, are highly inhibiting in this regard. An absence of agreed methodologies can lead to misunderstandings, helping to undermine confidence in the data, information and advice as seen during the COVID-19 pandemic (Box 2.1).

Box 2.1 COVID-19 – Adding Up the Toll

The COVID-19 pandemic which started in 2020 highlighted many of the challenges in establishing the dimensions of disasters, particularly with regard to numbers. Whether it is the number of cases of COVID-19, the number of related deaths, the numbers debilitated by after-effects called 'long COVID', or the R-number used to represent infection growth rates, different jurisdictions have chosen differing criteria to determine these numbers. These data are often compared and ranked internationally with little indication of the lack of standardisation in the data presented.

In the UK, for example, for the first few months, deaths were reported only if they occurred in hospitals and did not include those that occurred in care homes until 28 April 2020. In comparison, France and Germany were reporting deaths in both hospitals and care home settings. Moreover, these countries differed in how such deaths were framed. Germany counted deaths in care homes only if people tested positive for the virus. Belgium included in its tally any death in which a doctor suspected coronavirus was involved.

There are many indirect deaths that are not included in the UK official tally, including those who died from a lack of treatment of other health issues such as cancer, heart attacks, tumours, strokes and mental health concerns due to the UK's health systems being overwhelmed by COVID-19 cases, and heavy restrictions put into place as to what health services and operations could be conducted. This example highlights the question of 'whose deaths do we include' when looking at the dimensions of a disaster. How do we obtain data on deaths, and how confident can we be that the death was a direct impact of a hazard such as infectious disease?

The UK also reported two different death tolls: deaths within 28 days of a positive COVID-19 test and deaths with COVID-19 listed as the cause on the death certificate. Further confusions arose regarding how to tabulate people in hospital because of COVID-19 compared to people in hospital for non-COVID-19 reasons but who tested positive for COVID-19. Others entered hospices due to terminal illnesses, caught COVID-19 and then died, perhaps days or weeks earlier than they would have died without COVID-19.

The R-number is the number of people that one infected person will, on average, infect. Given that the moment of infection is hard to pinpoint, scientists work backwards to use data such as the number of people admitted to hospital or testing positive for the virus over time to estimate how easily the virus is spreading. If the R-number is higher than one, then the number of cases increases exponentially.

The R-number can only be as good as the recording of the cases across a location which is also linked to the number of tests conducted. Early in the pandemic in the UK, it was not possible to obtain tests for COVID-19 unless an individual was hospitalised. Meanwhile, many Pacific Island countries were said to be COVID-19-free – and North

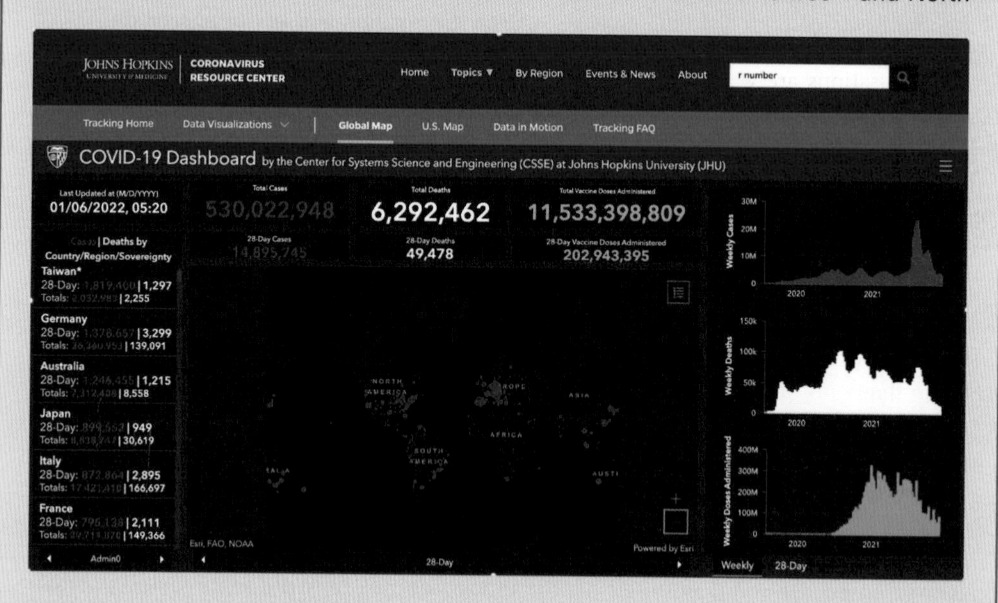

Figure 2.1 COVID-19 Dashboard by the Center for Systems Science and Engineering (CSSE) at Johns Hopkins University (JHU).

Source: COVID-19 Data Repository by the Center for Systems Science and Engineering (CSSE) at Johns Hopkins University (https://coronavirus.jhu.edu/map.html) taken from 1/05/2022

Korea declared itself to be so – even though tests were not being conducted. In many countries, people isolated at home due to COVID-19 symptoms which might have been COVID-19 or might have been other diseases such as cold and flu – or even hay fever.

The numbers of COVID-19 cases and deaths across countries were subsequently documented in global platforms that use aggregated datasets in a 'dashboard' format. One example is the COVID-19 Meta Dashboard by the Johns Hopkins Coronavirus Resource Center (CRC) (https://coronavirus.jhu.edu) (see Figure 2.1). Whilst the dashboard provides access to differing databases, the numbers among countries are hard to compare due to different methods of collection and reporting as well as absences of data. Countries have varying resources – or varying interests from ruling administrations – to monitor COVID-19 cases and record them. The COVID-19 pandemic demonstrates the challenges in using numbers to understand the impact of a disaster, both during and after.

B. DEFINING DISASTER

1. Reporting Disaster

Media, including social media, play a major role in shaping our views on what a disaster is and its impacts, including for organisations collecting disaster data. News organisations are well equipped to collect and transmit information continuously in times of crisis. Knowledge of a disaster spreads rapidly from the hazard zone to a global audience, often using imagery sourced by social media without extensive fact checking. News of a disaster can stimulate public interest and may also influence the flow of disaster aid.

Eisensee and Stromberg (2007) found that emergency relief decisions by the US government were strongly influenced by media reports of disasters. More aid was provided when US disasters were not crowded out of the headlines by unrelated but apparently more newsworthy items. For example, in order to have the same chance of receiving relief funds, a disaster during the Olympic Games required three times the number of casualties of a disaster than occurred otherwise.

Media reports during the early phases of an emergency tend to be patchy in cover and unreliable in content, due to uncertainties at the time. They can be shaped by and repeat biased tropes, often racist, sexist, classist, ableist, gendered and presuming violence and criminal activity (Monahan & Ettinger, 2018), thereby deliberately victimising those already experiencing disaster. Numerous studies describe the ways in which African-American residents of New Orleans who could not evacuate before or after Hurricane Katrina's 2005 landfall were reported on through racially biased stereotypes (Sonnett et al., 2015). In counterpoint, Spike Lee's documentary film, *When the Levees Broke: A Requiem in Four Acts* (2006), narrates the unequal impacts of the disaster, and the portrayal of those impacts, as a window onto the ongoing role of race/racism and social class/classism in determining who becomes vulnerable from environmental hazards.

Media coverage can enhance the work of warnings and can assist with the effective evacuation of people. As shown by Rashid (2011) for Canada, local (as opposed to national) news media can provide good in-depth reporting by providing context on the emergence and impact of disaster (Figure 2.2). Yet, local printed outlets such as newspapers have declined significantly, moving increasingly to a digital presence and competing with instant, brief snippets found in newer media sources. News editors can prioritise stories according to death totals and death proximity in addition to the availability of (often unconfirmed) graphic video material from social media. Disaster reports become entertainment as much as news, termed infotainment or rescuetainment. Short-term dramatic hazards (for instance, earthquakes, volcanoes, heatwaves and wildfires) attract a lot of news coverage, but chronic issues (for instance, drought, famine and emerging infectious diseases) are more likely to be dealt with much later. The drive for 'instantainment' leads to an absence of in-depth reporting that explains the real reasons for disasters, which are vulnerabilities (including exposures) rather than hazards.

When there is a dependence on revenue related to advertising or clicks, the media tend to concentrate on the prosperous target markets of commercial sponsors. This can lead to under-reporting of disaster impacts on poorer social groups in disadvantaged areas (Rodrigue & Rovai, 1995).

Figure 2.2 Media reports of the 1954 Hurricane Hazel disaster in Toronto, Canada, displayed on the contemporary memorial.

Source: Photo: Ilan Kelman

Disaster coverage is increasingly determined by geopolitics, driven largely by where the disaster happened and who it affects, further framing how stories are presented. Local, national and international media respond differently to crises, depending on the scale of impact and who is affected. Some media watchers call out racist tropes such as highlighting foreign citizen deaths in plane crashes and sexist tropes such as referring to 'women and children' as being both vulnerable and passive in violent conflicts.

The coverage of disasters on television news channels is much influenced by the visual impact of video and still images. Media reporting shows an excessive concentration on the emergency phase of disaster, focusing on what is happening during the emergency rather than why it happened or what goes on afterwards, especially if striking images of distressed affected people (presented as 'victims') are available. There is also an undue concentration on disasters affecting those seen to be like 'us'. Solman and Henderson (2019) reviewed five UK newspapers' reports of flooding in India and the UK during 2015 and found a striking imbalance in terms of the volume of coverage, with the UK floods receiving far more attention than the floods in India, even though the latter had a greater number of fatalities. Noting that this may be a matter of geographical proximity in terms of access to stories and interest among readership, Solman and Henderson (2019) also found bias in the details with UK flood reports containing more emotional, vivid and personal accounts while Indian flood reports tended to be detached and dispassionate.

Fiction and entertainment have long been interested in the narrative power of disasters and show similar biases as the media reporting what they deem to be news. Blockbuster films such as *Armageddon* (1998), *Twister* (1996), *Dante's Peak* (1997), *The Day After Tomorrow* (2004), *Contagion* (2011), *The Wave* (2015), *The Hurricane Heist* (2018) and *Moonfall* (2022) have been joined by a genre called 'climate fiction' or 'cli-fi' for short. Cli-fi deals with human-caused climate change, its impacts in the present and near future, and the responses of individuals and groups (Chapters 16 and 18). Meanwhile, disasters enter other movie series such as James Bond flicks: *No Time to Die* (2021) is about targeting a disease to specific individuals while *A View to a Kill* (1985) is about generating a Silicon Valley earthquake and flood.

Many disaster movies depict real catastrophes such as the 2004 Indian Ocean tsunami in *The Impossible* (2012) and *Deepwater Horizon* (2016) covering the BP oil rig that exploded and released oil in the Gulf of Mexico. Many documentaries cover disasters such as Hurricane Katrina in 2005, the 2011 Japan earthquake and tsunami and the COVID-19 pandemic along with movies that marry fiction and documentary genres such as *The Age of Stupid* (2009), purporting to be a look back from 2055 about how humanity wiped itself out. Other disaster movies claim to be, effectively, dramatisations of real life or docudramas, although separating fact from fiction in the details is not always easy, such as for *The Perfect Storm* (2000) about a 1991 weather system and *United 93* (2006) about the 11 September 2001 terrorist attacks.

Despite a long history of publications before it, Rachel Carson's book *Silent Spring* (1962) (Figure 2.3) is often said to be the first to highlight to a wide public the significant environmental impact from human activities. It documented, with varying levels of scientific accuracy, the adverse

Figure 2.3 Rachel Carson's *Silent Spring* and Ian McEwan's *Solar*.

Source: Photo: Ilan Kelman

environmental effects from indiscriminate pesticide use. Human-caused climate change has since become a key subject for popular nonfiction books communicating the various hazards said to comprise this planetary condition – again with varying scientific accuracy – such as Naomi Klein's (2014) *This Changes Everything: Capitalism vs the Climate*, Mark Lynas's (2015) *Six Degrees: Our Future on a Hotter Planet*, Andrew Guzman's (2013) *Overheated: The Human Cost of Climate Change* and Greta Thunberg's (2019) *No One Is Too Small to Make a Difference*.

As with film, fiction plays a significant role in capturing the emergence and many dimensions of disasters. Ian McEwan's *Solar* (2010) (Figure 2.3) is based in part on the author's participation in a voyage organised by the art and climate change organisation Cape Farewell. Science fiction has long provided a commentary on the present by situating crises in a future world. In her *Parable of the Sower* (1993), Octavia Butler situates climate change amidst racism, social disintegration and the resilience of Black culture, while Margaret Atwood's *The MaddAddam Trilogy* (2003–2013) observes the scope and impact of a catastrophic pandemic.

Social media – as a form of entertainment, but also as a means of disseminating news, creating news and adding commentary – has become a crucial factor in how disasters are recorded, understood and responded to. For those with the privilege – further accentuating inequalities and inequities – social media networks and smartphone technology can be the initial, preferred and predominant method for accessing information on disasters. This contributes to understanding disasters when the public captures and shares local disasters. Within minutes of the March 2011 Tohoku earthquake, Facebook recorded dozens of posts containing the word 'Japan' and, within an hour, there were almost 1,200 tweets (Twitter messages) per minute emanating from Tokyo (Spong, 2011). From friends and family to the emergency services, people share critical information (and, inadvertently or deliberately, misinformation and disinformation) on locations of hazards, evacuation centres and road closures; preparedness and response strategies; and to 'check in', letting the world know that they are safe when in a disaster-affected area (Bird et al., 2012).

Volunteers from around the world play 'crowdsourcing' roles in analysing satellite imagery, social media postings and other data to generate maps that are meant to guide relief efforts, as happened after Typhoon Haiyan (Yolanda) in the Philippines in 2013 (Butler, 2013). During periods of relative quiescence, social media is used for promoting disaster risk reduction, to share information in an effort to maintain public interest and trust (Sennert et al., 2018). Box 2.2 highlights the role of social media combined with GIS tools to facilitate community disaster response.

All media forms promise new kinds of visibility: new opportunities to monitor and predict disasters, to try to prevent them or to reduce their impacts, to organise aid and to connect to others. The question is balance. How much do different media better democratise the visibility of disasters, especially disaster risk reduction? How much do they inhibit, perpetuate myths and harm through false information? For social media, it is up to us regarding how we use it and push for its governance.

Box 2.2 2011 Christchurch Earthquake, New Zealand

On 22 February 2011 at 12:51 p.m. local time, a 6.2 Mw earthquake struck the Canterbury region of the South Island, New Zealand. At only 5 km depth and potentially an aftershock of the 7.1 Mw earthquake on 4 September 2010, the earthquake killed 185 people and caused extensive damage in New Zealand's second largest city, Christchurch, where buildings were already weakened from the prior earthquake. The peak ground acceleration of this shaking was very high, and both the vertical and horizontal ground movements resulted in parts of Christchurch's centre being destroyed. The city was still recovering over a decade later with profound disagreements on how and what to rebuild, including the iconic cathedral which partly collapsed in the shaking.

Following the earthquake, social media in the form of the image and video uploading platforms Flickr and YouTube became key tools with which those affected could report damage and request assistance. These videos were collated using GIS software by the Christchurch City Council to highlight key areas that required attention, along with locations where critical services were operating so as to help the rescue and recovery process (Figure 2.4).

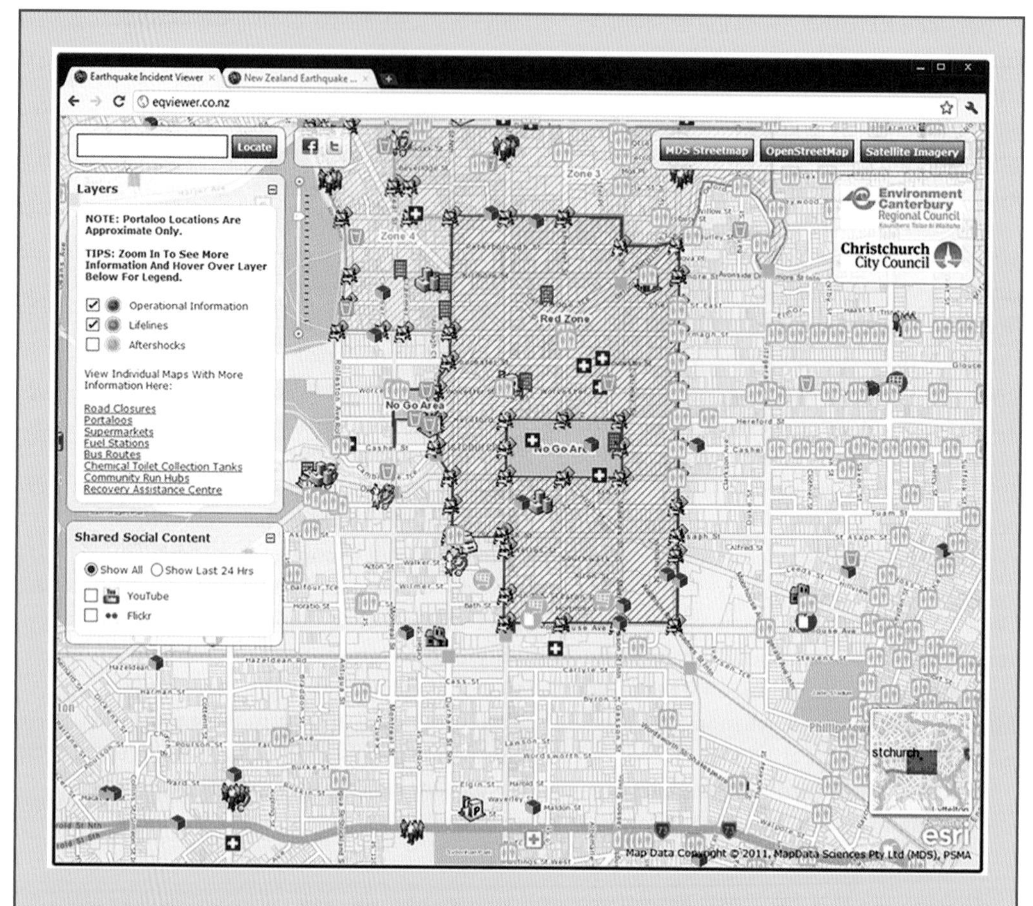

Figure 2.4 GIS software showing the social media content uploaded by residents.

Source: Christchurch City Counties and Environment Canterbury

Additionally, the GIS software company ESRI generated a social media response map that shared real time updates from a crowdsourcing tool Ushahidi on the condition of critical infrastructure such as energy, transportation and water (Figure 2.5). The power of volunteered geographic information (VGI) can be clearly understood following Christchurch's disaster.

Global social media helped manage the earthquake's aftermath. Google set up one of the first 'Crisis Response team' web pages providing key information for those affected by the earthquake, so that all official numbers and information would be available on one page. People used social media such as Twitter to highlight local facilities and resources using the hashtag #eqnz:

> RT @anthonybaxter: Google has people finder up for #eqnz #christchurch http://bit.ly/i0aAle please RT widely

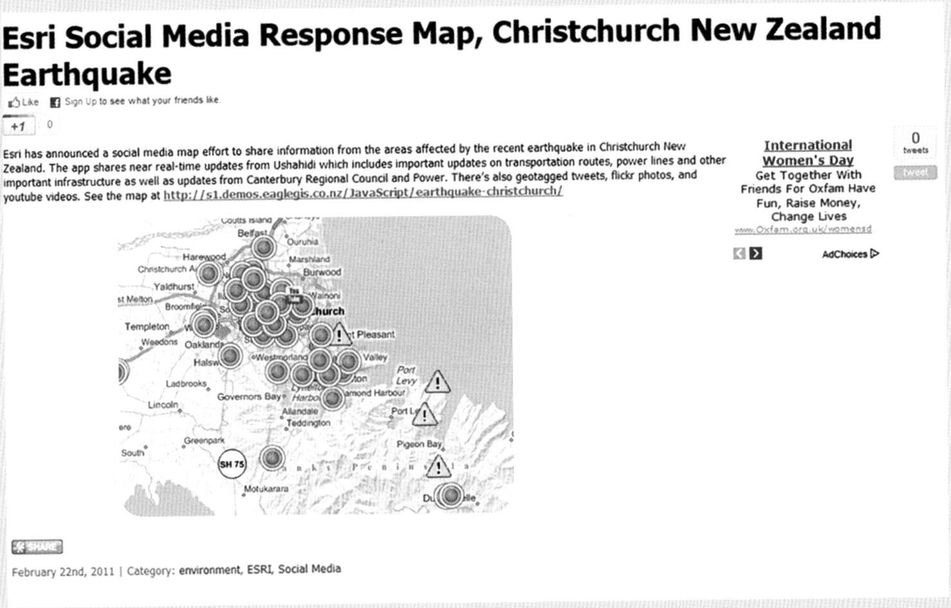

Esri Social Media Response Map, Christchurch New Zealand Earthquake

Esri has announced a social media map effort to share information from the areas affected by the recent earthquake in Christchurch New Zealand. The app shares near real-time updates from Ushahidi which includes important updates on transportation routes, power lines and other important infrastructure as well as updates from Canterbury Regional Council and Power. There's also geotagged tweets, flickr photos, and youtube videos. See the map at http://s1.demos.eaglegis.co.nz/JavaScript/earthquake-christchurch/

February 22nd, 2011 | Category: environment, ESRI, Social Media

Figure 2.5 ESRI social media response map, Christchurch New Zealand, 22 February 2011.

Source: ESRI

> RT @TelecomNZ: Please keep ALL calls nationwide to minimum to save capacity for emergency services. Txt instead if you can #eqnz

and to express relief when friends and family were found to be safe:

> RT @publicaddress: Just confirmed that [name redacted] is okay. Yay! #eqnz

Celebrities and organisations expressed concern and support online using hashtags that aggregated and amplified responses, giving the disaster and its impacts an increased presence. Stephen Fry's message of support, for example, read:

> RT @stephenfry: Oh dear, poor Christchurch. Another horrific earthquake. http://t.co/S5nL3lq #chch #eqnz http://t.co/plUcmcP

Using Twitter and Facebook in the aftermath enabled affected locals to identify locations with running water, functioning toilets, food and family or friends with whom to shelter (see also Dabner, 2012). Bruns and Burgess (2012) highlighted how social media enabled communities to self-organise their response and recovery whilst the information assisted emergency management services in receiving updates from locals, as well as in disseminating key information. Social media continues to play a role in providing individual and community disaster data for response but can also equally be harnessed for preparedness.

2. Reporting Disasters and Impacts

While media – both fiction and nonfiction – play a crucial role in shaping our understanding of what a disaster is, there are more strict definitions used by humanitarian organisations, civil relief agencies, insurance companies, international organisations and governments. An example is from the International Federation of Red Cross and Red Crescent Societies (IFRC) that a disaster is 'a sudden, calamitous event that seriously disrupts the functioning of a community or society and causes human, material, and economic or environmental losses that exceed the community's or society's ability to cope using its own resources' (2022).

According to this definition, losses have to be sufficiently large to disrupt the functioning of a community beyond its ability to cope. Quantifying the precise impact thresholds necessary for such disruption – whether for fatalities, numbers of people affected, economic damage or other losses such as environmental and intangible aspects – is challenging. Loss of human life is usually the prime indicator which might seem to be readily available, but it may not be so in countries lacking detailed population data. Homeless and undocumented people are often left out of these data. Information on other impacts such as personal injury, disease and resulting homelessness is more difficult to obtain. In the absence of standardised methods of survey and assessment – difficult anyway when efforts need to be put into saving lives – the collection of robust and comparable information on property damage and less direct economic losses is also unlikely.

As such, the IFRC's definition is not really in line with disaster research's understandings outlined in Chapter 1 that a disaster is much more than a sudden (hazard) event and that some disruptions to society (which enacting disaster risk reduction could be viewed as) are helpful for avoiding disaster. As noted in Chapter 1, disasters are effectively cascading and complex processes, thereby not easily classified and compared. To avoid double counting impacts, each loss category, such as deaths or damages, should be recorded once only and allocated to a specific cause, but practical problems arise. For example, when an earthquake triggers a landslide, should deaths be attributed to the earthquake (the trigger event/primary hazard) or the landslide (the direct cause/secondary hazard)? For the Mt St Helens eruption in the US in 1980, an earthquake triggered a landslide, weakening the slope and permitting a lateral blast from the volcano that killed people. Will deaths occurring days, weeks or months later from injuries sustained in a hazard be recorded? Should suicides, strokes or heart attacks in the days or years after losses be included? Generally speaking, when recording mass fatalities, the initial trigger is named. This means that the effects of 'secondary' hazards, such as landslides, are underestimated (see Chapter 9). For long-duration hazards, such as drought, there may be doubts about the exact start and end dates. The precise location of a hazard or disaster can also be difficult to identify, especially when hazards cross administrative boundaries. This poses significant challenges in providing warnings for low-duration and slow-onset hazards.

Many disaster reports lack loss reporting beyond the immediate, direct and tangible impacts (Figure 2.6). Losses (being deprived of something) are a measure (quantified or not) of the damage caused by a disaster. *Direct effects* are the first-order consequences that occur immediately after an event, such as deaths, injuries and economic loss caused by the damage of infrastructure, including buildings, power plants and transport, and the instant impact on the continuation of businesses. *Indirect effects* tend to emerge over time and are more difficult to attribute specifically. These include factors such as adverse consequences on mental health and well-being, cultural and community losses (such as damage to or separation from historically and culturally significant sites, as well as separation from sites of knowledge and wisdom), broader impacts on the well-being of society and the environmental impacts. Additionally, a disaster can have significant long-term social impacts on education, health and other social services, and livelihoods (also from business disruption). *Tangible effects* are those for which it is possible to assign monetary values, such as the replacement of damaged property and infrastructure, usually provided via insured losses, which underestimates losses since plenty of property and possessions are uninsured or underinsured – or insurance companies decline payouts. *Intangible effects*, although real, are difficult to be fully assessed in monetary or other quantitative terms, yet some health costs are calculated over the long term, and modelling can be used to provide an approximation of the cost of intangible effects.

As per Figure 2.6, *direct losses* are the most visible consequence of disasters due to the immediate impacts on life, livelihoods and well-being and damage, such as building collapse. These losses are comparatively quantifiable – such as number of people killed or damage to buildings,

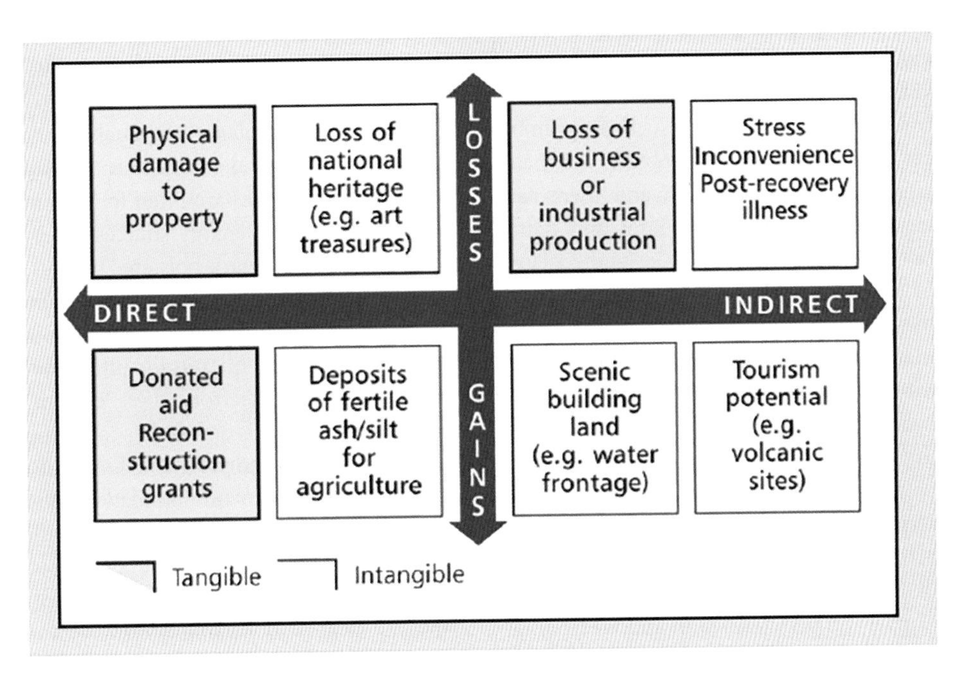

Figure 2.6 Possible losses (damage) and gains in disaster, both direct and indirect, with specimen examples of tangible and intangible effects. Some may be quantifiable, others non-quantifiable.

infrastructure and natural resources – within all the limitations discussed earlier in this chapter. Injuries can be physical such as cuts, abrasions and fractures as well as psychological such as stress, anxiety and depression (Marres et al., 2011; Norris et al., 2010). Direct losses tend to be monetised as 'economic losses' in that the destruction of infrastructure, crops and livestock – and even physical and mental health impacts are given a monetary value (Wouter Botzen et al., 2019).

These direct losses can be estimated using catastrophe models and measured using empirical data on losses such as insurance claims. Insurance must then be available and affordable to have been purchased, claims can be deliberately inflated and companies can deny or reduce payouts. Additionally, the extensive costs of many disasters are not accounted for by insurance losses (Poontirakul et al., 2017).

Direct gains represent benefits flowing to survivors after a disaster, including various forms of aid. People with skills in construction and trades such as plumbers and electricians may obtain well-paid employment post-disaster, and there can be improvements to the built and natural environments. On the Icelandic island of Heimaey, volcanic ash resulting from the 1973 eruption was used as foundation material to extend the airport runway while the lava flow into the ocean yielded a better harbour, albeit after significant efforts to not lose the harbour from the lava flow.

Indirect losses are the second-order consequences of disaster, like the disruption of economic and social activities that result in a decline in output or revenues; the impacts on people's health and well-being; and the disruption to the flow of goods and services. As property values may fall and prices may rise, consumers save rather than spend, business becomes less profitable and unemployment rises. Indirect losses and these wider disaster effects are rarely accounted for, especially for low-income and subsistence households and for marginalised people and communities. Ill-health effects often outlast other losses. Symptoms of psychological impacts can be shock, anxiety, depression, stress or apathy, possibly expressed through sleep disturbance, belligerence, substance use and self-harm. Again, data are incomplete and hard to collect. Information on the impacts of a disaster may eventually show up in metrics such as disability-adjusted life years (DALYs) and quality-adjusted life years (QALYs). As with many datasets, they rely on the scope and quality of information received and the ways in which particular health conditions are recognised and recorded.

Indirect gains are the long-term benefits from a hazard-prone location. Examples are water supply, transport routes, environmental amenities and fertile land from living in a river's floodplain. Earthquake faults in deserts can provide freshwater in arid lands. Volcanic slopes are often fertile and provide geothermal energy sources. Tourism for natural heritage is bolstered by landscapes in arid lands, volcanic areas, coastal and freshwater floodplains and hills prone to landslides. A key question is often who gains and who loses from these circumstances. Luxury hotels on beaches can bring low-paid, seasonal employment for locals, placing them in a tsunami inundation zone and removing their fishing livelihoods with its associated environmental knowledges, while reaping huge profits for the affluent, far-away owners and shareholders.

Fully accounting for all disaster-related losses and gains is thus challenging. Disaster reports and analyses might present only key impacts, skewing the perception of the severity – even for direct loss of life. The 12 January 2010 earthquake in Haiti had an initial death toll of 170,000, mostly caused by building collapse, as documented on 27 January. By the disaster's first anniversary, the toll had risen to 316,000 deaths, due to further recoveries of bodies, a significant cholera outbreak and poor healthcare facilities to assist the injured. Others put the death toll closer to 200,000. Whilst numbers can be misleading, and it is remarkable that a prominent disaster's death toll can differ by 100,000 people, the values are important when developing a political and financial case for disaster risk reduction. Work must continue to explore how indirect gains and losses can be better integrated, especially to support investment in avoiding disasters.

Many large businesses manage their risk via insurance. In turn, catastrophe reinsurance is 'insurance for insurance companies', whereby insurers transfer portions of their risk portfolios to other parties to reduce their exposure to the financial risks of a catastrophe. Catastrophe reinsurance allows the insurer to shift some or all of the risk associated with policies that it underwrites in exchange for a portion of the premiums that it receives from policyholders. In contrast, the aim of disaster microinsurance is to provide everyone with possibilities for affordable insurance irrespective of livelihoods and income which might cover life insurance, health insurance, liability insurance and others in addition to more standard disaster insurance. Small-scale losses for people with low incomes can be devastating, so microinsurance provides options for localised approaches to the insurance mechanism. A number of governments can also provide their own national insurance programs for citizens, making it more affordable.

C. ARCHIVING DISASTER

1. Archived Disaster Datasets

Disaster records are held by many, including international organisations, national government agencies, insurance companies and academic institutions. These bodies have different resources, reasons and validation approaches for data collection and reporting, and so the resulting information varies in content and quality. Datasets for environmental hazard data provide historical and real-time data (Table 2.2).

A quality framework for disaster databases was recommended by Below et al. (2010). Key features, updated here, are the following:

- *Prerequisites and sustainability.* Support and resources to maintain the dataset comprehensively without interruption.
- *Data accuracy, precision and reliability.* The information should be complete, with good and balanced geographical coverage alongside procedures to validate data and to test for bias and other faults.
- *Methodology.* The raw information should be processed according to clear concepts and definitions. Storage and backup must be part of the methodology.
- *Credibility.* There should be evidence of the expertise of the archiving body, including assurances on transparency, accountability and quality control.
- *Serviceability.* The information should be useful, convenient and straightforward to interpret, have perceived relevance and have timely dissemination.

Table 2.2 Examples of environmental hazard datasets and resources

Organisation	Dataset	Website
ReliefWeb	All hazards	https://reliefweb.int/disasters
Global Disaster Alert and Coordination System (GDACS)	All hazards	www.gdacs.org
United Nations Office for Disaster Risk Reduction (UNDRR), Science and Technology Advisory Group (STAG), working group on 'Data'	All hazards	www.undrr.org/publication/stag-data-working-group-report
WMO	Meteorology	https://public.wmo.int/en
USGS Seismology	Earthquakes	https://earthquake.usgs.gov
European-Mediterranean Seismological Centre (EMSC)	Earthquakes	www.emsc-csem.org
International Tsunami Information Center	Earthquakes, volcanic eruptions and tsunamis	http://itic.ioc-unesco.org/index.php?option=com_content&view=category&id=1072&Itemid=1072
Pacific Tsunami Warning Center (PTWC)	Earthquakes and tsunamis	www.tsunami.gov
USGS Volcano Hazard Program (VHP)	Volcanoes	www.usgs.gov/volcano
Global Volcanism Program (GVP)	Volcanoes	https://volcano.si.edu
Volcanic Ash Advisory Centers	Volcanic ash	www.ospo.noaa.gov/Products/atmosphere/vaac/index.html
EM-DAT	The International Disaster Database	www.emdat.be
DesInventar	Disaster loss data	www.desinventar.net
Famine Early Warning System (FEWS)	Famines	https://fews.net
WHO	Health data	www.who.int/data
CDC	All hazards, focused on health data	www.cdc.gov/datastatistics/index.html
Dartmouth Flood Observatory	Floods	https://floodobservatory.colorado.edu
NASA's Center for Near-Earth Object Studies	Near-earth objects	https://cneos.jpl.nasa.gov
UK Natural History Museum	Meteorites	www.nhm.ac.uk/our-science/data/metcat/search/bgintro.dsml
Munich Re's NatCatSERVICE	Insurance data	www.munichre.com/en/solutions/for-industry-clients/natcatservice.html
Swiss Re's sigma data	Insurance data	www.swissre.com/institute/research/sigma-research

- *Accessibility*. The data should be readily accessible to a wide variety of users. Contact details should be available for those seeking further information.

Comparing disaster data compiled by two European reinsurance companies and EM-DAT reveals major inconsistencies (Table 2.3). The EM-DAT records listed many more fatalities, whilst the two reinsurance companies recorded higher levels of economic loss. This difference reflects the priorities of, and definitions from, each organisation.

In a review of 31 databases covering natural and technological hazards related to

Table 2.3 Disaster data from three different databases for four countries from 2002

	CRED	Munich Re	Swiss Re
Honduras			
Number of disasters	14	34	7
Number killed	15,121	15,184	9,760
Number affected	2,982,107	4,888,806	0
Total damage ($US million)	2,145	3,982	5,560
India			
Number of disasters	147	23	1,220
Number killed	58,609	877	65,058
Number affected	706,722,177	2,993,281	16,188,723
Total damage ($US million)	17,850	112	68,854
Mozambique			
Number of disasters	16	23	4
Number killed	106,745	877	233
Number affected	9,952,500	2,993,281	6,500
Total damage ($US million)	27	112	2,085
Vietnam			
Number of disasters	55	101	36
Number killed	10,350	11,114	9,618
Number affected	36,572,845	20,869,877	2,840,748
Total damage ($US million)	1.915	3,402	2,681
Totals			
Total number of disasters	232	387	167
Total number killed	189,825	96,418	84,669
Total number affected	756,139,629	277,490,405	19,035,971
Total damage ($US million)	21,937	29,629	79,180

Source: After Guha-Sapir and Below (2002). The Quality and Accuracy of Disaster Data: A Comparative Analysis of Three Global Data Sets. Working Paper prepared for the Disaster Management Facility, World Bank, CRED (2002). Copyright CRED 2009.

disasters, Tschoegl et al. (2006) concluded that a lack of standardisation in definitions, differences in classifications, inadequate accounts of methodologies and variations in the availability of the sources undermined the usefulness of the information. This remains a challenge, with some groups such as the World Organisation of Volcanoes database (WOVOdat) working hard to try and standardise data so that it is possible to compare data between different volcanoes to assist prior and during volcanic unrest.

The quality of disaster data is under continuous scrutiny, largely to ensure that users understand the limitations of the information. Gall et al. (2009) identified several problems, updated here:

- *Hazard bias.* Users assume that every hazard is represented, whereas selective reporting of hazards may occur based on the priorities of the collectors and reporters.
- *Temporal bias.* Users assume that losses are comparable over time, whereas changes exist in collection and recording procedures, while baselines of population, livelihoods and wealth data are changing.
- *Threshold bias.* Users assume that all losses are counted, whereas a filtering process means that many losses go unreported, such as from smaller scales or marginalised peoples.
- *Accounting bias.* Users assume that all types of loss are included, whereas it is a highly variable feature of disaster loss estimation with indirect and non-quantifiable losses often downplayed.
- *Geographical bias.* Users assume that losses are comparable across geographical areas and units, whereas changing political and administrative boundaries, internal variations and variations in governance and resources make it inconsistent.
- *Systemic bias.* Users assume that losses are computed uniformly, whereas monetary losses may or may not be adjusted for price inflation/deflation, cost-of-living indices and currency exchange rates. Where a range of estimates exists, the archive may report the highest, the lowest, the mean or a combination.

Even with the highest-quality data, it can be difficult to draw totally valid conclusions. The widespread use of absolute impact thresholds obscures differences in the relative impact of disasters between – and within – individual countries. For example, a US$100 million loss would result from a much smaller and higher-frequency hazard in – say – California than in Bangladesh or in the richer areas of California and Bangladesh compared to the poorer areas. Generally, datasets of financial losses, like those produced by the reinsurance industry, tend to emphasise places where the exposed assets are higher. Conversely, datasets prioritising fatalities tend to be biased towards places where large numbers of vulnerable people live, often larger cities.

Aggregated datasets do not capture local impacts or horrendous lived experiences, including effects on marginalised and oppressed groups. The loss of ten people from a small, remote fishing hamlet would be far more devastating for that community's survival than the death of 1,000 people across a megacity. This in no way diminishes the loss of 1,000 people. Disaster impacts should be scaled according to national, regional or local population numbers, to examine proportional vulnerability, proportional impact and proportional losses – in addition to the absolute tallies. Both absolute and relative measures for damage and effects indicate different dimensions of disasters, so both should be recorded and reported.

2. Archived Lives

While archived datasets provide insight into some numerical dimensions of disasters, questions remain regarding how the experiences of impacted individuals and groups are recognised and recorded. There are numerous archives set up by universities, arts organisations, charities and libraries (amongst others) that are intended to provide a sense of how the dimensions of disasters are felt at the individual scale and how such experiences allow for a critical reflection on how disasters might be better avoided and prepared for.

The *Hurricane Digital Memory Bank* established in 2005, for example, collects, maintains and presents first-hand accounts, on-scene images, blogs and podcasts concerning Hurricane Katrina and Hurricane Rita, both of which moved through the Gulf of Mexico to make landfall in 2005. Organised by George Mason University's Roy Rosenzweig Center for History and New Media and the University of New Orleans, in partnership with the Smithsonian Institutions, National Museum of American History and other partners, the archive currently holds more than 25,000 freely accessible digital items and hopes 'to foster some positive legacies by allowing the people affected by these storms to tell their stories in their own words, which as part of the historical record will remain accessible to a wide audience for generations to come' (http://hurricanearchive.org/).

UNESCO has driven several initiatives to try and preserve cultural memories of hazards. In 2015, UNESCO published a booklet *Remembering the 1945 Makran Tsunami: Interviews With Survivors Beside the Arabian Sea* (see Figure 2.8). This report captures coastal residents' recollections of the 1945 disaster, from either persons born before 1945 or younger people who recounted what their elders had told them. Whilst many of the accounts are sketchy and can be regarded as hearsay, the striking details consistent with 'flashbulb memory', a psychological and neurological response to traumatic events, is invaluable. The project also produced videos and audio recordings of interviews. Further examples exist in the UNESDOC Digital Library (UNESDOC: UNESCO Digital Library – MED DIALOGUE). A memorial centre in Oslo to the 22 July 2011 terrorist attack included videos and text of people's first-hand accounts.

The range of archives being produced is well illustrated by collections recording the COVID-19 pandemic. UNICEF, for example, has undertaken the 'Stay Home Diaries' project, which asked children from around the world to document for YouTube their lives at home and provide 'a glimpse into their world and to inspire others' (Stay Home Diaries – YouTube). The Cities and Memories project (https://citiesandmemory.com/covid19-sounds/) built a crowdsourced sound map with contributions from 70 countries, aiming to show 'how our aural lives are shifting as lockdown affects the world and as urban environments evolve'. Archived sounds have included 'everything from the changing sounds of morning prayer from a Tibetan monastery in northern India to a Finnish woman reading out excerpts of Roald Dahl stories to her infant relatives in isolation' (Bakare, Tue 31 Mar 2020, *Guardian*). Similarly, the Museum of London launched its 'Collecting COVID' project, which captures sounds and photos from sites across London during lockdown that correspond with

Figure 2.7 The IOC-UNESCO publication of *Remembering the 1945 Makran Tsunami – Interviews With Survivors Beside the Arabian Sea.* This booklet captures nearly 100 accounts of the 1945 Makran disaster as told by eyewitnesses and second-generation survivors in Pakistan, Iran, Oman and India.

Source: IOC-UNESCO Secretariat for the Indian Ocean Tsunami Warning and Mitigation System (IOTWS) in Perth, Australia, and the Indian Ocean Tsunami Information Centre (IOTIC) in Jakarta, Indonesia

recordings made in 1928. This comparison presents another dimension of the COVID-19 pandemic, insofar as 'For almost a century, it would have been unfathomable to imagine what silent central London streets would sound like. Through the Collecting COVID project, a record of this extraordinary moment in time will be preserved for future generations' (Brazee, 2020; from the project: Recording London Landscapes, *Past and Present*) (www.museumoflondon.org.uk/discover/recording-london-sound-scapes-past-present).

Case studies of specific hazards and disasters, additional data and differing perspectives are also archived in reports published by NGOs, humanitarian agencies and other government or supranational governmental organisations. Examples are the *World Disasters Report* (IFRC), the *Global Risks Report* (World Economic Forum), the UN Global Assessment Report (GAR) on Disaster Risk Reduction, the *Views From the Frontline* (GNDR) report for local disaster risk reduction, and the Intergovernmental Panel on Climate Change (IPCC) synthesising and assessing the science of natural and human-caused climate change. Further case studies and data are continually being published across all media, scientific, popular science and non-scientific. A significant amount of data is published in 'grey' literature that rarely sees an audience beyond the library or office it sits in, yet can provide valuable insights into a particular hazard or disaster.

D. EXPLAINING DISASTER: TIME TRENDS

1. Changes in Data Recording Methods

Any changes in data collection and recording methods which lead to the inclusion of more hazards and disasters will clearly increase the numbers and impacts reported. Such changes may reflect the establishment of a new agency, a shift in policy within an agency, or wider influences on what is important to record. The example of the infectious disease smallpox (Box 2.3) is illustrative of the impact of data recording techniques on understanding disaster dimensions.

Similarly, the upper part of Figure 2.9 shows a long-term upward trend from 1790 to 1990 in annual volcanic eruptions, despite the fact that there is no physical reason for global volcanic activity to increase over this 200-year period. As shown by the lower graph in Figure 2.9, the number

Box 2.3 The Geographies of Smallpox

Smallpox is an infectious disease wherein the variola virus is transmitted from person to person when infected individuals develop sores in their mouth or throat and droplets of infected mucus are released by a cough or sneeze. People are the only known hosts of this disease. Prolonged exposure to an infected individual raises the risk of infection, and smallpox has become pandemic numerous times, devastating communities. One of the deadliest diseases to have impacted people, smallpox is also the only human disease to have been eradicated by vaccination.

Figure 2.8 from the Our World in Data project, using data collected by the World Health Organization, indicates the changing spread of smallpox – and hence the virus triggering it – over time. By the turn of the 20th century, smallpox was present across the globe, following migration and trade routes while also spreading via colonial conquest and settlement. Smallpox was spread to the Americas as part of the 'Columbian Exchange', which saw Spanish conquistadors introduce this disease, as well as measles, to the Caribbean in 1507. Smallpox had spread as far as Mexico by 1520, causing massive mortality in indigenous communities with an estimated one-quarter to one-third of those infected dying. In addition to the suffering caused, the depopulation of the Americas had profound environmental consequences, such as through the regrowth of forests.

From Figure 2.8, there appears to be a very rapid spread of smallpox in 1923 in South Asia. This actually reflects data collection since, from 1920 onwards, infectious disease statistics are collected and forwarded to the newly established Health Office of the League of Nations (a political precursor to the United Nations). This change in reporting made the scope of this particular disaster more visible.

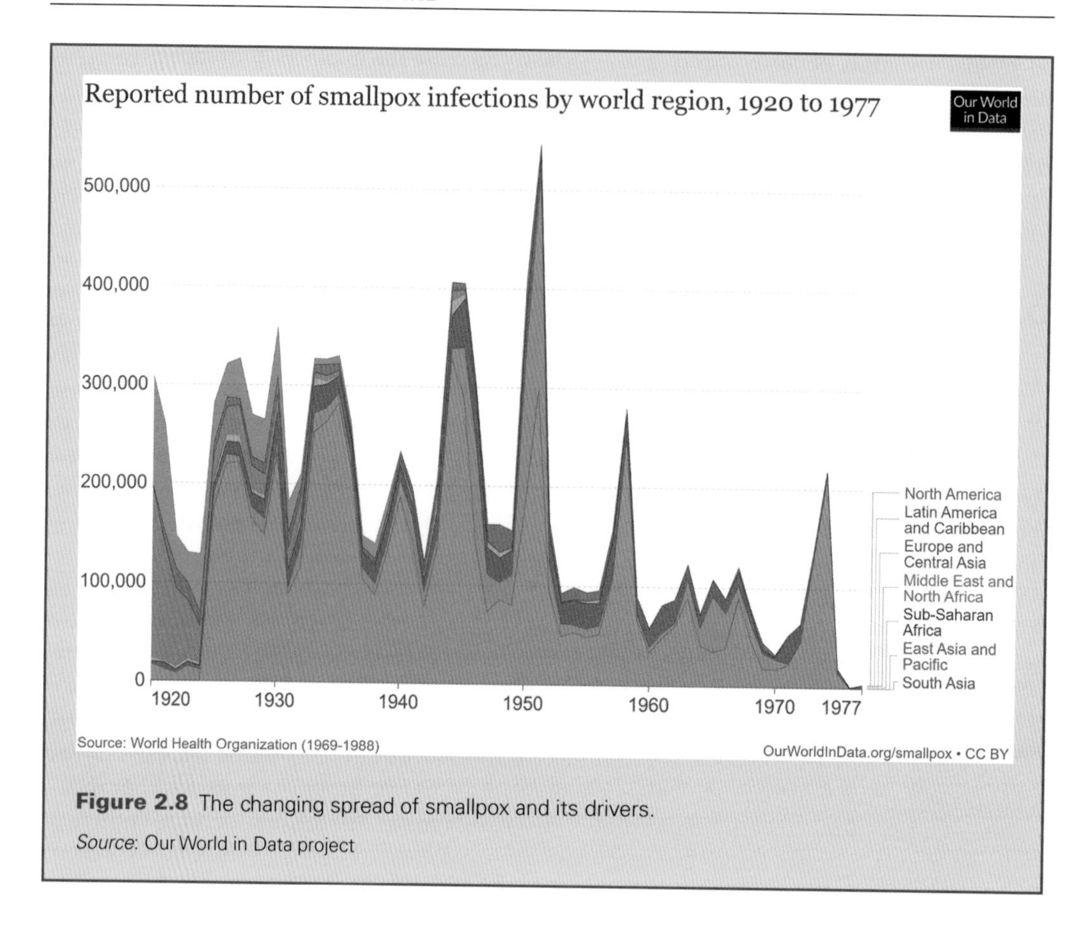

Figure 2.8 The changing spread of smallpox and its drivers.

Source: Our World in Data project

of large eruptions – those most likely to be reported and archived – remained fairly constant. Therefore, the apparent upward trend is really a measure of volcanic hazard awareness and improved monitoring and reporting over the period. When the world was preoccupied with alternative matters, such as World Wars I and II, volcanoes became less newsworthy, so the number of eruptions apparently declined due to less reporting rather than due to fewer eruptions.

These two examples show the importance of carefully evaluating how datasets, and the graphs from them, are constructed, so that changes in the data collection and reporting processes can be considered. Once these are taken into account, it is possible to examine how the hazards and disasters captured by these datasets are indeed changing due to factors such as social policies and programmes and environmental changes.

2. Disaster Trends

Since 1900, a rising trend in the recorded number of hazards and disasters is evident due to increased awareness of hazards, better scientific and technological monitoring, expanded media reporting and the development of organisations generating and disseminating databases. There is no clear evidence that the number of

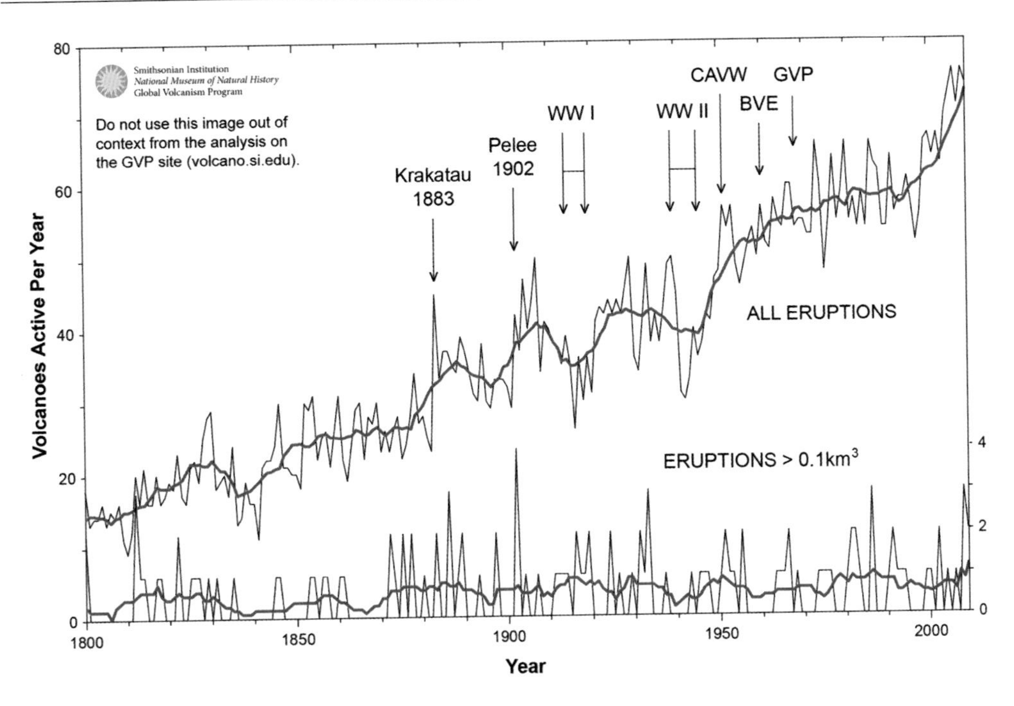

Figure 2.9 (a) Annual number of volcanic eruptions reported 1800–2010. This graph, for all reported eruptions, shows a clear upward trend, disturbed by two world wars. (b) Note the annual number of volcanic eruptions of size 0.1 km^3 or greater reported in the lower part of the graphs shows no overall trend but for the largest eruptions only.

Source: (a) Adapted from L. Siebert, T. Simkin and P. Kimberly, *Volcanoes of the World*, 2011.

hazards or disasters is actually rising. Changes in environmental processes that produce hazards, and which in turn can intersect with social processes to produce disasters, are influencing many hazards – especially meteorological, hydrological and climatological – with some increases and some decreases.

Often, a trend in declining mortality is touted in terms of absolute and proportional numbers, attributed to greater awareness of hazards and disaster risk reduction measures being implemented. For example, a Bangladesh cyclone in 1970 (when the country was geopolitically identified as East Pakistan) might have killed up to half a million people. Since then, warning systems, evacuation procedures and shelters were established and the mortality rates fell significantly to thousands and, lately, dozens per cyclone. Whilst such trends can indicate successes, they can also be misleading due to high-mortality events that do not recur frequently. Indian Ocean tsunami deaths were low until the disaster on 26 December 2004, while US hurricane deaths were clearly declining until Hurricane Katrina hit in 2005. A million-death earthquake has not yet occurred at the time of writing here, so when it does happen, trends in earthquake mortality will suddenly change.

Certainly, hazard parameters typically correlate poorly with disaster parameters. Two earthquakes in December 2003 with similar parameters shook California and

Iran leading to, respectively, two deaths and about 25,000 deaths (Box 1.9 in Chapter 1). Major North Sea storm surges affected eastern England in 1953 killing over 300 people on land and, with some higher water levels, in 2013 killing perhaps just two people. While hazards can be fairly similar, where they manifest will impact in different ways. Places lacking long-term disaster risk reduction typically display higher impacts including mortality, and mortality rates, than those that have put in a concerted effort to avoid disasters. Broader trends can hide important data needed to nuance the context of environmental hazards and disasters.

3. Changes in the Environment

Few databases contain records consistent and suitable enough for long-term analysis, although proxy geological and historical data can be used to extend directly observed time series. Understanding environmental changes, especially to hazards, is therefore not always straightforward. Tectonic activity is almost entirely natural, and discernible trends within a human lifetime are rare. Some climate- and weather-related trends, like tropical cyclones (referring to hurricanes, cyclones and typhoons, depending on the storm's location), show natural variations on numerous annual, decadal and multidecadal timescales. Many others, including local wind speeds and flood depths, are subject to multiple human influences across space and timescales. Land use and building construction alter these parameters, for instance by funnelling wind for a wind tunnel effect or augmenting running by paving over green areas for houses, roads and car parks.

As a result, identifying time trends in hazards and their parameters, as well as attributing causes to any apparent trends, is often difficult and changes with space and timescales. For instance, storm severity might show a trend over 10–20 years, but then it cannot be linked to natural or human-caused climate change because climate, by definition, is weather statistics over longer time periods. Even if a real trend towards more frequent, less frequent, more severe or less severe hazard parameters could be verified – for example, ash volume from volcanoes or maximum river flood flow rate – it is not inevitably possible to establish direct, one-to-one links with disaster dimensions including impacts. This is because disasters are usually caused by vulnerability, not environmental hazards. Any reported increase in disasters or disaster impacts can be attributed mainly to changes in society, not the environmental hazards. This reinforces the need to view disasters as complex (see Chapter 1, Section D6).

E. EXPLAINING DIMENSIONS: SPATIAL

While a dynamic physical environment continually triggers a host of potential hazards, the experience of disasters is mediated through vulnerabilities including poverty, resource distribution and exploitation, underdevelopment, conflict, oppression and inadequate governance. The current, entrenched global economic and resource systems thrive on political and economic inequalities and inequities, profoundly and detrimentally altering natural and human ecologies. Disasters have disparate consequences across spatial differences, depending on vulnerable populations and assets.

Many comparisons of disaster risk across spatial dimensions foreground factors such as (the flawed metric of) a country's GDP and (more robustly) a location's formal education trends. They then aggregate such values into higher, medium and lower ranges to categorise (a colonialist perspective of) 'developed' and 'developing' (and sometimes 'underdeveloped') countries, linking such labels to the assessment of disaster risk. Other approaches exist, which are always critiqueable but which might have some value.

One example, to be considered with respect to its advantages and limitations, is the World Risk Index (WRI) allowing aggregation and specificity to become visible, including the following (as taken from the World Risk Report, 2021, p. 6):

- *The vulnerabilities of island states*: A total of ten island states are among the 15 countries with the highest disaster risk. Their risk profile is increasingly determined by climate change induced sea-level rise.
- *Continent comparisons*: Oceania has the highest disaster risk. Africa, the Americas, Asia and Europe follow in descending order of disaster risk.
- *Societal vulnerability*: Africa has the highest societal vulnerability with 12 of the 15 most vulnerable countries in the world located there.
- *Disaster risk*: Europe has by far the lowest disaster risk of all continents, with a median of 3.27 across 40 countries. It is also in the most favourable position in all other components of the risk analysis.
- *Key trends*: Countries with low economic capacity and income tend to have higher vulnerability or lower coping capacities to avert disasters. In these countries, hazards often lead to further reductions in existing coping capacities. The examples of the Netherlands, Japan, Mauritius, and Trinidad and Tobago illustrate the principle that low or very low vulnerability can drastically reduce disaster risk.
- *Exposure and vulnerability*: Vanuatu is the most exposed, followed by Antigua and Barbuda, and Tonga. The most vulnerable country in the world is the Central African Republic, followed by Chad and the Democratic Republic of the Congo.

With appropriate critiques and further analyses, these findings (see also Figure 2.10) may contribute to formulating action recommendations.

F. TECHNOLOGICAL DIMENSIONS TO DISASTERS

The contribution of information technology and big data to addressing environmental hazards and disasters continues to grow. Prominence has been given to the use of near real-time data by emergency services and the public during emergency responses. More anticipatory forms of risk assessment emerge. Information technology and big data certainly have places in hazard monitoring and compiling disaster databases for later analysis and preparedness planning. Many applications of information technology are hazard- and disaster-specific – for example, developments in hazard forecasting and warning schemes – as described are reserved for Part 2 of this book. The section aims to provide a context in which to place more detailed treatments.

Increased computing capacity and speed continually offer fresh opportunities in project planning and real-time decision-making during hazards and disasters. By the early 1990s, relatively

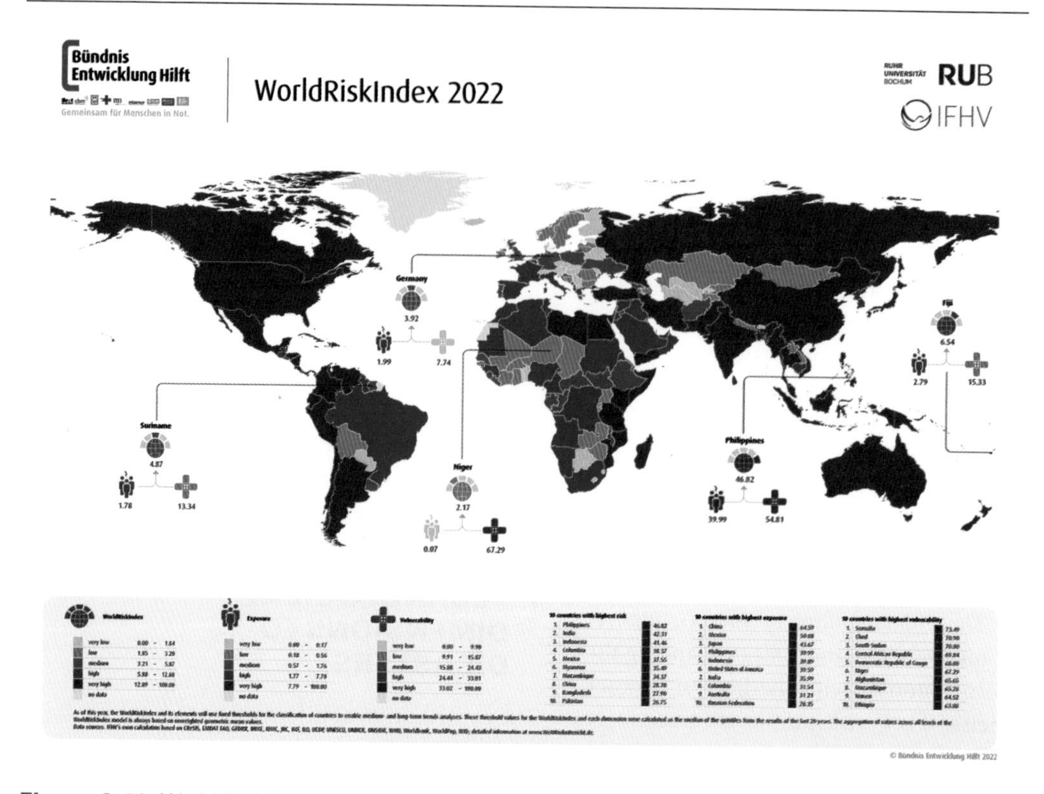

Figure 2.10 World Risk Index that states the risk of disaster for 181 of the world's countries in 2021. It is calculated on a country-by-country basis through the multiplication of exposure and vulnerability.

Data source: IFHV, based on the PREVIEW Global Risk Data Platform, Oak Ridge National Laboratory LandScan, CReSIS, CIESIN, NatCatSERVICE and global databases; detailed information at www.WorldRiskReport.org

powerful and networked desktop computer systems were an integral part of disaster management operations (Stephenson & Anderson, 1997). This time period saw improved telecommunications and simulation modelling and aiming for integration of the emerging technologies. In the 1990s to 2000s portable radio-based transmission systems permitted communication even when the ground-based infrastructure was destroyed, meaning that computers could be carried into remote or devastated areas where GPS technology permits instantaneous location fixing and tracking. By 2020 smartphones and tablets were key tools for inputting data to shared, cloud-based platforms (Usuda et al., 2019). This pivot to cloud storage raises issues around technology companies, data sovereignty, reliability and security.

Today whilst networked computer systems remain prone to power failure during disasters, the increasing reliability and speed of technology, as well as off-grid power sources for backup, can bring a number of significant benefits to dealing with environmental hazards and disasters, some already explored in this chapter. Sources such as Ushahidi, Twitter, Facebook and OpenStreetMap are discussed regarding their efficacy in assisting affected people in a disaster, with many local and contextual initiatives springing up during and after a disaster. Technologies can take observational data from real-time sensors such as river monitors or tsunami buoys and send it to car navigation systems and smartphones where it could be combined with GPS data from the vehicle or phone to immediately map out evacuation routes including the locations and flows of others, thereby preventing gridlock. Electronic road signs could be automatically updated and on-ramps closed to facilitate evacuations such as through reversing the flow of traffic away from an impending hazard. Countries have long used automatic systems to stop trains and shut down power plants and other operations as soon as a major nearby earthquake is detected. Telemetry from satellites, aircraft and drones (air, land, surface water and underwater) can be reoriented or specifically deployed for imaging and field survey work, helping to provide early warnings of famines, climate trends and out-of-sight but oncoming hazards, such as dust storms and flash floods. New technology can enable quicker, more reliable forms of communication, particularly via smartphones and satellite phones.

Expectations of where and how technologies could be implemented vary significantly. Often less 'smart' or less up-to-date technology has a lower entry barrier to access, enabling quick and efficient use while reducing costs and concerns about damage to or theft of the device. Decision-makers who need to use the technologies are often missing in the discussion around its development and application, leading to technological solutions without a problem having been identified. Training and alternative plans must be available for the failure of any technology.

1. Satellite Remote Sensing

Many forms of remote sensing have become integral to disaster risk reduction strategies and efforts, and technology continues to evolve at a rapid pace, making it imperative to be aware of recent developments (see Figure 2.11). In general, Earth observation satellites have supported pre-disaster preparedness through monitoring of physical processes, while information and communication technologies (ICT) have contributed to the dissemination of knowledge about hazard and risk and the coordination of response and rescue operations (Freeman et al., 2019). As already indicated, the specific application depends on the task and the hazard. For example, automated techniques are now in place for the detection of volcanic activity and wildfire by the satellite remote sensing of excess heat (Massimetti et al., 2020; Wooster et al., 2021). During the routine monitoring and land zoning of a volcanic cone, the imagery is unlikely to be time dependent, but during emergency operations, information is urgently required and must be available in all weather conditions.

The type of sensor used depends on the spatial or spectral resolution required. Radar data is needed when cloud obscures disaster areas, and a mixture of optical and infrared bands is best for wildfire detection. Integrated techniques are increasingly used – for example, the use of synthetic aperture radar (SAR) in assessing landslide and coastal erosion hazards (Lissak et al., 2020; Gebremichael et al., 2020). SAR technology is particularly useful for mapping the extent of surface floods. Other applications include lahar monitoring on the flanks of volcanoes (Cigna et al., 2020; Poland, 2022). Early detection of regional drought, before it grows into a disaster, is

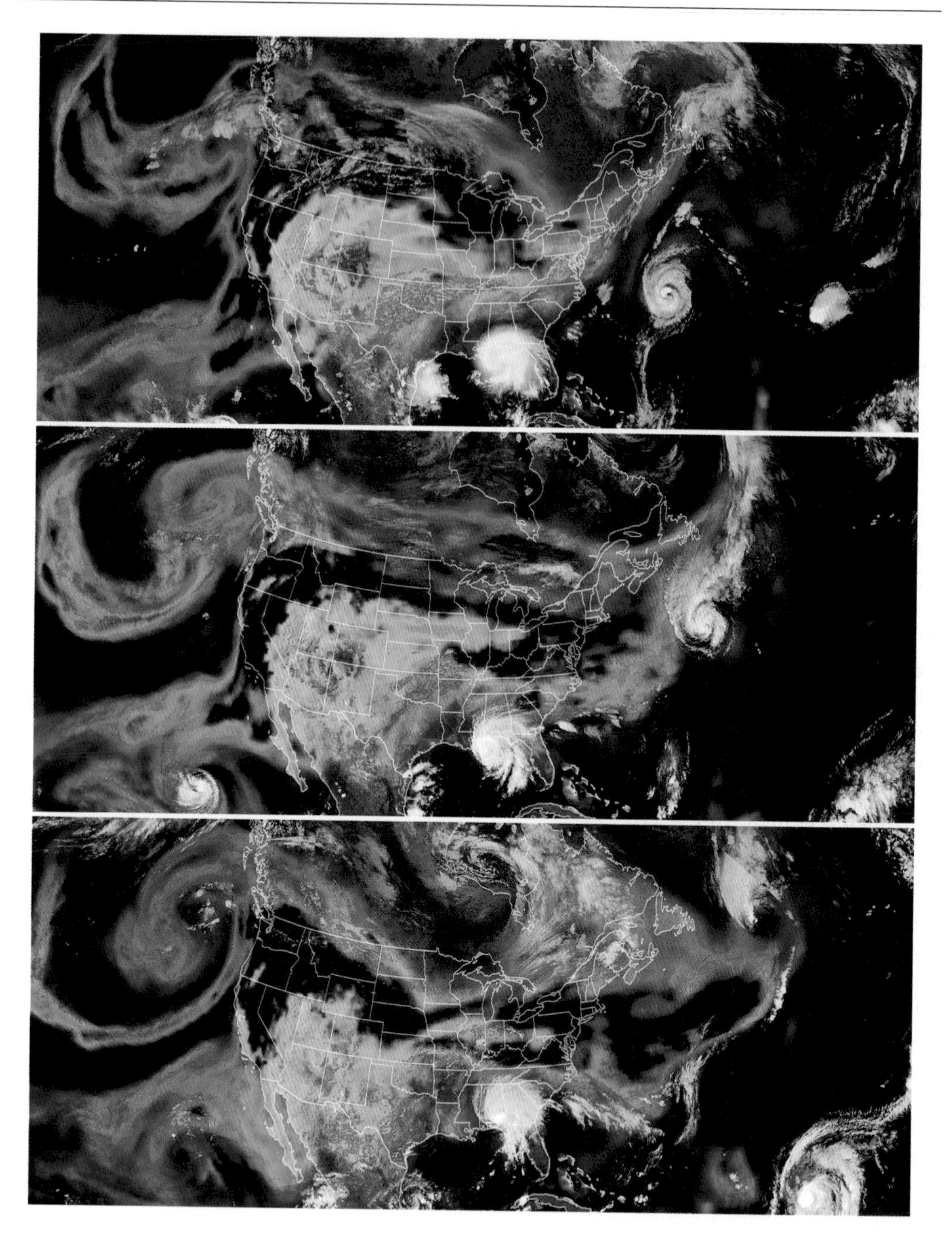

Figure 2.11 Three satellite maps from 14 (top), 15 and 16 September in 2020 show the gargantuan smoke plume from the West Coast wildfires brushing past hurricanes in the Gulf Coast and East Coast. In the middle image (15 September), Hurricane Paulette blocks the smoke from travelling further east over the Atlantic Ocean.

Source: Image: Joshua Stevens/NASA

also possible. Because of the high cost of development and launching, Earth observation satellites have not been deployed as widely as communication platforms, although developments in small-satellite and constellation-satellite technologies have increased their number and scope (da Silva Curiel et al., 2002; Gauthier et al., 2021; Marcuccio et al., 2019).

Satellites provide a relatively cost-effective, global coverage of volcanic activity through the detection of thermal anomalies and plume tracking. Similarly, large-scale drought monitoring is possible through changes in surface albedo and the application of a vegetation index (VI) that measures vegetation stress (Chen et al., 2019; Wang et al., 2020). This information can be used for a variety of purposes, ranging from a change in cropping patterns and irrigation practices early in a growing season to the late-season estimation of crop yields and their possible effects on food supplies. The mapping of flood-affected areas is also highly successful, due to differences in the spectral signatures for different types of inundation – standing water, submerged crops, areas of floodwater retreat and so on. In addition, the topographic information necessary for hazard zone mapping can be provided by instruments such as the SPOT and ERS satellites, which have stereo imaging for this purpose. Over the last two decades the availability of very high-resolution instruments, such as through the *Ikonos* and *Quickbird* satellite series, has allowed the identification of individual structures and even earthquake-induced cracks in the ground (Fan et al., 2019, Ghaedi, 2021). This is now permitting the assessment of damage to be undertaken remotely.

There are limitations. Remotely sensed imagery needs filtering and correction, a process that is expensive and time consuming. The very high-resolution satellites typically image each area every few days only. In addition, the instruments are optical in character, meaning that they cannot penetrate cloud cover. Radar instruments can do this, but the data resolution is often too poor to allow useful analysis for short-term damage assessment. High-resolution data is also very expensive to purchase. An attempt was made to address some of these issues by the establishment of the International Charter on Space and Major Disasters in 1999. Almost all of the satellite data providers are signatories to this charter, which allows member organisations (mostly government bodies, international agencies and the major NGOs) to acquire satellite data for disaster areas free of charge. In the aftermath of the 2005 Kashmir earthquake, the charter was used to allow the acquisition of data to assist with the relief operations in Pakistan and India. By 2018, the charter had been activated over 550 times and has provided data to 119 countries (Clark, 2018). However, the effectiveness of such voluntary systems remains limited by difficulties in analysing the remotely sensed data in a timely fashion and the problems of communicating the analyses to end users on the ground in an area with poor communication networks. Consequently, satellite data is not always available for real-time monitoring depending on the location of the hazard and event and the frequency of data coverage by the satellite.

2. Geographical Information Systems

GIS refers to the coming together of hardware, in the form of a physical computer system (which may be a desktop but increasingly includes tablets and smartphones), software in the form of algorithms (programs, rules or commands) for performing tasks as data input, storage, retrieval, analysis, query, output and updating, and people tasked with collecting and/or analysing and using the outputs of an analysis. Key to a GIS is the geolocating of a wide range of quantified datasets, such that spatiotemporal patterns within these datasets can be assessed, and these patterns in turn can be framed as a new, constructed dataset. In addition to organising, building and analysing datasets, GIS can be used to predict and project data trends, thus enabling decision-making and policy formation. What is more, GIS can be used to visualise cartographically both the initial datasets and those created via integration and analysis. Spatial maps can be produced, for example, that build on the integration and analysis of a series of datasets to produce various measures for 'risk' and 'vulnerability'. Indeed, there have emerged various aggregate measures of and mapping approaches to vulnerability (Bankoff et al., 2004; Cadag & Gaillard, 2012; Papathoma-Koehle et al., 2011). As Fekete (2012) cautions:

> Increasingly, this type of spatial assessment aims at providing 'policy-relevant' information. However, among scientists, little is known about how these maps or products are perceived by their users or the public. Vulnerability maps may be published and communicated with unanticipated and unintended reactions or even misinterpretation by their recipients.
>
> (Fekete, 2012, p. 1162)

Though GIS technology can be a major resource for disaster mitigation and emergency managers, there is a significant inequality in regard to what data is available where, what scale the data is at, and who can access and use this data (see Table 2.4). Many local government offices and other agencies do routinely hold geolocated archives of contours, rivers, geology, soils, highways, census data, phone listings and the areas subject to flooding, or other hazards, for their area. This archived information can then be integrated with dynamic layers of information on evolving hazards such as floods, vulnerabilities of particular groups and the impacts of disasters. Much of this information can be extracted from remotely sensed satellite data and/or drone data, as well as social media. However, there remains a lack of geolocated archives at the local level especially in some areas, which makes it difficult to map the upholding dimensions of hazard, risk and disaster and to integrate georeferenced data. When disasters do occur, accessible satellite imagery becomes a crucial input for a GIS assessment of the scope and impact of these.

Much deployment of GIS in hazard and disaster fields has aimed for improved warning and evacuation systems, planning for evacuation routes that minimise the distance people travel to reach shelters (No et al., 2020). Emergency managers in many countries now use GIS information to identify the areas to be evacuated when a hazard such as a hurricane, tornado, fire or flood is forecast. GIS can also be applied in the recovery period following disaster, when the need may be to devise alternative evacuation routes, plan the relocation of facilities and make other land use decisions. GIS and GPS technology has also been adopted in the humanitarian sector (Kaiser et al., 2003). Early applications were in the control of disease outbreaks, but later deployments involve large-scale vulnerability assessment, mortality

Table 2.4 Satellite data comprises spectral bands in visible (VIS), near infrared (NIR), short-wave infrared (SWIR), thermal infrared (TIR) and synthetic aperture radar (SAR), providing information on proxies for a series of environmental phenomena, as well as proxies for population density and land use.

Wavelength	Waveband	Applicable for	Sensors example
Visible (VIS)	0.4–0.7 mm	• Vegetation mapping • Building stock assessment • Population density • Digital elevation model	• SPOT; Landsat TM • AVHRR; MODIS; IKONOS • IKONOS; MODIS • ASTER; PRISM
Near infrared (NIR)	0.7–1.0 mm	• Vegetation mapping • Flood mapping	• SPOT; Landsat TM; AVHRR; MODIS; • MODIS
Short-wave infrared (SWIR)	0.7–3.0 mm	• Water vapour	• AIRS
Thermal infrared (TIR)	3.0–14.0 mm	• Active fire detection • Burn scar mapping • Hotspots • Volcanic activity	• MODIS • MODIS • MODIS; AVHRR • Hyperion
Microwave (Radar)	0.1–100 cm	• Earth deformation and ground movement • Rainfall • River discharge and volume • Flood mapping and forecasting • Surface winds • 3D storm structure	• Radarsat SAR; PALSAR • Meteosat; Microwave Imager (aboard TRMM) • AMSR-E • AMSR-E • QuikScat radar • Precipitation radar (aboard TRMM)

Source: Krishnamoorthi, Natarajan. 'Role of remote sensing and GIS in natural-disaster management cycle.' Imp J Inter Res 2.3 (2016): 144–154. Page 146.

surveys, the rapid identification of basic disaster needs (such as water, food and fuel) and the mapping of population movements. As Greenough and Nelson (2019) observe,

> Logisticians characterize areas most affected or insecure and optimize supply lines. Water and sanitation engineers delineate coordinates of water tables, envision distribution lines and plot water delivery points. Food security agencies develop models to streamline the route between food sources and affected-populations. Shelter managers digitize the ebb and flow of displaced populations, and health providers map facilities, disease incidence and care delivery capacities.
>
> (p. 1)

Tracking the movement of people has in recent years been enabled by the aggregation of large amounts of mobile phone data. Many mobile phone applications have a geolocation component, allowing the tracking of people before, during and after, for example, an evacuation (Hong & Akerkar, 2021).

GIS as an integrative technology that can be used to assess spatiotemporal patterns is dependent on the scope and quality of the data available. While environmental datasets may be derived from satellites, aircraft and drones, there are challenges to accessing social datasets, despite the massive uptick over the past decade in data brokering, wherein large amounts of information is aggregated, packaged and sold as a product. As Balsari et al. (2021) note in regard to assessing the dimensions of disaster health impacts via GIS:

> Remarkably, although comprehensive patient profiles are routinely constructed from electronic medical records, pharmacy data, and claims data – to be traded and sold commercially – they are not readily available to patients, hospitals, or response planners. Similarly, although technology companies have extensive data on user habits, location, and mobility patterns, relevant aggregated data sets or analyses are seldom shared with public health response agencies.
>
> (p. 1527)

3. Information and Communication Systems

The role of smart technologies, applications and social media platforms continues to expand, in both constructive and retrogressive ways. They are widely used for disseminating official information about hazards and disasters and general advice on emergency procedures. For example, many emergency and disaster agencies from governments, businesses and NGOs update online resources in real time, leave archived material and provide distant learning for certification courses in disaster-related fields. Less formal channels permit the rapid sharing of information, including images, advice for people in the affected area and platforms for matching needs with donations. These processes can stimulate and direct emergency action whilst also increasing public awareness. Handheld devices connected to data or wifi can be routinely programmed to receive automated alerts from an emergency operations centre. Cell networks may not always operate during a disaster, but they can be, so this facility will be more useful before a hazard manifests, perhaps to provide guidance on evacuation, or afterwards to provide information for recovery. Fire departments and other emergency personnel can communicate over wireless networks to exchange text-based information about a developing situation.

The use of social media to crowdsource information on hazards and disasters has become an area of interest for disaster management and humanitarian organisations, as well as emergency and relief services. According to Kankanamge et al.:

> The 2010 Haiti Earthquake was the tipping point for realising the value of crowd power, outsourcing, and volunteerism. . . . 640 volunteers contributed to creating road maps of Haiti in two weeks. Volunteers mapped the roads and humanitarian features including displaced persons camps of Haiti. Further, the Haiti government launched a mission, where people could send text requests for shelter, food, and medicines. Later, the tasks of translating, geocoding, and categorising of 4,636 messages were outsourced through social media showed the possibility of using social media for disease prevention after a disaster.
>
> (2019, p. 2–3)

The 2010 Haiti earthquake also illustrates limitations when applying ICTs for dealing with environmental hazards and disasters. First, quality control. Who triangulates, validates, corroborates or filters crowdsourced and voluntary information? With so much information appearing and spreading swiftly, what resources are put into checking it? How much of voluntary supplied information is really new or more useful than what would be provided by other channels?

Yet attempts at quality control demonstrate the second major limitation: power relationships. Power relationships are not limitations of technologies per se, rather being part of all human interactions. Who decides which information needs vetting and how to vet it? In particular, leading to a third major limitation, who decides about access to the information and the technology, in terms of issues such as devices, power supplies, internet access and languages for the information? How will all people with disabilities be served? What happens when companies force software or hardware upgrades or security bugs are uncovered?

These issues have not diminished with time. In January 2022, a volcanic eruption in Tonga led to a tsunami inundating many of the countries' settlements. In February 2022, Russia invaded Ukraine. In both disasters, a private company Starlink offered its satellite-based internet access through delivering kits with the router, cables and base station along with bandwidth. Who prioritises recipients? Who decides when access is reduced or cut off? Who deals with the electricity needs and repairs? All technologies have always raised ethical dilemmas with application of the technology generally outpacing resolution of moral conundrums.

Communication technologies and devices are increasingly entangled with other smart technologies that collect, analyse and distribute big data. As a 2019 report prepared for the European Parliament observes, these collections of smart, digital technologies are fundamentally transforming the humanitarian sector. Components of the 'digital turn' in this sector include:

> the availability and usage of mobile phones and social media platforms by people affected by [disaster], geospatial technologies and unmanned aerial vehicles (UAVs) to detect evidence of [disaster impacts], biometric identification to facilitate humanitarian support, and a shift to digital payments as relief provisions with e-vouchers and mobile money.
>
> (Capgemini Consulting, the Netherlands, 2019, p. 10)

These technologies are valued as providing access to more accurate information that aggregates fine-grained data, thus strengthening the capacity of organisations to act quickly and to anticipate and better prepare for disasters, as well as respond (see Figure 2.12). Digital technologies including blockchain/encryption and satellite/drone data – both of which can be engaged with using smartphones – have also become sites of experimentation and practice in well-established UN organisations such as the UN Agency for Refugees (UNHCR), World Food Programme (WFP) and the UN Children's Fund (UNICEF). As Searle notes, drones 'are beaming data directly to software programmes to produce real-time maps of disaster-affected areas and populations in extremis' (2017, p. 3). In addition to carrying high-resolution cameras and thermal imaging equipment that can sense a range of environmental phenomena and inform three-dimensional maps, drones can also host wifi, becoming makeshift towers. They can even be used to transport materials (including medical samples and treatments) from one location to another. Nevertheless, there are substantial concerns regarding the ways in which personal information is protected (or not) in the collection and distribution of big data and concerns around the manner in which an individual's privacy (that is, their capacity not to be observed and recorded) can be impinged upon.

Continued discussion, exchange, education, exchange, teaching and learning are essential. Not as so-called experts telling everyone else. Instead, conversations and dialogues to shore up everyone's limited knowledge with the best from others. At local and household levels, a need exists to understand capabilities, capacities, abilities, limitations and gaps regarding hazard mitigation and especially vulnerability reduction. This must be achieved through multiple forms of communication, including but not limited to brochures, maps, videos, seminars, workshops, training exercises, scenarios, focus groups, academic presentations and scientific publication. Arts must be part of this work: song, dance, paintings, photography, theatre, story telling, narration and many more (see examples on the Anticipation Hub: www.anticipation-hub.org/). All these apply at national, continental and global levels too. International organisations, relief agencies and scientific institutes are not immune to bias and require social, governance and technical support to improve, which would involve pooling resources and knowledge while learning from impacted communities.

Beyond environmental hazards, risks are usually spread unevenly among people, locations and groups. Estimating costs and benefits of action and inaction has major difficulties, particularly when considering lives, injuries, livelihood and intangibles gained or lost. When funds are allocated, institutional weaknesses, lack of technical and personnel expertise, the absence of regulations and the poor enforcement of existing legislation might constrain the effectiveness of the resources – including complaints about corruption, misuse, fraud, misappropriation,

Figure 2.12 Drone footage recorded over the Nepalese capital Kathmandu showing the scale of destruction following a major earthquake on 25 April 2015.

Source: Image captured from YouTube (Sky News, 2015)

siphoning and lack of accountability. These factors are significant, continual problems, and their persistence might lead to low aspirations about and low achievements of disaster risk reduction. All such dimensions of disaster need to be admitted and documented in order to overcome them.

KEY TAKE AWAY POINTS

- The scale of a disaster is typically measured in quantitative terms, such as loss of life and insurance payouts. Not all losses and impacts can be measured. Consider how all losses and impacts can be represented, as well as the many uncertainties around quantitative data itself.
- The public's perception of the scale of a disaster – and the potential for disaster – is informed by traditional and new social media and consists of fictional as well as documentary accounts. While fictional accounts – in film, TV and literature – might be dismissed as inaccurate, they nevertheless can convey the scope and magnitude of events, the complex social response to events and the emotional experience of those living through events.
- New social media technologies are enabling the crowdsourcing of information about hazardous conditions and disasters. This information is increasingly recognised by disaster risk managers and humanitarian organisations as helpful, while acknowledging limitations, as all data and information have.
- Satellite and drone imagery – particularly when used in conjunction with geographic information systems – is providing ever more detailed information on the unfolding of environmental hazards and indicators of vulnerability, including exposure. Much work remains to be done to ensure that the outputs of this work are shared with impacted communities and that rights to privacy and meaningful consent are accounted for in data collection and analysis.

BIBLIOGRAPHY

Atwood, Margaret. 2013. MaddAddam. New York: Doubleday.

Bakare, L. (2020). Art project captures sound of cities during coronavirus outbreak Tue 31 March. *Guardian*. www.theguardian.com/world/2020/mar/31/art-project-captures-sound-of-cities-during-coronavirus-outbreak

Balsari, S., Kiang, M. V., & Buckee, C. O. (2021). Data in crisis—Rethinking disaster preparedness in the United States. *The New England Journal of Medicine, 385*(16), 1526.

Bankoff, G., Frerks, G., & Hilhorst, T. (Eds.). (2004). *Mapping vulnerability: Disasters, development and people*. London: Earthscan.

Below, R., Vos, F., & Guha-Sapir, D. (2010). *Moving towards harmonization of disaster data: A study of six Asian databases.* Brussels: Centre for Research on the Epidemiology of Disasters.

Bird, D., Ling, M., & Haynes, K. (2012). Flooding Facebook: The use of social media during the Queensland and Victorian floods. *The Australian Journal of Emergency Management, 27*, 27–33.

Brazee, E. (2020). *Recording london landscapes, past and present.* www.museumoflondon.org.uk/discover/recording-london-soundscapes-past-present. Accessed 10 November 2021.

Bruns, A., & Burgess, J. (2012). Local and global responses to disaster:# Eqnz and the Christchurch earthquake. Proceedings of the 2012 Australian and New Zealand Disaster and Emergency Management Conference (pp. 86–103), AST Management Pty Ltd, Brisbane. https://snurb.info/files/2012/Local%20and%20Global%20Responses%20to%20Disaster.pdf

Butler, D. (2013). Crowdsourcing goes mainstream in typhoon response. *Nature* [Online]. www.nature.com/news/crowdsourcing-goes-mainstream-in-typhoon-response-1.14186. Accessed 26 October 2018.

Butler, Octavia E. Parable of the Sower. New York: Warner Books, 1993

Cadag, J. R. D., & Gaillard, J. C. (2012). Integrating knowledge and actions in disaster risk reduction: The contribution of participatory mapping. *Area, 44,* 100–109.

Capgemini Consulting, the Netherlands. (2019, May). Technological innovation for humanitarian aid and assistance. *European Parliamentary Research Service.* www.europarl.europa.eu/RegData/etudes/STUD/2019/634411/EPRS_STU(2019)634411_EN.pdf. Accessed 9 November 2022.

Chen, Y., et al. (2019). Application of Sentinel 2 data for drought monitoring in Texas, America. 2019 8th International Conference on Agro-Geoinformatics (Agro-Geoinformatics), IEEE, Danvers, MA.

Cigna, F., Tapete, D., & Lu, Z. (2020). Remote sensing of volcanic processes and risk. *Remote Sensing, 12*(16), 2567.

The Cities and Memories Project. (2022). https://citiesandmemory.com/covid19-sounds/. Accessed 10 November 2022.

Clark, N. E. (2018). Gauging the effectiveness of soft law in theory and practice: A case study of the International Charter on Space and Major Disasters. *Air and Space Law, 43*(1).

Dabner, N. (2012). 'Breaking ground' in the use of social media: A case study of a university earthquake response to inform educational design with Facebook. *The Internet and Higher Education, 15*(1), 69–78.

da Silva Curiel, A., Wicks, A., Meerman, M., Boland, L., & Sweeting, M. (2002). Second generation disaster-monitoring microsatellite platform. *Acta Astronautica, 51*(1–9), 191–197.

Eisensee, T., & D. Strömberg (2007). News Droughts, News Floods, and U. S. Disaster Relief. *The Quarterly Journal of Economics, 122,* 693–728.

Fan, X., et al. (2019). Earthquake-induced chains of geologic hazards: Patterns, mechanisms, and impacts. *Reviews of Geophysics, 57*(2), 421–503.

Fekete, A. (2012). Spatial disaster vulnerability and risk assessments: Challenges in their quality and acceptance. *Natural Hazards, 61,* 1161–1178.

Freeman, J. D., et al. (2019). Use of big data and information and communications technology in disasters: An integrative review. *Disaster Medicine and Public Health Preparedness, 13*(2), 353–367.

Gall, M., Borden, K.A., & Cutter, S.L. (2009). When Do Losses Count? Six Fallacies of Natural Hazards Loss Data. *Bulletin of the American Meteorological Society, 90,* 799–810.

Gauthier, D. J., Bollt, E., Griffith, A., & Barbosa, W. A. (2021). Next generation reservoir computing. *Nature Communications, 12*(1), 5564.

Gebremichael, E., et al. (2020). Flood hazard and risk assessment of extreme weather events using synthetic aperture radar and auxiliary data: A case study. *Remote Sensing, 12*(21), 3588.

Ghaedi, K., Gordan, M., Ismail, Z., Hashim, H., & Talebkhah, M. (2021). A literature review on the development of remote sensing in damage detection of civil structures. *Journal of Engineering Research and Reports, 20*(10), 39–56.

Greenough, P. G., & Nelson, E. L. (2019). Beyond mapping: A case for geospatial analytics in humanitarian health. *Conflict and Health, 13*(1), 1–14.

Guha-Sapir, D., & Below, R. (2002). The quality and accuracy of disaster data: A comparative analysis of three global data sets. Working Paper prepared for the Disaster Management Facility, World Bank, CRED, Copyright CRED 2009.

Guzman, Andrew T. Overheated: The Human Cost of Climate Change. Oxford University Press, USA, 2013.

Hong, M., & Akerkar, R. (2021). Victim detection platform in IoT paradigm. *Concurrency and Computation: Practice and Experience, 33*(3), e5254.

IFRC. (2022). *What is a disaster?* www.ifrc.org/what-disaster. Accessed 31 May 2022.

Intergovernmental Oceanographic Commission and Indian Ocean Tsunami Information Center. (2015). *Remembering the 1945 Makran tsunami: Interviews with survivors beside the Arabian Sea.* https://iotic.ioc-unesco.org/wp-content/uploads/2020/08/Remembering-the-1945-Makran-Tsunami.pdf. Accessed 31 May 2022.

Joyce, Karen E., et al. "A review of the status of satellite remote sensing and image processing techniques for mapping natural hazards and disasters." Progress in physical geography 33.2 (2009): 183-207.

Kaiser, R., Spiegel, P. B., Henderson, A. K., & Gerber, M. L. (2003). The application of geographic information systems and global positioning systems in humanitarian emergencies: Lessons learned, pro gramme implications and future research. *Disasters, 27*(2), 127–140.

Kankanamge, N., et al. (2019). Can volunteer crowdsourcing reduce disaster risk? A systematic review of the literature. *International Journal of Disaster Risk Reduction, 35*, 101097.

Lissak, C., et al. (2020). Remote sensing for assessing landslides and associated hazards. *Surveys in Geophysics, 41*(6), 1391–1435.

Lynas, Mark. "Six Degrees: Our Future on a Hotter Planet. London: Fourth Estate." (2007).

Marcuccio, S., Ullo, S., Carminati, M., & Kanoun, O. (2019). Smaller satellites, larger constellations: Trends and design issues for earth observation systems. *IEEE Aerospace and Electronic Systems Magazine, 34*(10), 50–59.

Marres, G. M., Leenen, L. P., deVries, J., Mulder, P. G., & Vermetten, E. (2011). Disaster-related injury and predictors of health complaints after exposure to a natural disaster: An online survey. *BMJ Open, 1*(2), e000248.

Marulanda, M. C., Cardona, O. D., & Barbat, A. H. (2010). Revealing the socioeconomic impact of small disasters in Colombia using the DesInventar database. *Disasters, 34*(2), 552–570.

Massimetti, F., et al. (2020). Volcanic hot-spot detection using SENTINEL-2: A comparison with MODIS: MIROVA thermal data series. *Remote Sensing, 12*(5), 820.

Monahan, B., & Ettinger, M. (2018). News Media and Disasters: Navigating Old Challenges and New Opportunities in the Digital Age. In: Rodríguez, H., Donner, W., & Trainor, J. (eds) *Handbook of Disaster Research. Handbooks of Sociology and Social Research*. Springer, Cham, pp. 479–495.

No, W., Choi, J., Park, S., & Lee, D. (2020). Balancing hazard exposure and walking distance in evacuation route planning during earthquake disasters. *ISPRS International Journal of Geo-Information, 9*(7), 432.

Norris, F. H., Sherrieb, K., & Galea, S. (2010). Prevalence and consequences of disaster-related illness and injury from Hurricane Ike. *Rehabilitation Psychology, 55*(3), 221.

Papathoma-Koehle, M., Kappes, M., Keiler, M., & Glade, T. (2011). Physical vulnerability assessment for alpine hazards: State of the art and future needs. *Natural Hazards, 58*(2), 645–681.

Poland, M. P. (2022). Synthetic aperture radar volcanic flow maps (SAR VFMs): A simple method for rapid identification and mapping of volcanic mass flows. *Bulletin of Volcanology, 84*(3), 1–11.

Poontirakul, P., Brown, C., Seville, E., Vargo, J., & Noy, I. (2017). Insurance as a double-edged sword: Quantitative evidence from the 2011 Christchurch earthquake. *The Geneva Papers on Risk and Insurance-Issues and Practice, 42*(4), 609–632.

Rashid, H. (2011). Interpreting flood disasters and flood hazard perceptions from newspaper discourse: Tale of two floods in the Red River valley, Manitoba, Canada. *Applied Geography, 31*, 35–45.

Reddick, Christopher. "Information technology and emergency management: preparedness and planning in US states." Disasters 35.1 (2011): 45-61.

Rodrigue, C. M., & Rovai, E. (1995). The 'Northridge' earthquake: Differential geographies of damage, media attention, and recovery. *National Social Science Perspectives Journal, 7*(3), 98–111.

Sakurai, Mihoko, and Yuko Murayama. "Information technologies and disaster management–Benefits and issues." Progress in Disaster Science 2 (2019): 100012.

Searle, M. (2017). *Humanitarian technology: New innovations, familiar challenges, and difficult balances.*

Sennert, S. S. K., Klemetti, E. W., & Bird, D. K. (2018). Role of social media and networking in volcanic crises and communication. In C. J. Fearnley, D. K. Bird, K. Haynes, W. J. Mcguire, & G. Jolly (Eds.), *Observing the volcano world: Volcano crisis communication* (pp. 733–743). Cham: Springer International Publishing.

Sky News (2015, 27th April). *Drone footage shows Nepal earthquake damage.* YouTube: https://www.youtube.com/watch?v=Wwlw1-voHKQ

Solman, P., & Henderson, L. (2019). Flood disasters in the United Kingdom and India: A critical discourse analysis of media reporting. *Journalism, 20*(12), 1648–1664.

Sonnett, J., Johnson, K.A., & Dolan, M.K. (2015). Priming Implicit Racism in Television News: Visual and Verbal Limitations on Diversity. *Sociological Forum, 30*, 328–347.

Spong, D. (2011). New media and the 2011 Tōhoku earthquake and tsunami [Online]. *Prezi*. http://prezi.com/sh2lm6f-pleyg/new-media-and-the-2011-tohoku-earthquake-and-tsunami/. Accessed 10 January 2012.

Stephenson, R., & Anderson, P.S. (1997) Disasters and the Information Technology Revolution. *Disasters, 21*, 305–334.

Tschoegl, L., Below, R., & Guha-Sapir, D. (2006). *An analytical review of selected data sets on natural disasters and impacts*. Brussels, Belgium: Centre for Research on the Epidemiology of Disasters.

UNDRR and International Science Council. (2020). *Hazard definition and classification review*. www.undrr.org/publication/hazard-definition-and-classification-review

UNESDOC Digital Library. https://meddialogue.eu/e-library/unesdoc-unesco-digital-library/

Usuda, Y., Matsui, T., Deguchi, H., Hori, T., & Suzuki, S. (2019). The Shared Information Platform for Disaster Management – The Research and Development Regarding Technologies for Utilization of Disaster Information. *Journal of Disaster Research*, 14, 279–291.

Wang, Z., et al. (2020). Integration of microwave and optical/infrared derived datasets from multi-satellite products for drought monitoring. *Water, 12*(5), 1504.

Wooster, M. J., et al. (2021). Satellite remote sensing of active fires: History and current status, applications and future requirements. *Remote Sensing of Environment, 267*, 112694.

World Risk Report (2021). World Risk Report. Bündnis Entwicklung Hilft Ruhr University Bochum – Institute for International Law of Peace and Armed Conflict (IFHV), Bochum, Germany.

Wouter Botzen, W. J., et al. (2019). Integrated disaster risk management and adaptation. In R. Mechler, L. Bouwer, T. Schinko, S. Surminski, & J. Linnerooth-Bayer (Eds.), *Loss and damage from climate change: Climate risk management, policy and governance*. Cham: Springer. https://doi.org/10.1007/978-3-319-72026-5_12

Yu, Manzhu, Chaowei Yang, and Yun Li. "Big data in natural disaster management: a review." Geosciences 8.5 (2018): 165.

FURTHER READING

Aronsson-Storrier, M., & Dahlberg, R. (Eds.). (2022). *Defining disaster: Disciplines and domains*. Cheltenham, UK: Edward Elgar. An up-to-date volume of the needs and perils of defining a disaster.

Bankoff, G., & Hilhorst, D. (2022). *Why vulnerability still matters: The politics of disaster risk creation*. Abingdon, UK: Routledge.

Mosquera-Machado, S., & Dilley, M. (2009). A comparison of selected global disaster risk assessment results. *Natural Hazards, 48*, 439–456. An attempt to compare two methods of quantitative disaster risk assessment.

Quarantelli, E. L. (1998). *What is a disaster?* New York, USA: Routledge. One of the original books compiling multiple understandings of disaster.

Scanlon, J., & Frizzell, A. (1979). Old theories don't apply: Implications of communications in crises. *Disasters, 3*(3), 315–319. A foundational article on media, communications, and disasters.

WEB LINKS

See Table 2.2 for individual hazard links and monitoring data.

Disaster Deaths Analysis. https://ilankelman.org/disasterdeaths.html

EM-DAT: The International Emergency Events Database Run by the Centre for Research on the Epidemiology of Disasters (CRED). www.emdat.be/

NASA Earth Data. https://earthdata.nasa.gov

UN Spider Network Knowledge Portal. www.un-spider.org/network

World Organization of Volcano Observatories (WOVO). www.wovodat.org

CHAPTER THREE

Vulnerability, Resilience and Sustainability

3

OVERVIEW

This chapter provides an overview of vulnerability and resilience as concepts used in the hazards and disaster field, emphasising why they are important when considering the different impacts a hazard will have on a community or area over time. While noting briefly which particular academic disciplines these terms have emerged from, as well as summary definitions by disaster-focused organisations, the key message of this chapter is that these concepts are contested in their definition and are used differently by different stakeholders involved with hazards and disasters. Vulnerability is a concept used in many disciplines, from engineering to psychology, each with their own particular approaches and concerns. Nevertheless, what these diverse uses of the concept tend to have in common is an awareness of the many different economic, political, and social relationships – as well as environments – that people are part of, exposing them to harm and possibly limiting their ability to both anticipate and respond to harm. Here, we focus on vulnerability and development; mapping vulnerability; and vulnerability and the environment. Resilience is a more recent concept in the hazard and disaster field, used by academics, NGOs, funding agencies and governmental organisations operating at all scales. Here, we note how resilience has been understood as the capacity to recover from disaster and explore some of the challenges as well as opportunities that emerge from this concept. We finish the chapter with a brief section on sustainability science.

A. INTRODUCTION

In Chapter 1 we introduced hazards by noting the physical phenomena and processes that trigger conditions that can prove dangerous to people, alongside the observation that many of the Earth's ecologies are overexploited and/or subject

DOI: 10.4324/9781351261647-4

to degradation and pollution, producing a suite of hazardous conditions such as loss of biodiversity and increased toxicity. We discussed disasters as the differential exposure of individuals and communities to these hazards, and the differential impacts that can emerge that are damaging for people. Economic insecurity, for example, was discussed as a term that foregrounds the fact that many disasters are felt more keenly by people dependent on subsistence livelihoods, many of whom continue to live with the political and economic legacies of colonialism and postcolonialism. Moreover, many people across the globe are experiencing economic insecurity because of austerity policies that have stripped back social welfare and public health infrastructures. Rapid-onset hazards such as earthquakes and flash floods, but also more creeping changes such as pandemics and toxic build-up within ecologies, can be devastating to those already vulnerable through lack of access to financial resources (including insurance) and insufficient welfare and public health infrastructures.

In this chapter we build on this work by focusing on vulnerability and resilience and by referencing the ever more important, broad debate around sustainability. We look at how vulnerability has been understood and researched as a differential exposure to and loss from disasters. Resilience broadly refers to the ways and means by which individuals and communities can recover from the impacts of disasters. And, sustainability as a concept broadly refers to ways of living within an environment that are reproducible in the future without significant, damaging impacts on healthy societies or ecologies. All three terms – vulnerability, resilience and sustainability – have many divergent definitions, interpretations, explanations and understandings.

As our discussion in Chapter 1 of the changing perspectives in hazard and disaster research makes clear, how these concepts are defined and used shifts according to which theoretical paradigm is adopted. Rather than view such theoretical diversity as a problem, the debates over what and how such terms emerge, what they are intended to communicate and how they are understood by various publics and policy-makers, practitioners and agencies, as well as academics trained in various fields, indicate a vibrant world of ideas and propositions.

B. VULNERABILITY

From Vulnerability as Loss to Vulnerability as Marginalisation

The concept of vulnerability has long roots. Very broadly understood as the socially produced and individualised characteristics of people that help determine their exposure or potential exposure to harm and that contribute to constraining their ability to anticipate, cope with and recover from harm, vulnerability is interpreted very differently by academics working in geography, environmental science, anthropology, political science, psychology, civil engineering, postcolonialism, feminism, indigenous studies and others. Some are more concerned with identifying measurable aspects of people's lives that are believed to be indicators of an increased susceptibility to risk and harm, in order to map these for future mitigation planning. Key debates tend to return to the question of what kinds of datasets are both appropriate and available and what kinds of data needs to be collected. Others are more concerned with delving into why these differences exist, and so look into the economic and political structures within which people are embedded. Debate here tends to turn on the question of how economic, political and other social structures are themselves to be understood, so that more just and caring policies and actions can be supported and enacted.

Early applications of vulnerability in the field of hazard and disaster focused on exposure of people and objects (such as possessions, land and dwellings, and other infrastructure) to potentially harmful environmental hazard conditions. In 1979, an influential definition of the concept was produced by the United Nations Disaster Relief Organisation (UNDRO). Vulnerability was explained as:

> the degree of loss to a given element at risk or to a set of such elements resulting from the occurrence of a natural phenomenon of a given magnitude and expressed on a scale from 0 (no damage) to 1 (total loss).
>
> (UNDRO, p. 5)

Similar framings of vulnerability have become used in civil engineering and geophysics especially, wherein vulnerability is framed as a dynamic condition of people and objects, including infrastructure. It is constantly in flux, due to changing interactions between geophysical and social processes through time and space. This condition might be expressed on a scale of 0 to 1 as earlier; or, it might be expressed in other quantitative terms as the expected degree of loss for an element at risk as a consequence of a certain event, measured on a metric scale (such as monetary units) that allows comparisons, or on an ordinal scale. The latter might rank losses according to perceptions and evaluations that are specific to particular events, phenomena and communities, and which thus cannot be used for direct comparison. For highly specific hazards such as slow-rise flood depth or volcanic tephra, this form of vulnerability assessment could comprise an evaluation of the parameters of the physical processes leading to the environmental hazard (such as mapping the environmental conditions, modelling the past and projecting the future) alongside an assessment of an infrastructure's designs, materials, techniques, placement and the state of maintenance of this infrastructure alongside the presence of and efficacy of hazard mitigation measures.

While the concept outlined previously hinges on both people and objects bearing a 'loss' of some (quantified or ranked) form, a more human-focused framing emerged in the analyses of scholars working within the 'human ecology' paradigm outlined in Chapter 1. Kates, for example, described vulnerability as the 'capacity to be wounded' (1985, p. 9) – a definition that draws on the Latin derivation of the term, which is *vulnerare*, 'to wound' – and produced a model that combines characteristics of what he called the 'natural event system' with a 'human use system' (Kates, 1970). This human use system includes socio-economic and demographic aspects, damageable wealth and the perception of hazards. Importantly for this model, perception of the riskiness of environmental hazards is key to how people behave to either reduce or exacerbate their susceptibility to hazards.

While this understanding of vulnerability remains a core component of hazard and disaster research, the concept has been expanded to encompass different kinds of hazard and to address the complexity of social and environmental relations within which people live (Lewis, 1979). It now rests at the interface between science and policy-making with respect to several themes, perhaps most prominently:

- The inequalities and inequities produced under economic, political and other social systems, which can build on as well as intersect with sexist, racist and ableist social structures, and have diverse impacts on people's mental as well as physical well-being. These inequalities and inequities differentially exacerbate exposure to and harm from particular risks and threats, as well as the lived experiences of disaster.
- Global environmental issues, including human-caused climate change, that pose particular challenges to sustainable livelihoods and which are managed and responded to in different ways.

An acknowledgement of the role of global capitalism – as well as other ideologies – in producing inequalities and inequities, entrenching poverty and so increasing the vulnerabilities of particular individuals and communities emerged in concert with the 'underdevelopment paradigm' outlined in Chapter 1. Building on empirical findings, vulnerability was reworked as a key component of underdevelopment, with particular attention paid to how economic and political structures marginalise particular groups, placing them at greater risk (e.g. Hewitt, 1983, 1997).

Although scales of investigation vary, factors that exacerbate marginality can include:

- *Age*. Disaster preparedness plans and alert/warning systems need to account for the needs of all ages, including young and old people who might have limited mobility and so need help to escape rapid-onset hazards.
- *Ableism*. Disaster preparedness plans and alert/warning systems need to account for the needs of people with disabilities who may not receive, or be able to respond to, ableist warning messages or other information.
- *Sexism*. Due to sexist societal structures that produce gendered job sectors and a gendered pay gap, women tend to have lower incomes and may be the sole carer for a large family in a single-parent household. Some sexes and genders might hesitate to heed warnings due to past experiences of discrimination, harassment, bullying and abuse.
- *Racism*: Many individuals and groups experience race-based economic, political and other social marginalisation and may face the challenge of negotiating alert/warning systems that are not expressed in plural languages or diverse mediums.
- *Insufficient Welfare and Public Health Infrastructures*. Chronic disease intersecting with limited access to formal health systems and services will hamper disaster recovery. Self-employed, occasional, zero-hour contract and gig workers may lose their job while lacking tools and other resources to support them in a disaster for recovery.

Vulnerability and Development

Vulnerability as a lived condition that needs to be tackled by development policies has long been adopted in research and practice. In 1986, for example, the International Relief Development Project (IRDP) was established with the aim of exploring the linkages between relief and development, such that agencies might respond better to disasters involving various environmental hazards. Reviewing many of the relief efforts addressed by the project, Anderson and Woodrow (1989, 1991) outlined three dimensions of vulnerabilities in a matrix (Figure 3.1) – material/physical, social/organisational and motivational/attitudinal – while noting the 'capacities' that people can bring to bear to manage these impacts. This point is reiterated by Wisner (2016), who writes:

> Exclusive focus on vulnerability (versus capacity) is misplaced and misleading for several reasons. The most obvious reason is that information about the ways that local people are proactive in protecting themselves does not come to light. The potential for building on such local knowledge and skill can be missed.
>
> (Wiser, 2016, p. 7)

Development is a highly contested term in and of itself (see also Hewitt, 1983, 1979; Lewis, 1979; Wisner et al., 2004). Nevertheless, a largely uncritical approach to development as a means of tackling poverty, and hence vulnerability, has been adopted by organisations such as the World Bank and the UN and humanitarian organisations such as the Gates Foundation. These have promoted development via economic growth, alongside increased aid and debt relief, as means of reducing

	VULNERABILITIES	CAPACITIES
PHYSICAL/ MATERIAL What productive resources, hazards, and skills exist?		
SOCIAL/ ORGANIZATIONAL What are the main features of community organization and the distribution of power?		
MOTIVATIONAL/ ATTITUDINAL How do members of the community view their ability to create change?		

Figure 3.1 Dimensions of vulnerability and capacity.

Source: Anderson and Woodrow (1991, p. 46)

poverty and thus general vulnerability. Various development interventions – ranging from investments in schools and public health infrastructure to microcredit financing – are assumed to help mitigate poverty and thus reduce the vulnerability of particular groups.

While arguing for greater state spending on health and education, the foreword to the World Bank's (2000) *World Development Report 2000–2001*, for example, firmly situates vulnerability as a dimension of poverty. This condition includes not only low income and low consumption, but also low achievement in education, health, nutrition and other areas of human development. The term 'vulnerability' appears 96 times in this 2001 report, with a section devoted to vulnerability as 'the risk that a household or individual will experience an episode of income or health poverty over time' (p. 9) as well as 'the resilience against a shock – the likelihood that a shock will result in a decline in well-being. . . . Vulnerability is primarily a function of a household's asset endowment and insurance mechanisms – and of the characteristics (severity, frequency) of the shock' (p. 139). More recent initiatives, such as the United Nations' Sustainable Development Goals, similarly link poverty and vulnerability, advocating for increased resources to be given to health and education.

From an underdevelopment perspective, these measures attempt to soften some of the consequences of social (including economic) structures rather than tackle how capitalism and other equally harmful economic systems operate to produce and maintain inequalities and inequities as part of their baseline. Moreover, development by many meanings and actions does not necessarily reduce vulnerability or poverty. The assumption that development interventions aimed at reducing poverty will translate into reduced vulnerability has been critiqued, as noted in Box 3.1, which focuses on vulnerability, development and precarity.

Box 3.1 Precarity and Vulnerability in Nepal

Development interventions aimed at reducing poverty, and hence vulnerability, have become a feature of policy-making at all scales. Working in Nepal, Rigg et al. (2016) argue that while specific development measures are geared towards what they call 'Vulnerability (inherited exposure)', such interventions can also produce new forms of vulnerability, or 'Precarity (produced exposure)'.

Nepal is poor by many metrics and experiences diverse environmental hazards including earthquakes, landslides and floods. Both the World Bank and Asian Development Bank have projects intervening to support economic growth via enhanced transport connectivity and improvements to the business environment. For Rigg et al. (2016, p. 65),

> while such interventions may indeed address some aspects of vulnerability, in so doing they change the texture of livelihoods. In other words, they alter not just the amount of production, income or return, but also the means and methods

by which these are generated and their social distribution and environmental consequences.

Debt-financed crop production may benefit families in rural Nepal in the short-term, but may also 'leave them open to market shocks in the medium and long-term, thereby raising the possibility of foreclosure' (Rigg et al., 2016, p. 66). Table 3.1 draws out this distinction between vulnerabilities that have accrued before intervention and the precarities that can emerge after intervention. This study shows that there is no simple link between poverty and vulnerability. In assuming that such a simple link exists, poverty reduction measures can actually increase different kinds of vulnerability.

Table 3.1 Drawing a line between vulnerability and precarity

Character of exposure		
	Vulnerability (inherited exposure)	**Precarity (produced exposure)**
Environmental	• Occupation of marginal, hazard-prone land • Small landholdings • Steep, poor quality lands in the hills which limits agricultural productivity • Absence of irrigation technologies	• Dispossession of (and from) land • Commercial logging and associated soil degradation • Resettlement on marginal lands • Chemicalisation of agriculture • Loss of biodiversity
Economic	• High dependency on agriculture • Lack of access to credit • Lack of access to markets	• Unsustainable levels of debt • Market dependencies • Growing inequalities between rich and poor • Out-migration
Political and socio-cultural	• Caste system and associated marginalisation • Gender divisions in society • Participatory exclusions • Lack of empowerment • Feudalism, lack of land reform	• Erosion of the community covenant (moral economy) • Falling fertility rates, ageing population • Emergence of multi-sited households and crisis of care for elderly • Left-behind children
Poverty	Old poverty	New poverty

Source: Rigg et al. (2016, p. 66)

Mapping Vulnerability

Policy-oriented research has also been directed at mapping vulnerability across space and time in a predictive way, with the intent of helping to shape how resources are directed (Bankoff et al., 2004). Indeed, measuring vulnerability has become one of the key aims of the subfield 'disaster risk reduction'. The World Conference on Disaster Reduction (WCDR) in Kobe 2005 promoted the need for standardised risk and vulnerability measures for multiple scales in the resulting call for action: the Hyogo Framework for Action 2005–2015: Building the resilience of nations and communities to disasters (UNISDR, 2005).

A challenge here is the production of measurable variables that will indeed capture vulnerability. Vulnerability indices are normally constructed from archived datasets of information, many of which will have been collected and assembled for other purposes. Some measures are single variables, such as the number of houses or the age of people. Others are less well-defined aggregate indices, such as the extent of local reliance on food aid or the importance of local customs and heritage sites.

The Disaster Impact Index (DII), devised by Gardoni and Murphy (2010), focuses on changes in individuals' capabilities. Here, the fulfilment of particular needs – such as being adequately nourished, sheltered and mobile – is assumed to be key to individual well-being and, hence, producing lower vulnerability. These needs are translated into quantitative indicators – such as the number of individuals without access to sanitary water following a disaster – that allow for international comparisons to be made. These comparisons in turn allow for international aid funding to be directed towards those deemed to be most vulnerable because of the impact of a disaster on their individual capability for health and well-being. Such a comparative approach is very much subject to the issues around preconstructed data collection noted in Chapter 1 as well as assumptions regarding the meaning of vulnerability.

By contrast, the Vulnerability and Capacities Index (VCI) from Mustafa et al. (2011) hinges on the *in situ* collection of data. Building on Anderson and Woodrow's (1991) matrix of vulnerabilities and capacities (Figure 3.1), the intent is to assess a range of indicators including income (which takes note of dependency on a local-level resource, such as fisheries or a small shop, as well as how sensitive that income is to local environmental or other hazards); educational attainment (which takes note of literacy, and weights this by gender); exposure (which takes note of location in a high impact area such as a floodplain); social networks (which is memberships in ethnic, caste, professional or religious organisations); extra-local kinship ties; infrastructure (which includes access to water, telecommunications, electricity, roads and health care); warning systems (both present and trusted); earning members in a household; membership of disadvantaged lower caste, religious or ethnic minority; sense of empowerment (which addresses participation in or access to leadership structure at any level); and knowledge of potential hazards. Comprehensive, extensive fieldwork is required to undertake such a survey. This allows for the collection of much more detailed information. As with the Disaster Impact Index, this approach assumes that academics observing (directly or indirectly) impacts on people have the most accurate and appropriate knowledge as to the disastrous or potentially disastrous nature of the hazard(s) involved.

Vulnerability and the Environment

For those working in the hazard and disaster field, a key consideration is how vulnerability is related to environmental conditions. A prominent framework for empirical analyses is pressure and release (PAR), following the work of Wisner et al. (2004). As an explanatory framework, PAR describes a disaster as the intersection of two opposing forces: (1) hazardous conditions, processes and events and (2) forces generating vulnerability. Vulnerability has three components:

- Root causes, which comprise the distribution and exercise of power within a society, both directly through the control of financial and political resources and indirectly through the curtailment of choices.
- Dynamic pressures, which comprise the work of institutions that can produce decisions that create, increase and maintain the vulnerability of groups and individuals.
- Unsafe conditions, which are the outcomes of root causes and dynamic pressures and comprise the specific, potentially harmful conditions within which people live and work.

PAR is applied to an example in Box 3.2, while Figure 3.2 unpacks various parts of the PAR framework, following from Wisner et al. (2012). On the right side of Figure 3.2 are various forms of hazard that can become disastrous: climatological, meteorological, geomorphological and geological, biological and ecological, and astronomical. On the left side are root causes in the form of social and economic structures, ideologies and historical and cultural factors. The latter, it must be stressed, do not treat culture and history as unique or homogeneous characteristics of societies. Rather, they combine global-scale formations such as colonialism and practices that have passed along from generation to generation as being significant and formative to the relationship between people and place.

Turner et al. (2003a, 2003b) argue for a coupled human–environment systems (CHES) approach in which society and the environment are so intertwined that they are inseparable. Other incarnations of this point include coupled human and natural systems (CHANS) and social-ecological systems/socio-environmental systems (SES, with many other variations for this acronym). These statements are, of course, a truism with many cultures having lived this experience since the beginning of humanity. Those advocating for CHES, CHANS and SES do not always acknowledge this human experience – or even the science on the topic from fields such as indigenous studies, anthropology, human ecology, and development studies.

Focusing on Turner et al. (2003b):

> vulnerability resides in the condition and operation of the coupled human – environment system, including the response capacities and system feedbacks to the hazards encountered.
>
> (2003b, p. 8080)

Despite the emphasis on case studies of place, significant processes are not presumed to exist purely at a local scale. As noted in Chapter 1, political and economic processes, as well as physical processes, extend across spatial and temporal scales, and so case studies must take into account how these processes are intersecting in place to produce vulnerability.

Turner et al. (2003a) detail that a CHES approach comprises a coupled system forged from the intersection of 'human' and 'environmental subsystems' that are

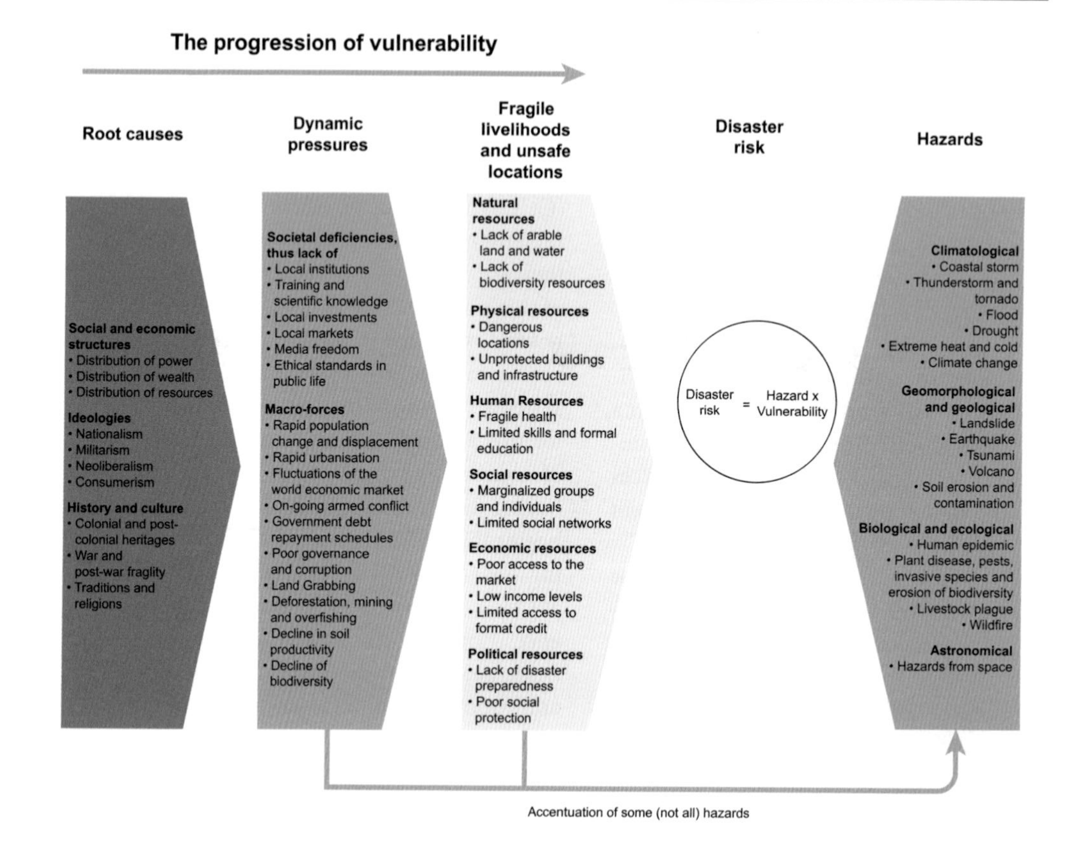

Figure 3.2 The pressure and release framework for explaining disaster.

Source: Wisner et al. (2012, p. 32)

Box 3.2 Applying the Pressure and Release Framework to River Flooding in Norway

River flooding in Norway has a long history (Figure 3.3) and typically emerges from heavy precipitation combined with or separate from snowmelt. Tsunamis or storm surges propagating upstream are also possibilities but have not contributed much to river flooding so far. Freshwater floods from tsunamis in lakes were involved in disasters in 1905, 1934 and 1936, with the risk remaining high. Most freshwater flooding, though, has tended to be slow-rise river or highly localised ponding in urban areas, rather than flash flooding. Rainfall-induced rock falls and landslides are common during the floods.

Figure 3.3 A history of flooding along the River Glomma at Elverum, Norway.

Source: Photo: Ilan Kelman

Applying the pressure and release framework to river flooding in Norway (Figure 3.4) leads to the identification of two main root causes: the political and economic systems of the country. Norway's political system creates dynamic pressures when national governance allows and supports municipalities to make land use and development choices irrespective of the impacts on river flood vulnerability, which tends to support increasing risk. Norway's economic system sits within this governance, producing another dynamic pressure via an insurance system that has typically encouraged the development of floodplains by municipalities without encouraging awareness of flood vulnerabilities. These pressures could be significantly released through regulatory and insurance mechanisms for flood risk reduction, rather than shifting blame among governance levels, and prioritising better flood response.

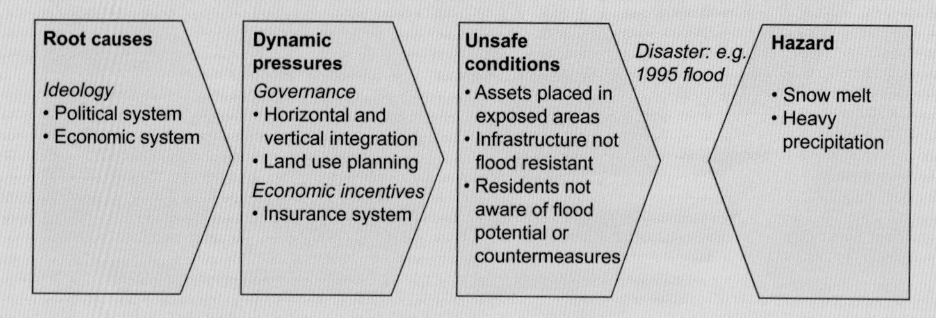

Figure 3.4 Applying the pressure and release framework to river flooding in Norway.

Source: Rauken and Kelman (2010)

structural but dynamic, cross-scalar but intersecting in place, with:

> (i) linkages to the broader human and biophysical (environmental) conditions and processes operating on the coupled system in question; (ii) perturbations and stressors/stress that emerge from these conditions and processes; and (iii) the coupled human – environment system of concern in which vulnerability resides, including exposure and responses (i.e. coping, impacts, adjustments, and adaptations). These elements are interactive and scale dependent, such that analysis is affected by the way in which the coupled system is conceptualized and bounded for study.
>
> (2003a, pp. 8076–8077)

In determining what a 'subsystem' comprises, the decision taken is to sort out human versus environmental phenomena and processes not because these somehow exist separately in the real world, but because these are said to indicate different kinds of phenomena and processes. The term 'coupling' is thus intended to indicate how these different kinds of processes and phenomena are indeed linked, while avoiding reducing an understanding of social dynamics to environmental ones and vice versa. In other words, coupling is what happens through every second and every epoch from microscale to planetary scale.

In CHES, CHANS and SES approaches, the actual harms caused by or averted during a disaster can be noted in their resilience component, alongside coping responses, adjustments and adaptations. Bearing in mind the close linkages foregrounded by the idea of coupling – also developed, theorised and grounded much earlier in Perrow's (1984, 1999) 'Normal Accidents' – these together comprise 'trade-offs' within and among both human and environmental subsystems and their interconnections, each impacting on the rest in various ways, positively and negatively. Some trade-offs will take the form of perturbations – an ecological term, not always suitable for society – within a subsystem, while others may produce enough stress that a 'tipping point' – another contested term especially as tipping points do not always exist in reality (Hillebrand et al., 2020) – is reached, and then the fundamental character of the subsystem is apparently altered. Attempts to manage a particular hazard such as coastal erosion, for example, will have specific repercussions on the ecosystems irrespective of the techniques used. Similarly, attempts to manage floods (Chapter 11) or food production will have diverse consequences through an environmental system, from a drop in pollinators to an increase in albedo. Only some of these consequences can be tracked via their economic cost or via a change in a physical metric. Some may become apparent very quickly, while others may take generations to manifest.

It is challenging to research, monitor and detail the numerous trade-offs among losses, coping responses, adaptations and adjustments for a given community entwined with the environment they inhabit. Nevertheless, a simplified account (perhaps as a model) can sometimes still provide important insights into the potential repercussions of specific, proposed policies and guidelines or suggest innovative responses to potential hazards drawing on lessons from elsewhere. Much of this work, though, emerged from and is framed within a global environmental change baseline rather than all environmental hazards per se.

In fact, vulnerability has long been a key concern for climate change scientists (Chapters 16 and 18), evidenced in the work of the Intergovernmental Panel on Climate Change (IPCC) from its inception in 1988 and First Assessment Report in 1990 and continuing as of 2022 with the Sixth Assessment Report. The IPCC is a UN group producing a synthesis and assessment of the science of climate change and what to do about it, as approved by member state governments. Its reports are meant to be policy neutral and policy relevant, but not policy prescriptive – perhaps an almost impossible task to fulfil. In its current form, the IPCC has three working groups of which Working Group II 'Impacts, Adaptation and Vulnerability' addresses 'socio-economic and natural systems' – which is yet another rendition of CHES, CHANS and SES approaches. On vulnerability, a key contributing cohort to Working Group II was the International Human Dimensions Programme on Global Environmental Change (IHDP), which ran in different forms from 1990 to 2014.

For the IPCC, vulnerability is 'The propensity or predisposition to be adversely affected. Vulnerability encompasses a variety of concepts and elements including sensitivity or susceptibility to harm and lack of capacity to cope and adapt' (IPCC, 2022, p. AII-47). This definition retains the sense of loss ('adversely affected') that can accrue to both social and environmental phenomena by referring to 'sensitivity or susceptibility to harm' that does not seem to change or to be changeable. 'Lack of capacity to cope and adapt' arguably retains a sense of passivity, not indicating why capacity might or might not be present. From an underdevelopment perspective, such a definition is conservative and disempowering, implying a commitment to current structures as 'normal' or 'business as usual', while marginalising any radical designs for the future.

Certainly in regard to large research programmes such as the IHDP and massive syntheses such as the IPCC, the focus on helping to design and shape policies that reduce vulnerabilities has often lost meaning to people directly affected, becoming imbued with typologies and complicated frameworks. Despite the best efforts to communicate extremely complicated data in visuals, tables and text, there is a lack of connection to the daily and decadal realities faced by people with few options made available for them to change their vulnerability dimensions and vulnerable circumstances for themselves. This sense of disconnect means that it is all the more important that different kinds of science communication organisations and events, from science news columns to science festivals, and mediums, from radio to theatre, are valued and supported.

C. VULNERABILITY'S ORIGINS

Vulnerability as a long-term societal process emerges from many influences, not all of which are clear or straightforward. Table 3.2 provides some examples of how aspects of vulnerability in disaster might be and frequently have been presented and portrayed. No list could ever be complete and underlying assumptions must always be challenged, especially when they become pejorative (such as classifying all women, minorities or elderly as 'vulnerable'). No one – not children, not women, not minorities, not people with disabilities, not homeless people and not specific groups – should be labelled or seen solely as 'victims'. Aside from groups not being homogenous but varying in vulnerability-linked traits within themselves, so much evidence provides support for how people are able to help themselves reduce their vulnerability when given the opportunity to do so. Examples are children (Wisner, 2006), people with disabilities (Kelman & Stough, 2015) and sexual and gender minorities (Gaillard et al., 2017).

Aspects of vulnerability must account for variations within groups, variations over time and people's interest in doing what they can with what they have to overcome the difficulties in which they are placed. Table 3.2 is thus presented as illustrative for discussion, not a final statement of aspects of vulnerability in disaster.

The dangers of drawing out simple linkages between the factors noted in Table 3.2 and vulnerability are illustrated by considering population numbers and density. The Caribbean island of Montserrat, for example, is a UK Overseas Territory and an active volcano. In 1995, eruptions began affecting the approximately 12,000 people who lived there. Gas and ash emissions continued regularly for the next two decades, wrecking most of the infrastructure around the island and leading to almost 100% of the population leaving

Table 3.2 Aspects of vulnerability in disaster

Aspect	**Influence on disaster vulnerability**
Socio-economic status – wealth/poverty and power	Limitations to power, wealth and resources tend to translate into limitations in overcoming vulnerability and reducing disaster risk.
Sex/Gender	Women tend to have lower wages and also tend to have family care responsibilities while being more discriminated against, harassed and subject to different forms of violence.
Race and ethnicity	Minority groups can be hampered in vulnerability reduction by language and cultural barriers, as well as societal marginalisation.
Age	Those younger and older might have more limited mobility and cognition than others, which can reduce awareness and response times.
Livelihood loss	Those with precarious or underemployed livelihoods have few resources to tackle vulnerability. Large numbers of unemployed and underemployed people enhance economic insecurity.
Residential property	Expensive homes are costly and time consuming to replace while mobile homes and poorly built properties are easily destroyed by hazards.
Infrastructure and lifelines	Poor-quality infrastructure (such as water, energy and transport services) impedes abilities to address vulnerabilities.
Renters	People who rent often lack financial resources to address their vulnerability, such as knowing that their dwelling could not withstand a tornado or earthquake, but the owner could be unwilling to retrofit.
Family structure	Families with large numbers of dependents (old and young) or single-parent households must account for everyone's needs.
Education	Lower educational achievements are likely to produce lower earnings and also limit personal access to and understanding of warnings, vulnerability reduction information and preparedness and response actions.
Community growth	Rapid community growth, especially when informal, can outstrip the provision of housing and essential social services (such as emergency services, education and health care), notably for newcomers to the area.
Medical services	Ready availability of and access to health care are important for preparing for disasters and are part of reducing disaster risks.
Specific populations	People infirm, institutionalised, incarcerated, undocumented, transient or homeless are at a high risk of being neglected for vulnerability reduction, as are many minorities such as with regard to sexuality.

their homes. Up to two-thirds left Montserrat at some point, many going to Antigua or the UK, although many returned later to continue living in the north of the island where new settlements were built. While many have been affected by stress and years of inhaling volcanic ash, the immediately attributable death toll was around 19 people on 25 June 1997 from hot gas and ash clouds, called pyroclastic flows or pyroclastic density currents.

This number of fatalities appears to be low as an absolute number and even 12,000 people might not appear to be a huge total affected when millions are evacuated elsewhere due to tropical cyclones.

Yet 100% of the island's population experienced major disaster-related impacts, having their homes destroyed, their livelihoods upended and their lives severely disrupted. Twelve thousands remains less than half those killed in the 2003 Bam, Iran, earthquake (Box 1.9). This comparison of numbers does not mean that Montserrat's vulnerability is less than Iran's or that Iran's disaster was worse than Montserrat's. Calculating by country/territory, the proportion of people who died in 1997 in Montserrat is around quadruple as much as the proportion killed in 2003 in Iran. This comparison of numbers does not mean that Iran's vulnerability is less than Montserrat's or that Montserrat's disaster was worse than Iran's.

Ultimately, the comparisons themselves are flawed. Both disasters were terrible, and both populations experienced immense suffering. In both cases the vulnerabilities and environmental hazards were known long before each disaster, and the adverse impacts could have been averted or, at minimum, reduced through action beforehand. Each disaster represents unnecessary circumstances exposing aspects of vulnerability based on a few demographic variables.

Similar complexities emerge for population densities affecting vulnerability. Do cities worsen disasters? Does urbanisation drive vulnerability? Do higher population densities mean more people affected and more infrastructure damaged by environmental hazards? Certainly, a hazard over a specific land area might affect more people and could likely cause more damage. These outcomes are all subject to disaster risk reduction to address impacts from environmental hazards before they manifest. Meanwhile, higher populations in urban areas provide skills and equipment that might be able to assist, if permitted and available. More infrastructure can produce more health and emergency services, including personnel with the most varied experience, best training and most equipment – provided all this was implemented before an environmental hazard. Higher population density means that people are spread over a more confined area, leading to possible advantages for logistics, planning, supply chains and transportation – provided these advantages are applied for disaster risk reduction.

Some influences of population numbers and densities directly affect environmental hazards. Cities typically have large paved areas, augmenting run-off and hence floods. Groups of high-rise buildings can lead to wind tunnels, speeding up wind. The urban heat island is called so for a reason. Yet techniques exist to counter these changes to hazards, with green spaces assisting with heat and floods (Figure 3.5) while permeable surfaces absorb water.

Considering age and gender, being younger and older can drive vulnerability, as does discrimination on the basis of sex, sexuality and gender. In the Bangladesh cyclone disaster of 1970, over half of all the deaths were children under 10 years of age, who comprised only one-third of the population (Sommer & Mosely, 1972). Disasters involving earthquakes have shown that survivors over 60 years of age and females are most likely to have severe physical injuries and to experience mental health impacts (Peek-Asa et al., 2002; Chen et al., 2001). These results are highly

Figure 3.5 Boulder, Colorado, has moved properties out of the flash floodplain reducing vulnerability, giving the river space and creating shaded recreational areas, improving quality of life.

Source: Photo: Ilan Kelman

contextual to the specific situations studied in the references, and they differ in other circumstances. Much more work is needed to adequately disaggregate the data on who is affected by environmental hazards and how, such as being killed or losing jobs, in order to explore more consistently and comparatively why people end up being more vulnerable or less vulnerable.

One such way in which people should be supported to tackle their vulnerability is by reducing inequalities and inequities. Vulnerability is driven by large gaps between rich and poor people because the latter do not have the resources to change their situation. Rich–poor gaps have been generally increasing, with more and more wealth concentrated in the hands of fewer and fewer individuals and corporations. The Gini coefficient (also called the Gini index or the Gini ratio) measures household inequality from 0 (where everyone has the same income) to 1 (where one person has all the income). Some of the most recent values (World Bank, 2022) have not been updated since 1992 for Trinidad and Tobago, although many countries have the coefficient calculated for 2019 or 2020. Values range from 23.2 for the Slovak Republic to 63.0 for South Africa, with over four dozen countries and territories not listing a value including New Zealand and Singapore. It is not straightforward to work out connections between vulnerability and a single measure of inequality based on a single metric of household income.

Inequality and inequity are key factors to consider in assessing the making and scope of disasters, but then decoupling and resolving the reasons for their importance with regard to particular events and particular individuals and groups is challenging. Relevant factors include the meanings and levels of poverty, the opportunities for livelihoods and the lack of effective support systems such as affordable, accessible and reliable insurance as well as promulgated, implemented, monitored and enforced building codes and planning regulations. Access to information, and the ability to mobilise support from outside the household, can be significant. Poor people may appear to have little to lose, but that hardly justifies them losing it and it does not support them in doing better following an environmental hazard. When Hurricane Mitch struck rural Honduras in 1998, the households in the lowest wealth quintile had their meagre assets reduced by 18%, as compared with average losses of 3% recorded for those in the upper quintile (Morris et al., 2002). As with the discussion on Montserrat and Iran, it demonstrates the importance of considering percentages, that is proportional numbers, as well as the actual numbers, that is absolute numbers.

Many social aspects work in tandem to influence vulnerability. During flash floods in July 1993 in a densely populated rice-growing area in southern Nepal, over 1,600 people died (Pradhan et al., 2007). A survey of more than 40,000 residents showed that the fatalities were concentrated in certain groups. The crude fatality rate for all household residents was 9.9 per 1,000, but those most likely to die were those identified as children or women, those of low socio-economic status and those living in thatched houses (Figure 3.6). Over 70% of houses were built of thatch and many were washed away. Those living in thatched houses were over five times more likely to die than those in a cement or brick home.
Violent conflict and other forms of violence are important drivers of vulnerability. People forcibly displaced typically have fewer options than others have for understanding and responding to environmental hazards. In mid-2022, for example, over 100 million were displaced from their homes, including millions

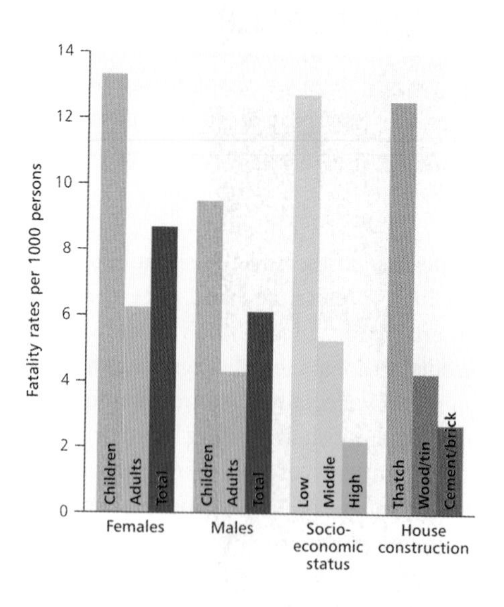

Figure 3.6 Socio-economic factors and fatality rates in flash floods during July 1993 in Nepal. Children were defined as 2–9 years of age, adults as 15 years or older. Socio-economic status was based on household ownership of land.

Source: Adapted from Pradhan et al. (2007)

each from Ukraine and Venezuela due to violence. History demonstrates how political upheavals can lead to greater vulnerability when people are not supported in dealing with the new situation. The collapse of the USSR in 1991 left already weakened social systems in tatters with ramifications around the world as Soviet support for governments and nongovernmental groups suddenly evaporated. Cuba, for instance, had depended on the USSR for oil and entered an energy crisis exacerbated by sanctions from the US, which also limited many other imports (Figure 3.7). Countries such as Afghanistan and the Democratic Republic of Congo have had decades-long violent conflicts. People in such situations lose opportunities to consider and address environmental hazards and so their vulnerability increases.

Yet these discussions about violence and the factors listed in Table 3.2 barely scratch the surface on the full story of vulnerability. Vulnerability as a process means that its drivers, diminishers and connections are not always overt. Violences and traumas can be short term and

Figure 3.7 Long-running US sanctions have led to a proliferation of old vehicles in Havana, even in 2019.

Source: Photo: Ilan Kelman

obvious, but these can also be insidiously complex and hard to capture in a comparative table. Indeed, different forms of violence, often not acknowledged or seen but which are foundations of vulnerability are:

1. 'Quiet violence' (Hartmann & Boyce, 1983) refers to exploitation, poverty and hunger – which can be intentional via austerity measures or inadvertent through those with power not caring or just not caring to be aware. It is typically deliberate in order to keep people in 'their place' and to actively deny them the possibility to overcome their own vulnerability.
2. 'Silent violence' (Watts, 1983) as a concept was developed to explain the social causes of famines, irrespective of environmental inputs into droughts (see more on drought as a social construction in Chapter 12).
3. 'Slow violence' (Nixon, 2011) covers the adverse consequences on people of environmental destruction including pollution, showing how these environment–society interactions happen slowly and continue to undercut people's efforts to maintain healthy lives and lifestyles.

Building on these violences, many of the other factors raised as aspects of vulnerability also degrade the environment and interfere with constructive people–nature interactions, leading to knock-on impacts on environmental hazards. In some countries, the vast majority of the population depends on agricultural livelihoods, typically either subsistence, marginal and precarious or else low-paid to support cash cropping. In the former, poverty then forces the adoption of unsustainable land use practices, such as deforestation, soil erosion and overcultivation. In the latter, overuse of fertilisers and pesticides can increase short-term output in exchange for long-term damage to the land, ecosystems and people.

Highly localised experiences and actions impact ecosystems and can lead to further vulnerabilities. For example, the Mangrove Ecosystems for Climate Change Adaptation and Livelihoods (MESCAL) project – applied to select Pacific islands from 2009 to 2013 – delved into how vulnerability-reducing actions can transform into vulnerability-creating actions. In Tonga (Figure 3.8), traditional waste management involved tossing much of the waste into the mangroves where it would not be visible. While this sounds like a damaging way to deal with garbage, much of it was local and organic, including coconut husks and fish bones. Not only do they decay rapidly and naturally, but they also represent compost, thereby fertilising the mangroves. As everyday materials shift in the wake of global market transformations, solid waste increasingly includes plastics, batteries and chemical containers, which end up leaching as they decompose slowly. Human–mangrove interactions remain the same in terms of behaviours, yet other influences have changed a vulnerability-reducing action to a vulnerability-creating action.

Governance and power have much more impact on creating, maintaining and reducing vulnerability than individual characteristics. This analysis does not deny the physicality of some vulnerabilities such as children and elderly people not being able to run as fast as others and women being more susceptible to osteoporosis than men are. It does indicate that giving people options and resources commensurate with their vulnerability

Figure 3.8 Mangroves in Tonga.

Source: Photo: Ilan Kelman

reduction needs, some of which are based on individual traits, is a major step forward in tackling vulnerability to environmental hazards and beyond.

D. RESILIENCE

Resilience has many different and overlapping meanings and interpretations, especially when used in tandem with other terms such as vulnerability and sustainability. References to resilience – including for environmental hazards – can be traced back through decades of academic literature (e.g. Timmerman, 1981). Broadly speaking, this interest in resilience has been a response to what advocates have considered a limitation to theorisations of 'vulnerability' that primarily address the production and distribution of loss and harm consequent to a hazardous event, process or condition. For many academics working on impact and policy, the concept of vulnerability needs to be complemented by a focus on resilience that:

- anticipates and redresses exposure, vulnerability and risk;
- manages hazard, vulnerability and risk through disaster risk reduction; and
- shapes the processes giving people and communities such capacities, capabilities and abilities.

In assessing the resilience of people and place, then, there are a diverse range of phenomena – from policy and planning priorities to emergency response technologies – to be looked at. For Twigg (2007), these 'aspects of resilience' can be organised thematically (Table 3.3).

Resilience as a concept in hazard and disaster research has drawn particularly on work in environmental science and ecology. In ecological terms, resilience refers to the amount and/or type of so-called disturbance an ecosystem can resist, absorb and accommodate and still remain within the same state. Assumptions include stability as normality and desirable, as well as a system being able to deal with an influence without change. Despite the importance of renewal and reorganisation to an ecological framing of systems, it is the interplay of (a positively perceived) persistence and (a negatively perceived) change that can be seen in the 2009 UNISDR definition of resilience:

> The ability of a system, community or society exposed to hazards to resist, absorb, accommodate to and recover from the effects of a hazard in a timely

Table 3.3 Aspects of resilience in disasters

Aspect	Influence on disaster resilience
Governance	• Policy, planning, priorities and political commitment • Legal and regulatory systems, formal and informal • Integration with development policies and planning • Integration with emergency response and recovery • Institutional mechanisms, capacities and structures • Allocation, monitoring and enforcement of responsibilities • Partnerships • Accountability and community participation
Risk assessment	• Hazards/risk data and assessment • Vulnerability and impact data and assessment • Scientific and technical capacities and innovation
Knowledge and education	• Public awareness, knowledge and skills • Information management and sharing • Education, skill development and training • Cultures, attitudes and motivations • Learning, teaching and exchange • Science and research
Risk management and vulnerability reduction	• Environmental and natural resource management • Physical and mental health and well-being • Sustainable livelihoods • Social protection • Financial instruments • Physical protection; structural and technical measures • Planning regimes
Disaster preparedness and response	• Organisational capacities and coordination • Warnings, including early warning systems • Preparedness and contingency planning • Emergency resources and infrastructure • Emergency response and recovery • Participation, consultation and voluntarism • Accountability and responsibility

Source: Twigg (2007, p. 9)

> and efficient manner, including through the preservation and restoration of its essential basic structures and functions.

This resilience approach might stress the strengths of communities, as well as their vulnerabilities. It can also give them false hopes regarding the nature and alleged finality of recovery. And it can reinforce a 'business as usual' stance by stressing the need for 'preservation and restoration of its essential basic structures and functions' – which were typically the poverty, oppression, poor governance, power imbalances and marginalisation that caused the disaster in the first place.

Studies of famine responses (de Waal, 1989) evidence the realities that people must deal with that are not especially reflected in such top-down and ecology-based definitions of resilience and their implied assumptions. Social cohesion and the networks of mutual support among families, friends and neighbours are important in dealing with difficulties. In turn, such links often depend on gender, religion, caste, ethnicity, race and other characteristics which may attenuate the vulnerability of some, as well as build the resilience of others. Indeed, individuals may well experience and contribute to a complex, intersecting set of relationships that marginalise and empower, producing vulnerabilities and resiliences simultaneously (Fordham, 1999). For example, Ayeb-Karlsson (2020) described the importance of gender roles in Bangladesh regarding cyclone survival, with those identified as men and women showing different vulnerabilities and resilience especially with respect to evacuation, safety, sheltering, trauma and dealing with the trauma.

Resilience may nonetheless be used as an inclusive concept, wherein the goal is to acknowledge and foster local capacities and resources so that, through greater and more diverse forms of political and social participation, communities can better anticipate and reduce aspects of vulnerability as the real drivers of disaster. For UNDRR, community competence is a matter of education and warning systems. UNISDR (2007), for example, featured 16 case studies illustrating a range of practices used by local NGOs to build up community strength. Essentially, these were microscale initiatives that showcased the role of education in many aspects of disaster reduction. Examples are hazard awareness and preparedness, first-aid training programmes, small-scale insurance schemes, hazard warning systems and strengthened buildings.

What is more, 'justice' concepts are becoming popular to represent how to build and perpetuate resilience. A key phrase, for example, is 'social justice', which is then applied to particular concepts such as 'disaster justice' and 'climate justice' (Lukasiewicz & Baldwin, 2020). Social justice is, in effect, the principle that all people should be heard on matters that affect them, whatever their position in a community. While it has many modern incarnations, the idea is far from new, and learning from past struggles and philosophies would be needed to inform disaster justice, climate justice and others before embarking on what many promote as a wholescale realignment of the meaning and application of resilience.

Numerous practical consequences also result from the justice turn in environmental hazards. First, hearing people does not necessarily mean listening to them, nor does it mean acknowledging and acting on their concerns. For human-caused climate change, fossil fuel companies are well aware of what 'climate justice' advocates are seeking through initiatives to remove investments from fossil fuel companies (called 'divestment'), seeking reparations for losses and damages purportedly attributed to climate change (typically called 'loss and damage'), and legal cases in which the companies are sued for the environmental harm their product has caused and continues to cause. Second, who decides what justice means for everyone affected by environmental hazards and who monitors and enforces these meanings? For court cases, it is the legal jurisdiction in which the case goes to court, so it varies widely around the world. Third, rather than adding another word – justice – to the already cluttered domain (including vulnerability, resilience, development and sustainability), should justice supplant or underpin some or all? What would be the implications, for instance, of focusing on 'justice' instead of 'resilience' – or, perhaps, resilient justice or just resilience?

Given the introduction of justice to large research projects such as the IDHP and the IPCC, it is unsurprising to find that academic publications take up and attempt to apply the concept, often without a proper definition or full understanding of its history and critiques. Certainly, 'climate justice' has become a buzzword in top-down processes attempting to address the causes and consequences of human-caused climate change, some of which do not necessarily transition to a just and green economy. The Bali Principles of Climate Justice from 2002, for instance, state in Article 18, 'Climate Justice affirms the rights of communities dependent on natural resources for their livelihood and cultures to own and manage the same in a sustainable manner, and is opposed to the commodification of nature and its resources'. Without any form of commodification of natural resources, most manufacturing – including for the book or device you are reading here – would grind to a halt.

The Bali Principles of Climate Justice do recognise their much earlier foundations by explaining that they were 'Adopted using the "Environmental Justice Principles" developed at the 1991 People of Color Environmental Justice Leadership Summit, Washington, DC, as a blueprint'. Environmental justice has a rich history and foundation in research, policy and practice (Althor & Witt, 2020), which disaster justice ought to learn from.

E. BEYOND VULNERABILITY AND RESILIENCE: TOWARD SUSTAINABILITY?

Will vulnerability reduction and resilience building achieve what we need for avoiding environmental hazards from becoming disasters? Arguably, more exploration is needed of the connections in Figure 1.18 with the overarching ethos of sustainability. Sustainability has various forms, including 'sustainable development', numerous definitions, a lot of criticism and critiques, and a long history (Costanza & Patten, 1995; Salas-Zapata & Ortiz-Muñoz, 2019). It should not come as a surprise that research, policy and practice do not always agree among or within each other. Multiple forms are presented and applied in multiple ways for the same concept.

The origins of one strand – sometimes termed 'sustainability science' – lie in the work of George Perkins Marsh, an American environmentalist who drew attention to the damaging effects of human activities on some of the world's ecosystems (Marsh, 1864). Almost a century later, Gilbert White's work revealed that engineering schemes alone were not the answer to flood losses and that strategies based on conservation could play a part (White, 1942/1945). Many top-down international endeavours continue to dominate scientific discussions and practical actions. The United Nations Conference on the Human Environment held in Stockholm, Sweden, on 5–16 June 1972 (UNCED, 1973) was hardly the beginning of these efforts, but it did coincide with landmark sustainability publications and phrases, notably 'limits to growth' (Meadows et al., 1972) and 'small is beautiful' (Schumacher, 1973). The Stockholm meeting also laid the foundation for the UNCED meeting in Rio de Janeiro in 1992 producing 'Local Agenda 21' (UN, 1992), among other outcomes. From this followed a long series of meetings including Stockholm+50 convened by the United Nations General Assembly in Stockholm on 2–3 June 2022.

In between, many initiatives and publications continued with the sustainability agenda. 'A Programme for Survival' (Brandt, 1980), also known as 'The Brandt Report', and 'Our Common Future' (WCED, 1987), also known as 'The Brundtland Report', solidified the meaning of 'sustainable development' with the baseline of aiming to meet legitimate needs of the current generation without undermining the possibility for future generations to meet their legitimate needs. Extensive discussion continues on the difference between needs and wants as well as the balance between rights and duties, responsibilities and obligations. The Millennium Development Goals ran from 2000 to 2015, followed by the Sustainable Development Goals within the 2030 Agenda for Sustainable Development (see also Chapter 1 and Figure 3.9).

Figure 3.9 The UN's 17 Sustainable Development Goals for 2015–2030.

Source: © Alamy Images/Randy Duchaine

Yet these examples are relatively modern and predominantly Anglophone, representing a minority perspective on human–environment relationships and practices. Aside from the perspectives and histories of sustainability science from many other countries and languages, many peoples around the world have been practising it for millennia. Without assuming that they have or should have all the answers, and recognising the environmental destruction caused by many human cultures throughout history, it remains important to accept, use and learn from indigenous, vernacular, local and traditional knowledges. Sustainability is fundamentally a human endeavour, so all human beings – whether scientists or not, and whether from a particular background or not – have plenty to offer and plenty to gain through sharing.

How useful is all this direction toward 'sustainability'? Is sustainability ever reached or is it merely a catch-all term for the never-ending process of doing better for humanity within our local and planetary environments? Is it really 'beyond vulnerability and resilience' or must vulnerability and resilience always play an integral part of seeking sustainability? There have been so many meetings, publications, goals, agendas, agreements, reports and initiatives, all seeking to achieve the nebulous ideal called 'sustainability', within which addressing environmental hazards sits. A proliferation of niche terminologies, concepts and acronyms pervade this work and material, and so a full discussion mentioning, describing and critiquing everything would not be possible here. The glossary for just one document, the IPCC's Working Group II of the Sixth Assessment Report published in 2022 (see section B of this chapter), runs to 46 pages of terminology and definitions.

Meanwhile, levels of success and failure are evaluated, discussed and disputed in yet more reports and meetings. They typically show many highly impressive successes, many horrendous failures and plenty in between, as will become much more evident in Part 2 of this book when specific environmental hazards are detailed. In the meantime, moving away from the generalities of sustainability to the specifics of what could and should be done for vulnerability and resilience in the context of environmental hazards is detailed further in the next chapter about assessing and managing risk.

KEY TAKE AWAY POINTS

- Though the terms vulnerability, resilience and sustainability are used by academics, governmental organisations, NGOs and funders, there is no one generally accepted definition of these. This lack of fixity indicates the dynamic debate taking place within and between these entities, but can also lead to miscommunication if it is not made clear how a term is being used.
- In some fields such as civil engineering and geophysics, vulnerability is a measure of loss from potential disasters that can be assessed via quantitative and qualitative methods of assessment, including a modelling of physical processes that can produce hazardous conditions. In other fields vulnerability is a concept that identifies the economic and political structures that marginalise particular groups, placing them at greater risk of exposure to and loss from environmental hazards. Though addressing different concerns, these approaches are not necessarily incompatible.
- Vulnerability has been linked to poverty in recent initiatives, such as the United Nations' Sustainable Development Goals.

The emphasis on increased resources for health and education as 'development' does ignore, however, the ways in which poverty is linked to dominant economic structures such as capitalism that hinge on the exploitation of people and ecologies and the work of different forms of violence.

- Resilience is a debated term that has its roots in ecological approaches to systems, their disturbance and the ability to rebound from these. More critical understandings emphasise the complex social relations within which people are embedded and the constraints and capacities for action that emerge from these, as well as the importance of 'justice'.
- Sustainability brings a welcome focus on the well-being of future generations, as well as the diverse ecologies that sustain life. Though general discussions of sustainability have proliferated, there is a need to identify – in an inclusive manner – the specifics of what could and should be done for vulnerability and resilience in the context of environmental hazards.

BIBLIOGRAPHY

Althor, G., & Witt, B. (2020). A quantitative systematic review of distributive environmental justice literature: A rich history and the need for an enterprising future. *Journal of Environmental Studies and Sciences, 10*, 91–103. https://doi.org/10.1007/s13412-019-00582-9

Anderson, M. B., & Woodrow, P. J. (1989). *Rising from the ashes: Development strategies in times of disaster*. Boulder, CO/Paris: Westview Press/Unesco Press.

Anderson, M. B., & Woodrow, P. J. (1991). Reducing vulnerability to drought and famine: Developmental approaches to relief. *Disasters, 15*(1), 43–54.

Ayeb-Karlsson, S. (2020). 'I do not like her going to the shelter': Stories on gendered disaster (im)mobility and wellbeing loss in coastal Bangladesh. *International Journal of Disaster Risk Reduction, 50*, article 101904.

Bankoff, G., Frerks, G., & Hilhorst, D. (Eds.). (2004). *Mapping vulnerability: Disasters, development and people*. London: Earthscan.

Brandt, W. (1980). North–South: A programme for survival. Report of the Independent Commission on International Development Issues, Boston, MA, MIT Press.

Chen, C. C., Yeh, T. L., Yang, Y. K., Chen, S. J., Lee, I. H., Fu, L. S., . . . Chen, L. Y. (2001). Psychiatric morbidity and post-traumatic symptoms among survivors in the early stage following the 1999 earthquake in Taiwan. *Psychiatry Research, 105*(1–2), 13–22.

Costanza, R., & Patten, B. C. (1995). Defining and predicting sustainability. *Ecological Economics, 15*(3), 193–196.

De Waal, A. (1989). Famine mortality: A case study of Darfur, Sudan 1984–5. *Population Studies, 43*(1), 5–24.

Fordham, M. (1999). The intersection of gender and social class in disaster: Balancing resilience and vulnerability. *International Journal of Mass Emergencies and Disasters, 17*(1), 15–36.

Gaillard, J. C., Gorman-Murray, A., & Fordham, M. (2017). Sexual and gender minorities in disaster. *Gender, Place & Culture, 24*(1), 18–26. DOI: 10.1080/0966369X.2016.1263438

Gardoni, P., & Murphy, C. (2010). Gauging the societal impacts of natural disasters using a capability approach. *Disasters, 34*(3), 619–636.

Hartmann, B., & Boyce, J. K. (1983). *A quiet violence: View from a Bangladesh village*. London, UK: Zed Books.

Hewitt, K. (1983). The idea of calamity in a technocratic age. In K. Hewitt (Ed.), *Interpretations of calamity from the viewpoint of human ecology* (pp. 3–32). Winchester: Allen & Unwin Inc.

Hewitt, K. (1997). *Regions of risk: A geographical introduction to disasters*. Harlow: Addison Wesley Longman Limited.

Hewitt, P. (1979). Making contingency studies for a nuclear disaster. *Occupational Health; A Journal for Occupational Health Nurses, 31*(5), 238–239.

Hillebrand, H., Donohue, I., Harpole, W. S., Hodapp, D., Kucera, M., Lewandowska, A. M., Merder, J., Montoya, J. M., & Freund, J. A. (2020). Thresholds for ecological responses to global change do not emerge from empirical data. *Nature Ecology & Evolution, 4*, 1502–1509.

IPCC. (2022). *Climate change 2022: Impacts, adaptation, and vulnerability: Contribution of working group II to the sixth assessment report of the intergovernmental panel on climate Change*. H.-O. Pörtner, D. C. Roberts, M. Tignor, E. S. Poloczanska, K. Mintenbeck, A. Alegría, M. Craig, S. Langsdorf, S. Löschke, V. Möller, A. Okem, & B. Rama (Eds.). Cambridge: Cambridge University Press. www.ipcc.ch/working-group/wg2

Kates, R. W. (1970). Natural hazard in human ecological perspective: Hypotheses and models. Working Paper 14, Department of Geography, University of Toronto, Toronto.

Kates, R. W. (1985). The interaction of climate and society. In R. W. Kates, J. H. Ausubel, & M. Berberian (Eds.), *Climate impact assessment: SCOPE 27* (pp. 3–36). New York: Wiley.

Kelman, I., & Stough, L. M. (Eds.). (2015). *Disability and disaster: Explorations and exchanges*. New York, USA: Palgrave Macmillan.

Lewis, J. (1979). The vulnerable state: An alternative view. In L. Stephens & S. J. Green (Eds.), *Disaster assistance: Appraisal, reform and new approaches* (pp. 104–129). New York: New York University Press.

Lukasiewicz, A., & Baldwin, C. (Eds.). (2020). *Natural hazards and disaster justice challenges for Australia and its neighbours*. Singapore: Palgrave Macmillan.

Marsh, G. P. (1864). *Man and nature: Or, physical geography as modified by human act*. New York: Scribner.

Meadows, D. H., Meadows, D. L., Randers, J., & Behrens III, W. W. (1972). *The limits to growth*. New York: Universe Books.

Morris, S. S., Neidecker-Gonzales, O., Carletto, C., Munguía, M., Medina, J. M., & Wodon, Q. (2002). Hurricane Mitch and the livelihoods of the rural poor in Honduras. *World Development, 30*(1), 49–60.

Mustafa, D., Ahmed, S., Saroch, E., & Bell, H. (2011). Pinning down vulnerability: From narratives to numbers. *Disasters, 35*(1), 62–86.

Nixon, R. (2011). *Slow violence and the environmentalism of the poor*. Harvard, CT, USA: Harvard University Press.

Peek-Asa, C., García, L., McArthur, D., & Castro, R. (2002). Severity of intimate partner abuse indicators as perceived by women in Mexico and the United States. *Women & Health, 35*(2–3), 165–180

Perrow, C. (1984). *Normal accidents: Living with high-risk technologies*. New York, NY: Basic Books.

Perrow, C. (1999). *Normal accidents: Living with high-risk technologies* (2nd ed.). Princeton, NJ: Princeton University Press.

Pradhan, E. K., West Jr., K. P., Katz, J., LeClerq, S. C., Khatry, S. K., & Shrestha, S. R. (2007). Risk of flood-related mortality in Nepal. *Disasters, 31*(1), 57–70.

Rauken, T., & Kelman, I. (2010). River flood vulnerability in Norway through the pressure and release model. *Journal of Flood Risk Management, 3*(4), 314–322.

Rigg, J., et al. (2016). Between a rock and a hard place: Vulnerability and precarity in rural Nepal. *Geoforum, 76*, 63–74.

Salas-Zapata, W. A., & Ortiz-Muñoz, S. M. (2019). Analysis of meanings of the concept of sustainability. *Sustainable Development, 27*(1), 153–161.

Schumacher, E. F. (1973). *Small is beautiful: A study of economics as if people mattered*. London: Blond & Briggs.

Sommer, A., & Mosley, W. (1972). East Bengal cyclone of November, 1970: Epidemiological approach to disaster assessment. *The Lancet, 299*(7759), 1030–1036.

Timmerman, P. (1981). *Vulnerability, resilience and the collapse of society: A review of models and possible climatic applications*. Environmental Monograph 1. Toronto: Institute for Environmental Studies, University of Toronto.

Turner, B. L., Kasperson, R. E., Matson, P. A., McCarthy, J. J., Corell, R. W., Christensen, L., Turner, B. L., Kasperson, R. E., Matson, P. A., McCarthy, J. J., Corell, R. W., Christensen, L., Eckley, N., Kasperson,

J. X., Luers, A., Martello, M. L., & Polsky, C. (2003a). Science and technology for sustainable development special feature: A framework for vulnerability analysis in sustainability science. Proceedings of the National Academy of Science (Vol. 100, No. 14, pp. 8074–8079).

Turner, B. L., 2nd, Matson, P. A., McCarthy, J. J., Corell, R. W., Christensen, L., Eckley, N., Hovelsrud-Broda, G. K., Kasperson, J. X., Kasperson, R. E., Luers, A., Martello, M. L., Mathiesen, S., Naylor, R., Polsky, C., Pulsipher, A., Schiller, A., Selin, H., & Tyler, N. (2003b). Illustrating the coupled human-environment system for vulnerability analysis: Three case studies. *Proceedings of the National Academy of Sciences of the United States of America, 100*(14), 8080–8085. https://doi.org/10.1073/pnas.1231334100.

Twigg, J. (2007). *Characteristics of a disaster-resilient community: A guidance note, version 1 (for field testing).* London, UK: Department for International Development's Disaster Risk Reduction Interagency Coordination Group.

UNCED. (1973). Report of the United Nations Conference on the Human Environment, Stockholm, 5–16 June 1972, New York.

UNISDR. (2005). *Hyogo framework for action 2005–2015: Building the resilience of nations and communities to disasters.* Geneva: United Nations Office for Disaster Risk Reduction. www.unisdr.org/files/1037_hyogoframeworkforactionenglish.pdf. Accessed 18 January 2015.

UNISDR. (2007). *Good practices and lessons learned.* Geneva: United Nations Development Program.

United Nations. (1992). Agenda 21, the Rio Declaration on environment and development. Adopted at the United Nations Conference on Environment and Development (UNCED) held in Rio de Janeiro, Brazil, June 3 to 14, UN for Sustainable Development, New York.

Watts, M. (1983). *Silent violence: Food, famine and peasantry in Northern Nigeria.* Berkeley, CA, USA: University of California Press.

WCED (World Commission on Environment and Development). (1987). *Our common future.* Oxford: Oxford University Press.

White, G. F. (1942/1945). Human adjustment to floods: A geographical approach to the flood problem in the United States. Doctoral Dissertation, University of Chicago (1942) and republished as Research Paper No. 29 (1945), University of Chicago, IL, Department of Geography.

Wisner, B. (2006). Let our children teach us. UNISDR, Bangalore.

Wisner, B. (2016). Vulnerability as concept, model, metric, and tool. *Oxford Research Encyclopedia of Natural Hazard Science.* https://oxfordre.com/naturalhazardscience/view/10.1093/acrefore/9780199389407.001.0001/acrefore-9780199389407-e-25. Accessed 9 June 2022.

Wisner, B., Blaikie, P., Cannon, T., & Davis, I. (2004). *At risk: Natural hazards, people's vulnerability and disasters* (2nd ed.). London: Routledge.

Wisner, B., Gaillard, J. C., & Kelman, I. (2012). Framing disaster: Theories and stories seeking to understand hazards, vulnerability and risk. In B. Wisner, J. C. Gaillard, & I. Kelman (Eds.), *The Routledge handbook of hazards and disaster risk reduction* (pp. 18–34). London: Routledge.

World Bank. (2000). *World development report 2000/2001: Attacking poverty.* The World Bank. https://elibrary.worldbank.org/doi/abs/10.1596/0-1952-1129-4

World Bank. (2022). *Gini index.* https://data.worldbank.org/indicator/SI.POV.GINI

FURTHER READING

Alexander, D. E. (2013). Resilience and disaster risk reduction: An etymological journey. *Natural Hazards and Earth System Sciences, 13*(11), 2707–2716. A detailed exploration of different meanings origins of "resilience".

Gaillard, J. C. (2010). Vulnerability, capacity, and resilience: Perspectives for climate and disaster risk reduction. *Journal of International Development, 22*(2), 218–232. An excellent history of understandings of vulnerability, resilience, and their links.

Hewitt, K. (2007). Preventable disasters: Addressing social vulnerability, institutional risk, and civil ethics. *Geographisches Rundscahu: International Edition, 3*(1),

43–52. A deep dive into vulnerability to explain processes by which disasters are created.

Kelman, I., Gaillard, J. C., Lewis, J., & Mercer, J. (2016). Learning from the history of disaster vulnerability and resilience research and practice for climate change. *Natural Hazards, 82*(S1), S129–S143.

Kelman, I., Gaillard, J. C., & Mercer, J. (2015). Climate change's role in disaster risk reduction's future: Beyond vulnerability and resilience. *International Journal of Disaster Risk Science, 6*(1), 21–27.

Pugh, J. (2014). Resilience, complexity and post-liberalism. *Area, 46*, 313–319.

Wisner, B. (1993). Disaster vulnerability: Scale, power and daily life. *GeoJournal, 30*, 127–140. https://doi.org/10.1007/BF00808129

WEB LINKS

Human Development Index. www.hdr.undp.org/en/content/human-development-index-hdi

US National Risk Index. https://hazards.fema.gov/nri

World Bank GINI Index. https://data.worldbank.org/indicator/SI.POV.GINI

World Happiness Report. https://worldhappiness.report

CHAPTER FOUR

Risk Assessment and Management

4

OVERVIEW

This chapter explores the complex concept of risk in the context of environmental hazards, drawing on a number of differing perspectives to provide an overview of how risk is understood, communicated and managed. We begin by exploring the relationship between risk and uncertainty. We then turn to risk perception, emphasising that how diverse risks are perceived is fundamental to how an individual and community understands, prepares for, and responds to a hazard. Risk perception is shaped by cultural context, alongside modes of knowledge dissemination such as social media, as well as the history of people's own exposures to various hazards and their impacts. For academics risk is typically approached from (1) a quantitative perspective that uses statistics, models, event trees and magnitude–frequency relationships to better understand a hazard and assist in the decision-making processes; and (2) a qualitative perspective that highlights the complexity of people's understanding of and management of risk, and which takes account of 'incertitude' as encompassing uncertainty, but also ambiguity and misinformation. We go on to introduce the important question of 'expertise' in risk assessment and response. While there are tools such as 'expert elicitation' that can be used to determine who has most credibility with regard to particular risks, the concept of the 'expert' has expanded significantly in recent decades to include a wide range of stakeholders, as well as the range of experiences, concerns and issues that expertise is called upon to address. This shift has implications for the sharing and co-production of knowledge of risk between different stakeholders and the effectiveness of risk management plans and initiatives.

A. THE NATURE OF UNCERTAINTY AND RISK

Two magnificently dressed young women sit upright on their chairs, calmly facing each other. Yet neither takes notice

DOI: 10.4324/9781351261647-5

Figure 4.1 Fortuna, the wheel-toting goddess of chance (left), facing Sapientia, the divine goddess of science (right).

Source: Gigerenzer (2014, p. 22)

of the other. Fortuna, the fickle, wheel-toting goddess of chance, sits blindfolded on the left while human figures desperately climb, cling to, or tumble off the wheel in her hand. Sapientia, the calculating and vain deity of science, gazes into a hand-mirror, lost in admiration of herself. These two allegorical figures depict a long-standing polarity: Fortuna brings good or bad luck, depending on her mood, but science promises certainty.

This sixteenth-century woodcut was carved a century before one of the greatest revolutions in human thinking, the 'probabilistic revolution,' colloquially known as the taming of chance. Its domestication began in the mid-seventeenth century. Since then, Fortuna's opposition to Sapientia has evolved into an intimate relationship, not without attempts to snatch each other's possessions. Science sought to liberate people from Fortuna's wheel, to banish belief in fate, and replace chances with causes. Fortuna struck back by undermining science itself with chance and creating the vast empire of probability and statistics. After their struggles, neither remained the same: Fortuna was tamed, and science lost its certainty.

(Gigerenzer, 2014, p. 22)

Risk is a complex word. According to Bernstein (1996), the term 'risk' is derived from the Italian *risicare*, which means to dare to throw oneself into the new and unknown. The Chinese term for risk, *wei ji*, combines the two characters meaning 'danger' and 'opportunity'. Another interpretation is 'precarious moment'. All of these meanings imply that risk is associated with the choices we make, which in turn depends on the freedom we have to make those choices. As the preceding vignette indicates, however, risk has increasingly been approached from a mathematical perspective and has become associated with the future probability of occurrence of a given event. In mathematical terms, a given probability will range between 0 (which equates to the event not occurring) and less than one. Though probabilities in and of themselves do not have a positive or negative value, the focus of interest has been on unwanted events, such as hazardous events and vulnerabilities leading to disasters. The practical application of probability calculations flourished with the rise of the insurance industry, as well as financial investment, but also the emergence of weather forecasting, aided by shipping fleets, observatories and the logistics of communication systems. Indeed, developments in business and natural science were closely connected, as 'economic forecasting' calculated the future profitability of investments in agricultural goods typically transported long distances by ship, and so was very much concerned with upcoming weather conditions.

Risks are by definition future-orientated, in that they are a calculator of what may or may not come to pass in an uncertain future. And, for some, this indicates a key difference between risk as a probability that can be quantified and those aspects of the future that cannot similarly be calculated. In 1921, for example, University of Chicago economist Frank Knight (1885–1972) distinguished risk and uncertainty in his work 'Risk, Uncertainty, and Profit' as:

> A measurable uncertainty, or 'risk' proper, as we shall use the term, is so far different from an unmeasurable one that it is not in effect an uncertainty at all. We . . . accordingly restrict the term 'uncertainty' to cases of the non-quantitative type.
>
> (Knight, 1921, p. I.I 26)

This statement indicates that unless the hazardous conditions that can emerge from an event risk are quantified then these are not in fact risks but are rather uncertainties.

Known risks created the field of *probability*. For example, the probabilities of rain can be measured based on the observation of historical and geographical patterns of rain. Probabilities have three different key aspects:

- Frequency: Probabilities are about counting. For example, we can count the number of days with rainfall divided by the total number of days over the examination period to obtain the frequency of days with rainfall.
- Physical design: Probabilities can be built into the physical make-up of objects. A fair six-sided die has a one-in-six chance of rolling a four, whereas other (ostensibly fair) dice have different probabilities of rolling a four (Figure 4.2). Casino machines might pay out a percentage of what is put in them. Here, probabilities are known because they have been designed that way.
- Degrees of belief: A probability reflects our degree of belief in a hypothesis. Probabilities are therefore subjective, and someone with different knowledge may have different probabilities. Thus, by collecting evidence, our degrees of belief change, as evidence changes probabilities.

Figure 4.2 Different kinds of dice exist, with different probabilities for rolling a specific number.

Source: Photo: Ilan Kelman

In counterpoint to the mathematical approach to risk there are social scientists who view risk as a social construct that is difficult to reduce to numbers given: (1) the diversity of apprehensions of risk and concerns over whose apprehensions are marginalised and not listened to by those in positions of power; (2) the irreducibility of contingency, insofar as simple taxonomies and cause-and-effect relationships between phenomena may be suitable means of describing physical events but are much more challenging – and arguably dehumanising – when used to address social events that emerge from and reshape particular contexts; and (3) the effect of structures such as economic and political formations on how knowledge of risk is produced, disseminated and responded to.

Here, risk is no longer a calculation but is instead a concept that can be used to undertake particular aims. What is more, risk can be a feeling, or even an atmosphere that we might sense but have difficulty articulating in thoughts and words let alone numbers. This multifaceted nature of risk is summed up by Garland who writes:

> Risk is a calculation. Risk is a commodity. Risk is a capital. Risk is a technique of government. Risk is objective and scientifically knowable. Risk is subjective and socially constructed. Risk is a problem, a threat, a source of insecurity. Risk is a pleasure, a thrill, a source of profit and freedom. Risk society is our late modern world spinning out of control.
>
> (Garland, 2003, p. 49)

The Risk of Certainty

Illusions of certainty exist. One is the 'zero-risk illusion', wherein the technology used to assess the presence of a threat is mistakenly assumed to be infallible. A false positive or false negative result for a viral disease exemplifies this kind of risk. Another illusion is the calculable risk illusion that mistakes uncertainty for known risks.

In practice, people generally strive to assess and manage their own risks in order to reduce any adverse consequences. *Risk assessment* involves evaluating the significance of a particular threat, by either quantitative or qualitative means. *Decision theory* distinguishes between so-called risky prospects, where the probabilities of possible outcomes are thought to be known, and less-certain outcomes, where such information is unavailable. Quantitative assessments are generally based on estimates of the probability of an event together with the magnitude of its known adverse consequences, often expressed in various forms such as:

Risk = Hazard Probability × Elements at Risk × Vulnerability
Risk = Hazard impact × Elements at risk × Vulnerability of elements at risk
Risk = Hazard × Vulnerability
Risk = Hazard × Exposure × Vulnerability
Risk = Probability × Consequences
Risk = Hazard × Vulnerability × Value threatened ÷ Preparedness
Risk = Hazard × Vulnerability ÷ Capacity
Risk = Hazard × [(Vulnerability ÷ Capacity) – Mitigation]

Note that including the operators, such as multiplication and division, is not always necessary. Sometimes, risk is expressed as a function of the variables – such as stating that risk is a function of hazard and vulnerability or risk is a function of hazard, vulnerability and capacity. Then, the prospect remains for different forms of the function as well as considering interdependencies among the variables rather than, for instance, presuming that hazard and vulnerability are independent. This approach is particularly important when including exposure, capacity and resilience, since none of these are separate from vulnerability; exposure especially works well as a subset enfolded within vulnerability.

Not all the variables in the equations are easy, consistent or appropriate to quantify. Even when risks have been quantified the level of uncertainty associated with estimates of loss may be high, with numerous assumptions and statistical models adopted to establish the estimates that are in essence subjective, rather than objective. Therefore, it is important that risk assessments are communicated in a transparent and accessible manner to the public and stakeholders, and that care is taken to explain the extent of uncertainty attached to any estimate. There are numerous examples where poor communication of risk assessments in percentages has resulted in adverse impacts.

In terms of disaster risk reduction, one practical goal is *risk management*, which aims to lower the known threats whilst maximising any related benefits. Potentially, almost every person, community and organisation has something to contribute to risk management, but achieving the optimum balance between risk and safety involves controversial value judgements and a balance of cost and benefit. There are challenges in answering even basic questions, such as: How do we know risk? What is an acceptable level of risk? Who benefits from risk assessment and management? How is it perceived? Who pays for the process? What is meant by success or failure in risk reduction policy? How can we best communicate and understand risk? And, how do people and institutions embody this knowledge?

Inherent risk is the risk present in any scenario where no attempts at mitigation have been made and no controls or other measures have been applied to reduce the risk from initial levels to levels more acceptable to the organisation. Residual risk is the risk that remains after efforts to identify and eliminate some or all types of risk have been made, typically via mitigation strategies (see Chapter 5).

Neither risk assessment nor risk management can be separated from choices and the freedom to make choices. These choices are conditioned and enabled by individual beliefs and circumstances, financial resources and wider attitudes in society. Since most people make decisions based on their personal assessment of a threat, *risk perception* has to be regarded as a valid element in risk management. Distinctions are often drawn between *objective* and *perceived* (subjective) risks. This is because the level of personal risk perceived by the individual concerned often differs from the results obtained by supposedly more 'objective' assessments. Care must be taken to ensure that an 'objective' risk analysis (as commonly used in financial models of costs and benefits) cannot be assumed to be correct, or to lead to better outcomes, than assessments based on perception. The objectivity of these risk assessments can be queried given they require choices about elements to include and how to value these elements.

When dealing with the perceptions of individual people, risks are often allocated into two categories:

- *Involuntary risks* happen to us without our prior knowledge or consent. As such, they are often seen as external to the individual. So-called acts of God, like being struck by lightning or a meteorite, are considered to be involuntary risks, as is exposure to some environmental contaminants. Occasionally, such risks are known to the individual, but then they are often seen as inevitable or uncontrollable, as in the case of earthquakes. Many of the hazards considered in this book represent involuntary risks to individuals or communities, due to their location in a hazard-prone environment.
- *Voluntary risks* are associated with activities that we elect to undertake, such as driving a car or smoking cigarettes. These risks, willingly accepted by individuals, are generally more common and controllable. Because they are undertaken on a personal basis, they also have less catastrophe potential. The control of voluntary risk is exercised either through modifications of individual behaviour (stopping smoking or ceasing participation in a dangerous sport) or by government action (introduction of safety legislation, such as the requirement to wear a crash helmet when riding a motorcycle).

In reality, the division between these two risk categories is less clear than it appears. For example, while cigarette smoking and mountain climbing are obvious cases of voluntary activities, the same cannot be so firmly stated for second-hand smoke from others smoking. Driving a car may be an essential form of transport for people in remote areas and the risk from it involves many others, from road and vehicle designers to other drivers. The alternative to working in a dangerous chemical factory may be unemployment. In other words, some risks are more voluntary than others and accepting risk might not be a foolish or ignorant decision. Some floodplain dwellers may have bought or rented a home near a river because it is cheaper than an equivalent property outside the floodplain – or because it is more expensive due to the view and greenspace amenities or because other hazards reduced the prices of homes elsewhere. Such decisions can be both voluntary and have rationality.

B. RISK PERCEPTION

A person's perception of risk is the result of complex interactions between the general attitudes taken by the community in which the person lives, their prior experience of dealing with the hazard in question, and the scope and effectiveness of knowledge around risks authoritatively distributed by administrations and organisations. The cultural environment provides the setting within which the risk is interpreted. For example, a person living in a community with strong religious beliefs may be more likely to view a hazard as an 'act of God', albeit one that needs to be prepared for and managed. Past experience is important because people with knowledge of the emergence and impacts of disasters tend to have a greater apprehension as to the probability of future occurrence. For example, people moving from rural to urban areas may be vulnerable to landslides on excavated and reconstructed ground because they are unaware of the prior history of land transformation here. Similarly, people moving to floodplains may be unaware of the history of water–land interactions here that produce a range of hazards, from inundation and mould to bank collapse.

When direct experience of a particular hazard is lacking, then perceptions of risk are influenced in other ways. The media, including social media, are a powerful source of information. Given the extent of in-built bias in reporting disaster news (Chapter 2) and an increasing reliance on information from non-official sources, especially social media, the public's hazard perceptions are likely to be shaped differently from more calculated (not necessarily objective) risk analysis outcomes. Nonetheless, media provide an opportunity for scientists to influence risk perceptions and to attempt to provide an evidence base for analysing and responding to risks.

For some academics working on risk perception, it is useful to broadly categorise people's approach to hazards and their potential impact, albeit with the recognition that these are by no means fully representative and that individual understandings are complicated. Here, taxonomies are often built on the degree to which people believe that the impact of a hazardous event is dependent upon fate (in that it is externally controlled) or their own actions (and is thus internally controlled). Clearly, a range of views exists surrounding what is usually described as the 'locus of control'. Within this spectrum, three types of perception have been identified. These are:

- *Determinism*: This pattern of behaviour, sometimes called the gambler's fallacy, exists when people erroneously believe that a certain random event is less likely or more likely to happen based on the outcome of a previous event or series of events.
- *Dissonance*: Although it takes many forms, the dissonant perception category captures a denial or a minimis

ation of risk. Here, an event is viewed as a freak occurrence unlikely to be repeated. In extreme cases the existence of a past event may even be denied. In an early study, Jackson and Burton (1978) suggested that people living in areas subject to high levels of seismic hazard did not consider the hazard to be troublesome, partly because of the practical difficulty of coping with the consequences of a large earthquake and partly due to the challenge of coming to terms with continuing vague threats. From this viewpoint, dissonance is an attempt to deal with ongoing risks on a bearable day-to-day basis.

- *Probabilism*: Probabilistic perception captures understandings that accept that disasters will occur and patterns are not always clear. It generally aligns with the views of officials charged with making decisions about risks. But in some cases the acceptance of risk is combined with a need to transfer the responsibility for dealing with the hazard to a higher authority, which may range from the government to God. Indeed, the probabilistic view has sometimes led to a fatalistic, 'acts of God' approach, whereby individuals feel no responsibility for hazard response and wish to avoid any actions or expenditure on risk reduction.

An important feature of risk perception is *social amplification*, wherein risk is ramped up or down with negative consequences for hazard response. Social amplification can happen when factors combine to create an exaggerated fear of a threat. It tends to occur when the threat is new to the individual, when people believe that the true magnitude of the risk is being hidden from them in some way, when there is a belief that the hazard cannot be controlled, when the individuals exposed to the hazard are considered to be highly vulnerable (e.g. if they are children) and when there is a feeling that experts do not understand the risks. In counterpoint, risk can be ramped down when individuals or groups are not able to relate directly to the hazard, when the level of media reporting about the hazard is limited or short term, when there are perceived benefits associated with the hazard, and when there is a belief that the hazard is well understood and that the responsible individuals are trusted.

Some of the factors that can increase or reduce public risk perception are listed in Table 4.1. Broadly speaking, risks are taken more seriously if they are life threatening, immediate and direct. This means that an earthquake tends to be rated more seriously than a drought, which can emerge more slowly and have impacts on the more vulnerable memes of society whose plight is not being picked up on by government or media. Indeed, the type of potential 'victim' of a disaster can be significant: awareness is heightened if children are at risk or if the people affected are another readily identifiable group who are viewed positively. Level of knowledge can be important, particularly when related to the degree of trust in the sources of hazard information. This is a common feature in the perception of technological risks, especially if a lack of scientific understanding is combined with disbelief of opinions expressed by those managing the technology.

Much weight is already given to hazards perceived to be more common, like road safety, which can be significant for some countries. In New Zealand, the death toll on the roads every two years has recently been exceeding the loss of life due to earthquakes throughout the country's recorded history (Abeling et al.,

Table 4.1 Twelve factors influencing public risk perception, with some examples of relative safety judgements

Factors tending to increase risk perception	Factors tending to decrease risk perception
Involuntary hazard (radioactive fallout)	Voluntary hazard (mountaineering)
Immediate impact (wildfire)	Delayed impact (drought)
Direct impact (earthquake)	Indirect impact (drought)
Dreaded hazard (cancer)	Common hazard (road crash)
Many fatalities per event (air crash)	Few fatalities per event (car crash)
Identifiable 'victims' (chemical plant workers)	Statistical 'victims' (cigarette smokers)
Processes not well understood (nuclear power plant disaster)	Processes well understood (snowstorm)
Uncontrollable hazard (tropical cyclone)	Controllable hazard (ice on highways)
Unfamiliar hazard (tsunami)	Familiar hazard (river flood)
Lack of belief in authority (donor-driven political manifestos)	Belief in authority (university scientist)
Much media attention (nuclear plant)	Little media attention (chemical plant)

Source: Adapted from Whyte & Burton (1982)

2020; Te Manatū Waka Ministry of Transport, 2023). A single earthquake, perhaps the 'Big One' that Wellington is preparing for and which could happen any day, could kill more people than New Zealand's total traffic-related deaths throughout the country's recorded history.

Slovic (1987) evaluated the risk of 30 common activities where people cited the factors in explaining why they believed activities to be more risky than the statistical risk of death would indicate (see Figure 4.3). His findings demonstrated that there is a social amplification of risk triggered by situations such as disasters, outbreaks and incidents. For example, whilst caffeine is something many of us consume daily, too much can result in death, yet the public do not generally worry about caffeine consumption. Nuclear weapons are highly dreaded, but the only time in human history (at the time of writing) that they have been detonated against an enemy was in 1945. Slovic (1991) suggested that such trade-offs between risks and benefits are not always made because perceived dangers influence attitudes more strongly than do perceived benefits. This situation is typified by so-called dread hazards like nuclear power. It is also the case that perceived levels of risk can change quickly over time. For example, the perceived risk posed by tsunamis increased greatly after the 2004 disaster around the Indian Ocean.

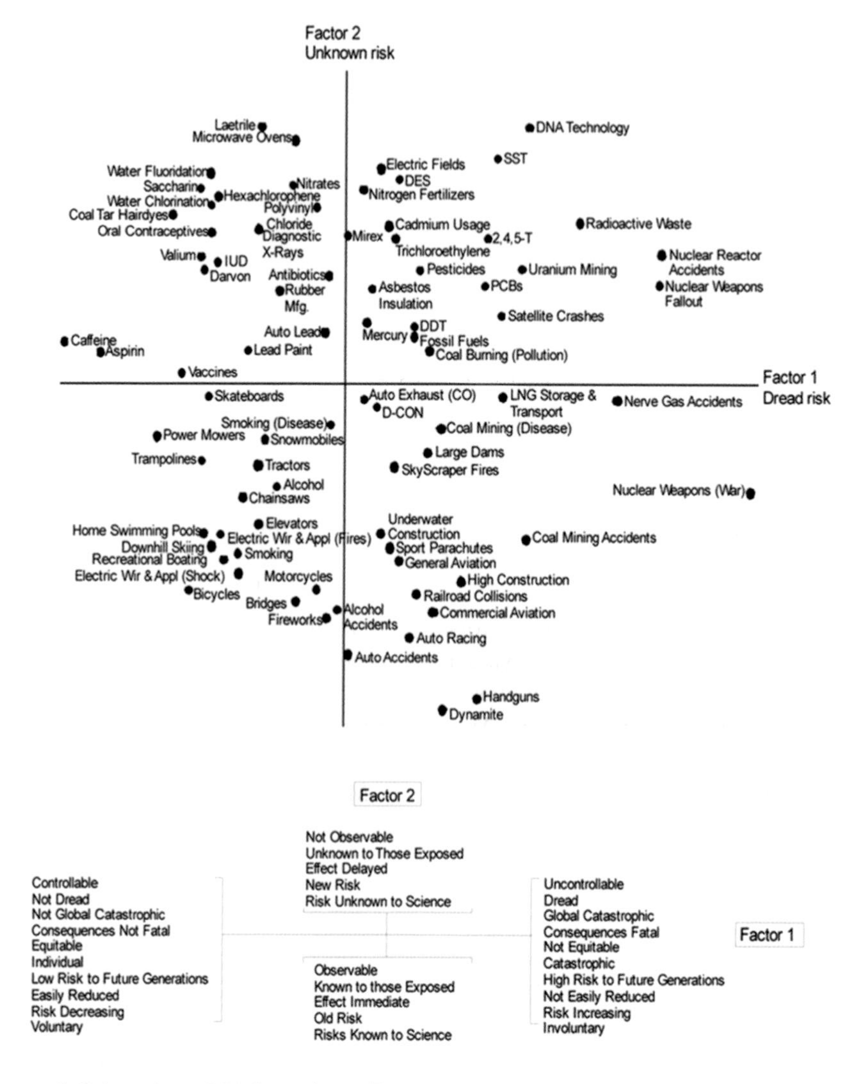

Figure 4.3 Location of 81 hazards on Factors 1 and 2 derived from the interrelationships among 15 risk characteristics. Each factor is made up of a combination of characteristics, as indicated by the lower diagram.

Source: Slovic (1987, p. 236)

The study of risk perception arose out of the observation that those labelled experts and laypeople often disagree about how risky various technologies and natural hazards are. There are two main approaches to this finding. On the one hand, a 'realist' approach seeks to bring perception as close as possible to the objective risk of an activity or an event. This approach can thus foreground outreach and education as means of filling a 'knowledge deficit'. On the other hand, a 'constructivist' approach notes not only the ways in which particular understandings are formed, but also how and with what impact some are designated as 'expert' in regard to particular risks. This approach can thus not only foreground the marginalisation of some views and the impact of this on people, but also place under scrutiny the assumptions held by experts and the wider import of their work.

Risk perception has numerous different but overlapping theoretical approaches, three of which are:

- The psychological approach uses heuristics and cognitive knowledge to understand how people process information. In particular the psychometric paradigm identifies numerous factors responsible for influencing individual perceptions of risk, including dread, novelty, stigma and other factors. Also people are influenced by the emotional state of the perceiver.
- The psychometric paradigm is based on studies of how people process information. It suggests that newly discovered risks, including 'dread' risks such as nuclear power plant disasters, greatly increase the seriousness with which the threats are perceived. Anthropological and sociological approaches integrate cultural theory by exploring how risk perceptions are produced by and supported by social institutions, cultural values and ways of life.
- Interdisciplinary approaches combine research in psychology, sociology, anthropology and communications theory to examine how communications of risk events pass from the sender through intermediate stations to a receiver, and in the process serve to amplify or attenuate perceptions of risk. All links in the communication chain – comprising individuals, groups and media fora – contain filters through which information is sorted and understood.

In summary, the perceived risk an individual holds is subjective and variable. The fact that people tend to tolerate substantially more risk when the threat is associated with voluntary behaviour has been explained as an unrealistic belief in personal control. In other words, individuals rarely have the degree of control over events that they assume or would wish. Cross-cultural studies have evidenced that people are influenced by their social context in ways that may restrict the scope for individual perception of hazards. Thus, variations in risk perception exist according to location, occupation and lifestyle, even between individuals of the same age and gender, as well as between nations (Rohrmann, 1994).

Moving the focus away from individualised perceptions, and their aggregation, some academics have drawn on critical social theory to address the political and economic relations that together produce particular kinds of risk and that comprise collective actions and reactions to hazards and disasters. The work of Giddens (1990) and Beck (1992) suggested that people today are faced with new and complex threats well beyond those posed by traditional 'natural hazards'. Many of these threats have been created by human activity, in particular the technological hazards that have emerged with industrialisation. The concept of the 'risk society' focuses on political and public concern about industrial practices, epitomised by low probability but high consequence disasters such as the nuclear power plant disasters at

Windscale (now called Sellafield, UK) in 1957, Three Mile Island (USA) in 1979, Chernobyl (USSR at the time) in 1985, and Fukushima (Japan) in 2011.

When people believe they are well informed, whatever the source or reliability of the information, they are more likely to attempt risk assessment for themselves. Consequently, with the spread of information via the internet and other media, there has been a decline in the level of public trust given to figures in authority, such as industrialists, politicians and experts of many kinds. In turn, this distrust has been accompanied by a rise in support for concepts such as the *precautionary principle* (designed to minimise potential risks) and *sustainable development* (designed to secure future livelihoods). People currently living in the risk society attach greater importance to safety, and to a secure future, than did previous generations. As people enjoy longer, healthier lives, they perceive a definite value in an extended life and become more risk averse. One obvious expression of this has been the marked growth of an enthusiastic health and well-being culture among affluent groups.

Given that absolute safety is impossible to achieve, determining the level of risk that is acceptable for any activity or situation is helpful for understanding possible consequences of environmental and other

Box 4.1 The Risk Society

Bringing together insights from geography, political science, sociology, psychology, economics and law, Anthony Giddens and Ulrich Beck have argued for a new kind of academic inquiry that directly addresses the social relations within which new technological risks are introduced, so that new institutions can be designed that can assess, monitor and regulate these risks. The social relations that Giddens and Beck point to are industrial economies that are loosely governed by democratic administrations; here, it is argued, the drive for wealth accumulation has been tempered by a rising awareness of the impact of new kinds of industrial disasters. This consciousness of being at risk permeates all levels of society, and even undermines a class consciousness of the different vulnerabilities to harm that different groups have.

The risk society concept picks up on everyday feelings of uncertainty, anxiety, insecurity and alienation that are a response to developments in nuclear, chemical and biomedical technologies. What it also draws attention to, however, are (1) efforts to anticipate and plan for the near and distant future as a response to this experience; and (2) a distrust of established sources of authority, including science, as a means of explaining and managing these new threats. Thinking about these two issues together, Beck argues that by now well-established political authorities tasked with producing knowledge on the presence and scope of risk need to be overhauled. These authorities are adept, he argues, at evading the question of who is responsible for the introduction of new kinds of industrial risk and who is responsible for mitigating and ideally removing harms to society. Without radical change, he concludes, our collective well-being and ultimately our survival will be compromised by an 'organised irresponsibility' that holds no one accountable.

hazards. *Acceptable risk* is the degree of loss that is perceived by the community, or the relevant regulatory authorities, to be most relevant when managing risk. It is a much misunderstood term. For example, it does not describe either the level of risk with which people are happy or even the lowest risk possible. Fischhoff et al. (1981) concluded that the word describes the 'least unacceptable' option because the associated risk is not really 'acceptable' in any absolute sense. As a result, the term *tolerable risk* is often used instead, in that it indicates the level of risk that is tolerated, rather than accepted. Tolerable risk is a dynamic concept because the actual level varies according to a wide range of factors. These include the severity of the risk itself, the nature of the potential impacts, the level of general understanding of the risk, the familiarity of the affected people with the risk, the benefits associated with the risk and the dangers and benefits associated with any alternative scenario.

It is important when specifying the level of acceptable or tolerable risk to be clear about the people *to whom* it is acceptable. Actual behaviour does not necessarily reflect the optimum choice. For example, in the case of a consumer buying a car, the act of purchase need not imply that the product is safe enough, just that the trade-off with other forms of transport is the best available. In this instance, the risk is tolerated rather than accepted. There are many factors that influence the consumer's choice of a car. Perhaps surprisingly, statistics on the safety of the vehicle are rarely prioritised, so, in most decisions, risk perception is just one element in the process.

To summarise, there is no fully objective approach to risk perception and decision-making. Since there is often uncertainty and ambivalence about the best way to manage hazards and risks, quantitative analysis is best viewed as a partial, rather than as a complete, approach.

C. QUANTITATIVE RISK ASSESSMENT

Risk assessment is a process comprising the identification of hazards that are likely to cause harm to people or damage to the environment or infrastructure. Typically it involves three steps:

- *Hazard analysis*: Identifying and analysing potential (future) events that may negatively impact society and the environment, potentially resulting in a disaster.
- *Risk evaluation*: Estimating the likelihood of such events, which can include calculating the probability that they will happen. Risk evaluation also comprises an analysis – typically via a risk matrix – of where and for how long individuals will be exposed to a potential hazard.
- *Establishing control measures*: Evaluating the consequences of the hazardous event(s) in terms of what the likely losses will be and assessing what effective controls can be put in place to mitigate against and reduce these impacts.

In reality the process is more complex because there is an additional need to understand the magnitude of the event and how it may affect risk outcomes. For example, the probability of occurrence of an avalanche is related to the volume of snow; large avalanches occur less frequently and are, therefore, less probable. But, the frequency–volume relationship may not be the full story because the threat posed could be influenced by the velocity of the avalanche flow and the nature of the snow, as well as the volume. The risk is therefore more complicated to assess.

Assuming that these problems can be overcome, the statistical analysis of risk is based on theories of probability whereby risk (R) is taken as a product of probability (p) and loss (L):

$$R = p \times L$$

If every event resulted in the same consequences, it would be necessary only to calculate the frequency of occurrence. But, as already indicated, environmental hazards have variable impacts. Therefore, an assessment of damaging consequences is required (Box 4.2). For many threats, especially technological hazards, the available data of past events are rarely adequate for a reliable statistical assessment of risk. In these cases *event* and *fault tree* techniques are used (Figure 4.5). These use a process of inductive logic whereby there may be some observations that can help develop some form of pattern or tentative theory or hypothesis, but it is not possible to know if this is correct, as there is no absolute certainty. Therefore, the quality of the idea or model or theory depends on the quality of the observations and the analysis conducted.

Induction can be thought of as probability based on the strength and quality of evidence. In contrast deduction is the process of having a theory that predicts an outcome and is tested by experimentation whereby observation confirms or disproves a theory. In the context of environmental and human-made hazards, deduction may be possible as it requires a sequential set of facts that are known to be true, for example the time it takes for seismic waves from an earthquake to travel from the hypocentre (in the crust) to the surface using quantitative models which are guaranteed to be true (or

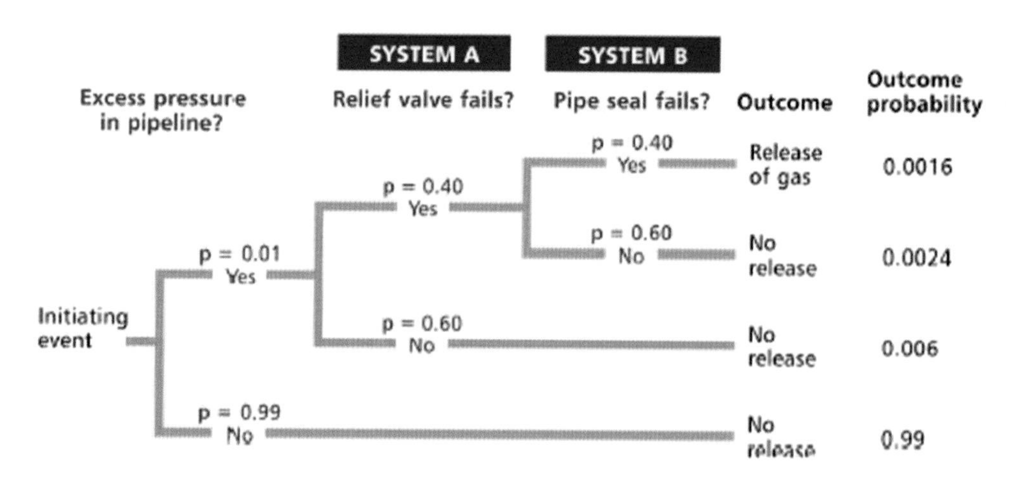

Figure 4.4 A probabilistic event tree for a hypothetical gas pipeline disaster. The performance of safety systems A and B determines the outcome probability of the initiating event.

Source: Diagram courtesy of Dr J. R. Keaton

at least accurate). However, when looking at how that earthquake may impact society, deductive approaches are limited and inductive approaches help deal with the partial information available about our world, such as the time and date of the earthquake, the impact on buildings and other hazards that may occur at the same time.

1. Magnitude–Frequency Relationships

Many natural hazards can be measured on scientific scales of magnitude or intensity, for example earthquakes (Mw and Mercalli scales), tornadoes (Fujita scale and variations) and hurricanes (Saffir-Simpson scale). Unfortunately, such scales tend to measure just one physical factor that influences disaster impact. For hurricanes, the Saffir-Simpson scale relates to the maximum sustained wind speed only, whereas most damage is due to extreme wind gusts, storm surge or intense precipitation (Chapter 10). Even if scientific scales could incorporate all aspects of the damaging phenomena, the event alone is a poor guide to disaster impact because of moderating effects by local environmental and societal conditions. More significantly, disaster impact severity reflects the level of vulnerability in the communities where hazard strikes. Time of day can be important. At night more people will be indoors: perhaps protected from strong winds and rain, but vulnerable to earthquakes if buildings collapse. Vulnerability is not static but varies over time as both human populations and the physical environment change (Meehl et al., 2000).

Box 4.2 Using Historical Data to Quantify Risk

Consider if, from experience, it is known that in different, mutually exclusive, events $E_1 \ldots E_n$ may occur. These events might be a series of damaging floods or urban landslides, but the effectiveness of quantifying risk depends heavily on the availability of a good database. Thus, this method is less satisfactory for rare or less observed hazards including some technological hazards, such as the release of radionuclides from nuclear facilities.

From historical data, it can be determined that event E_j will occur with probability p_j and cause a loss equivalent to L_j, where $_j$ represents any of the individual numbers $_{1 \ldots n}$ and $L_1 \ldots L_n$ are measured in the same units, e.g. a currency or lives lost. It is assumed that all the possible events can be identified in advance. Therefore, $p_1 + p_2 \ldots p_n = 1$.

After arranging the n events in order of increasing loss ($L_1 < \ldots < L_n$), the cumulative probability for an individual event can be calculated as $P_j = p_j + \ldots p_n$. This specifies the probability of the occurrence of an event for which the loss is as great as, or greater than, L_j, as shown in Table 4.2 – using clams as a universal currency (even if less used in locations far from water).

If we can categorise all possible events in terms of the property loss (expressed in clams), it may be possible to produce a risk analysis along the following lines:

Table 4.2 Basic elements of quantitative risk analysis – property loss

Property loss (clams)	Probability (p)	Cumulative probability (P) of exceedance
0	0.950	1.000
10,000	0.030	0.050
50,000	0.015	0.020
100,000	0.005	0.005

Table 4.3 Basic elements of quantitative risk analysis – event and probability

Event	Probability	Loss*	Cumulative probability
E_1	P_1	L_1	$P_1 = p_1 + \ldots + P_n = 1$
E_j	p_j	L_j	$P_j = p_j + \ldots + p_n$
E_n	p_n	L_n	$P_n = p_n$

Source: After Krewski et al. (1982)
Note: * Arranged in increasing order ($L_1 \leq \ldots \leq L_n$).

This theoretical example shows that there is a 95% chance of no property loss and only a 2% chance of a property loss of 50,000 clams or greater.

In some circumstances, it may be necessary or desirable to produce a summary measure of risk (R). This can be done by calculating the *total probable loss*:

$$R = p_1L_1 \ldots + p_nL_n$$

In this example, R would be 1,550 clams. Alternatively, the *maximum loss* could be calculated. This is a rather extreme summary that ignores the probability of occurrence and takes the risk to be equal to the maximum loss, which, in this case, would be 100,000 clams. Because of the skewed distribution, another way would be to take a given *percentile loss*, for example 98% level of loss.

The same methodology can be applied when damaging events cause loss of life. For the prior example, an appropriate tabulation might be:

Table 4.4 Basic elements of quantitative risk analysis – probability of number of deaths

Number of deaths	Probability	Cumulative probability
0	0.99	1.000
1	0.006	0.010
2	0.003	0.004
3	0.001	0.001

Source: After Krewski et al. (1982)

As noted earlier, the *magnitude* (size or intensity) of a hazardous process is usually inversely related to the *frequency* of its occurrence. For example, large earthquakes occur much less often than small ones and major disasters result from relatively rare, big events. The energy release from the 2004 Boxing Day earthquake, which killed about 250,000 people, was about 100 times that of the 2005 Kashmir earthquake, which resulted in 74,500 deaths. The five largest events during the 20th century were responsible for over half of all the earthquake-related deaths. When the magnitude of an event is plotted against the logarithm of its frequency, it normally exhibits the relationship shown in Figure 4.5a. The *recurrence interval* (or return period) is the time that elapses, on average, between two events that equal, or exceed, a given magnitude. A plot of recurrence intervals versus associated magnitudes (Figure 4.5b) produces a group of points that approximates to a straight line on a semi-logarithmic graph.

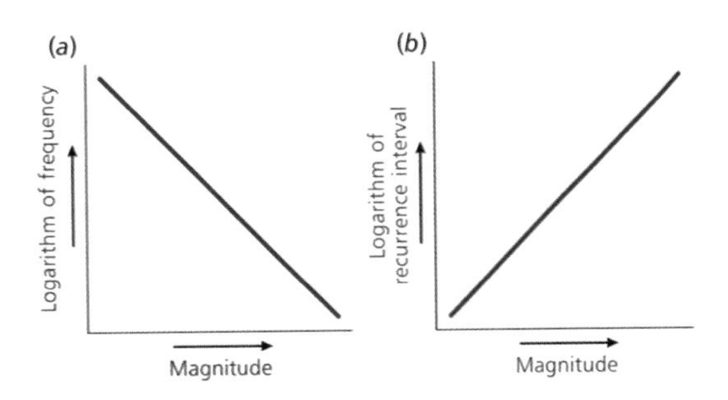

Figure 4.5 Generalised statistical relationships between the magnitude and (a) the frequency and (b) the return period for some hazards. A few very high-magnitude events are responsible for the majority of disaster losses.

Magnitude–frequency relationships are used in other areas of hazard management. For example, a mortgage lender or an insurer might wish to know the magnitude–frequency relationships of flood risk for new houses built on a floodplain during the average mortgage span of 30 years. Figure 4.6 shows the risks of an event being equalled or exceeded during this period. It can be seen that an event as high as, or higher than, the 50-year flood has a 45% probability of occurrence, but if the 100-year return period is chosen, the probability drops to 26%. This is valuable information. If the probability of a claim being made and the likely cost of that claim are known, then the insurance premium can be set appropriately for that business. If the estimate of probable losses is too high, the premium will be high and may prove to be uncompetitive in the insurance market. If the estimate is too low, the insurance company stands to take a loss from unexpected claims.

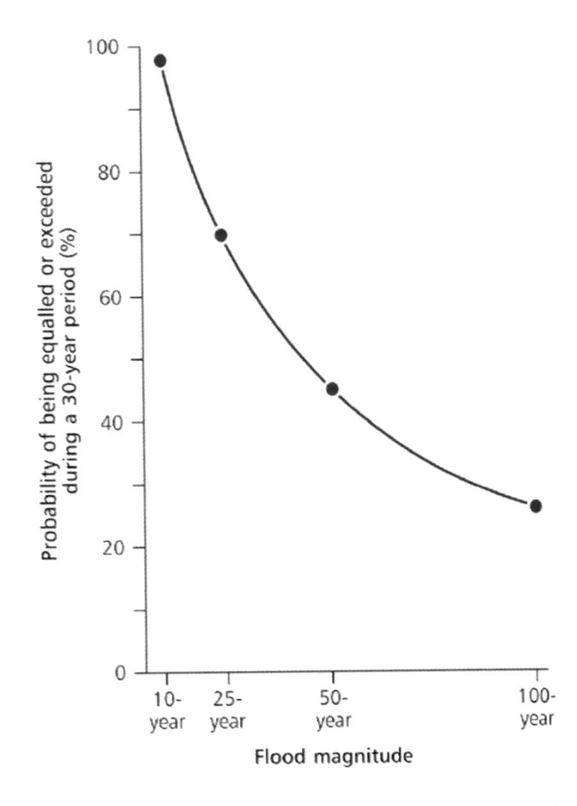

Figure 4.6 The probability of occurrence of floods of various magnitudes during a period of 30 years. This is the average duration of a standard property mortgage and the information on risk will be of interest to mortgage lenders and to property insurers.

2. The Analysis of Extreme Events

The analysis of extreme events by probability methods is arguably predicated partly on the assumption of *uniformitarianism*, which is the belief that past processes and events are a good guide for the future. Uniformitarianism as an approach can be traced back to the field of geology, such as James Hutton's (1794)

Investigation of the Principles of Knowledge. While this assumption is generally poor, as some hazards can substantially increase or decrease in intensity or cause new issues, it can have some reasonable results for hazards which tend to remain fairly constant on human lifetime scales, such as some global tectonic processes. Probability analysis is less suitable for environmental processes that are far more dynamic, particularly in the context of global warming. In recent years, there have been rapid and significant shifts in magnitude–frequency relationships for floods of a particular size if extensive deforestation has taken place over the drainage basin or if engineered measures have been built along a river or coast (Chapter 11). Setting such limitations aside, probability-based approaches can be used to estimate the size of floods that might be expected once every year, once every 10 years, and once every 100 years and so on. But, whilst a 100-year flood has a *probability* of 1:100 of happening in any one year and an estimated average return period of 100 years, in practice such a flood could occur next year, be exceeded several times in the next 100 years or not occur for 200 years. As such, these calculations are vague approximations rather than clear indications for action, although the use of a defined method yielding a specific number often gives them far more credibility than they should be accorded. This can create significant confusion with the public who often think that if a 1:100-year event has happened then they need not worry or prepare for the next one.

Despite such limitations, probability-based estimates are used to design and build key structures in hazard-prone areas. Engineers plan for a selected *design event*, which is often the magnitude of the hazardous process that a structure is built to withstand during its expected lifetime. The actual return period for the design event varies according to the nature of the hazard and the vulnerability of the elements at risk. As an example, large dams on major rivers are often built to withstand the 1:10,000-year flood because the consequences of failure would be catastrophic for many downstream communities. On the other hand, in the UK, railway bridges are generally designed to withstand the 1:100-year flood event because the potential consequences of failure are much less catastrophic. When the science is not accurate, disasters can form, as seen during the 2011 Tohoku tsunami (see Box 4.3).

Box 4.3 Understanding Hazard Events and Their Statistics

There were two significant failures in probability-based estimates for seismic and tsunami hazards in Japan. First, the pre-2011 national seismic hazard assessment for Japan indicated the largest earthquake from the northeastern Japan subduction zone to be magnitude Mw 8 with a maximum tsunami run-up of 10 m. This assessment was based only upon instrumental seismicity (since circa 1900) and made the big assumption that the subduction zone behaves in a consistent way and will not produce earthquakes bigger than in recorded history. It also ignored evidence of the 1896 Sanriku tsunami (that had a peak run-up of around 33 m) and older earthquake tsunamis with large inundations (AD 1611, AD 869) that imply earthquakes larger than Mw 8 can occur. Consequently, the mitigation strategies developed in Japan were intended to accommodate an Mw 8 earthquake. Unfortunately the Tohoku earthquake that struck on March 11th in 2011 was an Mw 9 quake. The extraordinary and expensive sea wall and vertical evacuation structures were not built to accommodate a tsunami generated by this large an earthquake and therefore the sea walls failed, and many people who fled to vertical evacuation structures were drowned (Koshimura & Shuto, 2015). In addition, the Fukushima nuclear plant was damaged (see Box 1.2 in Chapter 1).

These examples serve as examples of what can be classed as 'brittle' mitigation (Day & Fearnley, 2015), a strategy that works up to a limit of hazard intensity, and then fails catastrophically. In this case, mass casualties from decisions indirectly influenced by the national hazard assessment could have been avoided. First, the disaster planning, land use decision-making, tsunami defences and siting of critical facilities could have been made less sensitive to errors in the assessment. Second, the public could have been instructed to evacuate to high ground if at all possible, rather than to vertical evacuation shelters. Third, the potential for exceptionally large earthquakes and tsunamis should have been recognised – as they were by many scientists long before 2011 (Abe, 1973; Fukao & Furumoto, 1975). Personal experiences of the 1896 and 1933 tsunamis were also available alongside memorial stones and stories explaining where on the coastline it would be safe to build (Figure 4.7).

Figure 4.7 The grand tsunami monument in Aneyoshi, Miyako City, built after the Showa-Sanriku earthquake.

Source: Photographer: T. Kishimoto, Wikimedia Commons

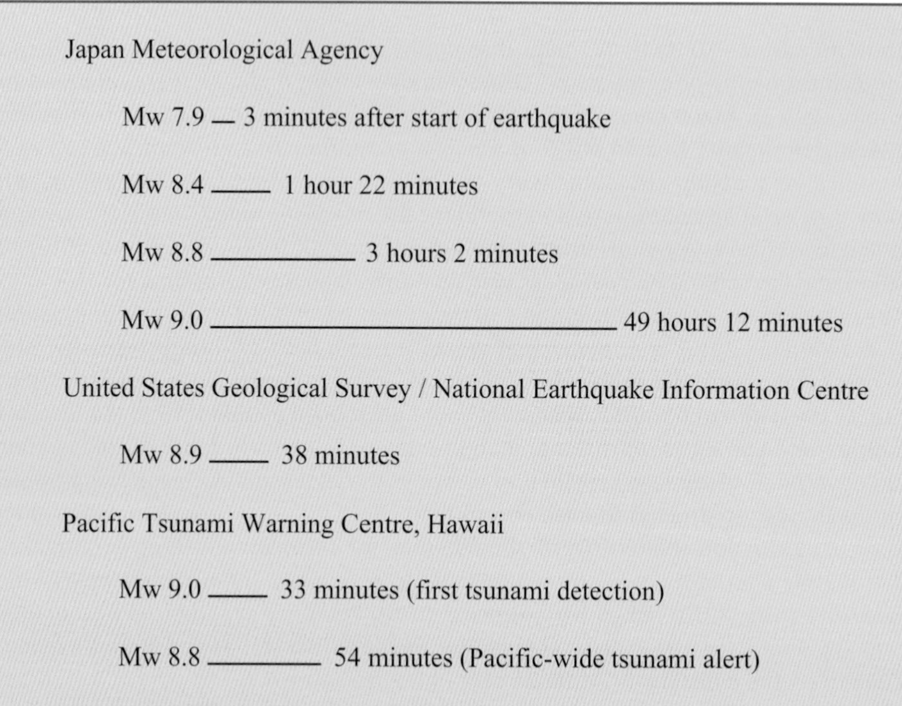

Figure 4.8 The relative issuance of earthquake magnitude over time following the Tohoku earthquake from differing sources.

Source: JMA, PTWC and USGS

The second failure in estimates occurred when the earthquake first struck, with the Japan Meteorological Agency establishing it as an Mw 7.9 earthquake, while the Pacific Tsunami Warning Center classified it as an Mw 9.0 earthquake, only 30 minutes later (see Figure 4.8), which may have been too late for some.

The initial (under-)estimates of the magnitude of the Tohoku earthquake resulted in the immediate response assuming a smaller earthquake than that experienced, and also a smaller tsunami, and no doubt had an influence on the response chosen by members of the public (Cyranoski, 2011). The diversity in models of the scale of the earthquake also created confusion for some in Japan who were receiving conflicting information from the USGS and PTWC that the earthquake and expected tsunami were larger.

In conclusion, hazard and risk assessments in most of the world are based on very short event catalogues, yet there is a critical need for geological studies to extend the records of past events to 10,000 years or so to help inform risk assessments and engineering designs. It is important that disaster plans make sufficient allowance for uncertainties and errors in hazard estimates. Additionally, there is a need to take more care to avoid 'brittle' mitigation strategies that fail catastrophically in the face of larger-than-expected hazard events.

When a dataset is too short to be representative it may be necessary to extrapolate a design event, despite the risk of substantial error. It is for this reason that efforts are made in earthquake engineering and flood hydrology to extend the instrumental records, using historical documents and other proxy records to estimate the frequency and size of unmeasured extreme events. The situation is most difficult for very rare hazards, such as large-scale volcanic eruptions, for which there is no statistically valid dataset. In these cases the only viable approach is to examine the geological record for proxy evidence in order to create modelling scenarios for such an event, as well as incorporating archaeological and anthropological studies to examine how prior civilisations coped with larger scale eruptions than witnessed in recent history. Inevitably there are numerous limitations in making assumptions around prior events, particularly in developing models.

A major conclusion must be that the reliability of results from probability-based analysis depends heavily on the quality of the database. Ideally, each event in the record should be drawn from the same statistical population, should be independent and should follow a known distribution curve. Other environmental phenomena are not necessarily independent. Earthquake occurrence is not random in time, as the magnitude of the event depends in part upon the amount of strain energy that is stored up in the Earth's crust. When a large earthquake occurs, at least part of the strain energy is released. This reduces the immediate likelihood of another large event on the same section of fault until the strain energy has built up again. On the other hand, the stress may have been transferred onto other local faults, increasing the chance of an earthquake on a nearby fault.

Whilst it is sometimes assumed that the statistics are best described by a normal distribution function, this is not always so. Daily rainfall data, for example, have a skewed, rather than a normal, statistical distribution, with resulting complications for probability analysis. Other problems arise, as mentioned in the previous section, when past records are used for prediction purposes on the assumption that there will be no change in causal factors. This assumption, known as *stationarity*, ignores the possibility of wider environmental change. Changes to physical systems can occur naturally over very long time periods, but changes resulting from human activities are often more important. In terms of near-surface geophysical processes, such as floods, the relevant systems have almost certainly been affected to some degree by human activity over the last century or so. The prospect of climate change (Chapters 16 and 18) means that the existing statistical distributions are less likely to provide a reliable estimate of future events.

The consequences of such changes, when expressed in statistical terms, are complex. Changes in the frequency of hazardous events can be expressed most simply as shifts in the mean and standard deviation of the dataset. Figure 4.9 illustrates a climate change situation in which the mean value remains constant but the variability, expressed by the standard deviation, increases. Thus, the frequency

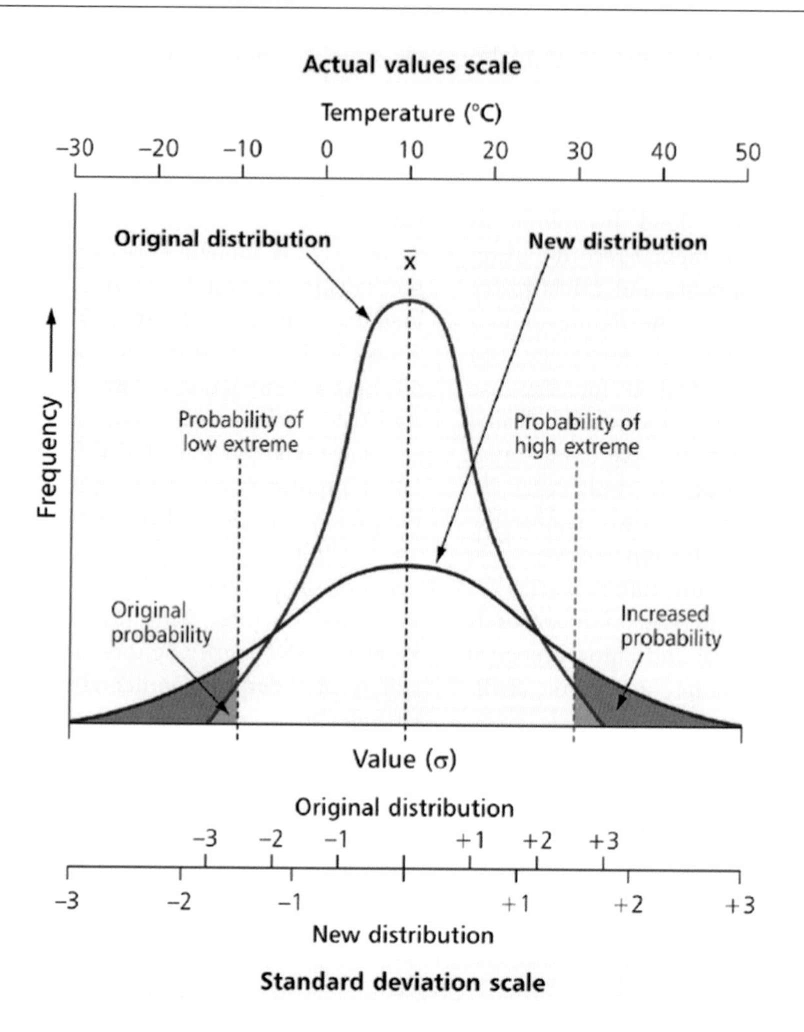

Figure 4.9 The effects of a change to increased variability on the occurrence of extreme events. Both upper and lower hazard-impact thresholds are breached more frequently as a result of the increased standard deviation, although the mean value remains constant. The example is provided on a temperature scale.

of both 'high' and 'low' extreme events increases relative to the thresholds which define the relevant social band of tolerance. This might simulate climate change that leads to both colder winters and warmer summers, as shown on the temperature scale. On the other hand, Figure 4.10 shows the consequences of an increase in the mean value but with no change in variability. This might simulate the effects of climate change in which a location undergoes net warming without a major change to the weather patterns. In this case the frequency of 'high' extremes relative to the impact threshold rises, whilst the incidence of 'low' extremes falls. Needless to say, this effect would be reversed with a lower mean value.

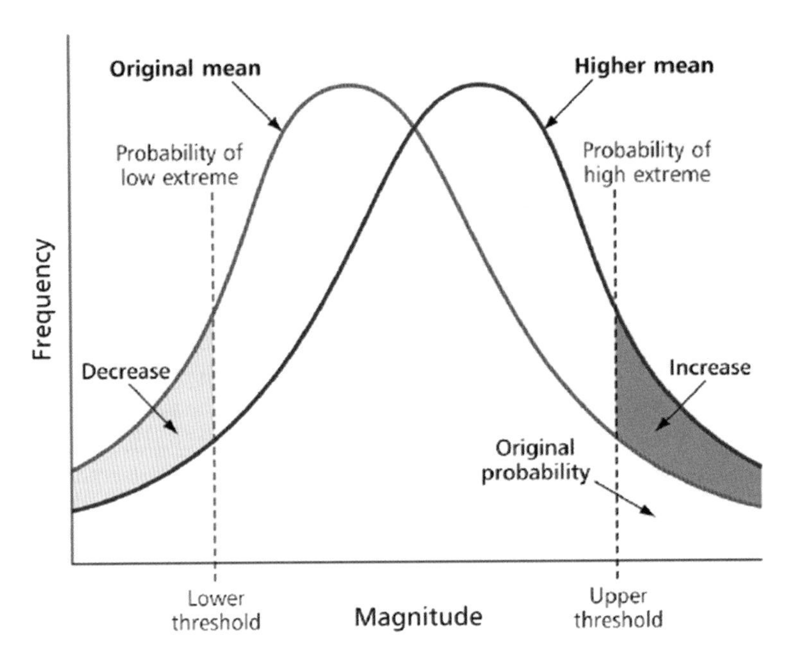

Figure 4.10 The effects of a change to an increased mean value on the distribution of extreme events. The shift results in an increased frequency of hazard impacts from 'high'-magnitude events, with a corresponding decrease in the frequency of 'low'-magnitude events.

In reality, environmental change might cause shifts in both the mean temperature and the variability. It might also alter the shape of the distribution. For this reason, accurate and precise projections of the effects of climate change on the occurrence of hazardous events can be difficult to achieve with existing models. One complication is the non-linear relationships that exist between driving factors and the hazards themselves, such as between sea-surface temperatures and the formation of tropical cyclones (see Chapter 10). The probability function for most hazardous processes is itself sensitive to changes in the mean value (Wigley, 1985). A shift in the mean value of only one standard deviation would cause an extreme event expected once in 20 years to become five times more frequent. Similarly, the return period for the one-in-a-100-year event would fall to only 11 years, an increase in probability of nine times. This is one reason why some researchers believe that the impacts of atmospheric hazards will increase significantly with only modest changes in climate. There are a number of risk modelling methods used to assess potential future losses that bring a wide range of data together to help disaster preparedness.

D. SITUATING RISK ASSESSMENT

Scientists commonly regard risk as the interaction between the hazard and society; more specifically this is understood as 'the magnitude of a potential loss – of life, property, or productive capacity – within the area subject to hazard(s)' (Wright & Pierson, 1992, p. 28). Historically, risk has been reviewed using risk assessments that are quantitative, despite often being based on qualitative measurements. Scientific methods of understanding risk include a range of 'quantitative and/or expert-based risk assessment techniques, involving varying forms of scientific experimentation and modelling, probability and statistical theory, cost-benefit and decision analysis, and Bayesian and Monte-Carlo methods' (Stirling, 2007, p. 309). These are intrinsically reductive processes, simplifying complex and contested realities into a discrete set of ordered categories.

Within mathematical and scientific disciplines risk assessment is usually conducted as shown in Figure 4.12, which demonstrates 'the unifying approach to risk and risk analysis is based on the idea that risk is a way of expressing uncertainty related to future observable quantities' (Aven, 2003, p. 47). Models in risk analysis require observable quantities that are used to develop a deterministic model that links the systems performance with these observations. Ultimately, it is possible to calculate the uncertainty distribution of the performance measures and determine a suitable prediction.

Without such quantities it is not possible to conduct a risk analysis, as there is no basis on which to calculate probabilities. Stirling states, 'reductive quantitative risk fails to recognise the intrinsic limitations and contradictions in the rational choice foundations that underlie risk assessment' (2003, p. 52). This suggests that models, such as Bayesian event trees and expert elicitation, are less useful because they

Box 4.4 Risk Modelling

Hazus is a nationally standardised risk modelling methodology operated in the US by the Federal Emergency Management Agency (FEMA) for estimating potential losses. It is a commercial off-the-shelf loss and risk assessment software package built on GIS technology and natural hazard analysis tools that helps identify areas with high risk for natural hazards and estimates physical, economic and social impacts of these hazards. Hazus models multiple types of hazards: flooding, hurricanes, coastal surge, tsunamis and earthquakes using three steps: first, calculating the exposure for a selected area; second, characterising the level or intensity of the hazard affecting the exposed area; and third, using the exposed area and the hazard to calculate the potential losses in terms of economic losses and structural damage, as exemplified in Figure 4.11. Hazus is used by communities to support risk assessments that perform economic loss scenarios for certain natural hazards, and to raise hazard awareness. You can learn more here: www.fema.gov/flood-maps/products-tools/hazus. Tools like these enable better understanding of the risks and enable key decision-makers to take appropriate action based on the available data.

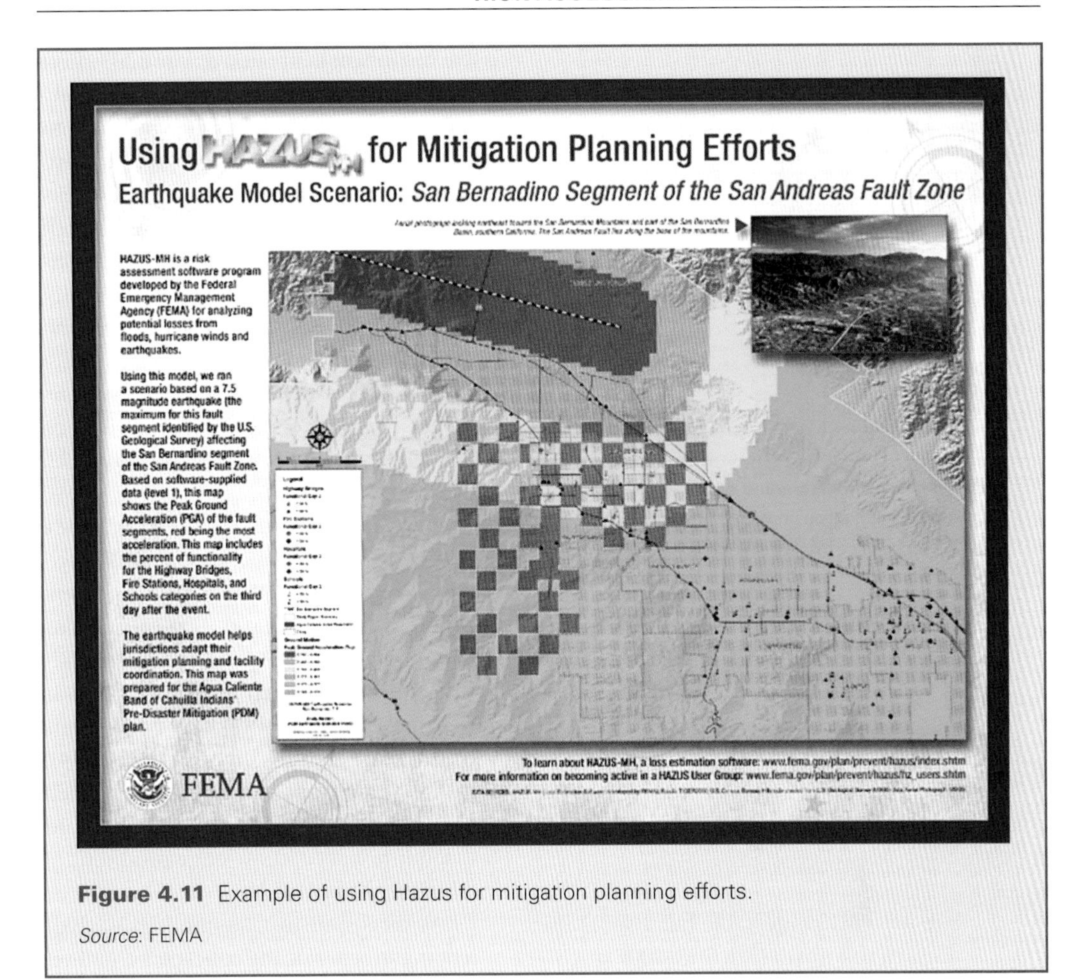

Figure 4.11 Example of using Hazus for mitigation planning efforts.

Source: FEMA

do not consider ambiguity and ignorance when attempting to quantify risk.

Traditional risk assessment uses a powerful set of methods when looking at risk, but as shown in this section they are not applicable under conditions of high uncertainties. Although numerous aspects of hazards and threats are uncertain, understanding uncertainty alone does not capture 'incertitude': a concept that combines uncertainty, ignorance, ambiguity and risk to provide a more holistic view of knowledge, or lack of (Stirling, 2007). This has practical implications for the robustness of conventional reductive risk assessment in decision-making, and it is argued that persistence in using these reductive methods under conditions other than a strict state of risk (as defined by Knight, 1921) are irrational, unscientific and potentially misleading. Uncertainty indicates that a single aggregated picture of risk is neither rational nor 'science-based'; ambiguity highlights that it is not rigorous or rational to provide a single 'sound scientific' picture of risk; and ignorance demonstrates that parameters

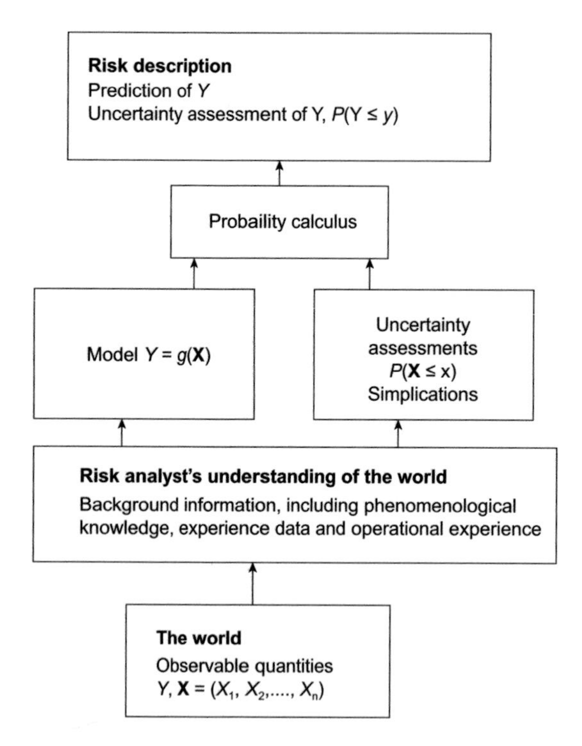

Figure 4.12 Basic elements of risk analysis.

Source: Adapted from Aven (2003, p. 49)

are not only contestable but also, at least in part, unknown (Stirling, 2007, p. 310). Ignorance can reflect our reliance on inductive knowledge as well as the 'unknown unknowns' for which we do not know what we do not know. These four elements of incomplete knowledge are presented in Figure 4.13 providing schematic examples and showing the relationship between knowledge of probabilities (risk and uncertainty) and the outcomes (ambiguity and ignorance).

Post-Normal Science

In the last 20 years there has been a growing recognition by philosophers and sociologists of science that when there are extensive scientific uncertainties, science can no longer be treated as 'normal', where normal assumes there is universal, objective and context-free knowledge (Gibbons et al., 1994). Science, it is argued, is 'post-normal' in that it has entered the polity and is no longer viable as 'normal' puzzle solving conducted in abstraction from the issues of who pays and why. Consequently, this suggests some natural hazard science is more akin to post-normal science (PNS). This is where 'facts are uncertain, values in dispute, stakes high and decisions urgent' (Funtowicz & Ravetz, 1993, p. 744).

A similar framework to PNS is 'Mode 2', a mid-20th century theory of knowledge production that is context-driven, problem-focused and interdisciplinary. It involves bringing together multidisciplinary teams for short periods of time to work on specific problems in the real world (Gibbons et al., 1994), rather than traditional research based on academic, investigator-initiated and discipline-based knowledge production, labelled Mode 1.

As discussed in Chapter 1, it is challenging to separate the concepts of 'nature', 'science' and 'society', therefore decisions of risk must consider various sorts of uncertainty and value commitments. Like Mode 2 knowledge, PNS brings together an extended peer community to enter into a dialogue about the uncertainty, ignorance, perspectives and values of each stakeholder, using their expertise. It is the values of relevant policy stakeholders therefore – derived from qualitative methods of data collection such as interviews, focus groups, scenario mapping and participant observation – that are foregrounded. In recent years, the principles and practices of PNS have been widely incorporated into some 'participation' science. PNS can be represented by

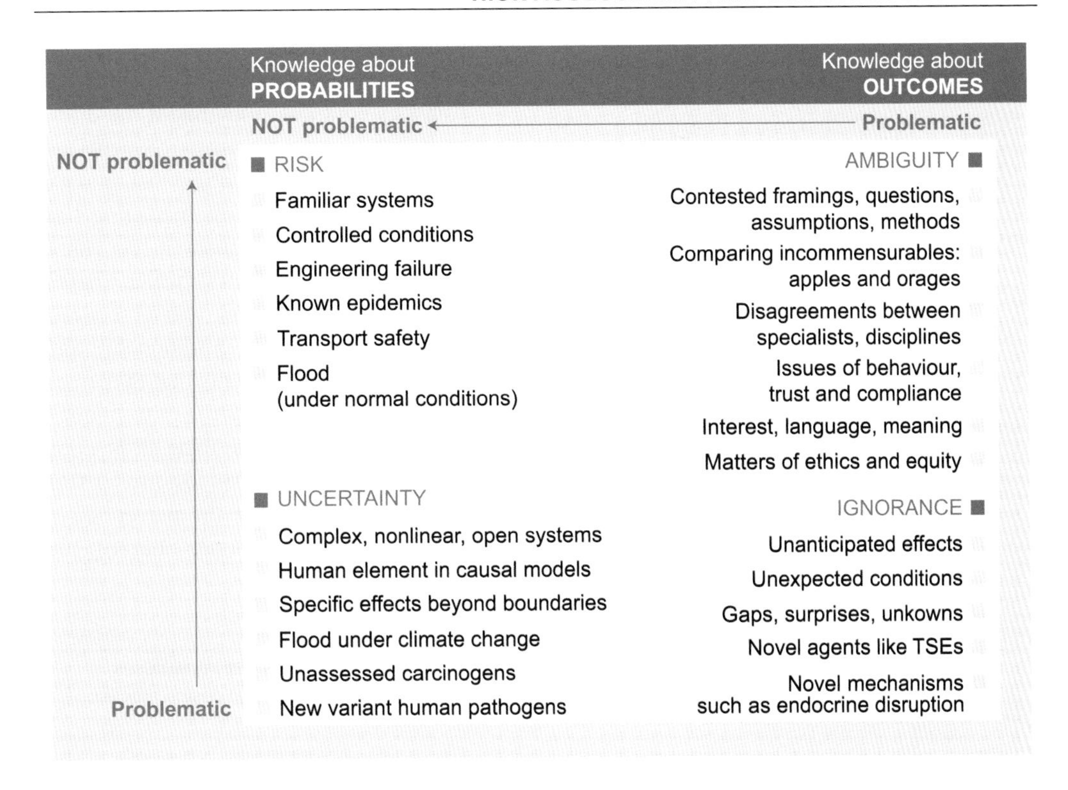

Figure 4.13 Contrasting states of incomplete knowledge, with schematic examples.

Source: Stirling (2007, p. 312)

Figure 4.14 that reviews increasing levels of uncertainty and decision-making stakes that indicate that a complex situation is within a PNS state.

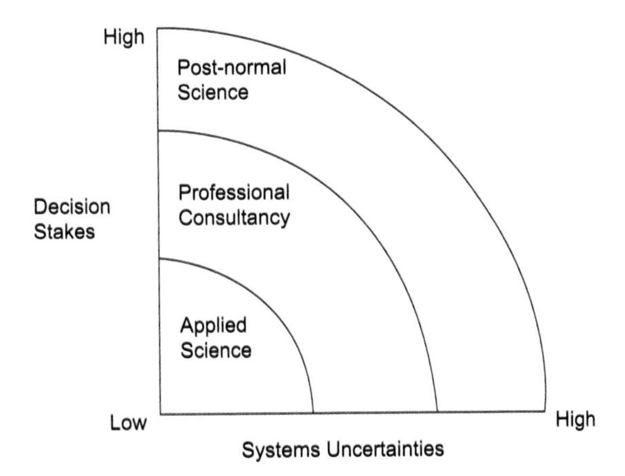

Figure 4.14 Post-normal science.

Source: Ravetz (2004, p. 354)

Qualitative Methods

There is a growing movement to place uncertainty at the centre of the science-policy and science-society interface, under the assumption that the public and other stakeholders have a role to play in policy development, and that top-down policy-making is no longer viable. Effective risk reduction requires that all actors are involved in the social learning process; adopting a bottom-up approach promotes local stakeholders to create a more resilient community (O'Brien, 2008; O'Brien & Read, 2005). However, there is a distinction between risk perceived by experts and laypersons as already outlined in this chapter. Stirling (2007) outlines a general framework for effectively articulating conventional risk assessment with other broader qualities and associated methods by using a criteria-based screening process to identify crucial attributes of scientific uncertainty, or social or political ambiguity as outlined in Figure 4.15. This framework highlights that there are better alternatives than just using traditional risk assessments.

E. WHO IS THE EXPERT?

Analyses continue to highlight the differences between expert risk assessments and lay risk perceptions, noting that this gap can lead to problems in managing hazards. A statistician might well rate voluntary and involuntary risks equally, whereas most non-experts show greater concern for involuntary risks. Additionally, high risk–low frequency events that take many lives can be seen as equivalent to frequent hazards that take only a single life at a time but which, over time, lead to a similar scale of loss. Conversely, in the perception of most laypeople, the dramatic hazards that

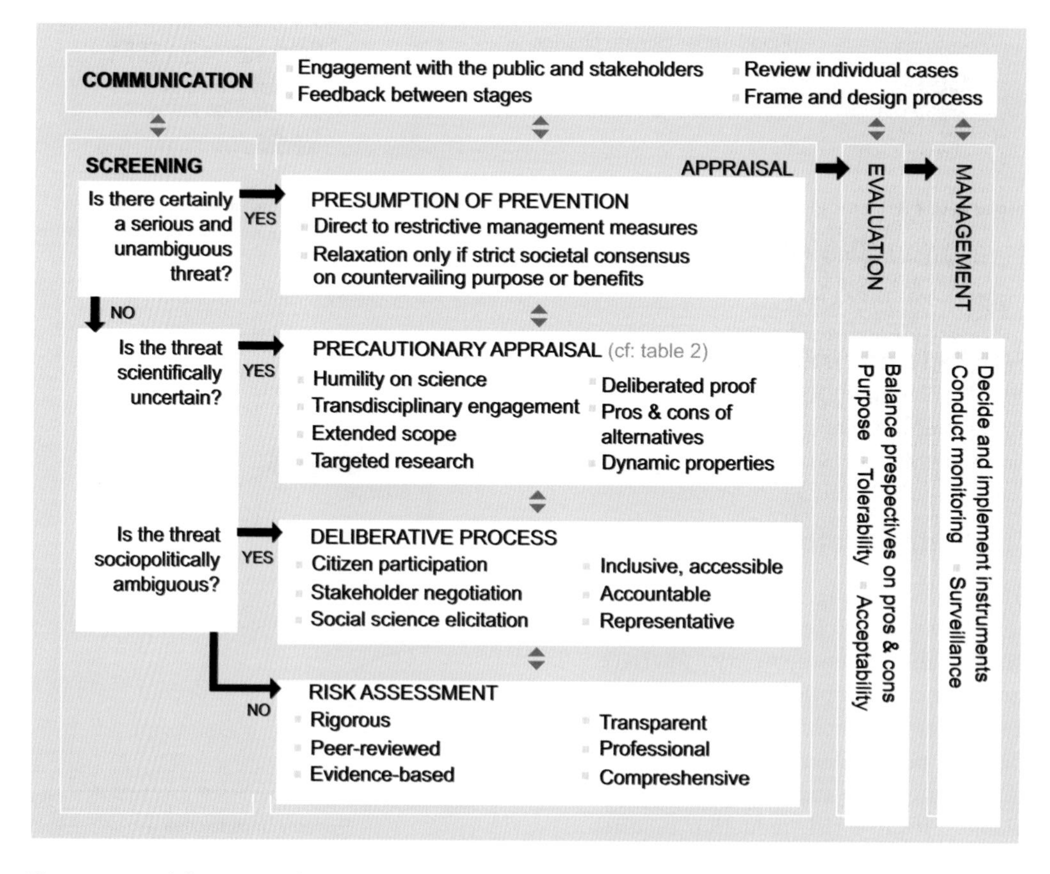

Figure 4.15 A framework for articulating precaution and risk assessment.

Source: Stirling (2007, p. 313)

take many lives at a time are more significant. For example, in the UK, on average more people die each day in road crashes than die each year in rail crashes, but railway crashes attract more media coverage. This is partly because rail crashes are considered the result of involuntary risk and the events tend to produce striking images. Vehicle crashes are perceived as the result of voluntary risk and the number of deaths per event tends to be low (Figure 4.16).

There has been a growing recognition of the need to engage with the 'lay' public who are in many cases the experts in understanding their environment and what needs to be done to prepare, mitigate and respond to potential hazards.

An expert is a person who has extensive skill or knowledge in a particular field, as opposed to a layperson who does not have specialised or professional knowledge of a subject. Scientists will differ in expertise determined by their research experience, skills training and ability to access and keep up to date with the latest debates, ideas, concerns and methods. This provides a degree of diversity within this cohort. If you were to provide ten world-leading seismologists a seismograph, you may receive ten different opinions about when they think a

Figure 4.16 Emergency response to a 2009 road traffic crash in Geneva, Switzerland. Transport risks tend to be acceptable to the public, despite the cumulative death rate in vehicle crashes.

Source: Photo: Ilan Kelman

volcano may erupt. Therefore, one of the key tools used within the scientific world to determine which expert is best to listen to/make a decision in a given situation is via structured expert elicitation. This process comprises the synthesis of opinions of authorities on a subject where there is uncertainty due to insufficient data, or when data is unattainable due to physical constraints or lack of resources. Expert elicitation is essentially a scientific consensus methodology. The process enables a level of accountability and transparency to the decision-making process, neutrality, fairness and quality control through its structure. A now common approach introduced by the Australian government is the IDEA protocol, which outlines four key steps: investigate, discuss, estimate and aggregate (see Figure 4.17).

Whilst there are tools to help identify who an 'expert' scientist is, they are still somewhat subjective, and the scientific input into managing risk is only one component. Bringing together the concepts outlined previously around post-normal science and Mode 2 knowledge, it is clear that within the context of natural and technological hazards no one person is truly an 'expert'. The management of hazards, threats and risks requires a range of experts to come together, along with the vulnerable groups.

One prompt for thinking about expertise beyond the scientific community emerged from the impact of the 1986 Chernobyl fallout and its impact on sheep farming in Cumbria in the UK during 1986. Adopting top-down narratives that ignored local expertise proved costly. Brian Wynne (1992, 1996) studied how government scientists claimed the problem would clear up in weeks, contradicting the farmers' own expertise about the contingencies of farming in the Lake District, taking no account of the idiosyncratic features of the area. This top-down approach undermined the status of local expert knowledges and created disillusionment with scientists' ability to predict and manage risks. This study highlights 'the forms of institutional embedding, patronage, organisation and control of scientific knowledge' (Wynne, 1992, p. 42) that often lead to a lack of action by the public in response to warnings.

Certainly there has been a gradual shift in the understanding of how to engage

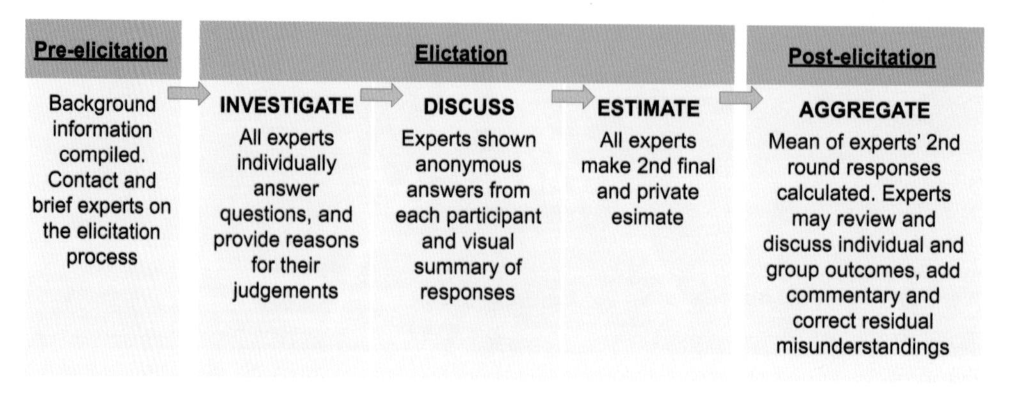

Figure 4.17 The IDEA protocol.

Source: Adapted from Burgman (2015)

with the public, at varying paces globally. In the post-war period some scientists and disciplines felt that if the public understood science better, then they would comply with what scientists thought was best for society. This 'deficit' model assumes that if you educate people they will be enlightened and understand the situation. Instead, the public needs to be included for science and scientific issues not at the end of these processes, but at the start of them. This approach is sometimes called 'upstream' engagement, so that the public influences (but does not have full say over) science and technological development. Public engagement contributes to addressing environmental hazards and risk management.

Public engagement can be defined as 'whether the public (whoever they are) should, can, or would want to understand (whatever that means) science (however defined)' (Gregory & Miller, 1998, p. 8). This influences science and risk education, communication and policy. The public are increasingly contributing to scientific research by doing different aspects of the research themselves. This can take place in a number of different ways, as highlighted in Figure 4.18.

Citizens can act as 'sensors' by collecting data, such as managing weather stations and sending that data to national meteorological agencies. This helps create engagement, knowledge and continued awareness around different types of hazards that give the community a sense of ownership in the hazard management process. At the other end of the spectrum, extreme citizen science is a bottom-up practice that takes into account local needs, practices and culture and works, with broad networks of people to design and build new devices and knowledge creation processes. In many instances this is the work facilitated by humanitarian agencies and other NGOs alongside academic partners to help identify problems

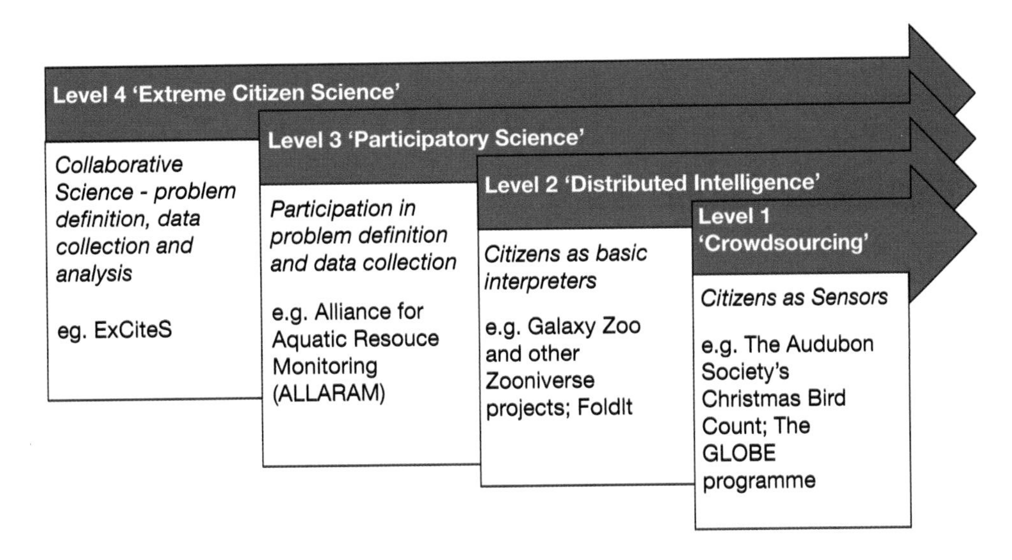

Figure 4.18 Web-based channels for science communication: A review of trends and platforms.

Source: Bultitude (2014, p. 238). 'Web-based channels for science communication: A review of trends and platforms'. In L. Tan Wee Hin, R. Subramaniam (Eds.), *Communicating Science to the Public: Opportunities and Challenges for the Asia-Pacific Region*. Netherlands: Springer.

and solutions for a community in relation to managing the risks posed by environmental hazards.

Notwithstanding the aforementioned ways in which 'expert' has been expanded, a profound and long-lasting problem in addressing environmental hazards has been the serial, often violent, dismissal of indigenous knowledges of place. It is increasingly recognised that different forms of knowledge can learn from each other (Mazzocchi, 2006). Supranational organisations such as UNESCO, and some states, are belatedly acknowledging that there is significant indigenous and traditional knowledge and practice around hazards and threats that can provide valuable insights into understanding, communicating and managing hazards and risks. Such knowledges can be customary, pragmatic, experiential and holistic. They often take the form of oral stories and traditions, which convey cross-generational knowledge of landscapes, waterscapes and icescapes. Participatory engagement aims to respect the risk perceptions held by local stakeholders and to work jointly towards the application of indigenous skills and coping measures. This can be a challenging and complex problem, particularly when Eurocentric assumptions as to the innate superiority of modern scientific knowledge becomes a reason for dismissing alternate understandings (see Box 4.5).

The valuing of indigenous knowledges can be seen in some institutional contexts,

Box 4.5 'There's No Plastic in Our Volcano': A Story About Losing and Finding a Path to Participatory Volcanic Risk Management in Colombia

On the fourth day of a week-long workshop designed to promote dialogue and co-operation between stakeholders at the base of Galeras, an active volcano just outside the city of Pasto in Nariño province, southern Colombia, an indigenous elder took the stage in an agitated state. He began to outline all the reasons that it was ridiculous to suggest that their volcano, on whose flanks they had lived and worked for several generations, could possibly contain plastic. The many stakeholders attending the University-led workshop included scientists, local and national government agencies, emergency services and at-risk communities. The indigenous elder stated that he knew there to be rocks, ash and all sorts of other materials for sure, but categorically not plastic. He spoke so fast and with such force that the official translators in the box at the rear of the auditorium gave up after the first few minutes. At this point, most of the foreign scientists left the auditorium for a tea break and the attending community members stated that they no longer wished to remain for the final days of the workshop. On discovering this, the scientists, both local and international, congregated outside to discuss how such a well-intentioned and carefully planned opportunity to create new collaboration and what was believed to be a participatory path to managing risk could have broken down so completely.

(Wilmshurst, 2018, p. 500)

What stands out from this excerpt is the lack of inclination by the scientists gathered from other countries to listen to the concerns being expressed by the elder and to ascertain how the communication of 'pyroclastic' could have been so badly handled. In leaving the workshop, there is no evident awareness of the challenges involved in communicating not only between languages, but also between different kinds of expertise. It becomes clear by the end of the excerpt how such a lack of valuing of other knowledges can spiral into a breakdown in relationships and the undermining of efforts to nurture a comprehensive and effective risk management strategy. Indeed, the need for such a strategy was well evidenced by the nine deaths caused by eruptions since the Galeras volcano became active again in 1989, of which seven were scientists and two were tourists who went up to the crater with them. Wilmshurst goes on to describe the subsequent, difficult process by which people came together from a place of mistrust and miscommunication to build a participatory approach to managing risks at Galeras volcano, which resulted in a 3-month facilitated process of dialogue and collaboration, that then continued independently which, after some initial issues, appears to continue to this day.

such as GNS Science (New Zealand's leading provider of Earth and geoscience research) and NIWA (National Institute of Water and Atmospheric Research) where Māori knowledge has been introduced. This includes knowledge captured via oral histories and traditions that capture past catastrophic events, place names that indicate areas of hazards and environmental indicators about the safety of activities. This is typically classed as Māori environmental knowledge (MEK). A 2006 report by NIWA highlighted, 'the detail contained in Māori nomenclature is a valuable and neglected area of information about the past which can help with hazard identification, management, and prevention in New Zealand' (King & Goff, 2006, p. iv).

The public are subsequently contributing more to public policy. Responsible research and innovation (RRI) takes up the challenge of listening, taking account of public perspectives and scrutinising the values of science that truly reflects wider social needs and values and how we can best do this. Local, traditional and international knowledge can all feed into policy development by a range of public engagement mechanisms (see Table 4.5).

The public can even go even further by changing the lines of inquiry pursued by scientists using grassroots activism. There have been a number of environmental cases won by Greenpeace where grassroots activism has assisted in reducing disasters, but one particularly interesting example is that of AIDS during the 1980s (Epstein, 1996), see Box 4.6.

It is vital to involve everyone, not just the vocal minority or presumed majority, when thinking about risk management. Multiple cross-cutting characteristics of individuals/social groups/communities include sex, gender, sexuality, age, race, ethnicity, caste, abilities (e.g. physical, mental and cognitive), religion, languages, communication forms and precarity (e.g. detained, undocumented, homeless, asylum status). Tailored and bespoke approaches are essential to ensure that those vulnerable are contributing to the development of risk

Table 4.5 Public participation methods: A framework for evaluation.

Participation method	Nature of participants	Time scale/ duration	Characteristics/ mechanism	Examples/ references
Referenda	Potentially all members of national or local population; realistically, a significant proportion of these	Vote cast at single point in time	Vote is usually choice of one of two options. All participants have equal influence. Final outcome is binding.	Biotechnology in Switzerland (Buchmann, 1995); waste repository in Sweden (af Wahlberg, 1997).
Public hearings/ inquiries	Interested citizens, limited in number by size of venue. True participants are experts and politicians making presentations.	May last many weeks/months, even years. Usually held during weekdays/ working hours.	Entails presentations by agencies regarding plans in open forum. Public may voice opinions but have no direct impact on recommendation.	Frequent mechanism in, for example, United States (Fiorino, 1990), Australia (Davison et al., 1997); review by Middendorf and Busch (1997).
Public opinion surveys	Large sample (e.g. 100s or 1,000s), usually representative of the population segments of interest.	Single event, usually lasting no more than several minutes.	Often enacted through written questionnaire or telephone survey. May involve variety of questions. Used for information gathering.	Radioactive sites in United States (Feldman & Hanahan, 1996); genetically modified food in the United Kingdom (Vidal, 1998); biotech surveys (Davison et al., 1997).
Negotiated rule making	Small number of representatives of stakeholder groups (may include public representatives).	Uncertain: strict deadline usually set: days/weeks/ moment.	Working committee of stakeholder representatives (and from sponsor). Consensus required on specific question (usually, a regulation).	Used by US Environmental Protection Agency (Hanson, 1984); method discussed by Susskind and McMahon (1985) and Fiorino (1990).
Consensus conference	Generally, ten to 16 members of public (with no knowledge on topic) selected by steering committee as 'representative' of the general public.	Preparatory demonstrations and lectures (etc.) to inform panellists about topic, then 3-day conference.	Lay panel with independent facilitator questions expert witnesses chosen by stakeholder panel. Meetings open to wider public. Conclusions on key questions made via report or press conference.	Used in Denmark and Netherlands on topic from food irradiation to air pollution (Joss & Durant, 1994; Grundahl, 1995); also used in United Kingdom on plant biotechnology (Ellahi, 1995).
Citizens' jury panel	Generally, 12 to 20 members of public selected	Not precise but generally involve meetings over	Lay panel with independent facilitator questions	Examples in Germany, United States and United

Participation method	Nature of participants	Time scale/ duration	Characteristics/ mechanism	Examples/ references
	by stakeholder panel to be roughly representative of the local population.	a few days (e.g. four to ten).	expert witnesses chosen by stakeholder panel. Meetings not generally open. Conclusions on key questions made via report or press conference.	Kingdom (e.g. Crosby et al., 1986; Coote et al., 1994; Lenaghan et al., 1996).
Citizen/public advisory committee	Small group selected by sponsor to represent views of various groups or communities (may not comprise members of true public).	Takes place over an extended period of time.	Group convened by sponsor to examine some significant issue. Interaction with industry representatives.	Particularly evident in United States, for example, in clean-up of waste sites (Lynn & Busenberg, 1995; Perhac, 1998); see Creighton (1993) for guidelines.
Focus groups	Small group of five to 12 selected to be representative of public; several groups may be used for one project (comprising members of subgroups).	Single meeting, usually up to two hours.	Free discussion on general topic with video/tape recording and little input/direction from facilitator. Used to assess opinions/ attitudes.	Guidelines from Morgan (1993); UK example to assess food risk (Fife-Schaw & Rowe, 1995).

Source: Rowe & Frewer (2000, pp. 8–9)

Box 4.6 Activism in AIDS in North America

The AIDS epidemic officially began in the USA on 5 June 1981, when the US Centers for Disease Control and Prevention (CDC) issued findings on the disease. It was shocking to those who were diagnosed with AIDS how little the US government and medical establishment was doing to address a crisis that, at the time, mostly afflicted gay men. Sadly, many governments globally also 'ignore' this pandemic. With thousands dead, no significant interest by pharmaceutical companies to look at cures or treatments and little said in relation to the disease, many were left disempowered. Desperate, and with their lives at stake, AIDS activists underwent a metamorphosis process becoming a new group of experts speaking credibly in the language of scientific researchers to engage fully with the project of biomed research and treatment. The grassroots activism movement, with groups such as AIDS Coalition to Unleash Power (or ACT UP), recast the status of the patient from victim to participant-expert. Those who learned the scientific language – including research protocols, drug testing and regulation systems, and the workings of pharmaceutical companies and government advisory committees – generated social acceptance, respect from doctors and recognition of their own expertise.

One of the key issues that was fought for (and won) by activists was the use of medicines in trials. When pharmaceutical companies were testing AIDS medications, they placed one group on placebos and the other on the active medication. With people's lives at risk, the pharmaceutical trials were unlikely to get clean data when dealing with a desperate population who understandably took a wide range of medicine, swapped drugs and broke open their pills to share so they could increase their chances of not being on a placebo. ACT UP used its 'inside-outside strategy' and deployed it over and over again with the National Institutes of Health, and then with pharmaceutical companies, eventually becoming full partners with key scientists until they revolutionised the very way that drugs are identified and tested. This included scrapping the prevailing practice of testing drugs on a small number of people over a long period of time in favour of testing a huge sample of people over a much shorter period. This significantly sped up the time it took to conduct drug trials. Similarly, ACT UP insisted that the researchers and pharmaceutical companies searching for a cure for AIDS also research treatments for the infections that were killing off AIDS patients while they waited for a cure.

In sum, the AIDS grassroots activity created a model for patient advocacy within the research system that had never existed before. Today, it is normal that people suffering from a disease (cancer, Alzheimer's or diabetes) should have a voice in how it is researched and treated. More details can be read in Steven Epstein's phenomenal book *Impure Science: AIDS, Activism, and the Politics of Knowledge* published in 1996.

Figure 4.19 Demonstrators from the organisation ACT UP protest in front of the headquarters of the Food and Drug Administration. The FDA opened up access to experimental drugs soon after.

Source: © Getty Images/Mikki Ansin

management and have their needs met. Depending on the communities involved, as well as the differences and similarities of the people and groups within those communities, prioritisations will be needed. Privileged individuals/social groups/communities will need to consider the interconnected nature of social categorisations such as race, class, gender and overlapping and interdependent systems of discrimination or disadvantage, often referred to as intersectionality (Zaidi & Fordham, 2021) after the ground-breaking work of Kimberlé Crenshaw (see Crenshaw, 1989, 2017).

Whilst the public increasingly have a vital and important role via engagement activities with scientists, government and other stakeholders, there have been numerous high-profile and false public health fears that have inhibited risk management. Misinformation disseminated by groups but also those considered experts by virtue of their qualifications can cause significant problems during a crisis. One classic and pervading example is the incorrect claims that children risk autism from the triple vaccine for measles, mumps and rubella (MMR) that occurred in 1998, which resulted in significant reductions in MMR vaccination rates and a significant increase in the incidence of two of the three diseases. Even today, issues around the impact of immunisations have pervaded, as witnessed by the anti-vax movement during COVID-19, including that COVID-19 vaccines include a microchip to track the person's movements. Such false statements are, ironically, often spread through social media which harvests the user's preference data, and are often posted or read on mobile devices that *do* track the user's movements. In a 'risk society' driven by echo chambers repeating unverified conspiracies, the public is increasingly cautious about accepting scientific views about hazards.

F. RISK MANAGEMENT

Risk management is a process whereby risk is evaluated in order to facilitate the targeted and effective introduction of hazard-reducing strategies. To some extent, there are differences between risk management for technological and natural hazards. For example, fault tree methods are more common for technological hazards, but, in both cases, an appreciation of the entire system involved is needed.

In most countries the prime responsibility for risk management lies with national governments, who create health and safety legislation, which, in turn, is implemented and enforced by regional authorities and specialised agencies, most commonly emergency managements agency, such as FEMA in the USA, and civil defence or protection, such as Badan Nasional Penanggulangan Bencana (BNPB) in Indonesia. Different legal systems determine the structure and style of the management of risk, whether risk management actions are enforced by the government or it is the responsibility of individuals to respond, and whether nations are likely to rebel against governments or be compliant. One of the biggest issues is whether decisions made on risk management fall to these specialised agencies or whether these are decisions made across agencies usually incorporating risk scientists as part of the process. Whatever administrative structure exists, successful risk management depends on the use of effective and transparent methods that are cost effective and acceptable to the stakeholders within the relevant community.

Management methods vary according to the type of risk and the needs of vulnerable cohorts. In an ideal world, there would be a clear set of agreed priorities for formal risk management, with the highest levels of risk addressed first. In order to develop such a priority list, a quantitative risk assessment of all relevant factors and consequences is often used, forming a national risk register. Often risk registers fail to capture risks that are not previously experienced, infrequent or existential (low risk but high impact events such as meteorite strikes). This goal is almost impossible to achieve, due to a lack of data, the need to balance risks between high- and low-frequency events, financial constraints and, as already discussed, the complexities attached to public perceptions of risk. In particular, top-down approaches frequently fail to capture the needs of some at-risk groups, and there is now focus on more community-based participatory approaches to risk management and disaster mitigation.

1. The Formal Approach

Recent experience of large disaster losses, combined with reduced financial resources, has encouraged governments in Europe and elsewhere to adopt more rigorous and systematic methods of risk management, as exemplified by the procedures introduced for natural hazards reduction in Switzerland (Bründl et al., 2009). As shown in Figure 4.20, this approach consists of a detailed, three-stage procedure that includes risk analysis, risk evaluation and risk management.

- *Risk analysis* is the first step when a general hazard appraisal takes place using archived information from maps,

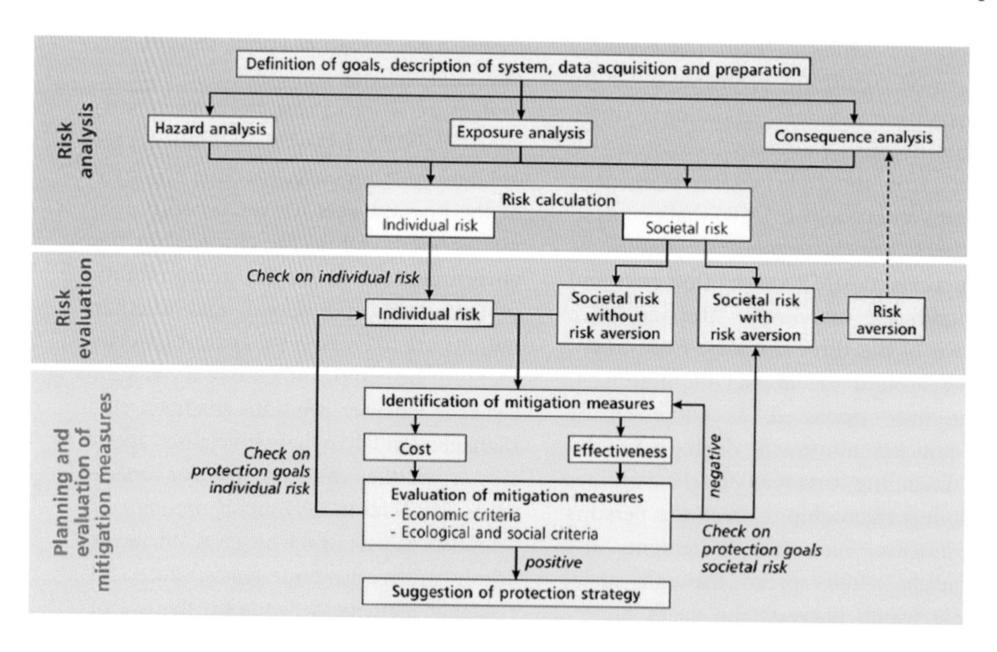

Figure 4.20 A sequential approach to natural hazard risk management as adopted in Switzerland. The process progresses through stages of analysis, evaluation and mitigation for individual and societal risks.

Source: After Bründl et al. (2009). Reproduced under Creative Commons Attribution 3.0 License.

terrain analysis, aerial photographs and satellite imagery. *Exposure analysis* then identifies and assesses the extent to which people or other local assets are at risk. *Consequence analysis* combines the hazard and exposure outcomes to provide estimates of expected damages or other losses from given events. *Risk calculation* is then conducted to determine the scale of expected losses for persons or socio-economic assets.

- *Risk evaluation* follows, so that the expected losses, expressed in terms of fatalities or economic damage, can be scaled against predetermined safety goals (e.g. with respect to fatalities) in order to determine what losses are acceptable and what are not. Intervention action is then prioritised accordingly.
- *Risk management* is the final phase of evaluation and planning, when a search is made for the most appropriate mitigation strategy based on a variety of economic, ecological and social criteria in use at the time.

2. The Participatory Approach

The notion that community risk assessment is best served by an inclusive approach is fundamental. As highlighted in this chapter, local people often have a good understanding of the risks that they face and through public engagement risk knowledge and management can be co-produced. In the immediate aftermath of a disaster, local residents are usually the first responders, and so training and reliance on family and community networks can improve the search-and-rescue phase. The recovery process following a disaster is crucial because this phase offers a 'window of opportunity' to adopt initiatives designed to correct prior faults in development policies, alongside weaknesses in infrastructure design and builds, so as to make communities more resilient in the future. Communities themselves have skills and sometimes the resources to take on these tasks, even though these assets are often neglected by the outside world.

Box 4.7 Frameworks of Risk Management

As previously indicated, the framework for risk management is usually set by government regulations operating at local, regional, national and international levels. For example, in the UK many everyday risks are managed through laws originating from international, European and British parliaments which are administered by agencies such as the Health and Safety Executive and the Environment Agency. Under the Civil Contingencies Act 2004, enforcement may then be undertaken by those bodies, by the police (with respect to the management of risk on the highways) or by local authorities. In addition, the British Standards Institute provides a set of codes of practice that, although not legally binding, provide appropriate guidance to enable organisations to comply with the legislative requirements. Finally, some specialised industries have their own legislative frameworks and enforcement systems. For example, the aviation industry is covered by a specific set of laws, agreements and frameworks which, in the UK, are enforced by the Civil Aviation Authority. In a crisis situation a gold–silver–bronze command structure is established that determines a command hierarchy used for major operations by the emergency services. Some practitioners use the term strategic–tactical–operational command structure instead, but the different categories are equivalent (see Figure 4.21).

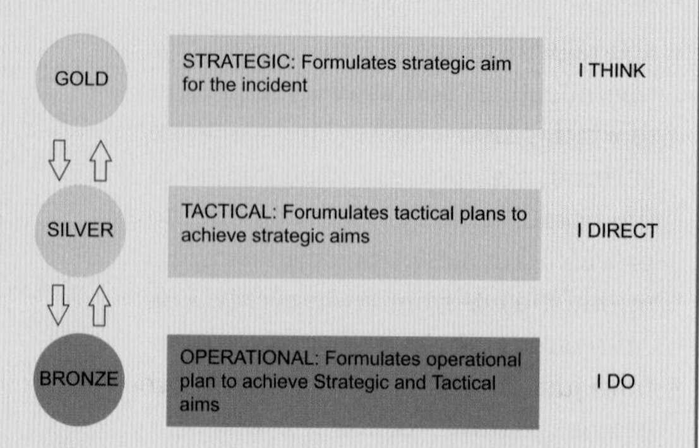

Figure 4.21 Command and control in an emergency in the UK.

Source: Town & Parish Council, Emergency Planning Conference, 22 February 2017, p. 5

In contrast, the USA National Incident Management System (NIMS) provides a systematic, proactive approach to guide departments and agencies at all levels of government, nongovernmental organisations and the private sector to work seamlessly to prevent, protect against, respond to, recover from and mitigate the effects of incidents, regardless of cause, size, location or complexity, in order to reduce the loss of life and property and harm to the environment (see Figure 4.22). The Incident Command System (ICS) is a standardised approach to the command, control and coordination of on-scene incident management that

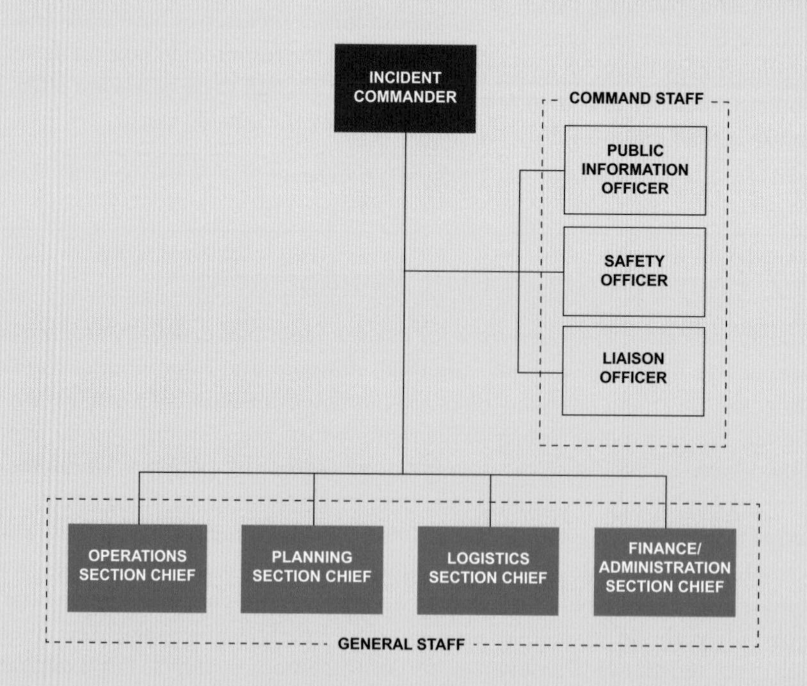

Figure 4.22 Example of an Incident Command System organisation with a single incident commander.

Source: Federal Emergency Management Agency, 2017. National incident management system. FEMA p. 24.

provides a common hierarchy within which personnel from multiple organisations can be effective. This specifies an organisational structure for incident management, and by using this for every incident it helps hone and maintain skills needed to coordinate efforts effectively and is used by all levels of government.

To compare the two systems, the UK approach is locally based but also hierarchical, with clearly defined roles and responsibilities embedded within the structure. The US approach adopts a community-based method, with responsibilities for the response delegated to the local level. These two systems are very different, and are just two examples in many used globally for the management of risk. However, many nations follow similar styles and structures of that implemented in the US and UK.

Another interesting example of risk management is the Indonesian approach, where there is more of a focus on the co-production of knowledge and decision-making that cuts across hazard and risk. Whilst Indonesia also uses the Incident Command System (ICS) in disaster response, the National Disaster Management Agency (BNPB) (Badan Nasional Penanggulangan Bencana) is the primary agency responsible for coordinating preparedness, response, prevention and mitigation, and rehabilitation and recovery all together. Figure 4.23 outlines the organisational structure and works from national to local levels, with input from a wide range of government departments via the steering committee. The chief of BNPB reports directly to the president.

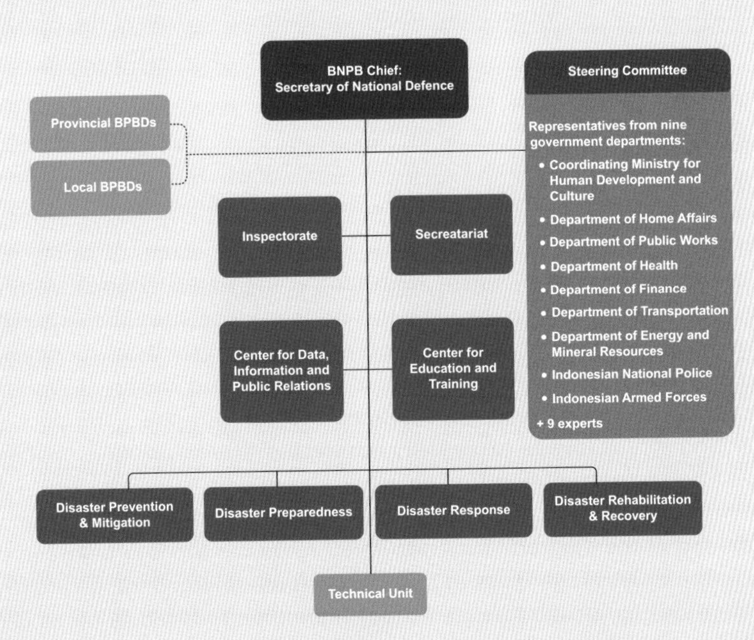

Figure 4.23 Organisation structure for disaster management in Indonesia.

Source: Indonesia: Disaster Management Reference Handbook, Center for Excellence in Disaster Management and Humanitarian Assistance, June 2018, p. 24 https://reliefweb.int/report/indonesia/indonesia-disaster-management-reference-handbook-june-2018

The legal framework for risk management is supported by a range of other measures, such as the use of public information programmes that inform people about hazards, the purposes and nature of the regulatory framework and the actions that people can take to minimise their own risks. This advice may be backed up by economic instruments such as financial subsidies and tax credits for compliance, combined with fines for non-compliance. As an example, the authorities in an urban area with a high level of seismic risk might reduce risk by the following:

- Enforcing a building code that requires all new structures to be able to withstand a specified earthquake risk. Ideally this building code would be enforced through legislation, with high penalties (including demolition in extreme cases) imposed for non-compliance, or by withholding insurance coverage.
- Providing tax incentives and subsidies to owners to encourage the retrofitting of existing buildings in order to meet building code standards.
- Educating the public about the building code and suitable measures for retrofitting buildings. Programmes that increase public awareness of the earthquake risk may be undertaken. Emphasis is often placed on teaching children how to react because this helps to protect some of the most vulnerable people in society and assists in the transfer of information to adults.

Despite its importance, risk management is only one of many goals in society. The resources required have to be balanced against other demands. In effect, the aim of risk management is to reduce threats to an acceptable level that is compatible with other socio-economic needs and is guided by some variant of the ALARP principle (Box 4.8 and Figure 4.24).

Box 4.8 The ALARP Principle

The ALARP principle – which maps out the severity and scope of safety risks and the costs of addressing these – arises from the UK legislative context. With roots in workplace casework dating back to 1949, this principle was specified as a regulatory requirement by the Health and Safety at Work (etc.) Act of 1974 (HSWA) following recommendations set out in the Robens Report on Safety and Health at Work in 1972. ALARP stands for 'as low as reasonably practicable' (or 'possible') and parallels with ALARA standing for 'as low as reasonably achievable'. The principle is applied to risk management on the assumption that society is faced with a hierarchy of risks from acceptable through tolerable to unacceptable (Figure 4.24). Risks in the unacceptable range (at the top of the diagram) are considered to be too great to bear and must be addressed, more or less regardless of cost. Risks in the tolerable range are then tackled using the ALARP principle that states they should be reduced as far as is feasible within a wider economic and social framework. Finally, the lowest category, negligible (acceptable) risk, is not addressed through risk management because it would represent a misuse of resources.

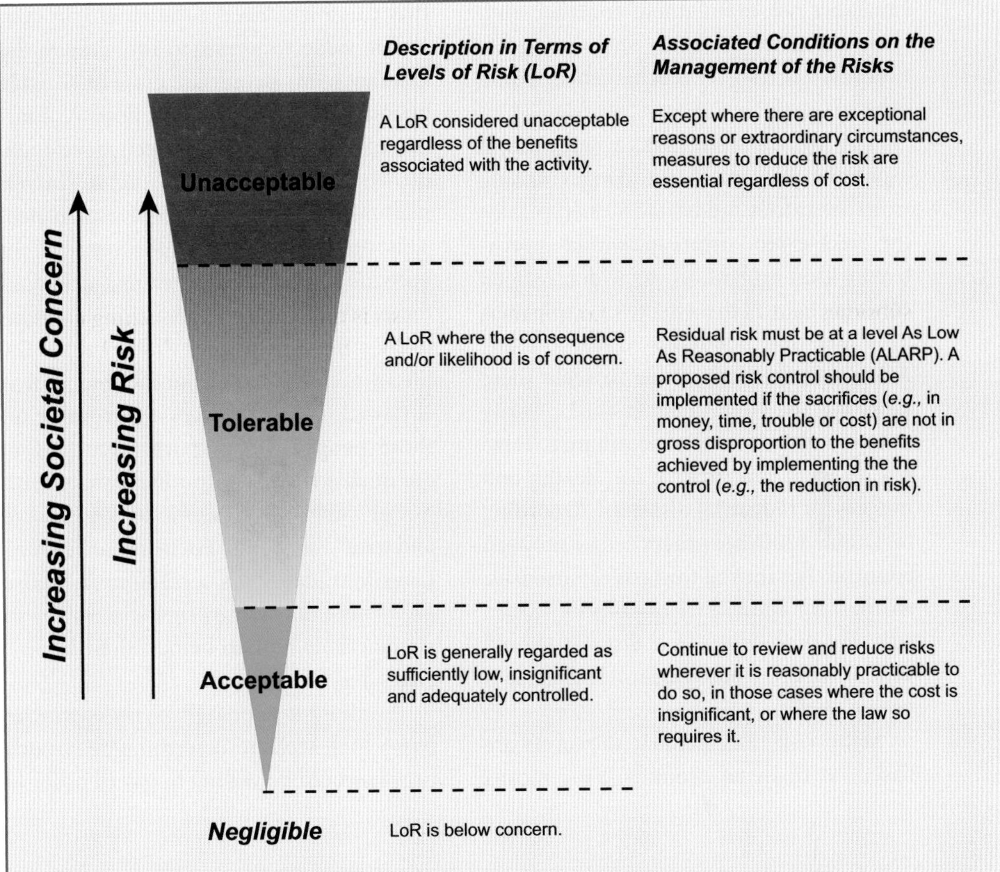

Figure 4.24 The ALARP approach to risk management. High-level (unacceptable) risks are at one end of the scale, lower-level (acceptable) risks are at the other end. The majority of tolerable risks should be managed and reduced as far as possible by practicable means.

Source: Clothier et al. (2013, p. 5)

In 2007 the UK's approach was challenged by the European Commission, with the charge that it was at odds with the EU's own 1989 European Framework Directive to ensure the safety and health of workers. This directive, which backgrounded the cost of mitigation in favour of taking action for safety, had been embedded in other European member states' legislation. The UK's approach was successfully defended, and this balance between safety and cost was to become adopted by other countries such as Malaysia and Australia.

Where the objective of ALARP risk is achieved, residual risk remains: the amount of risk remaining after risk management measures. The general formula to calculate residual risk is:

$$\text{residual risk} = \text{inherent risk} - \text{impact of risk controls}$$

To manage residual risk it is important that emergency response and recovery capacities are maintained, together with socio-economic policies such as safety nets and risk transfer mechanisms, among others.

The principle of risk acceptance is especially relevant to the danger arising from hazardous industrial sites. Typically, the risks extend from individual operatives within a plant, up to wider societal consequences if the effects of explosions or dangerous pollutants travel off-site. Attempts have been made to scale these risks and the consequent hazard severity in terms of their level of acceptance. Table 4.6 shows a risk matrix for deaths and injuries for industrial disasters. According to Duijm (2009b), practice in European countries has moved towards a consensus, so that, for individual or location-based risk, the probability of a disaster should be less than 10^{-6} per year, and this value is used globally.

Many risk management decisions are based on financial grounds. This means that there is a need to attribute an economic value to a human life, despite the fact that many people are quite reasonably uncomfortable with the notion. A number of approaches have been developed, of which the so-called *human capital* method is perhaps the best established. This method works on the basis of an individual's lost future earning capacity in the event of injury or death. It is a relatively simple principle, which values the life of a child at the highest level, but it is flawed in that it places a zero value on those people who, for whatever reason, are unable to work. Another flaw, as with 'social capital', is commodifying people and what they offer. Focusing on 'capital' plays into ideologies seeking to monetise everything and assign financial values to irreplaceable human beings, their knowledges and their skills.

A different approach is *willingness to pay*, which seeks to determine how much people would be willing to pay in order to

Table 4.6 An example of a risk matrix for industrial disasters

Frequency classification	Frequency per year	Disaster magnitude
Frequent Occurs several times during lifetime of installation	$1–10^{-2}$	Undesired event Minor material damage
Likely Will probably, but not necessarily, happen	$10^{-2}–10^{-4}$	Minor incident Minor occupational injuries on site
Not likely Could possibly happen	$10^{-4}–10^{-6}$	Serious incident Serious occupational injuries on site
Very unlikely Almost unthinkable	$10^{-6}–10^{-8}$	Major incident Fatalities on site, injuries to people off site
Extremely unlikely Frequency is under the limit of reasonability	$<10^{-8}$	Disaster Fatalities on and off site

Source: Adapted from Duijm (2009a). Used with permission.

achieve a certain reduction in their chance of a premature death (Jones-Lee et al., 1985). This is preferable because it measures *risk aversion*; that is, the value people place on reducing the risk of death and injury, rather than on more abstract, long-term concepts. Willingness to pay can be assessed by questionnaires which ask the respondents to estimate either the levels of compensation required for assuming an increased risk or the premium they would pay for a specified reduction in risk. Studies have found that the valuation of risk should include some allowance for the pain from, and aversion to, the potential form of death (e.g. high values for death by cancer) and that willingness to pay tends to decline after middle age, as the risk of mortality from natural causes increases.

G. RISK COMMUNICATION

This chapter so far has explored many different aspects of risk, including risk communication through challenging who is the expert, exploring different ways of risk management and exploring the decision-making processes. All of these elements require communication and understanding of risk and of various perceptions. There has been a gradual shift in environmental hazard disciplines away from a model of risk communication as an add-on to risk assessment process, toward more of an integrated and engaging approach to risk communication. This section looks more in-depth at how risk communication happens and how it could be improved.

Risk literacy is the basic requirement that people are able to deal with both natural and human-made threats, yet many are not risk literate. Risk can be communicated in different ways, as already discussed by using percentages, but it is important to understand what these percentages mean and how information communicated is understood in different ways by different audiences. In the 1980s Stuart Hall, a sociologist who worked within communication studies, proposed that audience members can play an active role in decoding messages as they rely on their own social contexts and are subsequently capable of changing messages themselves through collective action. Hall (1980) stated that not only are frameworks of knowledge, relations of productions and technical infrastructure important to understand when encoding a message (or warning), but also when decoding it. Hall's work on encoding and decoding challenged prior mass communication models by arguing that

i. meaning is not simply fixed or determined by the sender;
ii. the message is never transparent or clear; and
iii. the audience is not a passive recipient of information and meaning.

Consequently, there is a lack of fit between the two sides of communication where encoding and decoding are the points of entrance and exit to different systems of discourse. This is illustrated in Figure 4.25, which highlights Hall's model to show information production (encoding) and readings (decoding).

For example when the weather forecast says the probability of rain is 50% today, would you bring an umbrella with you? Whilst this may be more about your perception of risk, do you also know what 50% of rain means? It is important to clarify frames of reference when giving information, in this case for weather

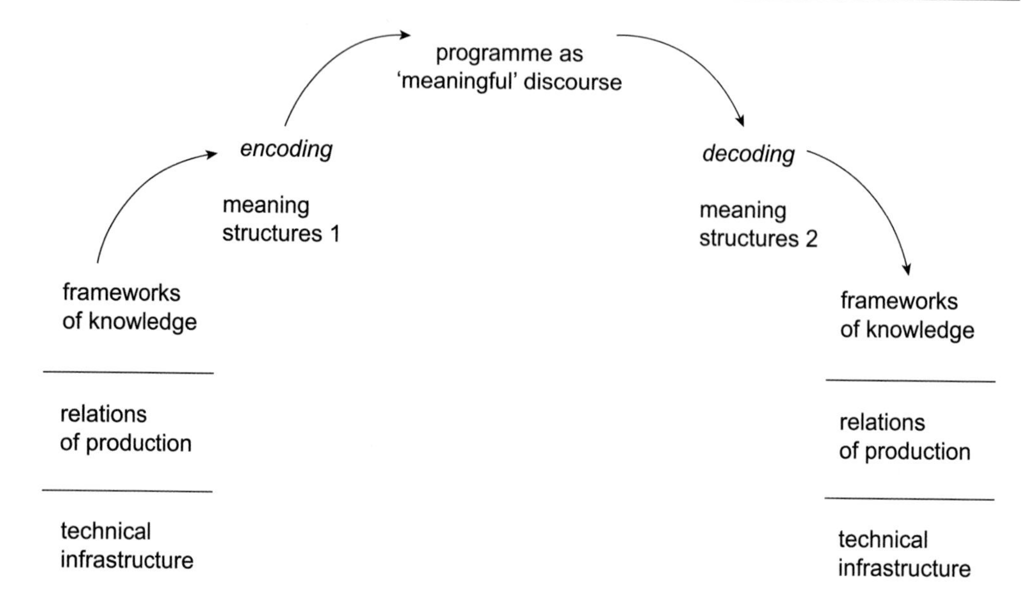

Figure 4.25 Encoding/decoding.

Source: Hall (1980, p. 130)

it means that it will rain on 50% of days for which the announcement is made. Whilst encoding and decoding can cause significant issues in risk communication, misunderstandings can also occur if terminologies are not explored and used well. This can come down to the way risk information is communicated in percentages. For example, in the 1960s there was an emergency announcement that women taking the contraceptive pill had a twofold increase in the risk of thrombosis. Twofold means a 100% increase, and this is a relative risk of 100%, but the absolute risk was moving from one in 7,000 women to two in 7,000 women. As a consequence some women decided to come off the pill, and in the years that followed there were an estimated increase of 13,000 additional abortions in England and Wales; pill sales fell considerably; for every abortion there was one extra birth, particularly teenage girls with some 800 additional conceptions; and it cost the NHS £4–6 million for abortion provision and resulted in a lack of trust in NHS (Furedi, 1999). If the results had been reported in a different way; that is, using absolute risk, the consequences would have been less severe. Therefore it is important that risk information is clearly communicated in a way that does not amplify the threat.

Whilst probabilities are one form of communication, there are a wide range of tools that can be used to communicate risk. In *Observing the Volcano World: Volcanic Crisis Communication*, over 42 chapters have been written to explore different aspects of crisis communications, whether in relation to different types of hazards (volcanic and associated), differing contexts and culture driven by the geographical, political, cultural and economic context of a crisis and by specific tool. These can range from learning about risks from previous civilisations, indigenous and traditional learnings, different sectors, statistics and insurance, mapping data, remote sensing data into maps, community learning and educational activities, and social media (see Fearnley et al., 2018). All of these aspects bring different types of communication together to help prepare for, prevent and reduce risks from natural hazards.

It is possible to break down the various processes and issues that make communication work in practice before and during a crisis, by highlighting commonly adopted tools. Using Cash et al.'s (2003) model (see Figure 4.26), it is possible to classify these tools under the need to:

i. ensure they are scientifically robust and are driven by the scientific credibility requirement;

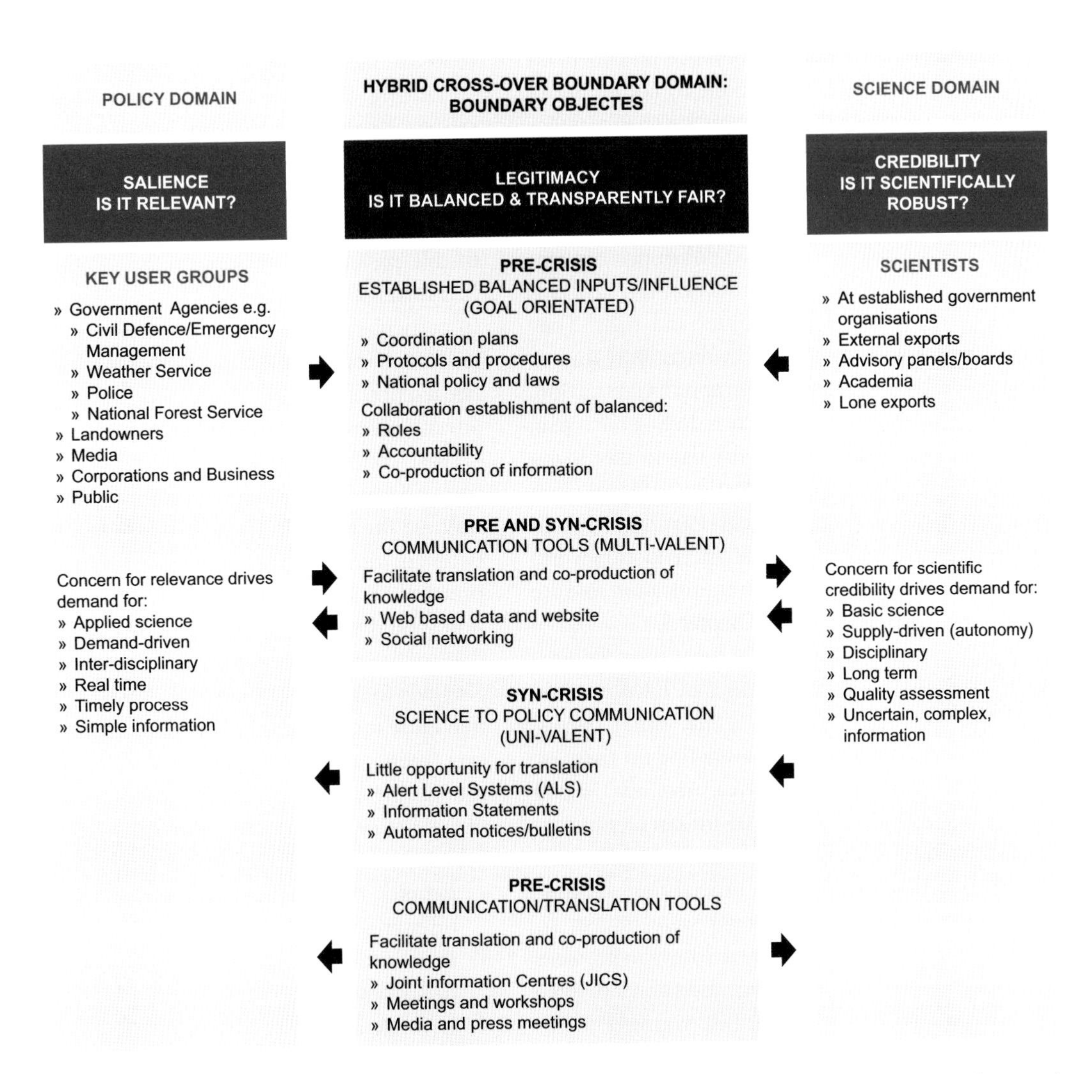

Figure 4.26 Mapping credibility, relevance and the generation of legitimacy to translate, communicate and mediate crisis information.

Source: Adapted from Fearnley & Beaven (2018, p. 11)

ii. generate salient knowledge relevant to the needs of decision-makers; and
iii. ensure that (i) and (ii) are balanced, in order for both processes and information to be perceived as legitimate.

These tools help create legitimacy between the scientists who are concerned about being credible and the politicians asking how this information is relevant. Being able to understand data and its relevance and to make effective decisions is done by communicating between the actors on the science policy divide and all the varying stakeholders.

Another key risk communication tool is warnings, often developed as part of an early warning system (EWS). Warnings contribute to disaster risk reduction for all forms of hazards, vulnerabilities and risks. They communicate hazard, vulnerability and/or risk and can generate appropriate action. Warnings save countless lives every year, can be used to support day-to-day living and vulnerability reduction, and are often operated by government organisations with legal remits for disaster risk reduction work. They and their risk communication role are discussed further in Chapter 5.

All too often there have been failures in risk communication that have resulted in considerable loss of life, either by not receiving warnings, not acting on them or being given poor advice. Around 250,000 people died in the 2004 Indian Ocean tsunami as the region lacked a formal tsunami warning system. Few official risk communications (via warnings) were issued, so the people who needed to evacuate were not informed, demonstrating a lack of integration between scientific warning information with the people directly affected (United Nations, 2006). Not understanding or acting on volcanic risk information presented in the City of Armero in Colombia in 1985 resulted in over 23,000 people dying. Despite solid science and warnings being issued, these were not accepted or acted on by local governors due to a lack of understanding of what the hazard and associated impact would look like (due to a lack of imagination) or simple denial (Fearnley & Kelman, 2021). In Box 4.9 the controversial and high profile L'Aquila trial resulting from an earthquake in 2009 that devastated central Italy demonstrates how poor risk communication can contribute to loss of life.

During contentious decision-making, serious breakdowns in trust between risk managers and the public can occur. In practice a balance must be achieved. Perceptions of risk can sometimes be driven by unjustifiable prejudices, which are often amplified through the media and by politicians. In particular, where risk perception is used to drive hazard management, great care is needed to ensure that the results do not disadvantage already marginalised groups.

Past conflicts, between technical assessors of hazard and risk and the public, demonstrate the need for good communication between the two interest groups. From a practical standpoint, improved communication should enable laypeople to understand the results of objective analyses of risk and also help to inform scientists about the risks that cause most concern to the public. However, there are problems attached to communicating complex technical assessments of risk to the public (Slovic, 2016) such as:

- People's initial perceptions of risk are often inaccurate.
- Risk information often frightens and frustrates the public.
- Strongly held, preconceived beliefs are hard to modify, even when the justification for those beliefs is incorrect.
- Naive or simplistic views are easily manipulated by presentation format. When it is stated that there is a 10%

Box 4.9 The 2009 L'Aquila Earthquakes

At a meeting held on 31 March 2009 at L'Aquila's government office, central Italy, scientists gathered to discuss recent earthquake swarms (several small earthquakes in a short period). After extensive discussion, the scientists did not rule out the possibility that a large earthquake could occur, but there was little discussion about what to do ahead of a major shaking, despite the vulnerability of many old buildings in the region. After the meeting, a press conference provided updates on the latest thinking for the earthquakes and Bernardo De Bernardinis, then vice-director of the Department of Civil Protection, said the situation in L'Aquila posed 'no danger, the scientific community continues to assure me that, to the contrary, it's a favourable situation because of the continuous discharge of energy', adding that local citizens should go have a glass of wine. The scientists themselves had very little engagement with the media or process of communication with the local communities who were understandably worried. They left town following the meeting leaving the Civil Protection to manage the media.

At 03:32 CEST on 6 April 2009, the main 6.3 moment magnitude shock occurred at 8.8 km depth. The earthquake was felt throughout central Italy and 308 people are known to have died, making it the deadliest earthquake to hit Italy since the 1980 Irpinia earthquake. Some residents did not leave their homes because they felt unable to make informed choices, despite evacuation being a traditional decision made by generations before.

A long, extensive trial followed the disaster, and on 22 October 2012, six scientists and one ex-government official were convicted of multiple manslaughter for downplaying the likelihood of a major earthquake six days before it took place. The media echoed the anger of numerous scientists that earthquakes are impossible to predict and that scientists should not go to jail for not being able to predict the impossible. Few drew attention to the trial not being about the failure of science to predict earthquakes, but rather about missing and incorrect information about the risk prior to the earthquake. At the start of the trial, Dr. Vincenzo Vittorini, who lost his wife and daughter in the quake, stated, 'Nobody here wants to put science in the dock. We all know that the earthquake could not be predicted, and that evacuation was not an option. All we wanted was clearer information on risks in order to make our choices'. The seven individuals were accused of giving inexact, incomplete and contradictory information about the danger of the tremors prior to the main quake. They were each sentenced to six years' imprisonment, but the verdict was overturned on 10 November 2014. Criticism was also applied to poor building standards that led to the failure of many modern buildings in a known earthquake zone.

The L'Aquila trial itself had impacts globally on risk communication, with some scientists concerned about focusing on the present processes rather than future probabilities. The consequences of the trial are long lasting, as scientists and key emergency managers face more strained relationships as scientists have arguably become overly cautious of providing information around the risks. Whilst expertise is needed to help predict and plan for risks to society in general, the public needs help to know choices they have to act as individuals. Scientists remain generally trusted sources of information in many contexts. As much as they share their expertise about risk with governments, they should also communicate with and educate individuals looking for advice and guidance.

Figure 4.27 Amatrice, Italy, was very close to the epicentre and faced significant destruction from the 6.3 earthquake.

Source:© Shutterstock/Antonio Nardelli

chance of an event occurring, rather than a 90% chance that it will not, opinions change.

When those with the power to manage risks to save lives do not, a hazard is often termed a 'surprise'. We could never have known, so how could we have acted? In popular parlance, they are often called 'black swans' which, allegedly, could not have been considered beforehand despite the large impacts they produce. The flaw in the black swan argument is that disasters are caused by vulnerabilities which are easy to know about and redress if the decision is made to do so. Vulnerabilities, and hence the disaster process, are easily observable as poverty, oppression, poor governance, marginalisation, inequities and inequalities. Certainly, many aspects of nature remain unknown, unpredictable and surprising. Environmental hazards beyond human scope and human survivability exist and some we might never before have known. These do not justify the common applicability of 'black swan' to hazards which are known, such as hurricanes, earthquakes and tsunamis.

The growth of online sources of information has made matters more complicated, especially helping the populism of misnomers such as 'black swan' and myths about numerous disasters being unprecedented and unpredictable. Many people have access to a vast range of information and can undertake their own independent research online and offline. Whilst this is a valid and empowering trend, the quality of the information accessed is sometimes poor and data may be selected to reinforce initial misconceptions. Risk assessment, management and communication faces and has always faced challenges and opportunities in terms of how and why people obtain their information and make their own interpretations.

KEY TAKE AWAY POINTS

- Risk is a complex and contested concept. There are different approaches to risk, whether qualitative or quantitative, but neither are objective approaches to risk perception and decision-making.
- Subsequently the management of risk is challenging, and the concept of incertitude can be used to bring together elements of risk, uncertainty, ambiguity and ignorance. These elements force us to move away from more traditional risk assessments to consider the non-scientific inputs when making decisions or establishing policy.
- The role of an 'expert' has become blurred as in the management of environmental hazards, there cannot be just one expert. It requires a diverse range of experts to work together, including the public.
- Risk is therefore becoming more of a co-production of knowledge and values between different stakeholders, using a wide range of communication processes. Balancing credibility and relevance is key to generating legitimacy.

BIBLIOGRAPHY

Abe, K. (1973). Tsunami and mechanism of great earthquakes. *Physics of the Earth and Planetary Interiors, 7*(2), 143–153.

Abeling, S., Horspool, N., Johnston, D., Dizhur, D., Wilson, N., Clement, C., & Ingham, J. (2020). Patterns of earthquake-related mortality at a whole-country level: New Zealand, 1840–2017. *Earthquake Spectra, 36*(1), 138–163. https://doi.org/10.1177/8755293019878190

Aven, T. (2003). Foundations of Risk Analysis. John Wiley and Sons, Chichester, UK.

Beck, U. (1992). *Risk society: Towards a new modernity*. London: Sage.

Bernstein, P. L. (1996). *Against the gods: The remarkable story of risk*. John Wiley and Sons.

Bründl, M., Romang, H. E., Bischof, N., & Rheinberger, C. M. (2009). The risk concept and its application in natural hazard risk management in Switzerland. *Natural Hazards and Earth System Sciences, 9*(3), 801–813.

Bultitude, K. (2014). Web-based channels for science communication. In *Communicating science to the public* (pp. 225–245). Dordrecht: Springer.

Burgman, M. A. (2015). *Trusting judgements: How to get the best out of experts*. Cambridge, UK: Cambridge University Press.

Cash, D. W., Clark, W. C., Alcock, F., Dickson, N. M., Eckley, N., Guston, D. H., . . . Mitchell, R. B. (2003). Knowledge systems for sustainable development. *Proceedings of the National Academy of Sciences, 100*(14), 8086–8091.

Clothier, R., Williams, B., Fulton, N., & Lin, X. (2013). ALARP and the risk management of civil unmanned aircraft systems. Proceedings of 2013 Australian System Safety Conference (pp. 3–13), Australian Computer Society.

Crenshaw, K. (1989). Demarginalizing the intersection of race and sex: A black feminist critique of antidiscrimination doctrine, feminist theory and antiracist politics. *University of Chicago Legal Forum, 1*, 139–167.

Crenshaw, K. W. (2017). *On intersectionality: Essential writings*. The New Press.

Cyranoski, D. (2011). Japan faces up to failure of its earthquake preparations. *Nature,*

471, 556–557. https://doi.org/10.1038/471556a

Day, S., & Fearnley, C. (2015). A classification of mitigation strategies for natural hazards: Implications for the understanding of interactions between mitigation strategies. *Natural Hazards, 79*(2), 1219–1238.

Duijm, N. J. (2009a). *Acceptance criteria in Denmark and the EU*. Danish Environmental Protection Agency.

Duijm, N. J. (2009b). Safety-barrier diagrams as a safety management tool. *Reliability Engineering & System Safety, 94*(2), 332–341.

Epstein, S. (1996). *Impure science: AIDS, activism, and the politics of knowledge* (Vol. 7). Berkeley: University of California Press.

Fearnley, C., & Kelman, I. (2021). *Enhancing warnings*. National Preparedness Commission.

Fearnley, C. J., & Beaven, S. (2018). Volcano alert level systems: Managing the challenges of effective volcanic crisis communication. *Bulletin of Volcanology, 80*, 1–18.

Fearnley, C. J., Bird, D. K., Haynes, K., McGuire, W. J., & Jolly, G. (Eds.). (2018). *Observing the volcano world: Volcano crisis communication*. Cham: Springer International Publishing.

Federal Emergency Management Agency. (2017). *National incident management system*. FEMA.

Fischhoff, B., Lichtenstein, S., Slovíc, P., Derby, S. L., & Keeney, R. L. (1981). *Acceptable risk*. Cambridge: Cambridge University Press.

Fukao, Y., & Furumoto, M. (1975). Mechanism of large earthquakes along the eastern margin of the Japan Sea. *Tectonophysics, 26*(3–4), 247–266.

Funtowicz, S. O., & Ravetz, J. R. (1993). Science for the post-normal age. *Futures, 25*(7), 739–755.

Furedi, A. (1999). Social consequences. The public health implications of the 1995 'pill scare'. *Human Reproduction Update, 5*(6), 621–626.

Garland, D. (2003). The rise of risk. *Risk and Morality, 1*, 48–86.

Gibbons, M., Limoges, C., Scott, P., Schwartzman, S., & Nowotny, H. (1994). The new production of knowledge: The dynamics of science and research in contemporary societies. *The New Production of Knowledge*, 1–192.

Giddens, A. (1990). *The consequences of modernity*. Cambridge: Polity Press.

Gigerenzer, G. (2014). *Risk savvy: How to make good decisions*. Penguin.

Gregory, J., & Miller, S. (1998). *Science in public: Communication, culture, and credibility*. Plenum Press.

Hall, S. (2019). Encoding: Decoding (1980). In *Crime and media* (pp. 44–55). Abingdon, UK: Routledge.

Hutton, J. (1794). *Investigation of the principles of knowledge*.

Indonesia: Disaster management reference handbook (June 2018): Indonesia. (2018, August 9). *ReliefWeb*. https://reliefweb.int/report/indonesia/indonesia-disaster-management-reference-handbook-june-2018

Jackson, E. L., & Burton, I. (1978). The process of human adjustment to earthquake risk. In *The assessment and mitigation of earthquake risk* (pp. 241–260). Paris: UNESCO.

Jones-Lee, M. W., Hammerton, M., & Philips, P. R. (1985). The value of safety: Results of a national sample survey. *The Economic Journal, 95*(377), 49–72.

King, D. N., & Goff, J. (2006). Māori knowledge in natural hazards management and mitigation. *NIWA Report*. https://niwa.co.nz/sites/niwa.co.nz/files/niwa_report_akl2006-055.pdf

Knight, F. H. (1921). *Risk, uncertainty and profit* (Vol. 31). Houghton Mifflin.

Koshimura, S., & Shuto, N. (2015). Response to the 2011 great East Japan earthquake and tsunami disaster. *Philosophical Transactions of the Royal Society A: Mathematical, Physical and Engineering Sciences, 373*(2053), 20140373.

Krewski, D., Clayson, D., Collins, B., & Munro, I. (1982). Toxicological procedures for assessing the carcinogenic potential of agricultural chemicals. In R. A. Fleck & A. Hollaender (Eds.), *Genetic toxicology: An agricultural perspective* (pp. 461–497). New York: Plenum.

Mazzocchi, F. (2006). Western science and traditional knowledge: Despite their variations, different forms of knowledge can learn from each other. *EMBO Reports, 7*(5), 463–466.

Meehl, G. A., Karl, T., Easterling, D. R., Changnon, S., Pielke Jr, R., Changnon, D., . . . Zwiers, F. (2000). An introduction to trends in extreme weather and climate events: observations, socioeconomic impacts, terrestrial ecological impacts, and model projections. *Bulletin of the American Meteorological Society, 81*(3), 413–416.

O'Brien, G. (2008). UK emergency preparedness: A holistic response? *Disaster Prevention and Management: An International Journal, 17*(2), 232–243.

O'Brien, G., & Read, P. (2005). Future UK emergency management: New wine, old skin? *Disaster Prevention and Management: An International Journal, 14*(3), 353–361.

Ravetz, J. (2004). The post-normal science of precaution. *Futures, 36*(3), 347–357.

Rohrmann, B. (1994). Risk perception of different societal groups: Australian findings and crossnational comparisons. *Australian Journal of Psychology, 46*(3), 150–163.

Rowe, G. and L.J. Frewer (2000). Public Participation Methods: A Framework for Evaluation. *Science, Technology, & Human Values, 25*, 3-29.

Slovic, P. (1987). Perception of risk. *Science, 236*(4799), 280–285.

Slovic, P. (1991). Beyond numbers: A broader perspective on risk perception. *Acceptable Evidence: Science and Values in Risk Management, 48.*

Slovic, P. (2016). *The perception of risk.* Routledge.

Stirling, A. (2003). Risk, Uncertainty and Precaution: Some Instrumental Implications from the Social Sciences. In: Berkhout, F., Leach, M., & Scoones, I. (Eds), Negotiating Change: new perspectives from the social sciences, Edward Elgar, Cheltenham, UK, pp. 33-76.

Stirling, A. (2007). Risk, precaution and science: towards a more constructive policy debate. *EMBO Reports, 8*, 309–315.

Te Manatū Waka Ministry of Transport (2023). *Te Marutau– Ngā mate i ngā rori | Safety – Road deaths.* (2023). www.transport.govt.nz/statistics-and-insights/safety-road-deaths

United Nations. (2006). *Global survey of early warning systems. Gaps and opportunities towards building a comprehensive global early warning system for all natural hazards.* Report prepared at the request of the Secretary-General of the United Nations.

Whyte, A. V., & Burton, I. (1982). *Perception of risk in Canada: Living with risk.* Toronto: Institute of Environmental Studies, University of Toronto.

Wigley, T. M. (1985). Impact of extreme events. [Impact of climate variations]. *Nature (London);(United Kingdom), 316.*

Wilmshurst, J. (2018). "There's no plastic in our volcano": A story about losing and finding a path to participatory volcanic risk management in Colombia. *Observing the Volcano World: Volcano Crisis Communication*, 499–514.

Wright, T. L., & Pierson, T. C. (1992). *Living with volcanoes* (No. 1073). US Geological Survey.

Wynne, B. (1992). Misunderstood misunderstanding: Social identities and public uptake of science. *Public Understanding of Science, 1*(3), 281.

Wynne, B. (1996). May the sheep safely graze? A reflexive view of the expert-lay knowledge divide. *Risk, Environment and Modernity: Towards a New Ecology, 40*, 44.

Zaidi, R. Z., & Fordham, M. (2021). The missing half of the Sendai framework: Gender and women in the implementation of global disaster risk reduction policy. *Progress in Disaster Science, 10*, 100170.

FURTHER READING

Ball-King, L. N., & Ball, D. J. (2016). Health and safety and the management of risk. *Routledge Handbook of Risk Studies, 143.*

Beck, U. (1996). *The reinvention of politics: Rethinking modernity in the global social order.* Cambridge: Polity Press.

Beck, U. (1999). *World risk society.* Cambridge: Polity Press.

Boustras, G., & Waring, A. (2020). Towards a reconceptualization of safety and security, their interactions, and policy requirements in a 21st century context. *Safety Science, 132*, 104942.

Douglas, M., & Wildavsky, A. (1983). *Risk and culture: An essay on the selection of technological and environmental dangers*. Berkeley: University of California Press.

Giddens, A. (1998). Risk society: The context of British politics. In J. Franklin (Ed.), *The politics of risk society order*. Cambridge: Polity Press.

Giddens, A. (1999). Reith lecture 2: Risk', Vol. 2000. news.bbc.co.uk/hi/english/static/events/reith_99/week2/week2.htm

Jones-Lee, M., & Aven, T. (2011). ALARP: What does it really mean? *Reliability Engineering & System Safety*, *96*, 877–882.

Keeney, R. L. (1995). Understanding life-threatening risks. *Risk Analysis*, *15*, 627–637. A clear statement on disaster-type risks.

Sjöberg, L. (2001). Limits of knowledge and the limited importance of trust. *Risk Analysis*, *21*, 189–198. An interesting development from Starr's early work.

WEB LINKS

Prevention Web. www.preventionweb.net

Radix: Radical Interpretations of Disasters. www.radixonline.org

Risk Reduction Education for Disasters. www.riskred.org

CHAPTER FIVE

Disaster Risk Reduction

5

OVERVIEW

Disasters are driven by human actions: these are the consequences of decisions and plans that shape our lives and to which not all are allowed to contribute. Disaster risk reduction (DRR) is the term used to bring together the activities that reduce vulnerabilities and disaster risks via the development and action of policies, planning, practices and actions. This chapter explores the role of DRR in managing disasters using disaster risk reduction phases as a broad guide and reviewing the various tools available at different stages of a disaster, including mitigation, preparedness, response and recovery. Mitigation is vital to minimise the adverse impacts of a disaster, and different elements of mitigation are explored from protection and damage mitigation (e.g. retrofitting), to impact mitigation (e.g. insurance), to adaptation (e.g. land use planning) and the role of mitigation strategies within a broader mitigation system. Preparedness is the effective anticipation and response of a current or potential disaster and typically involves developing preparedness plans, emergency operations and training, predictions, forecast and warnings. Response is the actions taken directly before, during and after a disaster and includes the challenging actions of evacuation, search and rescue, and the assessment and restoration of critical infrastructures and services. Recovery is the restoration or improvement of communities, infrastructure and livelihoods following a disaster, often involving domestic and international aid, rehabilitation and reconstruction. Finally, disasters must be evaluated after the event, which can be undertaken via a cost-benefit analysis. A more critical accounting would probe the political and economic context of a disaster and address the question of who gains and who loses from building back in particular ways.

DOI: 10.4324/9781351261647-6

A. MANAGING DISASTERS: DISASTER RISK REDUCTION

The environmental hazards involved in disasters are part of complex physical systems that cannot ever be fully under human control, nor should they be. In a single 24-hour day, the Earth's atmosphere receives enough solar energy to power 10,000 hurricanes, 100 million thunderstorms or 100 billion tornadoes. The amount of energy in our planet's systems is enormous (Figure 5.1). Expressed relative to solar energy receipt (i.e. taking the daily global solar energy receipt as 1 unit), a very strong earthquake releases 10^{-2} units; an average cyclone 10^{-3} units. During its lifetime, and taking all atmospheric processes into account, a single large hurricane could release energy equivalent to about 200 times the current electrical generating capacity on the planet. Tectonic processes also hinge on tremendous amounts of energy. The 1960 Chilean earthquake, for example, the largest yet recorded (with a magnitude Mw = 9.5), radiated energy of about 1.1 × 1,026 ergs, roughly equivalent to 2,600 megatons of TNT explosives, or the energy that would be released by about 130,000 atomic bombs.

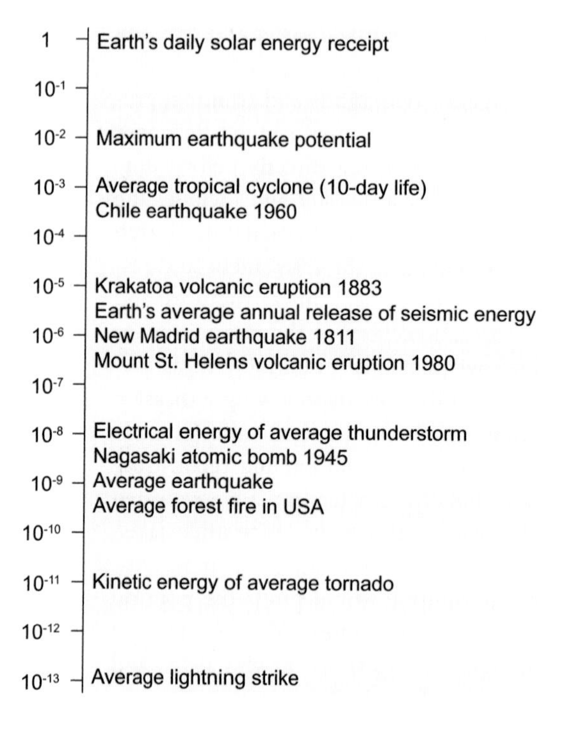

Figure 5.1 Energy release (in ergs) on a logarithmic scale for selected hazardous geophysical events compared with the Earth's daily receipt of solar energy.

It is also impossible to avoid all potentially dangerous locations on Earth because of the ubiquity, diversity and overlapping of hazards (Figure 5.2). While many physical processes are reasonably understood, a dearth of information remains for predictive modelling. As an example, reliable forecasting and warning schemes for earthquakes are reasonable spatially, in that we know the locations prone to high-magnitude shakings. Earthquake warning systems, as implemented in countries such as Japan and Mexico, are also useful over short timescales, sometimes giving up to a minute or two for action before the shaking strikes. Yet, they cannot provide warning long in advance of a specific tremor in a specific place.

The prevalence and diversity of environmental hazards, and the vulnerabilities that can turn them into disasters, make efforts to manage hazards challenging. There is frequent political inertia from governments around hazard awareness, risk reduction and disaster management, with short-term economic gains for a minority traded off against long-term sustainability for all. At an individual level, economic insecurity may impel people to attach an unduly low priority to environmental hazards as compared to day-to-day problems like inflation or unemployment – or simply getting to school and work and

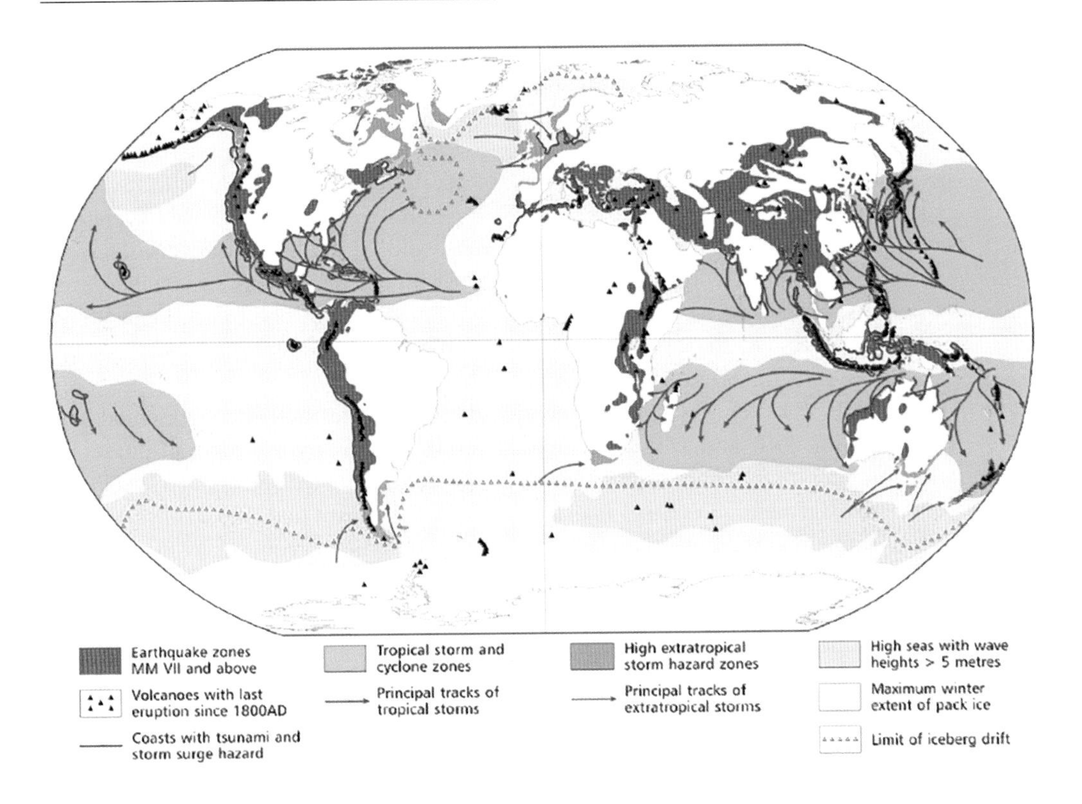

Figure 5.2 Simplified world map of selected natural hazards.

Source: After Munich Re at www.MunichRe.com (accessed 5 February 2012)

finding enough food and water to survive. Sometimes, the view is expressed that individuals should be free to assume whatever environmental risks they wish, as long as they accept the consequences of their decision. But lack of information, opportunities and resources, rather than a properly calculated choice, is the most common factor in forcing so many people to locate in hazardous areas. In fact, people can know exactly what the hazard situations are in the places they live and work and want to address them in order to live more safely. They may have no mechanisms by which to do so and might be actively prohibited from changing their predicament because of marginalisation, oppression and lack of power.

This chapter brings together the different elements of the prior chapters – focused on hazards in the environment; different dimensions of disaster vulnerability, resilience and sustainability; and risk assessment and management – to look at how disasters can be managed, most notably by avoiding them happening through steps that can be taken to reduce disaster risk. This is commonly referred to with the term disaster risk reduction (DRR), defined as 'preventing new and reducing existing disaster risk and managing residual risk, all of which contribute to strengthening resilience and therefore to the achievement of sustainable development' by the UNDRR (2017). A global, agreed policy of DRR is set out in the United Nations–endorsed Sendai Framework for Disaster Risk Reduction 2015–2030, adopted in March 2015. The expected outcome over the next 15 years is: 'The substantial reduction of disaster risk and losses in lives, livelihoods and health and in the economic, physical, social, cultural and environmental assets of persons, businesses, communities and countries' (UNISDR, 2015, p. 6).

DRR is sometimes referred to as, or linked with, disaster risk management (DRM). Both DRR and DRM are more wide ranging than conventional emergency management. Disaster management is sometimes presented as an activity limited to the emergency phase of the event, a period seen by the media and others as chaotic and crisis ridden. In reality, effective disaster management depends on the implementation of a carefully planned sequence of actions over many years that embraces pre-disaster protection, within-disaster action and post-disaster recovery (Figure 5.3).

This chapter addresses the four key components of DRR in turn discussing different actions that can be taken to try to reduce disaster risk in relation to (terms defined by the UNDRR):

- *Mitigation* – The minimising of the adverse impacts of a disaster.
- *Preparedness* – The knowledge and capacities developed by governments, response and recovery organisations, communities and individuals to effectively anticipate, respond to and recover from the impacts of likely, imminent or current disasters.
- *Response* – Actions taken directly before, during or immediately after a disaster in order to save lives, reduce health impacts, ensure public safety and meet the basic subsistence needs of the people affected.
- *Recovery* – The restoring or improving of livelihoods and health, as well as economic, physical, social, cultural and environmental assets, systems and activities of a disaster-affected community or society. It would preferably avoid or reduce future disaster risk, with phrases such as 'sustainable development' and 'build back better' often used to describe the process.

DRR activities are often presented as a cycle, where four temporal phases or stages are adopted and applied (see Figure 5.3). It is meant to define a continual process of actions required to manage risk and impacts in places prone to disasters and emergencies. It is not meant to be understood as a cycle where phases have a beginning and an end – as this framing has significant criticisms and alternatives proposed – but as a series of overlapping options and actions, where any one might dominate for a time but each is constantly reviewed and adapted as a result of the others.

There are a wide range of measures that can be taken in each or multiple phases to reduce disaster risk and manage hazards and threats. Table 5.1 outlines some examples.

These activities can take place pre-, synchronous with and post a hazard or crisis, depending on the event itself. DRR activities can be designed around a range of events including frequent small-scale events right through to existential risks, where an event could cause human extinction or permanently and drastically curtail humanity's potential (e.g. via an asteroid impact or supervolcanic eruption). In many cases, specific actions are taken in a top-down, technocratic fashion, viewing the environment as separate from and dangerous to society. The actions aim 'to protect' people against hazards and reduce

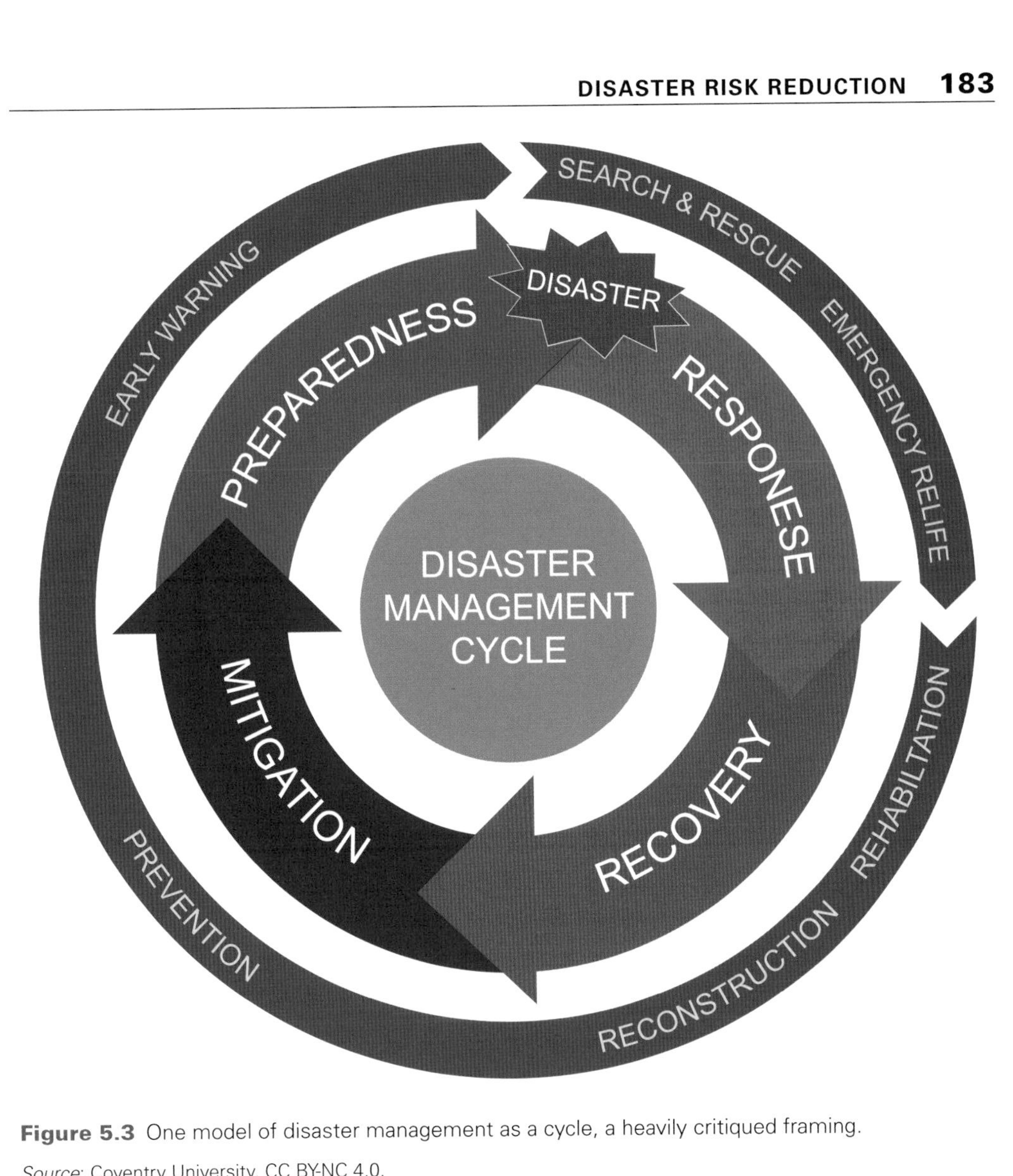

Figure 5.3 One model of disaster management as a cycle, a heavily critiqued framing.

Source: Coventry University. CC BY-NC 4.0.

Table 5.1 Examples of measures in each disaster management phase

Phase	Disaster			
	Earthquake	**Flood**	**Storm (cyclone, typhoon, hurricane)**	**Landslide**
Prevention/ migration	Seismic design Retrofitting of vulnerable buildings Installation of seismic isolation/ seismic response control systems	Construction of dike Building of dam Forestation Construction of flood control basins/ reservoirs	Construction of tide wall Establishment of forests to protect against storms	Construction of erosion control dams Construction of retaining walls

(*Continued*)

Table 5.1 (Continued)

Phase	Disaster			
	Earthquake	**Flood**	**Storm (cyclone, typhoon, hurricane)**	**Landslide**
Preparedness	Construction and operation of earthquake observation systems	Construction and operation of meteorological observation systems	Construction of shelter Construction and operation of meteorological observation systems	Construction and operation of meteorological observation systems
	Preparation of hazard maps			
	Food and material stockpiling			
	Emergency drills			
	Construction of early warning systems			
	Preparation of emergency kits			
Response	Rescue efforts			
	First aid treatments			
	Firefighting			
	Monitoring of secondary disaster			
	Construction of temporary housing			
	Establishment of tent villages			
Rehabilitation	Disaster-resistant reconstruction			
	Appropriate land use planning			

Source: Asian Disaster Reduction Center (2005, p. 5)

disaster impacts in this way. These actions start with identifying known 'threats' in order to input into a quantitative assessment of the degree of risk to life and property. Many technical tools are available, although they are fundamentally social processes since they involve people and since their implementation must involve politics.

Similarly, many other strategies and activities exist for DRR that are not covered by these technical dimensions. Examples are changing individual and collective behaviours to avoid forcing others into unwanted risks, establishing a culture of DRR and altering political processes to serve everyone rather than particular subsets (as discussed in Chapter 1). These social bases for effective long-term DRR are achieved by seeking education for all, physically and mentally healthy lifestyles, equity, equality and balanced resource distribution.

B. MITIGATION

Mitigation is defined by the UNDRR (2022) as the 'lessening or minimizing of the adverse impacts of a hazardous event'. Whilst the adverse impacts of hazards, in particular natural hazards, often cannot be prevented fully, their scale or severity can be substantially lessened by various strategies and actions. Mitigation measures include engineering techniques and hazard-resistant construction (such as tsunami barriers, river or tidal flood defences and seismically resilient buildings), as well as improved environmental and social policies and public awareness (land use and development planning and warning systems that foster education, evacuation plans and communication to enable mitigation actions at the time of hazard events or in anticipation of them). Note that in climate change policy, 'mitigation' is defined differently and refers to the reduction of greenhouse gas emissions that are the source of climate change (Chapters 16 and 18).

Mitigation strategies are policies or procedures that lead to more or less preplanned actions that operate before, during or after a hazard event to reduce its impact on vulnerable populations. Different types or classes of mitigation strategy require differing timescales and methods of implementation according to both the nature of the hazards and the vulnerabilities of the exposed societies. Multiple mitigation strategies are typically implemented in any nation, and mitigation strategies can be either developed and applied together or, more commonly, added in a progressive sequence reflecting technological and socio-economic developments rather than any systematic plan (one starting from a state of no mitigation). Although in principle one integrated system could provide for mitigation of multiple hazards (especially those that commonly occur in association such as earthquakes, landslides and tsunamis), mitigation strategies have historically targeted a single hazard. A mitigation system is a concept by Day and Fearnley (2015) that 'comprises all the mitigation strategies implemented together to ensure the lessening or limitation of the adverse impacts of hazards and related disasters'. Therefore a mitigation system is a set of interacting, interrelated or interdependent mitigation strategies implemented for the purpose of mitigating the effects of a particular hazard or group of hazards.

Mitigation depends on two key criteria that shape the actions that can be implemented prior to a hazard/threat (taken from Day & Fearnley, p. 1222):

1. *Timeframe*: When do the mitigative actions resulting from the strategy occur – does it produce actions (such as construction of flood defences) long before the hazard event whose effects are then permanently operative (or at least continuously operative over a specified time period), or are these actions triggered either by the successful detection of the source event that generates a hazard (or an early manifestation of the hazard such as first landfall of a tsunami) or by successful detection and interpretation of precursors to the hazard source event that are recorded and interpreted by observers or by some form of warning technology?
2. *Adaptability*: Does the strategy enable the actions resulting from it to be modified or adapted immediately before or during an unfolding hazard by optional decisions made following observations of the hazard or its precursors? Optional here means to emphasise that a potential decision-maker may choose not to intervene and rely upon decisions made by others. Or are these actions fixed in advance of the hazard event? The latter applies most obviously in the case of structural approaches such as dikes and walls, but also in the case of some automated responses as discussed later. The optional decisions involved in modifying or adapting the actions

may take many forms and may be made by different people, ranging from individuals within directly vulnerable populations through professional emergency managers to political leaders.

These two criteria highlight the possibility that once implemented, a mitigation strategy may require no further decision-making or actions unless and until public, political or scientific opinion seeks to change it as a result of social, economic, environmental or political changes, or changes in knowledge of the hazard, that have occurred since implementation of the strategy. This feature, which applies most obviously in the case of permanent physical defences against a hazard, are typically classed as technocratic categories of mitigation. Given the two criteria identified previously, we identify three classes or categories of mitigation strategy: permanent, responsive and anticipatory.

1. *Permanent mitigation*: Permanent mitigation strategies involve actions that are put in place long before hazard events and remain in place at all times, at least within specified periods (as for example in the case of an annual insurance policy). Their operation is not dependent upon human decisions or actions shortly before or at the time of the hazard event, and so cannot be adapted in the light of observations of impending or occurring hazard events, to mitigate those events. This typically involves permanently-in-place physical defences, reductions in exposure and compensatory reliefs (e.g. insurance), automated systems and land use restrictions. These methods are frequently both costly and 'brittle': actions work up to a design limit of hazard intensity or magnitude and then fail. An example is the 2011 tsunami in Japan that exceeded the design capacity of the sea walls (see Box 4.3).
2. *Responsive mitigation*: Responsive mitigation strategies involve decisions and actions in response to the detection and interpretation of a hazard event once the source event has occurred, taking advantage of the time gap compared to the time needed to implement changes to the plans previously devised in accordance with the mitigation strategy between the source event and its impact on vulnerable populations and environments. Responsive mitigation strategies therefore have a potential for adaptation of actions based on the observations of the hazard event, although the capacity for adaptation may be severely limited by the short duration of this time gap. They may be based upon complex technological subsystems or rely on simple direct sensory (non-instrumental) observations made by groups and individuals within vulnerable populations. These strategies require capacities to recognise the signs that a hazard-causing event has occurred or is about to impact vulnerable exposures; to interpret those signs in terms of the intensity of the hazard and if necessary communicate that interpretation to distinct decision-makers; to optionally decide what action to take to mitigate the hazard; and to carry out those mitigative actions in a timely fashion so that they have a positive effect. An example of this is the use of a warning system, for example that a tsunami has been generated and that

a population may have two minutes or 20 hours to prepare for the tsunami prior to it reaching the shoreline.

3. *Anticipatory mitigation strategies*: Anticipatory mitigation strategies involve actions that are implemented based on observations and interpretations of precursory phenomena to hazard-causing events, in the time period in which those precursors occur (which may vary from hours to days, in the case of many meteorological hazards, to months or even years in the case of volcanic eruption hazards). These actions may be adapted to allow for new information about the nature, intensity and extent of the anticipated hazard that is gained as a result of interpretation of the hazard precursors. Anticipatory mitigation strategies therefore depend fundamentally upon the knowledge base used to reliably identify precursory phenomena and to interpret those phenomena to enable prediction of the timing and intensity distribution of the resulting hazard in a form that provides a basis for mitigative actions (using multiple knowledge forms). Often, there is uncertainty about the nature of the hazards that are about to occur and the occurrence and intensity distributions of further hazards such as landslides, dam breaks and fires. As an example, volcano alert levels are typically used to raise awareness of a volcano's increasing behaviour in order to notify vulnerable populations and other authorities to take appropriate action. This action depends on the level of tolerance of uncertainty and risk by those affected, who must be prepared to accept the social, livelihood and reputational costs of false alarms.

The classification of mitigation strategies for different hazards may operate on different scales, so the details of the knowledge base differ widely. However, there are some common features of the knowledge base that apply to the different mitigation strategies, as shown in Table 5.2.

Whilst the boundary between responsive mitigation and anticipatory mitigation is clear for those hazards that have a clear onset time (most notably earthquakes and tsunamis), the distinction is less clear for those in which the hazard is caused by an evolving event such as a hurricane. Hurricane warnings and consequent mitigation activities, such as coastal evacuations, begin with many features of anticipatory mitigation – coping with uncertainties in the track and strength of the hurricane – and only gradually evolve into responsive mitigation as the hurricane approaches landfall.

Importantly, mitigation strategies can interfere with each other, within the broader mitigation system. Unexpected destructive interactions (or 'interferences') can damage the effectiveness of the strategies or even render particular combinations of strategies actively dangerous (see Box 4.3). Conversely, it is possible that constructive interaction may occur between mitigation strategies so that the whole system is more effective. This tends to require conscious design for constructive interactions within a mitigation system. Therefore, individual mitigation strategies should not be chosen and implemented in isolation but evaluated within the framework of the overall mitigation system.

There are three traditional technocratic categories of mitigation:

- *Protection/Damage Mitigation*: These responses seek to reduce hazard impacts by adjusting damaging events to people.

Table 5.2 Classification of mitigation strategies and the relative importance of different aspects of knowledge of hazard as a basis for different classes of mitigation strategy, as indicated by sizes of crosses in table

	Mitigation strategy classes and their defining characteristics		
	Permanent	**Responsive**	**Anticipatory**
Timing of implementation actions prescribed by strategy	Long before hazard event	In response to detection and characterisation of hazard event	In response to detection and interpretation of precursors to hazard event
Adaptability of implementation actions to characterisation of individual hazard events	None	Some (limited mainly by time available)	Some (limited mainly by uncertainty)
Key science tasks			
Long-term spatial frequency/ magnitude distribution	X	X	X
Understanding mechanisms and interpretation of precursors			X
Rapid detection, event quantification and communication		X	X
Accurate alert information and avoidance of false alarms		X	X

Source: Day & Fearnley (2015, p. 1227)

They apply measures to exert some control over hazardous processes, often by creating special structures or by strengthening existing infrastructure to resist the physical stresses of the hazard. The scale of intervention varies from *macro-protection* (large-scale defensive works designed to protect whole communities) to *micro-protection* (strengthening of individual buildings or rooms).

- *Impact Mitigation*: These responses aim to modify the loss burden in the immediate aftermath of disaster. *Emergency aid* is delivered through government agencies, the non-profit sector and the private sector. For some disaster-affected people, financial compensation will be available through *insurance schemes* administered through government agencies, the non-profit sector and the private sector. All these measures attempt to spread the financial burden of disaster beyond those suffering loss. To that extent, they are *loss-sharing* – rather than *loss-reducing* – devices, but the potential exists for encouraging more loss-reducing responses in the future. These measures are usually focused on tangible, quantifiable losses, since it can be hard to mitigate the impact on deaths, injuries, disruption and loss of intangible items such as photographs, computer files and heritage sites.
- *Adjustment/Adaptation*: These responses are technical measures seeking to adjust or adapt society and its activities to potential and actual hazards. Those covered here are formal, top-down mechanisms aiming to promote changes in human behaviour and attitudes so that hazards do not become

disasters or so that disasters have less impact. Associated protective methods, behavioural change, cultural change, relocation and altering routines and ways of living may be necessary, with the measures here trying to push people in such directions, with examples of activities being *preparedness*, *planning*, *forecasting*, *warnings*, *land use planning* and *legislation*.

Protection/Damage Mitigation

Physical protection from the elements is a basic human need, but few human-made structures have been designed, built and maintained to withstand the most extreme forces of nature. Vernacular architecture using traditional building methods and local materials, such as mud bricks, wood and thatch, typifies many parts of the world where experience has been built up over generations, leading to important and successful approaches in providing physical protection to hazards. For example, indigenous houses in Bali, Indonesia, which are lightly built with plant-matting walls and palm frond roofs to ensure ventilation, have successfully withstood earthquakes. Wooden, loft-style stilt houses, called *palafitos* in Chile, are common along the banks of tropical rivers in Southeast Asia, West Africa and South America. They provide accommodation for fishers above the seasonal flood flows but are not always successful.

Increasingly, the spread of manufactured products like concrete and modern construction methods has provided more opportunities for stronger buildings, but when implemented improperly or not accounting for detrimental consequences of new techniques, disaster damage can increase. For example, reinforced concrete for earthquakes might seem suitable for Bali, but it could reduce ventilation and so increase deaths related to heat and humidity. Similarly, if reinforced concrete is not produced and maintained properly, or the quality of steel used in construction is poor, then buildings will crumble in earthquakes.

Protection against hazards can be supported when structures are designed, erected and maintained according to approved and monitored standards, which can usefully combine local, vernacular and imported materials and methods. These approaches require skills from engineering, architecture, planning, sociology, participatory processes, law and many more in order to ensure compliance with appropriate building codes and other regulations which emerge from political initiatives and community acceptance.

1. Macro-Protection

Purpose-built structures are widely used to protect people and property against surface or near-surface flows of potentially damaging solid, fluid or mixed materials such as rock falls, lava flows, lahars, mudslides and avalanches, as well as floodwaters from rivers or coastal sources (storm surge and tsunami). These structures tend to act in two ways: either by holding back excess material (storing floodwater in reservoirs; Figure 5.4) or by diverting flows away from vulnerable sites or stopping its movement (avalanche control walls and fences; Figure 5.5). They can also be built to withstand the direct impact from the material's forces, such as roofs strengthened to avoid rocks punching through. They can exist at point locations such as individual builds or can take a linear or areal form such as embankments, dams with reservoirs, levees and artificial channels.

Long embankment (levee or dyke) systems run alongside many of the world's

Figure 5.4 Singapore's Marina Barrage stopping some inundation in the city with a reservoir and recreational area behind it.

Source: Photo: Ilan Kelman

rivers. Embankments run over 1,500 km along the Red River in North Vietnam, for example, and over 4,500 km along the Mississippi River in the USA. Huge dams store floodwaters upstream, with tens of thousands in operation around the world. Many were constructed to supply irrigation water or electricity, but many are also used to provide some flood control. The Three Gorges Dam on the River Yangtze in China is claimed to reduce major floods downstream from a frequency of 1:10 years to 1:100 years, yet it also forcibly displaced large groups of people and might not remain functional for its expected lifetime due to issues such as siltation. In attempts to reduce coastal flooding from the North Sea, the 1,000 km long coastline of the Netherlands has been transformed by a mix of stabilised dunes, beach nourishment, sea walls, concrete embankments and tidal barrages – all of which have encouraged more development and settlement in the almost one-third of the country that lies below sea level (Figure 5.6; Govarets & Lauwerts, 2009). An estimated 40% of Japan's coastline has concrete sea walls and breakwaters against storm surge and tsunami. Smaller structures have been erected against other similar types of hazard, and Box 5.1 shows how deflecting dams in Iceland have been

Figure 5.5 Fences against drifting, blowing and sliding snow in Liechtenstein.

Source: Photo: Ilan Kelman

Figure 5.6 An example of small-scale flood structures in the Netherlands.

Source: Photo: Ilan Kelman

Box 5.1 Avalanche-Deflecting Dams in Iceland

Avalanches threaten many communities in Iceland. On 26 October 1995 an avalanche containing about 430,000 m^3 of snow struck the village of Flateyri in northwestern Iceland and killed 20 people in an area previously thought to be safe (Jóhannesson, 2001). The avalanche was created by strong northerly winds blowing large quantities of snow from the plateau into the starting zones of the two avalanche paths of Skollahvilt and Innra-Bæjargil above Flateyri. Following this event two large deflecting dams, connected by a short catching dam, were built to divert future flows away from the settlement and into the sea (Figure 5.7). Each earth dam is about 600 m long and 15–20 m high and designed to intercept avalanche flows at angles of 20°–25°. The purpose of the central catching dam, which is about 10 m high, is to retain snow and other debris that might spill over from two deflectors in a large event. The total holding capacity of the structure is around 700,000 m^3.

The dams were completed in 1998. Since then they have successfully deflected two separate avalanches (February 1999 and February 2000), each with snow volumes over 100,000 m^3, impact velocities of 30 m s^{-1} and estimated return periods of 10–30 years. Estimated outlines of the avalanche run-out paths in the absence of the dams show that the Skollahvilt flow would have caused little loss, largely because houses destroyed in 1995 in this part of the village have not been rebuilt. But the 2000 avalanche from Innra-Bæjargil would have destroyed several houses. Although these two events are smaller than the design capacity of the deflecting dams, they provide a good example of the use of structures against moderately sized hazards.

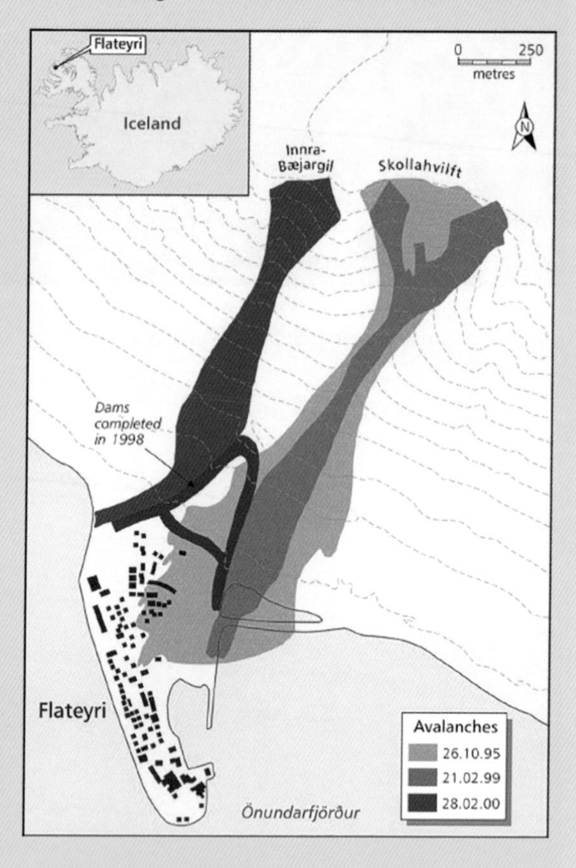

Figure 5.7 The effectiveness of deflecting dams in steering snow avalanches in 1999 and 2000 away from the township of Flateyri, northwest Iceland. The extent of the damaging 1995 avalanche that led to the construction of the dams is also shown.

Source: After Jóhannesson (2001). Reprinted from the *Annals of Glaciology* with permission of the International Glaciological Society.

effective in protecting property against snow avalanches.

Macro-scale engineering for flood protection has come under increasing scrutiny on grounds of financial, social and environmental acceptability, but also long-term effectiveness. What are the consequences of building structures that encourage settlement in highly flood-prone locations? Alternative approaches can be placed into two main categories, with overlaps as well as the usual advantages and limitations.

Floodplain Management

Mixed approaches, especially using non-structural measures such as warning, evacuation, land use planning and insurance, focus on reducing damage and casualties from floods. Certain activities or structures might be removed from floodplains or settlements designed to channel floodwater away from properties. Green spaces and floodways can be built to generate floodplains that store excess water until it can drain. Chapter 11 provides more details.

Living With Floods

Local communities are encouraged to live safely with floods in more connected ways. More comprehensive land use and livelihood considerations are applied to ensure that people use both the land and water. It might entail accepting a certain amount of regular damage and disruption from floods as well as building infrastructure that gets wet with limited damage. Chapter 11 provides more details.

2. Micro-Protection

In most countries the design and construction of key infrastructure like dams, bridges, power supplies, water supplies and gas supplies are governed by legally enforceable regulations, although questions remain regarding how much and how consistently they are enforced. The same is true for large industrial sites and government facilities. Such centralised regulation sometimes also applies to residential buildings. Yet no amount of regulations, enforcement or technical effort guarantees or could guarantee a structure's survival in a hazard. The formal standards can apply to a specified event magnitude occurring during the lifetime of the structure. Figure 5.8 shows the hypothetical example of a building designed to cope with a wind stress that is estimated to occur on average once in 100 years (1% probability). Wind speeds just outside the design limits have little effect, but progressive stress beyond the performance envelope causes structural failure. Many buildings remain in use longer than their expected lifetime of, say, 50 years and therefore are exposed to

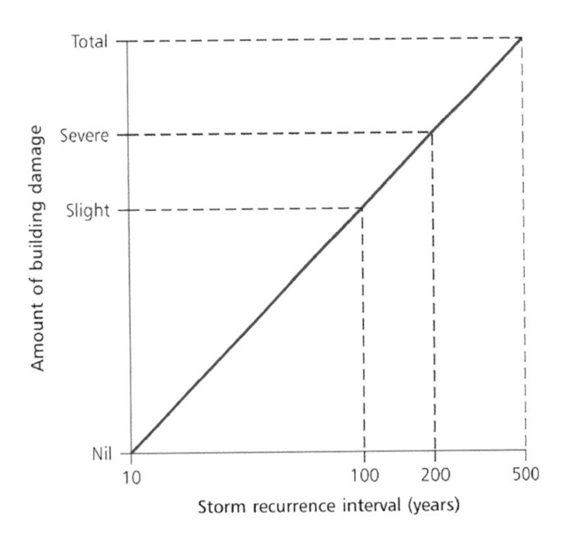

Figure 5.8 A theoretical illustration of the resistance of an engineered building to wind stress from storms of different return periods. It is important that the associated building codes are enforced if such performance standards are to be achieved.

higher risk. It follows that detailed structural inventories of the building stock should be maintained and routinely updated, but such information is rarely available because of a lack of qualified surveyors and the costs involved.

The construction methods used to reduce risk, whether for newly built or retrofitted structures, tend to be both hazard- and property-specific. Common features include:

- *Earthquakes* (Chapter 6). Soft-storey collapse occurs when the ground floor is too weak to support the upper levels. This can be prevented by introducing beam columns or other types of support. Single or two-storey wood-frame domestic buildings need bolting to their concrete foundations so they cannot be shaken free or displaced. Where soft soil exists beneath the structure, deep pilings can be driven down to underlying bedrock. Brick chimneys can be reinforced and braced onto structural elements to prevent collapse. Some *unreinforced masonry buildings* (URMs) are difficult to retrofit, although walls can be strengthened and tied to adequate footings. In any structure, closets and heavy furniture and appliances can be strapped to the walls to protect contents and people.
- *Floods* (Chapter 11). To reduce flood damage, walls can be made watertight, and flood-resistant doors and windows can be fitted. Another priority is to raise items and utilities, especially electricity and boilers, above the expected maximum water level. Some smaller wood-frame houses can be lifted above ground level to protect contents and basic fittings such as electrical supplies and heating boilers. Special attention is given to protect basements. Check valves ensure that the drains do not back up into the property, and small walls may be erected around low-lying properties.
- *Wind, including from hurricanes and tornadoes* (Chapter 10). Here, the priority is to maintain the structural integrity of the property and prevent major damage when strong gusts create differential air pressures between the inside and the outside of a building. Measures include storm-resistant shutters to protect glass windows and reinforcement of doors. Care is needed to make roofing materials secure, such as by bracing roof gables so that the roof is not torn off. Most failures occur because of poor design, construction and maintenance practices. A lack of financial resources and technical expertise often prevents safe design and building work.

Corruption is a recurrent, global problem and includes the abuse of building permits, illegal construction methods and non-observance of land use controls (Lewis, 2003; Chen & Lee, 2022). Public sector corruption can lead to the construction of substandard housing units, business buildings and infrastructures that ultimately fail in the face of major events such as earthquakes, leading to a greater death toll than would occur in the absence of such corruption. In the 2001 Bhuj earthquake in India, over 450,000 traditional rural houses of rubble masonry construction were destroyed. Although anti-seismic codes of practice existed for modern buildings, non-enforcement of regulations and inadequate inspection methods also led to the failure of 179 high-rise reinforced concrete buildings in Ahmedabad, about 230 km away from the epicentre of the quake (Benson & Twigg (2007).

It is important to note that corruption is more complex than a 'pay off'. Garrett and Sobel (2003) offer evidence that both a US president's declaration of a disaster area and the flow of Federal Emergency Management Administration (FEMA) money to such areas are highly correlated with the political importance of a particular state to the sitting president and to the state's congressional representation on FEMA oversight committees.

Establishing building codes is not enough; there is a need to consider the compliance of these codes. Whilst this may appear a key issue in more low-income countries, it can also affect those countries with the highest earthquake building standards. The 22 February 2011 earthquake in Christchurch, New Zealand, killed 185, of which 115 were in one building, the CTV building. This concentrated loss of life prompted two investigations, including a Royal Commission, as to why the building had collapsed (see Figure 5.9). Following the 4 September 2010 Darfield earthquake, the CTV building was deemed safe (green), but then suffered some damage during the 26 December2010 (Mw 4.9) aftershock. The investigations found that in 1986 the Christchurch City Council had granted a building permit despite concerns about

Figure 5.9a The CTV building in Christchurch, New Zealand, before the earthquake.

Source: Phillip Pearson via Wikipedia

Figure 5.9b The CTV building in Christchurch, New Zealand, following the earthquake at 12:51 p.m. on 22 February 2011.

Source: © Alamy Images/Newscom

structural design issues. Moreover, the authority did not insist on structural analyses of the building after the initial earthquake of 4 September 2010 (Seibel, 2022). It was determined that the failure of the CTV building was due to fundamental design errors that hinged on incorrect calculations by the chief designer (The Canterbury Earthquakes Royal Commission, 2012).

Nwadike and Wilkinson (2020) further investigated the root cause of these errors, comprising the complexities of building codes, a lack of capacity building among the relevant stakeholders, lack of training needs assessment, irregular building code update and lack of awareness. Most of these issues could be addressed by proactive training, raising awareness and provision of resourceful technical supports to the code users.

Often, protection is prioritised for public buildings and facilities expected to remain operative during emergencies (such as hospitals, police stations, pipelines). Schools, offices and factories have also been strengthened in the belief that they will provide shelter for local residents. This focus on so-called critical infrastructure can mean that little attention is paid to private homes. The defects of this policy were exposed when tropical cyclone 'Tracy' struck Darwin, northern Australia, in 1974. This storm destroyed 5,000 houses, comprising about 60% of all homes in the city. Faced with the loss of electricity and other basic services, three-quarters of the population was evacuated to cities further south (Walker & Minor, 1980). This disaster demonstrated that residential housing is of equal importance to public buildings in most communities. When cyclone warnings are issued, public buildings close down and people seek shelter in their own homes.

There are examples of good practice and of better construction methods. In 1977, following a cyclone in the coastal area of Andhra Pradesh, India, 1,500 new houses were built using concrete block walls and reinforced slab roofs (Benson & Twigg, 2007). Of these properties, 1,474 survived an even stronger cyclone in 1990. Similarly, 450 housing units were constructed in the Philippines after the passage of typhoon 'Sisang' in 1987, using concrete footings and steel post-straps bolted onto the core frame. These houses also withstood later typhoons with little damage. Developing buildings that can withstand multiple hazards is commonly done in earthquake- and tsunami-prone areas where both hazards can be mitigated against without weakening the structure (see Figure 5.10).

3. *Retrofitting*

In order to provide a safer environment, hazard resistance has to extend to existing infrastructure. Retrofitting is the process of reinforcing all types of structures and their contents against loss. Most existing structures can be strengthened, but homeowners might be reluctant to invest in protective measures, due to underestimation of the risks and a preoccupation with short-term returns or because they cannot afford the work required or distrust that it will be done properly.

Owners often do not implement cost-effective improvements for risk reduction, despite the incentives of reduced insurance premiums, quicker and cheaper post-disaster occupancy and most notably less chance of dying. They might not pay because they cannot afford it; they think that any losses will be compensated by insurers or government aid; or they might underestimate the risk. Some local authorities resist the introduction of building codes on the grounds that the costs of compliance and inspection will hamper inward investment and economic development. If commercial insurance against hazards becomes harder to obtain and as some taxpayers increasingly complain about paying for property owners who take no responsibility for hazard reduction, better design, code enforcement and

Figure 5.10 Office buildings in Concepción, Chile, designed and purpose-built to withstand earthquakes and tsunamis. These structures are elevated 10 metres above street level in order to allow a tsunami wave to pass below without causing damage.

Source: Photo: Walter D. Mooney, USGS

retrofitting will become more important (Nwadike et al., 2019).

All too often there is a lack of information about the most appropriate retrofit measures available. In Istanbul, Turkey, where earthquake risk is high, over half of all residents questioned had no knowledge of seismic strengthening (Eraybar et al., 2010). About one-third lacked plans for a safer home other than by purchasing a new, earthquake-resistant house. Without more information and some financial stimulus from the government, property owners might not take action. Some local authorities do have schemes for identifying and strengthening (or demolishing) existing hazardous buildings, although low-value public housing can often be left out. Many jurisdictions have special regulations for preserving buildings of historic or cultural significance, even if unsafe.

A diversity of examples from around the world calculating the costs and benefits of retrofitting residential buildings found that high returns on investment could be obtained (by Sharma et al., 2011):

- Strengthening doors and windows of middle-income homes to resist hurricane forces in St Lucia (Figure 5.11).
- Elevating high-income homes in Jakarta.
- Strengthening apartment buildings against earthquake risk in Istanbul.
- Building, and even replacing, brick homes on a firm plinth in Uttar Pradesh.

Figure 5.11 In St Lucia, how could you tell which homes have hurricane-resistant measures and which ones do not?

Source: Photo: Ilan Kelman

Costs are reduced if hazard resistance is factored in at the outset. Simple building changes could improve the cyclone resistance of *kutcha* (temporary non-masonry) houses in Bangladesh at an additional cost of only 5%, whilst the introduction of comprehensive anti-seismic construction principles at the planning stage, including optimum layout and better design of structural connections, would increase building costs by less than 15%. For the most tornado-affected US states, Simmons et al. (2020) estimated that better tornado protection would add to construction costs US$16–27 per square metre, which was easily recouped within most 30-year mortgages, in addition to protecting life.

Common barriers to enacting these measures include not being concerned about hazards because not much has happened recently; the owners of rented properties not worrying about the safety of the renters; and multi-occupancy structures such as high rises sometimes requiring all owners to agree on any measures proposed. Other complications are legislative frameworks; for example, in England, the owner of a building does not necessarily own the land on which the building sits, so disputes can arise between the two owners when seeking major changes.

C. INSURANCE

Disaster insurance exists when an asset is perceived to be at risk and the owner pays a fee (the premium), usually on an annual basis, to buy a contract (the insurance policy) that transfers the risk to a financial partner (the insurer). The insurer – usually a private company, but insurers can also comprise a not-for-profit service or a government agency – guarantees to meet specified costs when a trigger occurs, which is often loss or damage to the asset. Another approach to insurance is when a hazard over a certain parameter manifests, and insurance money can be paid which might cover expected losses but which could also be used to prepare for the hazard and to mitigate damage.

By such means the policyholder is able to spread the cost of a potentially unaffordable disaster over several years. A commercial insurer assumes either that no loss will occur during the term of the policy – that, over time, the sum paid will total less than the premiums received – or that they take a loss on their insurance business, but they collect and pool the premiums which they invest to make an overall profit. Commercial insurers should also pay claims out of investments and profits; government insurers pay claims out of tax revenues or loans.

1. Commercial Disaster Insurance

Private companies cover (underwrite) property such as buildings against flood, storm or other specified environmental perils. Policy underwriters should try to ensure that the properties they insure are varied and spread over diverse geographical areas so that only a fraction of the total liability is at risk from a single hazardous event. By this means, the cost of payouts to claimants is distributed across all policyholders. If premiums are set appropriately, they would cover costs, but the insurance company must also price premiums so that customers will buy them, meaning that premiums might be priced at a loss for the insurance company. The insurance company then makes profits largely by investing the money received from premiums.

Insurance claims after earthquakes, tropical cyclones and other sudden, focused hazards cluster within short timescales and small areas compared to the globe. Claims can be highly focused in certain time slots. For example, in 1994 the insurance industry in California collected about US$500 million in earthquake premiums, but it paid out over US$15 billion over a period of more than four years for damage caused by the 1994 Northridge earthquake. Unless a company accumulates a large catastrophe fund, it may not survive such demands and will go out of business (Born & Viscusi, 2006). Another problem is *adverse selection*, which occurs when the policyholder base is narrow and dominated by bad risks. For example, perhaps mainly floodplain or coastal dwellers take out flood insurance, leading to a geographical clustering of risk. After Hurricane Andrew in 1992, nine insurance companies became insolvent and others attempted to quit the market in Florida.

The insurance industry can increase profitability in various ways:

- *Raising the premiums*. This is the most obvious method but is unpopular with customers and so product purchases could decline – which might be advantageous for the company if high-risk customers purchase less insurance.

- *Re-rating the premiums*. The detailed setting of premiums in line with local levels of risk is possible. Insurers then place individual policyholders in different bands of hazard exposure and charge premiums appropriate to the risk.
- *Restricting the cover*. The size of claims can be restricted by applying a policy deductible (excess), by capping policies at a maximum payout or by providing different forms of cover for different locations. In Japan, the risk of huge losses from earthquakes in urban areas has led to limits on any one claim. Above an agreed threshold, the government shares costs with the policyholder. As a last resort, a company can refuse to sell insurance cover in high-risk areas, although this is unpopular with the public and with governments. Following the Eyjafjallajökull eruption in 2010 in Iceland, flight disruption due to volcanic ash became a popular add-on for tourists as part of their holiday insurance.
- *Widening the policyholder base*. This is done by spreading liability through a basket of cover rather than limiting cover to a single hazard. In the UK, the industry offers policies that include storm, flood and frost damage, as well as fire and theft. By this means, the uptake of household insurance – which is a requirement of all mortgage lenders – is high. Any losses arising from a specific hazard event are subsidised by all policyholders, including those with lower risk to that hazard.
- *Reinsurance*. Companies join together, or with the government, to pass on part of the risk. For example, the primary company might agree to pay the first US$10 million of claims; for losses in excess of this sum the company would be reimbursed – perhaps up to 90%–95% – by its partners. The reinsurance market is international, and very high risks are spread through the world markets, although the rising cost of claims, and fears about factors such as human-caused climate change, can make it difficult to obtain all the reinsurance that is required. Reinsurance firms typically conduct research on hazards and threats to aid their clients.
- *Reducing the vulnerability*. Insurers can offer lower premiums to policyholders who reduce their risks by retrofitting against hazards. Cover for new properties can depend on the use of approved construction techniques such as anchoring the structure to the foundations to prevent slippage or using wind-resistant roofing and walling materials. These measures are not widely deployed and hard to monitor without government support through legislation and enforcement.

Some advantages and disadvantages of commercial insurance are shown in Box 5.2. Disaster claims are rising while other factors need to be accounted for in comparisons, such as inflation and currency exchange fluctuations. Nonetheless, as a rough guide, before Hurricane Alicia in 1988, no single disaster had cost the insurance industry more than US$1 billion (Clark, 1997). According to Munich Re (2006), the costly year of 2005 brought six major disasters that contributed overall losses of US$170 billion (out of a global total of US$212 billion) and insured losses of US$82 billion (out of a global total of US$94 billion). In 2005, Hurricane Katrina alone created total economic losses estimated at US$125 billion, with US$45 billion covered in the private market, and became the most expensive disaster to that date. Then, loss estimates for the Great Tōhoku

Box 5.2 Advantages and Disadvantages of Commercial Insurance

Advantages

- It guarantees the disaster survivor compensation after loss (as long as the insurance company survives and does not dispute the claim). This can be more reliable than disaster relief and appeals to those opposed to government regulation because it depends on individual choice and the private market – which also means that it depends on affordability.
- It provides an equitable distribution of costs and benefits, provided that property owners pay a premium that fully reflects the risk and that insurance payments compensate the insured loss according to the policy purchased.
- Insurance can be used to reduce vulnerability. Provided that residents in hazardous areas pay the full-cost premium, there should be a financial disincentive to locate in such areas. The difficulty is that most residential development is by speculative builders and, until insurance premiums become high enough to make new hazard-prone properties impossible to sell, it is unlikely that developers will be deterred.
- Existing homeowners can be encouraged to reduce their vulnerability, and enjoy lower insurance premiums, by strengthening their property and lowering the risk of loss.

Disadvantages

- Private insurance may be unobtainable in very high-risk areas. Even when available, landslide insurance might cover the cost of structural repairs to property only and not that of permanent slope stabilisation, because of the potential high costs.
- People might want insurance but not be able to afford it or might be misled by inappropriate sales techniques, leaving them without compensation when they need it.
- Insurers can decide to stop selling annual policies in a location or for a hazard, leaving property owners with uninsurable and so difficult-to-sell buildings.
- There is frequently a low voluntary uptake of hazard insurance. Only 10% of the buildings damaged in the 1993 Midwest flood in the US were covered by flood insurance. Fewer than 20% of the US$500 billion losses sustained in the USA between 1975 and 1994 were insured. Such underinsurance may benefit the industry when a major disaster strikes, and Japanese insurers survived the Kobe earthquake largely because only 3% of affected homeowners had earthquake cover. Then, governments often step in to garner political support which adds a further disincentive for people to purchase insurance themselves, because they assume that the government will step in.
- Even when insurance policies are taken out, a significant proportion of policyholders will be underinsured for the full value of property at risk and are, therefore, unlikely to be fully reimbursed in the event of a claim.
- Unless premiums are scaled directly to the risk, hazard-zone occupants do not bear the cost of their location. From the 1960s until the 2000s, many UK insurance companies tended to charge a flat-rate premium of buildings cover for all houses.

This amounts to a subsidy from the low-risk to the high-risk property owners. Even if some link is attempted between premium and risk, the most hazardous locations will likely benefit from cross-subsidisation through the company charging higher premiums than necessary in less hazardous areas – although different areas might have higher risk for different hazards, so some form of balancing out could occur.

- Although insurance can be employed to reduce losses, the existence of moral hazard can often limit effectiveness. Moral hazard arises when insured persons reduce their level of care and thus change the risk probabilities on which the premiums were based. For example, some people may not move furniture away from rising floodwater if they know they will be compensated for any loss. They thus replace their old possessions with new ones, funded by the insurance company.

earthquake and tsunami in Japan (in 2011) started at US$250–300 billion, with insured property costs of US$20–30 billion.

When property owners do not have insurance due to lack of affordability, they might also lack the resources to restore their assets after a disaster. They might instead fall back on poorly targeted public assistance which typically is not provided equitably. Poverty and marginalisation can be reinforced, leading again to lack of affordability of and access to insurance, with the same people ending up in the same post-disaster circumstances as previously. Incentives such as long-term retrofit loans and subsidies for low-income households are techniques which can make insurance more financially attractive and more of a long-term endeavour, as long as the insurance provider will offer the required insurance.

2. Government Disaster Insurance

Creating government-run disaster insurance funds can address some flaws associated with commercial insurance, but has its own disadvantages. If made compulsory, government-run cover not only widens the policyholder base but also can be used to raise public awareness of hazards and the need to implement disaster risk reduction techniques. In theory, this would enable premiums to be related more closely to the risk. For example, the government could legislate that properties not built to specific standards would not be eligible for government insurance. The National Flood Insurance Act (1986) was an early attempt by the US government to adopt this approach and shift some federal costs to state governments and the private sector.

Some countries dictate disaster insurance cover through partnerships between government and the insurance industry. Spain has had such a scheme since 1954. In France, property and motor insurance has included mandatory disaster cover since 1982, financed by a surcharge on private premiums and state reinsurance. New Zealand introduced government cover for earthquakes through the Earthquake and War Damage Act (1944). The scheme, operated by the Earthquake

Commission (EQC), was extended to cover storms, floods, volcanic eruptions and landslides. Claims were financed by a surcharge on all fire insurance policies of 5 cents per NZ$100 of insured value. The EQC could rate premiums according to risk and refuse claims on poorly maintained properties, although political pressure can force many claims to be met.

Some governments aim to shift towards persuading individuals to accept more responsibility for their disaster losses. In Turkey, government payout for earthquake loss has been replaced by a mandatory insurance scheme, although this becomes operative only when a property is sold and the new owner becomes liable. One of the most radical changes in public–private risk sharing occurred in New Zealand when the state scheme was reformed in 1993 (Hay, 1996). From 1996 onwards, the EQC withdrew disaster cover for non-residential property and, although insurance for residential property remained automatic for owners with fire insurance, the extent of cover was limited. The EQC retains a large fund as a first call on disaster claims and has reinsurance arrangements, but the New Zealand government

Box 5.3 Microinsurance

There are an increasing number of insurance and financial tools to assist governments in their insurance schemes. For example the World Bank has a 'Disaster Risk Financing and Insurance Program' (DRFIP) that helps countries ensure that their populations are financially protected in the event of a disaster. Microfinancing – which actually formed the backbone of some of today's global commercial insurance companies, emerging in the 1800s as a mutual protection scheme for low-income workers – has increasingly been offered as a contemporary tool for the economic security of those people unable to get coverage from established commercial insurance companies, such as those working in the informal economy (Churchill, 2007). Microinsurance is the 'protection of low-income people against specific perils in exchange for regular premium payments proportionate to the likelihood and cost of the risk involved' (Churchill, 2007, p. 402). Most models of disaster relief aid are reactive (as discussed later under 'Response'), where financial assistance is pledged during and after the event. The money pledged is often unpredictable and slow to materialise. Microinsurance can challenge the humanitarian status quo to present a wide range of financial products that provide viable alternatives to aid, cash transfers and emergency lending. However it must be affordable, accessible, appropriate, timely and consistent (Yore & Faure Walker, 2019). To achieve this there needs to be more of focus on how consumers can access and experience the products and accompanying services. Figure 5.12 highlights how sustainability of microinsurance must be sought beyond the typical reliance of funding from donors and external sources. For low-income countries, premium subsidisation is important in supporting the social security function that can form a part of state assistance programmes, for which external funding becomes necessary.

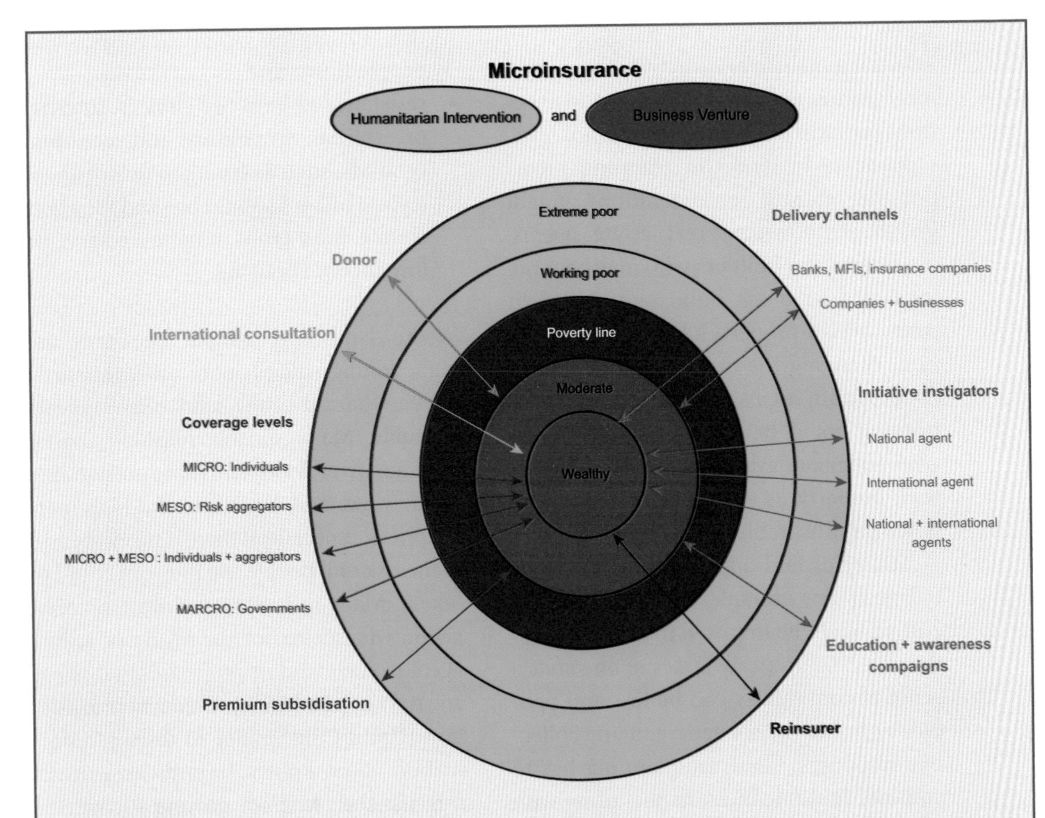

Figure 5.12 Markets for insurance, from the wealthiest in society to the poorest. The segments with vertical lines signify the target markets that often benefit from and are appropriate for microinsurance as a form of humanitarian and development assistance. The segments with horizontal lines indicate the markets for which insurance as a business venture is often better suited. The two overlap among the populations hovering around the poverty line (depicted by the check pattern), suggesting that the humanitarian model, the business model or a combination of both can be suitable, depending on the context. The microinsurance characteristics showing significance in this study are overlaid in colour, intersecting the segments where we suggest they are most appropriate. Where the coloured lines are dashes, the characteristics are considered potentially helpful but more discretionary for that market segment.

Source: Yore & Faure Walker (2019, p. 26)

remains responsible for any shortfall in disaster payments.

D. ADAPTATION – LAND USE PLANNING

Land use planning can zone and control hazardous areas to assist existing communities in dealing with hazards while new development can be made safer. Land use changes driven by competition, a desire for profits or only market-based considerations tend to augment hazard-related harm because disaster costs are not incorporated into the calculations. Instead, land use can be regulated at all scales combining multiple purposes, from national guidance or directives through to local zoning ordinances down to instructions for individual plot and subplot level. Opponents to land use planning incorporating hazard considerations might include landowners, real estate agents, developers and builders who may be more interested in quick build and sale than safety over a property's lifetime. In addition, changing the designation of hazard zones within areas already developed for housing could be opposed by residents who anticipate a loss in market value of their property, as well as the possibility of it being uninsurable or unsellable.

Strategies and approaches need to be balanced. For example, while low-density zoning might be imposed in order to limit the potential property losses in one area, a sense of community can be lost and people might end up isolated, reducing collective responses. Land use planning has to consider multiple interests including with regards to environmentally sensitive areas, such as with rare or endangered species; recreational uses and quality of life, such as bike and walking paths through river ravines; site contamination or ongoing pollution, from factories, noise or light; the transportation situation such as regular public transit or designing for every individual using a private vehicle; and balancing residential, commercial, industrial and agricultural uses.

Some land use planning limitations are:

- lack of knowledge about the hazard potential of smaller areas, such as individual building plots;
- existing development and use;
- the infrequency of many hazardous events and their changing nature;
- costs of detailed hazard mapping, including detailed inventories of existing land use, structures, occupancy rates and people flows; and
- resistance to land use choices, internal or external, on political and economic grounds, such as trying to determine if a piece of land should be used for a golf course, food crops, a nature reserve or low-income housing.

Where land development pressures are high, zoning will be more difficult, partly because hazard-prone land often appears desirable. Many slidable hillsides, coastal areas and river floodplains have outstanding scenic views and can command high market prices, especially if there is no awareness of the risk or if people feel they can afford to take it. Many jurisdictions impart no obligation whatsoever for the owner or selling agent of such land or properties to disclose any risks or past history. Even where legislation requires the vendor to disclose information about known hazards at an early stage in the purchase process so that the potential buyer can make an informed decision, delay is common and the purchaser may disregard the information supplied. Disclosure is generally unpopular with local commercial interests, and local authorities may not adopt regulations, in the belief that they will lose economic initiatives to more lenient communities nearby.

1. Macro-Zonation

Macro-zonation (regional planning) shapes broad land use policy. For example, a probabilistic map of seismic shaking hazards in California could be used to identify areas suitable for retrofitting buildings with anti-seismic measures or introducing building codes for new development (Figure 5.13). Peak ground acceleration (PGA) values are relevant because they indicate the maximum forces that

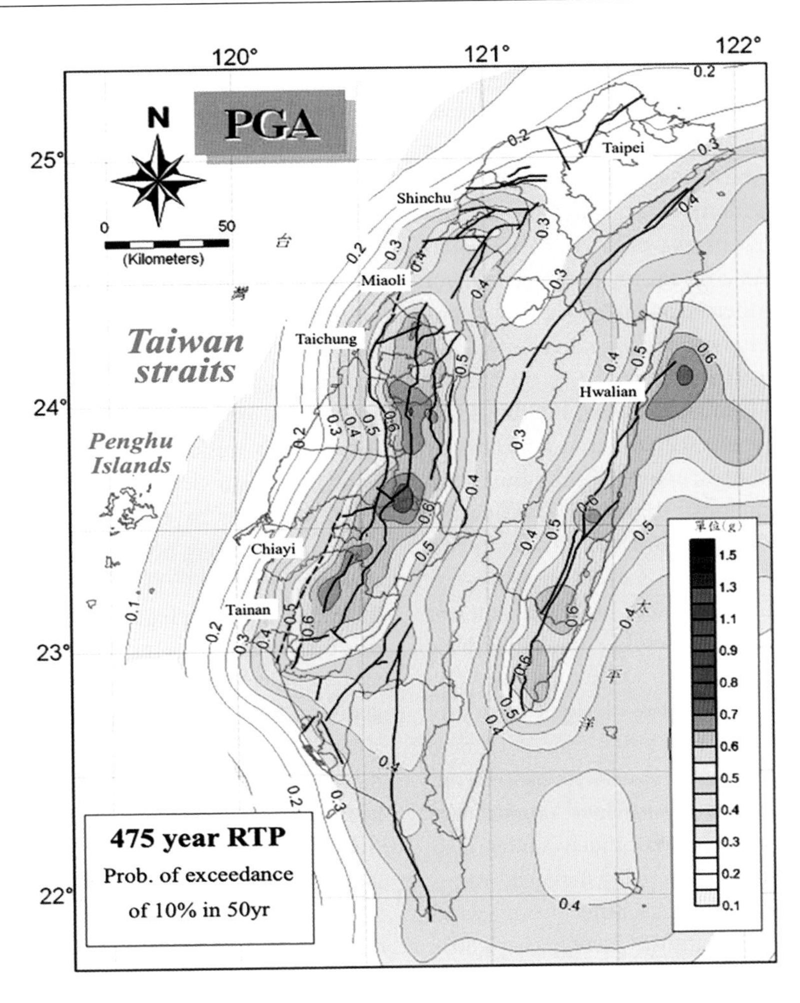

Figure 5.13 Map showing seismic shaking hazards from earthquakes in Taiwan relevant to the design of hazard-resistant buildings and the formulation of building codes. Shaking is measured by the peak ground acceleration (PGA) expected in firm rock with a 10% probability of being exceeded in 50 years. The unit 'g' is the acceleration of gravity.

Source: Cheng et al. (2007, p. 26)

buildings should be able to withstand. Of course, the performance of individual buildings will vary for the same PGA value – especially regarding maintenance, but also because the direction of earthquake shaking can make a difference.

The sort of information required for detailed land use regulation can be illustrated for the island of Martinique in the West Indies, which suffered a major disaster when the Mount Pelée volcano erupted in 1902. Planning policy here depends on estimating and mapping the maximum extent of hazards arising from a possible future eruption and the likely effect on buildings, so that development can be limited in the most dangerous areas (Leone & Lesales, 2009). Scientific views on future eruptions might have to blend with political priorities that can include 'housing for all', as well as different perceptions of and interests in 'an acceptable level of risk'. For example, the residents of Martinique have rated lava flows as a greater threat than the eruptive history of Mount Pelée would suggest. As a result, further development in the northern part of the island is a controversial issue. Four planning zones have been suggested (Figure 5.14): no permanent construction allowed; buildings can be located in exceptional circumstances; hazard-proof structures can be built; and no restrictions. Effective community engagement is required to carry forward plans like this.

Coastal areas are of particular interest since much of the world's population lives there and long-term viability is questionable under scenarios of significant sea-level rise due to human-caused climate change. Policies are sometimes referred to as *Integrated Coastal Zone Management* (ICZM), although how much is integrated and how much is actually managed varies significantly. Settings range from island countries like Samoa, where perhaps 80% of the population and most of the infrastructure is located along the coast (Daly et al., 2010), to huge countries such as Canada in which the biggest two cities are well inland.

2. Micro-Zonation

Zoning by-laws and local ordinances implement regional planning at the scale of communities and building lots. These local regulations can be used to control development through the provision of detailed reports on aspects such as soils, geological conditions, grading specifications, drainage requirements and landscape plans as well as specific hazard threats. Relatively large-scale maps (at least 1:10,000) are usually required for zoning in high-risk urban areas. Other regulations may apply when applications are made for development at the building-plot level. For example, subdivision regulations ensure that the conditions under which land may be subdivided are in conformity with the general plan.

Micro-zonation is most successful for relatively frequent surface flows of material guided by topography which can be located with sufficient precision to permit land regulation down to individual plot level. The Swiss have developed a generic approach that has been adopted in several other countries (Röthlisberger et al., 2016). As shown in Figure 5.15, it is based on the magnitude–frequency principle and combines the intensity and probability of the event to draw up three chief colour-coded zones with differing potentials for land development. For example, in relation to a river flood, the intensity would be measured by the expected maximum flood depth or the depth × velocity, and the risk levels for an urban area would be typically interpreted as the following:

- *Red zone (high risk) – prohibiting*. New houses are prohibited, existing buildings are allowed to be used, lives are at risk even in buildings.
- *Blue zone (medium risk) – restraining*. New houses are allowed but must have significant measures for hazards, more detailed regulation may be required by the local authority.
- *Yellow zone (low risk) – warning*. Key public buildings (hospitals, schools) must be strengthened against hazard impact, residents to be warned and possibly evacuated.
- *Other (residual risk only)*. No specific building or land use controls, public

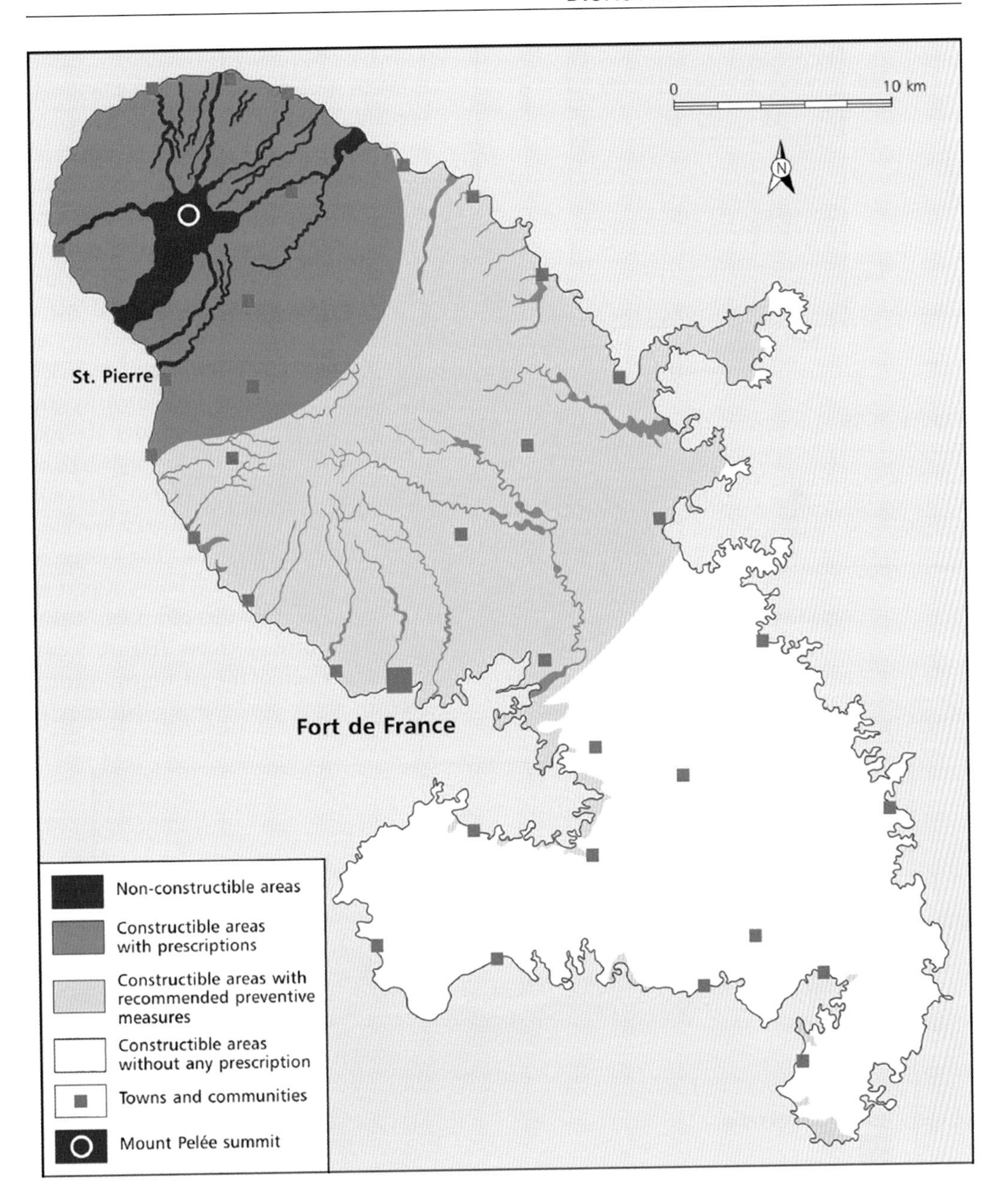

Figure 5.14 Proposed regulation map for volcanic risk reduction around Mount Pelée, Martinique. Such maps are intended to help planners and disaster agencies reduce future losses by controlling land use and may also contribute to improving risk perception by the general public.

Source: Adapted from *Journal of Volcanology and Geothermal Research* 186 (3–4), F. Leone and T. Lesales, The interest of cartography for a better perception and management of volcanic risk: from scientific to social representations (2009, pp. 186–194)

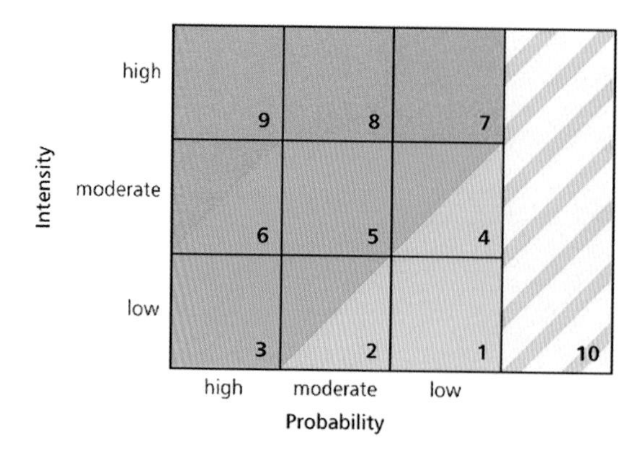

Figure 5.15 A matrix of the Swiss hazard zoning system. The system employs four main hazard zones scaled into ten risk classes. Red – high risk; blue – moderate risk; yellow – low risk; hatched white/yellow indicates high intensity but very low probability.

Source: After Röthlisberger et al. (2016)

organisations (schools and hospitals) should have response measures and emergency plans ready in the event of disaster.

This type of zoning is suitable for debris flows in steep terrain that threaten development on the limited flat land available. In 1998, Andorra – a country in the Pyrenees between France and Spain – adopted an Urban Land Use and Planning Law prohibiting new building development in zones exposed to natural hazards (Hürlimann et al., 2006). Geo-hazard maps have been produced of debris flows at the scale of 1:2,000. The maps are based on a matrix analysis of flow intensity and estimated annual probability for the following recurrence intervals: high hazard <40 years, medium hazard 40 to 500 years, low hazard >500 years and very low hazard lacking flow evidence. The village of Llorts in the Pyrenees is built on part of a debris fan at the outlet of a 4 km^2 basin drained by three torrents (Figure 5.16). The highest part of the catchment reaches 2,600 m above sea level; the fan apex is

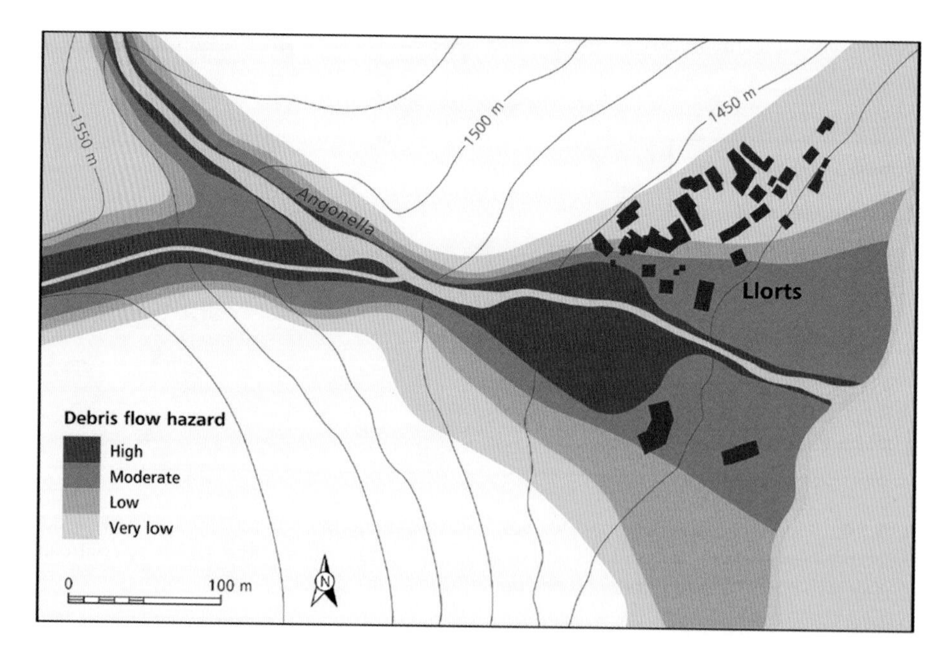

Figure 5.16 Debris flow hazard map of the alluvial fan at Llorts in the Pyrenean Principality of Andorra. Most of the village lies in a safe area, but some existing development is exposed to moderate risk.

Source: After *Geomorphology* 78 (3–4), M. Hürlimann et al. Detailed debris flow hazard assessment in Andorra: a multi-disciplinary approach, 359–372

250 m long with a steep slope averaging 12°. Debris flows enter the apex of the fan and a high-hazard zone was identified in this presently undeveloped area. Although most of the village lies within the safe area, some buildings are exposed to moderate and low-level hazard.

Earthquake micro-zonation might sometimes be less precise, but it remains important due to its potential to input into planning. Identifying active fault lines is crucial to determine how to manage the seismic risk at a highly local level, perhaps by imposing a setback corridor running alongside the fault. Sahdev et al. (2020) illustrate micro-zonation as one criterion within six for assessing earthquake vulnerability for Delhi, India.

For high-risk areas, the public acquisition of hazardous land is perhaps the most direct option for local governments. Once acquired, the land can be managed to protect public safety and meet other community objectives, such as ecosystem restoration or recreational facilities. Any movement of structures or occupants or the demolition of unsafe buildings is usually difficult, expensive and controversial. Voluntary buyouts add expense, and a single detractor can scuttle a project. Mandatory possession by governments impinges on freedoms and leads to protests, although it is done for developments such as airports and shopping malls, with owners often undercompensated. Forced relocation away from an area may destroy any potential the land might have to promote growth and generate local tax revenues – but, then again, that is often the point of the action. The purchase and demolition of buildings with historical or architectural importance would also generate opposition from pressure groups.

Since effective hazard-reduction strategies depend on the understanding and cooperation of the community as a whole, public information is essential. Simple methods – like the posting of warning notices – highlight the threat. Publicity may be disseminated through conferences, workshops, press releases and the publication of hazard zone maps. Financial measures can also be applied. These work indirectly by altering the relative advantage of building at a dangerous location. For example, local governments may decide to restrict investment in public facilities, such as roads, water mains and sewers, to hazard-free areas zoned for development. Any national scheme that provides grants, loans, tax credits, insurance or other types of financial assistance to encourage development can be adapted for local use. For example, tax credits can be offered on hazard-prone land that remains undeveloped or development funds can be approved for low-density development only. In the USA, land conversion is deterred by withholding federal grants and benefits from flood-prone communities not participating in the National Flood Insurance Program.

E. PREPAREDNESS

Disaster preparedness consists of measures undertaken by governments, organisations, communities or individuals to better respond and cope with the immediate aftermath of a disaster, whether natural or human-made. There are two main aims:

> to help people avoid impending disaster threats, and to put plans, resources and mechanisms in place to provide adequate assistance. The main elements of disaster preparedness are forecasting events and issuing warnings; taking precautionary measures;

> and improving response through timely and effective rescue, relief and assistance.
>
> (Twigg, 2015, p. 301)

Table 5.3 sets out the main components of disaster preparedness.

Whilst the categories in Table 5.3 should not be seen as a linear progress, there is a logical sequence whereby planning must be preceded by understanding the hazard/threat, risk and vulnerability, this leads to the establishment of an institutional framework, that enables setting up warnings and putting preparedness mechanisms in place, then testing them. Part of each stage of this is public engagement, education and training.

The objective of disaster preparedness is to reduce loss of life and livelihoods. Simple initiatives can go a long way, for instance in training for search and rescue, establishing early warning systems, developing contingency plans and stockpiling equipment and supplies. Disaster preparedness plays an important role in building the resilience of communities and can enable a prompt and effective response to disaster. In theory, it involves the planning – and testing – of damage reduction actions for hazards covering all timescales, ranging from seconds and minutes (earthquake or tsunami warnings) to decades and centuries (better land planning or measures to combat climate change). Greater risk awareness helps hazard zone occupants to identify threats

Table 5.3 Main components of disaster preparedness

1. Vulnerability, hazard and risk assessment	**2. Planning**	**3. Institutional framework**
Starting point for planning and preparation, linked to longer-term mitigation and development interventions as well as disaster preparedness.	Disaster preparedness plans agreed and in place, which are achievable and for which commitment and resources are assured.	Well-coordinated disaster preparedness and response system at all levels, with commitment from relevant stakeholders. Roles and responsibilities clearly defined.
4. Information systems	**5. Resource base**	**6. Warning systems**
Efficient and reliable systems for gathering and sharing information between stakeholders (e.g. forecasts and warnings, information on relevant capacities, role allocation and resources).	Goods (e.g. stockpiles of food, emergency shelter and other materials), services (e.g. search and rescue, medical, engineering, nutrition specialists) and disaster relief funding (e.g. for items not easily stockpiled or not anticipated) available and accessible.	Robust communication systems (technologies, infrastructure, people) capable of transmitting warnings effectively to people at risk.
7. Resource mechanisms	**8. Education and training**	**9. Rehearsals**
Established and familiar to disaster response agencies and disaster victims (may include evacuation procedures and shelters, search-and-rescue teams, activation of emergency lifeline facilities, reception centres and shelters).	Training courses, workshops and extension programmes for at-risk groups and disaster responders. Knowledge of risk and appropriate response shared through public information and education systems.	Evacuation and response procedures practised, evaluated and improved.

Source: Adapted from R. Ken, *Disaster Preparedness* (New York/Geneva: UNDP/DHA, 1994), in Twigg (2015, p. 302)

and take appropriate actions, although there will always be a gap between what people are advised to do, what they say they will do and what they actually do at any given point.

Preparedness arrangements differ widely. In some countries the task is devolved to existing bodies, like the armed forces or the police, or to dedicated emergency management or civil protection agencies. These agencies depend heavily on volunteer bodies, like local Bushfire Brigades, plus the police, fire and ambulance services. NGOs and international agencies always have an important role to play, often filling gaps which the government cannot or will not.

Proactive planning for disaster saves numerous lives. Key elements include warning, evacuation and sheltering. Such measures have to be well organised and developed long before a hazard strikes, especially to focus on people's needs and approaches so that they can respond effectively in their own way without waiting for outside interventions. Preparedness measures depend on the hazard, and whether it is rapid or slow onset, as the time leading up to an event is variable and affords different preparedness actions and opportunities.

It is important to note that all decisions are made within a political context. All planning and policy decisions by organisations are essentially 'political' rather than technical ones, taking into context the wider context of events, societies and institutions. The policies of governments, donors and other institutional actors shape the context in which the work will take place and should be analysed as part of the planning process (see Figure 5.17).

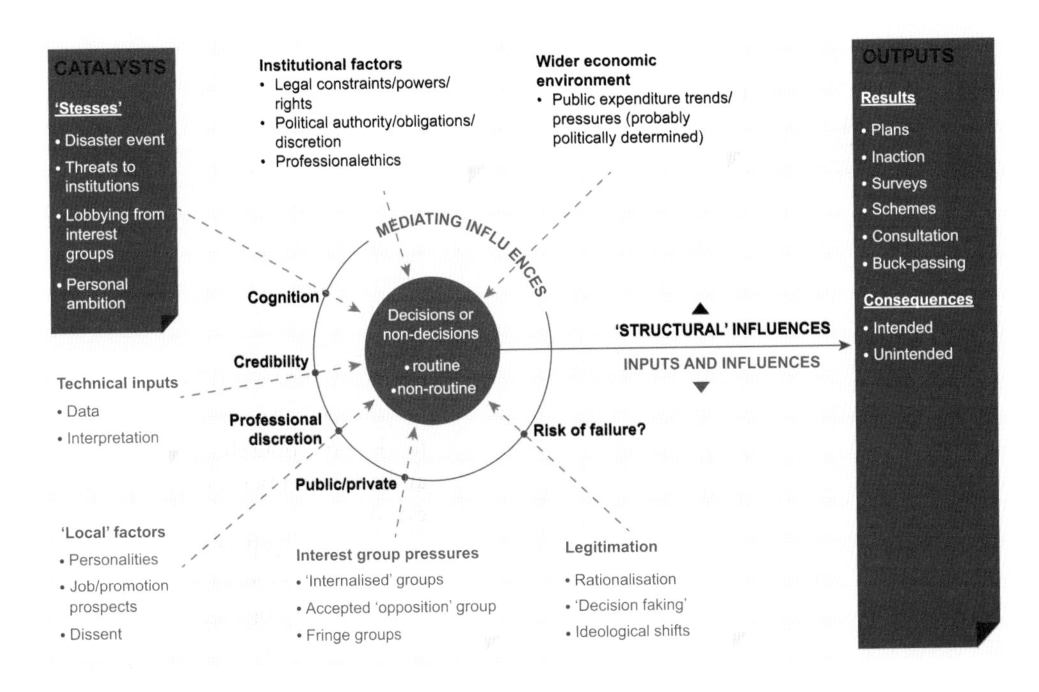

Figure 5.17 Decision-making in a political context.

Source: Mitigating Natural Disasters: Phenomena, Effects and Options – A Manual for Policy Makers and Planners (New York: Office of the United Nations Disaster Relief Coordinator, 1991), p. 18, taken from Twigg (2015, p. 31)

1. Preparedness Plans

Preparedness planning evolves over time. In California there is a long history of raising earthquake hazard awareness. Some 40 years ago residents were poorly prepared to face a damaging event, but, following newspaper publicity in the San Francisco Bay Area, residents responded positively (Table 5.4). Once practical advice was available, a clear majority of those surveyed stored emergency equipment, together with food and water supplies. The proportion of those taking other steps, such as strapping water heaters to walls or purchasing earthquake insurance, was also encouraging, given that many residents were in apartments that precluded certain responses (Mileti & Darlington, 1995). Ongoing work is needed to determine the effectiveness of such campaigns today compared to social media, word-of-mouth through apps, individual influencers, pranking and crowdsourcing compared to traditional media awareness and education campaigns. For earthquakes in particular, ShakeOuts – large-scale exercises (e.g. www.shakeout.org) involving the public to respond to a disaster scenario – have been used from California to the Philippines to New Zealand, raising awareness of what could happen, how to prepare and how to respond. Examples of actions that residents can take against hazards can be roughly categorised as the following:

Table 5.4 San Francisco Bay Area residents preparedness.

Preparedness action	Pre-publicity (%)	Post-publicity (%)	Increase
Stored emergency equipment	50	81	31
Stockpiled food and water	44	75	31
Strapped water heater	37	52	15
Rearranged breakable items	28	46	18
Bought earthquake insurance	27	40	13
Learned first aid	24	32	8
Installed flexible piping	24	30	6
Developed earthquake plan	18	28	10
Bolted house to foundation	19	24	5

Source: Adapted from Mileti and Darlington (1995).

Note: The postal sample in this survey consisted of 1,309 households, and 806 usable questionnaires were obtained.

- *Learning* – use available official sources to identify the risks and consider options, incorporating hazard learning into school curricula.
- *Planning* – discuss cooperative emergency responses with social networks, have a family plan of action, have off-site storage of important documents – perhaps with family on another continent, in paper or on hard drives in safety deposit boxes, or electronically on email or in the cloud – while managing their security.
- *Training* – practise physical and mental health first aid techniques, know how to shut off household water supplies and other utilities and know where to evacuate to.
- *Organising* – collect and store necessary equipment and supplies – first aid kit, a wind-up flashlight and radio (or battery-operated with spare batteries), non-perishable food (and a tin opener if using tins), bottled water, sandbags, shovels, basic rescue equipment and perhaps a generator (while being aware

of accompanying dangers such as electrocution, fire and poisonous gas).

- *Securing contents* – raise key items above flood level or be ready to put them in fire-resistant boxes, strap down heavy appliances and store hazardous materials safely.
- *Securing structures* – raise floor levels, electrics and plumbing for floods; make walls, windows, doors and roofs secure; use non-flammable and heat-resistant materials; and landscape around a property (within the context of neighbouring properties) to channel floodwater away or to reduce the chance of fire ignition.

2. Emergency Operations and Training

Loss-reducing measures include the activation of temporary evacuation plans and the distribution of stockpiled medical aid, food supplies and emergency shelters. It is important to pre-designate an operational control centre because many services – transportation, water and communications – are unlikely to be fully available. Most importantly, training in self-help techniques – first aid, search and rescue and firefighting – should be given to communities at risk. Most disaster survivors who are trapped are rescued in the immediate aftermath by other survivors, rather than by aid workers. Following the 1999 earthquakes in Turkey, about 50,000 people were rescued from damaged buildings; local people saved 98% of them (Walter, 2002).

In areas of high-population density, emergency sheltering might suddenly be needed to cope with large numbers of people arriving with very few possessions – but often with pets and service animals. This requires usable buildings; the massive stockpiling of food, medical supplies and sanitation equipment; the skills to attend to the various needs of different people and animals; and measures to prevent harassment, assault and other crimes. Facilities and equipment must be available for people dealing with babies (feeding and nappy changes) or menstruation and for those with disabilities. It is easy for the authorities to underestimate the physical devastation of a major hazard and the hourly challenges for survivors, including attempting to communicate with family who are missing or outside of the disaster zone while dealing with in-your-face media attention and other survivors live streaming.

One prevalent challenge is ensuring that schemes are compatible with prevailing social and cultural conditions. In the past in many places, disaster planning has tended to follow military lines, with a stress on communications, logistics and security. The military often have the most rapid deployment of basic resources and are able to transport materials and aid effectively. These are important requirements, but the 'command and control' model, represented by a top-down approach, is not always appropriate. External aid can be perceived – often legitimately so – as an act of foreign policy on behalf of distant colonial, ex-colonial or postcolonial powers. In addition, forces with authority, such as military or paramilitary, can lack sensitivity when assisting disaster-affected people, such as operating displaced person settlements, supporting people who have experienced violence, providing safe environments for women and children and trying to find missing relatives. Military assistance may nonetheless be vital, such as for transporting and protecting relief supplies and accessing dangerous locations, but such support must never forget the human element in post-disaster work, which ranges from the need to grieve and heal through to mental health and well-being as much as physical. Bringing together all these elements to work in tandem cannot be done when people are desperately in need but must be set up, practised and tested long before.

The fostering of local preparedness is an ever-present need and process. In the coastal areas of Tamil Nadu, India, community-based preparedness has been pioneered with preliminary risk assessments by focus groups comprising village leaders, NGOs, active youth organisations and many others. Special care was taken to ensure that marginalised and underprivileged people were included. This approach is a considerable achievement, given the hierarchical caste structures and different socio-economic backgrounds found throughout the state. The practical outcome has been a Village Disaster Risk Management Training model (VDRMT). As shown in Figure 5.18, this framework encourages a comprehensive participatory disaster response owned by local people. It should embed sustainability for the future as local leaders are trained to incorporate disaster preparedness into village-level planning which, in turn, should be complemented by local government plans for development.

Disaster preparedness plans can take several forms, ranging from a broad mitigation and preparedness strategy to detailed contingency plans for responding to a particular threat. Operational priorities generally have the aim of saving lives, meeting people's emergency needs (principally medical care, food and shelter) and restoring essential facilities (hospitals, water and sanitation, power and transport). A written plan is not a final product; it requires well-functioning managerial and operational systems, structures and procedures. Constant practice, review and dialogue between partners are required for this. In some cases certified crisis awareness courses have been developed to work across and link emergency responders. Contingency planning involves planning for specific situations by developing scenarios, deciding appropriate objectives for response and working out how to achieve them. Often contingency planning is conducted with key stakeholder agencies or the public via training events, presentations and workshops (which are necessary for grassroots preparedness), with educational resources supporting the process. This helps identify key issues that may not have been immediately obvious from drafting up a plan and enables a learning opportunity to make plans more effective.

For example, a tsunami warning plan for Galle in Sri Lanka identified five vulnerable population groups (women, children, people with disabilities, fishermen and workers in densely populated areas) when drawing up priority routes for evacuation from this coastal city (Villagrán de León et al., 2013). These groups are included in the first-priority escape routes shown in Figure 5.19.

3. Predictions, Forecast and Warnings

Disaster impacts can be reduced by forecasting the approach of many threats – from volcanic eruptions to drought and malaria outbreaks – and then warning those at risk – provided that actions described in previous sections have been enacted long before the forecast is made. Much prominence is given to sophisticated forecasting and warning systems,

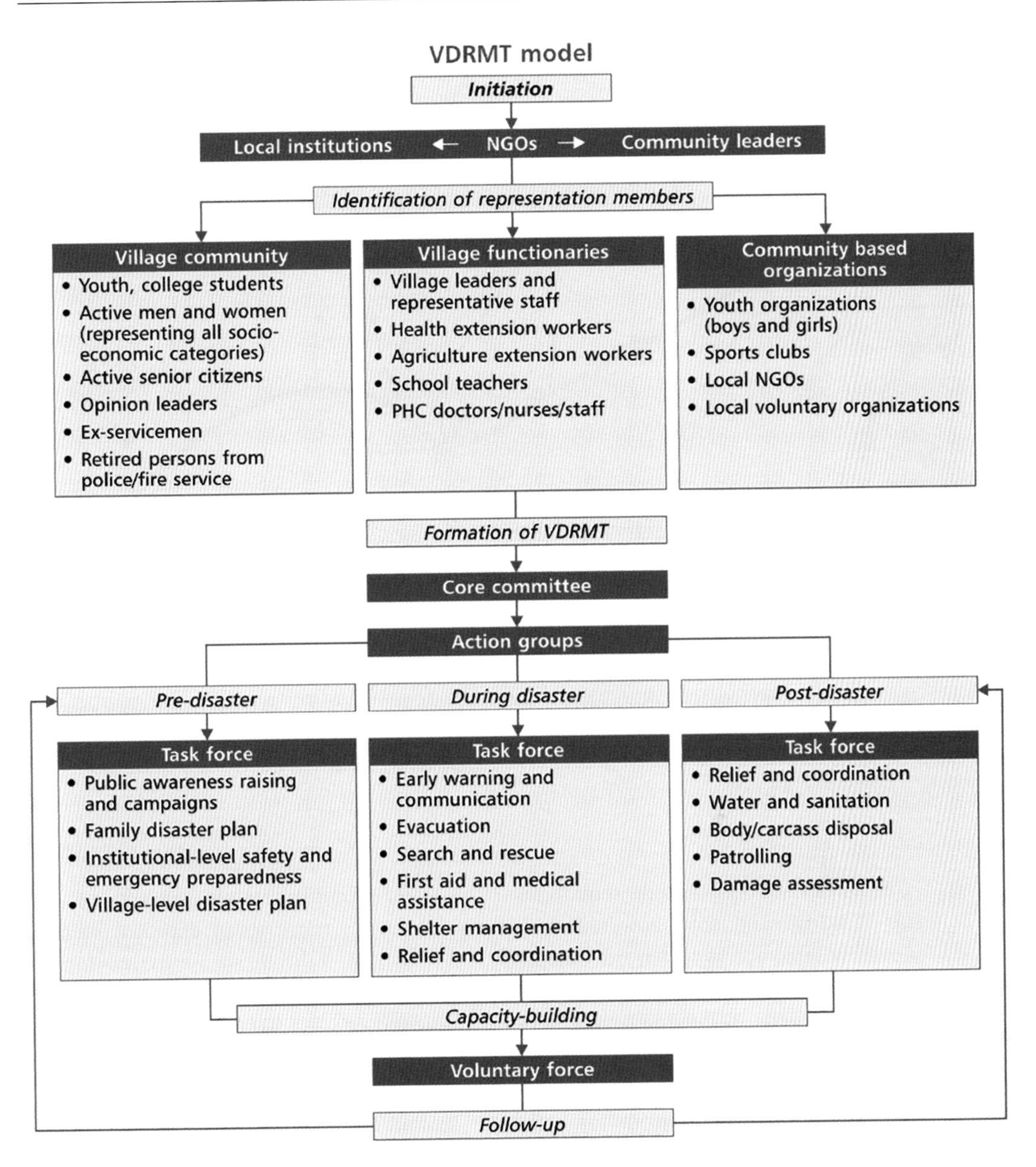

Figure 5.18 The Village Disaster Risk Management Training model (VDRMT). This conceptual model was developed as a capacity-building and disaster-reducing tool for rural communities in India, based on a participatory approach.

Source: After Anonymous (2007)

due to advances in understandings of the physical processes involved, such as tornado formation or an explosive volcanic eruption, real-time Earth observations from telemetry for ground-based instruments to satellites, and information and communications technology, which means that many people can be continually online wherever they are through their handheld devices.

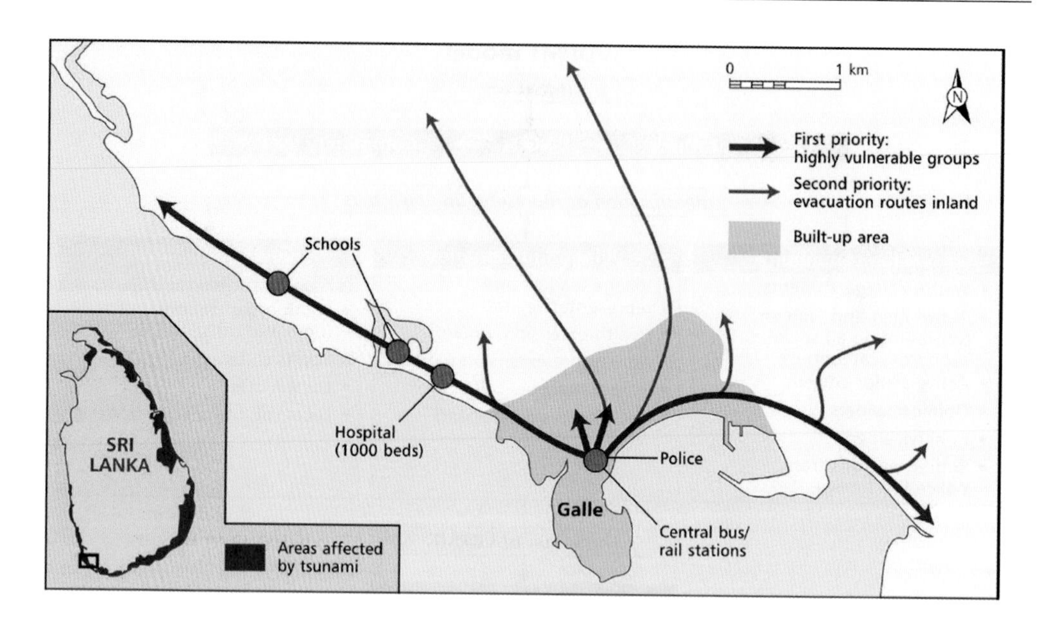

Figure 5.19 Evacuation map for Galle City, Sri Lanka, showing the first and second priority routes recommended for police enforcement during a tsunami emergency. This project was conducted within the UN Flash Appeal Indian Ocean Earthquake – Tsunami programme.

Source: After Villagrán de León et al. (2013). Reproduced with permission of UN HIC DSU Galle, Technical Committee – Early Warning DMC United Nations University for Environment and Human Security UNISDR and the Platform for the Promotion of Early Warning.

Simpler methods, though, remain useful. Wind-up and solar-powered radios permit reliable access to hazard warnings for some of the world's poorest or most isolated households. Following severe floods in Mozambique during 2000 which killed hundreds, the local Red Cross Society integrated *Freeplay Lifeline* radios within a community-based early warning system for cyclones and floods (IFRCRCS, 2009).

Differentiating among predictions, projections, forecasts and warnings is helpful:

- *Predictions* and *projections* are based on combinations of statistical theories, the known record of past events and different forms of modelling for the future. Predictions tend to aim for precision and accuracy, while projections are typically more about expected scenarios that indicate an approximation of potential futures, although there is no fixed or set differentiation between the two. They can be expressed in many different forms with varying indications and levels of confidence of specifically when and where an event may occur and at what magnitude. For instance, it is known where most high-magnitude earthquakes will strike, but so far not when. Conversely, tropical cyclones tend to form in specific months depending on the location, but it is rarely clear exactly where they will make landfall until a few days beforehand. Powerful tornadoes in the southeast US can form in any month, and their path can take them anywhere in the region, often with just minutes of notice.

- *Forecasts* depend on the detection and evaluation of a potentially hazardous event as it evolves. If the event can be monitored, it may be possible to specify the timing, location and likely magnitude of an impending hazard strike. Strictly speaking, forecasts are scientific statements and offer no advice as to how people should respond. They tend to be short term, and the limited lead-time can limit the effectiveness of warnings.
- *Warnings* are messages advising people at risk about a hazard and the steps that should be taken to reduce losses. Warnings are most useful when based on predictions, projections or forecasts. Until recently, many agencies – including those responsible for national weather services – issued forecasts and warnings with only limited advice on safety and hazard limitation. Now, much more advice tends to be included with such messages leading to discussions about whether too much is being provided, possibly diluting action-oriented messages, especially if uncertainties are detailed. Some parts of the meteorological world are moving toward impact-based warnings.

(Golding, 2022)

Warnings are often just regarded as a siren or phone alert, which is a very technocratic approach. In reality, warnings are a long-term social process that is a carefully crafted, integrated system of preparedness involving vulnerability analysis and reduction, hazard monitoring and forecasting, disaster risk assessment, and communication. Together, these activities make a warning system that is defined by the UNDRR (2017) as 'An integrated system of hazard monitoring, forecasting and prediction, disaster risk assessment, communication and preparedness activities systems and processes that enables individuals, communities, governments, businesses, and others to take timely action to reduce disaster risks in advance of hazardous events'. Warnings are represented via different iconographies and communicated via different mediums that usually express some form of threshold or tipping point. These vary enormously contingent on the hazard and the social, political and economic context of the warning.

Figure 5.20 depicts what the UN highlights as the four key elements of a warning system:

1. Disaster risk knowledge based on the systematic collection of data and disaster risk assessments.
2. Detection, monitoring, analysis and forecasting of the hazards and possible consequences.
3. Dissemination and communication, by an official source, of authoritative, timely, accurate and actionable warnings and associated information on likelihood and impact.
4. Preparedness at all levels to respond to the warnings received.

These four interrelated components need to be coordinated within and across sectors and multiple levels for the system to work effectively and to include a feedback mechanism for continuous improvement. Failure in one component or a lack of coordination across them could lead to the failure of the whole system. The links and integration between the four elements is vital to understand (Garcia & Fearnley, 2012).

Warning systems incorporate many aspects of DRR as they tie into mitigation

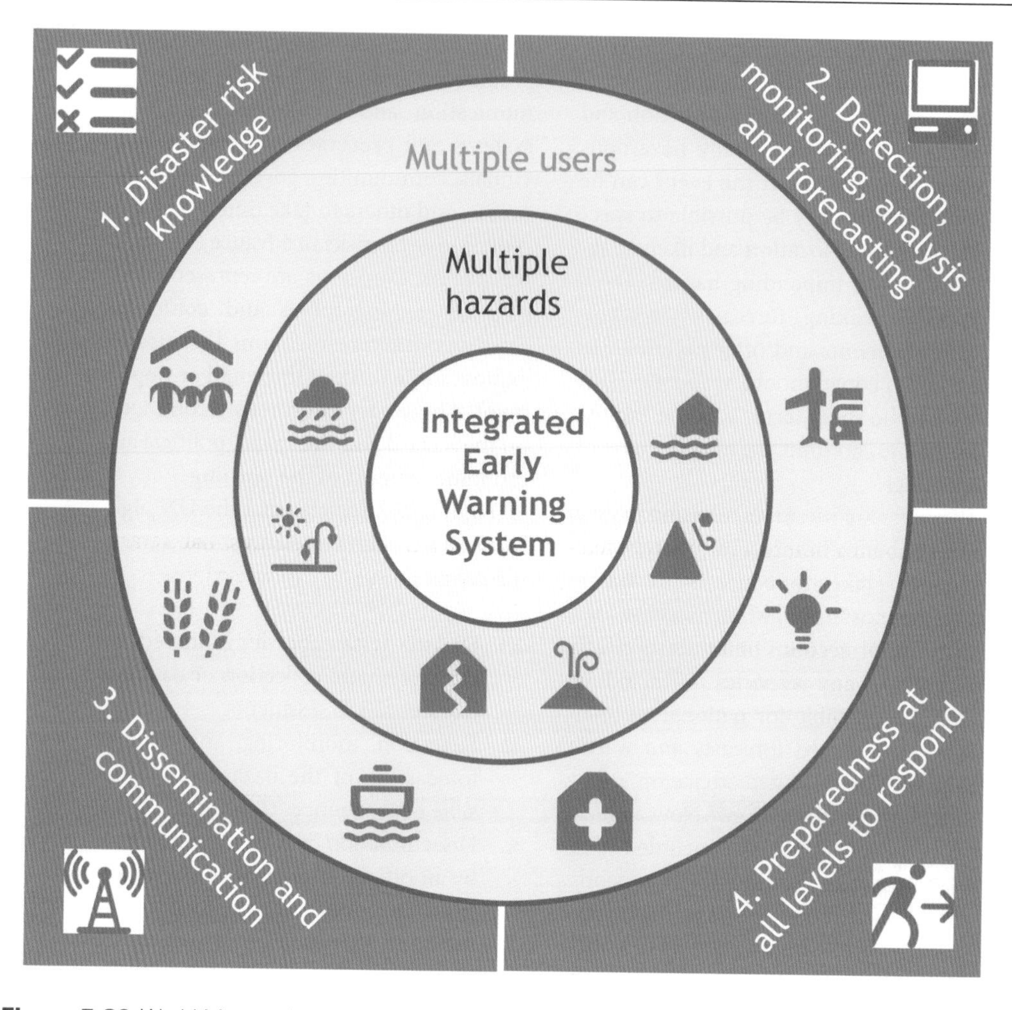

Figure 5.20 World Meteorological Organization model of an integrated early warning system.

Source: Cumiskey et al. (2019)

strategies, preparedness, response capacities and recovery processes (via ongoing warnings) and therefore sit across all the components of DRR.

For warnings to be effective, they need to be accurate, flexible, timely and transparent. In particular, receiving a warning when you only have minutes to evacuate is not sufficient, therefore if a warning cannot be effective, then other mitigation actions are needed. For example, for a near-shore tsunami, a warning may only provide 10–20 minutes for action. If there is no high ground, then vertical evacuation structures need to be built to enable people to reach safety or land use development in high-risk areas should be prohibited. Hence warnings have to be integrated as part of the mitigation system and as part of the DRR measures in place. Warnings can be classified into three different types: permanent, anticipatory and response (as per Day & Fearnley, 2015), that highlight the capability for warnings to work in a timely manner given the lead-time and spatial location of hazard events, as summarised in Figure 5.21.

Paramount to warnings, governance is establishing the leaders, decision-makers and disseminators. Several institutional frameworks exist for warnings at differing levels: local levels via community-based or community-established warnings via groups such as the Red Cross and Red Crescent; national and subnational governments, for example the USGS; and global or regional organisations, for example the United Nations, the World Health Organization and the World Meteorological Organization. It is important that warning systems include clarification on legal responsibility, have clearly defined roles and responsibilities and are integrated into everyday life via various mechanisms.

To be effective, all forecasting and warning should involve the people directly affected, from the initial concept and design stage right through to the operations of warnings. A realistic perception of the risk and prior knowledge of suitable actions will increase response quality. An understanding of local contexts and settings is as important as the precision and accuracy of the information, because of the usual gaps between the technical capacity of the forecast and the ability of a community to respond. Earlier adoption of long-term measures, like retrofitting, and integration of hazard knowledge into regular life will improve acceptance of manifesting threats with more effective responses and actions due to connection

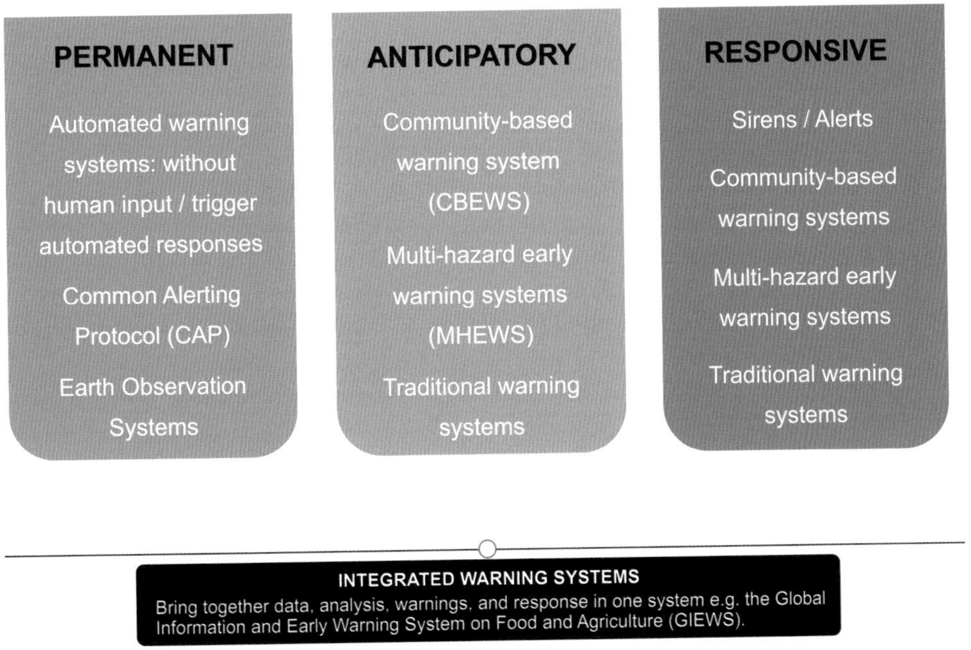

Figure 5.21 Different types of warnings depending on whether they are permanent, anticipatory, responsive or integrated.

Source: Fearnley (2022)

with typical life. Often there is a reluctance to evacuate, with many reasons explaining the logic behind such choices, such as lack of affordability or specific destination, messages and prior exercises failing to specify this action or how to enact it, people preferring to be in familiar surroundings, and because they fear (rightly or wrong) theft and looting of their property.

Decision-making throughout the forecast and warning process is not always easy. Forecasters might have to make quick decisions and can be caught between the choices of issuing a false warning or issuing no warning at all. For instance, tornadoes and flash floods can appear suddenly and be highly localised – or there can be hours-to-days' notice of a high likelihood of a major event across a wide swathe. Earthquake warnings are typically issued automatically after a major tremor, so there is no forecasting decision to be made. They can give people only up to a minute or two to take immediate life-saving action, which is little time to think and decide, meaning that continual training and readiness are essential.

Official forecast agencies occasionally assume little responsibility for their outputs after they have been issued, an attitude sometimes reflecting a wish to avoid legal liability if accused of offering inadequate advice or information. Mistakes can be costly. The erroneous prediction of an eruption of the Soufrière volcano in Guadeloupe in 1976 led to the evacuation of 72,000 people for several months and, for long afterwards, affected the livelihoods of the people who were displaced.

Alert level systems are tools used for a range of different purposes, but in essence they provide a framework to public and civil authorities that can be used to gauge and coordinate response to a developing emergency. They can be ascribed as tools to provide public awareness about both escalating and de-escalating crises. Alert level systems are typically tiered warnings, using either a traffic light system, numbers or terminologies such as 'advisory', 'watch', 'alert' and 'warning' and are typically employed when a hazard or disaster is emerging (i.e. not earthquakes). Pre-planning by everyone should ensure that basic procedures are in place long before they are needed, such as how to disseminate information, identifying the people most at risk and what specific actions to take. There should also be multiple means to distribute messages in case electricity, phone service or the internet go out.

Warnings can take place using a range of tools:

- The Common Alerting Protocol (CAP): implemented globally to provide an international standard format for emergency alerting and public warning via technology and radio (see Figure 5.22).
- Earth Observation Systems (EOS): numerous warning systems for hazards rely on deployed sensors on or near the Earth's surface or on satellites to track changes in the environment, be it a short-term volcanic eruption with volcanic ash, or longer-term impacts such as deforestation or sea-level rise.
- Integrated warning systems: These systems bring together data, analysis, warnings and response in one system as seen in the Global Information and Early Warning System on Food and Agriculture (GIEWS).
- Community-based warning systems (CBEWS): These systems empower people by involving them in the data collection and analysis processes, with communities leading and operating them.

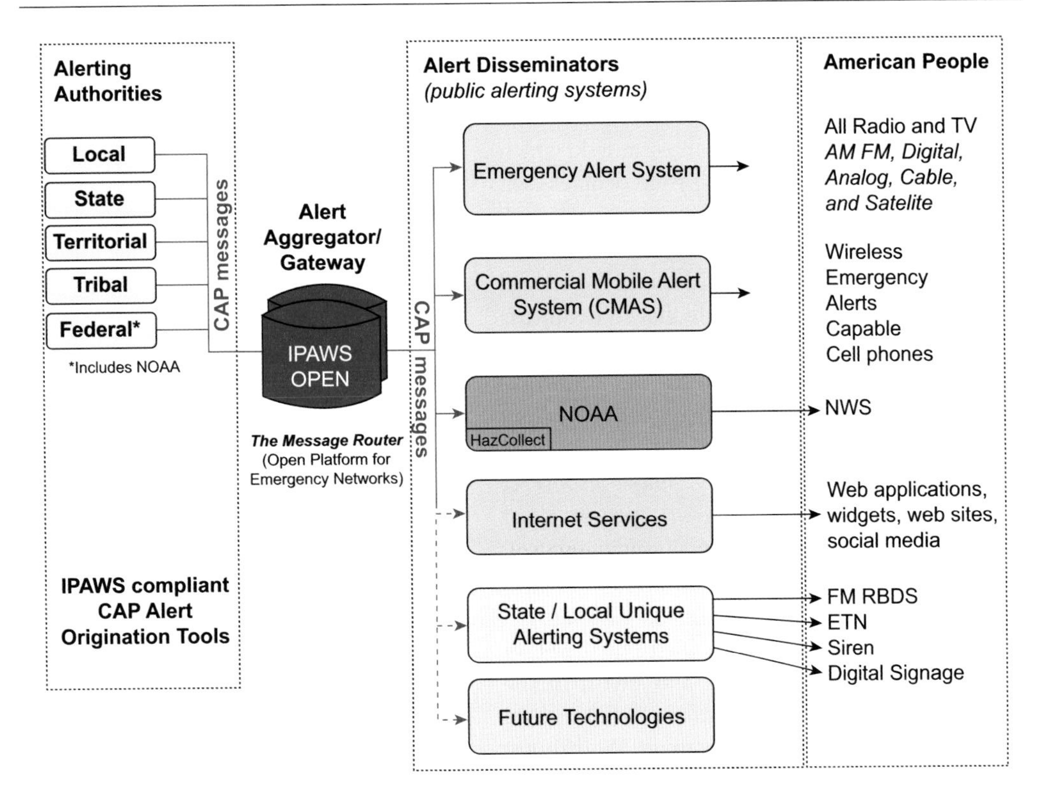

Figure 5.22 The Common Alerting Protocol as implemented in the USA.

Source: FEMA

- Traditional warning systems: These systems incorporate traditional knowledge and observations, often through storytelling, songs and regular conversations about the local environment.
- Multi-hazard early warning systems (MHEWS): These systems facilitate coordination and consistency of warnings for multiple hazards occurring simultaneously or in succession, as differing actions may be required.
- Automated warning systems: Once established these technologically based systems operate without human input to provide warnings based upon pre-assigned criteria and may trigger automated responses (e.g. bridge closures).

Real-time feedback within the system, including an accuracy check with the forecasters and a response check on those being warned, is important. This is because the onward transmission of the message may be unnecessarily delayed, or even halted, by operators seeking confirmation on some aspect. Most warning messages should contain a sense of the urgency, estimate the time before impact, indicate the scale of the event and provide specific instructions for action, including the need for spectators/tourists to stay clear of the hazard zone (Gruntfest, 1987; Mileti & Sorensen, 1990). Continuing advice on the evolving situation, such as weather, transportation and other possible threats, plus notification of the next warning update, is also usually helpful.

The best warnings make the content personally relevant to those at risk, which can often mean coming from people's own social networks, whom they trust already. Although warnings via the mass media are likely to be believed, the initial message may simply alert people that something is wrong, so stay tuned for further information. People almost always seek confirmation of the first warning received before action is taken, hence the advantage of tiered warnings. They might also react adversely to trite phrases such as 'don't panic' and 'everything is under control', so messages tailored to specific audiences can be helpful.

Advice needs to be distributed well in advance, both widely and often, to the public through multiple conduits, since people trust different sources. Workshops, pamphlets, social media, videos and other materials are important tools, especially in combination to ensure that everyone is reached, irrespective of languages, literacy and abilities to hear and see. In addition, public bodies and private sector companies can build hazard awareness into existing health and safety schemes, but any introduction at household level is difficult to monitor.

Warnings tend to fail when they are not integrated, do not consider the context they operate in and are not flexible to manage differing circumstances. Too often, warning failures are seen as needing to close the so-called last mile representing gaps between the origin of warning information and users. Instead, the warning process must prioritise the 'first mile' representing a beginning with users to listen to and respond to their needs and contributions. Good practice examples of integrating many such elements, including the first mile, are hurricane warnings and evacuations for the US and Cuba. Despite very different governance systems, both countries have long had successful monitoring and projection of hurricane tracks followed by millions evacuating to safer ground.

G. RESPONSE

Disaster/emergency response provides immediate assistance to maintain life, improve health and support the morale of the affected population. This can range from temporary shelter and food to establishing semi-permanent settlements in

Box 5.4 Bangladesh Cyclone Preparedness Programme

In Bangladesh, a Cyclone Preparedness Programme (CPP) began in 1973 in order to benefit the millions living in the coastal region. The aim was to ensure that people are adequately alerted about an approaching storm and can move to safety in cyclone shelters or other buildings (see Figure 5.23). Around 2010, the set-up was that the Bangladesh Meteorological Department transmitted warnings to six zonal offices and 31 subdistrict offices over high frequency (HF) radio. This message was passed on to some 274 village level unions by very high frequency (VHF) radio. Team leaders, typically responsible for one or two villages accommodating 2,000 to 3,000

Figure 5.23 Ever since the 1991 cyclone catastrophe, numerous cyclone shelters have been built to provide a safe haven for local people. These buildings are built at a height of 7 metres above ground and are made to withstand a potential storm surge. There is a need for community and government contribution for the construction of flood-proofed access routes to cyclone shelters. The construction of coastal embankments might also serve as a potential protection against the storm surges.

Source: https://people.uwec.edu/jolhm/eh2/molnar/mitigation.htm

Source: © Alamy Images/Neil Cooper

people, then spread out on bicycles to disseminate cyclone warnings door-to-door using megaphones, hand sirens and public address systems. In total, the scheme at that point covered about 3,500 villages through the use of 34,140 trained volunteers, including 5,000–6,000 females. Disaster-awareness training was given by the Red Cross and Red Crescent movement as part of the CPP programme. The system at that point could evacuate over 300,000 people in 48 hours, with demonstrable and substantive improvements. Whilst at least 300,000 people were killed in a 1970 cyclone, a similar tempest in 1991 claimed a much-reduced (although still catastrophic) 140,000 fatalities, but then the reported fatalities in 2007 from Cyclone Sidr were 4,000–5,000. In 2020 and 2021, cyclones killed only dozens each, assisted not only by the now-ingrained culture of cyclone warning, evacuation and sheltering, but also by the advent and prevalence of mobile phones permitting swift communication of cyclone-related information via text message, emails and apps. Given how frequently people trust their social networks more than governments, swift changes in information dissemination and availability can be used – also being aware of the power of misinformation and disinformation which spreads rapidly, is highly trusted and is difficult to counter.

Today the International Federation of Red Cross and Red Crescent Societies (IFRC) provides excellent field guides in relation to community-based early warning systems via their guiding principles brochure (www.ifrc.org/document/community-early-warning-systems-guiding-principles) and their field guide (www.ifrc.org/document/community-early-warning-systems-cews-training-toolkit-field-guide).

camps and other locations, to health and medical supplies, to initial repairs or diversions to key damaged infrastructure. The main focus is to keep people safe, prevent or mitigate against any further imminent disaster (such as aftershocks following an earthquake), and provide the basic needs of the people until more permanent and sustainable solutions can be found.

The emergency period is classed as the first 24 hours to three weeks following a disaster. During this time it is usually the government or governments in whose territory the disaster has occurred that are responsible to address the needs of the affected populations. For more severe events the government may be overwhelmed and declare a state of emergency, or simply request additional aid. Typically, humanitarian organisations are strongly present in this phase, particularly in countries where the government lacks the resources or political will to respond adequately to needs. The Office for the Coordination of Humanitarian Affairs (OCHA) of the UN Secretariat is responsible for coordinating responses to emergencies. It does this through the Inter-Agency Standing Committee, whose members include the UN system entities most responsible for providing emergency relief. OCHA emphasises,

> response efforts will focus on the immediate provision of quality life-saving humanitarian supplies, including ready-to-eat rations and food baskets, basic relief items for the most vulnerable households, including light hygiene and dignity kits, and a series of initial – and largely mobile – emergency protection interventions. Delivery of basic services will be supported through the reinforcement of available service providers relating to sectors including health, WASH, shelter, protection and education in the areas hosting those newly displaced.

Depending on the hazard or threat event, the first element of response by most agencies and the public involved is a warning. This is usually via a national warning system that will deliver messages using a variety of methods. It is imperative that the public know what to do in response to this warning, such as where to evacuate, how to get there and how to communicate with immediate family or vulnerable persons. Commonly, shelter is sought for safety. Once the event has struck, emergency operations respond mostly via search and rescue to try to save the lives of those trapped or affected by a hazard.

1. Evacuation

Evacuation is one of the best ways of ensuring public safety for a majority of disasters. This is dependent on the evacuation being efficiently managed (which involves significant preparedness) and on the hazard. For some hazards like infectious emerging diseases (like COVID-19) evacuation is not effective, in fact the opposite where people are required to stay at home and isolate as required. The process of evacuation is typically triggered by emergency managers based on the threat and risk, and then released to the public. This notification process takes time, and the public also need time to prepare; there is also a response time from when people depart and arrive outside the area that is dangerous or to a designated evacuation centre (that also needs to be set up). Therefore, evacuations ideally require a fair amount of time to implement. The final stage of the evacuation is to return once the emergency is over. This is not possible in some cases where buildings and infrastructure have been completely destroyed (e.g. landslides, volcanic eruptions, where the landscape has been significantly changed).

Evacuations can be enforced by a government or they can be voluntary, although incentives are usually required for the latter; this largely depends on the legal requirements of the disaster management policies in place. Evacuation is not an easy decision to make. Many people worry about their property or business and whether it will be safeguarded. In urban areas, the owners of property and shops might fear looting by those who remain. Farmers may be reluctant to leave their cattle behind as they are their family's livelihood. Some countries such as Costa Rica have devised schemes that evacuate cattle with farmers so as to ensure the safety of the farmers who otherwise might stay behind. Typically, tourists are more willing and able to evacuate (or return home early) than local populations. This also reduces the burden on the local population, especially if it is a tourist destination.

When evacuating, people might be stressed, fatigued, in unfamiliar circumstances or locations, or feeling uncertain, so they drive more poorly. People might compensate by being more careful. People might also drive faster to get away from a hazard-impacted area, plus the amount of traffic can vary, being heavier if many are evacuating simultaneously or being lighter if others are staying put or if they had evacuated earlier – all affecting driving speed and hence collision consequences. In a number of evacuations, buses are provided for those without cars. For tropical cyclones (Chapter 10), many locations open up both sides of a road to traffic travelling in the evacuation direction, providing more space to vehicles, although it can also be disconcerting driving on the 'wrong' side of the road.

The process of evacuation can be distressing. In the case of Hurricane Katrina, many families were split. Six weeks after Hurricane Katrina hit New Orleans and surrounding areas, a Red Cross survey highlighted that at least 100,000 families were still not reunited, including a large number of parents still not reunited with their children (Moore, 2005). Many people suffered hardships such as no access to water or food for a day, no shelter, the loss of a loved one or a pet, injury, subjected to crime and fearing for their life. It is easy in hindsight to ask why people did not evacuate when an evacuation order is issued, but often, as found in the same Red Cross study, people misjudged the event or did not anticipate it would be as bad as it was, had no means of leaving (i.e. money, transport, shelter) or had health issues.

Evacuations can take place in different forms, and typically this is done in a staged approach, evacuating those nearest to the potential hazard first (see Box 5.5).

Evacuees can seek shelter in various forms, whether temporary shelters established by the government or humanitarian agencies, staying with relatives and friends or, for the wealthier, going to second homes or hotels/rental properties. There are challenges, however, in evaluating particularly vulnerable groups of people. For example evacuating hospitals is exceptionally challenging, so it is better to build hospitals in areas that are low risk. Other groups that require more

Box 5.5 Volcanic Evacuation

In volcanic eruptions as seen during the eruption of Mt Pinatubo, the Philippines, in 1991, considerable work was done by the scientists and local authorities involved to evacuate populations. At first, Mt Pinatubo had not been identified as an active volcano by contemporary scientists, so considerable effort was invested in communicating what a volcano is, what hazards they produce and the impact on populations near the volcano. The video titled *Understanding Volcanic Hazards and Reducing Volcanic Risk* used extensive footage caught by Maurice and Katia Krafft to communicate just how deadly volcanic hazards are and to encourage people to evacuate should they need to. An evacuation map was devised as per Figure 5.24 that defined different

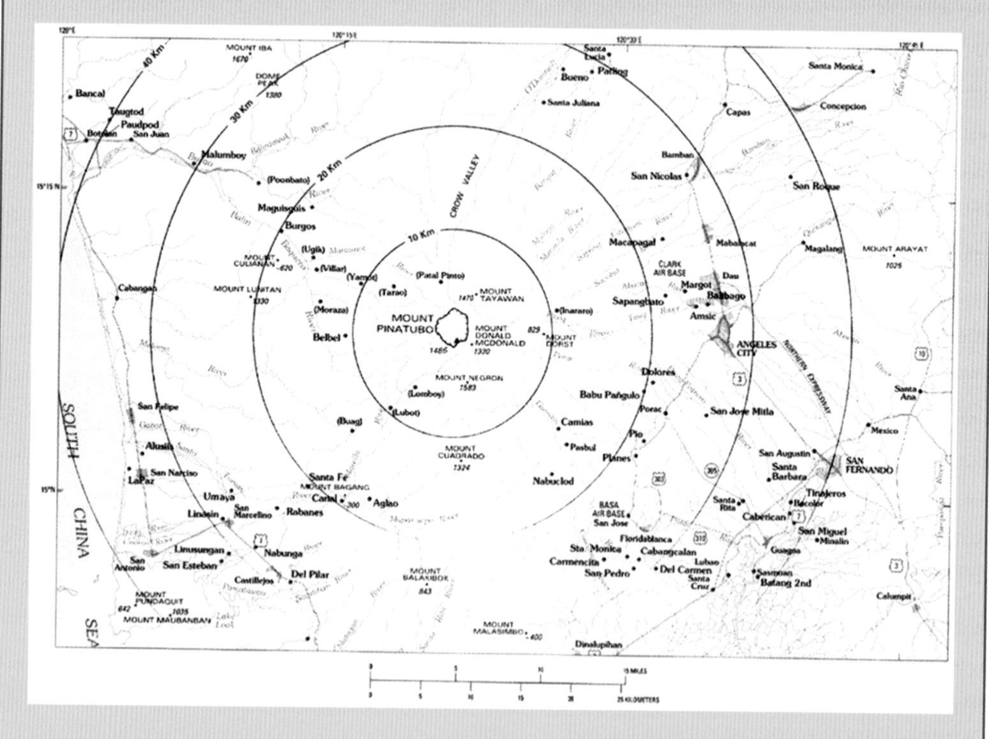

Figure 5.24 Map of the Mount Pinatubo area showing recommended evacuation zones ('danger zones') of various radii, and barangays cited in the text.

Source: Newhall & Punongbayan (1996)

zones around the volcano to be evacuated in stages as the volcanic eruption became more imminent, starting with those within closest proximity to the volcano.

Table 5.5 highlights how the alert level and volcanic activity led to evacuation of the different zones during what was the second largest volcanic eruption last century.

During the cataclysmic stage of the volcanic eruption, Typhoon Yunya also hit the main island of Luzon where Mt Pinatubo was erupting. In the end, hundreds died through direct results of Mt Pinatubo hazards, mostly from roof collapse due to ash mixing with rain, although thousands of indigenous people might have died through inadequate temporary settlements after they were evacuated from the volcano's proximity. The major city of Olongapo within the 40 km zone was not evacuated in the end, as the mayor decided the risk of evacuation was greater than staying put. Whilst this may have been the best decision, it is notable that most of the initial deaths occurred in the city, in the 40 km zone, where over 15 cm of ash fell. The crisis affected over 1 million people, many of whom were saved by the evacuations but did not necessarily escape suffering through inadequate post-eruption support.

Table 5.5 Summary of evacuation stages for Mt Pinatubo

Date	Alert	Volcano activity	Action
5 June	3	Many earthquakes and explosions	10,000 Aetas were taken to refugee camps
7 June	4	Many earthquakes, steam and ash explosions	–
9 June	5	Large amounts of sulphur dioxide gas, small pyroclastic flows	People evacuated within 20 km of Mt Pinatubo
10 June	5	–	Clark Air Base began evacuation of personnel and equipment
12 June	5	Violent eruption with 20 km high columns of ash, pyroclastic flows	People evacuated within 30 km of Mt Pinatubo
15 June	5 Typhoon warning	Typhoon Yunya passed 75 km NE of Mt Pinatubo; Cataclysmic eruption	People evacuated within 40 km of Mt Pinatubo

time and planning include the elderly (in retirement homes or at home), children (whose parents often want to pick their child up before doing anything else) and prisoners (who need to be under tight security). Environment is a big factor in the capability to evacuate. Small islands frequently face difficulties when evacuating all or part of an island as it can require significant logistical support. During the eruption of Mt Soufrière in Montserrat, where over 75% of people on the island ended up abandoning their home, only the northern part of the island remained safe for habitation (see also Chapter 3).

Evacuation lasting longer than a few days is extremely concerning for people. They may fear dispossession and loss of title, or be concerned about livestock or pets, as well as their economic security.

In the case of Christchurch in New Zealand, even 10 years after the destructive earthquake, some buildings, roads and areas are still not reconstructed. This has made it incredibly difficult for the local population, some of whom moved away, while others reside in more temporary housing. Cargo ship containers became used for housing and also shopping areas (e.g. Re:START Mall, see Figure 5.25) to provide temporary infrastructure.

Challenges arise when people are evacuated permanently because either the landscape has changed significantly and is no longer habitable or the infrastructure is irreparable. In 2008 the Wenchuan earthquake in China, for example, devastated 80% of the buildings of the city Beichuan, which was subsequently abandoned. The old city now stands as a memorial, with survivors moved to Yongchang Town. Evacuations generally can increase the number of homeless people and have a severe social, economic and cultural impact on society. For some cultures it is not culturally acceptable to evacuate, or there are issues around gender mixing or mixing of different classes. Some of the biggest challenges of evacuation is when key infrastructure is damaged and not replaced by the government. This can be either due to lack of resources or because a government wishes to use the land in a different way

2. Search and Rescue

Search and rescue is the search for and provision of aid to people who are in

Figure 5.25 Re:START in Christchurch, NZ, opened in October 2011 following the devastating earthquake.

Source: © Alamy Images/Nigel Spiers

Figure 5.26 Rescue workers recover a body from rubble created by the Mw = 8.0 magnitude earthquake at Hangwan, Sichuan Province, China, in 2008. About 5 million people were made homeless in this disaster.

Source: Photo: Panos/Qilai Shen QSH01973CHN

distress or imminent danger. These include services such as mountain rescue (e.g. avalanches, debris flows); ground search and rescue (e.g. earthquakes), including the use of search-and-rescue dogs; urban search and rescue in cities (e.g. landslides, flooding, cyclones); combat search and rescue in areas of conflict; and air-sea rescue over water (e.g. for tsunami or storms). The International Search and Rescue Advisory Group (INSARAG) is a UN organisation that promotes the exchange of information between national urban search-and-rescue organisations.

One of the most common tasks of search-and-rescue teams is the search for people injured or trapped following the collapse of a building due to a hazard event. Search-and-rescue teams have developed a standard international system for marking houses that have been searched following a disaster to help speed up and streamline the process. The 'golden hour' is the period immediately after a disaster or injury during which there is the highest likelihood that prompt medical and surgical treatment will prevent death. Whilst initially this was defined as an hour, the exact period depends on the injury and environmental conditions such as temperatures and weather conditions. After an hour, survival rates fall off dramatically. Given the limited timeframes, search and rescue have devised a scheme to optimise recording whether a building has been searched, as shown in Figure 5.27 and then exemplified in Figure 5.58.

The UN usually decides to call off search-and-rescue attempts between five and seven days after a disaster, once no one has been found alive for a day or

Figure 5.27 FEMA search-and-rescue markings.

Source: UrbanSurvivalNetwork.com

two. However, people have been known to be rescued alive beyond this point. In May 2013 a woman was pulled from the ruins of a factory building in Bangladesh 17 days after it collapsed. She was found by using video and audio detection equipment to locate her exact position after someone heard her crying out 'please save me'. Coordination between the various search-and-rescue teams and the agencies working on relief is complex and typically managed using the UN's cluster systems (see Figure 5.29). The cluster approach was applied for the first time following the 2005 earthquake in Kashmir.

New technologies have been critical in assisting search and rescue, as well as assessing and restoration following an event. The use of unmanned aerial vehicles (UAV) – drones – are exceptionally useful for assessing the damage and identifying where people may be trapped and need urgent assistance. This also helps the emergency managers coordinate their response. Increasingly people are being found by people using their mobile phone to call for help, or via their GIS in their phone. There are a number of apps that can be downloaded to help contact people during a crisis, including key phone numbers, alongside smart watches that have an emergency function. More broadly the use of the Facebook 'Safety Check' function enables family and friends afar to check on loved ones, without having to clog up phone lines and satellite data that could otherwise be used to provide the immediate response.

Figure 5.28 FEMA markings show a home that was searched twice and let emergency services see at a glance that the home was searched, when it was searched, by whom and what they found.

Source: Photo: Ilan Kelman

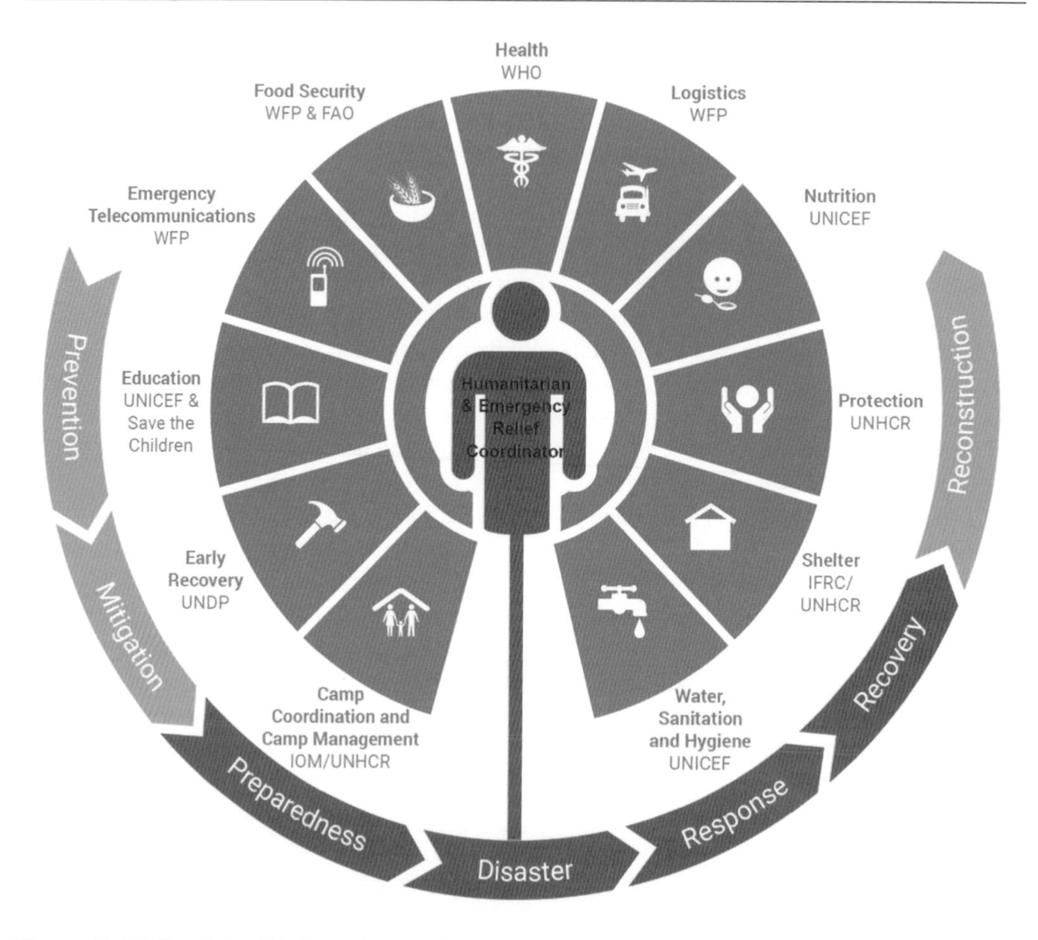

Figure 5.29 The United Nations cluster scheme.

Source: OCHA (2020)

3. Assessing and Restoration

Following a disaster, damage assessments record the extent of damage and what can be replaced, restored or salvaged. They may estimate the time required for repair, replacement and recovery. Damage assessments are needed to facilitate effective and efficient response by government agencies and other organisations, helping with response and relief operations. GIS and maps are frequently used during this process to provide large-scale perspectives on the damage, alongside available resources. More detailed assessment can be followed by various groups such as the Earthquake Engineering Field Investigation Team (EEFIT), a joint venture between industry and universities, conducting field investigations following major earthquakes (see more at www.istructe.org/get-involved/supported-organisations/eefit/). Amongst a number of tasks they aim to 'assess the effectiveness of earthquake protection methods, including repair and retrofit, and to compare actual performance with designer expectations'.

During the response, most effort is invested in the continued assistance required to get people to safety and the immediate restoration or construction of infrastructure. This includes the building of temporary bridges or storm drains, diversion dams and roads to enable relief to be distributed. Following the 2004 earthquake and tsunami in Indonesia, for example, numerous populations were cut off due to the destruction of railways, bridges and roads, making the distribution of vital aid exceedingly challenging.

For some crises the response is ongoing, with populations unable to move towards a recovery stage. COVID-19 presents a number of long-term challenges with various waves of the virus and differing strains emerging that meant a period of 'normalisation' of living under COVID-19. To manage such ongoing crises, alert levels can be used to change the various mitigating and responsive actions required by the public, moving between higher and lower levels as required. Alert levels like these are employed for health issues, volcanic alerts, wildfires and famine/desertification.

H. RECOVERY

Disaster recovery involves a set of policies, tools and procedures to enable the recovery or continuation of society following a natural or human-made threat. Most commonly, recovery is achieved via disaster aid, either nationally or internationally, to assist the affected areas. Typically, disaster management addresses four stages:

- *Relief.* This period covers the first hours or days after the disaster. The rescue of survivors is followed by the distribution of basic supplies (food, water, clothing, shelter and medical care) to ensure no further loss of life.
- *Rehabilitation.* This happens in the first few weeks or months, when the priority is to start the community functioning again. An early and expensive task is the removal of disaster debris, such as building rubble blocking roads or food spoiled due to power failure.
- *Reconstruction.* This is a much longer-term activity designed to recreate 'normality' after devastation. It includes improved disaster preparation for future resilience, e.g. construction of hazard-resistant buildings and better land use planning.
- *Learning review.* This process allows the dissemination of greater disaster awareness amongst all stakeholders in the community and the training of emergency workers for the future.

Yet, reconstruction can itself produce new forms of risk. Disaster areas are seen to be recoverable in particular ways, such as by actively removing business regulations so as to allow for investment. But, of course, such an approach sets up other forms of hazardous conditions, such as pollution, worker exploitation and further inequalities, inequities or discrimination.

1. Disaster Aid

The main purpose of disaster aid is to prevent further mortality and morbidity while providing shelter, social support, subsistence and longer-term security for the affected population. Full disaster recovery rarely happens quickly and might never really be achieved if major changes have resulted or if the disaster is used as an opportunity to enact major changes. Stages of recovery are neatly partitioned into phases on paper that rarely occur in practice as per Table 5.6. Instead, they overlap with similar aspects continually weaving in and out, and unfortunately many key tasks are not fulfilled or are completed superficially.

In order to prevent additional loss of life, clinical support and medical supplies must be delivered to disaster-affected people within the 'golden hour' immediately after the hazard. This period can be extremely short. Almost 90% of earthquake casualties brought out alive from collapsed buildings are rescued in the first 24 hours. International donations and teams typically arrive too late to help the most needy. A classic example was the Guatemala City earthquake of 1976, where the peak delivery of medical supplies came two weeks after the disaster, when most casualties had been treated and hospital attendance had fallen back to typical levels (see Figure 5.30).

After initial search and rescue while attending to immediate needs of injured people, including for mental health, the priority is for shelter, food, water and more sustained medical care and welfare. Some hazards, such as floods, can be associated with diarrhoeal, respiratory and infectious diseases, whereas earthquakes often yield fracture and crush injuries. The use of local medical teams is preferred because they can be mobilised quickly and are culturally integrated with the survivors. This can be achieved only by preparedness. The rebuilding of lives and livelihoods includes everything from psychological counselling to practical help using family and other social networks. It is important to monitor community morale, especially for those transferred to temporary collective shelter such as camps or common buildings,

Table 5.6 Chronological disaster aid that can vary in timescales significantly

Emergency period:	Search-and-rescue operations (golden hour)
24 hours–3 weeks	Emergency shelter
	Community care (medicine, food, water)
	Evacuation of survivors from unsafe areas
	Identification of the dead and dealing with the bodies
Relief period:	Debris removal and refuse disposal
2 weeks–6 months	Clean-up of any toxic materials
	Restoration of water and energy supplies
	Demolition of unsafe buildings
	Return of evacuees to homes declared safe
	Distribution of emergency funds
	Assessment of overall damage
Recovery period:	Return of livelihoods and other socio-economic functions
5 weeks–10 years or more	Permanent repairs to infrastructure
	Preservation of heritage sites
	Evaluation and update of local plans
	Long-term planning for sustainability
	Implementation of hazard-resistant measures

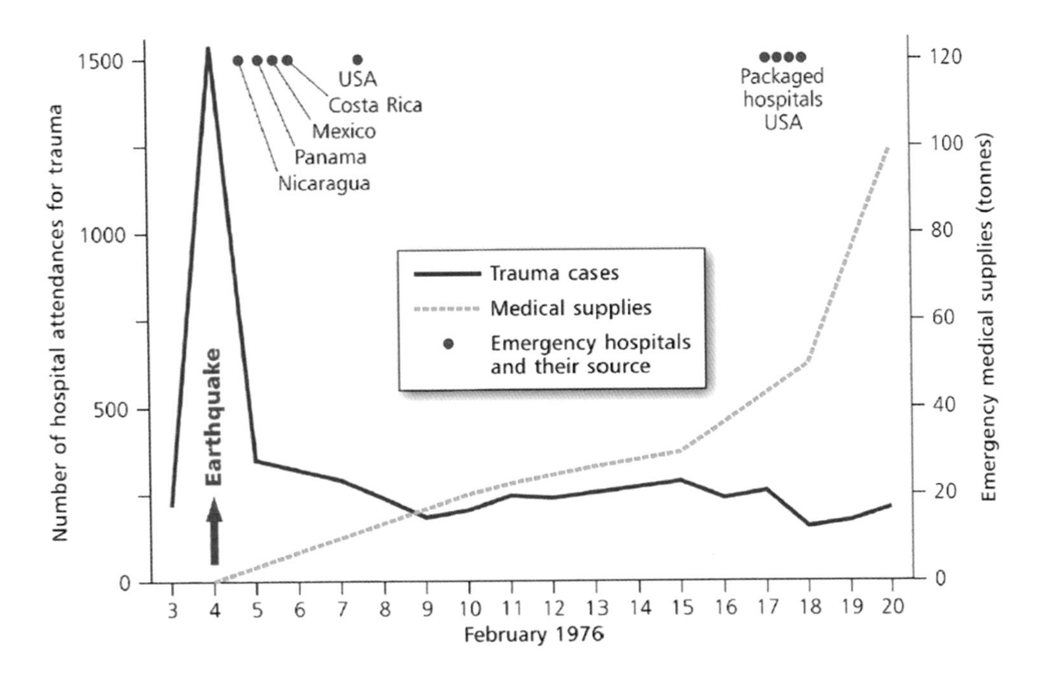

Figure 5.30 The daily number of disaster-affected people attending hospitals in Guatemala City in relation to the arrival of medical supplies and emergency hospitals after the 1976 earthquake.

Source: After J. Seaman et al., Epidemiology of natural disasters. In M.A. Klingberg (ed.) *Contributions to Epidemiology and Biostatistics*, vol. 5 (1984), 145–156. Copyright 1984 S. Karger, Basel. Reproduced with permission.

and to ensure that survivors are empowered with roles in decision-making for the future. Special attention should be given to addressing everyone's needs, including children, the elderly, people with animals, people with disabilities and typically marginalised groups such as many minorities, people who are homeless, prisoners and undocumented people.

The relief phase begins a dynamic process of recovery with no clear end point. Emergency aid donations enable some people to return home, provide limited water and energy supplies, restore links to the outside world and enable the restart of urban business functions. This phase typically merges into a much longer period of rehabilitation. The basic aim is to replace lost housing and infrastructure so that former livelihoods and conditions can not only be restored, but also improved.

Shortages of people, skills, building materials and other factors can impede recovery. Progress can occur more quickly if a pre-disaster response plan exists and adequate external finance is available. However, time is required to involve residents and other stakeholders in the many decisions to be taken, especially if the opportunity is grasped to plan for post-disaster improvements.

Over a protracted recovery period, emergency aid and development assistance blend together and it becomes difficult to measure the real effectiveness of disaster appeals and responses. In addition, disasters often form part of more complex failings that require long-term solutions in health care, education and social welfare. Many aid donors recognise such needs. Donors might try to make strategic investments that foster what they see as democratic and representative institutions, in order to support local action. Charitable bodies increasingly stress the need for ongoing disaster prevention rather than short-term aid.

Most humanitarian aid is raised by emergency appeals following disaster, which begs the question why people are more willing to donate when it is too late than to support fundamental DRR. This remains a challenge as it has been proven time and time again that investing one clam (or any other currency) in preparedness and risk reduction saves many times over that amount, and usually quickly (Shreve & Kelman, 2014).

As illustrated in Figure 5.31, various forms of support travel through a complex network of sources and intermediaries before they reach the recipients. *Governments* at several levels provide funding through the UN, the International Red Cross and Red Crescent Movement, NGOs, public–private partnerships and the government of the affected country. The *United Nations*

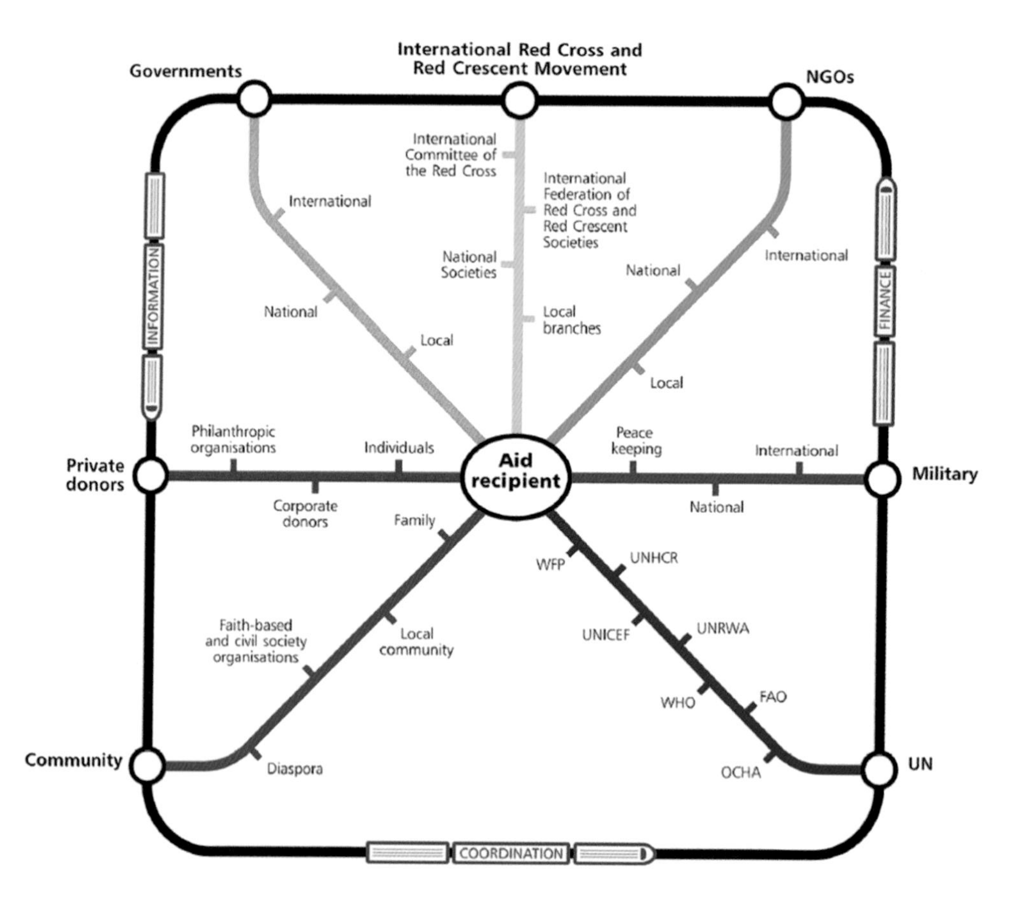

Figure 5.31 An overview map of the aid players involved in humanitarian emergencies. The diagram does not intend to suggest linear connections or funding relationships.

Source: Scott, R. (2014, p. 8)

manages some funding, provides direct support to aid recipients and also channels money to other delivery agencies (including NGOs) who are responsible for project management. *NGOs* provide direct support to aid recipients either through their international arms or via national or local offices. Individual charities sometimes group together to optimise disaster appeals. For example, the Disasters Emergency Committee (DEC) formed in 1963 as an umbrella organisation which currently coordinates the efforts of 15 UK aid agencies. *Private donors* may provide funds to all these bodies or pledge support directly to affected communities and individuals. *Military assistance* is a specific form of government aid, required when logistical transport or a peacekeeping role is required in the post-disaster period.

2. Domestic Aid

The US illustrates some of the challenges associated with domestic disaster aid. The president can issue formal disaster declarations (PDD) which should be accompanied by needs or damage assessments, but this procedure can be short-circuited in the interests of political expediency, especially when media pressure exists (Sylves, 2007). A PDD routinely releases enough federal aid to cover up to 75% of the costs of repairing or replacing damaged public and non-profit facilities, although this proportion has reached 100% in some cases. As shown in Figure 5.32, the number of PDDs has risen steadily over time. The procedure has been criticised as unduly political in nature and for failing to meet real needs. Schmidtlein et al. (2008) demonstrated a weak statistical relationship between the

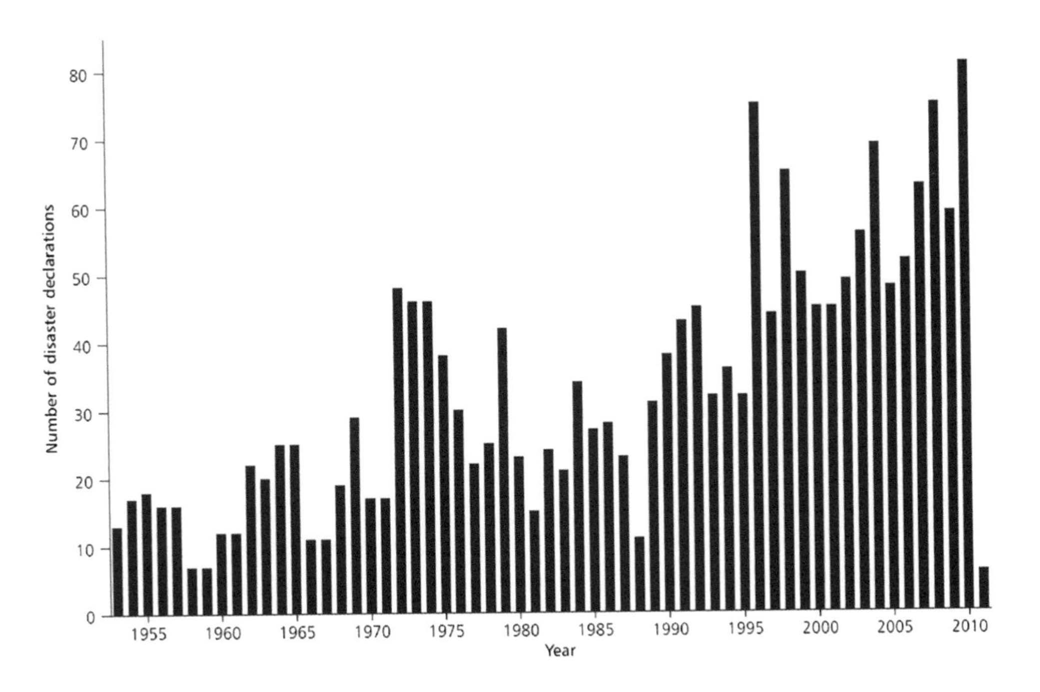

Figure 5.32 The annual number of presidential disaster declarations in the USA 1953–2011. There has been a progressive rise in declarations over the past half century.

Source: Data from FEMA at www.fema.gov/news/disaster-totals-annual.fema (accessed 2 March 2011)

annual number of PDDs and major hazard events across the country and claimed significant regional inequalities in the distribution of aid. It was suggested that more emphasis should be paid to the relative impact of events and that only severe events should qualify for support.

This upward trend has fuelled a debate on the extent to which federal funds should provide disaster assistance. In debates mirrored around the world, the efficacy of disaster aid from all sectors has long been questioned (Campbell & Cuny, 1983) given that it is:

- *Expensive* because of the rapid rise in appeals and payouts and because post-disaster costs are typically far higher than expenditures for loss prevention and mitigation.
- *Inefficient* because it allows those responsible to avoid a fair share of the costs, e.g. through government failure to enforce building codes, private sector failure to adhere to regulations, or insurers (for-profit, not-for-profit and government) not running schemes with disaster risk reduction incentives.
- *Inconsistent* because equivalent losses are not always treated in the same way, e.g. localised damage may not attain disaster-area status, thereby depriving disaster-affected people of assistance.
- *Inequitable* because it allows a misallocation of resources, e.g. when wealthy disaster-affected people are compensated by the general taxpayer or end up better off while already marginalised populations receive little.
- Consequently, many continue to question the success of disaster aid in terms of really helping people in need to deal with and to prevent disasters. Practical reforms examined and implemented include refusing disaster payouts to households over higher-income thresholds; tying aid to the enactment, monitoring and enforcement of building codes and planning regulations; ensuring that a proportion of emergency aid goes to prevention; and using the aid to push for wider social reforms such as gender equity, including people with disabilities, and ending violent conflict. No single formula or approach works, leading to continued examination of humanitarian ethics (Ahmad & Smith, 2018) not just for government aid, but also for international aid.

3. International Aid

Before the creation of the International League of Red Cross and Red Crescent Societies (now the International Federation of the Red Crescent, IFRC) in 1922, the transfer of aid was largely *bilateral* (directly government to government or indirectly through NGOs). As charitable bodies became more interested in overseas work, more agencies were set up, including the UN Children's Fund (UNICEF) in 1946 and the FAO World Food Programme (WFP) in 1963. In 1972, the UN established the Disaster Relief Organization (UNDRO), based in Geneva. This initiative was reinvented several times because of underfunding, internal rivalry with other UN agencies and criticisms from some member countries. In 1992, a new Department of Humanitarian Affairs (DHA) replaced UNDRO; in 1997 the DHA was itself replaced by the Office for the Coordination of Humanitarian Affairs (OCHA).

Key operating principles for international aid have been shifting from a viewpoint that external charity can and should save the afflicted to a positioning that it is only a small part of what is needed for giving people opportunities and resources to help themselves. In particular:

- Responsibility for people affected by emergency lies first and foremost locally, scaling up as needed when the smaller scale proves to be inadequate.
- Those receiving aid facilitate the work of responding organisations but are not subservient to them, instead having significant say, but not entire control, over what forms of aid are accepted, how much and distributed to whom.
- Humanitarian assistance often adopts broad principles such as humanity, neutrality, independence and impartiality, although the meaning, realism and appropriateness of these and other principles is continually questioned.

(Terry, 2002)

Much financial support from country governments comes from their Official Development Assistance (ODA) budget, although it can be masked in other forms of assistance such as trade, equipment, personnel such as consultants, and cultural or sports exchanges. Disaster relief as a proportion of ODA has varied significantly year to year and sometimes is included entirely separately from ODA. In many circumstances, the forms post-disaster assistance and ODA do not seem amenable to these tasks, such as providing military aid or giving tax breaks to multinational corporations to establish themselves in disaster-affected locales. Earlier assumptions were that aid is transferred from 'richer' or 'developed' countries to 'poorer' or 'less developed' countries, but this standard model has long been overturned as countries such as China, India and Russia take more leadership and donor roles, as well as the rise of individual and corporate philanthropists.

One of the largest global ODA projects at the moment, with some links to disaster aid, is China's Belt and Road Initiative (One Belt, One Road) seeking to gain legitimacy on the international stage and gain loyalty from countries around the world. After Fidel Castro took over Cuba in 1959, the country became an international donor of medical personnel, expertise and equipment, often using this medical diplomacy to gain allies and to trade for needed goods such as oil (Figure 5.33). During the COVID-19 pandemic, China and Russia made a point of providing aid in the form of medical equipment and then their own vaccines. They later became quieter, especially as their vaccines had less effectiveness than claimed, with Russia's invasion of Ukraine in February 2022 curtailing Russia's ability to be a donor.

Such responses to individual disasters vary greatly. The Indian Ocean tsunami of 26 December 2004 created an unprecedented disaster aid response, especially from voluntary bodies and private donors. India, despite having some devastated coastlines, at first tried to decline all offers of international assistance, instead positioning itself as a donor to other affected countries. In the end, India relented and accepted foreign aid. The tsunami disaster attracted great media interest, partly because the affected areas were familiar to donors through holiday experiences in that part of Asia and because so many international tourists were affected.

In the prior example, emergency relief was not really limited by finance, and a lot of the money was ostensibly reserved for reconstruction projects extending over several years. As is typical, many such projects did not always happen or

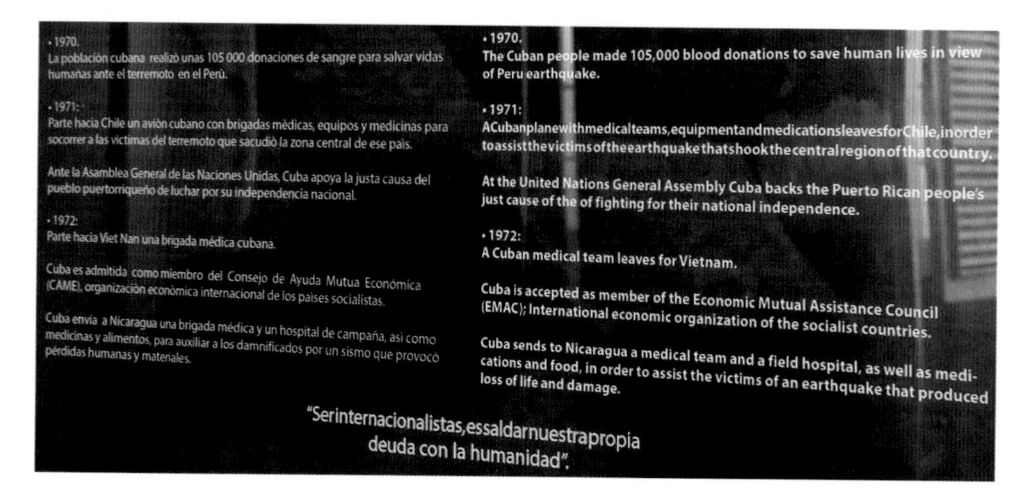

Figure 5.33 Cuba promoting its medical diplomacy in Havana's Museum of the Revolution (Museo de la Revolución).

Source: Photo: Ilan Kelman

hurt the disaster-affected people even more, such as forcibly grabbing their land and homes to permit the construction of luxury resorts while undermining local livelihoods such as fishing. Another key issue in Indonesia during the tsunami aftermath was the logistical challenges of distributing significant amounts of aid that were stockpiling at key airport facilities, unable to be distributed to those who needed it the most. The aid response to the 2004 Indian Ocean tsunami was unprecedented for a natural hazard, with a colossal $6.25bn donated to a central UN relief fund assisting 14 countries.

Comparing disasters, and evaluating the efficiency of disaster aid, is difficult, especially when substantial funding occurs outside the formal appeal system. There is no central monitoring arrangement that covers all donations and tracks money through the system. Greater transparency is required on the delivery of pledges, but the most significant omission is the lack of feedback on what has actually been received directly by people, from whom, and when. Frequently, aid pledges made with great fanfare are not delivered or the aid boomerangs back to the donor country to hire their own consultants for more assessments, reports and plans which are never used.

The World Bank commits significant volumes of crisis response funding within 18 months of a crisis emerging. Across six crises in 2021, the World Bank committed US$1.9 billion, representing 43% of the overall humanitarian and development funding for the response. However, World Bank crisis funding is disbursed much more slowly than funding from other sources, impacting speed of recovery.

Figure 5.34 shows the donor response during the first 17 weeks for four disasters that occurred between 2004 and 2010. The large increase for the Indian Ocean tsunami on the 75th day was due to the coincidental reporting of US$1.3 billion donations, mainly from private sources (Kellett & Sparks, 2012). It is impossible

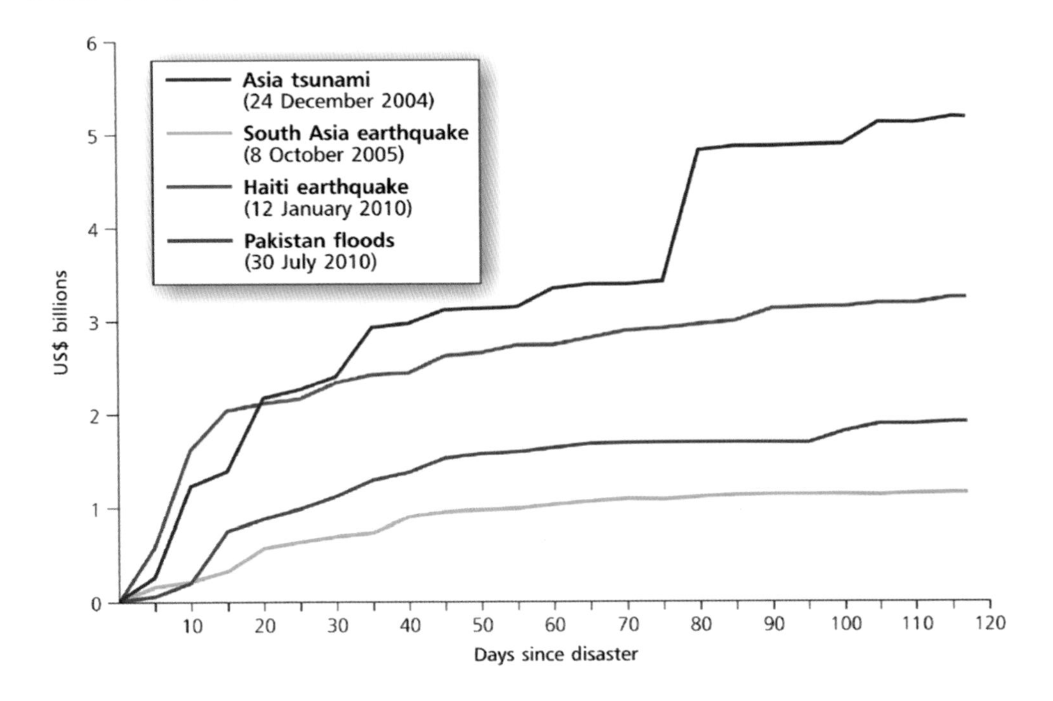

Figure 5.34 Cumulative donor response to appeals for aid in the periods following four major disasters. Generally the rate of funding declines with time.

Source: Data from OCHA FTS. Values in US$ billion. Global Humanitarian Assistance at www.globalhumanitarianassistance.org (accessed 27 March 2011). Reproduced with permission from Global Humanitarian Assistance.

to know how well these funds met local requirements or, more generally, how much money per person is necessary for equitable disaster relief. Even when funds are measured against the number of people affected by disaster, that term tells little about the degree of assistance required. The differences can be striking. The aid donated for the 2004 Indian Ocean tsunami amounted to over US$2,500 per person affected compared with US$110 and US$878 for each person affected by the 2010 Pakistan floods and the 2010 Haiti earthquake, respectively. This does not mean that the money actually reached the people affected in these amounts or at all.

Olsen et al. (2003) discuss how the scale of donations is affected by the intensity of media coverage, the degree of political interest and the strength of the international relief agencies in the country concerned. Sudden-onset hazards, like earthquakes and tropical cyclones, tend to attract more funds than slow-onset ones, like droughts and famine, irrespective of the number of people needing assistance. Journalists usually neglect the 'hidden' crises that arise from poverty, disease and discrimination in order to concentrate on events with large body counts and good photographic opportunities (Ross, 2004; International Federation of Red Cross, & Red Crescent Societies, 2006). Despite the promotion of principles such as neutrality and impartiality, disaster aid is highly political, depending on the priorities of the aid agencies and on the funds they receive from donors or the public (see Figure 5.34). Drury et al. (2005) showed that the most important long-term influence on the allocation of US aid for overseas disasters was foreign policy. In European countries, disaster aid is raised most readily for former colonies, with a subtext of continuing postcolonial influence.

Food aid began as a means of offloading surplus agricultural production in North America and Europe. Many food donations, unacceptable for local religious or dietary reasons, have been sent as aid along with outdated and clinically unusable drugs or entirely inappropriate material such as winter clothes for purely tropical climates. The US has even used disaster relief to try to force food with genetically modified organisms onto countries which had banned them. Over-generous donations of food aid can lower market prices and disrupt the local economy in the short-term while, in the longer term, deflect the receiving government from developing the local agricultural economy and make populations shift toward dependency on external supplies. Logistical difficulties, such as poor roads and lack of suitable transport, hinder supply distribution while other delays occur through bureaucracy and corruption, including syphoning. In some famines, food aid has rotted in depots as people starved nearby, because duties were not paid, certain people did not get their cut or the authorities were using the food aid as a weapon in a long-standing conflict over rival groups. Recurrent dilemmas for aid workers are whether to distribute more supplies to fewer numbers of affected people or fewer supplies to high numbers of affected people, and how much to look the other way when aid disappears or is partly withheld because at least some is reaching those in need.

How can disaster relief become more effective and really help people? Better local information and analysis would enable food aid to be delivered more efficiently to those with the greatest need. Quick and careful identification of at-risk groups is essential. This could be achieved through early warning systems, especially for disasters involving food shortage. More training for staff and volunteers, plus the retention of experienced aid operatives, would help, especially for quicker and more effective responses and especially soon after a major incident. Above all, donors and aid agencies need to move away from stereotypes of 'victims' and 'failed states' in order to be prepared to give more ownership of disaster relief to regional agencies and local communities, with external oversight to avoid many of the problems mentioned. This is partly an issue of respect, but it can also be highly practical when outsiders lack expertise surrounding local needs.

There are alternatives to the traditional, resource-driven distribution of aid in the form of commodities like food, blankets, medicine and shelter materials. So long as safeguards are in place to restrict corruption and other crime, there is growing support for better decision-making regarding *cash-based aid* allowing people to buy the goods they want compared to goods and services being provided directly (Castillo, 2021). Cash-based interventions can provide more dignity and flexibility for disaster-affected people and also liberate aid from donor-driven priorities that may distort receiving economies. *Cash-for-work schemes* that employ otherwise unskilled and semi-skilled workers provide disaster recovery employment to improve the urban and natural environments, and to generate livelihoods, also represent an option, as tried in the Philippines after Typhoon Haiyan (Yolanda) in 2013 (Eadie, 2019).

I. EVALUATING DISASTERS

This chapter has provided an overview of DRR and has highlighted numerous challenges. This final section reviews how to determine if DRR is beneficial, effective and worth the investment.

As discussed in Chapter 1, disasters can cause huge economic losses, and frequently the case for DRR measures is made in economic terms, using a cost-benefit analysis. This enables the justification of investment in different types of DRR work. On top of this, agencies and donors want evidence that their projects and programmes are delivering value for money. Whilst there is no standard methodology for conducting cost-benefit analysis, they are increasingly being used. Often cost-benefit analysis is used at the start of a project to try to analyse which methods would be the most beneficial within the financial constraints available. Whilst some projects deliver exceptional benefits for the investment (Shreve & Kelman, 2014), it is difficult to capture consistently and comparatively the true costs and savings involved. Typically, savings are expressed in monetary values, but it is possible to try to incorporate qualitative aspects by reviewing local populations' perceptions on the costs and benefits to their livelihood. Quantifying these aspects is hard and might not be useful, especially when looking at social aspects, including mental health and well-being issues, and environmental issues.

The concept of placing monetary value on lives, livelihoods, health and well-being raises ethical concerns. Calculations based on physical structures (such as buildings, infrastructure) and economic aspects (employment, savings, agriculture) are fairly straightforward. Data obtained can have all the challenges as discussed in Chapter 2, making it unrealistic, particularly in low-income countries. Also, cost-benefit analysis works better for short-term outcomes than longer-term trends. Knowing how much a particular DRR activity has saved during a crisis, should it be successful, is also challenging as we would not know what their impact would have been aside from using models and existing data. In summary, whilst cost-benefit analysis can be useful, it must be taken with caution as it is not possible to consider all the costs or benefits or the challenge of multiple hazards occurring at the same time.

The issue of devising multi-hazard DRR actions is of increasing global interest, particularly for early warning systems. The World Meteorological Organization (WMO) has produced a checklist on how to develop multi-hazard early warning systems and there have been a series of recent conferences focused on the topic, but the reality is that whilst such systems are necessary as many hazards can occur at the same time, it is very difficult to implement and operate due to existing silos between different monitoring agencies, emergency managers and varying stakeholders, and a current understanding by the public of particular systems. There is still considerable work needed to develop better multi-hazard DRR action but their potential remains significant.

Investing in disasters is challenging for reasons outlined in Chapter 1, but sometimes investment following a disaster can be manipulated. Klein (2007) describes the brutal tactic of using the public's disorientation following a collective shock from a disaster to declare a moment of 'extraordinary politics' for pushing through controversial and questionable policies (often radical pro-corporate measures), often called 'shock therapy', while citizens are too distracted (emotionally and physically) to engage and develop an adequate response and resist effectively. An example is during the 2004 Indonesian tsunami when coastal villages hit by the Indian Ocean tsunami are being cleared under the guise of safety, only to make way for tourist development, according to relief agencies operating in the areas. Not only does this action affect the displaced population, but also it places a whole new set of people at risk.

Nonetheless, caution is needed in the articulation of 'shock doctrine'. First, the shock doctrine and window of opportunity exist, but there are layers and nuances deserving of a more academic approach to its articulation. Overall, disasters in general do not typically create new change. They sometimes catalyse ongoing, existing processes where the preconditions are right.

Other political and ideological concepts produce disasters as much as capitalism does:

- *Disaster socialism*: The practice of taking advantage of major disasters to promote socialist policies that might be less acceptable without a disaster. Disaster socialism notes the transformative potential of highly visible and deeply felt disasters and the opening up of an opportunity for people to recognise the need for radical societal change if a repeat disaster is to be avoided. Radical change could be the introduction of socialist policies such as expanded public health programmes, as well as land redistribution. Disaster capitalism is arguably the counterpoint, being the retrenchment of unequal economic relations. What makes the phrase 'disaster socialism' challenging, however, is that, as with so many concepts addressed in this book, it has been reinterpreted by many.
- *Disaster communism*: Sometimes, this phrase refers to the efforts of communities hit by disaster that do not wait for the state, capital or other external intervenors to take the initiative, but instead 'negotiate with their hands', rebuilding their own communities and 'healing themselves', resulting in communities that are stronger. This phenomenon might be better referred to as 'disaster communalism', which is a long-evidenced point in disaster research that the first people to enact rescue, recovery and reconstruction are those directly affected. Disaster-affected people do not wait around as helpless 'victims' waiting for others to turn up. Instead, they typically do as much as they can with what they have to help themselves and each other. At other times, disaster communism is the political ideology of communism producing and failing to stop disasters, such as the millions killed by famines created by Joseph Stalin in the USSR and Mao Tse-tung in China.
- *Disaster centrism*: It is not just political extremes on the left, right or other political axes that can lead to disasters involving environmental hazards. Policies from centrist and cross-spectrum coalitions can create and maintain similar vulnerabilities as other political ideologies.

The key is to examine how policies and practices do and do not impact vulnerabilities as well as environmental hazards, rather than assuming that a particular political stance must either inevitably succeed or inevitably fail.

Whilst DRR measures can be hugely life saving, they always occur in political contexts. There may be numerous agendas behind their financing, implementation and planning. DRR remains a political field where significant research is required to understand better how multiple hazards can be managed at the same time, and how differing strategies can be designed for DRR across space and time scales.

KEY TAKE AWAY POINTS

- Disaster risk reduction (DRR) is a vital process that brings together the numerous and diverse range of activities to reduce vulnerabilities and disaster risks. These are typically classed under the four categories of mitigation, preparedness, response and recovery.
- These four categories relate to one another in a non-linear manner and require a holistic perspective to provide a system of strategies that are more effective.
- Warnings are a critical tool that bring many elements of DRR into practice, and warnings have been highlighted as one of the single most beneficial elements of DRR to help reduce loss of life and socio-economic impact. However, warnings must engage the users as part of the first mile and facilitate multi-directional communication to make sure warnings are received, understood, believed and acted on.
- Cost-benefit studies are useful to justify increasing investment in anticipatory or early action, yet are challenging to conduct. However, examples consistently demonstrate that preparedness and early action are key to the successful reduction of disasters, rather than the costly and often untimely response to disasters that can generate significant suffering.

BIBLIOGRAPHY

Ahmad, A., & Smith, J. (2018). *Humanitarian action and ethics*. London: Zed Books. https://www.bloomsbury.com/uk/humanitarian-action-and-ethics-9781786992703/

Benson, C., & Twigg, J. (2007). Tools for mainstreaming disaster risk reduction. *Guidance Notes for.*

Born, P., & Viscusi, W. K. (2006). The catastrophic effects of natural disasters on insurance markets. *Journal of Risk and Uncertainty, 33*(1), 55–72.

Campbell, J. R., & Cuny, F. (1983). *Agricultural development and disaster preparedness*. Honolulu, HI: Pacific Islands Development Program, East-West Center.

The Canterbury Earthquakes Royal Commission. (2012). Volume 6: Canterbury television building (CTV) (with foreign language translations). *Royal Commission of Inquiry into Building Failure* [Online]. https://canterbury.royalcommission.govt.nz/Final-Report-Volume-Six-Contents. Accessed 25 July 2022.

Castillo, J. G. (2021). Deciding between cash-based and in-kind distributions during humanitarian emergencies. *Journal of Humanitarian Logistics and Supply Chain Management, 11*(2), 272–295.

Center, A. D. R. (2005). *The concept of total disaster risk management*. http://cidbimena.desastres.hn/docum/crid/Abril2006/CD2/pdf/eng/doc16293/doc16293-2a.pdf

Chen, C., Liu, C., & Lee, J. (2022). Corruption and the quality of transportation infrastructure: Evidence from the US states. *International Review of Administrative Sciences, 88*(2), 552–569. https://doi.org/10.1177/0020852320953184

Cheng, C. T., Chiou, S. J., Lee, C. T., & Tsai, Y. B. (2007). Study on probabilistic seismic hazard maps of Taiwan after Chi-Chi earthquake. *Journal of GeoEngineering, 2*(1), 19–28.

Churchill, C. (2007). Insuring the low-income market: Challenges and solutions for commercial insurers. *Geneva Pap Risk Insur Issues Pract, 32*, 401–412. https://doi.org/10.1057/palgrave.gpp.2510132

Clark, K. M. (1997). Current and potential impact of hurricane variability on the insurance industry. In *Hurricanes* (pp. 273–283). Berlin, Heidelberg: Springer.

Cumiskey, L., Dogulu, N., Landaverde, E. R. M., Ali, J., & Sai, F. (2019). The role of young professionals in driving the integration of early warning systems. *WMO Bulletin, 68*, 38–44. https://public.wmo.int/en/resources/bulletin/role-of-young-professionals-driving-integration-of-early-warning-systems

Daly, M., Poutasi, N., Nelson, F., & Kohlhase, J. (2010). Reducing the climate vulnerability of coastal communities in Samoa. *Journal of International Development: The Journal of the Development Studies Association, 22*(2), 265–281.

Day, S., & Fearnley, C. (2015). A classification of mitigation strategies for natural hazards: Implications for the understanding of interactions between mitigation strategies. *Natural Hazards, 79*(2), 1219–1238.

Drury, A. C., Olson, R. S., & Van Belle, D. A. (2005). The politics of humanitarian aid: US foreign disaster assistance, 1964–1995. *The Journal of Politics, 67*(2), 454–473.

Eadie, P. (2019). Typhoon Yolanda and post-disaster resilience: Problems and challenges. *Asia Pacific Viewpoint, 60*(1), 94–107. https://onlinelibrary.wiley.com/doi/10.1111/apv.12215 https://doi.org/10.1016/j.ijdrr.2014.08.004

Eraybar, K., Okazaki, K., & Ilki, A. (2010). An exploratory study on perceptions of seismic risk and mitigation in two districts of Istanbul. *Disasters, 34*(1), 71–92.

Garcia, C., & Fearnley, C. J. (2012). Evaluating critical links in early warning systems for natural hazards. *Environmental Hazards, 11*(2), 123–137.

Garrett, T. A., & Sobel, R. S. (2003). The political economy of FEMA disaster payments. *Economic Inquiry, 41*(3), 496–509.

Golding, B. (2022). *Towards the 'perfect' weather warning: Bridging disciplinary gaps through partnership and communication.* https://link.springer.com/book/10.1007/978-3-030-98989-7

Govarets, A., & Lauwerts, B. (2009). Assessment of the impact of coastal defence structures. Biodiversity Series: OSPAR Commission.

Gruntfest, E. (1987). *What we have learned since the Big Thompson flood.* Boulder, CO: Natural Hazards Research and Applications Information Center. https://digitalcommons.usf.edu/nhcc/60

Hay, C. (1996). From crisis to catastrophe? The ecological pathologies of the liberal: Democratic state form. *Innovation: The European Journal of Social Science Research, 9*(4), 421–434.

Hürlimann, M., Copons, R., & Altimir, J. (2006). Detailed debris flow hazard assessment in Andorra: A multidisciplinary approach. *Geomorphology, 78*(3–4), 359–372.

International Federation of Red Cross, & Red Crescent Societies. (2006). World disasters report: Focus on neglected crises. Red Cross Red Crescent, Switzerland.

International Federation of Red Cross, & Red Crescent Societies. (2009). *World disasters report: Focus on early warning, early action.* Red Cross Red Crescent, Switzerland.

Jóhannesson, T. (2001). Run-up of two avalanches on the deflecting dams at Flateyri, northwestern Iceland. *Annals of Glaciology, 32*, 350–354.

Kellett, J., & Sparks, D. (2012). Disaster risk reduction. *Spending Where It Should Count, 2.*

Klein, N. (2007). *The shock doctrine: The rise of disaster capitalism.* Toronto: Alfred A. Knopf Canada.

Klingberg, M. A. (Ed.). (1995). *Contributions to epidemiology and biostatistics.* Cambridge: Cambridge University Press.

Leone, F., & Lesales, T. (2009). The interest of cartography for a better perception and management of volcanic risk: From scientific to social representations: The case of Mt. Pelée volcano, Martinique (Lesser Antilles). *Journal of Volcanology and Geothermal Research, 186*(3–4), 186–194.

Lewis, T. L. (2003). Environmental aid: Driven by recipient need or donor interests? *Social Science Quarterly, 84*(1), 144–161.

Mileti, D. S., & Darlington, J. D. (1995). Societal response to revised earthquake probabilities in the San Francisco Bay area. *International Journal of Mass Emergencies & Disasters, 13*(2), 119–145.

Mileti, D. S., & Sorensen, J. H. (1990). Communication of emergency public warnings: A social science perspective and state-of-the-art assessment (No.

ORNL-6609). Oak Ridge National Lab., TN (USA).

Moore, D. (2005). At least 100,000 Katrina victims still separated from families. *GALLUP* [Online]. https://news.gallup.com/poll/19225/least-100000-katrina-victims-still-separated-from-families.aspx#:~:text=Some%20of%20the%20major%20findings,families%20are%20still%20not%20reunited. Accessed 25 July 2022.

Munich Re. (2006). www.MunichRe.com. Accessed 5 March 2020.

Newhall, C. G., & Punongbayan, R. (Eds.). (1996). *Fire and mud: Eruptions and lahars of Mount Pinatubo, Philippines* (p. 1126). Quezon City: Philippine Institute of Volcanology and Seismology.

Nwadike, A., & Wilkinson, S. (2020). Challenges facing building code compliance in New Zealand. *International Journal of Construction Management*. DOI: 10.1080/15623599.2020.1801336

Nwadike, A., Wilkinson, S., & Clifton, C. (2019). Improving disaster resilience through effective building code compliance. 9th International i-Rec Conference 2019.

OCHA. (2020). www.humanitarianresponse.info/en/coordination/clusters/what-cluster-approach

Olsen, G. D., Pracejus, J. W., & Brown, N. R. (2003). When profit equals price: Consumer confusion about donation amounts in cause-related marketing. *Journal of Public Policy & Marketing, 22*(2), 170–180.

Ross, K. (2004). Sex at work: Gender politics and newsroom culture. In M. de Bruin & K. Ross (Eds.), *Gender and newsroom culture: Identities at work* (pp. 143–160). Creskill, NJ: Hampton Press. [Google Scholar].

Röthlisberger, V., Zischg, A., & Keiler, M. (2016). Spatiotemporal aspects of flood exposure in Switzerland. In E3S Web of Conferences (Vol. 7, p. 08008). EDP Sciences.

Sahdev, S., Kumar, M., & Singh, R. B. (2020). Seismic vulnerability assessment of NCT of Delhi using GIS-based multiple criteria decision analysis. In *Geoecology of landscape dynamics* (pp. 11–20). Singapore: Springer.

Schmidtlein, M. C., Deutsch, R. C., Piegorsch, W. W., & Cutter, S. L. (2008). A sensitivity analysis of the social vulnerability index. *Risk Analysis: An International Journal, 28*(4), 1099–1114.

Scott, R. (2014). Imagining More Effective Humanitarian Aid: A Donor Perspective. OECD (Organisation for Economic Co-operation and Development), Paris, France.

Seibel, W. (2022). Erosion of professional integrity: The collapse of the Canterbury TV building in Christchurch on 22 February 2011. In *Collapsing structures and public mismanagement* (pp. 87–128). Cham: Palgrave Macmillan.

Sharma, R. B., Hochrainer-Stigler, S., & Mechler, R. (2011). *Impact assessment of disaster microinsurance for pro-poor risk management: Evidence from South Asia.* https://pure.iiasa.ac.at/id/eprint/9750/1/XO-11-059.pdf

Shreve, C. M., & Kelman, I. (2014). Does mitigation save? Reviewing cost-benefit analyses of disaster risk reduction. *International Journal of Disaster Risk Reduction, 10*, 213–235. https://doi.org/10.1016/j.ijdrr.2014.08.004

Simmons, K. M., Kovacs, P., & Smith, A. B. (2020). State-by-state analysis of benefits to cost from wind-enhanced building codes. *Natural Hazards Review, 21*(2), 04020007.

Sylves, R., & Búzás, Z. I. (2007). Presidential disaster declaration decisions, 1953–2003: What influences odds of approval? *State and Local Government Review, 39*(1), 3–15.

Terry, F. (2002). *The paradox of humanitarian action: Condemned to repeat.* London: Cornell.

Twigg, J. (2015). *Disaster risk reduction.* London: Overseas Development Institute.

UNDRR. (2017). *Sendai framework terminology on disaster risk reduction.* https://www.undrr.org/hazard. Accessed 11 May 2022.

UNDRR. (2022). www.undrr.org/terminology

UNISDR, U. (2015, March). Sendai framework for disaster risk reduction 2015–2030. Proceedings of the 3rd United Nations World Conference on DRR (Vol. 1), Sendai, Japan.

Villagrán de León, J. C., Weerawarnakula, S., & Chandrapala, L. (2013). Elements to develop a tsunami-early warning plan for the city of Galle in Sri Lanka. Paper for Workshop on Human Impact of Tsunami and Disaster Risk Reduction in Bangkok, 16–17 June 2006. https://www.unisdr.org/2006/ppew/tsunami/highlights/TEWS-in-SL-UNU-EHS-v3.pdf

Walker, G. R., & Minor, J. E. (1980). Cyclone Tracy in retrospect: A review of its impact on the Australian community. *Wind Engineering*, 1327–1337.

Walter, J. (2002). *World disasters report 2002: Focus on reducing risk.* Switzerland: Kumarian Press.

Yore, R., & Faure Walker, J. (2019). Microinsurance for disaster recovery: Business venture or humanitarian intervention? An analysis of potential success and failure factors of microinsurance case studies. *International Journal of Disaster Risk Reduction, 33*, 16–32.

FURTHER READING

Alexander, D. E. (2002). *Principles of emergency planning and management.* Oxford: Oxford University Press on Demand.

Bosher, L. L. (Ed.). (2008). *Hazards and the built environment.* London, UK: Taylor and Francis. A good overview of hazards interacting with the built environment.

Kelman, I. (2020). *Disaster by choice: How our actions turn natural hazards into catastrophes.* Oxford: Oxford University Press.

Kelman, I., Mercer, J., & Gaillard, J. C. (Eds.). (2017). *The Routledge handbook of disaster risk reduction including climate change adaptation.* London: Routledge.

Mileti, D. (1999). *Disasters by design: A reassessment of natural hazards in the United States.* Joseph Henry Press.

Olsen, G. R., Carstensen, N., & Høyen, K. (2003). Humanitarian crises: What determines the level of emergency assistance? Media coverage, donor interests and the aid business. *Disasters, 27*, 109–126. This remains a complex and often unanswered question.

Paul, B. K. (2011). *Environmental hazards and disasters: Contexts, perspectives and management.* John Wiley & Sons.

Spence, R., & So, E. (2021). *Why do buildings collapse in earthquakes?* Chichester, UK: Wiley. An up-to-date explanation of infrastructure's earthquake vulnerability and mitigation measures.

Wisner, B., Gaillard, J. C., & Kelman, I. (2012). *Handbook of hazards and disaster risk reduction.* Abingdon, UK: Routledge.

WEB LINKS

Directorate-General for European Civil Protection and Humanitarian Aid Operations (DG ECHO). www.ec.europa.eu/echo/index

International Committee of the Red Cross (ICRC). www.icrc org

International Federation of Red Cross and Red Crescent Societies (IFRC). www.ifrc.org

United Nations Office for the Coordination of Humanitarian Affairs (UN OCHA). www.unocha.org

United Nations Refugee Agency (UNHCR). www.unhcr.org

Part Two

THE EXPERIENCE AND REDUCTION OF HAZARD

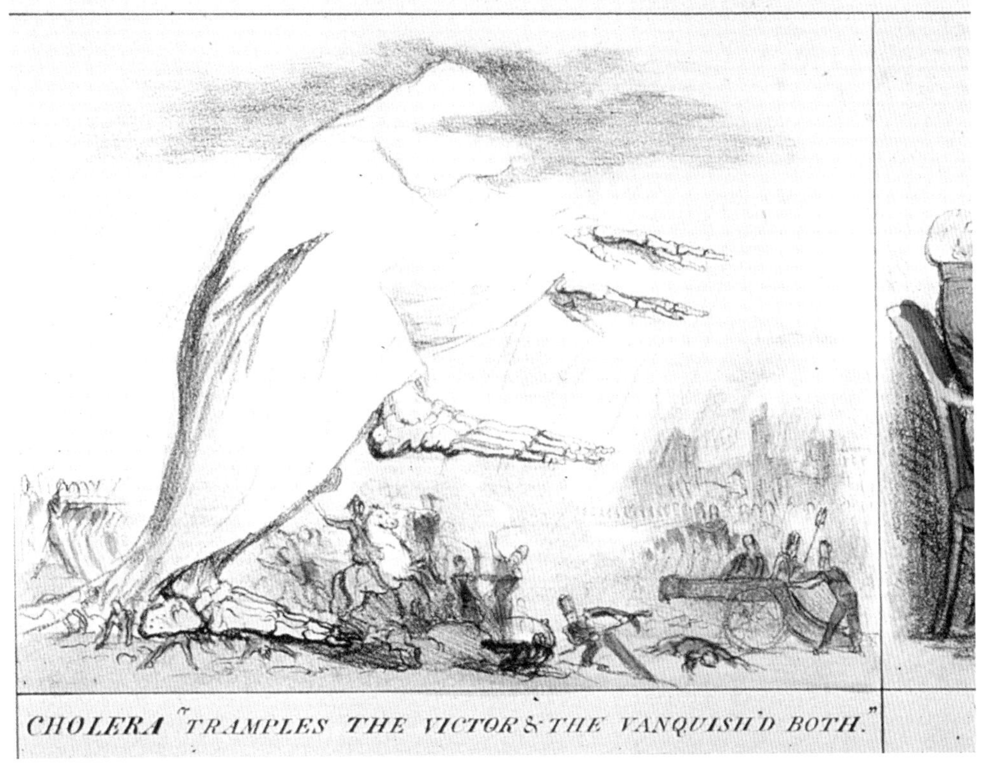

Cholera 'Tramples the victors & the vanquished both' by Robert Seymour, 1798–1836.

Source: Cholera Online, 1817 to 1923. Images from the History of Medicine, The National Library of Medicine.

Part 2:

I shall witness all the variety of appearance, that the elements can assume – I shall read fair augury in the rainbow – menace in the cloud – some lesson or record dear to my heart in everything.

– Mary Shelley, The Last Man

DOI: 10.4324/9781351261647-7

CHAPTER SIX

Earthquake Hazards

6

OVERVIEW

This chapter provides an overview of earthquakes, how they occur and how they are measured and mitigated against. Earthquakes also produce a range of secondary hazards including soil liquefaction, landslides, rock and snow avalanches, tsunamis and fires. Key risk management approaches are via preparedness in the development of buildings that are earthquake resistant, adhering to seismic safety codes, effective land use planning and the use of insurance. Most of these preparedness measures require investment and resources that may not be available to the poorest, therefore it has become typical that disaster aid becomes a key component of the response to earthquakes, providing shelter, safety and key supplies and looking for those trapped in buildings. Community preparedness exercises and actions can help avoid disasters involving earthquakes and, therefore, education plays a significant role in reducing earthquake impacts. Many lessons have been learnt from significant past earthquakes, but there are still challenges in protecting people from the dangers of earthquakes that are not necessarily large. Shallow earthquakes can also be devastating.

A. THE NATURE OF EARTHQUAKES

The Earth has a concentrically layered structure consisting of a core, mantle and crust, as seen in Figure 6.1. Earthquakes and associated hazards occur within the crust. This comprises oceanic crust (typically 6–11 km thick and composed of basalt) and continental crust (25–90 km thick and highly variable in composition).

The global distribution of earthquakes is far from random. About two-thirds of all large earthquakes are located in the so-called Ring of Fire around the Pacific Ocean. This pattern is closely related to the geophysical activity associated with plate tectonics (Rothery, 2010). The Earth's crust is divided into more than 15 lithospheric plates (Figure 6.2), with seven of these being major plates.

DOI: 10.4324/9781351261647-8

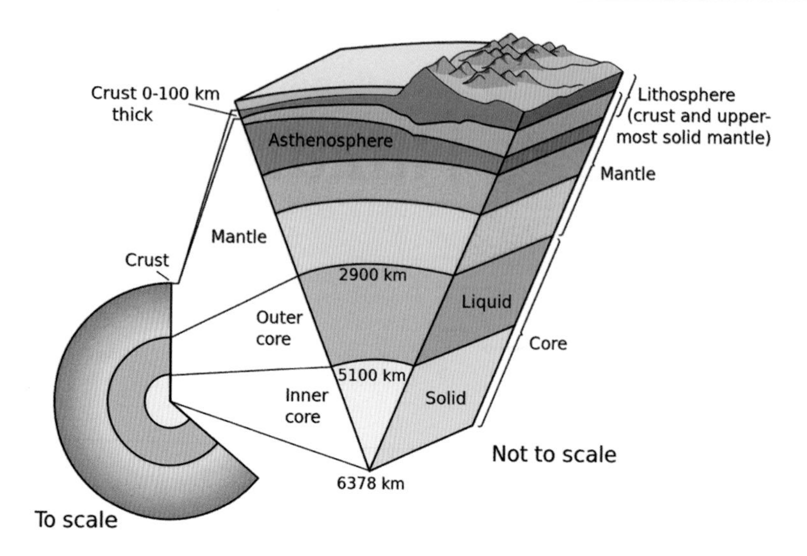

Figure 6.1 Cutaway views showing the internal structure of the Earth. Left: To scale drawing shows that the Earth's crust is very thin. Right: Not to scale, more detail of three main layers (crust, mantle, core).

Source: USGS

The plates move across the globe at speeds up to 180 mm yr^{-1}, carried along by convection currents in the mantle. Each plate is in contact with its neighbours and, due to convection currents within the Earth's mantle, these plates move relative to one another. Most earthquakes occur when two plates collide in a destructive plate boundary; this is typically in *subduction zones* where one plate is forced beneath another. Sometimes earthquakes occur at weak points within plates. Although these intra-plate earthquakes account for less than 0.5% of global seismicity, they are usually unexpected and pose a significant threat.

Earthquakes are caused by sudden movements in rock, relatively near to the Earth's surface, along a zone of pre-existing geological weakness called a fault. These movements are preceded by the slow build-up of tectonic strain that progressively deforms the crustal rocks and produces stored elastic energy. When the accumulated stress exceeds the resisting strength of the fault, it creates rock fractures. This is akin to holding two ends of a plastic ruler and bending it; eventually it will snap, releasing stress (the snap sound) from the brittle failure of the ruler. Stress is related to the forces deforming a region of land, and strain (which can be measured) is the amount of deformation generated by that stress.

The sudden release of energy produces seismic waves that radiate outwards. It is the fracture of the brittle crust, followed by elastic rebound on either side of the fracture, which is the cause of ground shaking. The point of rupture (or hypocentre) can occur anywhere between the Earth's surface and a depth of 700 km. The point on the earth's surface vertically above the point of rupture of an earthquake is called the epicentre (see Figure 6.3). The rupture propagates along the fault line, radiating seismic waves along the fault plane as well as from the hypocentre.

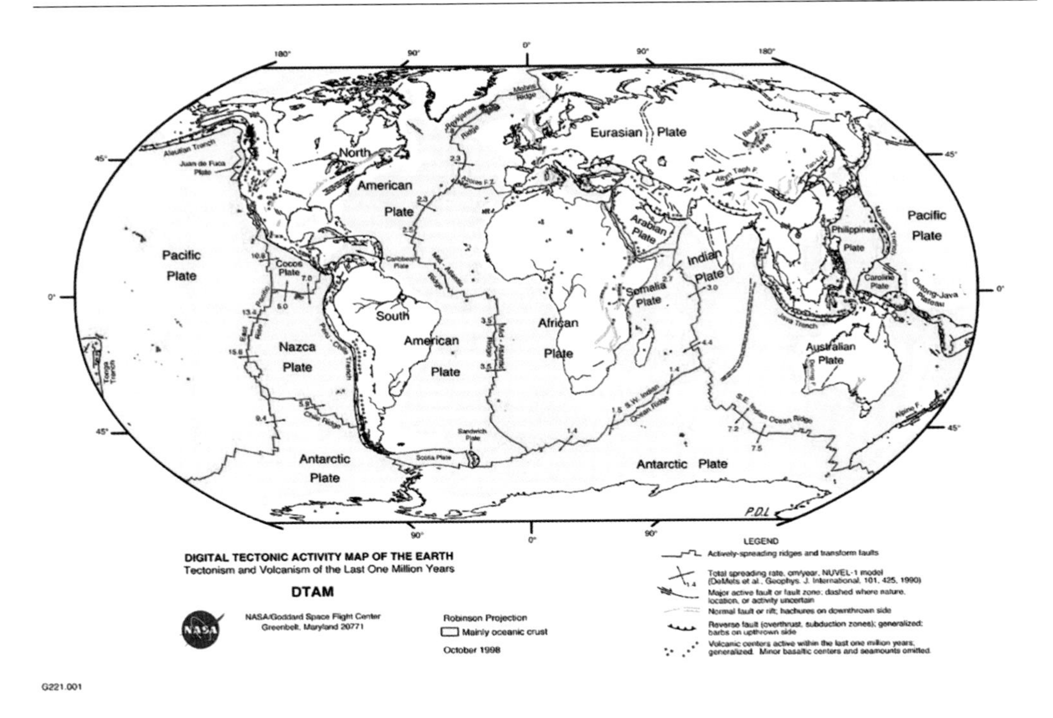

Figure 6.2 Digital tectonic activity map of the Earth. Note the different types of tectonic boundaries and locations of volcanoes and the oceanic crust.

Source: NASA/Goddard Space Flight Center

The size of an earthquake depends upon the amount of physical displacement on the fault. Generally speaking, larger fault movements – either vertically or laterally – and longer-length ruptures lead to bigger earthquakes. The 2004 Sumatra earthquake, for example, was very large because a major displacement of the fault (~15 m) took place over a long distance (1,600 km). The most damaging earthquakes – accounting for about three-quarters of the global seismic energy release – are shallow-focus earthquakes occurring <70 km below the surface. For example, the 2011 Christchurch earthquake in New Zealand was a moderate magnitude of Mw 6.2 but only occurred 5 km (3.1 miles) below the surface, resulting in significant damage to infrastructure and loss of life. Deep focus earthquakes occur 70–300 km, where the mantle eventually becomes too fluid-like to fracture.

Readjustments can occur at the site of the fault causing further slips that are typically classed as aftershocks and can continue from days to years after the initial events, often causing significant damage to already compromised buildings and infrastructure. Foreshocks are earthquakes that precede larger earthquakes in the same location, but often an earthquake cannot be identified as a foreshock until after a larger earthquake in the same area occurs. In some cases an earthquake swarm can occur, which is a sequence of seismic events occurring in a local area within a relatively short period of time. A swarm can last days, months or years with no single earthquake being the main shock.

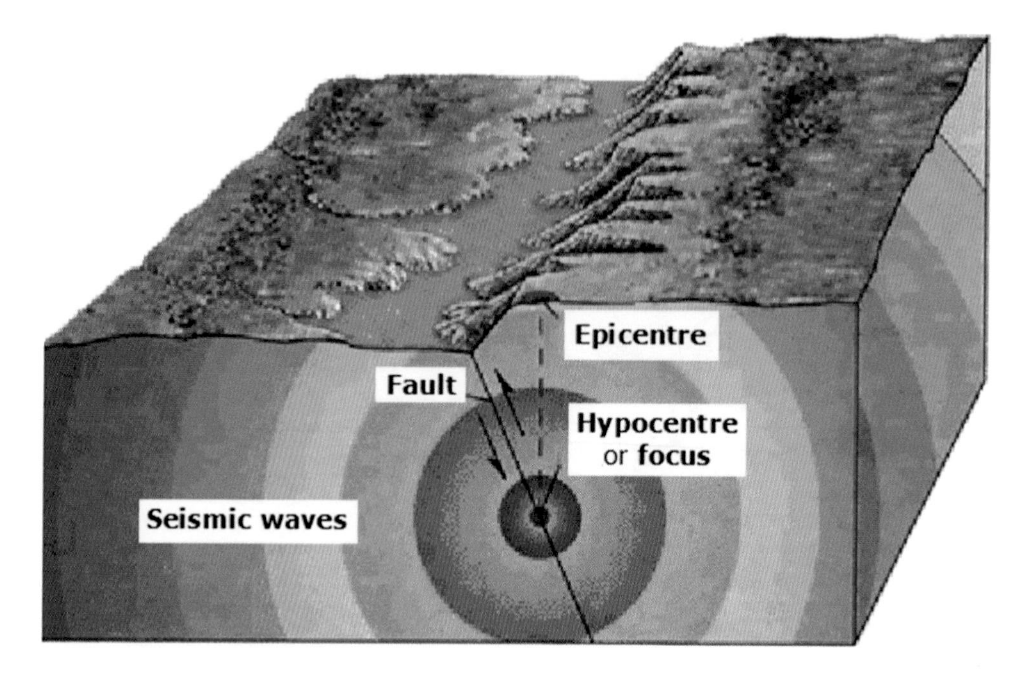

Figure 6.3 Elements of an earthquake.

Source: Jimdo

There are three main types of fault that are distinguished by the type of displacement of the ground. These are transcurrent or strike-slip fault (displacement is sideways); thrust faults where one area of ground is pushed over the other causing a reverse fault if it dips steeply; and normal faults in areas of extension and the fault is included to the vertical (see Figure 6.4).

Key plate boundaries are usually a zone of deformation, rather than a single fault. These zones form a fault system typically with a main fault with subsidiary faults that splay off it. A famous example is the San Andreas Fault in California, USA (see Figure 6.5). Areas of compression and extension develop as ruptures propagate along the fault (typically in a manner of unzipping).

1. Earthquake Magnitude

Earthquakes are measured at the epicentre, and the magnitude is calculated on scales based on the work of seismologist and physicist Charles Richter (1900–1985). These scales describe the total energy released by the earthquake in the form of seismic waves radiating from the fault plane. The energy is calculated from seismographs that record the amplitude of ground motion during the earthquake. The original system, still known as the Richter scale, measures the *local magnitude* (M_L) of the earthquake. This scale is unsuitable for very large earthquakes. Today, seismologists employ a slightly different measurement based on what is termed *moment magnitude* (Mw). This method takes into consideration both the area of the fault that has ruptured and the amount of movement that has occurred. It is thus a more reliable measurement of the total energy released. The moment magnitude scale has been refined so that the resultant values are reasonably close to those

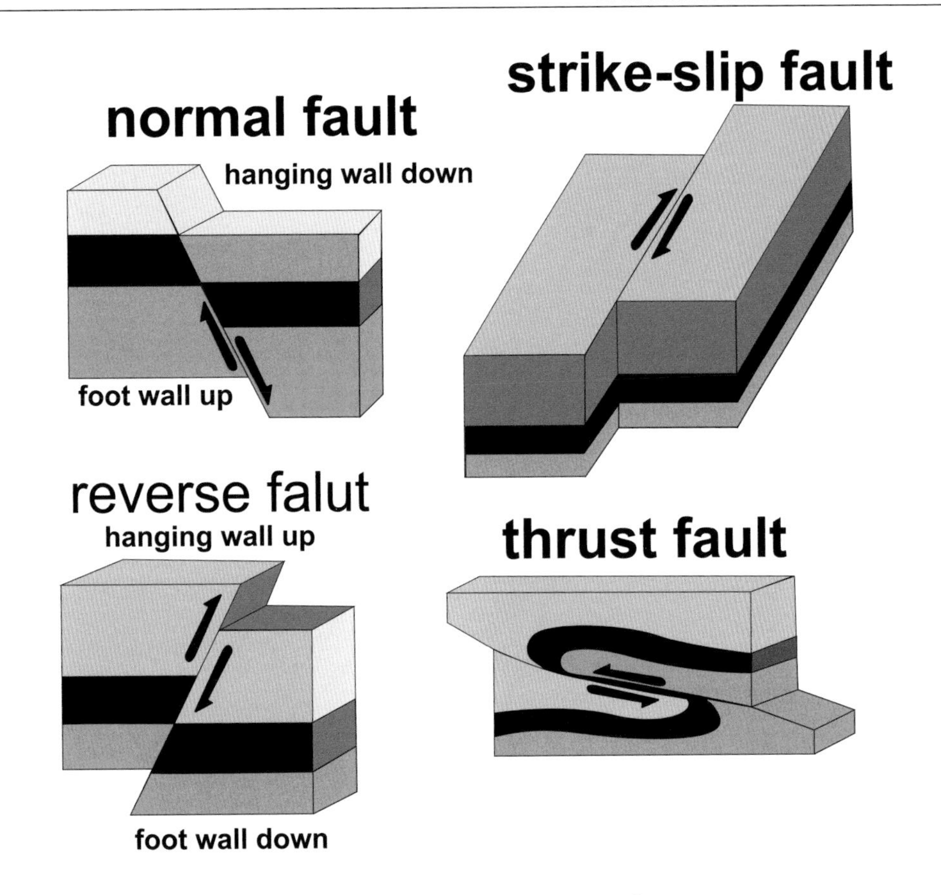

Figure 6.4 Types of fault line: normal, reverse, thrust and strike-slip.

Source: Robo Spark, Spark Preparedness (n.d.) http://sparkpreparedness.weebly.com/blog/earthquakes-happen-on-fault-lines-what-are-faults

of the original local magnitude scale, to ease comparisons.

It is important to understand that these scales are not linear. In Richter's original system, each point on the M_L scale indicated an order of magnitude increase in measured ground motion (see Table 6.1). Thus, an $M_L = 7.0$ earthquake produces about ten times more ground shaking than an $M_L = 6.0$ event, and around 1,000 times more ground shaking than an $M_L = 4.0$ event. Approximate energy–magnitude relationships show that, as the magnitude increases by one whole unit, the total energy released increases by about 32 times. The moment magnitude (Mw) scale measures this energy release. Thus, an Mw = 6.0 event releases about 32 times more energy than does an Mw = 5.0 event; an Mw = 9.0 event will release over 1 million times more energy than an Mw = 5 event. The scale has no theoretical upper limit but empirical evidence suggests that most shallow earthquakes need to attain a magnitude of at least Mw = 4.0 before damage is recorded on the surface (Bollinger et al., 1993). Such events occur several times each day worldwide, but few cause significant loss. The largest recorded earthquake was the 22 May 1960 Valdivia earthquake in Chile, which was an astonishing 9.4–9.6 Mw. Thankfully large earthquakes happen

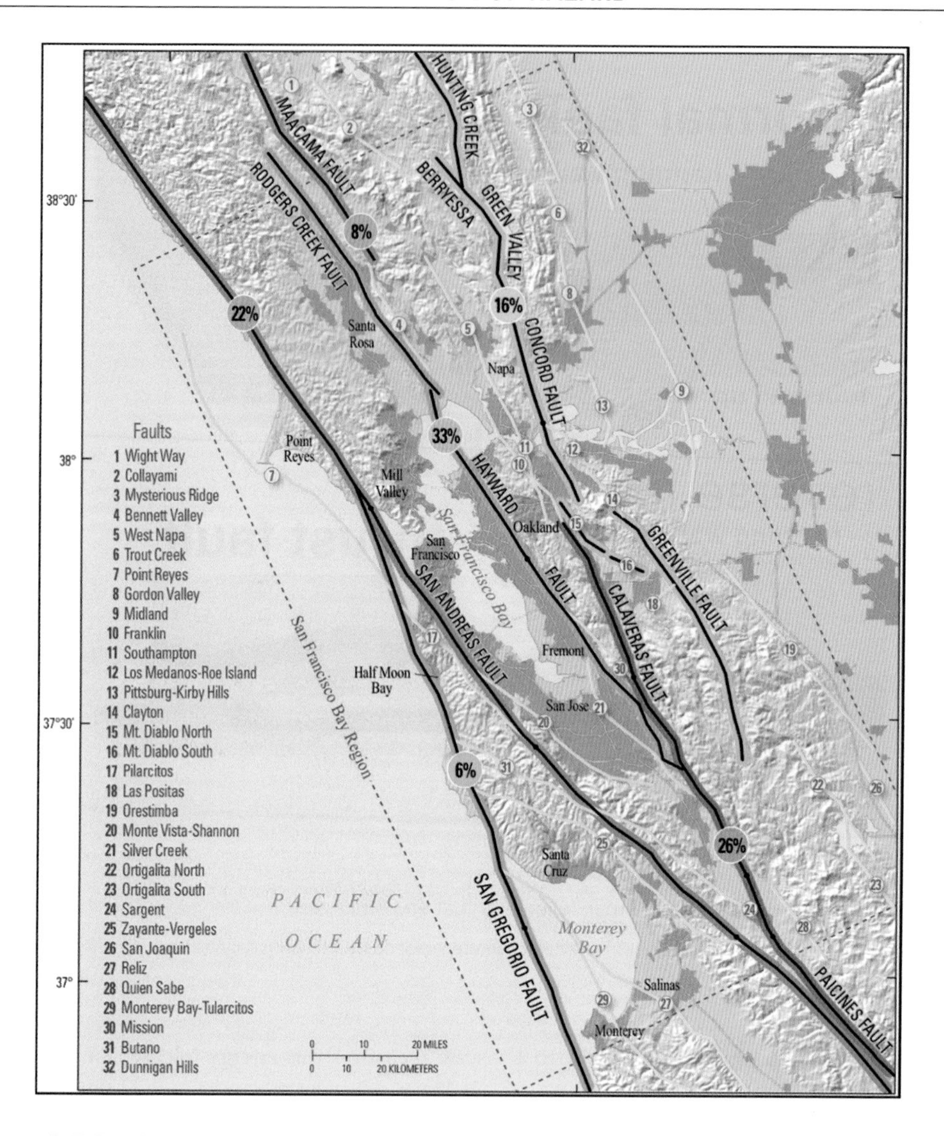

Figure 6.5 The San Andreas Complex Fault region.

Source: USGS map showing faults that span the Pacific–North America plate boundary

less frequently than smaller ones (much like volcanic eruptions).

2. Earthquake Intensity

Earthquake intensity is a measure of ground shaking that correlates better with disaster losses than with magnitude. Intensity is estimated on the Modified Mercalli (MM) scale, which allocates a numerical value to observations of the nature and extent of physical damage after the event (see Table 6.2). The scale ranges from MM = I (not felt at all) to MM = XII (widespread destruction). At first glance, the MM scale is less 'scientific' than other scales because it relies

Table 6.1 Annual frequency of occurrence of earthquakes of different magnitudes based on observations since 1900

Descriptor	Magnitude (Mw)	Annual average	Hazard potential
Great	8 and higher	1	Total destruction, high loss of life
Major	7–7.9	18	Serious building damage, major loss of life
Strong	6–6.9	120	Large losses, especially in urban areas
Moderate	5–5.9	800	Significant losses in populated areas
Light	4–4.9	6,200	Usually felt, some structural damage
Minor	3–3.9	49,000	Typically felt but usually little damage
Very minor	Less than 3	9,000 per day	Not felt, but recorded

Source: After US Geological Survey at http://neic.usgs.gov (accessed 16 January 2003)

Table 6.2 The Modified Mercalli earthquake intensity scale

Average peak velocity (cm s–1)	Intensity value and description of impacts		Average peak acceleration
	I.	Not felt except by a very few under especially favourable circumstances.	
	II.	Felt only by a few persons at rest, especially on upper floors of buildings. Delicately suspended objects may swing.	
	III.	Felt quite noticeably indoors, especially on upper floors of buildings, but many people do not recognise it as an earthquake. Standing automobiles may rock slightly. Vibration like a passing truck.	
1–2	IV.	During day felt by many, outdoors by few. At night some awakened. Dishes, windows, doors disturbed; walls make creaking sound. Sensation like heavy truck striking building. Standing vehicles rock noticeably.	0.015g–0.02g
2–5	V.	Felt by nearly everyone, many awakened. Some dishes, windows and so on broken; cracked plaster in a few places; unstable objects overturned. Disturbance of trees, poles and other tall objects sometimes noticed. Pendulum clocks may stop.	0.03g–0.04g
5–8	VI.	Felt by all, many frightened and run outdoors. Some heavy furniture moved; a few instances of fallen plaster and damaged chimneys. Damage slight.	0.06g–0.07g
8–12	VII.	Everybody runs outdoors. Damage negligible in buildings of good design and construction; slight to moderate in well-built ordinary structures; considerable in poorly built or badly designed structures; some chimneys broken. Noticed by persons driving cars.	0.10g–0.15g

(*Continued*)

Table 6.2 (Continued)

Average peak velocity (cm s–1)	Intensity value and description of impacts		Average peak acceleration
20–30	VIII.	Damage slight in specially designed structures; considerable in ordinary substantial buildings with partial collapse; great in poorly built structures. Panel walls thrown out of frame structures. Fall of chimneys, factory stacks, columns, walls and monuments. Heavy furniture overturned. Sand and mud ejected in small quantities. Changes in well water. Persons driving cars disturbed.	0.25g–0.30g
45–55	IX.	Damage considerable in specially designed structures; well-designed frame structures thrown out of plumb; great in substantial buildings with partial collapse. Buildings shifted off foundations. Ground cracked conspicuously. Underground pipes broken.	0.50g–0.55g
>60	X.	Some well-built wooden structures destroyed; most masonry and frame structures destroyed with foundations; ground badly cracked. Rails bent. Landslides considerable from riverbanks and steep slopes. Shifted sand and mud. Water splashed, slopped over banks.	>0.60g
	XI.	Few, if any, (masonry) structures remain standing. Bridges destroyed. Broad fissures in ground. Underground pipelines completely out of service. Earth slumps and land slips in soft ground. Rails bend greatly.	
	XII.	Damage total. Waves seen on ground surface. Lines of sight and level distorted. Objects thrown into the air.	

Note: g is gravity = 9.8 m s^2.

upon qualitative human descriptions rather than instrumental measurements. However, it captures important elements of earthquake impact. Another advantage is that in using written accounts of past events MM intensities can be assigned to earthquakes that occurred prior to the introduction of direct measurements. This allows the earthquake record to be extended back in history.

B. PRIMARY EARTHQUAKE HAZARDS

During an earthquake the extent of ground shaking is measured by strong motion seismometers. These instruments are activated by strong ground tremors and record both horizontal and vertical ground accelerations caused by the shaking (Box 6.1). Analysis of the data collected from these instruments shows that earthquakes produce four main types of seismic wave (Figure 6.6):

- *Primary waves* (P-waves) are vibrations caused by compression, similar to a shunt through a line of connected rail coaches. They spread out from the earthquake fault at a rate of about 5–7 km s^{-1} and are able to travel through

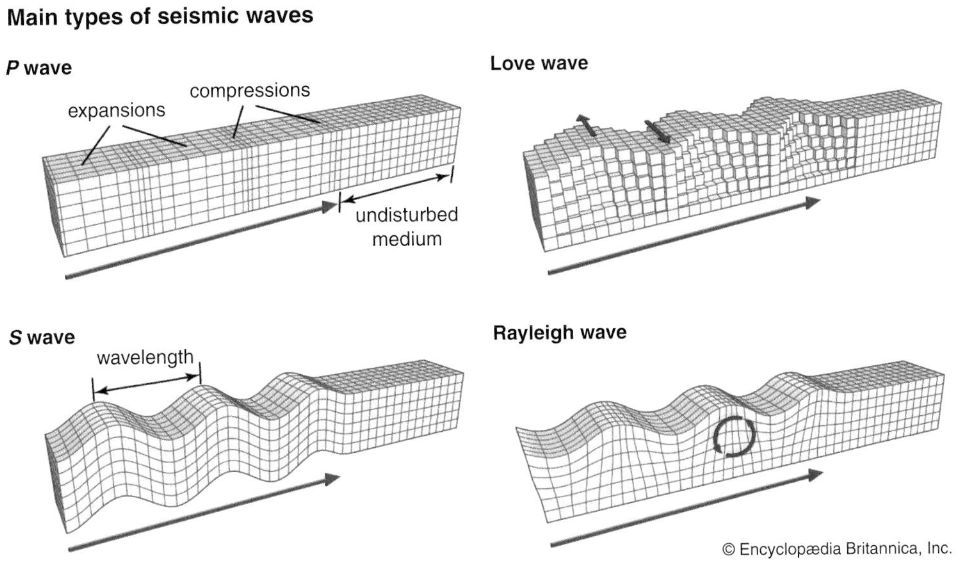

Figure 6.6 Schematic illustration of the four main types of earthquake waves: P-waves, S-waves, Love waves and Rayleigh waves. The secondary and Love waves tend to have the largest amplitude and destructive force.

Source: Encyclopedia Britannica: www.britannica.com/science/earthquake-geology/Shallow-intermediate-and-deep-foci

both solid rock and liquids, including the oceans and the Earth's liquid core.

- *Secondary waves* (S-waves) move through the Earth's body at about half the speed of primary waves. These waves vibrate at right angles to the direction of travel, similar to a wave travelling along a flexed rope held between two people. S-waves, which cannot travel through liquids, are responsible for much of the damage caused by earthquakes, as it is difficult to design structures that can withstand this type of motion. They travel 3–4 km s^{-1} in the crust.
- *Rayleigh waves* are surface waves in which particles follow an elliptical path in the direction of propagation and partly in the vertical plane, much like water moving within an ocean wave.
- *Love waves* (L-waves) are similar to Rayleigh waves but with vibration solely in the horizontal plane. They travel 2–4.4 km s^{-1} in the crust.

The overall severity of an earthquake is dependent on the amplitude and frequency of these wave motions. The S- and L-waves are more destructive than the P-waves because they have a larger amplitude and force. In an earthquake the ground surface may be displaced horizontally, vertically or obliquely, depending on the wave activity and the local geological conditions (Box 6.1).

C. SECONDARY EARTHQUAKE HAZARDS

1. Soil Liquefaction

An important secondary hazard associated with loose sediments is soil liquefaction. This is the process by which

Box 6.1 Ground Shaking in Earthquakes

Information on ground motion is necessary to understand the behaviour of buildings in earthquakes. Ground acceleration is usually expressed as fractions of the acceleration due to gravity (9.8 m s^{-1}). Thus, 1.0g represents an acceleration of 9.8 m s^{-2}, whilst 0.1g = 0.98 m s^{-2}. If an unsecured object experienced an acceleration of 1.0g in the vertical plane it would, in effect, become weightless and could leave the ground. Values as large as 0.8g have been recorded in firm ground from earthquakes with magnitudes as small as Mw = 4.5, whilst the 1994 Northridge earthquake had localised peak ground motions of nearly 1.82g. Even very strong structures struggle to deal with such high vertical accelerations. The greatest damage is often generated by the Love waves, which cause horizontal shaking. Some unreinforced masonry buildings (URMs) may be unable to cope with horizontal accelerations as small as 0.1g.

Local site conditions influence ground motion. Significant wave amplifications occur in steep topography, especially on ridge crests. Ground motions in soil are enhanced in both amplitude and duration, compared to those recorded in rock. As a result, structural damage is usually most severe for buildings founded on unconsolidated material. In the Michoacan earthquake of 1985, the recorded peak ground accelerations in Mexico City varied by a factor of five. Strong-motion records obtained on firm soil showed values of around 0.04g. This compared with observations from urban areas located on a dried lakebed where peak ground accelerations reached 0.2g. Similar effects were noted in the San Salvador earthquake of 1986. This had a very modest size (Mw = 5.4) but destroyed thousands of buildings as well as causing 1,500 deaths. The reason was rooted in layers of volcanic ash, up to 25 m thick, which underlie much of the city. As the three-second-long earthquake tremor passed upwards through the ash, the amplitude of ground movement was magnified up to five times.

The scale of destruction also depends on the frequency of the vibrations and the fundamental period of the structures at risk. The frequency of a wave is the number of vibrations (cycles) per second measured in units called Hertz (Hz). High-frequency waves tend to have high accelerations but relatively small amplitudes of displacement. Low-frequency waves have small accelerations but large velocities and displacements. During earthquakes, the ground may vibrate at all frequencies from 0.1 to 30 Hz. If the natural period of a building's vibration is close to that of the seismic waves, *resonance* can occur. This causes the building to sway. Low-rise buildings have short natural wave periods (0.05–0.1 seconds) and high-rise buildings have long natural periods (1–2 seconds). The P- and S-waves are mainly responsible for the high-frequency vibrations (>1 Hz) that are most effective in shaking low buildings. Rayleigh and Love waves are lower frequency and more effective in causing tall buildings to vibrate. The very lowest-frequency waves may have less than one cycle per hour and have wavelengths of 1,000 km or more.

water-saturated material can temporarily lose strength and behave as a fluid when subjected to strong shaking. Poorly compacted sand and silt situated at depths less than 10 m below the surface is the principal medium. In the 2001 Bhuj earthquake, many reservoir dams were damaged by soil liquefaction in the water-saturated alluvial foundations. According to Tinsley et al. (1985), four types of ground failure commonly result:

- *Lateral spread* involves the horizontal displacement of surface blocks as a result of liquefaction in a subsurface layer. Such spreads occur most commonly on slopes between 0.3° and 3°. They cause damage to pipelines, bridge piers and other structures with shallow foundations, especially those located near river channels or canal banks on alluvial floodplains.
- *Ground oscillation* occurs if liquefaction occurs but the slopes are too gentle to permit lateral displacement. Oscillation is similar to lateral spread but the disrupted blocks come to rest near their original position; lateral-spread blocks can move significant distances. Oscillation is often accompanied by the opening and closing of surface fissures. In the 1964 Alaskan earthquake, cracks up to 1 m wide and 10 m deep were observed.
- *Loss of bearing strength* occurs when a shallow layer of soil liquefies beneath a building. Large deformations within the soil mass can cause structures to settle and tip. In the 1964 Niigata, Japan, earthquake, four apartment buildings tilted as much as 60° in unconsolidated alluvial ground. Loss of bearing strength was a key reason for the high death toll in the 1985 Mexico City disaster, in which about 9,000 people died, even though the city was nearly 400 km from the fault rupture. Similar failures readily damage port facilities built on land reclaimed by dredged material like sand and silt.
- *Flow failure* is associated with the most catastrophic form of liquefaction because the slope fails at the surface as well as at depth. Flow failures can be very large and rapid, displacing material by tens of kilometres at velocities of tens – or even hundreds – of kilometres an hour. Such failures can happen on land or underwater. The devastation of Seward and Valdez, Alaska, in 1964 was largely caused by a submarine flow failure at the marine end of the delta on which these settlements were built. The harbour area was carried away and created water waves which then swept back onshore, causing further damage.

2. Landslides, Rock and Snow Avalanches

Severe ground shaking causes natural slopes to weaken and fail. The resulting landslides and rock and snow avalanches are major contributors to earthquake disasters because many destructive earthquakes occur within mountainous areas. For example, more than half of all deaths recorded after large-magnitude (Mw > 6.9) earthquakes in Japan have been attributed to landslides (Kobayashi, 1981). Correlations between event magnitude and landslide distribution show that landslides are unlikely to be triggered by earthquakes of less than Mw = 4.0. However, the maximum area likely to be affected by earthquake-related landslides increases rapidly thereafter, to reach 500,000 km^2 at Mw = 9.2 (Keefer, 1984). Central America and the Himalayas are regions where landslides frequently occur after earthquakes (Bommer &

Rodríguez, 2002). During the 25 April 2015 earthquake in Nepal, landslides played a major role in the approximately 9,000 deaths, thousands of injuries and hundreds of thousands of structures damaged or destroyed (Collins & Jibson, 2015). It also led to the most lethal Mount Everest environmental hazard to date when earthquake-triggered avalanches swept through base camp, killing at least 22 people.

There is considerable spatial variation in risk, due to differences in topography, rainfall, soils and land use. The greatest hazard of this form exists when high-magnitude events (Mw ≥ 6.0) create rock avalanches. These are large (at least 1 million m^3) volumes of rock fragments that can travel for tens of kilometres from their source at velocities of hundreds of kilometres per hour. A notable mass movement occurred when an offshore earthquake (Mw = 7.7) triggered a massive rock and snow avalanche from the overhanging face of the Nevados Huascarán mountain, Peru, in 1970 (Plafker & Ericksen, 1978). At an altitude of 6,654 m, Huascarán is the highest peak in the Peruvian Andes and its steep slopes have been the source of many catastrophic slides. In the 1970 disaster, a turbulent flow of mud and boulders, estimated at 50–100 × 10^6 m^3, passed down the Rio Shacsha and Santa valleys. It formed a wave 30 m high and travelled at an average speed of 70–100 m s^{-1} in the upper 9 km of its course (Figure 6.7). The towns of Yungay and Ranrahirca, plus several villages, were buried under debris 10 m deep. At least 18,000 people were killed within four minutes after the original slope failure high on the mountain.

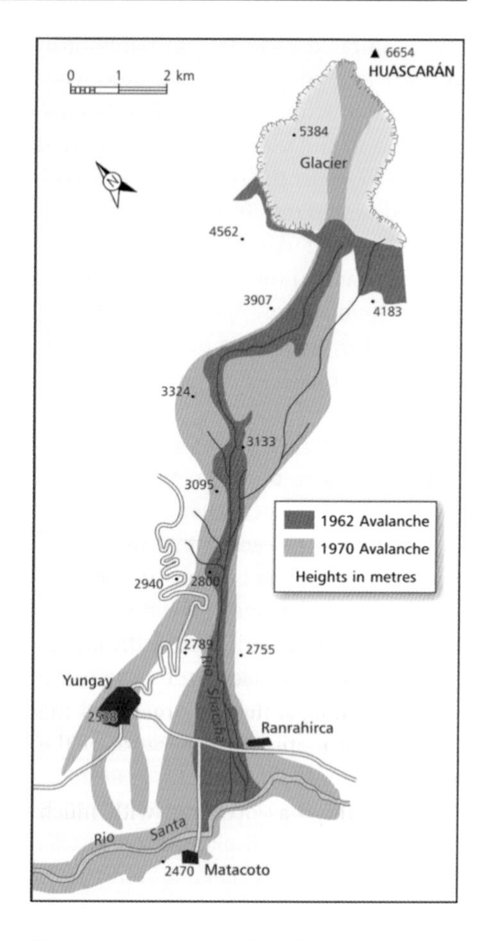

Figure 6.7 Map of the Mount Huascarán rock avalanche disasters in the Peruvian Andes in 1962 and 1970 showing the wide extent of the debris deposited in the 1970 event.

Source: After Whittow (1980)

3. Tsunamis

The most distinctive earthquake-related hazard is the seismic 'sea wave' or tsunami. A tsunami may be generated if (1) an earthquake rupture occurs under the ocean or in a coastal zone; (2) the hypocentre is not deep within the Earth's crust; or (3) the magnitude is large enough to create significant vertical displacement. The result is a series of ocean waves that spread out from this point carrying very large volumes of water. When the waves reach land they are charged with debris and can flow inland for up to several kilometres over beaches and along estuaries, with considerable destructive force (see Chapter 8).

D. IMPACTS AND CONSEQUENCES

Approximately 2 million earthquake-related deaths were recorded in the 20th century. Most resulted from a few high-magnitude events with at least 50,000 fatalities each. 'High-magnitude' is a relative term, but, as a guide, the ten largest earthquakes since 1900 were all above Mw = 8.5. Four were in the early years of the 21st century. Some relationships exist between earthquake magnitude and disaster impact, but factors other than size are needed to explain the pattern of deaths and other losses as can be seen in Figure 6.8. The Mw = 9.1 Sumatra earthquake of 2004 is amongst the ten largest and the ten deadliest earthquakes because of the many lives lost in the associated tsunami. High mortality has occurred with much lower-magnitude events. For example, the 1976 Tangshan earthquake in China had an official death toll of 255,000, although some estimates were of 655,000 or 750,000 deaths, whilst the 1920 Ningxia (Gansu) earthquake, also in China, is thought to have killed about 200,000.

The amount of destruction caused by an earthquake depends on various factors, including:

- *Duration of shaking* – in general, longer periods of shaking lead to more damage for the same magnitude of an event.
- *Distance from the fault* – as earthquake waves radiate outwards, their energy reduces with distance. Locations farther from the fault tend to experience lower levels of shaking.
- *Local conditions* – many local conditions affect the nature of shaking. For example, soil and rock properties alter the characteristics of the earthquake waves, whilst topographic effects can be significant.
- *Population density* – if the population density is high, more people will be at risk from an earthquake.
- *Building quality* – much depends on the quality of local building. Poorly built, unreinforced structures with heavy roofs are most prone to collapse; people may survive better in lightweight buildings.

The greatest losses of life and infrastructure exist when the intense seismic energy released along an earthquake fault coincides with high levels of hazard exposure and vulnerability. Most mortality associated with earthquakes is due to severe ground shaking and the collapse of buildings. Timing and location of the earthquake are often key to mortality rates; for example, a 1992 earthquake at Erzinçan, Turkey, claimed only 547 lives because it struck in the early evening, when many residents were at worship in mosques that were comparatively earthquake resistant. In contrast, the 2005 Kashmir earthquake killed over 19,000 children alone, largely because it occurred during school hours and many poorly constructed schools collapsed in the shaking.

The time of day is often significant in determining human mortality. In urban areas, fire due to the rupture of gas and water pipes is an important secondary peril. Over 80% of the property damage in the San Francisco earthquake of 1906, when about 3,000 people died, was due to fire. The 1923 Kanto earthquake killed nearly 160,000 people in Tokyo and Yokohama. This earthquake occurred at a time when over a million charcoal braziers were alight in wooden houses to

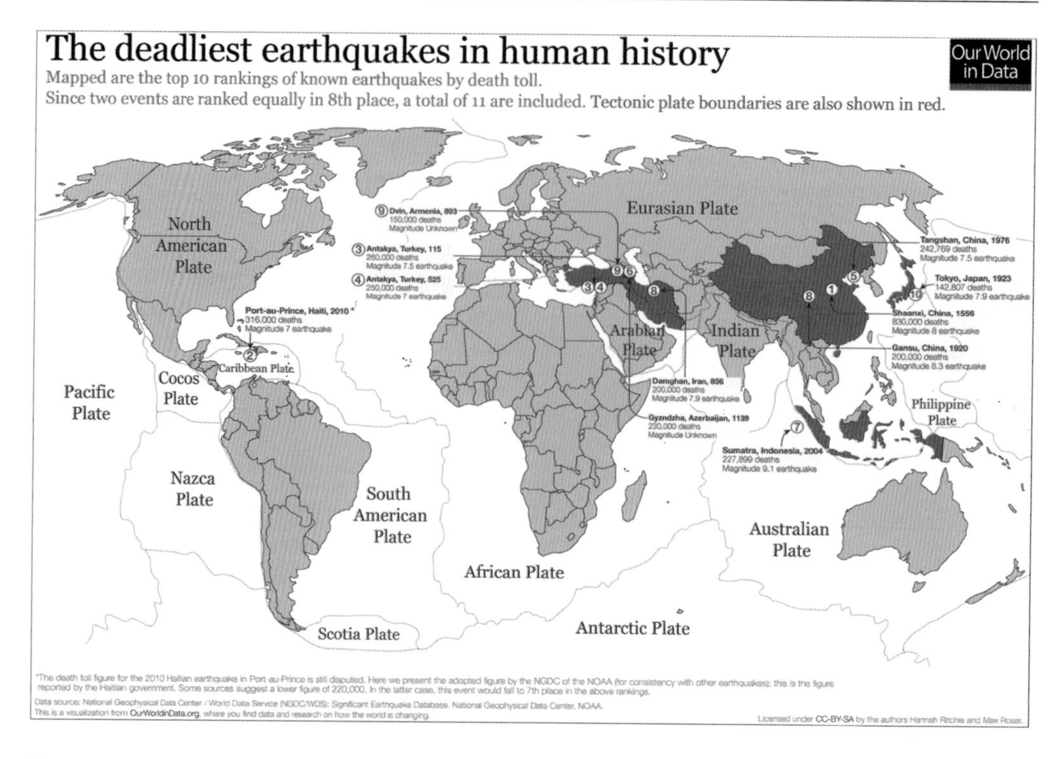

Figure 6.8 The top ten deadliest earthquakes in human history.

Source: Our World in Data

cook the midday meal. The resulting fires destroyed an estimated 380,000 dwellings. However, of equal importance is the economic status and socio-cultural development of the affected area. The 2010 Haiti earthquake experienced large loss of life and related impacts due to extreme levels of poverty and low socio-economic development (Box 6.2).

Large-magnitude earthquakes have great disaster potential because they shake the ground more severely, for a longer duration and over more extensive areas than do smaller events. But event impacts are influenced by local conditions. Geological factors, like steep slopes that cause landslides and alluvial soils that liquefy, enhance the effects of ground shaking. Most of the deaths in the 1920 Ningxia, China, earthquake arose from slope failures when loess deposits (periglacial or aeolian/windborne sediments) collapsed and buried entire towns.

Energy release alone is an imperfect guide to the socio-economic impact of earthquakes. For example, the 1995 Kobe, Japan, earthquake was a moderate (Mw = 6.8) event. Huge losses were experienced because the shock waves reached a densely populated industrial port where buildings near the shore were founded on soft soils and landfill that induced severe shaking. In

Box 6.2 The 2010 Haiti Earthquake: A Long-Term Disaster

On 12 January 2010, an Mw = 7.0 earthquake struck Haiti. The epicentre was 25 km west of the coastal capital of Port-au-Prince, home to over 2 million people. Due to the shallow 13 km depth of the earthquake, the resulting ground shaking was unusually severe and about 3.5 million people were subjected to Modified Mercalli scale intensities of VII to X, a range that can cause moderate to heavy damage to well-constructed properties. A total of 188,383 buildings collapsed. Of these, 105,000 were completely destroyed, including eight hospitals, while a further 22 hospitals were seriously damaged. The direct death toll was 222,570. About 2.3 million people, almost one-quarter of the national population, immediately fled their homes; 1.8 million were subsequently declared homeless. At the peak of the emergency, there were 1.5 million refugees in over 1,300 spontaneous settlements or camps. Damaged communication and transport systems affected the emergency response, together with confusion over lead responsibilities, failures to prioritise relief flights and delays in the distribution of aid supplies. Later in the recovery period, unpredictable movements of the survivors, disputes over land rights and fluctuating government policies further restricted progress. Later still, outbreaks of cholera spread to all ten departments of the country.

This disaster would have taxed the resources of any country. However, the scale of devastation and the patchy recovery show that the earthquake exposed long-term socio-economic problems that accumulated over time and greatly amplified the disaster impact. Haiti is the poorest country in the western hemisphere, with about 80% of its population living below the poverty line. More than two-thirds of the labour force lacks formal employment and is engaged in agriculture. The country is ranked about 150 out of more than 180 countries in the world on the Human Development Index scale and has been classed as economically vulnerable by the UN. Since the country gained independence from France in 1804, it has been under dictatorships for much of its history. It is also exposed to natural hazards other than earthquakes, including tropical cyclones, flooding and landslides.

Much of the damage was sustained in the capital, where about 85% of the population was living in urban slums, mostly tightly packed, inadequately built concrete buildings. Half of these people had no access to toilet facilities and only one-third had piped water available. Haiti has no properly approved construction code and, before the earthquake, it was estimated that about 60% of the houses were unsafe for occupation in normal circumstances. Following the extensive building collapse, there was 19 million m^3 of rubble in the streets of Port-au-Prince. The long-term challenge is to reinstate the capital and other areas in a more sustainable fashion. Previous attempts to improve the urban environment have faltered, due to a lack of public participation, low staff capacity and weaknesses in accountability.

addition, there was much wooden housing built to withstand tropical cyclones rather than earthquakes. The heavy, clay-tile roofs, typically weighing two tonnes, collapsed, and fires were readily started in the wooden structures. Over 90% of the 6,400 fatalities occurred in these areas of suburban housing. This exemplifies the challenges in dealing with multiple hazards simultaneously.

E. RISK ASSESSMENT AND MANAGEMENT

There is little immediate prospect of preventing earthquakes at source. The US Geological Survey conducted a long-term experiment starting in 1985 to identify precursory phenomena near the town of Parkfield, California, a 25 km stretch of the San Andreas Fault. This fault is intensively studied because it slips with a fairly short recurrence interval of about 22 years to give moderate (Mw = 6.0) earthquakes and also because over 120,000 households are at risk from seismic activity in the area. Despite significant research findings that have aided the study of earthquakes, subsequent analysis of the data collected revealed no precursory indications in time and therefore no temporal predictive elements could have been identified (Park et al., 2007). Disaster reduction, therefore, must focus on lowering human vulnerability and limiting secondary hazards. This is done via key approaches: monitoring, prediction and warnings, and preparedness via resistant design and community actions.

1. Monitoring

Detection of earthquakes is made possible by the use of seismographs, which are instruments used to record the motion of the ground during an earthquake. They are installed in the ground throughout the world and historically consisted of a pendulum or a mass mounted on a spring that moved with the ground, recording the movement on a seismogram (Figure 6.9). Today seismometers can also be high-tech electronic sensors, amplifiers and recording devices that operate as part of a seismographic network. The Global Seismographic Network collates data from over 150 seismic stations located around the globe and is supported by the United States Geological Survey and other non-government research institutions. This network locates and monitors earthquakes, but also studies the deep interior of the Earth and can locate underground nuclear explosions. Using just three seismometers it is also possible to triangulate the focus/hypocentre of the earthquake.

2. Forecasting and Warning for Earthquakes

A reliable forecast or prediction should specify that an earthquake of a given magnitude is likely in a certain area within a stated time-window. At the present time, such predictions are not possible temporally, and it is not even clear how such information could be applied for the maximum benefit. However, it is possible to provide a very short-term warning system. Figure 6.9 demonstrates the recording of seismic waves over time with the arrival of the P-wave, S-waves and surface waves over time. It is possible using high-tech seismological systems to issue a warning once the P-waves are first detected (that do not cause any significant shaking) and

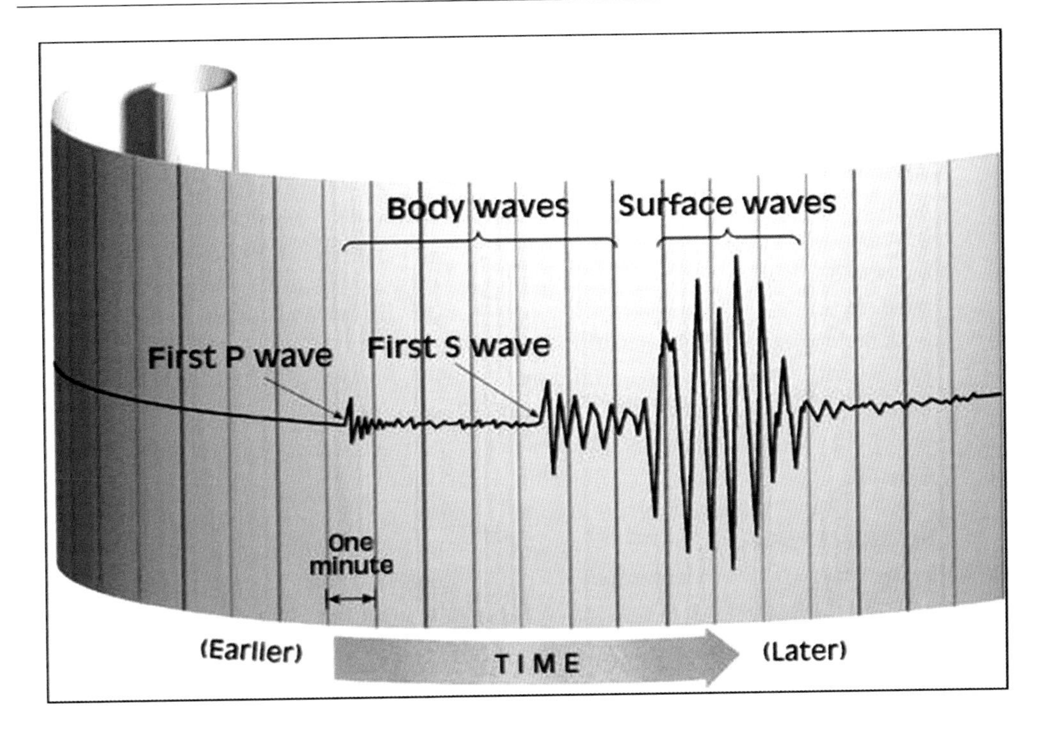

Figure 6.9 A typical seismograph.

Source: SMS Tsunami Warning (n.d.)

prior to the far more destructive S-waves arriving at the surface. Using the P-waves, an estimation of an earthquake's size and location can be established, and an integrated warning system can relay electronic warnings around 50 seconds ahead of the oncoming S-waves. Whilst this may not seem a lot of time, it is enough to provide a warning to the population, via text or other notification and TV/radio automated messages, and to trigger automated shut-downs of computer servers, gas pipelines, nuclear power plants, bridges, railways and elevators. This can make a significant difference to the impact, but it comes at a significant investment and reliance on technology. Examples include systems operated in Mexico, Japan, Taiwan and Chile, with a newer system in California, USA. For many nations, however, this type of warning is just not economically viable, and a nation must either use probabilistic or deterministic methods to forecast earthquakes and prepare as best as possible.

3. Probabilistic Methods

A record of the frequency of large earthquakes in any area can be used to estimate the future likelihood of similar events. In a country like New Zealand, where the pattern of earthquake activity does not correlate well with the surface geology, the historical record is useful (Smith & Berryman, 1986). Figure 6.10a employs data on shallow-focus earthquakes of $M \geq 6.5$ from 1840 to 1975 to map the return periods for intensity MM VI and greater, the level at which significant damage begins. Figure 6.10b shows the intensities with a 5% probability of occurrence within 50 years. This type of regional zoning has clear limitations because it is based on periods that are short on geological scales and fails to account for local ground conditions.

A major assumption in statistical analysis is that earthquakes occur randomly through time. This may not happen, because fault lines can interact with each other. This point was illustrated following the tsunami-generating 26 December2004 Sumatra earthquake, which led to increased stress concentrations on the adjacent fault lines. By February 2008, there had been seven subsequent earthquakes of Mw ≥ 7.0 in the same area, including the March 2005 Mw = 8.7 Nias earthquake, which killed 1,300 people; the May 2006 Mw = 6.2 Java earthquake, which killed 5,800 people; and the July 2006 Mw = 7.6 Java earthquake, which killed 660 people. Further large events seem likely on unruptured sections of the fault.

Seismic behaviour varies between individual fault segments. For example, the San Andreas Fault in California consists of both locked and creeping segments. Locked segments allow sufficient strain to build up to trigger major earthquakes, whilst creeping segments are characterised by more continuous sliding

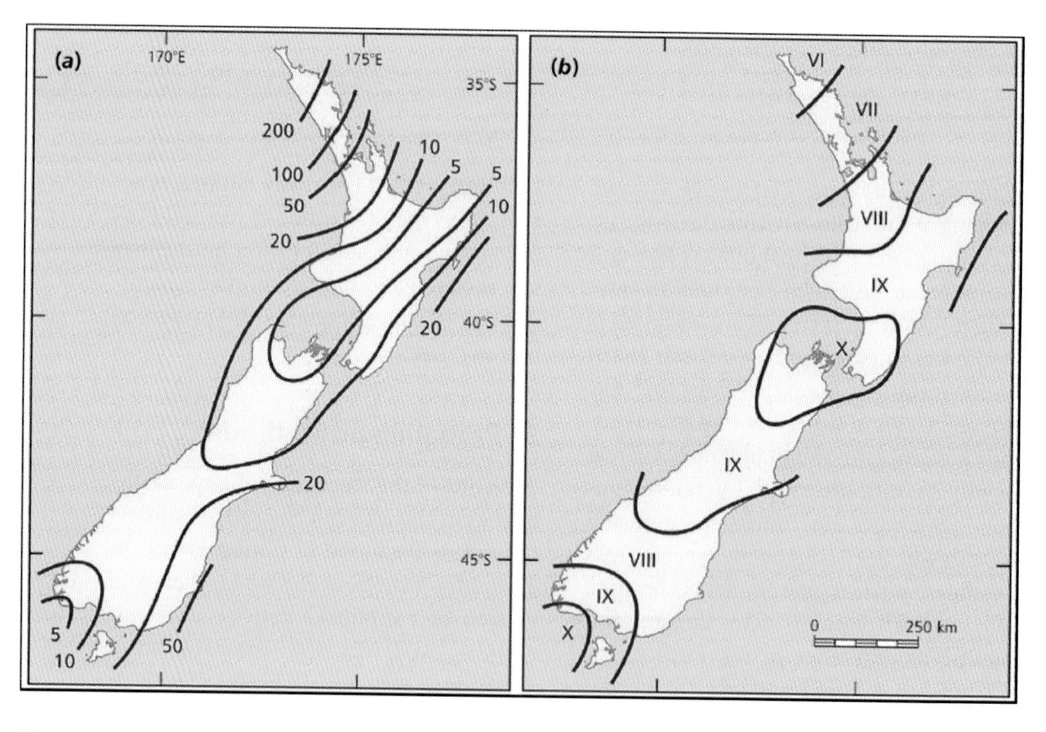

Figure 6.10 Earthquake prediction in New Zealand. (a) Return periods (years) for earthquakes of MM VI and greater; (b) intensities with a 5% probability of occurrence within 50 years.

Source: After Smith and Berryman (1986)

processes. Such creep appears to result from the presence of finely crushed rock and clay with a low frictional resistance, which limits the build-up of stress. However, more competent rocks at greater depth may accumulate stress. Dolan et al. (1995) claimed that, in the Los Angeles region, far too few moderate earthquakes had occurred during the last 200 years to account for the observed accumulation of tectonic strain. It is possible that the historic record reflects a period of unusual quiescence, but damaging strain may be accumulating. The US Geological Survey has calculated that the probability of at least one earthquake (Mw = 6.7 or more) striking between 2000 and 2030 in the San Francisco Bay area is 70% (twice as likely as not).

4. Deterministic Methods

Deterministic methods rely on the detection of earthquake precursors near the active fault. A number of different phenomena have been employed, including:

- *Seismicity patterns.* Some researchers have suggested that characteristic changes occur in background seismicity during the period before an earthquake, primarily due to shifts in the stress state of the fault as failure starts to develop (Wyss et al., 2012).
- *Electromagnetic field variations.* Others have claimed that the development of a fault rupture might lead to variations in the Earth's magnetic field that can be detected (Gershenzon et al., 1993).
- *Weather conditions and unusual clouds.* A few scientists maintain that distinctive cloud patterns can be observed along the line of earthquake faults prior to rupture. The reasons for this are unclear (Ikeya, 2004).
- *Radon emissions.* Post-earthquake analysis of borehole and soil-gas sensors indicate altered radon concentrations prior to an earthquake event. This is thought to result from the occurrence of cracking in the rock mass as the earthquake rupture begins and releases radon gas trapped within the rock (Barkat et al., 2017).
- *Groundwater level.* There is some evidence that groundwater levels change prior to an earthquake, probably because of the cracking process outlined previously (Hosono et al., 2020).
- *Animal behaviour.* Anomalous animal behaviour has been widely observed and reported from some countries prior to large earthquakes (see Woith et al., 2018).

The reliability of these techniques remains unproven because of inadequate scientific evidence, but significant progress has been made in recent years to demonstrate there is value to exploring each of these approaches further. It remains difficult to distinguish normal variations in the preceding parameters from the conditions associated with an impending earthquake. For example, the water level in wells falls and rises in response to atmospheric pressure changes and rainfall amounts.

Longer-term warnings or forecasting for earthquakes are possible by identifying regions or areas along a fault line that are more susceptible to fracture. By studying active faults, identifying recent epicentres and where earthquakes have not occurred, along with the build-up of strain along a fault line, it is possible to forecast where the next earthquake may occur, although it is not possible to pin down the exact location of a future rupture. Figure 6.11 demonstrates how applying this process to Indonesia in 2008 helped identify a

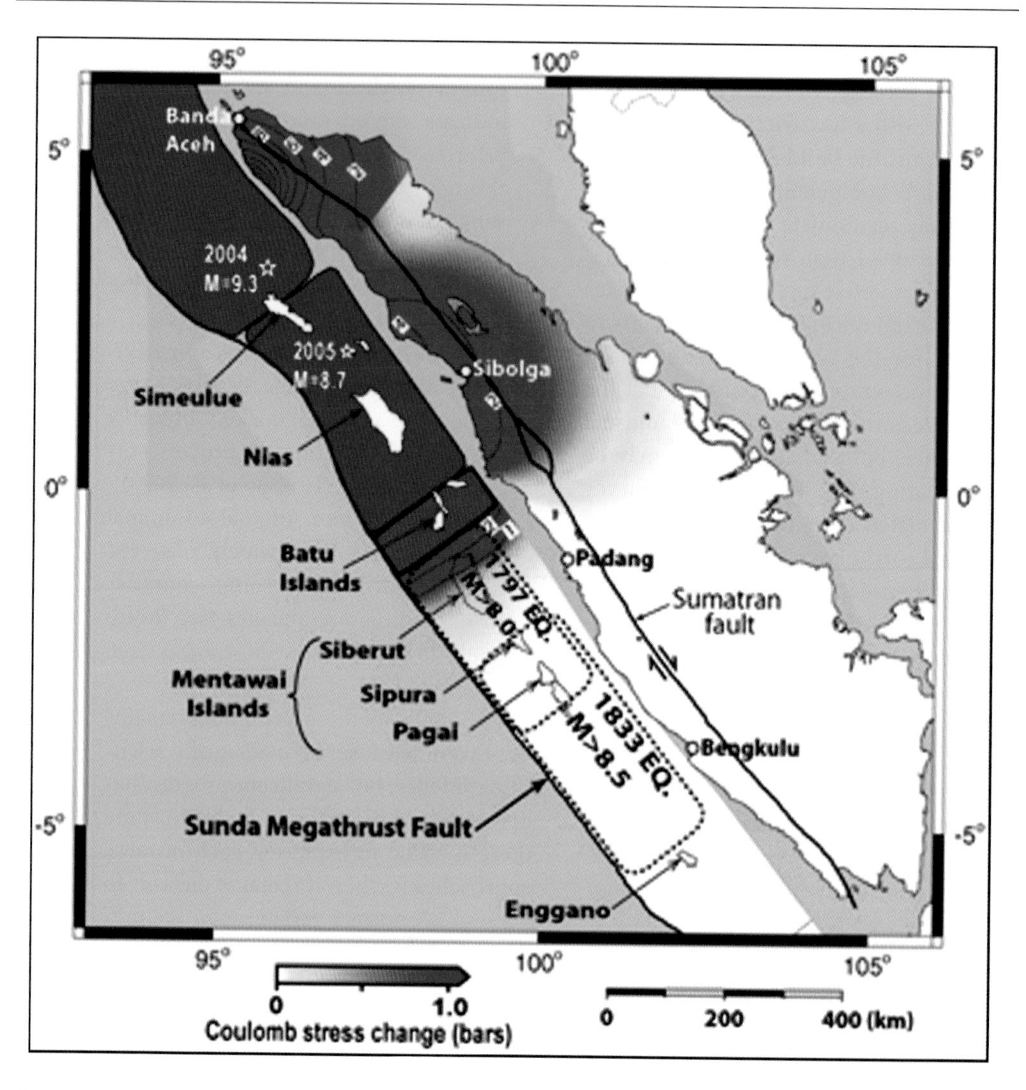

Figure 6.11 Historical earthquakes and current interaction stresses on the Sunda megathrust. Solid lines indicate the traces of important faults. Dotted lines indicate outlines of main historical earthquake ruptures as estimated by paleogeodetic studies (ancient measurements of the Earth's geometric shape, orientation in space and gravity field.) Stars indicate epicentres and dark blue areas indicate extent of recent great earthquake ruptures. Current interaction stresses, shown in the yellow–red colour scale, have been calculated for this study and include coseismic stress and both the effect of after-slip on the Simeulue–Nias source region and aseismic slip under the Batu islands. Colours indicate the stress, calculated at a depth of 10 km, resolved onto the appropriate mechanism (right-lateral strike slip for on-land Sumatra and thrust for the subduction zone).

Source: McCloskey et al. (2008, p. 63)

high risk of an earthquake in Padang as it became clear the fault line was 'unzipping'. In 2009 an Mw 7.6 earthquake struck just 45 km west-northwest of Padang, and whilst there were still over 1,115 dead and 135,000 houses severely damaged, preparedness work implemented to help the populations living in and around Padang helped to reduce the earthquake's devastating impacts (McCloskey, 2011).

5. Preparedness

In the absence of being able to predict, or warn for earthquakes, more than a very short time in advance, a useful investment is via preparedness and early action through developing earthquake-resistant buildings and preparing the community.

Resistant Design

According to Booth and Key (2006), about 60% of all earthquake-related deaths are caused by the failure of unreinforced masonry structures (URMs) in rural areas. URM buildings behave poorly under earthquake loadings due to their highly rigid and low-ductile behaviour (Mazumder et al., 2021). The most vulnerable buildings are constructed from adobe material (sun-baked clay bricks). Adobe construction is common in arid and semi-arid regions because it is cheap, easily worked and widely available. In Peru, an estimated two-thirds of all rural dwellers live in adobe houses; and in the 2003 Bam earthquake many such buildings collapsed, killing over 26,000 people (see Chapter 1, Box 1.9). Houses built of rubble masonry are also prone to collapse. In the Maharashtra, India, earthquake of 1993 the pucca houses – with thick granite walls and heavy timber roofs – led to more deaths than either the thatched huts or buildings poorly reinforced with concrete frames.

Some societies have deliberately employed 'weak' construction as a defence against earthquakes. Throughout tropical Asia, housing is typically lightly built with plant-matting walls and palm-frond roof (Leimena, 1980). During the Mw = 8.7 Nias earthquake in Indonesia in 2005, 1,300 people died in collapsed buildings, but the native wood-framed longhouses survived mostly undamaged. Other countries favour houses of wood-frame construction. Such buildings account for about 80% of all houses in the USA. There is a clear seismic benefit because these structures tend to flex, rather than fail, when subjected to ground shaking, although they carry a high fire risk – an important secondary hazard in earthquakes.

Over time, numerous URM buildings have been supplemented by high-rise, reinforced concrete structures – mainly apartment blocks – built to accommodate the growing population, particularly in urban areas or megacities (population greater than 30 million). This means that most urban areas present a complex array of risk. Figure 6.13, based on the effects of the Kobe earthquake, shows general relationships between earthquake intensity and building damage for different types of structures. It also emphasises how the risk of collapse rises with an ageing stock of buildings.

The only way to achieve safe buildings in seismically active urban areas is through detailed risk assessments and the adoption of best construction practices, usually by earthquake engineers (Elnashai & Di Sarno, 2015). The first requirement is for a geotechnical engineer to assess the suitability of the location. Other things being equal, buildings on solid rock are less likely to suffer damage than those built on clays or softer foundations. This initial assessment should include geological investigations to avoid building near faults or where the bearing strength of the foundation materials is inadequate (Box 6.3).

Seismic building codes, now adopted in over 100 countries, stipulate the minimum

Figure 6.12 Examples of structural collapse in different types of buildings. (a) A 100-year old adobe building in Talca, Chile, after the Mw = 8.8 earthquake in February 2010. Adobe construction using clay, sand and straw is no longer authorised in Chile. (b) A multi-storey residence in Port-au-Prince, Haiti, in January 2010. The weight of concrete floors and roof proved too great for the strength of the supporting columns.

Source: Photo (a): Walter D. Mooney, USGS

construction standards required to minimise the risk of collapse. To be effective, a seismic code needs full legal status and procedures for updating the criteria over time. Above all, the code should ensure that the construction methods for the planned building are adequate to withstand the selected design earthquake and

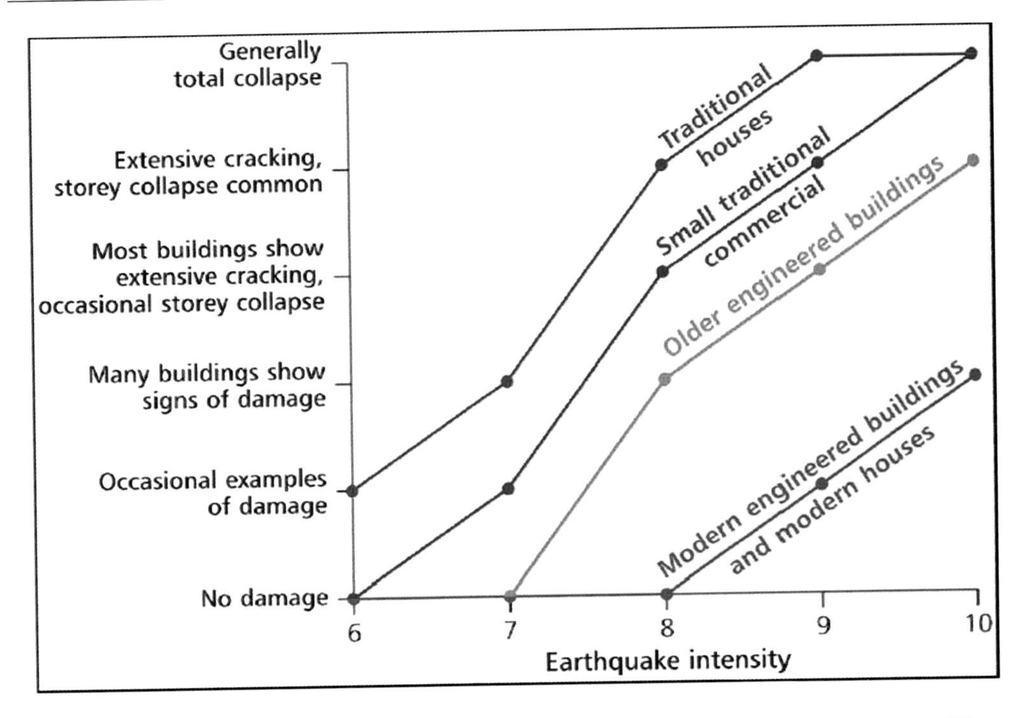

Figure 6.13 Relationships between earthquake intensity (Mercalli scale) and the damage for different types of building construction, based on the 1995 Kobe earthquake. Sharp difference in performance occurred between traditionally constructed and modern engineered buildings.

Source: After Alexander Howden Group Ltd and Institution of Civil Engineers (1995)

Box 6.3 Earthquake Safety and Buildings

The key to earthquake resistance is the appropriate choice of building design and construction methods. In this context, strong, flexible and ductile materials are preferred to those that are weak, stiff and brittle. For example, steel framing is a ductile material that absorbs a lot of energy when it deforms. Indeed, the spread of well-designed, steel-reinforced concrete buildings has been the primary factor in increasing earthquake safety for many decades. Glass, on the other hand, is a very brittle material that shatters easily. In practice, both types of material have to be incorporated into structures. Some otherwise well-designed structures collapse because of the failure of a single element which lacks sufficient strength or ductility. For example, buildings with flexible frames will often fail if the frames are in-filled with stiff, masonry brickwork.

The shape of a building will influence its seismic resistance. A stiff, single-storey structure (Figure 6.14a) will have a quick response to lateral forces, while tall, slender, multi-storey buildings (Figure 6.14b) respond slowly, dissipating the energy as the waves move upward to give amplified shaking at the top. If the buildings are too close

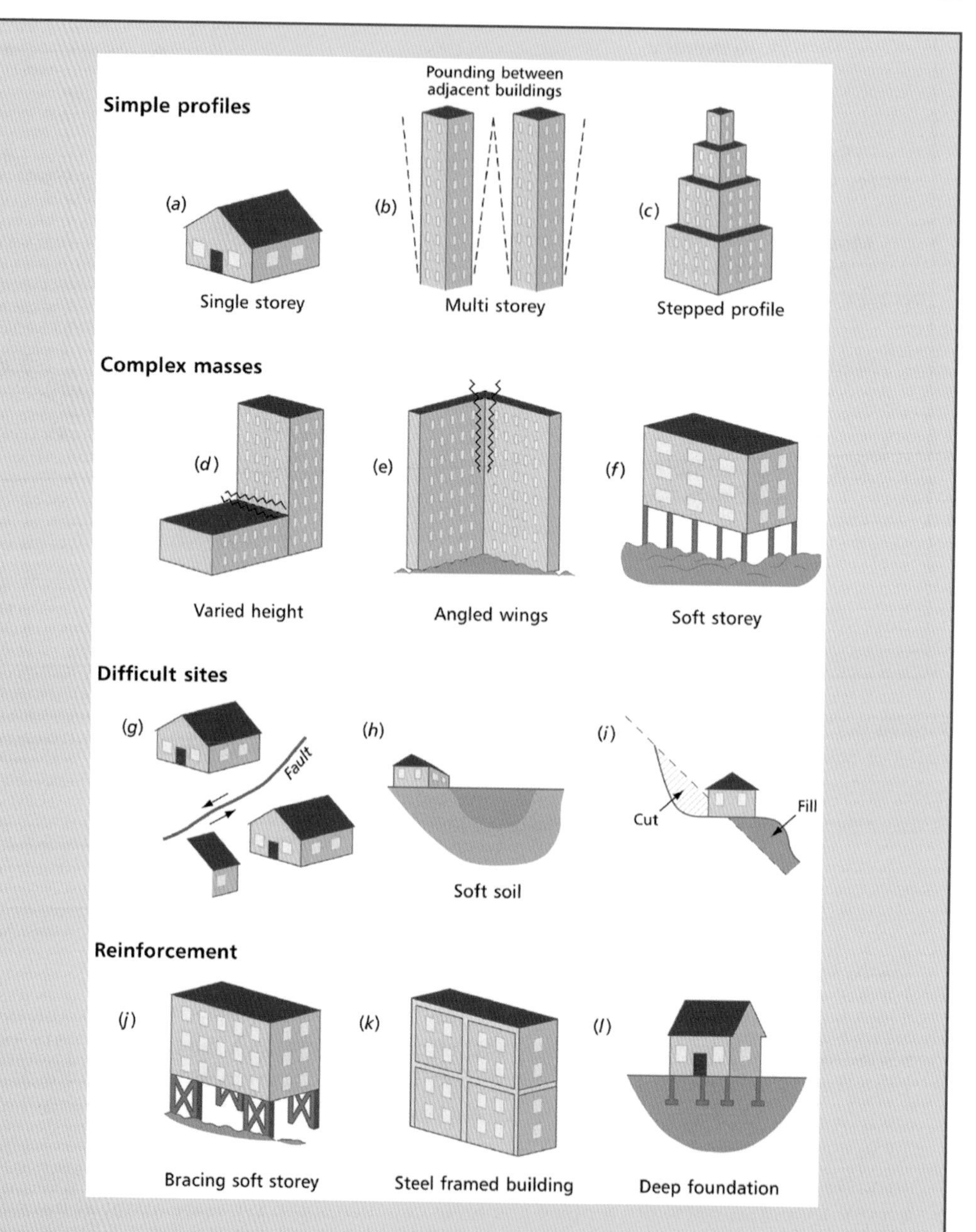

Figure 6.14 Schematic illustration of the effects of ground shaking on various types of buildings and examples of constructed measures designed to resist earthquake hazards: (a–c) simple building profiles, (d–f) complex building masses, (g–i) coping with difficult sites, (j–l) methods of building reinforcement.

together, pounding induced by resonance may occur between adjacent structures and add to the destruction. The stepped profile of the vertical mass of the building in Figure 6.14c offers stability against lateral forces. Most buildings are not symmetrical and form more complex masses (Figures 6.14d and 6.14e). These asymmetrical structures will experience twisting, as well as the to-and-fro motion. Unless the elements are well joined together, such differential movements may pull them apart. High-rise structures will be vulnerable if they do not have uniform strength and stiffness throughout their height. The presence of a *soft storey*, which is a discontinuity introduced into the design for architectural or functional requirements, may be the weak element that brings down the whole structure. Figure 6.14f shows a soft ground-floor storey, perhaps introduced to ease pedestrian traffic or car parking.

The weakest links in most buildings are the connections between the various structural elements, such as walls and roofs. Connections are important in the case of precast concrete buildings, where failure often results from the tearing-out of steel reinforcing bars or the breaking of connecting welds. In the 1994 Northridge earthquake a number of multi-storey car parks failed when vertical concrete columns were cracked by lateral ground shaking to the point where they became unable to support the horizontal concrete beams holding up the different floors. Exterior panels and parapets also need anchoring firmly to the main structure in order to resist collapse. Architectural style can contribute to disaster if features like chimneys, parapets, balconies and decorative stonework are inadequately secured.

Difficult construction sites (Figures 6.14g and 6.14h) include localities near to geological faults and soft soils that amplify ground shaking. As far as possible these should be avoided or built up at low densities so that, for example, buildings cannot collide as a result of downward movement on slopes. Some slopes may have to be reformed by cut and fill to limit the threat from earthquake-related landslides (Figure 6.14i). Methods of building reinforcement include the cross bracing of weak components, placing the whole structure in a steel frame and the installation of special, deep foundations on soft soils (Figures 6.14j–l). Adequate footings are important. High-rise buildings on soft soils should have foundations supported on piles driven well into the ground. Wood-framed houses should be internally braced with plywood walls tied to anchor bolts linked into foundations 1–2 m deep. Some new buildings can be mounted on isolated shock-absorbing pads made from rubber and steel which prevent most of the horizontal seismic energy being transmitted to the structural components. The technique is expensive but provides maximum protection for the loose contents of buildings, thus making it attractive for hospitals, laboratories and other public facilities. In addition, base-isolated buildings need less structural bracing to withstand lateral forces, so that the reduction in construction materials offsets the extra cost of the isolation system. Whilst earthquake-resistant buildings can be very high tech and expensive, it is possible to use cheap local materials to develop earthquake-resistant buildings using basic engineering principles that can be far more affordable.

should also require regular site inspections throughout the construction period. Unfortunately, building codes are frequently neglected and bypassed, due to a lack of resources, imperfect technical knowledge and local corruption. In the 1999 Marmara earthquakes in northwest Turkey, 20,000 people were killed and 50,000 were injured, despite the fact that building codes have existed there since the

1940s. Most deaths were blamed on non-compliance with the codes, largely due to financial corruption.

The majority of building stock in most areas will pre-date the building codes. Retrofitting old buildings to meet new standards costs money and can be equally difficult to undertake.

Even when properly applied, building codes cannot prevent all seismic damage. The codes may be based on an incomplete knowledge of local geology, and calculations regarding the performance of buildings under stress may be flawed, especially where regulations are based on experience imported from elsewhere. Administrators and decision-makers are not always convinced of the threats and may fail to see any financial advantage in the need for investment in earthquake security. At worst, building codes can lead to new buildings in hazardous areas if they create a false sense of security.

Much importance is attached to the earthquake resistance of public structures such as hospitals, dams, nuclear power stations and factories with explosive or toxic substances. Urban lifelines for transport, electric power, water supply and sewerage also need priority. Some commercial organisations take separate precautions, especially when insurance cover is unavailable unless there is compliance with standards in the building code. For example, the IBM manufacturing plant at San José, California, was subjected to an early retrofit programme (Haskell & Christiansen, 1985). As a result, it was able to get all but one of its Santa Clara buildings back into full operation the day after the Loma Prieta earthquake of 1989. Dams are also critical to adhere to the highest building standards as a failure can release a devastating flood.

Community Preparedness

Community preparedness and disaster recovery planning is a necessity in the face of not being able to receive timely warnings. Community preparedness is best developed at the local level within a framework established by state or national government. In some cases, it may be difficult to identify the areas at greatest risk. The 2005 Kashmir earthquake in Pakistan was the first large earthquake to affect that country in living memory. As a result, preparedness was inadequate. The emergency services were not trained for search-and-rescue operations and no contingency plan was in place to bring in assistance from outside the earthquake-affected area. Many people who survived the initial ground shaking were trapped under the rubble of their houses, and injured survivors died because of a shortage of hospital beds and specialist treatment (Rossetto & Peiris, 2009).

The creation of community preparedness is neither a short-term nor a simple process. The 'California at Risk: Reducing Earthquake Hazards 1987–1992' programme devised a basic checklist as shown in Table 6.3. Such complex programmes require multiple pieces of legislation, implementation and maintenance.

To be successful, preparedness must engage the public and involve household preparedness, where people can afford it.

If properly undertaken, emergency simulations can provide practical information on first aid and household evacuation as well as raise general hazard awareness. In Japan, legislation first introduced in 1961 includes emergency drills in disaster prevention. A comprehensive exercise is organised annually in designated areas on *Disaster Prevention Day* (1 September), with clear emphasis on raising

Table 6.3 Pre-earthquake planning for post-earthquake rebuilding (PEPPER) – a planning checklist

Existing development
- inventory hazardous buildings
- strengthen critical facilities
- reinforce hazardous buildings
- reduce non-structural hazards
- regulate hazardous materials

Emergency planning and response
- determine earthquake hazards and risks
- plan for earthquake response identify resources for response
- establish survivable communications system
- develop search-and-rescue capability
- plan for multijurisdictional response
- establish and train a response organisation

Future development
- require soil and geologic information
- update and improve safety element
- implement Special Studies Zones Act
- restrict building in hazardous areas
- strengthen design review and inspection
- plan to restore services

Recovery
- establish procedures to assess damage
- plan to inspect and post unsafe buildings
- plan for debris removal
- establish programme for short-term recovery
- prepare plans for long-term recovery

Public information, education and research
- work with local media
- encourage school preparation
- encourage business preparation
- help prepare families and neighbourhoods
- help prepare elderly and disabled
- encourage volunteer efforts
- keep staff and programmes up to date

Source: After Spangle and Associates Inc. (1986)

the awareness of children and other citizens. In 2001 the Tokai earthquake drill involved 1.5 million people from Tokyo and the surrounding area. Figure 6.15 demonstrates advice given in relation as to what to do when an earthquake strikes, and it is important that people understand the differences when in or outside a building. This highlights the value of education and sharing knowledge, and the vulnerable person being able to apply it.

Globally there is an International ShakeOut Day on the third Thursday of October, where individuals, businesses, schools, faith-based organisations, community groups, government agencies and others participate in the ShakeOut to get prepared for earthquakes and to share what they're doing so others can do the same. The website www.shakeout.org/index.html provides resources to help plan, promote, run and educate during the event, as well as find events in one's own country.

Land Use Planning

Land planning can guide earthquake-prone communities to a safer future. Land zoning based on seismic risk not only informs engineers about the design and construction methods to be employed to produce safer buildings, bridges and roads in a particular locality but also promotes other hazard-reducing measures such as building codes, retrofit priorities and building insurance rates.

There is always some resistance to planning restrictions, especially by local bodies seeking to pass disaster costs on to a higher level of administration. Even when hazard information is available, it may not be applied to best effect. Californian state law, for example, requires that estate agents (realtors) advise all potential purchasers if residential properties are located near mapped fault lines (see a typical brochure provided in 2020: https://ssc.ca.gov/wp-content/uploads/sites/9/2020/08/20-01_hog.pdf). In practice, such information may not be disclosed until sale

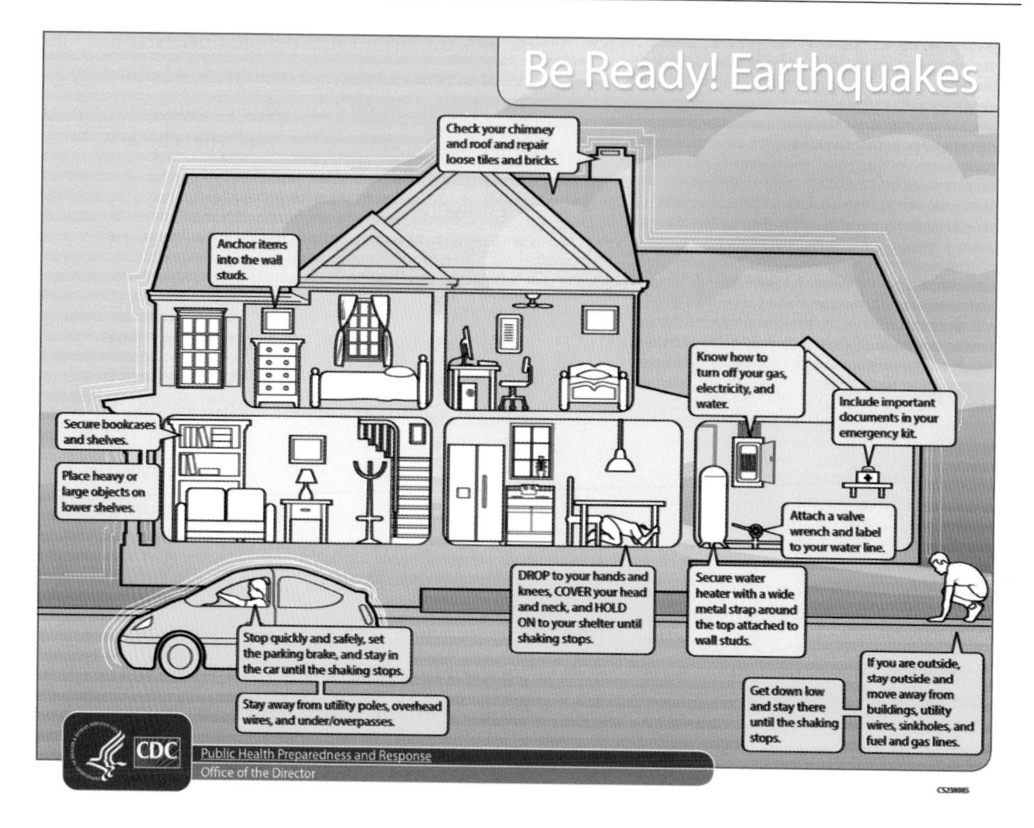

Figure 6.15 Be Ready! Earthquakes.

Source: CDC.Gov (n.d.) (accessed 2022)

negotiations are well advanced. In any case, earthquake hazards are unlikely to be a major factor in decision-making if other attributes – like an attractive view, schools, shops and investment potential – are important to buyers, especially if they intend to relocate in a few years' time.

Before land can be zoned and developed safely, the seismic risks have to be accurately mapped. A seismic hazard map shows the pattern of potential ground-shaking hazard in future earthquakes. Probabilistic maps of seismic shaking hazard (see Figure 6.16) are the basis for more detailed mapping that allows contours of expected ground motion, typically measured by peak ground acceleration (PGA), to be viewed at scales suitable for local planning decisions. Whilst these maps can be useful, they need refinement because the shaking depends on the detail of the surface deposits. The highest seismic hazard exists above geologically young deposits or human-made landfill, with the potential to liquefy into a sandy fluid during an earthquake. Such deposits exist in the low-lying sections of the San Francisco Bay area in the form of soft mud and sands and gravels. Figure 6.16 indicates the enhanced shaking in these deposits, relative to the bedrock, and the likely consequences for built structures.

In California, setback ordinances are a major tool, where there is a minimum

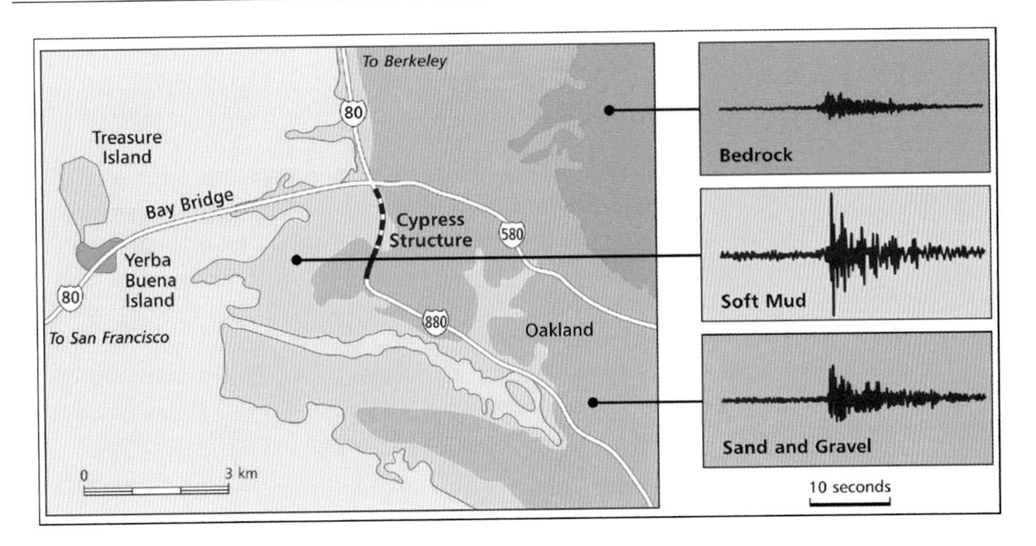

Figure 6.16 An illustration of variations in ground shaking due to surface geology. Deposits of mud and gravel shake to a much greater extent than bedrock. During the 1989 Loma Prieta earthquake, part of the elevated Cypress Structure freeway in the San Francisco Bay area collapsed, due to strong shaking in the soft muds on which it was constructed.

Source: After USGS Earthquake Hazards Programs at http://earthquake.usgs.gov/regional/nca (accessed 31 March 2011)

distance which a building or other structure must be set back from a street or road, a river or other stream, a shore or floodplain, or any other place which is deemed to need protection. Building setbacks are recommended where proposed development crosses known or inferred faults, and slope stability setbacks can be established where unrepaired active landslides, or old landslide deposits, have been identified.

6. After an Earthquake

Disaster Aid

Earthquake disasters readily attract emergency donations because of the sudden loss of life, heavily publicised by television imagery and social media. However, aid issues extend beyond the immediate crisis.

The 'golden hour' after an earthquake are crucial for the recovery of survivors trapped in fallen buildings (see Chapter 5). Landslides following earthquakes in more remote mountainous areas also cause problems for search-and-rescue operations and the delivery of aid (Box 6.4). Over 17 specialist search-and-rescue teams are kept on permanent stand-by around the world with the capability to reach a disaster zone within hours of a request from the relevant country. These teams carry enough food and water supplies to remain self-sufficient for up to two weeks. This activity is highlighted by the donor authority and in the media, but there is little reliable data on the practical value of such external support. Rom and Kelman (2020) argue that international search-and-rescue teams arrive too late to make a significant contribution to lives saved and that it is an expensive activity relative to training and preparing locals. Therefore, they argue that investing in

Box 6.4 Problems of Aid Delivery in the Aftermath of a Major Earthquake

The Mw 7.6 earthquake of 8 October 2005 had a devastating impact on a large area of Kashmir. According to official government statistics, the earthquake killed over 73,000 people in Pakistan. A further 1,360 persons died in India and a handful of deaths were reported for Afghanistan. The earthquake left over 100,000 people with injuries. Over 780,000 buildings were damaged beyond repair, the vast majority (97%) of which were houses, leaving perhaps 3 million people homeless.

A massive relief operation was initiated by the Government of Pakistan and by international agencies, such as the International Committee of the Red Cross and the World Food Programme. However, all assistance was hindered by two factors:

1. A lack of preparedness meant that little planning had been undertaken for the relief operation.
2. The Kashmir area is highly mountainous, with limited communication routes.

One of the main areas affected was the valley of the river Neelum. This area is accessed by a single road only, which crosses the fault in a zone with a steep river gorge. Landslides due to the earthquake blocked this road over a 20 km stretch and, in the aftershocks, landslides continued to occur on an hourly basis. Additionally, many of the most seriously damaged villages were located on the high slopes of the mountains, accessed only by small roads traversing very steep hillsides. In almost all cases, these roads were destroyed (Peiris et al., 2006). Aid delivery was extremely difficult. Even an assessment of the needs of the population was problematic, as the evaluation teams could not travel through the affected areas without helicopters, which were in short supply. Although the government mobilised 12 brigades of the engineer corps of the Pakistan Army, most of the major highways were not reopened for a month and reopening the Neelum Valley road took six weeks. Some minor roads remained closed three years later and are regularly damaged by landslides.

Logistical problems included restrictions on the supply of emergency medical care, food and shelters. The lack of shelter was a serious issue because winter in Kashmir is very cold. As a result, two unusual steps were taken in Kashmir. First, there was a major effort to move people into camps close to the main roads in the valley floors. In most cases, this meant moving people away from their home villages. This is undesirable because it increases disaster trauma and delays the rebuilding process, but, given the limitations of the transport system and the intense cold in the mountains, it was the only realistic solution. Second, there was much reliance on helicopters to deliver assistance. Helicopters were deployed from around the world – for example, the UK sent three RAF heavy-lift Chinook helicopters and the United States sent a further 12 to assist the large numbers used by the Pakistan Air Force. Additional assistance came from civilian agencies; the World Food Programme alone deployed 14 helicopters.

These humanitarian efforts undoubtedly saved lives, yet prevention remains the most effective action.

equipping and training local teams in high-risk, vulnerable earthquake-prone areas may help save more lives than traditional search-and-rescue efforts.

Once the emergency response phase is over, new decisions are required in the recovery (as discussed in Chapter 5). The basic aim is to provide temporary support for as short a time as possible to avoid dependency. This means working closely with local food vendors and water providers, sourcing goods and other services in a way that minimises competition with indigenous activities. In other words, a major question facing aid agencies is how best to manage the shift from short-term humanitarian support to longer-term intervention. In the first six months following the 2010 Haiti earthquake, the UK-based Disasters Emergency Committee spent about half of its appeal funds more or less equally between supplying water and sanitation and providing shelter. Other early needs included food and health care.

In the longer term, specific problems emerged regarding housing in Port-au-Prince, (Clermont et al., 2011). During recovery in urban settings, the safe return of tented residents to permanent accommodation within their own neighbourhoods is a priority. If this is done on a well-planned, participatory basis, it is preferable to the creation of new settlements – often little more than camps – in areas lacking public services or job opportunities. However, this policy may conflict with the needs of the most vulnerable victims, who depend for longer on short-term shelters and require more health care and other support. Some NGOs favour an intermediate move into prefabricated transitional 'T' shelters because it often fits the timeframe of their budgets and provides an earthquake- and hurricane-proof structure, but these shelters are almost always better suited to rural situations. These issues were prominent in Haiti, where it was difficult to erect 'T' shelters, due to disputes over land ownership and availability.

Long-term reconstruction rarely conforms to the original timetable. The 1988 Armenian earthquake made 514,000 people homeless and resulted in the evacuation of nearly 200,000 persons. Following the Soviet government's decision to accept international aid, over 67 nations offered cash and services amounting to over US$200 million. A reconstruction programme was announced to rebuild the cities within a two-year period on sites in safer areas, with building heights restricted to four storeys. But, during the first year, only two of the 400 buildings due for construction in Leninakan were completed, and many people lived as evacuees many months after the disaster. Continuing aid and support is also required for less tangible purposes, such as the treatment of post-traumatic stress following earthquake disasters (Karanci & Rüstemli, 1995).

Insurance

Most earthquake risk is uninsured, largely because few private companies have the capacity to absorb the potential costs. Some countries have national insurance schemes that guarantee the householder a payout for earthquake-damaged property (see Chapter 5). Part of the central fund is invested to provide capital to cover claims and part is used to purchase reinsurance. Any shortfall is underwritten by the government. Take-up globally generally remains low.

The lack of accurate forecasting and warnings for earthquakes beyond a minute require precautionary action. Therefore, earthquakes remain a significant risk that is mostly mitigated by adopting good engineering approaches that make buildings safer. This does not necessarily cost a lot, and with some simple preparedness and investment in developing and enforcing building codes, lives and communities can be saved.

KEY TAKE AWAY POINTS

- Though the geography of earthquakes can be broadly mapped according to geological conditions, there is a high degree of uncertainty as to when earthquakes will occur. This means that short-term warning systems and preparation are all the more important in reducing harm.
- The ground-shaking movement of earthquakes can cause a series of secondary hazards, with mass movements occurring in a range of materials. These hazards can become disasters when infrastructure collapses.
- Search and rescue in the aftermath of an earthquake and building collapse is key and is most effectively carried out by equipped and trained local teams in high-risk, vulnerable earthquake-prone areas. It is better to build earthquake-proof infrastructure, which does not need to be expensive or high tech if resources are limited.

BIBLIOGRAPHY

Barkat, A., Ali, A., Siddique, N., Alam, A., Wasim, M., & Iqbal, T. (2017). Radon as an earthquake precursor in and around northern Pakistan: A case study. *Geochemical Journal, 51*(4), 337–346.

Bollinger, G. A., Chapman, M. C., & Sibol, M. S. (1993). A comparison of earthquake damage areas as a function of magnitude across the United States. *Bulletin of the Seismological Society of America, 83*(4), 1064–1080.

Bommer, J. J., & Rodríguez, C. E. (2002). Earthquake-induced landslides in Central America. *Engineering Geology, 63*(3–4), 189–220.

Booth, E. D., & Key, D. (2006). *Earthquake design practice for buildings*. London: Thomas Telford.

CDC.Gov. (n.d.). *Stay safe during an earthquake*. www.cdc.gov/disasters/earthquakes/during.html. Accessed 1 August 2022.

Clermont, C., Sanderson, D., Sharma, A., & Spraos, H. (2011). *Urban disasters: Lessons from Haiti*. London: Disaster Emergency Committee.

Collins, B. D., & Jibson, R. W. (2015). *Assessment of existing and potential landslide hazards resulting from the April 25, 2015 Gorkha, Nepal earthquake sequence* (No. 2015–1142). USA: US Geological Survey.

Dolan, J. F., Sieh, K., Rockwell, T. K., Yeats, R. S., Shaw, J., Suppe, J., . . . Gath, E. M. (1995). Prospects for larger or more frequent earthquakes in the Los Angeles metropolitan region. *Science, 267*(5195), 199–205.

Elnashai, A. S., & Di Sarno, L. (2015). *Fundamentals of earthquake engineering: From source to fragility*. Chichester: John Wiley & Sons.

Gershenzon, N. I., Gokhberg, M. B., & Yunga, S. L. (1993). On the electromagnetic field of an earthquake focus. *Physics of the Earth and Planetary Interiors, 77*(1–2), 13–19.

Haskell, R. C., & Christiansen, J. R. (1985). Seismic bracing of equipment. *Journal of Environmental Sciences, 28*(2), 67–70.

Hosono, T., Yamada, C., Manga, M., Wang, C. Y., & Tanimizu, M. (2020). Stable isotopes show that earthquakes enhance permeability and release water from mountains. *Nature Communications, 11*(1), 1–9.

Ikeya, M. (2004). *Earthquakes and animals: From folk legends to science*. World Scientific.

Karanci, A. N., & Rüstemli, A. (1995). Psychological consequences of the 1992 Erzincan (Turkey) earthquake. *Disasters, 19*(1), 8–18.

Keefer, D. K. (1984). Landslides caused by earthquakes. *Geological Society of America Bulletin, 95*(4), 406–421.

Kobayashi, Y. (1981). Causes of fatalities in recent earthquakes in Japan. *Natural Disaster Science, 3*(2), 15–22.

Leimena, S. L. (1980). Traditional Balinese earthquake proof housing structure. *Disasters, 4*(2), 247–250.

Mazumder, R. K., Rana, S., & Salman, A. M. (2021). First level seismic risk assessment of old unreinforced masonry (URM) using fuzzy synthetic evaluation. *Journal of Building Engineering, 44*, 103162.

McCloskey, J. (2011). Focus on known active faults. *Nature Geoscience, 4*(8), 494–494.

McCloskey, J., Antonioli, A., Piatanesi, A., Sieh, K., Steacy, S., Nalbant, S., Cocco, M., Giunchi, C., Huang, J. D., & Dunlop, P. (2008). Tsunami threat in the Indian Ocean from a future megathrust earthquake west of Sumatra. *Earth and Planetary Science Letters, 265*(1–2), 61–81.

Plafker, G., & Ericksen, G. E. (1978). Nevados Huascaran avalanches, Peru. In *Developments in Geotechnical Engineering* (Vol. 14, pp. 277–314). Elsevier.

Park, S. K., Dalrymple, W., & Larsen, J. C. (2007). The 2004 Parkfield earthquake: Test of the electromagnetic precursor hypothesis. *Journal of Geophysical Research: Solid Earth, 112*(B5).

Peiris, N., Rossetto, T., Burton, P., & Mahmood, S. (2006). EEFIT mission: October 8, 2005 Kashmir earthquake. Published Report. The Institution of Structural Engineers, London.

Robo Spark, Spark Preparedness. (n.d.). *Earthquakes happen on fault lines: What are faults?* http://sparkpreparedness.weebly.com/blog/earthquakes-happen-on-fault-lines-what-are-faults. Accessed 1 August 2022.

Rom, A., & Kelman, I. (2020). Search without rescue? Evaluating the international search and rescue response to earthquake disasters. *BMJ Global Health, 5*(12), e002398.

Rossetto, T., & Peiris, N. (2009). Observations of damage due to the Kashmir earthquake of October 8, 2005 and study of current seismic provisions for buildings in Pakistan. *Bulletin of Earthquake Engineering, 7*(3), 681–699.

Rothery, D. (2010). *Volcanoes, earthquakes and tsunamis: Teach yourself.* Hachette UK.

Smith, W. D., & Berryman, K. R. (1983). Revised estimates of earthquake hazard in New Zealand. *Bulletin of the New Zealand Society for Earthquake Engineering, 16*(4), 259–272.

Smith, W. D., & Berryman, K. R. (1986). Earthquake hazard in New Zealand: Inferences from seismology and geology. In W. I. Reilly & B. E. Harford (Eds.), *Recent crustal movements of the Pacific region*, Royal Society of New Zealand, Bulletin 24 (pp. 223–243).

SMS Tsunami Warning. (n.d.). *Seismograph.* www.sms-tsunami-warning.com/pages/seismograph#.YufawuzMI_Y. Accessed 1 August 2022.

Spangle, W. (1986). Pre-earthquake planning for post-earthquake rebuilding (PEPPER). *The Journal of Environmental Sciences, 29*(2), 49–54.

Tinsley, J. C., Youd, T. L., Perkins, D. M., & Chen, A. T. F. (1985). Evaluating liquefaction potential. Evaluating Earthquake Hazards in the Los Angeles Region: An Earth Science Perspective: US Geological Survey Professional Paper, 1360, 263–316.

USGS. (n.d.a). *Cutaway views showing the internal structure of the Earth.* www.usgs.gov/media/images/cutaway-views-showing-internal-structure-earth-left. Accessed 1 August 2022.

USGS. (n.d.b). *Hayward fault.* https://en.wikipedia.org/wiki/Hayward_Fault_Zone#/media/File:122-38HaywardFault.jpg. Accessed 1 August 2022.

Whittow, J. B. (1980). *Disasters.* USA: University of Georgia Press.

Woith, H., Petersen, G. M., Hainzl, S., & Dahm, T. (2018). Can animals predict earthquakes? *Bulletin of the Seismological Society of America, 108*(3A), 1031–1045.

Wyss, M., Nekrasova, A., & Kossobokov, V. (2012). Errors in expected human losses due to incorrect seismic hazard estimates. *Natural Hazards, 62*, 927–935.

Wyss, M., Shimazaki, K., & Ito, A. (Eds.). (1999). *Introduction: Seismicity patterns, their statistical significance and physical meaning* (pp. 203–205). Basel: Birkhäuser.

FURTHER READING

Lewis, J. (2003). Housing construction in earthquake-prone places: Perspectives, priorities and projections for development. *Australian Journal of Emergency Management*, *18*(2), 35–44.

Mika, K. (2019). *Disasters, vulnerability, and narratives: Writing Haiti's futures*. London: Routledge.

Sorkhabi, R. (Ed.). (2014). *Earthquake hazard, risk and disasters*. Academic Press.

Spence, R., & So, E. (2021). *Why do buildings collapse in earthquakes? Building for safety in seismic areas*. Chichester: John Wiley & Sons.

WEB LINKS

Earthquake Engineering Field Investigation Team (EEFIT). www.istructe.org/get-involved/supported-organisations/eefit

Earthquake Engineering Research Institute (EERI). www.eeri.org

European-Mediterranean Seismological Centre. www.emsc-csem.org

Global Seismographic Network. www.usgs.gov/programs/earthquake-hazards/gsn-global-seismographic-network

Great Shakeout. www.shakeout.org/

USGS Earthquake Program. https://earthquake.usgs.gov/

CHAPTER SEVEN

Volcanic Hazards

7

OVERVIEW

Volcanoes are diverse in terms of eruptive styles, the types of hazards produced and the scale of their impact, the timeframe of activity and the scale of area affected. Volcanoes have the potential to cause mass extinction events over geological timescales or cause global cooling that can significantly impact life on Earth; however, these events are rare. This chapter explores how volcanoes form, the different types of volcanic eruptions and styles, and how they are classified. A range of both primary and secondary volcanic hazards are discussed, including lava flows, pyroclastic flows (the deadliest forces on Earth), lahars and gases. Given that it is difficult to forecast volcanic activity accurately, monitoring and precautionary and anticipatory action are all the more important in reducing harm. Effective land planning, robust warnings and communication tools, along with engaging with local communities, can have significant impacts of reducing loss of life and economic impact.

A. THE NATURE OF VOLCANOES

There are some 500 active volcanoes in the world, and about 50 erupt in an average year. The distribution of volcanoes is controlled by the global geometry of plate tectonics forming in three tectonic settings (see Figure 7.1 and Figure 6.1):

- *Subduction volcanoes* are located in the zones of the Earth's crust where one tectonic plate is thrust and consumed beneath another. They comprise about 80% of the world's active volcanoes and are the most explosive type. They are characterised by a composite cone and multiple hazards.
- *Rift volcanoes* occur where tectonic plates diverge and spread. They are generally less explosive and more effusive, especially when they occur on the deep ocean floor.
- *Hot spot volcanoes* exist in the middle of tectonic plates where a crustal weakness allows molten material to penetrate from the Earth's interior. The Hawaiian

DOI: 10.4324/9781351261647-9

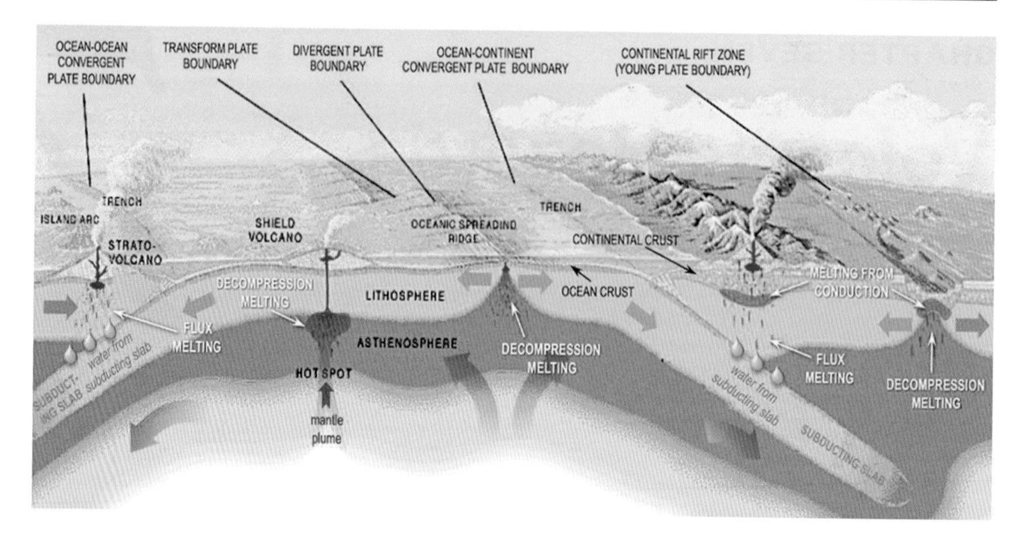

Figure 7.1 Plate tectonic settings of volcanism. Volcanoes along subduction zones are the result of flux melting (lowering the melting point by adding water). Decompression melting produces volcanoes along divergent margins (ocean spreading centres and continental rift zones), as well as above mantle plumes. Contact between hot mafic partial melts and felsic rocks can trigger partial melting of the felsic rocks (melting from conduction).

Source: José F. Vigil from This Dynamic Planet – a wall map produced jointly by the US Geological Survey, the Smithsonian Institution and the US Naval Research Laboratory, 1999

Islands, in the middle of the Pacific plate, are an example; another famous one is the supervolcano Yellowstone in the USA.

Not surprisingly, seismic activity is often associated with volcanic eruptions, although most volcanic-related earthquakes are small (Zobin, 2001).

All volcanoes are formed from molten material (*magma*) within the Earth's crust. Magma is a complex mixture of silicates, which contains dissolved gases and, very often, crystallised minerals in suspension. As the magma moves towards the surface the pressure decreases and the dissolved gases come out of solution to form bubbles. The bubbles expand and drive the magma further into the volcanic vent until it breaks through weaknesses to reach the surface.

The type of volcanic hazards produced depends on the tectonic setting, along with the chemistry of the magma (i.e. the mix of oceanic and continental plate compositions), the temperature of the magma and the pressure it is under. It is rather like the process of cooking; with an egg you can boil it, fry it or poach it, changing the end result of the egg. Changing the ingredients and the 'cooking' process results in very different eruptive behaviour and hazards. As such, each volcano is different and has differing characteristics that can even change during a volcano's lifetime, with eruptions varying in scale and types of behaviour. Hence, understanding a volcano's past behaviour may be useful in preparing for future activity, but must be interpreted with caution.

For a moderately large eruption the total thermal energy released lies in the range 10^{15}–10^{18} joules, which compares with the 4×10^{12} joules liberated by a one kilotonne atomic explosion. There are many ways to measure and classify volcanic eruptions,

Table 7.1 Selected criteria for the volcanic explosivity index (VEI)

VEI	Ejecta volume	Classification	Description	Plume	Frequency	Tropospheric injection	Examples
0	< 0.00001 km³	Hawaiian	effusive	< 100 m	constant	negligible	Kilauea
1	> 0.001 km³	Hawaiian/ Strombolian	gentle	100–1000 m	daily	minor	Nyiragongo (2002)
2	> 0.001 km³	Strombolian/ Vulcanian	explosive	1–5 km	weekly	moderate	Mount Sinabung (2010)
3	>0.01 km³	Vulcanian/ Peléan	severe	3–15 km	few months	substantial	Nevado del Ruiz (1985), Soufrière Hills (1995)
4	> 0.1 km³	Peléan/ Plinian	cataclysmic	10–25 km	≥ 1 yr	substantial	Mount Pelée (1902). Eyjafjallajökull (2010)
5	> 1 km³	Plinian	paroxysmal	20–35 km	≥ 10 yr	substantial	Mount St Helens (1980)
6	> 10 km³	Plinian/ Ultra-Plinian	colossal	> 30 km	≥ 100 yr	substantial	Krekatoa (1883), Mount Pinatubo (1991)
7	> 100 km³	Ultra-Plinian	super-colossal	>40 km	≥ 1,000 yr	substantial	Tambora (1815)
8	> 1,000 km³	Supervolcanic	mega-colossal	> 50 km	≥ 10,000 yr	substantial	Yellowstone (640,000 BP), Toba (74,000 BP)

Source: Adapted from Newhall and Self (1982)

but the most frequently used to measure the size of eruptions is via the *volcanic explosivity index* (VEI). The VEI combines the total volume of ejected products, the height of the eruption cloud, the duration of the main eruptive phase and several other items into a basic 0–8 scale of increasing hazard (Table 7.1). On average, an eruption with VEI = 5 occurs every 10 years, and a VEI = 7 every 100 years.

The disaster potential of volcanic eruptions depends on the effervescence of the gases and the viscosity of the magma. High effervescence and low viscosity lead to the most explosive episodes. Subduction zone volcanoes, for example, draw on magmas that are a mix of upper-mantle material and melted continental rocks rich in feldspar and silica. These *felsic* (acid) magmas produce thick, viscous lavas containing up to 70% silicon dioxide (SiO_2) and thus lead to violent eruptions. On the other hand, rift and hot-spot volcanoes draw on magmas high in magnesium and iron but low (< 50%) in silica content. Such *mafic* (or basic) lavas are fluid, retain little gas and erupt less violently.

These characteristics allow a broad recognition of volcanic eruptions by type. Common examples include the *Plinian type*, which produces the most violent upward expulsions of gas and other materials associated with very viscous magmas (dacite and rhyolite). In the 1991 eruption of Mount Pinatubo, Philippines, a plume of tephra was ejected more than 30 km into the atmosphere. The *Peléan type* is also dangerous because the rising magma is trapped by a dome of solid lava and then forces a new opening in the volcano flank. This can produce powerful lateral blasts such as occurred during the eruption of Mount St Helens, USA, in 1980, when much of the mountain top was destroyed and 57 people were killed. In contrast, the relatively low-hazard *Hawaiian type* of eruption jets highly fluid basaltic lava into the air from vents and fissures on the flank of the volcano to create *fire fountains*. The resulting surface flows of lava can travel miles from their source before cooling and hardening. This produces the low topographic profile associated with such volcanoes.

Another form of classification for volcanoes is via their structure (see Figure 7.2). Fissure volcanoes are most often associated with diverging plate boundaries and erupt large quantities of very fluid basaltic lava. This can be seen in Iceland, which sits atop the Mid-Atlantic Ridge. Shield volcanoes are built up over time after many eruptions have poured lava from a single vent, such as Mauna Loa, Hawaii. A dome volcano is created when lavas are erupted that are silica rich (and therefore viscous) and do not spread far. Cinder cone volcanoes are built up over time from layers of ejecta (solid fragments) released from the volcano during an eruption. Composite volcanoes are found along subduction zones, created by alternating layers of lava and ejecta (ash and rock fragments) that form after each eruption. These composite volcanoes typically form high, snow-capped peaks such as Mt Fuji in Japan. Caldera volcanoes are a vast crater created by eruptions that are so explosive and powerful that the magma chamber is partially emptied and the volcano summit collapses into itself, as seen at Toba volcano in Indonesia.

Volcanoes can occur for varying durations, with most eruptions lasting 10–1,000 days. Fewer than 20% are over within three days, and the median is seven weeks. In terms of volume over time a supervolcano erupts a vast amount of materials (>1000 km^3) within hours that can cause global

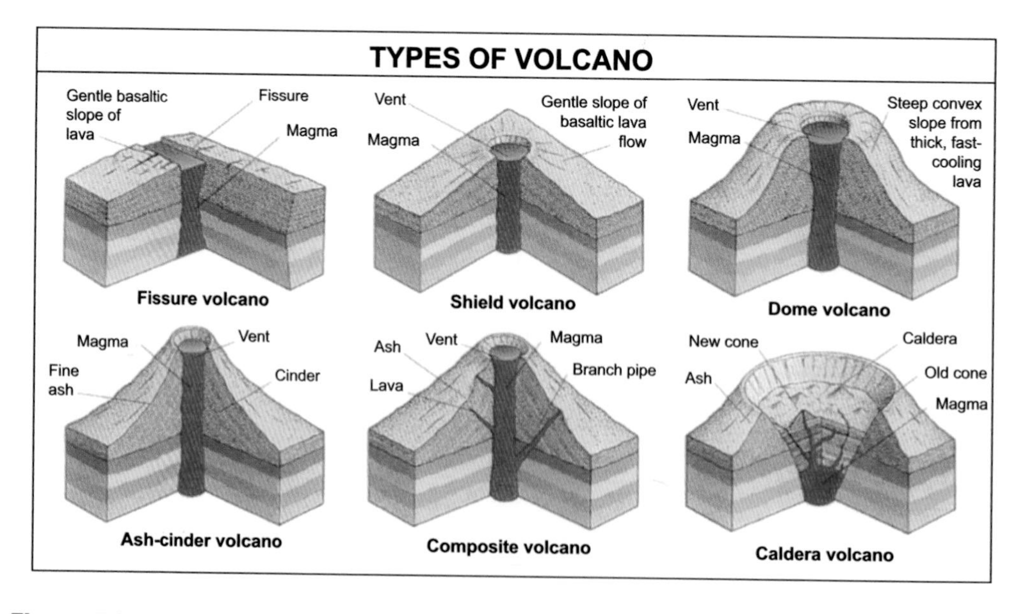

Figure 7.2 The six main styles of volcanoes found around the Earth; which type of volcano forms depends on the type of eruptions and the underlying geology.

Source: http://extremeenvironmentsguide.weebly.com/volcanoes.html

cooling (see Section B), whereas a continental flood basalt can erupt for up to a million years creating vast deposits called traps (such as the Deccan Traps in India). Whilst this chapter focuses predominantly on less extreme volcanism, it is important to note that continental flood basalts correlate with most significant extinction events and also five mass extinctions. Along with the destructive power that a supervolcano can have on the atmosphere, volcanoes at either end of the spectrum can ultimately have a devastating impact on humans and many other species.

Volcanoes produce an extensive diversity of hazard types. Figure 7.3 outlines some of the most common primary volcanic hazards.

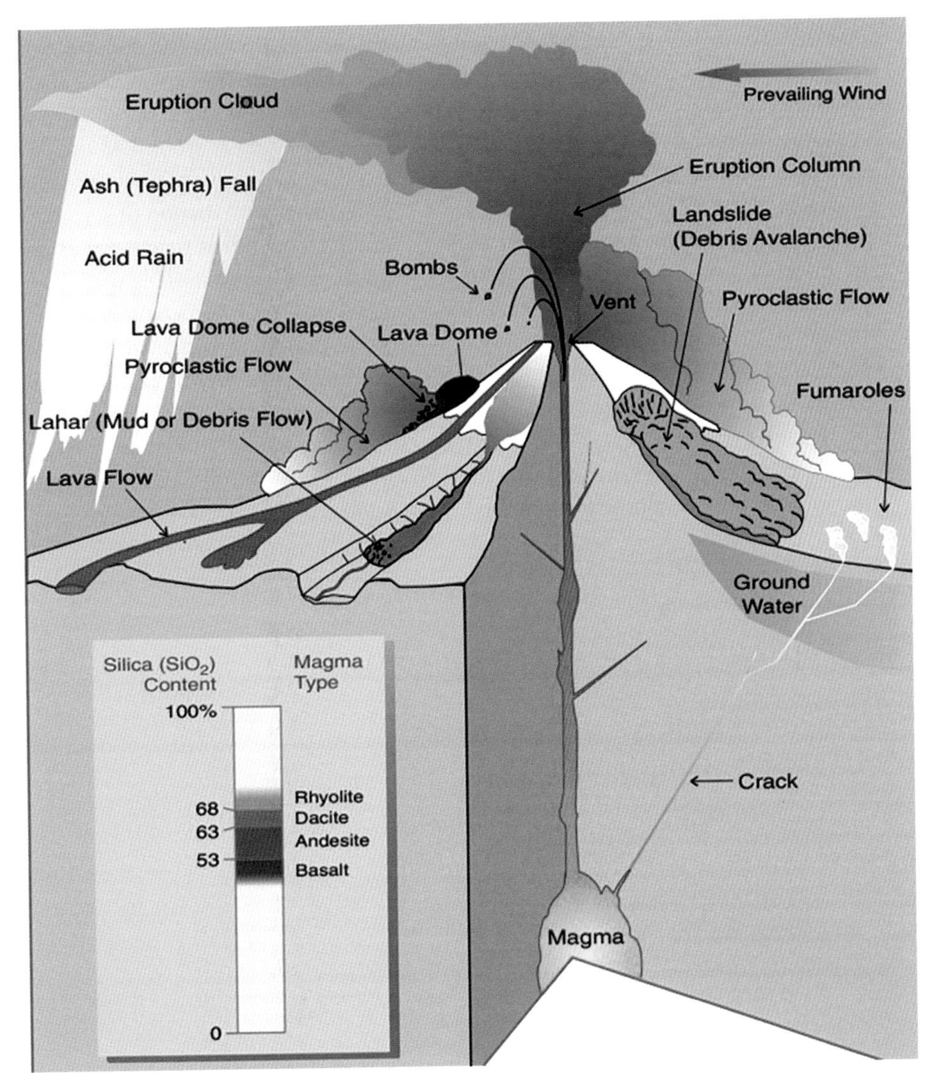

Figure 7.3 Section through a composite volcanic cone showing a range of possible hazards. Some hazards (pyroclastic and lava flows) occur during eruptions; other hazards (lahars) are more likely to occur after the event.

Source: Reproduced from Myers, B., Brantley, S.R., Stauffer, P., Hendley, J.W., 2008. What are volcano hazards? US Geological Survey Fact Sheet 002–97. Revised March 2008, two sheets (https://pubs.usgs.gov/fs/fs002-97/fs00297.pdf).

Primary Volcanic Hazards

These result from the products ejected by the volcanic eruption. A significant hazard feature is their long geographical reach away from the source (Figure 7.4).

Pyroclastic Flows

These have been responsible for most volcanic-related deaths to date and are one of the deadliest forces of nature. These flows were sometimes called *nuées ardentes* ('glowing clouds'), and they result from the frothing of molten magma in the vent of the volcano. The gas bubbles then expand and burst explosively to eject a turbulent mixture of hot gases and pyroclastic material (volcanic fragments, crystals, ash, pumice and glass shards). Pyroclastic bursts surge downhill because, with a heavy load of lava fragments and dust, they are appreciably denser than the surrounding air, and the energy required to keep the plume moving up into the atmosphere is too great given the volume of material.

The surges that collapse down the sides of the volcano can get up to 1,000°C and can travel up to 800 km h^{-1} which is the speed of a jet plane, although more typically is at 100 km h^{-1}. They pose the highest risks when blasts are directed laterally (Peléan type) close to the ground, but have enough energy that they can flow over mountains or valley edges (see Figure 7.5). During the Mont Pelée disaster in 1902, the town of St Pierre, some 6 km from the centre of the explosion, experienced a surge temperature about 700°C in a blast travelling at around 33 m s^{-1}. People exposed to these surges are immediately killed by a combination of severe external and internal burns together with asphyxiation. The surge itself is preceded by air pressures with sufficient force to topple some buildings. Modern volcanic hazard planning, as at Vesuvius, Italy, anticipates the emergency evacuation of people and extensive building damage within a radius of 10 km from the volcanic vent (Petrazzuoli & Zuccaro, 2004).

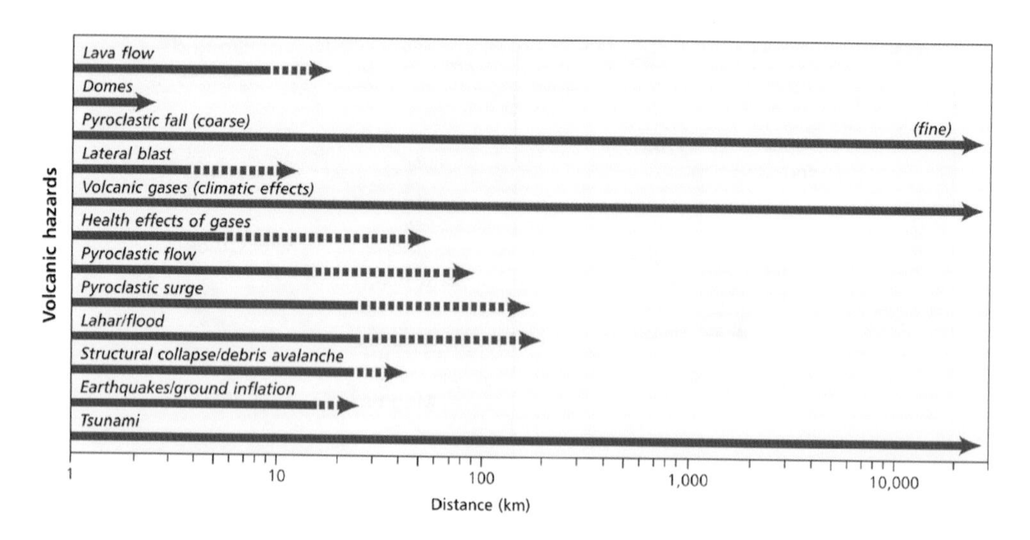

Figure 7.4 The influence of distance on hazardous volcanic phenomena. Many hazards are restricted to a 10 km radius of the volcano, but the effects of fine ash, gases and tsunami waves can extend beyond 10,000 km.

Source: Adapted from McGuire (1998) and reprinted from Chester et al. (2000, p. 97)

Figure 7.5 Eruption of the Soufrière Hills volcano on the island of Montserrat in August 1997. Pyroclastic flows claimed several lives and forced the evacuation of more than two-thirds of the residents.

Source: Photo: © Panos Pictures/Andy Johnstone AJH00072MSR

Air-Fall Tephra

Tephra comprises all the fragmented material ejected by the volcano that subsequently falls to the ground. Most eruptions produce less than 1 km^3 volume of material, but the largest explosions eject several times this amount. The particles range in size from so-called bombs (>32 mm in diameter) down to fine ash and dust (<4 mm in diameter). The coarser, heavier particles fall out close to the volcano vent. A dry layer of ash 10 cm deep weighs about 65–100 kg m^2, whilst wet ash weighs twice as much. As a result, flat-roofed, unreinforced buildings can collapse when ash accumulation approaches 1 metre. In some cases the tephra will be sufficiently hot to start fires on the ground. Depending on wind conditions, the finer dust may be deposited far away. Within six hours of the modest eruption (VEI = 5) of Mount St Helens in 1980, ash clouds had drifted 400 km downwind.

Although ashfalls account for fewer than 5% of the direct deaths associated with volcanic eruptions, they create other problems. Heavy falls of *scoria* (cinder) blanket the landscape; even light falls of ash contaminate farmland and create disruption and building damage in urban areas. The eruption of Mount Pinatubo in 1991 disrupted the livelihood of 500,000 farmers, as agricultural land up to 30 km distant was covered in ash. Wet ash conducts electricity and can cause the failure of electronic components, especially with high-voltage circuits and transformers. Fine ash can clog air filters and damage vehicle engines; roads and airport runways become impassable if made slippery by ash deposits. Airborne dust reduces atmospheric visibility and poses a risk to air transport (Box 7.1).

Volcanic ash can be hot enough to ignite fires and can cause buildings to collapse if too much ash accumulates on the roof. The density of compacted wet ash can be 1.6 tonnes m^3 causing further issues dependent on the weather and environment. Ash can severely impact visibility problems and can affect magnetic communication methods. It also has a habit of going everywhere. In Kagoshima in Japan, community members frequently sweep up volcanic ash from nearby Sakurajima volcano (see Figure 7.6).

Figure 7.6 Several plastic bags of volcanic ash lie on the roadside in Kagoshima, Japan.

Source: © Alamy Images/Wirestock, Inc.

Box 7.1 Airborne Volcanic Ash and Aviation

Volcanic ash consists of minute particles of pulverised rock and natural glass. It is extremely abrasive, is insoluble in water and conducts electricity when wet. The plume of ejected material can travel long distances in the atmosphere. In 1982, a British Airlines Boeing 747 lost all four engines flying though the volcanic plume of Mt Galunggung in Indonesia, falling over 7 kilometres before the engines were restarted (Miller & Casadevall, 2000). In 1989, during the eruption of Mt Redoubt in Alaska, another 747 airplane also had full engine failure; these were restarted only minutes before a collision would have occurred with mountains (Brantley et al., 1990). Since these near-disasters, the aviation industry has worked to develop internationally standardised protocols and procedures to prevent aircraft encountering ash (Casadevall, 1994). A policy of 'if ash, no fly' was adopted internationally by the International Civil Aviation Organisation (ICAO).

The main threat to planes arises when the small (< 2 mm diameter) fragments are ingested into a jet engine (see Figure 7.7). This material can erode the turbine blades and can melt at the high operating temperature of the engines and adhere to critical parts, thus causing engine failure. Forward-facing aircraft surfaces, including the cockpit windows, are also likely to be abraded, and the ash may interfere with navigation and other

Figure 7.7 The Eyjafjallajökull volcano in southern Iceland erupts and sends plumes of ash into the atmosphere just before sunset on 16 April 2010. This material was transferred southwards to create air traffic confusion over much of Western Europe.

Source: Photo: © Shutterstock/Petra Schneider

electronic systems on board (Neal & Guffanti, 2010). Ash falling onto runways, particularly when wet, will reduce the safety of aircraft movements, especially the braking performance. Ash accumulations greater than ~1 mm require complete removal for an airport to resume normal operating capability. Given that jet engines cost in excess of $10–$40 million and that ash also has a number of abrasive and destructive aspects to aircraft and its electronic requirement, the aviation industry is highly sensitive to these risks.

In 1982, the International Civil Aviation Organization (ICAO) created the International Airways Volcanic Watch programme (IAVW). Nine Volcanic Ash Advisory Centres (VAACs) exist, with the aim of improving the detection of airborne ash worldwide and issuing warning messages to air traffic controllers, dispatchers and pilots. Some countries, like the US, have made additional arrangements for their own air space. However, the warning system is not wholly reliable, as illustrated by the eruption of Eyjafjallajökull volcano in Iceland.

In April 2010 Eyjafjallajökull volcano in Iceland produced an ash cloud that led to the cancellation of 95,000 flights over Western Europe, with a cost to the airlines of US$1.7 billion and lost productivity in the region estimated at over US$600 million per day. This was a relatively small eruption, and the ash plume remained at comparatively low altitudes. This plume was dispersed by northerly winds over several countries and disrupted airport operations as far away as Moscow and Athens. British air space was closed for five days by the Civil Aviation Authority (CAA), acting on advice from the Meteorological Office. This policy of closing air space was much criticised at the time as an over-reaction. However, subsequent tests showed that the particles of explosive ash that reached Europe were unusually sharp over their entire size range and that the edges remained sharp after two weeks of artificial abrasion in the laboratory (Gíslason et al., 2011). Scottish airports were partially closed once more by an ash cloud from another Icelandic volcano in May 2011.

According to Donovan and Oppenheimer (2011), such seemingly unusual incidents have to be placed in context and highlight the importance of multidisciplinary studies of environmental hazards. It is evident that aviation policy and operational responses need greater clarification. This applies not only to the detection and forecasting of ash cloud movements but also to the development of techniques to determine the type and degree of aviation risk. Significant research has been invested in improving ash particle modelling, understanding the impact of volcanic ash on jet engines and enhancing forecasts for volcanic ash (see Beckett et al., 2020; Mastin et al., 2022). In turn aviation policy in relation to volcanic ash has changed globally, particularly in relation to who is responsible when flying through ash.

Lava Flows

Lava flows threaten human life when they emerge rapidly from fissure, rather than from central-vent, eruptions. As noted earlier, the fluidity of lava is determined by its chemical composition, especially the proportion of silicon dioxide (SiO_2). If silicon dioxide forms less than about half the total, the lavas are mafic and very fluid. This difference has led to the recognition of two types of lava flow:

- *Pahoehoe lava* flows, which are the most liquid, and where cooling leaves a relatively smooth, wrinkled surface.
- *Aa lava*, which moves downhill in a blocky, slow-moving manner and which leaves behind a rough, irregular surface.

Flowing lava typically has a temperature above the ignition point of many materials at 750°C to 1100°C. The crust forms at the top of the lava flow as it cools and, depending on the viscosity of the lava (driven by the amount of silica composition), the flows can range in thickness from a few centimetres to tens of metres. Often lava tubes can form when the crust of the flow cools, providing a hard crust, and the lava flows underneath at high speed.

On steep slopes low-viscosity pahoehoe lava can stream downhill at speeds approaching 15 m s^{-1}. In the 1977 eruption of Nyiragongo volcano, Zaire (now the Democratic Republic of Congo), five fissures on the flanks of the volcano released a wave of lava that killed 72 people and destroyed over 400 houses. In January 2002 a stream of lava from Nyiragongo devastated about one-third of the city of Goma, killing at least 45 residents and forcing 300,000 others to flee across the border to Rwanda. Many evacuees soon returned but had little to eat for several days because the approach roads were blocked by lava. More than one month later 30,000 people remained dependent on aid in temporary camps. Nyiragongo erupted again on 22 May 2021 and the lava flows cut off key highways, caused multiple electricity cuts across the area and eventually reached the outskirts of the city of Goma. Many people fled the area, with 3,000 people crossing into Rwanda from Goma and another 25,000 fleeing to the northwest in Sake (UNICEF).[1] Following a number of strong aftershocks that affected the area, on 25 May, at 11:03 a.m., a 5.3 Mw earthquake struck destroying several buildings. As of 27 May, 37 people were missing and presumed dead; many of these deaths were the result of car crashes during the evacuation. By 7 June, activity had subsided and evacuees returned to their homes, where possible.

Around Mount Etna, Sicily, aa-type lava flows have done much damage to property and agriculture in the past. The city of Catania was partially destroyed in 1669. The greatest lava-related disaster in historic times occurred in 1783, when lava flowed out of the 24 km long Lakagigar fissure in Iceland for more than five months (Thorarinsson, 1969). There was little direct mortality but more than 10,000 people – over one-fifth of Iceland's population at the time – died in the resulting famine (Grattan et al., 2005).

Volcanic Gases

Numerous gases are released from explosive eruptions and lava flows. The complex gaseous mixture commonly includes water vapour, hydrogen, carbon monoxide, carbon dioxide, hydrogen sulphide, sulphur dioxide, sulphur trioxide, chlorine and hydrogen chloride in variable proportions. Monitoring of the gas composition is difficult due to the high temperatures near an active vent and because the juvenile gases interact with the atmosphere and each other. This constantly alters their composition and proportions. Carbon monoxide has caused deaths because of its toxic effects at very low concentrations, but most fatalities have been associated with carbon dioxide releases. Carbon dioxide is dangerous because it is a colourless, odourless gas with a density about 1.5 times greater than air. When it accumulates in low-lying places, disasters can occur.

The release of carbon dioxide from previous volcanic activity can also create a highly unusual threat. In 1984 a cloud of gas, rich in carbon dioxide, burst out of the volcanic crater of Lake Monoun, Cameroon, and killed 37 people by asphyxiation (Sigurdsson, 1988). In 1986, a similar disaster occurred at the Lake Nyos crater, also in Cameroon. This time 1,746 lives were lost, together with over 8,300 livestock; 3,460 people were moved to temporary camps. The outburst of gas created a fountain that reached over 100 m above the lake surface before the dense cloud flowed down two valleys to cover an area over 60 km^2 (Kling et al., 1989). These hazards are very rare. They are a function of high levels of carbon dioxide in the waters of these lakes, probably built up over a long period of time from carbon dioxide-rich groundwater springs flowing into the submerged crater. Under normal circumstances the dissolved carbon dioxide would remain trapped below the water surface. In the case of Lake Monoun, the sudden gas release could have been due to disturbance of the water by a landslide originating on the crater's rim, or a small earthquake underwater, or a small eruption. Unfortunately, there is no concrete evidence for the exact trigger at Lake Nyos. Controlled de-gassing of lakes Nyos and Monoun, in Cameroon, began in 2001 and 2003, respectively (Kling et al., 2005). The success of this technique depends on the balance between artificial gas removal and natural recharge rates. Without the installation of more pipes and an increased abstraction rate, it seems likely that hazardous amounts of gas may remain within the lakes for some time.

On a bigger scale volcanic gases can cause significant global impact by influencing the climate. During large explosive eruptions, volcanoes emit vast volumes of volcanic gases, aerosol droplets and ash high into the stratosphere. The sulphur dioxide and hydrochloric acid react to form a sulphate aerosol that acts as a mirror, deflecting the sun and reducing the sunlight reaching the Earth's surface, resulting in global cooling (see Figure 7.8b). For example, the year following the 1815 eruption of Tambora, Indonesia – with a VEI = 7 – was called 'the year without a summer' throughout the northern hemisphere. The eruption of Mount Pinatubo in 1991 lowered surface air temperatures in parts of the northern hemisphere by up to 2°C in the summer of 1992, and, during the winters of 1991–92 and 1992–93, raised temperatures by as much as 3°C. These temperature variations have significant implications for weather-sensitive activities like agriculture.

Following a supervolcano, significant cooling may occur, causing a volcanic winter. However, for volcanoes that erupt into the troposphere, carbon dioxide can build up and as a greenhouse gas can result in global warming, as seen following the 1783 Lakagígar eruption in Iceland (see Figure 7.8a).

Very large eruptions also emit debris into the lower stratosphere, some 20–25 km above the Earth's surface, and can form a 'dust veil' over the planet. The maximum risk is from volcanoes in lower latitudes. After the 1883 eruption of Krakatau, Indonesia, for example, an aerosol cloud spread round the globe within two weeks. As shown in Table 7.2, the climatic effects can last from periods of a single day (by reducing the diurnal temperature cycle) to up to 100 years if a series of eruptions raises the mean optical depth of the atmosphere enough to cause decadal-scale cooling. Important changes in atmospheric chemistry, especially ozone depletion, can also occur.

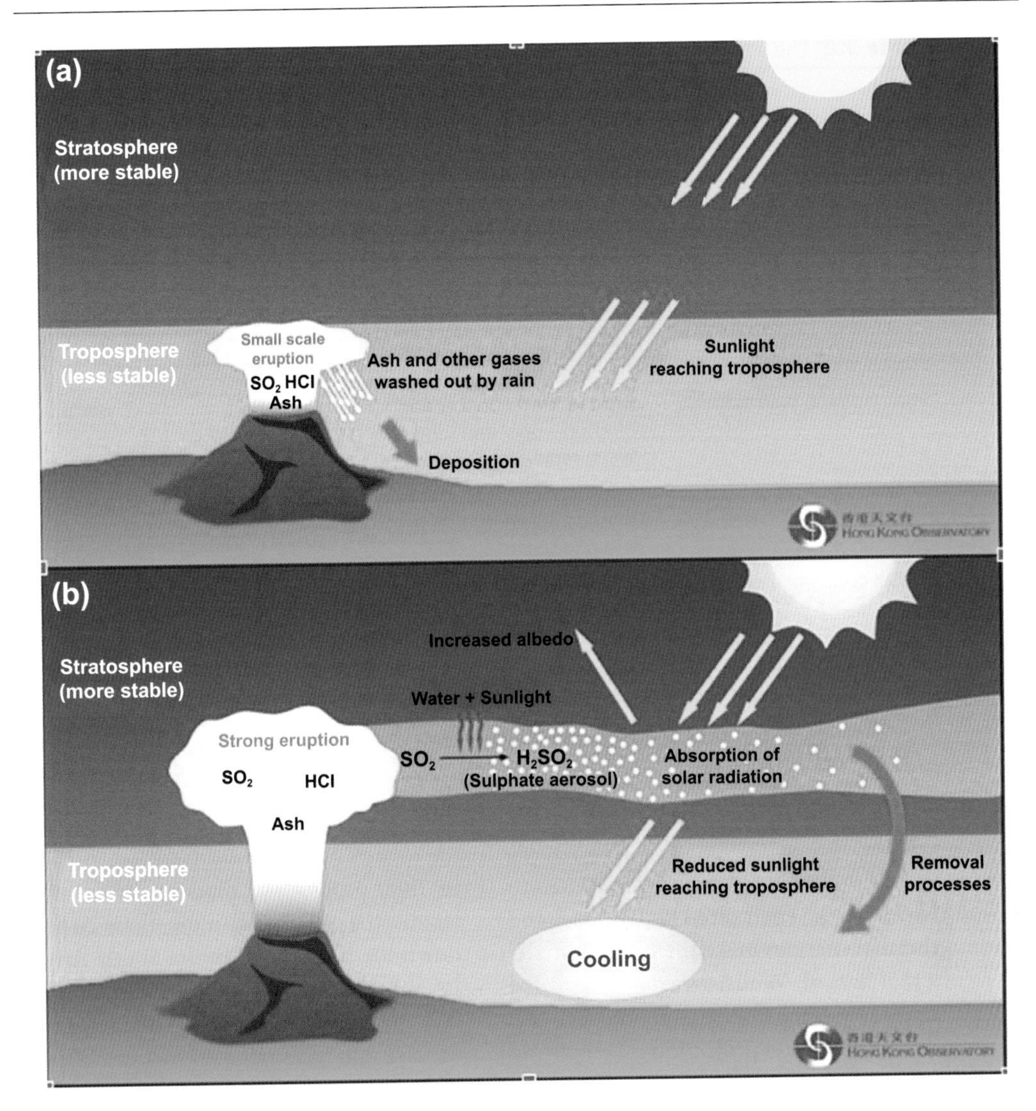

Figure 7.8 (a) Emitting volcanic gas into the troposphere that generates global warming, and (b) volcanic gas emitted into the stratosphere generating global warming.

Source: Hong Kong Observatory

Chinese: 『此資料由「香港特別行政區政府香港天文台」提供』

English: 'Courtesy of the Hong Kong Observatory of HKSAR' www.hko.gov.hk/tc/blog/00000107.htm

Ballistics

Volcanic ballistic projectiles (VBPs) are the main hazard to life and infrastructure from Strombolian-style eruptions. Ballistics are globally the most common cause of volcanic fatalities for tourists, who often venture too close to the volcanic vent. Ballistics are fragments of molten lava or solid rock, ranging from a few centimetres to tens of metres in diameter, that can travel tens to hundreds of metres per second up to ~10 km from

Table 7.2 The effects of large explosive volcanic eruptions on weather and climate

Effect	Mechanism	Begins	Duration
Reduction of diurnal cycle	Blockage of short-wave and emission of long-wave radiation	immediately	1–4 days
Reduced tropical precipitation	Blockage of short-wave radiation, reduced evaporation	1–3 months	3–6 months
Summer cooling of northern hemisphere, tropics and subtropics	Blockage of short-wave radiation	1–3 months	1–2 years
Stratospheric warming	Stratospheric absorption of short-wave and long-wave radiation	1–3 months	1–2 years
Winter warming of northern hemisphere continents	Stratospheric absorption of short-wave and long-wave radiation	6 months	1 or 2 winters
Global cooling	Blockage of short-wave radiation	immediately	1–3 years
Global cooling from multiple eruptions	Blockage of short-wave radiation	immediately	10–100 years
Ozone depletion, enhanced UV	Dilution, heterogeneous chemistry and aerosols	1 day	1–2 years

Source: After Robock (2000). Reproduced by permission of American Geophysical Union.

vent, although they are usually limited to within 5 km. Their often high impact and thermal energies make them a potentially lethal hazard. Whilst wearing a hard hat may help, when ballistics are the size of a bus there is little projection a helmet can offer (see Fitzgerald et al., 2020).

Secondary Volcanic Hazards

Ground Deformation

Ground deformation occurs as volcanoes grow from within by magma intrusion and new layers of lava and pyroclastic material accumulate on the surrounding slopes. The deformation may eventually lead to a catastrophic failure of the volcanic edifice and to mass movement hazards. For example, structural failure of the north flank of Mount St Helens in 1980 produced a massive debris avalanche that advanced more than 20 km down the North Fork of the Toutle River. In 2000 the almost total collapse of a new lava dome at the Soufrière Hills volcano, Montserrat, generated pyroclastic flows and a number of lahars and debris avalanches in the surrounding valleys (Carn et al., 2004). According to Siebert (1992), on average major structural failures of volcanoes have occurred worldwide four times per century over the last 500 years. Such instability is found on large polygenetic volcanoes, like Mauna Loa and Kilauea, Hawaii. Volcanoes like Etna are also prone to instability because of their complex construction of inter-bedded lavas and pyroclastic deposits lying on steep slopes, although relatively few deaths have resulted so far.

Lahars

After pyroclastic flows, lahars present the greatest risk to human life. They are often defined as volcanic mudflows composed of sand–silt sized sediments, although other volcanic material, such as pumice, can be transported too. Lahars consist of volcanic ash and rock up to at least 40% by weight and create dense, viscous flows that can travel faster than clear-water streams. They occur widely on steep volcanic flanks in the wet tropics, and the term is of Indonesian (Javanese) origin. Lahars are very rapid-onset, short-duration events, lasting between 30 min and 2 h 30 min, and have average velocities of 5–7 m s^{-1} at elevations of 1,000 m (Lavigne et al., 2000). Reliable forecasting of lahars is impossible because of the short lead-time and variations in rainfall intensity during the monsoon season, but monitoring is a necessary step to the better understanding of these hazards.

The degree of hazard varies greatly but, generally, the destructive potential tends to rise with flows containing the larger-size sediments, as shown at Popocatépetl volcano, Mexico, by Capra et al. (2004). Most lahars result from heavy rainfall and excess water at the volcano surface. At Mount Pinatubo, on the island of Luzon in the Philippines, lahars can transport and deposit tens of millions of cubic metres of sediment in a day and threaten a local population estimated at 100,000 people (Newhall & Punongbayan, 1996). Hot mudflows can emerge from sources below ground, and cold lahars are also possible. In May 2006 a 'mud volcano' in Indonesia caused several fatalities and the temporary displacement of about 25,000 people.

Lahars can be classified as:

- *primary* when they occur during a volcanic eruption and freshly fallen tephra is immediately mobilised by large quantities of water – sometimes resulting from the collapse of a crater lake – into hot flows.
- *secondary* when they are triggered by high-intensity rainfall between eruptions and old tephra deposits on the volcano slopes are reactivated into mudflows.

Some of the most destructive primary events arise from the rapid melting of snow and ice. This happens when hot lava fragments fall on snow and ice lying at the summits of the highest volcanoes. The water mixes with soft ash and volcanic boulders to produce a debris-rich fluid, sometimes at high temperatures, which then pours down the mountain side at speeds typically of 15 m s^{-1} that may reach >22 m s^{-1}. This is a particular hazard in the northern Andes, where at least 20 active volcanoes straddle the equator, from central Colombia to southern Ecuador (Clapperton, 1986). The highest peaks are capped with permanent snow and ice and are structurally weak because of the action of hot gases over time. During an eruption in 1877 so much ice and snow was melted that enormous lahars, 160 km long, discharged simultaneously to the Pacific and Atlantic drainage basins.

According to Tuffen (2010), ice thinning due to global warming may cause more explosive eruptions in the future because the removal of 100 m or more of ice thickness would likely reduce the load pressure on magma chambers. Any increase in magma–water interactions would create additional tephra hazards, including the possible collapse of the edifice of the volcano. Other sources of lahars are from breaching a crater lake.

The second-deadliest volcanic disaster so far recorded resulted from lahars generated by the 1985 eruption of the Nevado del Ruiz volcano, Colombia, the most northerly active volcano in the Andes. This disaster illustrates the need to understand the history of a volcano, because large lahars had previously been recorded, notably in 1595 and 1845, when the surrounding population was relatively low (Wright & Pierson, 1992). More recent volcanic activity began in November 1984, but the main eruption did not occur until one year later. Large-scale glacier melting produced a huge lahar that rushed down the Lagunillas valley, sweeping up trees and buildings in its path (Sigurdsson & Carey, 1986). The town of Armero, 50 km downstream, was overwhelmed with a mudflow deposit 3–8 m deep that destroyed over 5,000 buildings and killed more than 23,000 people within minutes. A preliminary hazard-zone map had been completed one month before the 1985 eruption (Figure 7.9). This indicated fairly accurately the vulnerability of Armero to mudflows, as well as the likely extent of ash fall, but sadly, in the words of Barry Voight,

> the catastrophe was not caused by technological ineffectiveness or detectiveness, nor by an overwhelming eruption, or by an improbable run of bad luck, but rather by cumulative human error – by misjudgment, indecision and bureaucratic shortsightedness. Armero could have produced no victims, and therein dwells its immense tragedy.
>
> (Voight, 1990, p. 151)

The accumulation of ash on volcanic flanks results in an increased threat of river flooding and sediment re-deposition, especially in countries subject to tropical cyclones or

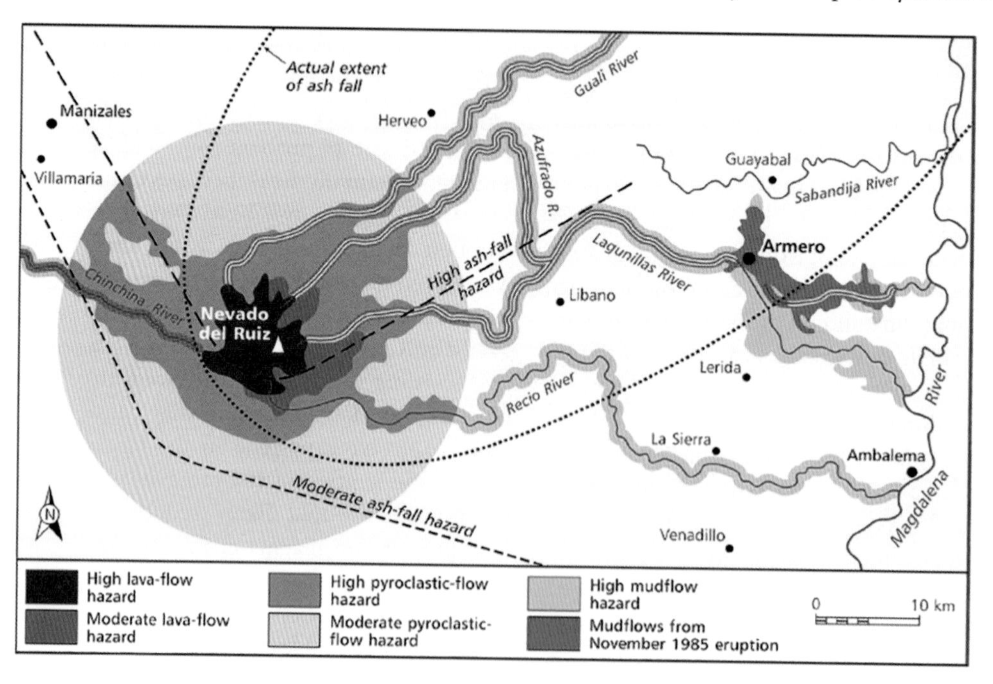

Figure 7.9 Hazard zone map for the Nevado del Ruiz volcano, Colombia, showing the risk of mudflows in the valley occupied by the town of Armero. The circle denotes a 20 km radius from the cone.

Source: After Wright and Pierson (1992)

monsoon rains. For example, over 5,000 people were killed in a mudflow following the eruption of the Kelut volcano, Java, in 1919. The most destructive lahars are those containing up to 90% by volume of debris, often created by the breaching of dams holding back temporary lakes along channels and tributary valleys filled with sediment. These lake outbursts occur without warning, even when there is no rainfall. Figure 7.10 shows the lahar deposits that cover over 280 km^2 at Merapi volcano, Java. Most of these deposits lie in river channels and are 0.5–2.0 m thick, although some have depths greater than 10 m (Lavigne et al., 2000). The sediments are quickly remobilised by tropical rainfall and eventually reach lowland rivers, where they reduce channel capacity and increase the risk of rivers migrating across floodplains. Over 1 million people live on the slopes of the Merapi volcano, central Java, and

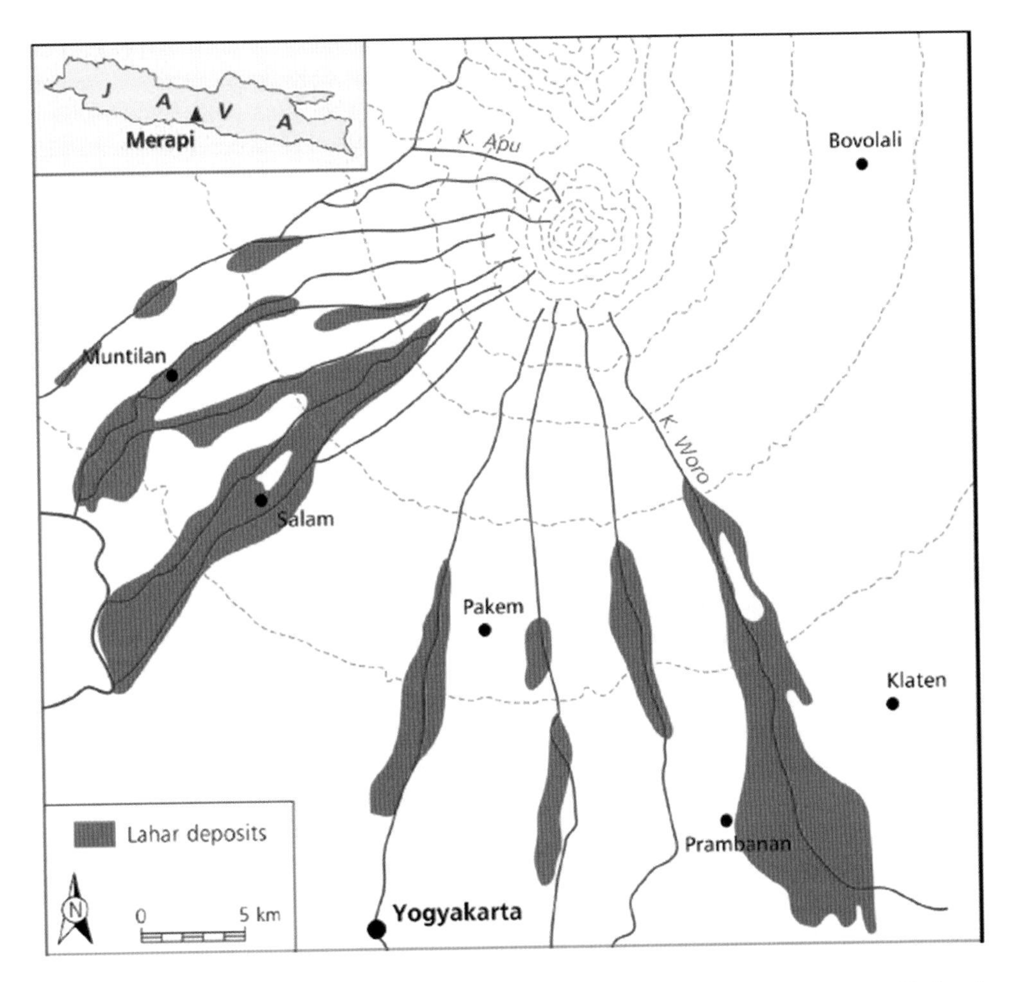

Figure 7.10 The distribution of lahar deposits on the slopes of Merapi volcano, Java. All the river courses shown have produced active lahars during historic times.

Source: After Lavigne et al. (2000). Reprinted from *Journal of Volcanology and Geothermal Research* 100, F. Lavigne et al. Lahars at Merapi volcano, central Java, 423–65.

secondary lahars are triggered by rainfall of about 40 mm in just two hours.

Landslides

Landslides and debris avalanches are a common feature of volcano-related ground failure. They are particularly associated with eruptions of siliceous (dacitic) magma of relatively high viscosity with a large content of dissolved gas. This material can intrude into the volcano, as happened in May 1980 at Mount St Helens, USA. Swarms of small earthquakes (Mw = 3.0) and minor ash eruptions were followed by ground uplift on the north flank of the volcanic cone. Before the main eruption the bulge was nearly 2 km in diameter and large cracks appeared in the cover of snow and ice (Foxworthy & Hill, 1982). On 18 May, when the surface bulge was 150 m high, an earthquake shook a huge slab of material from the oversteepened slopes and triggered a debris avalanche containing 2.7 km^3 of material (see Figure 7.11).

Tsunamis

Tsunamis can occur after catastrophic eruptions. The structural failure of volcanic islands, creating potential debris flows in excess of 1 million cubic metres, can produce 'super' tsunami hazards. The most-quoted example is that of the island volcano of Krakatoa, between Java and Sumatra, in 1883 (VEI = 6). A series of enormous explosions, audible at a distance of almost 5,000 km, produced an ash cloud that penetrated to a height of 80 km into the atmosphere and was carried round the

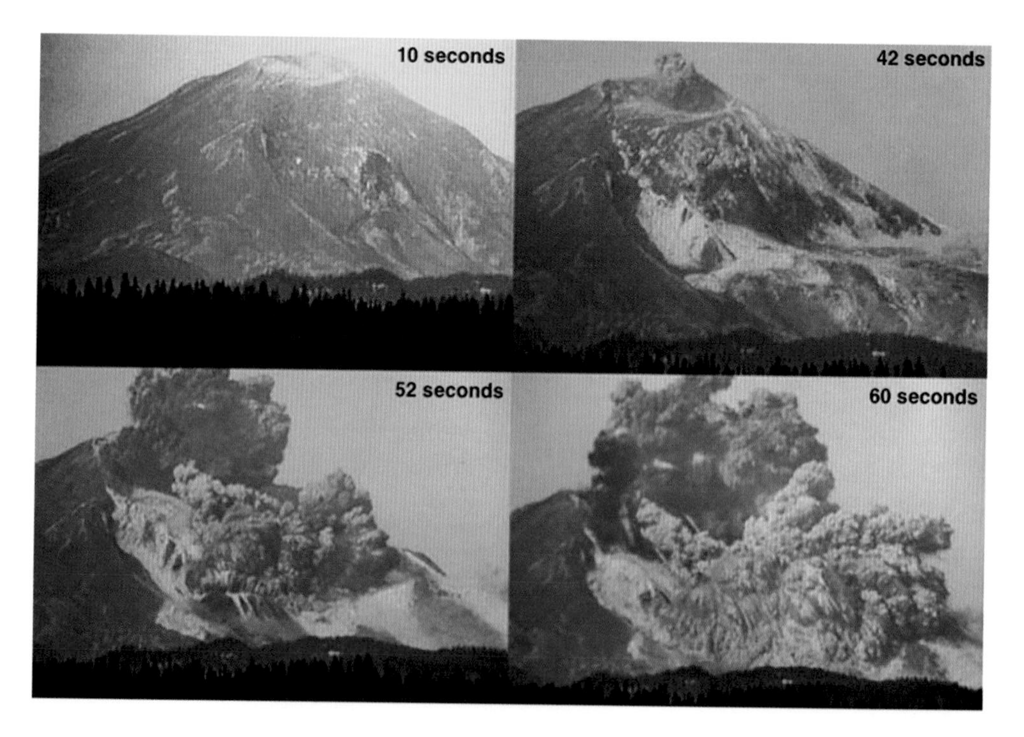

Figure 7.11 Sequence of Mount St Helens photos of the colossal landslide and ensuing lateral blast following the Mw 5.1 earthquake, 18 May 1980. Timestamps indicate the time following the earthquake.

Source: USGS, Gary Rosenquist https://volcanoes.usgs.gov/volcanic_ash/mount_st_helens_1980.html

world several times by upper-level winds. Such was the force of the eruption that the volcanic cone collapsed into the caldera. The resulting tsunamis swept through the narrow Sunda Straits, with onshore waves over 30 m high in places. It has been estimated that over 36,000 people drowned. See more in Chapter 8.

Jökulhlaups

Another common secondary hazard can be formed on glaciers such as seen in Iceland and Chile. A jökulhlaup, an Icelandic term, is a type of glacial outburst flood, commonly triggered by geothermal heating and occasionally by a volcanic subglacial eruption, but it is now used to describe any large and abrupt release of water from a subglacial or proglacial lake/reservoir. Discharges up to 300,000 $m^3 s^{-1}$ have been estimated for the jökulhlaup that was triggered by the 1918 eruption of the Katla volcano in Iceland. People in southern Iceland are at risk from such a potential outburst because the highly active Katla volcano, lying beneath the Myrdalsjökull icecap, has erupted about twice per century since the year 874. Survey interviews and a full-scale evacuation exercise in 2006 showed that many residents living in southern Iceland are concerned about a Katla eruption and the risk a jökulhlaup poses to their communities and livelihoods (Bird, 2009; Jóhannesdóttir & Gísladóttir, 2010). Tourists visiting the region, however, were unaware of the jökulhlaup risk and were unlikely to know what to do if a warning was issued (Bird et al., 2010; Bird & Gísladóttir, 2020).

B. IMPACTS AND CONSEQUENCES

More than half of volcanic deaths recorded in the 20th century occurred in just two events. The 1902 eruption of Mont Pelée, on the island of Martinique in the West Indies, killed 29,000 people in the port of Saint Pierre, leaving only two known survivors, whilst the 1985 eruption of Nevado del Ruiz in Colombia claimed a further 23,000 lives. These numbers do not fully account for wide-ranging deaths, such as from changes to the climate or ash interfering with crops and breathing far from the volcano's eruption. Even though volcanoes are typically classified as active, dormant or extinct, the infrequency of major eruptive events is dangerous. For instance, in 1951 Mount Lamington killed 5,000 people in Papua New Guinea, despite being considered extinct (Chester, 1993). To be prudent, all volcanoes that have erupted within the last 25,000 years should be regarded as potentially active. Most are located in geologically unstable areas and prone to multiple threats (Malheiro, 2006). On the other hand, volcanic terrain provides important natural resources, including geothermal energy, building materials and opportunities for tourism.

Witham (2005) noted that volcano-related fatalities have been underestimated and drew attention to other important impacts, such as the large number of people evacuated in volcanic emergencies (Table 7.3). Historically, most volcano-related deaths have been due to indirect causes, like famine arising from crop failure caused by falling ash. Today, pyroclastic flows are the chief cause of death, whilst lahars (volcanic mudflows) are the chief cause of injuries and ash fall accounts for most people made homeless, and less explosive eruptions create more localised hazards. However, there is concern about the increasing human exposure to volcanic hazards, especially in cities (Chester et al., 2000). The volcanic complex west of Naples, Italy, is now one of the most densely

Table 7.3 Best estimates of the human impacts of volcanic hazards in the 20th century (1900–99)

Human impacts	Number of events	Number of people
Killed	260	91,724
Injured	133	16,013
Homeless	81	291,457
Evacuated/affected	248	5,281,906
Any incident	491	5,595,500

Source: After Witham (2005)

Note: Each event may have had more than one consequence.

Table 7.4 Seven largest incidents in terms of loss of life (not including indirect or seismicity-related fatal causes). These total 125,000 fatalities (about 58% of total fatalities).

Volcano name	Year	Number of fatalities	Fatal cause
Krakatau, Indonesia	1883	36,000	Tsunami
Pelée, Martinique	1902	28,000	PDCs
Nevado del Ruiz, Colombia	1985	24,000	Lahars
Tambora, Indonesia	1815	12,000	Multiple (PDCs, Tephra, Tsunami)
Unzendake, Japan	1792	10,139	Tsunami
Kelut, Indonesia	1586	10,000	Lahars
Kelut, Indonesia	1919	5110	Lahars

Source: Brown et al. (2017, p. 7)

Note: The largest incidents are those with over 5,000 fatalities.

populated areas of active volcanism in the world, where some 200,000 people are at risk (Barberi & Carapezza, 1996). Many residents are vulnerable because they are poor, marginally employed and live in old, weakened buildings or unplanned housing.

Volcanoes attract human settlement, and it is the increase in exposed risk, rather than the frequency of eruptions, that explains the doubling of fatal eruptions from the 19th to the 20th century (Simkin et al., 2001) (see Table 7.4). According to Small and Naumann (2001), 10% of the world's population live within 100 km of a volcano active in historic times. The highest population densities at risk are in Southeast Asia and Central America. In Europe, the Etna region contains about 20% of Sicily's population, with rural population densities of 500–800 per km^2. Countries like Indonesia, located at the junction of three tectonic plates and with a population of over 150 million, face the greatest threat, and this nation has suffered two-thirds of all volcano-related deaths (Suryo & Clarke, 1985). The 1815 eruption of the Tambora volcano directly killed 12,000 people and a further 80,000 persons later perished through disease and famine. The rare supervolcano eruption at what is now Lake Toba in Indonesia some 75,000 years ago released an estimated 2,800 m^3 of debris, although whether it caused significant human deaths remains unconfirmed (Gathorne-Hardy & Harcourt-Smith, 2003). A repeat event would have a catastrophic impact on the world today.

Volcanoes affect livelihoods in many ways. The 2021 eruption of Cumbre Vieja volcano on La Palma (Canary Islands) had a devastating impact both on the local populations and on tourists. The eruption lasted 85 days and resulted in the evacuation of 7,000 people, and the lava flow covered over 1,000 hectares, including burying the town of and areas around the town of Todoque. The lava flow reached the sea, destroying more than 3,000 buildings, cutting access to the essential coastal highway and creating a new peninsula. The total damage caused by the eruption is calculated by the Canary Islands government as 843 million euros. La Palma is a popular tourism destination, and the ash generated caused the airport to partially close. Tourists visiting the volcano were required to wear protective goggles, and in some cases gas masks. The impact of the event was devastating for the island's economy, during an already challenging time of COVID-19, and many people who were affected were not familiar with the risk the volcano posed.

Volcanoes themselves can be a tourist attraction. A good example is that of Whakaari/White Island, an active andesite stratovolcano in the Bay of Plenty, New Zealand (Lim & Flaherty, 2020). The island is the peak of a much larger submarine volcano that is highly active. Whakaari has erupted continually from December 1975 until September 2000, and also erupted in 2012, 2016 and 2019. It has, however, become a popular tourist destination where guided tours enable people to get up close to a fairly active volcanic vent. Normally this type of activity would not be permitted by a government as the risk is too high, but the island is privately owned by the Buttle Family Trust and access to the island is allowed only as a member of a tour run by a registered tour operator. An eruption occurred on 9 December 2019 when 47 people were on the island. The eruption resulted in 22 fatalities; in addition 25 survivors were seriously injured, and three survivors suffered minor injuries. This was a devastating tragedy and legal investigations continue, highlighting that sometimes the best action is to not go near the volcano. This is not the first event to catch tourists out. Mount Ontake in Japan erupted suddenly on 27 September 2014, killing 63 people. It was a phreatic eruption where magma heats ground water or surface water causing a sudden explosion, with large amounts of ballistics.

C. DISASTER RISK REDUCTION

With over 10% of the world's population living within 100 km of a volcano believed to have been active during the last 10,000 years, DRR measures are essential to help protect populations, infrastructure, resources and livelihoods. The wide diversity of volcanic hazards means that each requires monitoring, warnings and mitigation strategies over differing spatialities and temporalities, with some occurring without an eruption. Key tools used to manage volcanic eruptions include monitoring the volcano, producing forecasts and warnings, building engineering structures, developing good land use planning and providing aid following an event.

Monitoring

Most volcanic eruptions are preceded by a variety of environmental changes that accompany the rise of magma towards the surface. UNDRO (1985) classified some of the physical and chemical phenomena that are observed before eruptions (Table 7.5). Unfortunately, such phenomena are not always present, and the most highly explosive volcanic eruptions are particularly difficult to forecast using the data. Over 80 volcano observatories across the globe are tasked with monitoring and communicating timely and useful information about the behaviour of a volcano (see the WOVO website and Newhall et al., 2017).

Table 7.5 Precursory phenomena that may be observed before a volcanic eruption

Seismic activity
- increase in local earthquake activity
- audible rumblings

Ground deformation
- swelling or uplifting of the volcanic edifice
- changes in ground slope near the volcano

Hydrothermal phenomena
- increased discharge from hot springs
- increased discharge of steam from fumaroles
- rise in temperature of hot springs or fumarole steam emissions
- rise in temperature of crater lakes
- melting of snow or ice on the volcano
- withering of vegetation on the slopes of the volcano

Chemical changes
- changes in the chemical composition of gas discharges from surface vents (e.g. increase in SO_2 or H_2S content)

Source: After UNDRO (1985)

Typically, real-time measurements of volcanic processes are made with GPS technology and satellite imagery. In some cases, local rainfall measurements may be useful (Barclay et al., 2006). These monitoring programmes provide the best hope of developing reliable data about what the volcano is doing. Without 'eyes and ears' on the ground, it is very difficult to know what is happening at a volcano, yet only 10% of the world's volcanoes have monitoring equipment to provide data (Scarpa & Gasparini, 1996). There have been advances in remote sensing to help provide data on volcanoes with no ground monitoring equipment, notably the observations made by ten Earth Observing System (EOS) satellites that monitor changes in volcanic activity involving features such as thermal anomalies, plume chemistry and lava composition (Ramsey & Flynn, 2004). Monitoring of volcanoes typically required a lot of ground data collection, which was dangerous on a volcano about to erupt. As such it is no surprise that being a field volcanologist is one of the most dangerous jobs to hold. However, the use of satellite and remote technologies have enabled scientists to obtain data remotely in safety.

Specific monitoring techniques apply to the following processes (see more at McGuire, 1998; Scarpa & Tilling, 2012):

- *Earthquake activity* is common near volcanoes and, for predictive purposes, it is important to gauge any increase in activity in relation to background levels. This requires a study of records going back many years. During a period of high alert, these records can be supplemented by data from portable seismometers. Most volcano-related earthquakes are less than Mw = 2 or 3 and occur less than 10 km beneath a volcano. A prior seismic signature has been incorporated into a tentative earthquake swarm model for the prediction of volcanic eruptions. As shown in Figure 7.12, the onset and subsequent peak of a 'swarm' of high-frequency earthquakes reflects the fracture of local rocks as magmatic pressure increases. This phase is followed by a relatively quiet period, when some of the pressure is relieved by cracks in the Earth's crust, before a final tremor results in an explosive eruption. Seismic signals are typically the single most useful form of monitoring data.
- *Ground deformation* is sometimes the forerunner of an explosive eruption, although the information is not easy to fit into a forecasting model. Movement is typically detected using survey equipment and *tiltmeters* that operate like a woodworker's level. They are very sensitive but can record changes in slopes only over short distances. Interferometric synthetic aperture radar, abbreviated InSAR, is a radar

technique used to map ground deformation using radar images of the Earth's surface from satellites and can provide data on ground deformation over a wide area by analysing changes in images over time.

- *Electronic distance meters* (EDMs) can measure the distance between benchmarks placed on a volcano in order to pinpoint when magma is rising and displacing the ground surface. This technique is usually less available and requires a series of visible targets on the volcano. There is a wide range of capability for these instruments, but short-range (<10 km) to medium-range (<50 km) EDMs are typically employed. Short-range EDMs can measure distances with an accuracy of about 5 mm, but the trick is to place benchmarks in the right places and make frequent observations.
- *Global Positioning Systems* (GPS) rely on about 24 satellites that orbit the Earth twice each day at altitudes around 20,000 km and constantly transmit information that allows the position of ground-based receivers to be determined. Normally five to eight of these satellites are in view from any point on Earth, provided that there is an unobstructed sky view. Since a ground-level accuracy of a few centimetres is necessary to detect the build-up of pressure from rising magma, GPS receivers are commonly placed at several locations on the volcano to minimise any variations in signal transmission through the atmosphere to individual receivers.
- *Thermal changes* might be expected as magma rises to the surface and increases the surface temperature. Many volcanoes have erupted without any detectable shift in temperature, but this approach has proved useful where a crater lake exists. An early example occurred at the Taal volcano, Philippines, when the temperature increased from a constant 33°C in June 1965 to reach 45°C by the end of July. The water level also rose during this period. In September 1965, a violent eruption occurred. Such ground observations can be supplemented and confirmed by thermal imaging from satellites. In addition, a thermal camera uses infrared-sensitive sensors to detect the IR energy and converts it into a thermal image. Colours are assigned to the differing temperatures to indicate variations in temperatures. Often infrared enables scientists to see lava flows that may not be visible to the naked eye.
- *Geochemical changes* can be detected in the composition of the gases issuing from volcanic vents. In addition, the rate at which a volcano releases gases is related to variations in the amount of magma within the subsurface systems. Direct field sampling of gases escaping from surface vents is the usual method employed, but remote sensing has been an important tool for the surveillance of volcanic activity for over 40 years. Juvenile gases show considerable variation over short periods and distances, and it is difficult to judge how gas samples represent more general conditions in the volcano. Larger-scale visual observations of steam emissions or ash clouds depend on meteorological conditions. Sulphur dioxide injected high into the atmosphere can be measured by on-board satellite instruments, and the behaviour of volcanic plumes can be monitored by weather satellites. Equipment can also be mounted on an aircraft to determine the quantity of gas being emitted on a daily basis and then also be used on the ground.

- *Lahars* have been monitored for years via visual sightings by local people and, more recently, with the aid of video cameras located in remote valleys. An automated detection system now exists using a series of acoustic-flow (AFM) monitoring stations placed downstream from the volcano (see USGS http://volcanoes.usgs.gov/activity/methods/hydrologic/lahardetection.php). At each location, a seismometer detects ground vibrations from an approaching lahar. The amplitude of vibration is sampled every second and information, including an emergency message, can be transmitted to a base station further down the volcano's flank and closer to a population centre. Short-term warnings and emergency evacuation are then possible.
- *Drones* (an unmanned aerial vehicle, UAV) enable close-up imagery of activity that can be sought without the need for a full video surveillance system and can catch footage once the volcano has erupted.

Forecasting and Warning

To date, the volcanological community has been unable to generate accurate and reliable predictive models for use on a single volcano, let alone for the many types of volcano and styles of eruption that occur. Scientists (typically geochemists, geophysicists, geologists, volcanologists and mathematicians) have exploited a range of quantitative theoretical and statistical models of volcanic monitoring data to develop better prognoses of a volcano's behaviour and to deal with the high levels of uncertainty involved (Kilburn, 2003).

However, in the face of such complexity a volcano early warning system is

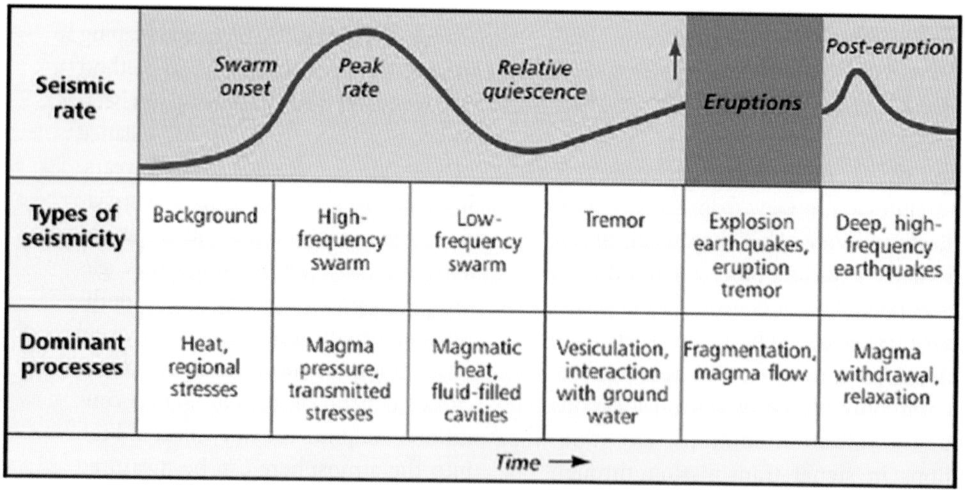

Seismic rate	Swarm onset	Peak rate	Relative quiescence		Eruptions	Post-eruption
Types of seismicity	Background	High-frequency swarm	Low-frequency swarm	Tremor	Explosion earthquakes, eruption tremor	Deep, high-frequency earthquakes
Dominant processes	Heat, regional stresses	Magma pressure, transmitted stresses	Magmatic heat, fluid-filled cavities	Vesiculation, interaction with ground water	Fragmentation, magma flow	Magma withdrawal, relaxation

Time →

Figure 7.12 The stages of a generic volcanic-earthquake-swarm model. The precursor earthquake swarm, which can be used as a warning sign, reflects the fracturing of rocks in response to growing magmatic pressure.

Source: After McNutt (1996). With kind permission from Springer Science+Business Media: S.R. McNutt, Seismic monitoring and eruption forecasting of volcanoes: a review of the state of the art and case histories in R. Scarpa and R.I. Tilling (eds) *Monitoring and Mitigation of Volcano Hazards* (1996), 99–146. Copyright Springer 1996. Reproduced with permission.

used to provide the best information to the civil protection/emergency managers and the vulnerable public so they can prepare. This requires bringing together a volcano's history (Chester, 2005) and the monitoring data and establishing a warning system. A volcano alert level system (VALS) is a series of levels that correspond generally to increasing levels of volcanic activity (Gardner & Guffanti, 2006). Volcanoes have the most diverse range of ALS of any hazard, varying by hazard (lahar, ash, VOG, tsunami); scale (for each volcano, regional, national, international); and geography (each nation generally has their own standardised systems) defined and shaped by the social, cultural, economic and political contexts in which they operate. Alerts provide a general awareness of increased hazard and/or risk around a volcano that may then result in evacuation prior to the eruption (see Figure 7.13).

Alert levels and warnings are used in conjunction with other communication techniques, such as information statements, reports and maps typically issued by a volcano observatory, and are often decided at meetings and issued by the scientific monitoring agency and/or emergency managers (Fearnley & Beaven, 2018). *Volcanic crisis communication* is the term used to encompass all forms of communication during a volcanic crisis

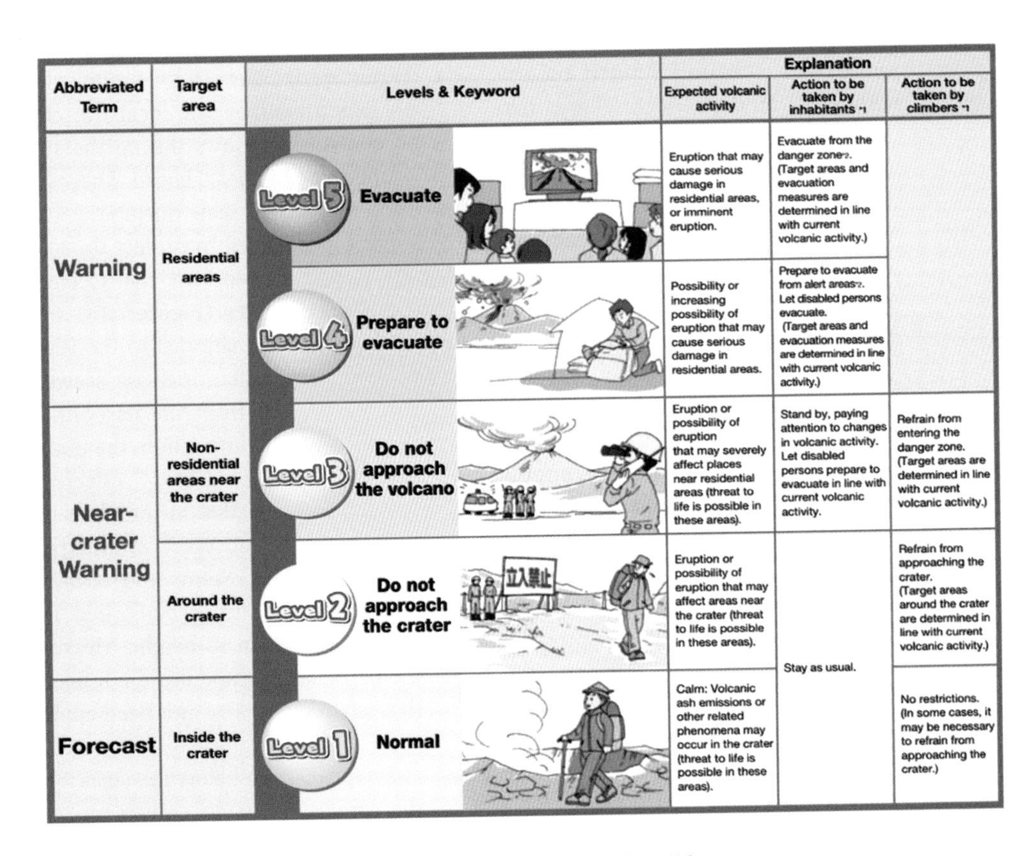

Abbreviated Term	Target area	Levels & Keyword	Explanation: Expected volcanic activity	Explanation: Action to be taken by inhabitants *1	Explanation: Action to be taken by climbers *1
Warning	Residential areas	Level 5 Evacuate	Eruption that may cause serious damage in residential areas, or imminent eruption.	Evacuate from the danger zone*2. (Target areas and evacuation measures are determined in line with current volcanic activity.)	
		Level 4 Prepare to evacuate	Possibility or increasing possibility of eruption that may cause serious damage in residential areas.	Prepare to evacuate from alert areas*2. Let disabled persons evacuate. (Target areas and evacuation measures are determined in line with current volcanic activity.)	
Near-crater Warning	Non-residential areas near the crater	Level 3 Do not approach the volcano	Eruption or possibility of eruption that may severely affect places near residential areas (threat to life is possible in these areas).	Stand by, paying attention to changes in volcanic activity. Let disabled persons prepare to evacuate in line with current volcanic activity.	Refrain from entering the danger zone. (Target areas are determined in line with current volcanic activity.)
	Around the crater	Level 2 Do not approach the crater	Eruption or possibility of eruption that may affect areas near the crater (threat to life is possible in these areas).	Stay as usual.	Refrain from approaching the crater. (Target areas around the crater are determined in line with current volcanic activity.)
Forecast	Inside the crater	Level 1 Normal	Calm: Volcanic ash emissions or other related phenomena may occur in the crater (threat to life is possible in these areas).		No restrictions. (In some cases, it may be necessary to refrain from approaching the crater.)

Figure 7.13 Japanese Meteorological Agency Volcano Alert Level System.

Source: Japanese Meteorological Agency

(Fearnley et al., 2018), from oral traditions to social media, to GIS, to dance and art.

Environmental Control

There is no way of preventing major volcanic eruptions, but attempts have been made to control the surface flow of lava. Engineered structures that can mitigate or protect against volcanic hazards (Ikeya, 2008; Williams & Moore, 1983) include the following:

- *Explosives* can be used in two situations. First, aerial bombing of fluid lava high on the volcano may cause the flow to spread and halt the advancing lava front by depriving it of supply. This method was first tried in 1935 on Hawaii with only limited success, although modern technology may achieve better results Second, control of *aa* flows has been attempted by breaching the walls which form along the edges of the flow so that lava floods out and starves the advancing front of material. This method was used on the *aa* flow of Mauna Loa's 1942 eruption and in the 1983 eruption of Etna, when it proved possible to divert some 20%–30% of blocky flow (Barberi et al., 1993). These methods are not without risk and possible collateral damage. They would be used with great caution today.
- *Artificial barriers* can divert lava streams away from valuable property if the topographic conditions are suitable and local landowners agree. Barriers must be constructed from massive rocks, or other resistant material, with a broad base and gentle slopes. The method works best for thin and fluid lava flows that exert a limited amount of thrust. It is doubtful if diversion would work with the powerful blocky flows that attain heights of 30 m or more. In the Krafla area of northern Iceland, land has been bulldozed to create barriers to protect a village and a factory from flowing lava. During the 1955 eruption of Kilauea, Hawaii, a temporary barrier diverted the flow from two plantations but later flows took different paths and destroyed the property. Such uncertainty raises the possibility of legal action if lava is deliberately diverted onto a property that otherwise would have escaped. Permanent diversion barriers have been proposed to protect Hilo, Hawaii. Such walls would need to be 10 m high with channels capable of containing a flow about 1 km wide.
- *Water sprays* were first employed in an experiment by the local fire chief to control lava flows during the 1960 eruption of Kilauea, Hawaii. They were used on a larger scale during the 1973 eruption of Eldfell to protect the town of Vestmannaeyjar on the Icelandic island of Heimaey (Figure 7.14). Special pumps were shipped to the island so that large quantities of seawater could be taken from the harbour. At the height of the operation, the pumping rate was almost 1 $m^3 s^{-1}$, effectively chilling about 60,000 m^3 of advancing lava per day. The exercise lasted for about 150 days. Soon after spraying began, the lava front congealed into a solid wall 20 m in height. Measurements of lava temperature confirmed that where water had not been applied the lava temperature was 500–700°C at a depth 5–8 m below the surface. In the sprayed areas an equivalent temperature was not attained until a depth of 12–16 m below the lava surface (UNDRO, 1985).

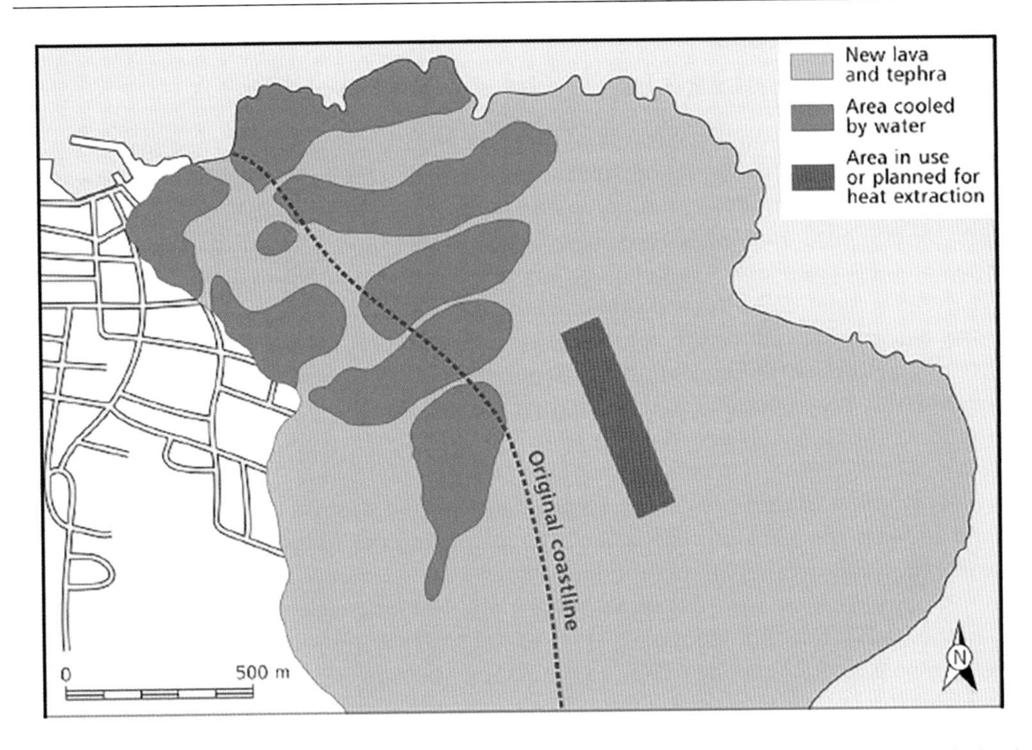

Figure 7.14 Simplified map of the eastern edge of the fishing port of Vestmannaeyjar, Heimay, Iceland, after the eruption of Eldfell volcano in 1973, showing the new lava field beyond the earlier coastline and areas cooled by pumping seawater between March and June 1973. Planned heat-extraction zones are a reminder that gains, as well as losses, can result from disasters.

Source: After Williams and Moore (1983, p. 17)

Physical protection against lahars and pyroclastic flows is possible via SABO dams, which depend on the construction of sediment traps and diversion barriers similar to those for other fluid flows. These structures are expensive and silt up with material over time. They can be located only where lahar and/or pyroclastic paths are well defined and do not work for major, destructive flows in deep valleys. Proposals for the diversion of lahars into wetland areas used for seasonal floodwater storage and fishing in the Philippines have proved controversial. The most ambitious attempts to control lahars at source have been undertaken on Kelut volcano on the island of Java (Box 7.2 and Figure 7.15).

Hazard-Resistant Design

If buildings remain intact during an explosive eruption they can provide a temporary refuge. Even after a large eruption, some buildings beyond 2–3 km of the volcanic vent have resisted collapse from the pressure of pyroclastic flows (Petrazzuoli & Zuccaro, 2004). According to Spence et al. (2004), the buildings most likely to survive are those of recent masonry construction or with reinforced concrete frames, so long as the door and window openings do not fail and allow the entry of hot gas and ash. Ash falls can cause the collapse of unstrengthened buildings, especially those with flat roofs. This is most likely if the ash is wet, because, whilst

Box 7.2 Crater-Lake Lahars in the Wet Tropics

Kelut volcano, eastern Java, is one of the most deadly volcanoes in Indonesia. This is due to lahars produced by releases from the large crater lake, which in 1875 was estimated to contain 78 × 10^6 m^3 of water. In the year 1586 about 10,000 lives were lost, and in 1919 an explosive eruption threw some 38.5 × 10^6 m^3 of water out of the crater lake and lahars travelled 38 km in less than one hour to claim 5,160 lives. To avoid a repetition of this disaster, Dutch engineers began a tunnel nearly 1 km long designed to reduce the volume of stored water from about 65 × 10^6 m^3 to 3 × 10^6 m^3. In 1923, with the existing crater already half full (22 × 10^6 m^3), the plan was changed to seven parallel tunnels that would progressively lower the water level (Figure 7.15). This work was completed in 1926 and the lake volume was reduced to < 2 × 10^6 m^3[30].

An eruption in 1951 created no large lahars but did destroy the tunnel entrances and added to the water storage capacity by deepening the crater by some 10 m. Even with repair of the original lowest tunnel the lake soon accumulated a volume of 40 × 10^6 m^3 and became a serious threat once more. The Indonesian government started another low tunnel but stopped it short of the crater wall in the hope that seepage would help to drain the lake. This did not happen because of the low permeability of the volcanic cone. At the time of the 1966 eruption the lake volume was about 23 × 10^6 m^3. Lahars killed hundreds of people and damaged much agricultural land. After this event a new tunnel, completed in 1967, was constructed 45 m below the level of the lowest existing tunnel and the lake volume was reduced to 2.5 × 10^6 m^3. Several sediment dams were also installed. When the 1990 eruption occurred no primary lahars were recorded, although at least 33 post-eruption lahars were generated which travelled nearly 25 km from the crater. By 2003 the lake was some 33 m deep and the 1.9 × 10^6 m^3 volume represented the lowest risk of primary lahar generation for many years (www.ulb.ac.be/sciences).

A similar, but larger, problem emerged at Mount Pinatubo after the 1991 eruption created a 2.5 km wide crater over 100 m deep and capable of holding over 200 × 10^6 m^3 of water. As a result, about 46,000 residents in the town of Botolan, 40 km northwest of the volcano, are at risk from a massive lahar, together with the people who have returned to the upper slopes since 1991. By August 2001, the threat of a breach in the crater wall, as a result of increasing water levels in the rainy season, persuaded the authorities to dig a 'notch' in the crater rim to drain water away from populated areas. The outfall started operating in September, accompanied by the short-term evacuation of Botolan as a safety measure. Despite this attempt at drainage the lake level continued to rise. In July 2002, part of the western wall of the crater collapsed, with the slow release of about 160 × 10^6 m^3 of water and sediment.

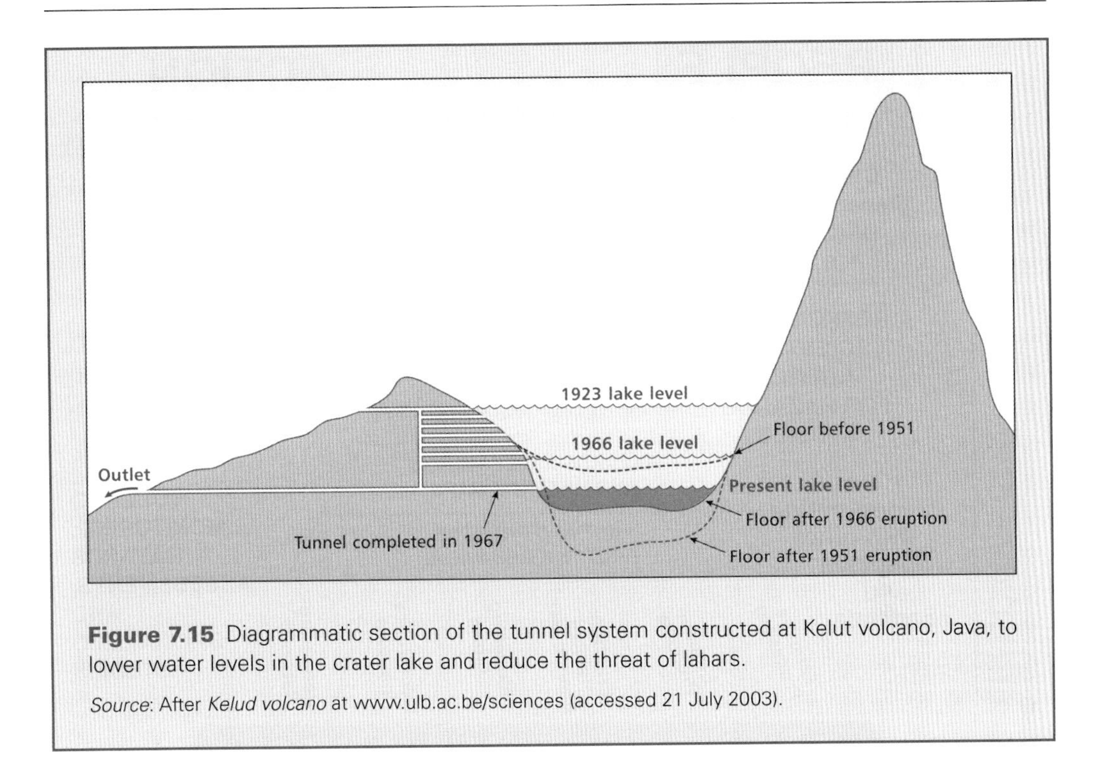

Figure 7.15 Diagrammatic section of the tunnel system constructed at Kelut volcano, Java, to lower water levels in the crater lake and reduce the threat of lahars.

Source: After *Kelud volcano* at www.ulb.ac.be/sciences (accessed 21 July 2003).

the bulk density of dry ash ranges from 0.5 to 0.7 t m^{-3}, that of wet ash may reach 1.0 t m^{-3}. Therefore, when warning times are too short for evacuation, people should be advised to seek shelter indoors, preferably with a sloping roof.

Community Preparedness

The cost of monitoring volcanoes and pre-disaster planning is small compared to the potential losses. The main elements of reducing volcanic risk are shown in Figure 7.16. The length of time available for a warning differs widely. Some volcanic activity may start several months before an eruption; in other cases, only a few hours may be available. For effective evacuation, it is essential that the population at risk is advised well in advance about evacuation routes and the location of refuge points. These directions have to be flexible, depending on the scale of the eruption (which will influence the pattern of lava flow) and the wind direction at the time (which will influence the pattern of ash fall). Some local roads may be destroyed by earthquake-induced ground failures and steep sections of highways can become impassable with small deposits of fine ash that make asphalt very slippery.

Unfortunately, the infrequency of volcanic activity induces poor hazard awareness and low levels of community preparedness.

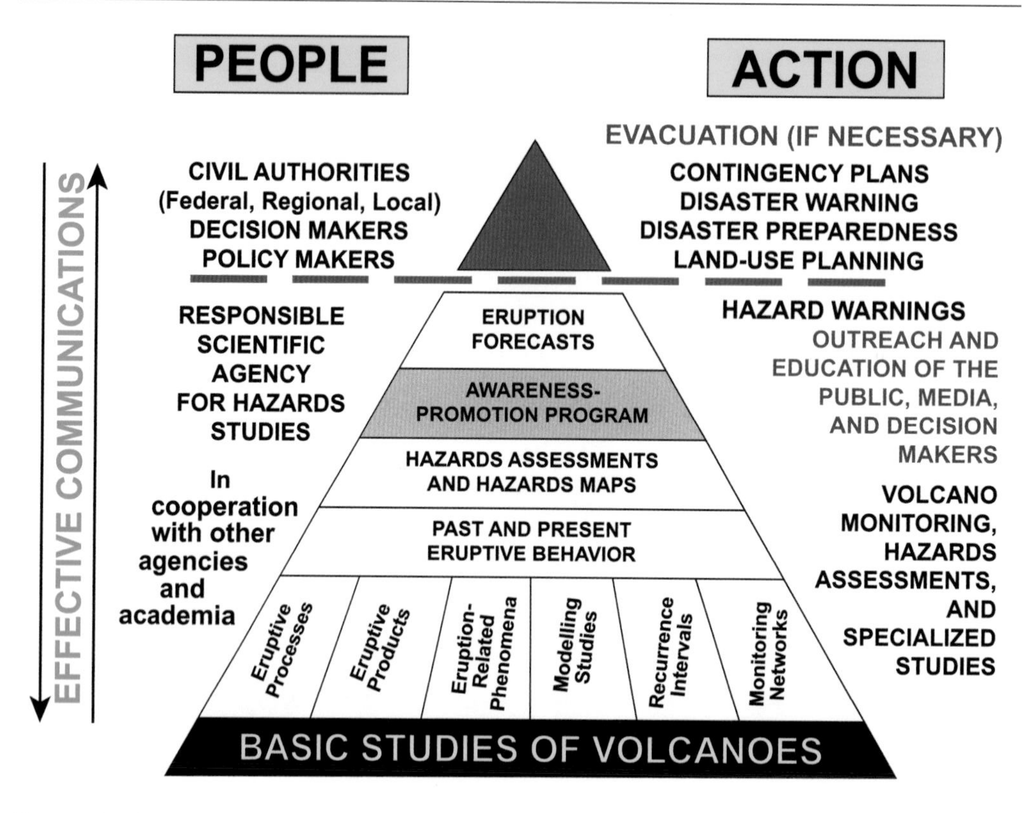

Figure 7.16 Schematic diagram of an idealised programme to reduce volcanic risk. The apex is separated from the rest of the triangle to emphasise that volcano scientists, while responsible for providing the best possible scientific information and advice, do not typically have knowledge of other key factors (e.g. socio-economic, cultural, political) and rarely have the authority to make final decisions regarding mitigation measures, including possible evacuation.

Source: Modified from Tilling (1989, p. 242)

Surveys of residents on Hawaii (Gregg et al., 2004) and Santorini (Dominey-Howes & Minos-Minopoulos, 2004) revealed a relatively poor understanding of volcanic risk and, in the case of the latter island, no emergency plan existed at the time.

All too often emergency planning for volcanic hazards occurs after a disaster. Before the Nevado del Ruiz disaster in Colombia in 1985, for example, there was no national policy for volcanic hazards. Similar policy failures occurred at the Galeras volcano, Colombia, where hazard mapping was held up by the unwillingness of the authorities to accept either the concept of disaster or the cost of mitigation (Cardona, 1997). The successful evacuation of densely populated areas requires adequate transportation. Evacuees need support services, including medical treatment (especially for dust-aggravated respiratory problems and burns), shelter, food and hygiene.

Because volcanic emergencies can last for months, the 'temporary' arrangements for displaced people sometimes function from one crop season to another. For example, a total of 26,000 people were evacuated in 1999 from the slopes of Tungurahua (Ecuador) because of the possibility of an eruption; some remained in special accommodation for over one year. Prolonged relocation is profoundly challenging for those in economic insecurity. During the Tungurahua emergency, the evacuated population of Banos, a town heavily reliant on tourist revenue, organised a return to their homes while the town remained under an evacuation order so that they could regain their livelihoods. Hazard mitigation specialists tend to assume that the key issue is communicating with vulnerable communities that they are at risk. Public education programmes, including field trips and evacuation exercises, help to raise awareness, as seen during the 2020–2021 eruption of La Soufrière volcano, St Vincent (see Box 7.3). However, this response does not tackle economic insecurity and concerns around loss of income.

Land Use Planning

Land zoning of high-hazard volcanic areas, plus the selection of safe sites for emergency evacuation and new development, depends on probability assessments of dangerous activity. To determine the likelihood of an event, all previous eruptions require dating on geological timescales using techniques such as radiocarbon

Box 7.3 Communicating Volcanic Risk During a Pandemic

After 40 years of low activity, La Soufrière, a stratovolcano on the Caribbean island of Saint Vincent in Saint Vincent and the Grenadines, began an effusive eruption on 27 December 2020. On 9 April 2021 there was an explosive eruption, and the volcano continued to erupt explosively over the following days resulting in pyroclastic flows, prompting the evacuation of at least 18,000 people. The COVID-19 pandemic made evacuation more difficult, requiring COVID-19 precautions to avoid outbreaks. The University of the West Indies (UWI) Seismic Research Centre (SRC) and the St Vincent & the Grenadines National Emergency Management Organization (NEMO) are the official sources of information on La Soufrière and, despite the challenges, were able to successfully manage the evacuation without a single death.

The key to their successes was to develop risk communications to target different learning styles, media platforms and preferences that were based on best practice and evidence. This included visual, print and audio products, live media interviews and social media posts. The St Vincent-based SRC scientists engaged with the local communities via daily activity updates on local television and radio stations and provided cabinet briefings and updates to key decision-makers, working with multiple stakeholders. A novel approach was taken by the scientists who, faced with the limitations of COVID-19 where the normal 'town hall' community meeting events were not possible, developed virtual and drive-through community meetings for populations living in the Red Zone (north part of the island). By adopting a wide breadth of communication tools, for a range of stakeholders, as outlined in Figure 7.17, it was possible to help manage the uncertainty involved during the crisis, counter misinformation and rumours and respond to the inability of people to mix and communicate in person. By producing clear and easily accessible resources, it was possible for international scientists to support the crisis management plan and provide the same messages to their own media channels.

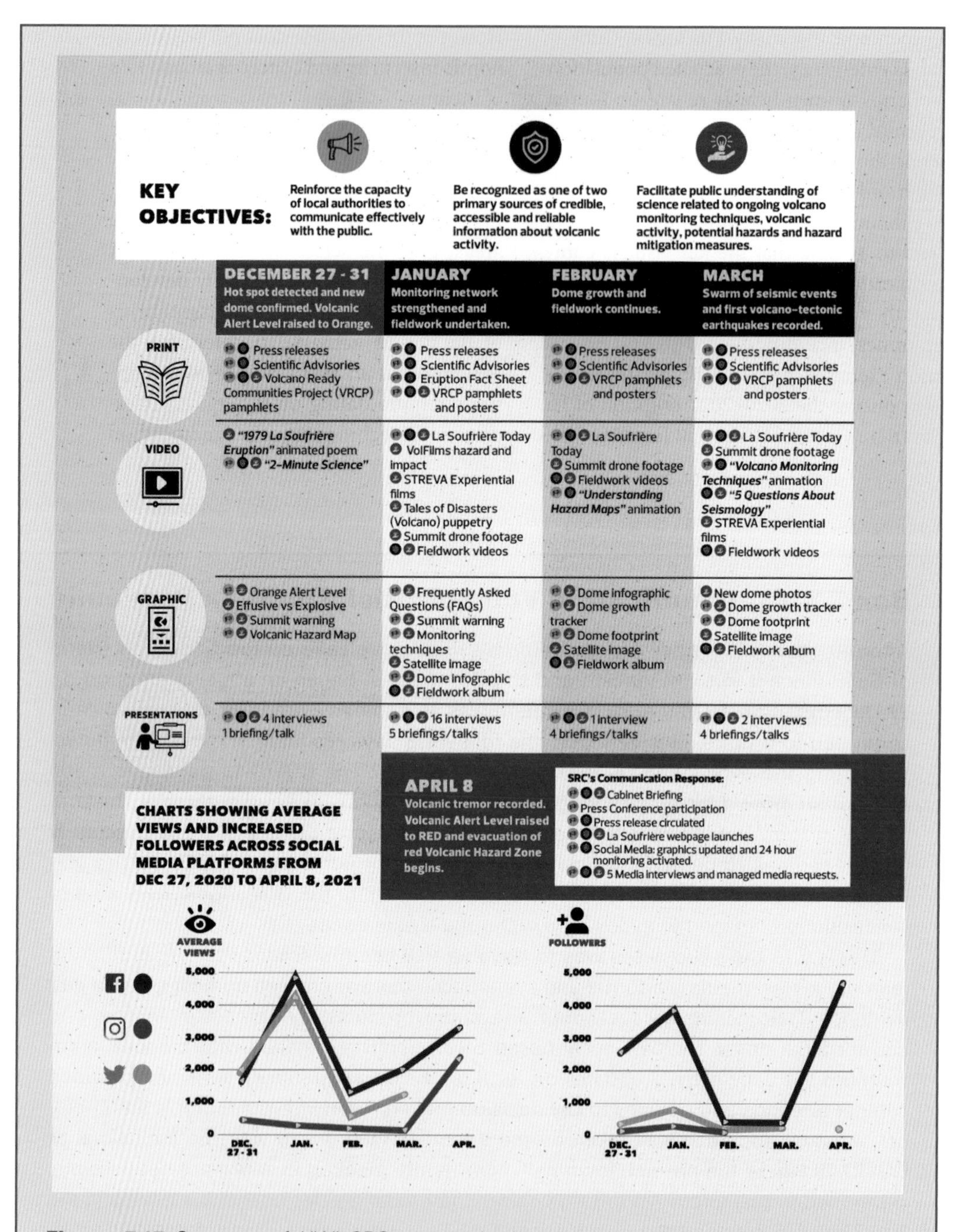

Figure 7.17 Summary of UWI SRC communication strategy and response throughout the 2020–2021 eruption of La Soufrière, St Vincent eruption.

Source: Joseph et al. (2022)

dating, tree-ring analysis, lichenometry and thermoluminescence. Volcanic-hazard maps can then be prepared to show the likely area at risk (Thompson et al., 2017). Such mapping is hampered by limited knowledge of the size of future eruptions and the extent of pyroclastic surges or lahars. Environmental conditions at the time of eruption will be important; the amount of seasonal snow cover will affect lahar and avalanche hazards, whilst the speed and direction of the wind will determine the airborne spread of tephra.

Because of these challenges many zoning maps are restricted to just one or two volcanic hazards. For example, the island of Hawaii has nine hazard zones ranked on the probability of land coverage by lava flows, based on the location of volcanic vents, the topography of the volcanoes and the extent of past flows. Zones 1–3 are limited to the active volcanoes of Kilauea and Mauna Loa, while zone 9 consists of Kohala volcano, which last erupted over 60,000 years ago (Figure 7.18). This map ignores hazards other than lava flows. In

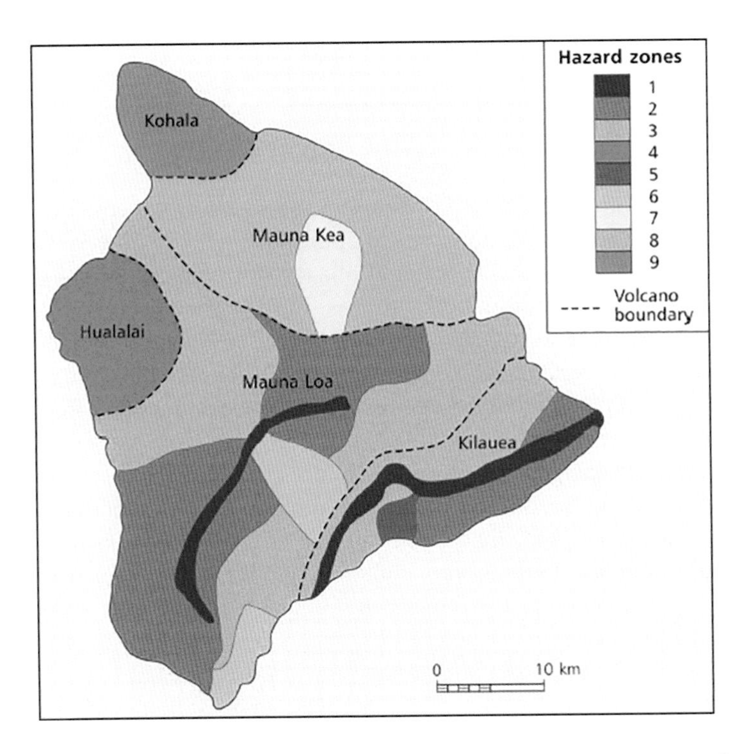

Figure 7.18 The island of Hawaii zoned according to the degree of risk from lava flows. Zone 1 is highest rated, Zone 9 is lowest rated and the change between zones is gradual rather than abrupt. All property destroyed in the last 30 years was in Zone 2 located within 12 km of the vent of Kilauea.

Source: After US Geological Survey at http://pubs.usgs.gov/gip/hazards/maps.html (accessed 26 February 2003)

comparison, the combined lava flow and ash fall risk assessment available for the island of Tenerife provides a more comprehensive picture (Araña et al., 2000). Most land planning maps undergo revision and updating and can be used for insurance purposes, for building development permission and to evaluate the location of key infrastructure such as hospitals and school facilities.

Disaster Aid

An eruptive spell can continue over months or years, creating a need for support that blurs the distinction between emergency aid and longer-term development investment. Evacuation is a common response. For example, the 1982 Galunggung, Indonesia, emergency was created by no fewer than 29 explosive phases occurring over a six-month period and led to the evacuation of over 70,000 people. Volcanic disasters on small islands tend to overwhelm local resources and pose extra difficulties for evacuation and disaster management (see Box 7.4).

Volcanoes remain one of the most challenging hazards to manage, with multiple hazards and risks that can occur over different timescales and geographies and have significant impacts on entire regions,

Box 7.4 Emergency Response in Montserrat

In July 1995, the Soufrière Hills volcano, located in the south of the small Caribbean island of Montserrat, began a prolonged phase of eruptive activity that continued for over a decade. It was characterised by several phases of dome building and subsequent collapse, accompanied by multiple hazards, including extensive ash fall and lahar deposits. Only 19 lives were lost, but most of the island's infrastructure was destroyed, with economic losses of about £1 billion. By December 1997 almost 90% of the population of over 10,000 people had been forced to relocate and more than two-thirds had been evacuated from the island.

Montserrat is an Overseas Territory of the UK. In the absence of an emergency plan the UK government and the Government of Montserrat had to work together on disaster response. An initial emergency plan was prepared in the first few days. Following various evacuations of people to temporary shelter in public buildings, such as schools and churches, relief rations were distributed to 4,000–5,000 displaced people, with many suffering due to delays.

The chronology of repeated evacuations from Plymouth, the island's capital, illustrates the uncertainty regarding when to evacuate. People were first evacuated on 21 August 1995 and allowed to return 7 September; a second evacuation on 2 December 1995 was followed by a return on 2 January 1996; and a final evacuation was begun on 3 April 1996. Later that month a state of public emergency was declared and residents were offered voluntary relocation off the island – either to a neighbouring Caribbean island or to the UK with full rights to state benefit and accommodation for a two-year period. By August 1997 about 1,600 displaced people were still in temporary shelter and, even in late 1998, 322 people were still housed in these conditions.

Figure 7.19 The spire of a church protrudes from ash, mud and rock deposited by pyroclastic flows that destroyed Plymouth, the former capital of the island of Montserrat, during multiple eruptions of the Soufrière Hills volcano in the 1990s.

Source: Photo: Panos Pictures/Steve Forrest SFR00093MSR

In June 1997, pyroclastic flows resulted in fatalities and the volcanic risk was reassessed, with more than half the island being placed in an exclusion zone. Around this time, more permanent arrangements began to appear, many aimed at securing the longer-term future of the island. These included the construction of an emergency jetty to aid evacuation; new directly built housing, aided by subsidised soft mortgage schemes; strengthening of the scientific capability of the Montserrat Volcano Observatory and the publication of a draft Sustainable Development Plan. Inevitably, there were further delays and, by November 1998, only 105 out of 255 planned houses had been built. In 1999 the UK provided an assisted return passage scheme for those who had left the island. Up to March 1998, the UK government had spent £59 million in emergency-related aid, with an estimated total expenditure of £160 million over a six-year period.

This emergency highlighted a need for pre-disaster planning and swifter investment decisions in volcanic crises, especially where governance is shared between different authorities. More regional cooperation, in this case between the Caribbean countries, would help with monitoring of volcanic activity and the raising of awareness and preparation for future events.

if not the whole globe via atmospheric impacts. It is vital that monitoring is established, relationships are forged with vulnerable communities around volcanoes and warning systems help populations gear up their awareness with increased activity. This remains challenging when volcanoes do not tend to erupt very often and many vulnerable communities are worried about everyday risks such as economic insecurity, health, crime and access to food security (Christie et al., 2015).

KEY TAKE AWAY POINTS

- Volcanoes have a wide diversity in terms of eruptive styles, types of hazards, scale of impact and timeframes of eruption that makes their management incredibly challenging. Consequently, volcanic hazards can have a significant impact on society.
- Volcanoes cause multi-hazard events that can be primary and secondary and can occur whether a volcano is active or not.
- Volcanoes have the potential to cause mass extinction events over geological timescales or cause global cooling that can significantly impact life on Earth; however, these events are rare.
- Warning systems are a key tool used to help facilitate precautionary and anticipatory action prior to volcanic eruptions, typically via the use of alert level systems.
- A key challenge is raising awareness and getting people to safety or to evacuate, especially if for an unknown length of time. This often requires precautionary actions, which places a lot of pressure on generating accurate forecasts and often results in authorities having to make difficult uncertain decisions.

NOTE

1 www.aljazeera.com/news/2021/5/23/dr-congo-volcano-eruption-leaves-death-and-trail-of-destruction

BIBLIOGRAPHY

Araña, V., Felpeto, A., Astiz, M., Garcıa, A., Ortiz, R., & Abella, R. (2000). Zonation of the main volcanic hazards (lava flows and ash fall) in Tenerife, Canary Islands: A proposal for a surveillance network. *Journal of Volcanology and Geothermal Research*, *103*(1–4), 377–391.

Barberi, F., & Carapezza, M. L. (1996). The problem of volcanic unrest: The Campi Flegrei case history. In *Monitoring and mitigation of volcano hazards* (pp. 771–786). Berlin, Heidelberg: Springer.

Barberi, F., Carapezza, M. L., Valenza, M., & Villari, L. (1993). The control of lava flow during the 1991–1992 eruption of Mt. Etna. *Journal of Volcanology and Geothermal Research*, *56*(1–2), 1–34.

Barclay, J., Johnstone, J. E., & Matthews, A. J. (2006). Meteorological monitoring of an active volcano: Implications for eruption prediction. *Journal of Volcanology and Geothermal Research*, *150*(4), 339–358.

Beckett, F. M., Witham, C. S., Leadbetter, S. J., Crocker, R., Webster, H. N., Hort, M. C., . . . Thomson, D. J. (2020). Atmospheric dispersion modelling at the London VAAC: A review of developments since the 2010 Eyjafjallajökull volcano ash cloud. *Atmosphere*, *11*(4), 352.

Bird, D. K. (2009). The use of questionnaires for acquiring information on public perception of natural hazards and risk mitigation: A review of current knowledge and practice. *Natural Hazards and Earth System Sciences*, *9*(4), 1307–1325.

Bird, D. K., & Gísladóttir, G. (2020). Enhancing tourists' safety in volcanic areas: An investigation of risk communication initiatives in Iceland. *International Journal of Disaster Risk Reduction*, *50*, 101896.

Bird, D. K., Gisladottir, G., & Dominey-Howes, D. (2010). Volcanic risk and tourism in southern Iceland: Implications for hazard, risk and emergency response education and training. *Journal of Volcanology and Geothermal Research*, *189*(1–2), 33–48.

Brantley, S. R. (1990). The eruption of Redoubt Volcano, Alaska, December 14, 1989–August 31, 1990.

Brown, S. K., Jenkins, S. F., Sparks, R. S. J., Odbert, H., & Auker, M. R. (2017). Volcanic fatalities database: Analysis of volcanic threat with distance and victim classification. *Journal of Applied Volcanology, 6*(1), 1–20.

Capra, L., Poblete, M. A., & Alvarado, R. (2004). The 1997 and 2001 lahars of Popocatépetl volcano (Central Mexico): Textural and sedimentological constraints on their origin and hazards. *Journal of Volcanology and Geothermal Research, 131*(3–4), 351–369.

Cardona, O. D. (1997). Management of the volcanic crises of Galeras volcano: Social, economic and institutional aspects. *Journal of Volcanology and Geothermal Research, 77*(1–4), 313–324.

Carn, S. A., Watts, R. B., Thompson, G., & Norton, G. E. (2004). Anatomy of a lava dome collapse: The 20 March 2000 event at Soufrière Hills Volcano, Montserrat. *Journal of Volcanology and Geothermal Research, 131*(3–4), 241–264.

Casadevall, T. J. (Ed.). (1994). *Volcanic ash and aviation safety: Proceedings of the first international symposium on volcanic ash and aviation safety* (No. 2047). US Government Printing Office.

Chester, D. K. (1993). *Volcanoes and society.* London.

Chester, D. K. (2005). Theology and disaster studies: The need for dialogue. *Journal of Volcanology and Geothermal Research, 146*(4), 319–328.

Chester, D. K., Degg, M., Duncan, A. M., & Guest, J. E. (2000). The increasing exposure of cities to the effects of volcanic eruptions: A global survey. *Global Environmental Change Part B: Environmental Hazards, 2*(3), 89–103.

Christie, R., Cooke, O., & Gottsmann, J. (2015). Fearing the knock on the door: Critical security studies insights into limited cooperation with disaster management regimes. *Journal of Applied Volcanology, 4*(1), 1–15.

Clapperton, C. M., & Vera, R. (1986). The Quaternary glacial sequence in Ecuador: A reinterpretation of the work of Walter Sauer. *Journal of Quaternary Science, 1*(1), 45–56.

Dominey-Howes, D., & Minos-Minopoulos, D. (2004). Perceptions of hazard and risk on Santorini. *Journal of Volcanology and Geothermal Research, 137*(4), 285–310.

Donovan, A. R., & Oppenheimer, C. (2011). The 2010 Eyjafjallajökull eruption and the reconstruction of geography. *The Geographical Journal, 177*(1), 4–11.

Fearnley, C. J., & Beaven, S. (2018). Volcano alert level systems: Managing the challenges of effective volcanic crisis communication. *Bulletin of Volcanology, 80*(5), 1–18.

Fearnley, C. J., Bird, D. K., Haynes, K., McGuire, W. J., & Jolly, G. (Eds.). (2018). *Observing the volcano world: Volcano crisis communication.* Cham: Springer International Publishing.

Fitzgerald, R. H., Kennedy, B. M., Gomez, C., Wilson, T. M., Simons, B., Leonard, G. S., Matoza, R. S., Jolly, A. D., & Garaebiti, E. (2020). Volcanic ballistic projectile deposition from a continuously erupting volcano: Yasur Volcano, Vanuatu. *Volcanica, 3*(2), 183–204.

Foxworthy, B. L., & Hill, M. (1982). *Volcanic eruptions of 1980 at Mount St. Helens: The first 100 days* (Vol. 1249). USA: US Government Printing Office.

Gardner, C. A., & Guffanti, M. C. (2006). *US Geological Survey's alert notification system for volcanic activity* (No. 2006–3139). USA.

Gathorne-Hardy, F. J., & Harcourt-Smith, W. E. H. (2003). The super-eruption of Toba, did it cause a human bottleneck? *Journal of Human Evolution, 45*(3), 227–230.

Gíslason, S. R., Alfredsson, H. A., Eiríksdóttir, E. S., Hassenkam, T., & Stipp, S. L. (2011). Volcanic ash from the 2010 Eyjafjallajökull eruption. *Applied Geochemistry, 26,* S188–S190.

Grattan, J., Rabartin, R., Self, S., & Thordarson, T. (2005). Volcanic air pollution and mortality in France 1783–1784. *Comptes Rendus Geoscience, 337*(7), 641–651.

Gregg, C. E., Houghton, B. F., Johnston, D. M., Paton, D., & Swanson, D. A. (2004). The perception of volcanic risk in Kona communities from Mauna Loa and Hualālai volcanoes, Hawai'i. *Journal of Volcanology and Geothermal Research, 130*(3–4), 179–196.

Ikeya, H. (2008). The Heisei eruption of Mt. Unzen-Fugendake and measures against volcanic disasters. *Journal of Disaster Research, 3*(4), 276–283.

Jóhannesdóttir, G., & Gísladóttir, G. (2010). People living under threat of volcanic hazard in southern Iceland: Vulnerability and risk perception. *Natural Hazards and Earth System Sciences, 10*(2), 407–420.

Joseph, E. P., Camejo-Harry, M., Christopher, T., Contreras-Arratia, R., Edwards, S., Graham, O., Johnson, M., Juman, A., Latchman, J. L., Lynch, L., & Miller, V. L. (2022). Responding to eruptive transitions during the 2020–2021 eruption of La Soufrière volcano, St. Vincent. *Nature Communications, 13*(1), 1–15.

Kaneko, T., Maeno, F., & Nakada, S. (2016). 2014 Mount Ontake eruption: Characteristics of the phreatic eruption as inferred from aerial observations. *Earth, Planets and Space, 68*(1), 1–11.

Kilburn, C. R. (2003). Multiscale fracturing as a key to forecasting volcanic eruptions. *Journal of Volcanology and Geothermal Research, 125*(3–4), 271–289.

Kling, G. W., Evans, W. C., Tanyileke, G., Kusakabe, M., Ohba, T., Yoshida, Y., & Hell, J. V. (2005). Degassing Lakes Nyos and Monoun: Defusing certain disaster. *Proceedings of the National Academy of Sciences, 102*(40), 14185–14190.

Kling, G. W., Tuttle, M. L., & Evans, W. C. (1989). The evolution of thermal structure and water chemistry in Lake Nyos. *Journal of Volcanology and Geothermal Research, 39*(2), 151–165. https://doi.org/https://doi.org/10.1016/0377-0273(89)90055-3

Lavigne, F., Thouret, J. C., Voight, B., Suwa, H., & Sumaryono, A. (2000). Lahars at Merapi volcano, Central Java: An overview. *Journal of Volcanology and Geothermal Research, 100*(1–4), 423–456.

Lim, Z., & Flaherty, G. (2020). Fiery eruptions: Travel health risks of volcano tourism. *Journal of Travel Medicine, 27*(6), taaa019.

Malheiro, A. (2006). Geological hazards in the Azores archipelago: Volcanic terrain instability and human vulnerability. *Journal of Volcanology and Geothermal Research, 156*(1–2), 158–171.

Mastin, L., Pavolonis, M., Engwell, S., Clarkson, R., Witham, C., Brock, G., Lisk, I., Guffanti, M., Tupper, A., Schneider, D., & Beckett, F. (2022). Progress in protecting air travel from volcanic ash clouds. *Bulletin of Volcanology, 84*(1), 1–9.

McGuire, W. J. (1998). Volcanic hazards and their mitigation. *Geological Society, London, Engineering Geology Special Publications, 15*(1), 79–95.

McNutt, S. R. (1996). Seismic monitoring and eruption forecasting of volcanoes: A review of the state of the art and case histories. In R. Scarpa & R. I. Tilling (Eds.), *Monitoring and mitigation of volcano hazards* (pp. 99–146). Springer Science & Business Media.

Miller, T. P., & Casadevall, T. J. (2000). Volcanic ash hazards to aviation. In H. Sigurdsson (Ed.), *Encyclopedia of volcanoes*. Academic Press.

Myers, B. (1997). *What are volcano hazards?* (Vol. 2, No. 97). USA: US Geological Survey, Cascades Volcano Observatory.

Neal, C., & Guffanti, M. (2010). *Airborne volcanic ash: A global threat to aviation*. US Department of the Interior, US Geological Survey.

Newhall, C. G., Costa, F., Ratdomopurbo, A., Venezky, D. Y., Widiwijayanti, C., Win, N. T. Z., Tan, K., & Fajiculay, E. (2017). WOVOdat: An online, growing library of worldwide volcanic unrest. *Journal of Volcanology and Geothermal Research, 345*, 184–199.

Newhall, C. G., & Punongbayan, R. (Eds.). (1996). *Fire and mud: Eruptions and lahars of Mount Pinatubo, Philippines* (p. 1126). Quezon City: Philippine Institute of Volcanology and Seismology.

Newhall, C. G., & Self, S. (1982). The volcanic explosivity index (VEI) an estimate of explosive magnitude for historical volcanism. *Journal of Geophysical Research: Oceans, 87*(C2), 1231–1238.

Petrazzuoli, S. M., & Zuccaro, G. (2004). Structural resistance of reinforced concrete buildings under pyroclastic flows: A study of the Vesuvian area. *Journal of*

Volcanology and Geothermal Research, *133*(1–4), 353–367.

Ramsey, M. S., & Flynn, L. P. (2004). Strategies, insights, and the recent advances in volcanic monitoring and mapping with data from NASA's Earth Observing System. *Journal of Volcanology and Geothermal Research*, *135*(1–2), 1–11.

Robock, A. (2000). Volcanic eruptions and climate. *Reviews of Geophysics*, *38*(2), 191–219.

Scarpa, R., & Gasparini, P. (1996). A review of volcano geophysics and volcano-monitoring methods. *Monitoring and Mitigation of Volcano Hazards*, 3–22.

Scarpa, R., & Tilling, R. I. (2012). *Monitoring and Mitigation of Volcano Hazards*. Springer Science & Business Media.

Siebert, L. (1992). Threats from debris avalanches. *Nature*, *356*(6371), 658–659.

Sigurdsson, H. (1988). Gas bursts from Cameroon crater lakes: A new natural hazard. *Disasters*, *12*(2), 131–146.

Sigurdsson, H., & Carey, S. (1986). Volcanic disasters in Latin America and the 13th November 1985 eruption of Nevado del Ruiz volcano in Colombia. *Disasters*, *10*(3), 205–216.

Simkin, T., Siebert, L., & Blong, R. (2001). Volcano fatalities: Lessons from the historical record. *Science*, *291*(5502), 255–255.

Small, C., & Naumann, T. (2001). The global distribution of human population and recent volcanism. *Global Environmental Change Part B: Environmental Hazards*, *3*(3), 93–109.

Spence, R. J., Baxter, P. J., & Zuccaro, G. (2004). Building vulnerability and human casualty estimation for a pyroclastic flow: A model and its application to Vesuvius. *Journal of Volcanology and Geothermal Research*, *133*(1–4), 321–343.

Suryo, I., & Clarke, M. C. G. (1985). The occurrence and mitigation of volcanic hazards in Indonesia as exemplified at the Mount Merapi, Mount Kelut and Mount Galunggung volcanoes. *Quarterly Journal of Engineering Geology*, *18*(1), 79–98.

Thompson, M. A., Lindsay, J. M., & Leonard, G. S. (2017). More than meets the eye: Volcanic hazard map design and visual communication. In *Observing the volcano world* (pp. 621–640). Cham: Springer.

Thorarinsson, S. (1969). The Lakagigar eruption of 1783, Bull. *Volcanol*, *33*, 910–927.

Tilling, R. I. (1989). Volcanic hazards and their mitigation: Progress and problems. *Reviews of Geophysics*, *27*(2), 237–269.

Tuffen, H. (2010). How will melting of ice affect volcanic hazards in the twenty-first century? *Philosophical Transactions of the Royal Society A: Mathematical, Physical and Engineering Sciences*, *368*(1919), 2535–2558.

UNDRO. (1985). Volcanic emergency management. *United Nations, New York*, 1–86.

Voight, B. (1990). The 1985 Nevado del Ruiz volcano catastrophe: Anatomy and retrospection. *Journal of Volcanology and Geothermal Research*, *42*(1–2), 151–188.

Williams, R. S., & Moore, J. G. (1983). *Man against volcano: The eruption on Heimaey, Vestmannaeyjar, Iceland*. USA: US Department of the Interior, Geological Survey.

Witham, C. S. (2005). Volcanic disasters and incidents: A new database. *Journal of Volcanology and Geothermal Research*, *148*(3–4), 191–233.

Wright, T. L., & Pierson, T. C. (1992). *Living with volcanoes* (No. 1073). USA: US Geological Survey.

Zobin, V. M. (2001). Seismic hazard of volcanic activity. *Journal of Volcanology and Geothermal Research*, *112*(1–4), 1–14.

FURTHER READING

Chester, D. (1993). *Volcanoes and society*. London: Edward Arnold.

Fearnley, C. J., Bird, D. K., Haynes, K., McGuire, W. J., & Jolly, G. (Eds.). (2018). *Observing the volcano world: Volcano crisis communication*. Cham: Springer International Publishing. https://link.springer.com/book/10.1007/978-3-319-44097-2

Francis, P., & Oppenheimer, C. (2004). *Volcanoes*. Oxford: Oxford University Press.

Marti, J., & Ernst, G. (2005). *Volcanoes and the environment*. New York: Cambridge University Press.

Oppenheimer, C. (2011). *Eruptions that shook the world*. Cambridge: Cambridge University Press.

WEB LINKS

IAVCEI. www.iavceivolcano.org/

International Volcanic Health Hazard Network. www.ivhhn.org/

Smithsonian Institution – Global Volcanism Program. https://volcano.si.edu

Volcanic Ash Advisory Centre. www.ssd.noaa.gov/VAAC/vaac.html

Volcano Hazards, US Geological Survey. www.usgs.gov/programs/VHP

WOVO. https://wovo.iavceivolcano.org/

CHAPTER EIGHT

Tsunami Hazards

8

OVERVIEW

Tsunamis are secondary hazards and can be triggered by earthquakes, landslides or submarine slides, volcanic activity or extraterrestrial events, most of which cause significant damage and loss of life. Whilst 90% of tsunamis are generated by earthquakes, knowledge of the mechanisms for tsunami generation by other hazards is an important research area. The impacts of tsunami are wide ranging and significant given the vast area of coastline that can be affected. Consequently, effort has been invested in reviewing tsunami impacts and exploring diverse forms of risk management. These largely comprise warning systems, whether highly technical or via self-warning or oral traditions, as often getting to high ground is the most effective way of being safe. This chapter explores the role of tsunami warnings, modelling, engineering approaches, land use management and community preparedness as means of mitigating the impact of tsunamis.

A. THE NATURE OF TSUNAMIS

The word 'tsunami' comes from two Japanese words, *tsu* (port or harbour) and *nami* (wave or sea), indicating the impact of these ocean waves as they inundate low-lying bays and coastal areas. In the past, tsunamis were sometimes referred to as 'tidal waves' by the general public and as 'seismic sea waves' by the scientific community. The term tidal wave is a misnomer, as tsunamis are unrelated to the tides, and seismic sea wave is also misleading because tsunamis can also be caused by a non-seismic events, such as landslides, volcanic eruptions and meteorite impact that displaces large amounts of water. Furthermore, a tsunami is not dangerous in the open sea.

A tsunami is a water-based, basin-wide result of energetic shock to a significant water column. Almost all tsunamis contain a number of waves (typically 3–20 major wave crests). Tsunamis are a 'secondary' effect of a number of hazards,

DOI: 10.4324/9781351261647-10

including earthquakes, volcanic collapse/eruption, landslides (either submarine or subaerial) and extraterrestrial (e.g. meteorite impacts).

Around 80% of the recorded damaging tsunami events have been triggered by earthquakes, and traditionally tsunamis have been studied by seismologists. Since the late 1980s more geoscientists have become interested in developing knowledge of tsunamis via the development of new marine mapping technology (multibeam) that has allowed detailed mapping of seabed morphology. This has enabled scientists to map the impacts of tsunamis by examining the sediments deposited, both historic and prehistoric.

There are three key processes to a tsunami. First is the tsunami generation, followed by the propagation of the wave through the ocean and then finally the run-up on land causing inundation. Generation and propagation from earthquakes is generally well understood. The run-up is dependent on local offshore and coastal morphology, and thus is site specific (Figure 8.1).

A tsunami is generated by a single event that displaces water, creating long waves that have a wavelength (λ) that can exceed 200 km, in comparison to a normal ocean wave that has a wavelength of about 100 m. A common analogy is that used when a pebble is dropped in a lake and a series of ripples dissipate in all directions. A tsunami wave in deep water can travel up to 720 km hr^{-1}, and is determined by:

$$\text{Speed of wave} = (\text{gravity} \times \text{depth}) \times 1/2$$

Tsunami waves cannot be seen or felt on ships in the deep ocean, but near to shore the water depth decreases. The speed of the wave is dependent on water depth. Typically the front of a wave is 100–200 km ahead of the back of the wave, so the

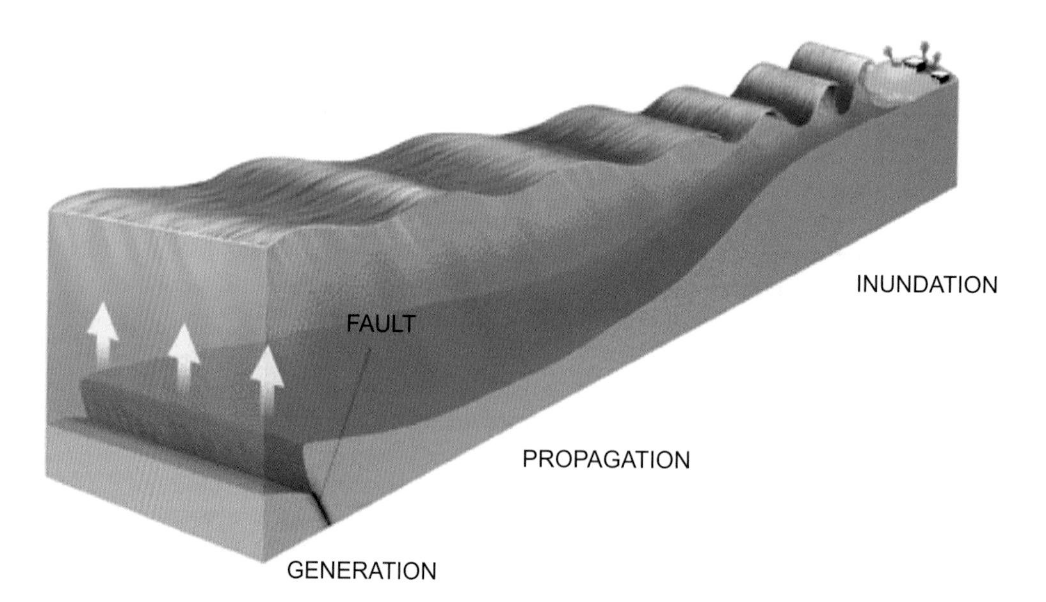

Figure 8.1 The three key steps to tsunami generation.

Source: González (1999, p. 61)

front meets much shallower water first and slows down. The rear of the wave catches up and the amplitude increases, resulting in a large amplitude, relatively long, wavelength (1–10 km) wave, moving at 30–100 km^{-1} (see Figure 8.2). This is why the first wave may not be the largest in the series of waves.

A tsunami wave crest has three general appearances from shore: as a fast-rising tide, as a cresting wave and as a step-like change in the water level that advances rapidly (called a bore). The tsunami wave is refracted and focused by the ocean floor topography (bathymetry). Most tsunamis come in a series of waves that may last for several hours. The outflow of water back to the sea between waves can cause more damage than the original incoming wave fronts.

The maximum height that a tsunami wave reaches on shore is called run-up: this is the vertical distance between max height reached by water on shore and mean sea level surface (see Figure 8.3). Flooding by individual waves lasts from 10 to 30 minutes, although the danger period can last for hours. The run-up at point of impact on a coastline depends on:

- how energy is focused/travel path of tsunami waves;
- coastal configuration/offshore topography;
- leading-depression wave: an initial tsunami wave is a trough, causing a drawdown of water level; and
- leading-positive wave: initial tsunami wave is a crest, causing a rise in water level, also called a leading-elevation wave.

In the last 400 years, 92 instances of a run-up of more than 10 m have been recorded; 39 of these have occurred with a run-up of more than 20 m, and 14 with more than 50 m (see Figure 8.4; Scheffers & Kelletat, 2003).

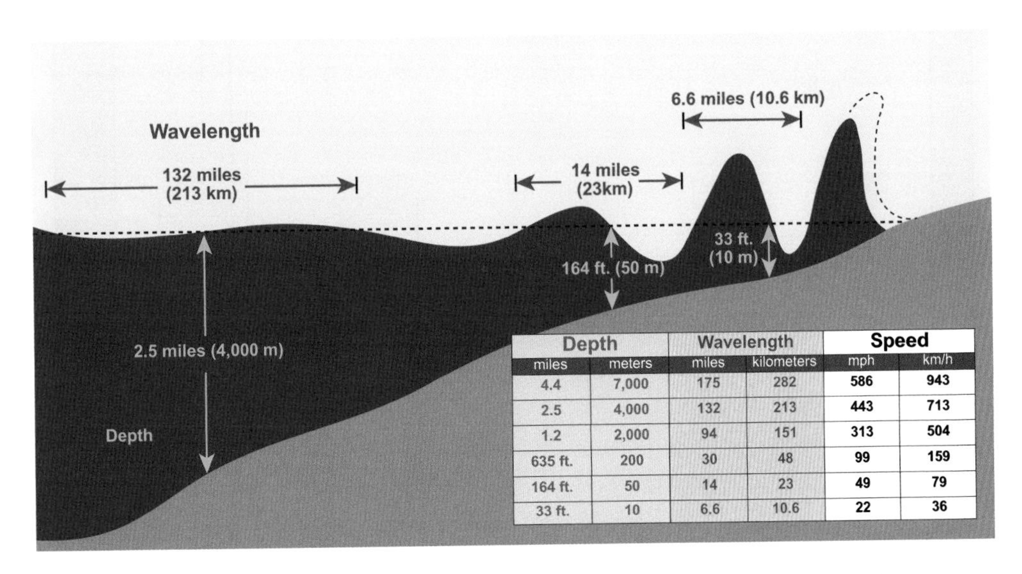

Depth		Wavelength		Speed	
miles	meters	miles	kilometers	mph	km/h
4.4	7,000	175	282	586	943
2.5	4,000	132	213	443	713
1.2	2,000	94	151	313	504
635 ft.	200	30	48	99	159
164 ft.	50	14	23	49	79
33 ft.	10	6.6	10.6	22	36

Figure 8.2 Cross section of a tsunami as its long waves move through the ocean and compress as they approach the coast.

Source: NOAA (n.d.b)

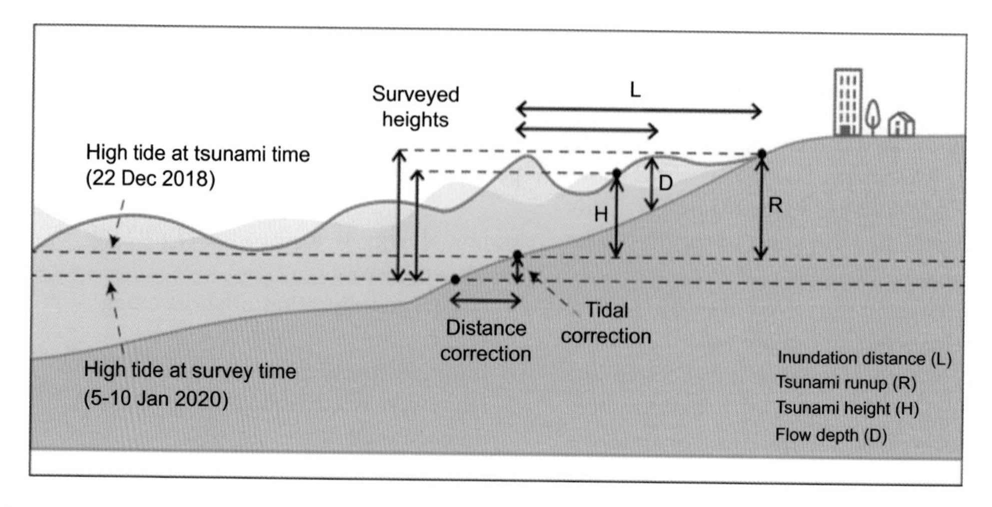

Figure 8.3 Sketch showing the method applied for surveys of tsunami heights and tidal corrections on the measured tsunami heights, tsunami run-ups and inundation distances in order to obtain actual heights and inundations.

Source: Heidarzadeh et al. (2020, p. 4579)

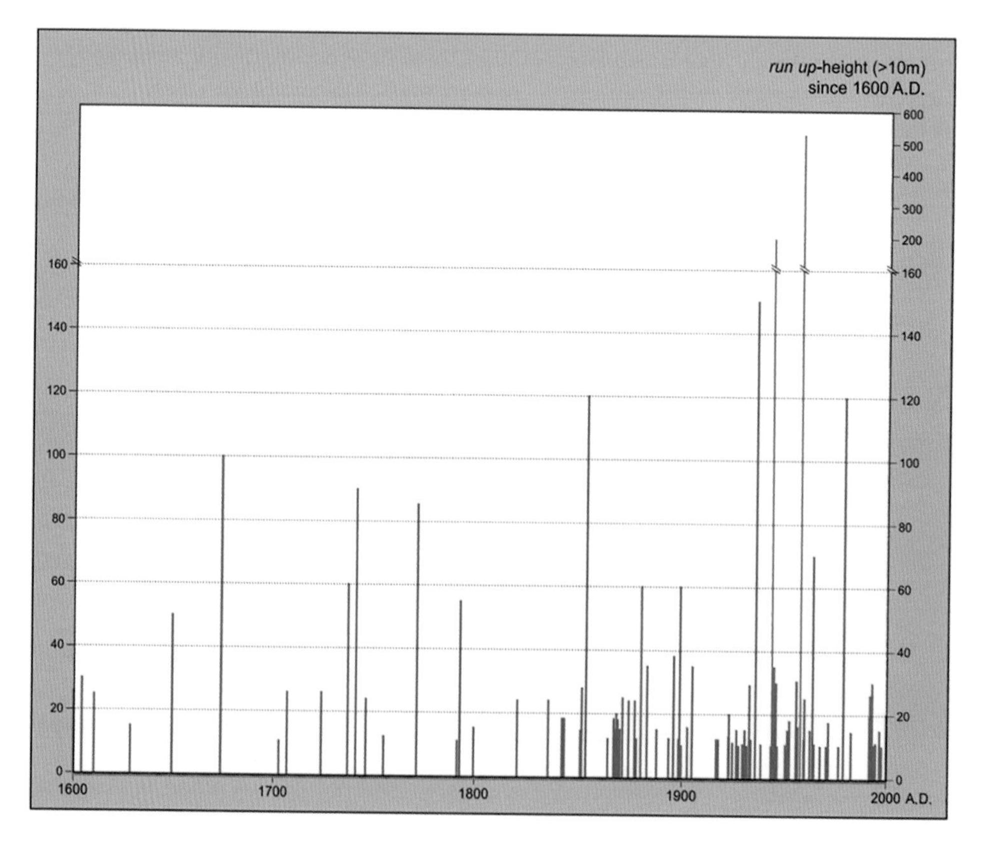

Figure 8.4 High run-up values during the last 400 years.

Source: Scheffers & Kelletat (2003, p. 176, adapted from NGDC, 2001)

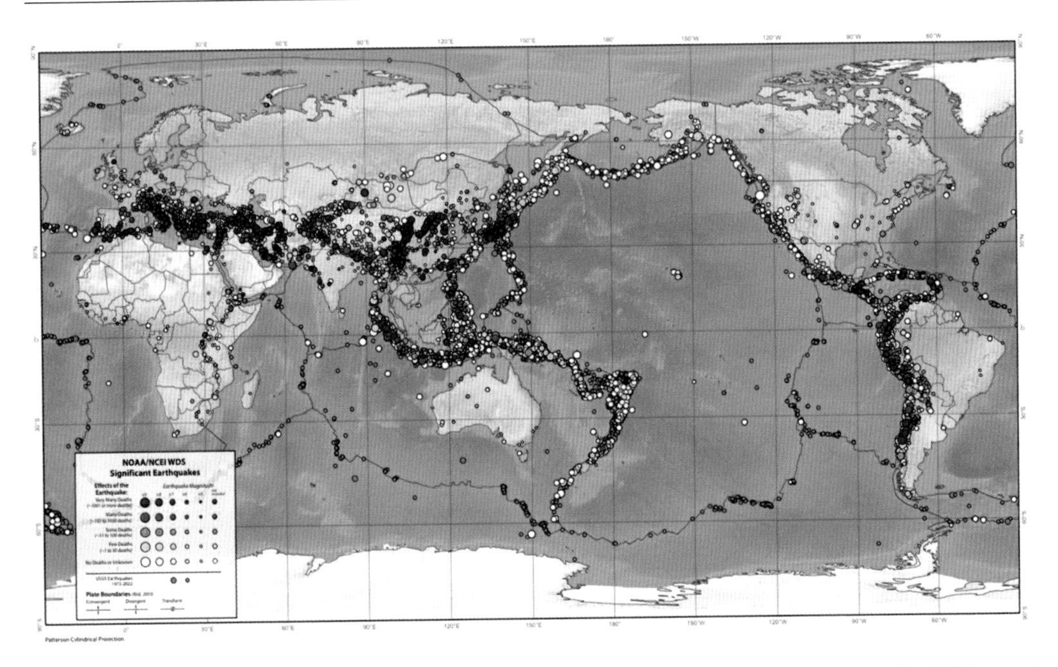

Figure 8.5 Tsunami sources 1610 BC to AD 2022 from earthquakes, volcanic eruptions, landslides and other causes.

Source: International Tsunami Information Centre (n.d.a, n.d.b)

Between 1610 BC and AD 2022 there have been over 1,400 confirmed tsunamis, of which 264 have been deadly. Sixty-nine per cent of these tsunamis occurred in the Pacific Ocean, and 80% were caused by earthquakes (see Figure 8.5 for locations).

Earthquake-Generated Tsunamis

Not all earthquakes generate tsunamis. The earthquake epicentre must be underneath or near the ocean and cause vertical movement of the sea floor (up to several metres) over a large area (up to 100,000 km^2) that creates displacement of the water. The earthquake needs to be of a large magnitude (> 7.5 Mw) and a shallow focus (< 70 km). The primary source for tsunamis are major thrust faults associated with subduction zones (thrust and normal faults), and the more steeply dipping, the more effective they are at producing tsunamis. Larger areas of seabed displacement tend to produce longer-period tsunamis (Figure 8.6).

Tectonic displacement of the seabed is associated with large, shallow-focus earthquakes. These are most common when two tectonic plates collide and the oceanic plate slides beneath the continental plate to form a deep ocean trench. Large subduction zones exist along the island arcs and coastlines of the Pacific Ocean; in the Indian Ocean, the Indo-Australian plate is subducted beneath the Eurasian plate. When rapid vertical movement of the sea floor occurs, it displaces the water column above. In 2004, for example, an accumulation of seismic stress

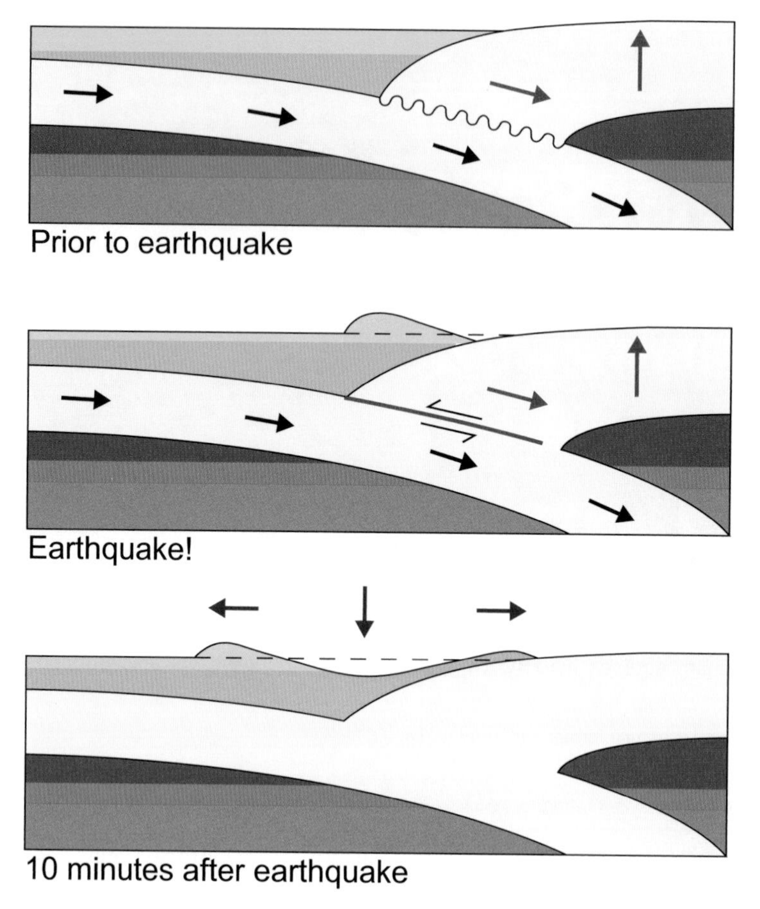

Figure 8.6 Mechanism for tsunami generation in an oceanic subduction zone.

Source: Thomas et al. (2007, p. 7)

near Sumatra caused the Earth's crust to deform downwards. When the fault ruptured, the crust rebounded upwards about 5 m. This raised a huge volume of water which flowed outwards to equilibrate sea level. The leading wave travelled quickly, arriving at the Indian and Sri Lankan coasts just 90 minutes after the earthquake and reached the coast of Somalia after about seven hours.

The first visible indication of an approaching tsunami may be recession of water (drawdown). When a tsunami is generated, a crest and trough make up the shape of the wave. Sometimes the trough reaches land first; sometimes the crest does. If the trough reaches land before the crest, observers will see a recession of the ocean water at the shore. This concept is shown in Figure 8.7(a) showing the generation of the earthquake, and Figure 8.7(b) the collision with the coast with the trough of the wave leading. In this tsunami, the areas to the east of the earthquake epicentre experienced recession. Figure 8.7(c) shows the extent to which the ocean receded between tsunami waves at Kalutara Beach in Sri Lanka during the 2004 Indonesian Boxing Day tsunami.

Landslide-Generated Tsunamis

Major landslides on coastal land or below sea level can create tsunamis with locally devastating powers. These events are

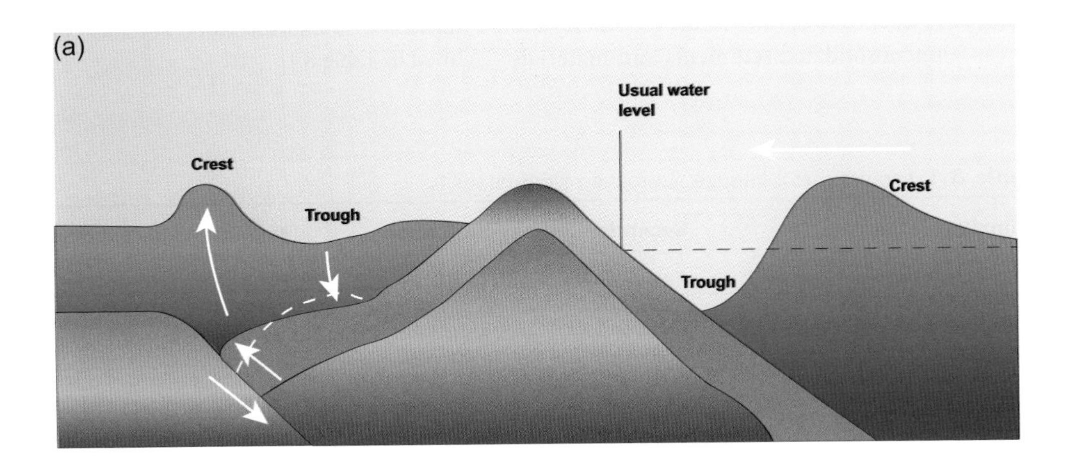

Figure 8.7 (a) On the trail of destruction and (b) the tsunami strikes the Sri Lankan coast, receding nearly 400 meters.

Sources: (a) Schiermeier (2005, p. 352); (b) NASA Earth Observatory (2004)

often caused by earthquakes and result in the displacement of ocean water due to large rock falls and debris slides into confined bays or lakes. The submarine landslides following the Prince William Sound earthquake in 1964 generated a tsunami that killed some 80 people in Alaska. However, wherever there are unconsolidated sediments and materials there is a potential for a submarine landslide, including open continental slope and rise areas, submarine canyon/fan systems, fjords, active river deltas, volcanic islands and convergent margins (Figure 8.8).

There are a number of factors that influence submarine slope stability, outlined in Table 8.1:

Table 8.1 Factors that influence submarine slope stability

Primary	Secondary (preconditioning)	Instantaneous triggers
• Sediment delivery to the continental margins; its rate, volume and type • Locations of depo-centres; particularly slope vs. shelf • Sediment thickness • Thickness	• Changes in groundwater flow • Changes in seafloor conditions, which can influence hydrate stability and the possible generation of free gas	• Earthquakes, wave, tide loading, salt movement • Questionable ones such as sediment overpressuring and hydrate disassociation

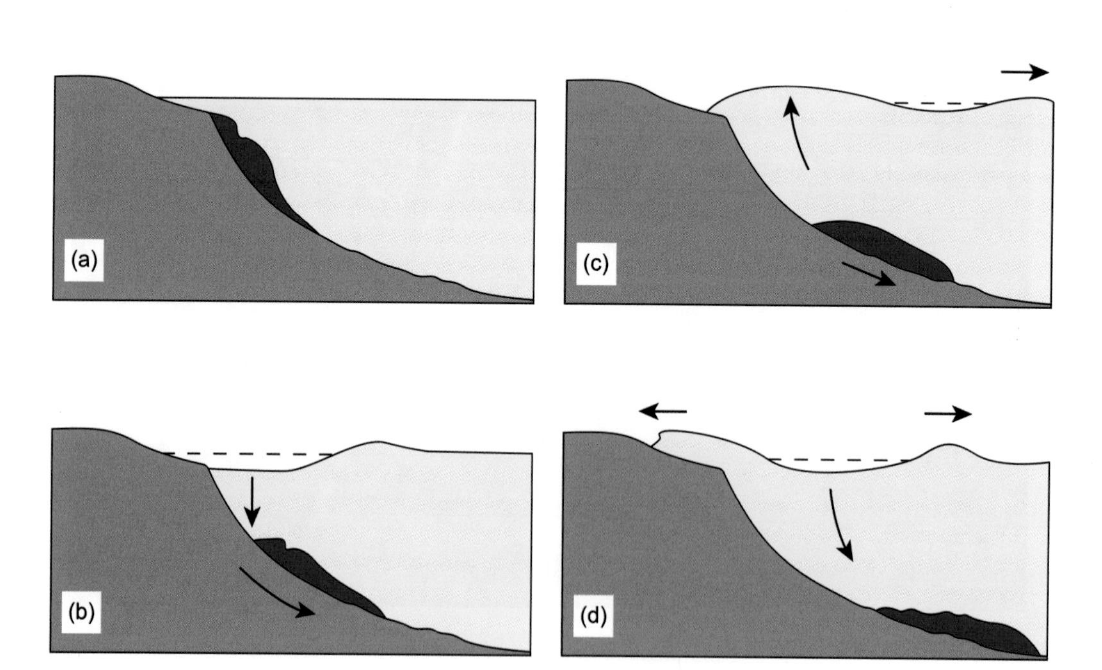

Figure 8.8 How submarine landslides generate tsunamis. (a) Prefailure, (b) initial slip and surface drag down above the rear of the submarine mass failure (SMF), (c) positive and negative wave generation, and (d) continued wave propagation.

Source: Tappin (2017, p. 192)

Using seismic technology and depth sounding (using sonar), the seabed and sub-seabed structure can be mapped, which enables the identification of past landslides. Whilst numerous landslides have been mapped using extensive mapping of continental margins, evidence for landslides from prehistoric events remains lacking as they may not be well preserved. Additionally, large areas of the seabed remain unmapped. Submarine landslides or mass movements may have dimensions greater than 200 km in length and 50 km in width, with depths ranging from several metres to 2 km (for debris avalanches) or up to 10 km (for slumps). Submarine slides may cover surfaces of several thousand square kilometres and range in volume from 20 km^3 to more than 1,000 km^3 (Figure 8.9).

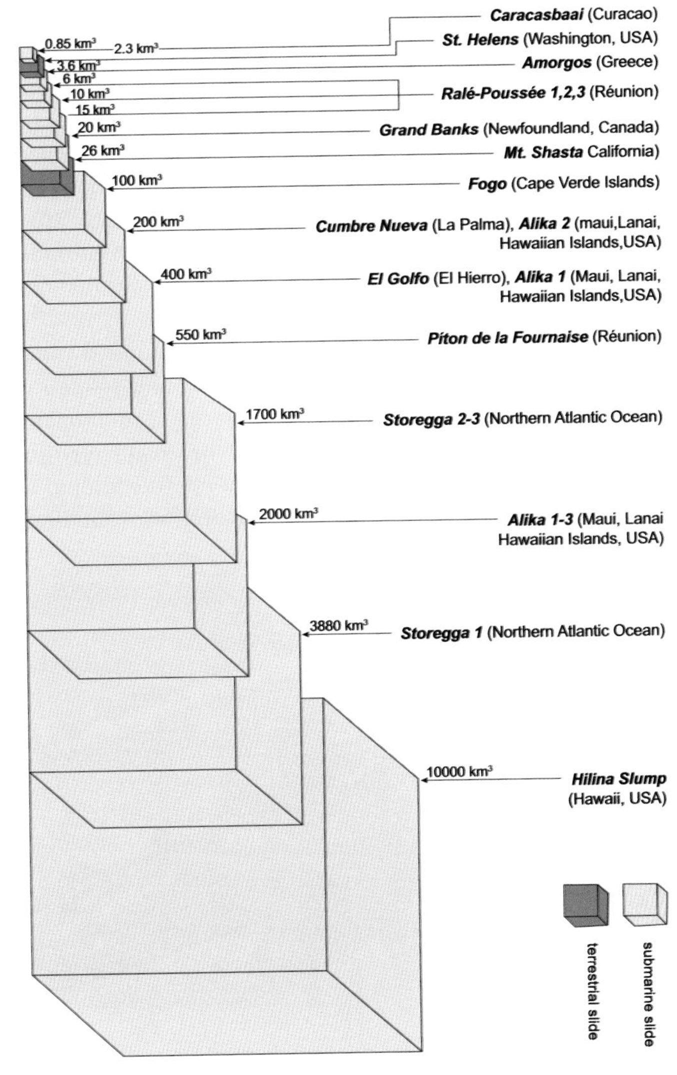

Figure 8.9 Illustration of volumes of submarine slides. Volumes are displayed as cubes for a simplified comparison.

Source: Whelan & Kelletat (2003, p. 208)

It was not until 1998 – when a slump on the seabed offshore of northern Papua New Guinea caused a tsunami wave up to 15 m high and killing over 2200 people (Box 8.1) – that the significance of submarine landslides in tsunami generation was realised by disaster and risk scientists. Significant historical landslide events include the Storegga slides (in the Norwegian Sea, 6225–6170 BCE), the Grand Banks earthquake or Larentian Slope earthquake (Newfoundland, in 1929), and in Japan (2011) the Tohoku earthquake. There are numerous other locations globally where the potential hazard remains high.

Box 8.1 The Papua New Guinea Landslide Tsunami

Following an earthquake, a tsunami struck Papua New Guinea on 17 July1998, killing over 2,200 people. The tsunami was 10 m high along over 10 km of coast. The maximum run-up of 15 m overwhelmed a sand bar where several small villages were built some 1–3 m above sea level. All wooden buildings within 500 m of the shore were swept away (González, 1999). Yet, the earthquake magnitude (Mw 7.0) did not seem sufficient to explain the elevation and impact of the tsunami. In addition, there was a 20-minute delay between the earthquake and the tsunami: the thrust fault was a shallow dip, and the epicentre was west of the peaked run-up.

Scientists working on five major offshore expeditions (funded mainly by Japan) used multibeam bathymetry, seismic sediment sampling and seabed video and still photography to investigate the tsunami generation. The results demonstrated evidence of a submarine landslide mechanism for the tsunami, whereby the earthquake had triggered a landslide slump that then generated the tsunami. This finding was critical for the development of research on the generation of tsunamis from submarine landslides.

Figure 8.10 Digital elevation model of sea-floor relief with bathymetric contours offshore of northern Papua New Guinea looking south (vertical exaggeration ×4). The box shows the slump location; the red star is the associated earthquake.

Source: Reproduced from Tappin (2017, p. 193)

Volcanic-Generated Tsunamis

Volcanic eruptions both above and underwater create a number of hazards that cause the displacement of water, triggering tsunamis that can in turn lead to high mortality rates. A quarter of volcanic-related tsunamis have no discernible origin, but the triggers of these tsunamis can be associated with (and as illustrated in Figure 8.11):

- 16% tectonic earthquakes
- 20% pyroclastic flows into water
- 14% submarine eruptions
- 7% collapse and caldera formation
- 5% landslides/avalanches of cold material
- 4% landslides/avalanches of hot material
- 3% lahars

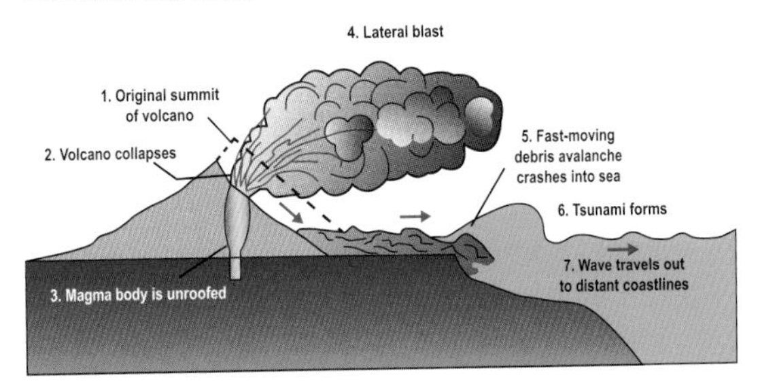

Figure 8.11 (a) Scheme of tsunami generation by a volcanic eruption and flank collapse above water.

Source: UNESCO

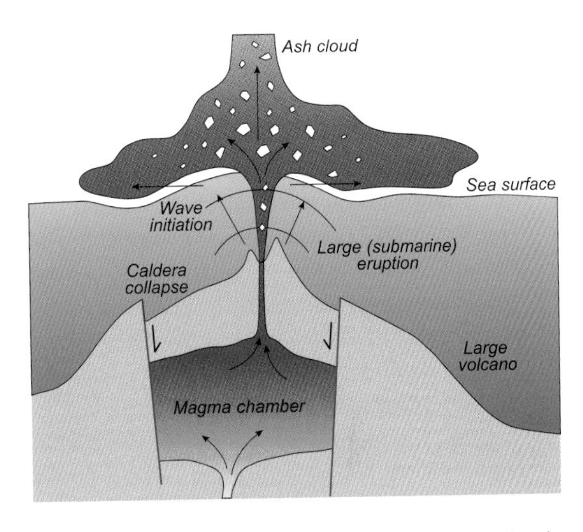

Figure 8.11 (b) Tsunami generation by a volcanic eruption underwater.

Source: Geological Survey of Canada

It is believed that about 5% of all tsunamis are due to volcanic activity. Perhaps 1% of the total, including some of the largest events, are related to the collapse of volcanic ocean islands. According to Keating and McGuire (2000), there are 23 distinct processes capable of destabilising volcanic islands. The highest risk of failure lies in the island arc volcanoes around the Pacific Ocean, due to their explosive nature and steep slopes. Mount Unzen, Japan, for example, created tsunamigenic landslides in 1792 that killed 14,500 people. Ward and Day (2001) postulated that a 500 km^3 block slide could potentially collapse at La Palma in the Canary Islands, generating a tsunami that would devastate shores across the whole of the Atlantic basin, with waves 10–25 m high possible along the coasts of the Americas. This would be an extremely rare event, but incredibly devastating.

Volcanogenic tsunamis can be deadly. Five thousand people were killed by tsunami waves from the 1792 debris avalanche at Unzen in Japan as it entered the sea. Tsunamis can also occur in crater lakes. For example, the volcano Taal that emerges from a lake within a caldera erupted in 1965, killing over 200 people when tsunami waves generated by the eruption hit the lake and capsized boats with people fleeing the island. One well-known volcanogenic tsunami was from Krakatau in 1883 and more recently in 2018 (see Box 8.2).

Today, populations living along coastlines are significantly larger, and large tsunamis can cause destruction thousands of kilometres away. The collapse of volcanoes presents a significant tsunami

Box 8.2 Krakatoa – the Big Bang

Krakatau was the first volcanic eruption to gain international reputation when it erupted in 1883, producing the loudest recorded explosion, heard over 4,000 km away, and generating a tsunami thought to be 37 m high and killing an estimated 36,417 people. Some reports state that up to 120,000 died. Around 90% of the mortality was from tsunamis produced by the volcanic eruption. Over 165 villages and towns were destroyed and 132 seriously damaged.

Located in Indonesia, at the time a Dutch colony, Krakatau was a large volcanic island around 800 m high situated in the Sundra Strait between Sumatra and Java. The Sundra Strait was a vital shipping shortcut between Asia and Europe. On 26–27 August 1883, a VEI 6 eruption (see Chapter 7) caused an explosive eruption that ripped apart the island, leaving a caldera and three islands, one of which became Anak Krakatau ('child of Krakatau').

In the aftermath Rogier Verbeek (see also Verbeek, 1884), a Dutch geologist and natural scientist, was sent with a small team to examine what happened using personal observation and eyewitness testimony. Verbeek spent over a year and a half investigating the volcanic eruption and submitted an official report of 550 pages of text, charts and images to the government of the Dutch East Indies, which they published 1884 as *The Krakatau Eruption* (the Dutch spelling of the volcano). The book not only was the definitive study of the Krakatoa event, but also arguably marked the beginning of modern volcanology as taught in European universities. It also presented a mystery. The images and data collected by Verbeek showed the coastlines of the main islands with trees that were clearly affected by pyroclastic flows. The questions were: how was the tsunami triggered? And, how did the pyroclastic flows travel over the sea to the land?

Over 100 years later scientists were able to investigate further the major mechanisms for tsunami generation, which could have been due to the caldera collapse, phreatomagmatic explosion and pyroclastic flow. Studies using numerical simulations concluded that a pyroclastic flow entering the sea is the most plausible mechanism of the 1883 Krakatau tsunami (Maeno & Imamura, 2011). It was also evidenced that the upper, dilute parts of pyroclastic flows travelled over the sea surface to traverse 40 km of the Sunda Straits and reach the southeast coast of Sumatra and kill approximately 2,000 people. As the pyroclastic flows hit the water, the heat caused a layer of steam that reduced friction, and thus enabled the upper, more gaseous part of the pyroclastic flow to travel over the water (Carey et al., 2000). There was significant concern that should this eruption repeat, and with so many more people living on the surrounding shores, the impact would be deadly.

On December 2018, during another small eruption at Anak Krakatoa, one of the flanks collapsed into the ocean generating a major tsunami, thought to have been at least 100 m high. This time, 437 were left dead, nearly 32,000 were injured and more than 16,000 people were displaced. The lack of a tsunami warning during the first few minutes after the generation of the first wave led to a significant number of human casualties in both of the affected areas. No warning was issued because the warning systems set up for

Figure 8.12 Drawing of Krakatau and the island with pyroclastic flow damage.

Source: The explosion of Krakatoa in 27 August 1883, engraved by Parker & Coward from photograph. Image © The Eruption of Krakatoa and Subsequent Phenomena, British Royal Society, 1888.

tsunamis in Indonesia relate to seismically generated tsunamis using large-scale DART buoys (see warnings later). A key lesson learned was that there is a need to focus on better monitoring for non-seismic tsunamis, such as landslides and volcanic-generated tsunamis (Luthfi et al., 2020). Krakatoa continues to pose a significant threat.

Figure 8.13a Anak Krakatau before the tsunami-producing volcanic eruption in December 2018. Photo taken August 2018.

Source: © Shutterstock/Azis Purwanto

Figure 8.13b Anak Krakatau after the tsunami-producing volcanic eruption in December 2018. Photo taken January 2019.

Source: GFX.com

An even larger set of tsunamis are thought to have been produced during the caldera-forming eruption of Sanorini, in the Aegean sea around 1620 BC. Also said to be a VEI 6 eruption, the tsunami produced is thought to have been 50 m high and a significant contributor to the downfall of the Minoan culture based in Crete.

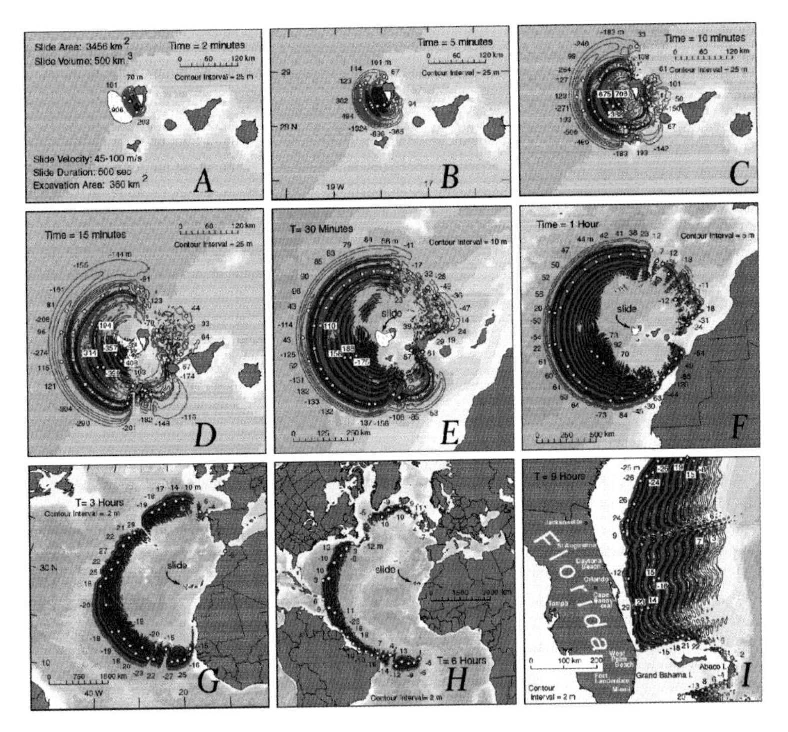

Figure 8.14 Model of evolution of the La Palma landslide tsunami from two minutes (a, upper left) to nine hours (i, lower right). Red and blue contours cover elevated and depressed regions of the ocean respectively, and the yellow dots and numbers sample the wave height, positive or negative, in metres. Note the strong influence of dispersion in spreading out an original impulse into a long series of waves of decreasing wavelength. See also that the peak amplitudes generally do not coincide with the first wave. Even after crossing the Atlantic, a lateral collapse of Cumbre Vieja volcano could impose a great sequence of waves of 10–25 m height on the shores of the Americas.

Source: Ward & Day (2001, p. 3399)

threat, being essentially composed of unconsolidated rock. This was seen at the volcanic island Hierro in the Canary Islands that had a large proportion of the volcano collapse into the sea over 120,000 years ago, generating a devastating tsunami on the east coast of the USA up to 20 km inland. This threat continues today as the volcano Cumbre Viejo, on the nearby island of La Palma, is already showing signs of considerable slipping. Ward and Day (2001) modelled the scale of the waves generated by such a collapse, and a worst case scenario would comprise a 500 km^3 block slide and an initial wave that could reach 600 m high, 20 km inland in the US (see Figure 8.14). Whilst this is highly unlikely, volcanic islands demonstrating slip must be monitored carefully.

Extraterrestrially Generated Tsunamis

The Earth receives over 10,000 tonnes of cosmic debris every year, most of which are small chunks that burn up in the atmosphere and travel around 20 km per second. Occasionally objects that are larger than 100 m across (e.g. comets and asteroids) can reach the surface intact, creating a crater that is around 20 times its own diameter. The impact causes a blast/shock wave and triggers local earthquakes, and if the object lands in the sea it generates a crater on the ocean floor. The result is a tsunami that is generated not just by the water displaced by the object, but also by the movement of the water surrounding the new impact crater moving to refill the crater (see Figure 8.15). There has not been an impact-generated tsunami documented in historic times.

Whilst a devastating impact-generated tsunami is less common than that caused by an Mw 9.0, over geological timescales such events are inevitable, and have indeed already occurred (see

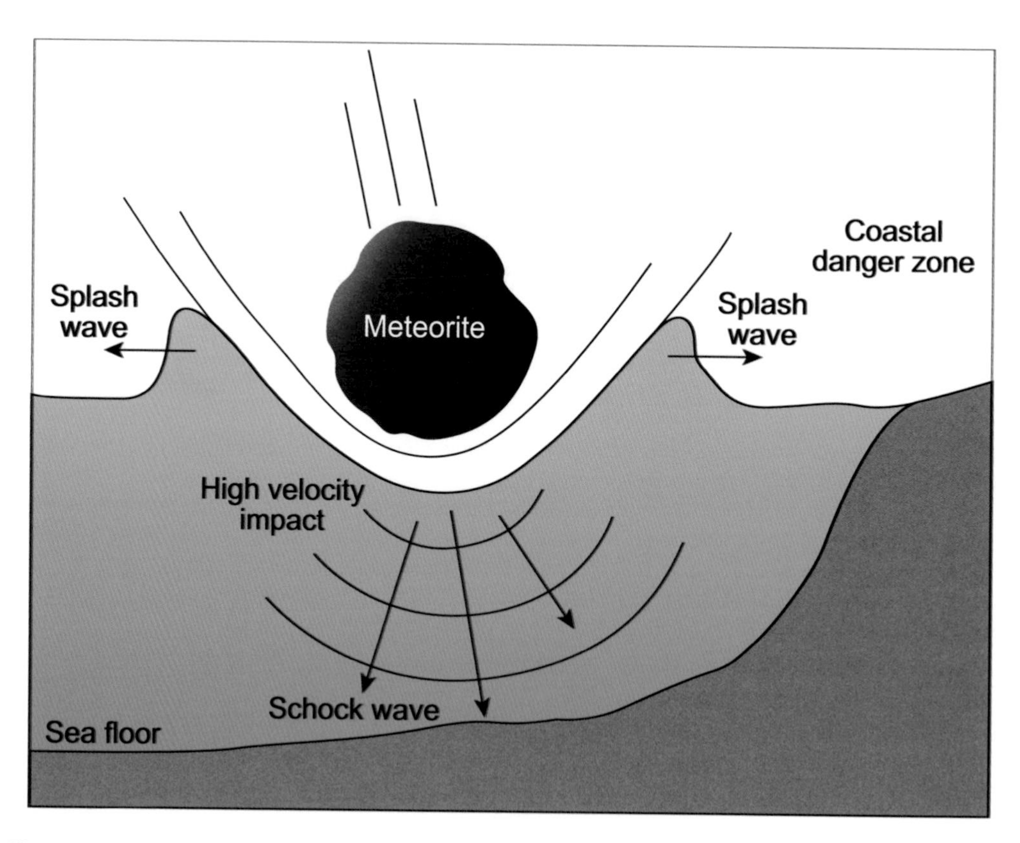

Figure 8.15 Tsunami generated by a meteorite.

Source: Geological Survey of Canada

Table 8.2 Regional and local tsunamis causing deaths since 1980

Date				
Year	**Month**	**Day**	**Source location**	**Estimated dead or missing**
1981	9	1	Samoa Islands	Few
1983	5	261	Nashiro, Japan	100
1988	8	10	Solomon Islands	1
1991	4	22	Limon, Costa Rica	3
1992	9	2	Off coast Nicaragua	170
1992	12	12	Flores Sea, Indonesia	1,169
1993	7	12	Sea of Japan	208
1994	6	2	Java, Indonesia	238
1994	10	8	Halmahera, Indonesia	1
1994	11	4	Skagway Alaska, USA**	1
1994	11	14	Philippine Islands	*81
1995	5	14	Timor, Indonesia	11
1995	10	9	Manzanillo, Mexico	1
1996	1	1	Sulawesi, Indonesia	9
1996	2	17	Irian Jaya, Indonesia	110
1996	2	21	Northern Peru	12
1998	7	17	Papua New Guinea	1,636
1999	8	17	Izmit Bay, Turkey	155
1999	11	26	Vanuatu Islands,	5
2001	6	23	Southern Peru	26
2003	9	25	Hokkaido, Japan	2
2004	12	26	Banda Aceh, Indonesia	*^227,899
2006	3	14	Seram Island, Indonesia	4
2006	7	17	Java, Indonesia	802
2007	4	1	Solomon Islands	50
2007	4	21	Southern Chile	8
2007	8	152	Southern Peru	3
2009	9	9	Samoa Islands	192
2010	1	12	Haiti	7
2010	2	27	Southern Chile	156
2010	10	25	Mentawai, Indonesia	431
2011	3	11	Tohoku, Japan	*^18,428
2013	2	6	Solomon Islands	10

(Continued)

Table 8.2 (Continued)

Date				
Year	**Month**	**Day**	**Source location**	**Estimated dead or missing**
2015	9	16	Central Chile	8
2017	6	17	Greenland**	4
2018	9	28	Sulawesi, Indonesia	*4,340
2018	12	22	Anak Krakatau, Indonesia***	437
2020	10	30	Aegean Sea	1
2022	1	15	Tonga Islands***	4
Total				256,723

Source: IOC (2016, p. 9)

* May include earthquake deaths
** Tsunami generated by landslide
*** Tsunami generated by volcanic eruption
^ Includes dead/missing near and outside source region

Table 8.2). The Chicxulub crater is 200 km wide and formed when a 10 km wide asteroid impacted Earth around 65 million years ago. It is widely accepted that this bolide impact was a significant contributor to the last mass extinction event during the Cretaceous–Tertiary (K–T) geological boundary and responsible for the demise of the dinosaurs, amongst other life globally. The energy from the impact would be equivalent to an Mw 11 earthquake, and research by Range et al. in 2018 used models to estimate that the tsunami wave heights in the Pacific and Atlantic basins would have been as large as 14 m. As these waves approached land and slowed down, they would have become even larger.

B. IMPACTS AND CONSEQUENCES OF TSUNAMIS

A total of 264 confirmed deadly tsunamis have resulted in over 544,000 known (or confirmed) deaths (see Table 8.2). The death total may include deaths from the generating event (e.g. earthquake), as it is not always possible to separate deaths from the different causes, and in many cases the actual number of fatalities is not known. The reporting of deadly tsunamis is not homogeneous in space or time, particularly for events prior to 1900. Over 60% of all tsunamis originate around the Pacific Ocean and more than 80% of damaging wave run-ups occur along the coasts around this basin.

The impact of tsunamis are devastating, and typical impacts include:

Loss of life: Many tsunami deaths are from drowning, as tsunamis are difficult to escape. For a far-field tsunami, where the tsunami wave has several hours before hitting the coastline, there is plenty of time to evacuate and prepare. Yet, in 1960 an earthquake-generated tsunami in Chile killed 2,000 people locally, but still killed over 226 people around the rest of the Pacific, despite warnings providing ample time before the waves hit the US, Japan and China. For tsunamis that are near-field – that is, the tsunami is generated near the vulnerable population – the warning times may vary from just a few minutes to half an hour, which is not a long time to enable people to evacuate. Whilst an earthquake also causes disruption and deaths, it can provide a warning for a tsunami, but for those tsunamis caused by landslides and volcanoes there may be no warning signs.

Damage to infrastructure and deaths: The energy and water contained in a huge tsunami can cause extreme destruction when it strikes land (see Box 8.3). Flooding, the collapse of buildings, electrocution and explosions from gas, damaged tanks and floating debris all contribute to the death toll and injuries caused. Whilst the initial wave of a tsunami can be tall (Chapter 8), extensive damage is from the huge mass of water behind the front, with the sea level rising fast and flooding powerfully onto the coastal areas. Destruction is caused by the force of a wall of water travelling at high speed, the force of breaking waves, the force of a large volume of water draining off the land and taking all the debris with it, and other flood forces (Chapter 11). These forces destroy buildings, and large objects such as ships and boulders can be carried several kilometres inland before the tsunami subsides. Tsunamis also do not just stop at the shoreline; they can travel inland for several kilometres, and the horizontal measurement of the path of the tsunami is called inundation. This causes significant damage, and small islands can be particularly vulnerable. The end result can be fires from ruptured tanks or gas lines, loss of key community infrastructure (police, fire and medical facilities) and a lack of electricity or transportation means.

Spread of disease: Often sewage and freshwater supplies are affected in the areas inundated, and following a tsunami it is common to not have access to clean water for weeks after. The contamination of drinking water can cause disease to spread in the tsunami-hit area, especially malaria when water is stagnant and contaminated. People struggle to stay healthy and hydrated and so infections, illnesses and diseases can spread very quickly, causing further secondary deaths.

Damage to the environment: Tsunamis have a devastating effect on insects, animals, plants and natural resources. A tsunami changes the landscape completely, uprooting trees and destroying animal habitat. Damaged human-made facilities can also cause significant danger to the environment.

Land animals are killed by drowning and sea animals are killed by pollution if dangerous chemicals are washed away into the sea. Solid waste and disaster debris are a major environmental problem faced by

Table 8.3 Tsunamis causing deaths greater than 1,000 km from the source location

Date				Estimated dead or missing		
Year	Month	Day	Source location	Local	Distant	Distant locations that reported casualties
1700	1	27	Cascadia Subduction Zone, USA		2	Japan
1755	11	1	SW Iberian Margin, Portugal	50,000	3	Brazil
1837	11	7	Southern Chile	0	16	USA (Hawaii)
1868	8	13	Northern Chile**	*25,000	7	New Zealand, Samoa, Southern Chile
1877	5	10	Northern Chile	277	2,005	Fiji, Japan, Peru, USA (Hawaii)
1883	8	27	Krakatau, Indonesia	34,417	1	Sri Lanka
1901	8	9	Loyalty Islands, New Caledonia	0	Several	Santa Cruz Islands
1923	2	3	Kamchatka, Russia	2	1	USA (Hawaii)
1945	11	27	Makran Coast, Pakistan	*300	15	India
1946	4	1	Unimak Island, Alaska, USA	5	163	Marquesas Islands, Peru, USA (California, Hawaii)
1957	3	9	Andreanof Islands, Alaska, USA	0	2	USA (Hawaii, indirect deaths from plane crash during tsunami reconnaissance)
1960	5	22	Central Chile	2,000	226	Japan, Philippines, USA (California, Hawaii)
1964	3	28	Prince William Sound, Alaska, USA	106	18	USA (California, Oregon)
2004	12	26	Banda Aceh, Indonesia	*175,827	52,072	Bangladesh, India, Kenya, Madagascar, Maldives, Myanmar, Seychelles, Somalia, South Africa, Sri Lanka, Tanzania, Yemen
2005	3	28	Sumatra, Indonesia	0	10	Sri Lanka (deaths during evacuation)
2011	3	11	Tohoku, Japan	*18,426	2	Indonesia, USA (California)
2012	10	28	Haida Gwaii, Canada	0	1	USA (Hawaii, death during evacuation)
2022	1	15	Tonga Islands	4	2	Peru

Source: IOC (2016, p. 9)

* May include earthquake deaths
** Local and regional deaths in Chile and Peru
*** Local and regional deaths in Indonesia, Malaysia and Thailand

tsunami-hit locations. Hazardous materials and toxic substances can be inadvertently mixed up with ordinary debris, along with the contamination of soil (affecting agriculture) and water (both fresh and human-stored). The rapid clean-up of affected areas can result in inappropriate cleaning or disposal methods, including air burning and open dumping, that can lead to other secondary impacts on the environment. In a worst case scenario there may be radiation resulting from damage to nuclear plants, as seen in Japan during the Tohoku earthquake and tsunami in March 2011, with long-lasting damage and impact (see Chapter 1, Box 1.2).

Livelihoods: Tsunamis affect people, communities and ecosystems in a variety of ways. Many coastal populations rely on the land and sea for their livelihoods and food security, in particular the fishing industry, tourism sector and local ecosystems that may include mangroves that are used for building materials and fuel. Those affected by a tsunami may well be affected by the primary hazard that triggered the tsunami, such as an earthquake or volcanic eruption, and have to manage multiple hazards and multiple impacts on their community that can hamper the recovery process. The impacts of a tsunami can be devastating in the medium term as seen by the 2009 tsunami on the Samoan archipelago. A total of 148 people were killed, and those that survived relocated to the highlands of the island's interior due to the coastal damage. Fishing equipment was damaged, and people were focused on rebuilding their life. Consequently, fishing activities died down and people had to find food from elsewhere. Government, NGOs and international organisations not only donated food for those affected to eat, but also initiated a clean-up programme to help encourage people to access the sea again and resume fishing with donations of fishing equipment.

It is no surprise that many victims of tsunami events suffer psychological problems that can last for days, years or a lifetime. Many victims suffer from post-traumatic stress disorder (PTSD) and grief and depression as loved ones, homes, businesses and entire communities were taken from them so rapidly, often with no warning. The cost of a tsunami is huge, not just financially, but also to the victims and survivors who need immediate help and ongoing support to rebuild.

Given the lack of any warning system people did not have any official warnings, and the devastation and trauma resulting from the event was harrowing (see Figure 8.18). The following quotes taken from an Oxfam report (Oxfam, 2005, p. 3) of interviews with survivors highlight some of the trauma experienced by the local populations affected.

> 'In rural coastal areas, many men who were fishing far out at sea survived, as the giant waves passed harmlessly under their small boats. When the waves hit the shore, they flattened coastal communities and killed many of the women and children, most of whom were at home on that Sunday morning. In agricultural areas, men were often working out in the fields or doing errands away from the house, or were taking produce to markets'.
>
> 'The sheer strength needed to stay alive in the torrent was often also decisive in determining who survived. Many women and young children, unable to stay on their feet, or afloat, in the powerful waves, simply tired and drowned.

Box 8.3 The Andaman Sumatra Earthquake and Indian Ocean Tsunami

The Andaman Sumatra earthquake was the second largest earthquake recorded on a seismograph at 9.1–9.3 Mw and generated a wide-scale Indian Ocean tsunami that killed over 225,000 people in about a dozen countries on 26 December 2004 at 07:58:53 local time. Its significance has been addressed throughout Part 1, but one of the key reasons why this event was so deadly was due to the lack of a warning system.

The earthquake was generated at the subduction zone between the India and Burma tectonic plates, called the Sunda trench. It was the longest duration of faulting ever observed at 8.3–10 minutes. The fault rupture propagated to the northwest from the epicentre (60 km off the coast from Banda Aceh), with a width of 100 km and an average displacement on the fault plane of 20 m. The waves generated were up to 30 m high and caused the planet to vibrate as much as 1 cm and continue to reverberate for several weeks later. The average recorded velocity of the 2004 tsunami wave from Sumatra was close to 640 km hr^{-1}, probably reflecting a slowing down of the wave as it approached coastal regions. Crossing the ocean, the wave was only about 60 cm in

Figure 8.17 Seawater splashes in the air as the first tsunami waves hit Ao Nang, Krabi Province, Thailand, on 26 December 2004.

Source: © Alamy Images/David Rydevik/Matteo Omied

height and posed no hazard. But as it neared the Banda Aceh shore, where much of the damage took place, the wave reached a maximum height of over 30 m.

Over 125,000 people were injured and 1.69 million were displaced, and the event had massive impacts. International donations of over $7 billion followed the event. The disaster was particularly devastating as most people did not know what to do in the face of the tsunami, a situation made worse since many of the hotels were full of international tourists who were not familiar with the local landscapes or with tsunamis. Some people recognised the feature of the sea withdrawing out to sea as an indication of danger and managed to escape. Others were attracted to the seashore by this unusual sight and were exposed to the full force of the wave when it made landfall (see Figure 8.17). In particular, the withdrawal of the ocean exposed lots of fish that locals then went to collect.

Figure 8.18 A file photo taken on 5 January 2005 of the devastated district of Banda Aceh in Aceh province located on Indonesia's Sumatra island in the aftermath of the massive 26 December 2004 tsunami.

Source: © Alamy Images/Mark Pearson (Choo Youn-Kong/AFP/Getty Images)

> Women clinging to one or more children would have tired even more quickly'.

One of the key issues during the 2004 Boxing Day tsunami was that there was no warning system established in the Indian Ocean to detect tsunamis or to warn the general populace living around the ocean. Despite a lag of up to several hours between the earthquake and tsunami in some locations (see Box 8.4), people were taken by surprise. What is even more tragic is that the Pacific Tsunami Warning Center (PTWC) detected the earthquakes and were aware that a large tsunami was likely to have been generated, but they had no capacity to inform the locations to provide a warning. Following the event an international review and survey of early warning systems was conducted by the United Nations (UN, 2005). As a consequence, the Indian Ocean Tsunami Warning System became active in late June 2006 following the leadership of UNESCO. The system consists of 25 seismographic stations relaying information to 26 national tsunami information centres, as well as three deep-ocean sensors. UNESCO warned that further coordination between governments and methods of relaying information from the centres to vulnerable populations at risk are required to make the system effective.

There is much to learn still about the science behind volcanoes and also the impact they have on populations. The volcano Hunga Tonga–Hunga Ha'apai erupted

Box 8.4 Chronology of Events (All Times Local)

07:58: M=9.3 Mw earthquake occurs off the northwest coast of Sumatra

08:07: Australia contacts the PTWC with initial earthquake parameters (stating 8 Mw)

08:10: Tsunami hits Andaman and Nicobar Islands

08:14: PTWC sends an earthquake information bulletin to Pacific participants stating no tsunami risk in the Pacific

08:20: Tsunami hits Banda Aceh province coastline in Indonesia, killing 170,000 people

08:28: PTWC target time for issuing Pacific-wide warnings (1/2 hr)

08:04: PTWC revises magnitude to 8.5

10:00: Tsunami hits Phuket, Thailand, killing 5,400 people (2,000 foreign tourists)

10:20: Tsunami hits eastern Sri Lanka, killing 30,000 people

10:30: Internet newswire reports from Sri Lanka provide first evidence of destructive tsunami

10:50: Tsunami hits eastern India, killing more than 16,000 people

14:30: Tsunami hits eastern Africa, killing 300 people in Somalia, Tanzania and Kenya

The resulting damage across the affected countries was analysed to prepare for future large-scale events. Variable degrees of damage were observed along the coasts along with large variations in inundation and water levels. Where water levels exceeded 4 m, the potential for near-shore life loss increased dramatically, and damage decreased with distance from shore except in the presence of waterways/rivers. There was a close correlation between locations of greatest life loss, high water levels and high building damage. There are many similar tectonic regions globally where a near-field tsunami could have a significant and rapid impact, demonstrating the need for preparedness, risk reduction and warning systems.

on 15 January 2022, sending a plume of ash soaring into the upper atmosphere and triggering a tsunami. This combination of both a shallow underwater eruption and a high eruption plume reaching 30 km high has not before been seen, suggesting a new mechanism of eruption (Witze, 2022). One of the significant issues faced in managing environmental hazards is noting and then investigating unexpected phenomena.

Whilst this eruption impacted the globe through reverberations that circled the globe multiple times, the impact for the Tongans has been significant. The eruption happened just 65 km from the Tongan capital of Nukuʻalofa, killing three people and affecting the islands' population of 100,000 people. A thick layer of ash has covered the island, and the crop damage was estimated at nearly 39 million Tongan paʻanga (US$17 million). Tragically, the islanders were exposed to their first identified cases of COVID-19 after relief ships arrived from other countries providing aid.

The tsunami itself impacted Tonga, Fiji, American Samoa and Vanuatu and also generated damaging tsunamis further afield in New Zealand, Japan, the United States, the Russian Far East, Chile and Peru. On many of the islands tsunami warnings were issued, but one of the biggest challenges was that the submarine cable that connects Tonga to Fiji was severed during the eruption, cutting off international communications. This hampered the response efforts significantly, and relatives living abroad suffered for days not knowing if their loved ones were safe.

C. RISK ASSESSMENT AND MANAGEMENT

There are a number of actions that can be taken to try to reduce the risk of tsunamis. Where possible, warnings can provide a significant impact on people's ability to get to safety, but warnings cannot always be timely. Consequently, it is important to understand what the tsunami risk is, prepare for it by building appropriate tsunami-resistant buildings or building vertical evacuation structures, and considering the role of land use planning, for example making sure nuclear facilities are built in safe locations. Community preparedness is vital, and education plays a key role in making sure people know to move inland and to high ground when they see the signs of a tsunami. Many residents already have a very good understanding of the hazards, which must be recognised and valued. These tools and resources all come together to produce a tsunami-resilient community as per Figure 8.19.

Forecasting and Warning

Without a warning system there is little capability for people to respond to a tsunami. A warning system can be and must include local and indigenous knowledges, as well as an international, coordinated warning lexicon. Whilst tsunami detection is difficult and requires sophisticated technology to detect it, setting up the communications infrastructure to issue timely warnings across a region is an even bigger problem, particularly across countries in conflict.

A tsunami warning system uses seismic data to monitor earthquake activity and integrates this with pressure sensors that sit on the ocean bottom and measure the weight of the water column above them to try to detect if a tsunami has been generated. A string of buoys are located in the ocean, each one costing tens of millions of dollars to install and are part of the Deep-ocean Assessment and Reporting

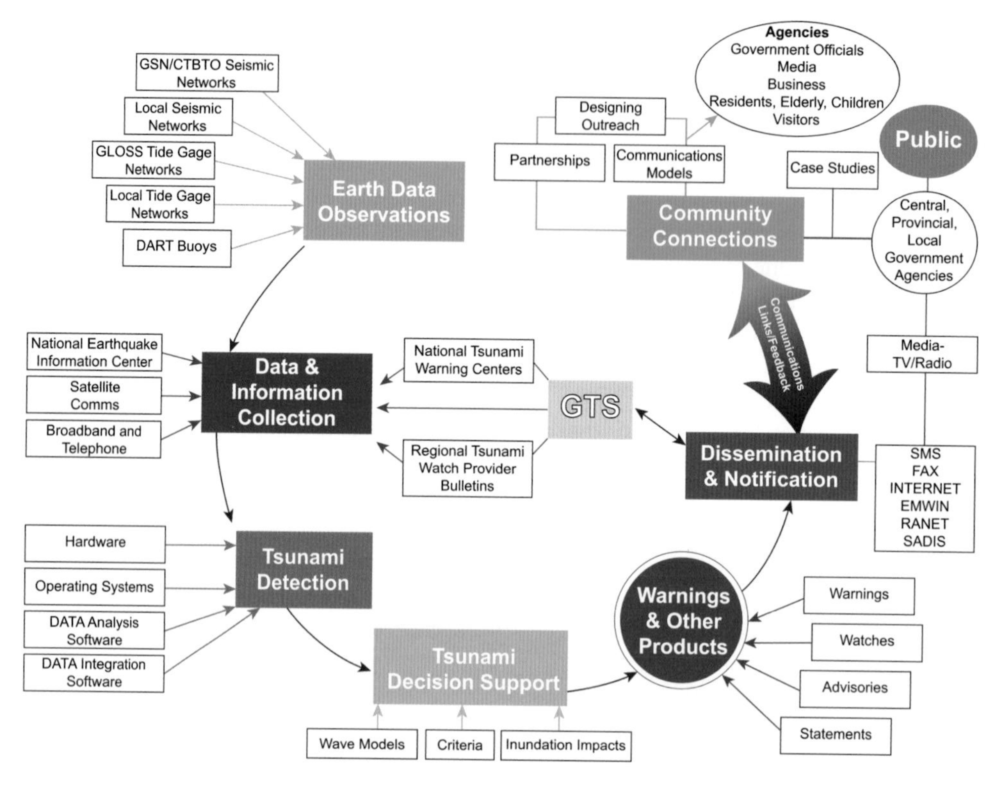

Figure 8.19 Key operational components of the tsunami warning centres.

Sources: US Indian Ocean Tsunami Warning System Program, 2007

of Tsunamis (DART) programme that forms part of an almost worldwide early warning system (see Figure 8.20). Should a pressure sensor measure an unexpected change in the weight of water above it a signal is sent to a buoy sitting on the sea surface. The buoy sends a signal to a satellite, which in turn alerts an early warning centre where warnings are typically issued via TV, radio, even a telephone. Many experts note that the biggest challenge is to establish an effective response infrastructure, which can reach everybody-no matter how remote.

Once an earthquake has been detected by the DART network of instruments and its epicentre located, it becomes possible to predict with some accuracy the arrival time of the tsunami wave on a distant shoreline. Figure 8.21 is a computer-generated map of the passage of the 2004 Sumatra tsunami across the Indian Ocean from the epicentre. Such simulated travel times are normally accurate to within one hour of the actual arrival.

The Pacific Tsunami Warning Center was set up in 1949 and is managed by the US's National Oceanic and Atmospheric Administration (NOAA). The first tsunami warning system was installed in 1949 for the Pacific Ocean, using a network of seismic stations to relay information to a warning centre near Honolulu, Hawaii. Following the tsunamigenic Alaska earthquake of 1964, the West Coast/Alaska Tsunami Warning Center was set up in

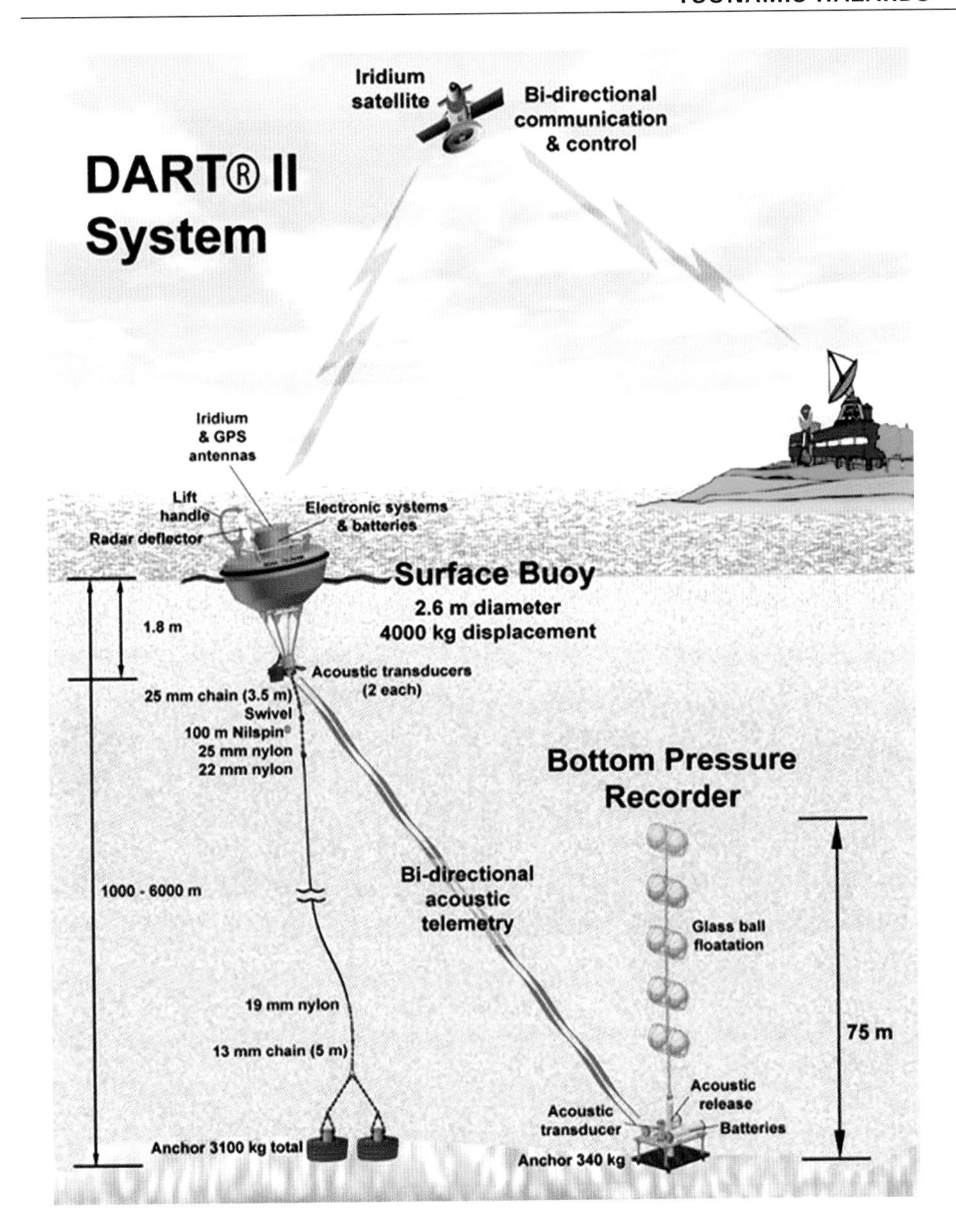

Figure 8.20 Deep-ocean Assessment and Reporting of Tsunami (DART®) stations.

Source: NOAA National Data Buoy Centre (2022)

1967 to provide more localised warning for Alaska, British Columbia, Washington, Oregon and California. In 1996 the USA approved the National Tsunami Hazard Mitigation Program (NTHMP), a collaborative state–federal partnership designed to reduce tsunami hazards along US coastlines.

The Pacific Tsunami Warning and Mitigation System (PTWS) relies on about 150 high-quality seismic stations around the world (Figure 8.22a) to locate and

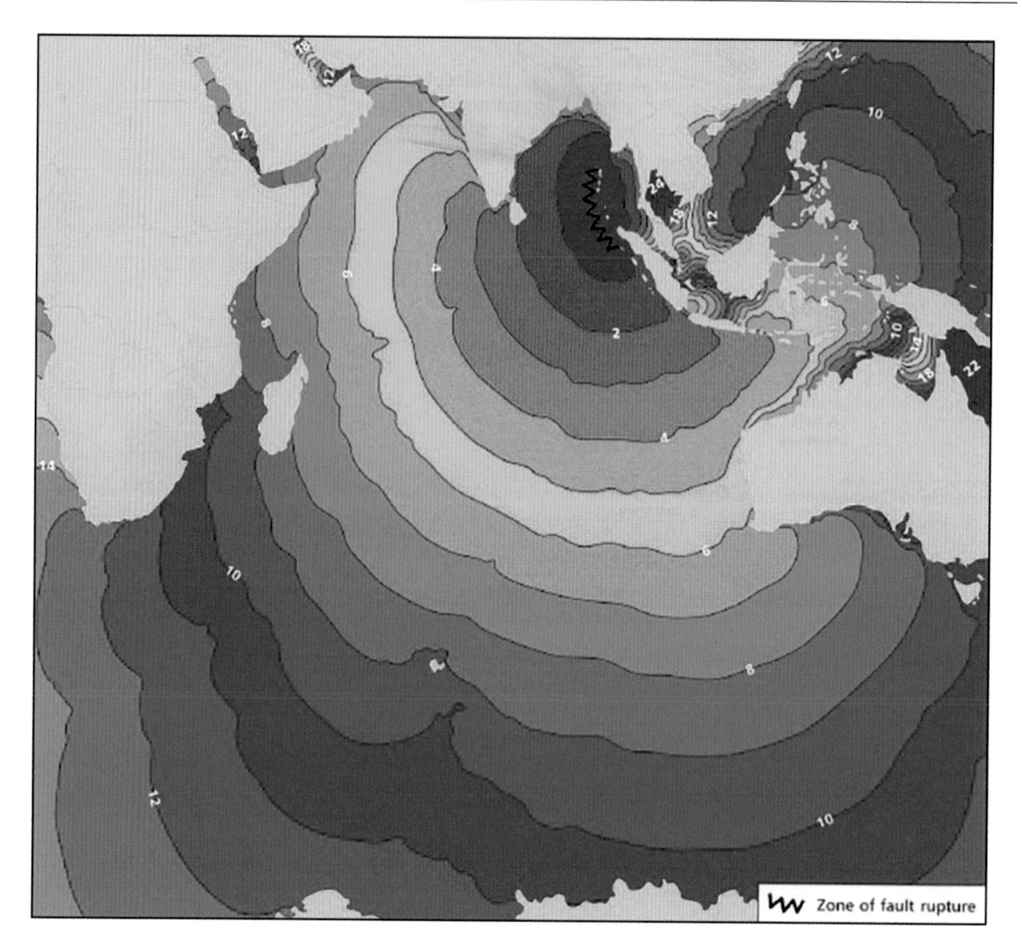

Figure 8.21 Progress of a tsunami wave across the Pacific Ocean after the Mw = 9.1 Sumatra earthquake at the Sunda trench on 26 December 2004. Isolines indicate one-hour time intervals. More than 4 million Indonesians live in tsunami-prone areas and need an effective monitoring system capable of delivering very short-term warnings.

Source: After NOAA at https://www.ngdc.noaa.gov/hazard/26dec2004.html (US government material not subject to copyright protection in the United States)

assess earthquakes with tsunami potential. Data are also accessed from about 100 sea-level stations (Figure 8.22b) to verify the generation of a tsunami and to assess the scale of the threat. The PTWS disseminates information and warning messages to over 100 designated offices in countries bordering the Pacific Ocean. Two sub-centres provide regional alerts to the US west coast, Alaska, Canada and the northwest Pacific and South China Sea areas.

The PTWS operates on two levels. The first level supplies warnings to all Pacific nations of large, destructive tsunamis likely to be ocean-wide. Following a high-magnitude earthquake ($M \geq 7.0$), tide stations near the epicentre are alerted to observe unusual wave activity. If this is detected, information is issued in the

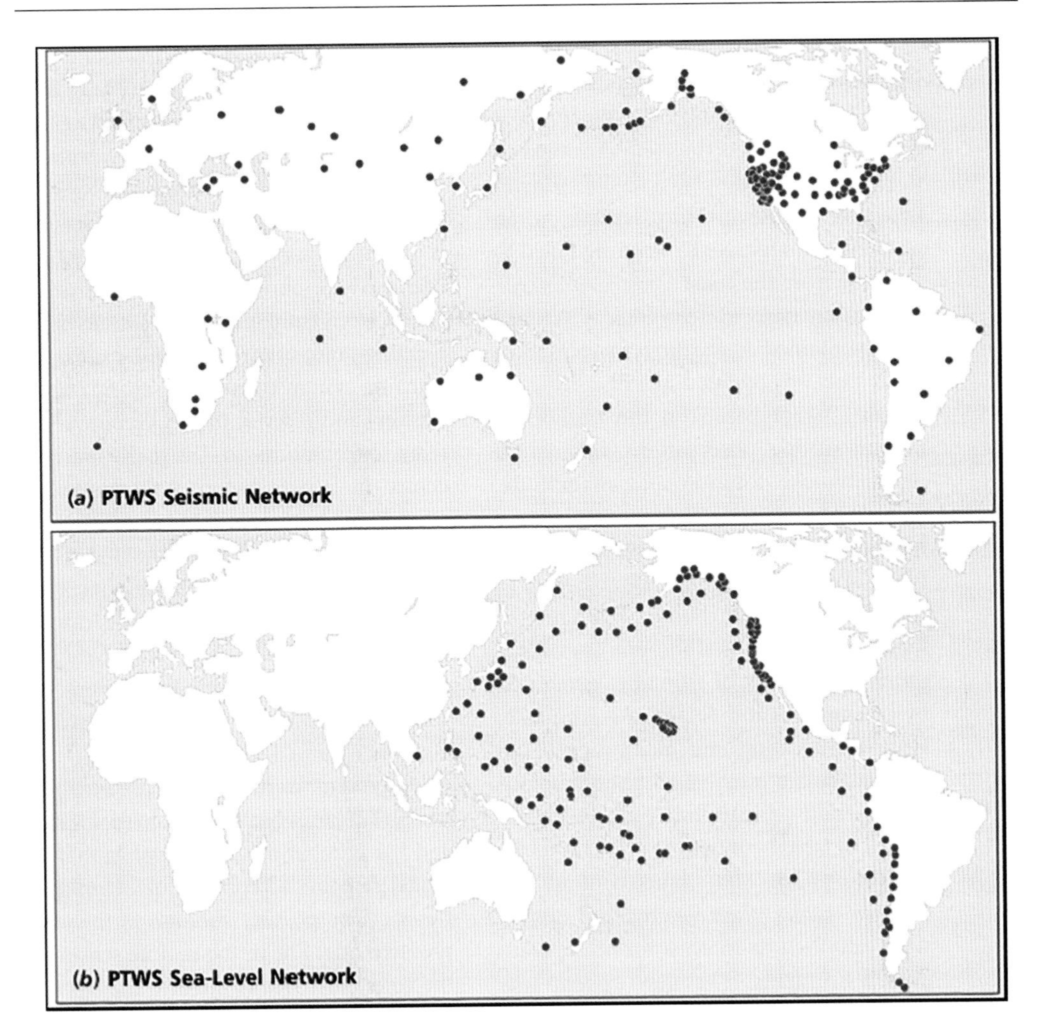

Figure 8.22 The Pacific Tsunami Warning System (PTWS) showing the network of monitoring stations for (a) seismic activity and (b) sea-level changes. The PTWS uses over 150 seismic stations and about 100 sea-level stations around the globe to evaluate the severity of a tsunami. Warning messages can be disseminated to authorities in more than 100 locations across the Pacific.

Source: After the International Tsunami Information Center at www.tsunamiwave.info (accessed 5 January 2012). US government material not subject to copyright protection in the United States.

form of watches and advisories and full tsunami warnings. The primary aim is to alert all at-risk coastal populations, within one hour, about the arrival time of the first wave, with an accuracy of ±10 minutes. More distant populations have longer to react and prepare for evacuation from the coast. The second level of cover is based on warning systems serving specific areas. Local tsunamis can pose a greater threat than ocean-wide events because they strike onshore very quickly. These systems rely on local data obtained in real time and typically aim to issue warnings within minutes for areas 100–750 km distant from the earthquake source.

Whilst Pacific tsunami forecasting and warning systems are well established, frequent false alarms and occasional threat denial have limited their effectiveness in the past. For example, a 1960 tsunami killed 61 people in Hilo, Hawaii, when residents who received a warning failed to evacuate from the coast.

The Japanese Meteorological Agency has maintained a warning service since 1952. This was updated in 1999. Previous warnings were based on the traditional method of tidal observations, calculation of earthquake location and magnitude, empirical estimation of tsunami wave heights, followed, if necessary, by the issue of a warning message for 18 coastal segments each several hundred kilometres long. The present method employs computer simulations of tsunamis generated by various sizes and depths of earthquake. Once the location and magnitude of an earthquake are established, then tsunami heights and arrival times are retrieved from a database containing about 100,000 simulations for 600 points around the Japanese coast. Wave heights and arrival times can then be forecast for 66 separate coastal segments. The new system was tested in the 2011 event and provides:

- an initial tsunami advisory or warning three minutes after an earthquake;
- the issue of maximum wave heights and arrival times within about five minutes after the earthquake; and
- the subsequent issue of the times of high tides and continuing updates about the hazard situation.

The impact of the 2004 Indian Ocean tsunami stimulated interest in tsunami disaster risk reduction for the Indian Ocean region (Jin & Lin, 2011). With over 225,000 lives lost, if a system similar to the PTWS had been in place, the death toll might have been reduced to about 15,000 (Wang & Li, 2008). At a United Nations conference held in January 2005, the German government agreed, under the leadership of UNESCO, to supply such a system based in Jakarta, Indonesia. Indonesia consists of 17,000 islands, with about 60% of the coastline at risk from tsunamis because it lies at the meeting point of three tectonic plates. The warning system consists of seismographic stations and a series of DART buoys that relay information to the early-warning centre in Jakarta before transmission to 11 regional hubs across Indonesia (Strunz et al., 2011). The pilot version of the scheme was launched in November 2008; full ownership passed to the government in Jakarta in March 2011. One of the key challenges in maintaining this system – as with all such systems – is making sure the equipment is well maintained.

At present, warning times are very limited for tsunamis generated in coastal waters, and the number of false warnings that are issued remains a problem for ocean-wide threats. One possible solution to the latter problem might lie in the remote sensing of the cold water that is lifted to the sea surface from deep in the ocean by submarine earthquakes. According to Lin et al. (2011), the arrival of the cold water at the surface emits an infrared radiation signal that was detected by satellite in both the 2004 Indian Ocean and the 2011 Tōhoku events. If this thermal anomaly proves to be a reliable signature of a tsunami, its detection could augment the DART system with a warning scheme that involves little time delay and is less likely to produce false warnings.

Risk Assessment

To gauge the near-shore impact of tsunami inundation, engineers and scientists can use three key tools to assess the risk of a tsunami:

- a field survey of past events;
- physical experimentation in a laboratory; and
- numerical modelling of tsunami.

Models help simulate the properties of the tsunami (e. g. earthquake strength, wave height and period) and the coastal profile (e. g. slope, roughness), enabling better risk information about how a coastline may be affected by tsunamis of different scales. It remains incredibly challenging to model coastlines, as they are so variable over time due to erosional processes. Detailed maps of future tsunami flooding (inundation) are needed for delineation of evacuation routes and long-term planning in vulnerable coastal communities. Computer models are used to develop the inundation maps used for coastal community planning, as shown in Figure 8.24. In addition, probabilistic tsunami hazard analysis (PTHA) can be used to quantify the probability of exceeding a specified inundation intensity at a given location within a given time interval. PTHA provides additional guidance for tsunami risk analysis and risk management, including coastal land use planning and the development of warning systems.

Engineering

Civil engineering techniques are also used to limit the impact of tsunami hazards. Typically, buildings and infrastructure vulnerable to tsunamis are prone to earthquake hazards and therefore additional protection is necessary against tsunamis. These engineering solutions can vary in scale from houses to extensive sea walls. Typical actions adopted to develop suitable engineering solution include:

- determining and understanding the community tsunami risk;
- avoiding new development in tsunami run-up areas;
- locating and configuring new development in the run-up areas to minimise future tsunami losses;
- designing and constructing new buildings to minimise tsunami damage;
- protecting existing development through redevelopment, retrofitting and land reuse plans and projects; and
- taking special precautions in locating and designing infrastructure and critical facilities.

A range of constructional mitigation measures can be designed to avoid or attenuate tsunami impact on the coastline and structures. Figure 8.23 outlines five key approaches that prevent direct wave impact or dissipate the tsunami impact energy. For smaller-scale tsunamis, they can help prevent or mitigate adverse impacts. For large-scale tsunamis, it may be a case of focusing on evacuation. Three structural options for preventing/mitigating the risks of damage or loss are available: *structural* (protecting; Figure 8.23a, b, c, e); what used to be called 'retreating' and is now *realigning* (accommodating, Figure 8.23d); and *non-structural* measures.

Following the 1933 Sanriku tsunami off the northeast coast of Japan, government grants were provided for the relocation of some fishing villages to higher ground (Fukuchi & Mitsuhashi, 1983). This policy was ineffective due to the limited availability of nearby land for resettlement and the desire of fishermen to remain close to the shore. Efforts then shifted to the

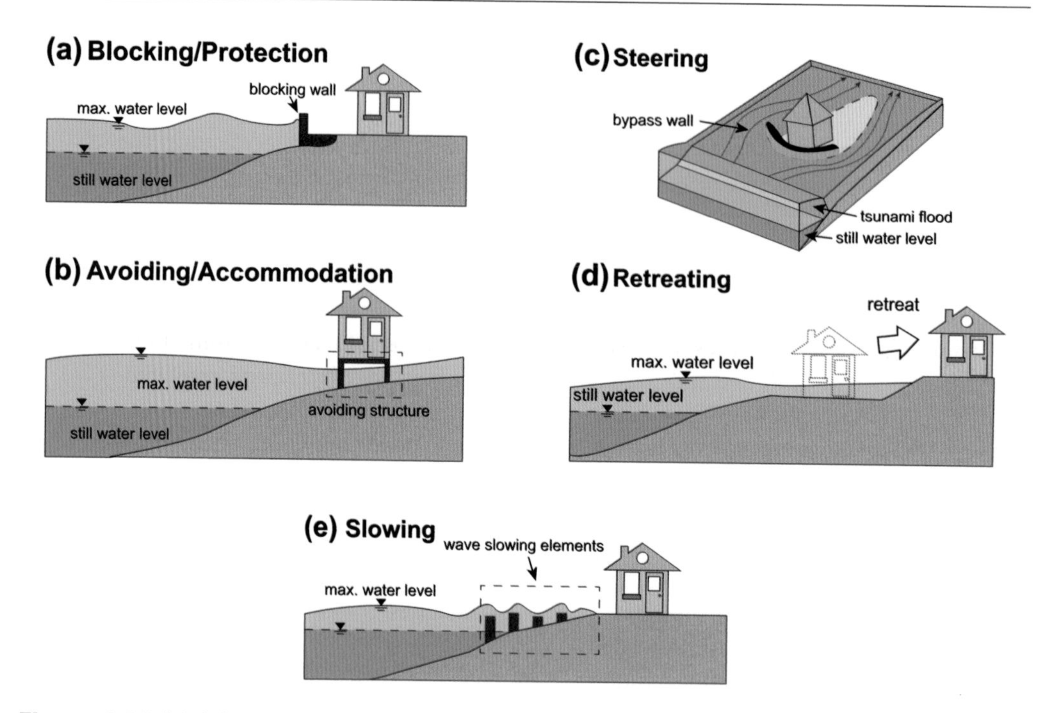

Figure 8.23 (a)–(e) Basic strategies to reduce tsunami risk following NOAA (2001, modified), with 'realigning' now preferred to 'retreating'.

Source: Oetjen et al. (2022)

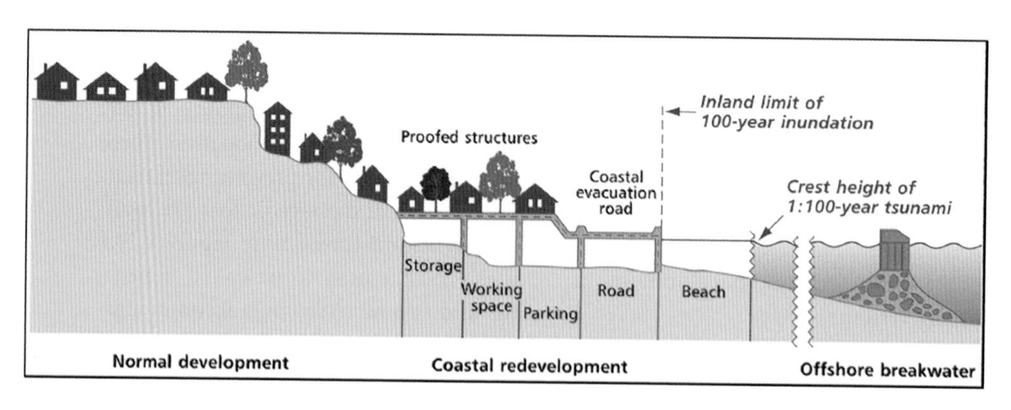

Figure 8.24 Depiction of tsunami protection works typical of the Sanriku coast, Japan, showing an offshore breakwater and some raised onshore development, including an emergency evacuation route.

construction of onshore tsunami walls, although these failed to prevent further losses in 1960. After this event, the government passed a special law to subsidise construction costs up to 80% for sea walls designed to cope with wave heights equivalent to the 1960 event. Since then, engineers have continued to protect against even larger tsunamis, and the highest walls stand up to 16 m above tidal levels.

Japan employs tsunami breakwaters in addition to sea walls. Breakwaters do not take up valuable land and provide shelter for shipping, but they are expensive and can interfere with tidal circulations and damage the local fishing industry. More recently, some elaborate onshore tsunami walls designed to protect coastal property and transport links have been used, although these structures are visually intrusive. Figure 8.24 shows a tsunami defence scheme, including an offshore breakwater and an onshore tsunami wall combined with coastal redevelopment. These projects are expensive and give coastal communities the appearance of fortified towns, with limited views of the ocean for residents.

The town of Taro has both inner and outer sea walls; the inner one is 10 m high and stretches 2.5 km across the bay. These defences were overtopped by the 2011 tsunami, which destroyed much of the town (see Box 4.3 and Figure 8.24). About 80 km to the south, what is probably the world's deepest breakwater was completed in 2009 after a 30-year construction period and at a cost of US$1.5 billion, to protect the city of Kamaishi. These defences also failed in 2011. The whole issue of coastal engineering is currently under fresh scrutiny as Japan searches for solutions to tsunami hazards.

Engineering solutions can be used to provide evacuation shelters for those trying to get to safety. Typically a vertical evacuation structure is built, especially where there is no high ground nearby, to provide protection. These structures can consist of a building or earth mound that is specially designed to resist earthquake and tsunami forces, along with the debris that is picked up in the tsunami waters. Its height allows people to evacuate above the level of tsunami inundation, as long as they are built to appropriate scientific criteria to ensure safety (see Box 4.3). These structures can take form as a building, a special purpose-built tower or a platform.

Additionally, some newer options have been considered such as a tsunami floating shelter where people can seek safety by floating along with the flow of turbulent waters to escape from direct exposure to the elements (see Figure 8.25). Most of these options are expensive. Vertical evacuation structures need to be several storeys high to remain above the tsunami run-up and, if not built properly, might increase the possibility of earthquake damage (Chapter 6) or could collapse when the tsunami undermines or breaks them. A simpler and cheaper option is the creation of paths and steps leading up hillsides away from the shore, but the local topography may be unsuitable (or flat), and some people with disabilities will be unable to use this type of escape route.

Coastal Zone Planning

The need for tsunami mitigation to be integrated into planning procedures along low-lying coastlines was emphasised by Preuss (1983) and by the Intergovernmental Oceanographic Commission (2008). In terms of tsunami protection, engineered sea walls remain relevant but, if the tsunami run-up height is over 4 m, they are of little benefit.

These structures can be complemented with 'bioshields' or coastal tree plantations. Although natural forest of mangrove and casuarinas is often severely damaged by tsunami waves, trees can provide a protective buffer for the land behind. Along the Sanriku coast of Japan, for example, the planting of spruce forest is widely used to

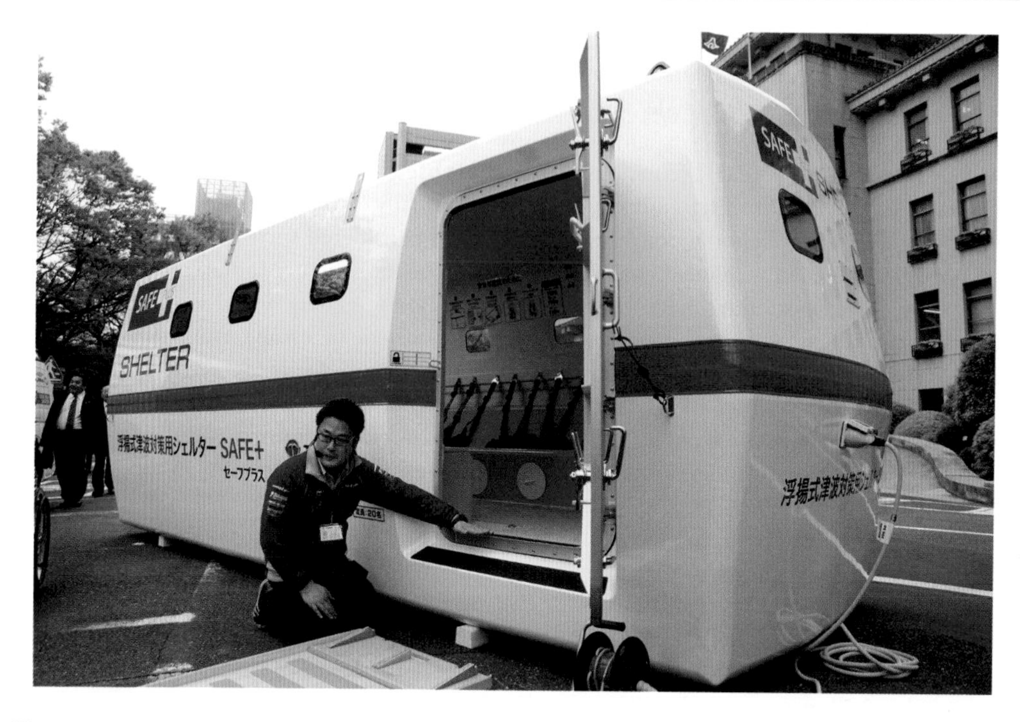

Figure 8.25 Tsunami & Flood Floating SAFE+ (Plus) Shelter from outside and inside perspectives; source: Tajima Motor Corporation, 2022.

Source: © Alamy Images/Newscom

dissipate wave energy and filter out floating debris. Otherwise, residential areas, public facilities and transportation should be relocated to higher ground. In particular key infrastructure that is high risk should not be placed in high-risk areas, such as the Fukushima Nuclear Power Plant (Box 1.2).

Figure 8.26 is an illustration of the measures that could be incorporated into a comprehensive anti-tsunami scheme, including physical structures and the provision of a coastal evacuation route. The presumption is always that new development will not be exposed to risk, but achieving a balance between safety and land function can be controversial, as in the case of the proposed Marine Education Centre for Wellington, New Zealand (Garside et al., 2009).

Transportation planning is a difficult issue in view of the expected traffic volumes generated by mass evacuation after a tsunami warning. Many existing roads follow low-lying shores or valley sides, which are likely to be inundated and can become congested with traffic.

Despite these problems, an important element of tsunami land planning is the identification of evacuation zones. Figure 8.27 is part of the tsunami evacuation zone map for the city of Hilo, Hawaii, showing the evacuation area with the position of roadblocks at its edge and the location of the emergency response agencies, civil defence, fire service, police and the public works department just outside the zone. Wherever possible, people are encouraged to

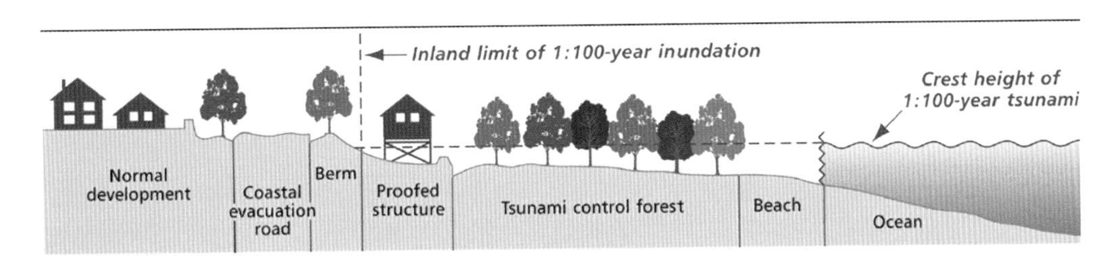

Figure 8.26 Typical example of coastal land planning for tsunami hazards. The beach and forest zones dissipate the energy of the onshore wave. Building development and the evacuation route are located above the predicted height of the 1:100 year event.

Source: After Preuss (1983)

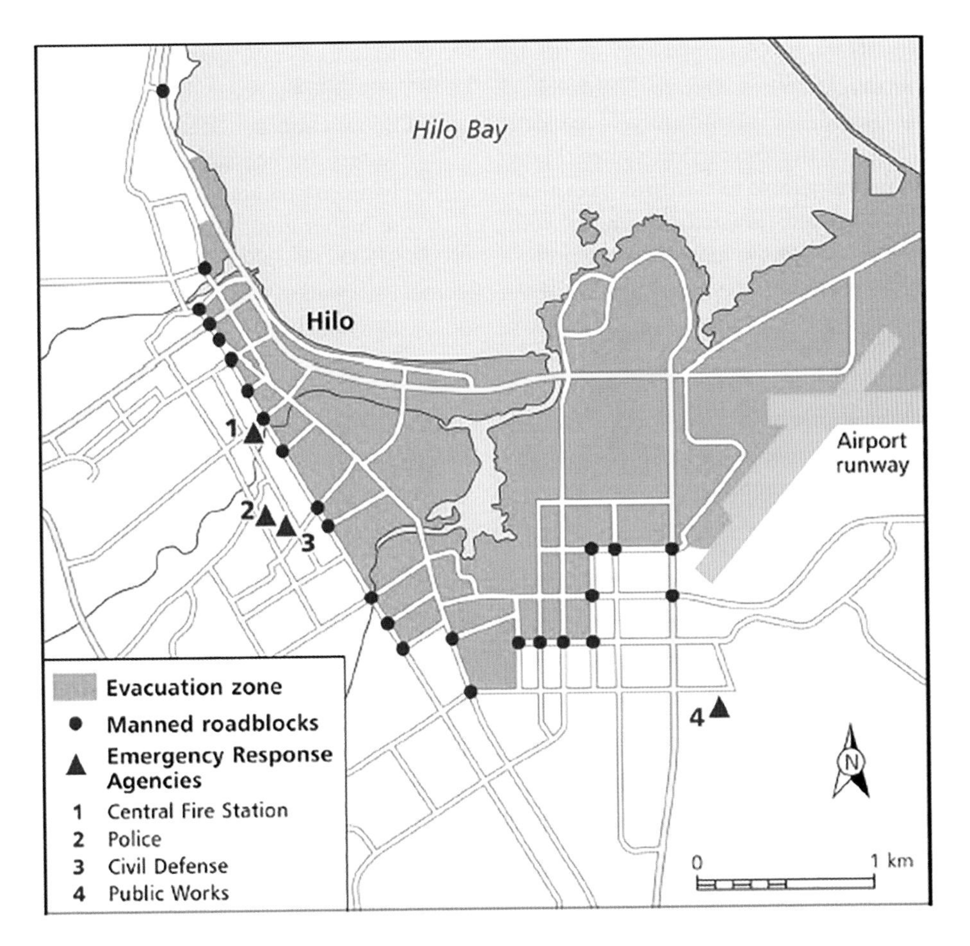

Figure 8.27 Part of the tsunami hazard map of the city of Hilo, Hawaii, showing the scheduled evacuation area and the location of the emergency response agencies at the outer edge of this zone. Note the proximity of the airport to the evacuation zone.

Source: PDC. *Tsunami Evacuation Zones (Hawaii)* at https://static.pdc.org/tsunami/index.html

evacuate such areas on foot so as to limit traffic congestion and to follow signed routes where these are indicated.

Community Preparedness

Many of the tools required to aid the management of risk involve complex and costly technology, modelling and engineering. However, there are a number of cheap and effective tools that communities can use to help their preparedness, warnings and actions to tsunamis. As seen with earthquakes, there are annual drills that take place globally on World Tsunami Awareness Day, on 5 November, that provide an opportunity to educate and to practice evacuation routes. Many nations vulnerable to tsunamis have dedicated tsunami evacuation routes with appropriate signs to help those unfamiliar with the location. In some areas like Wellington in New Zealand, blue lines have been painted on the roads marking out a tsunami-safe zone.

Educational programs focus on giving people the knowledge and tools to identify a tsunami themselves and to be able to evacuate without waiting for a formal warning that may not be timely enough to enable evacuation to safety. This can be classed as education for self-warning and voluntary evacuation (SWAVE) (Ranger & Day, 2007). Lori Dengler from Humboldt State University stated that for the 2004 tragedy:

> Even without a warning system, even in places where they didn't feel the earthquake, if people had simply understood that when you see the water go down, when you hear a rumble from the coast, you don't go down to investigate, you grab your babies and run for your life, many lives would have been saved.
>
> (*New Scientist Magazine*, January 15, 2005)

Guidance on how to sense a tsunami is:

TOUCH Strong local earthquakes may cause tsunamis.

FEEL the ground shaking severely? Immediately evacuate low-lying coastal areas and move inland to higher ground!

SIGHT As a tsunami approaches shorelines, water may recede from the coast, exposing the ocean floor and reefs.

SEE an unusual disappearance of water? Immediately evacuate low-lying coastal areas and move inland to higher ground!

SOUND The abnormal ocean activity, a wall of water, and approaching tsunami waves create a loud 'roaring' sound similar to that of a train or jet aircraft.

HEAR the roar? Immediately evacuate low-lying coastal areas and move inland to higher ground. Don't wait for official evacuation orders.

For many communities a knowledge of tsunamis is well established. For example, communities on the island of Simeulue, located 60 km from the epicentre of the 2004 Indonesian Sumatran earthquake, had a low death toll of a handful of residents, with about 70,000 people reaching safety. This is largely due to a local knowledge called 'smong' to warn the community to run to a higher and hence safer place when large earthquakes occur. This story was transferred from generation to generation through a lullaby called Buai-buai, the poem of Nandong,

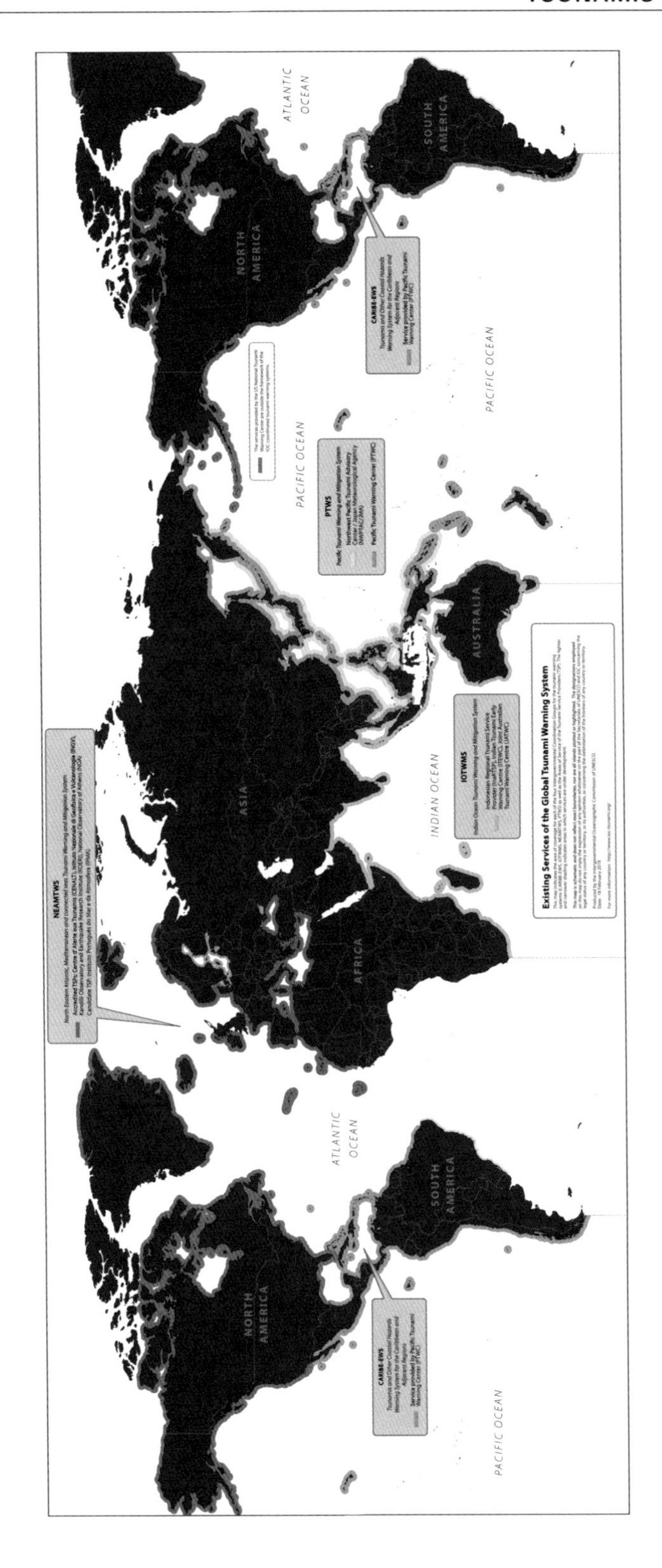

Figure 8.28 Existing services of the Global Tsunami Warning System.

Source: Intergovernmental Oceanographic Commission of UNESCO (February 2018)

and daily conversation (Gaillard et al., 2008; Rahman et al., 2017).

Global tsunami warning systems are increasing coverage over the globe, but there are still gaps, and significant work is yet to be done to integrate these warning systems with community knowledges to enhance awareness and preparedness. Tsunamis force international collaboration, which can be challenging, and significant investment is needed to find low-cost but accurate tsunami monitoring tools so as to provide more data and warnings. Focus will need to shift to managing more landslide-generated tsunamis in the future that provide no warning.

KEY TAKE AWAY POINTS

- Tsunamis can be generated by a range of hazards but typically affect large areas involving a number of nations. The most commonly generated tsunamis are those from earthquakes.
- Their highly destructive nature requires the use of early warning systems to help provide timely warnings for vulnerable populations to reach high ground. This may also require the building of vertical evacuation structures to provide safety in flat, low-lying landscapes.
- Tsunami early warning systems can be highly technical and complex technological tools requiring significant investment and maintenance. However, warnings via self-warning or oral traditions are often the most effective and cheapest way of being safe.
- Modelling tsunamis is key to exploring the damage a tsunami may create over a wide area and calculating time to arrival, alongside potential damage for insurance companies. However, with constantly changing coastlines specific and detailed data for inundation, rates are challenging to establish.
- Significant research and technological development is required to better monitor and provide warnings for tsunamis generated by landslides that can be significant in scale and provide no primary warning from shaking, as seen in earthquake-generated tsunamis.

BIBLIOGRAPHY

Board, O. S., & National Research Council. (2011). *Tsunami warning and preparedness: An assessment of the us tsunami program and the nation's preparedness efforts*. Washington, DC: National Academies Press.

Carey, S., Sigurdsson, H., Mandeville, C., & Bronto, S. (2000). Volcanic hazards from pyroclastic flow discharge into the sea: Examples from the 1883 eruption of Krakat these work effectively when there is more than 30 minutes warning before landfall, as less than this does not provide ample time for effective evacuation in most cases. Au, Indonesia. Special Papers-Geological Society of America, 1–14.

Fukuchi, T., & Mitsuhashi, K. (1983). Tsunami countermeasures in fishing villages along the Sanriku coast, Japan. In K. Iida & T. Iwasaki (Eds.), *Tsunami: Their science and engineering* (pp. 389–296). Tokyo, Japan: Terrapub.

Gaillard, J. C., Clavé, E., Vibert, O., Denain, J. C., Efendi, Y., Grancher, D., Liamzon, C. C., Sari, D. R., & Setiawan, R. (2008). Ethnic groups' response to the 26 December 2004 earthquake and tsunami in Aceh, Indonesia. *Natural Hazards, 47*(1), 17–38.

Garside, R., Johnston, D., Saunders, W., & Leonard, G. (2009). Planning for tsunami evacuations: The case of the Marine Education Centre, Wellington, New Zealand. *The Australian Journal of Emergency Management, 24*(3), 28–31.

González, F. I. (1999). Tsunami!. *Scientific American, 280*(5), 56–65.

Heidarzadeh, M., Putra, P. S., Nugroho, S. H., & Rashid, D. B. Z. (2020). Field survey of tsunami heights and runups following the 22 December 2018 Anak Krakatau volcano tsunami, Indonesia. *Pure and Applied Geophysics, 177*(10), 4577–4595.

Intergovernmental Oceanographic Commission (IOC). (2016). *Tsunami glossary* (3rd ed.), IOC Technical Series, 85 (IOC/2008/TS/85 rev.2). Paris: UNESCO. http://itic.ioc-unesco.org/index.php

Intergovernmental Oceanographic Commission, & International Tsunami Information Center. (2008). *Tsunami: The great waves.* UNESCO Intergovernmental Oceanographic Commission (IOC).

International Tsunami Information Centre. (n.d.a). *Global and regional hazard maps, tsunami sources.* http://itic.ioc-unesco.org/index.php?option=com_content&view=article&id=1672&catid=1075&Itemid=1075. Accessed 6 August 2022.

International Tsunami Information Centre. (n.d.b). *Intergovernmental Coordination Group for the Pacific Tsunami Warning and Mitigation System (ICG/PTWS).* http://itic.ioc-unesco.org/index.php?option=com_content&view=category&id=1153&Itemid=1153. Accessed 6 August 2022.

Jin, D., & Lin, J. (2011). Managing tsunamis through early warning systems: A multidisciplinary approach. *Ocean & Coastal Management, 54*(2), 189–199.

Keating, B. H., & McGuire, W. J. (2000). Island edifice failures and associated tsunami hazards. *Pure and Applied Geophysics, 157*, 899–955.

Kornei, K. (2018). Huge global tsunami followed dinosaur-killing asteroid impact. *Eos,* 99. https://doi.org/10.1029/2018EO112419. Published on 20 December 2018.

Lin, F. C., Zhu, W., & Sookhanaphibarn, K. (2011). Observation of tsunami radiation at tohoku by remote sensing. *Science of Tsunami Hazards, 30*(4).

Liu, J. Y., Chen, C. H., Lin, C. H., Tsai, H. F., Chen, C. H., & Kamogawa, M. (2011). Ionospheric disturbances triggered by the 11 March 2011 M9. 0 Tohoku earthquake. *Journal of Geophysical Research: Space Physics, 116*(A6).

Luthfi, M., Suppasri, A., & Comfort, L. K. (2020). The 22 December 2018 Mount Anak Krakatau volcanogenic tsunami on Sunda Strait coasts, Indonesia: Tsunami and damage characteristics. *Natural Hazards and Earth System Sciences, 20*(2), 549–565.

Maeno, F., & Imamura, F. (2011). Tsunami generation by a rapid entrance of pyroclastic flow into the sea during the 1883 Krakatau eruption, Indonesia. *Journal of Geophysical Research: Solid Earth, 116*(B9).

NASA Earth Observatory. (2004). *Tsunami strikes Sri Lanka.* https://earthobservatory.nasa.gov/images/5125/tsunami-strikes-sri-lanka?src=on-this-day. Accessed 6 August 2022.

National Data Buoy Center. (2022). Deep-ocean assessment and reporting of tsunamis (DART®) description. Accessed 23 July 2021.

NOAA. (n.d.a). *National data buoy centre.* www.ndbc.noaa.gov/dart/dart.shtml. Accessed 6 August 2022.

NOAA. (n.d.b). *Tsunami propagation.* https://www.noaa.gov/jetstream/tsunamis/tsunami-propagation. Accessed 7 March 2020.

Oetjen, J., Sundar, V., Venkatachalam, S., Reicherter, K., Engel, M., Schüttrumpf, H., & Sannasiraj, S. A. (2022). A comprehensive review on structural tsunami countermeasures. *Natural Hazards*, 1–31.

Oxfam Briefing Note. (2005). *Gender and the tsunami.* www.oxfamamerica.org/newsandpublications/publications/briefing_papers/briefing_note.2005-03-30.6547801151. Accessed 4 January 2006.

Preuss, J. (1983). Land management guidelines for tsunami hazard zones. Tsunamis: Their Science and Engineering, Reidel, Boston, MA, 527–539.

Rahman, A., Sakurai, A., & Munadi, K. (2017, March). Indigenous knowledge management to enhance community resilience to tsunami risk: Lessons learned from Smong traditions in Simeulue island, Indonesia. IOP Conference Series: Earth and Environmental Science (Vol. 56, No. 1, p. 012018), IOP Publishing Ltd.

Ranger, R., & Day, S. J. (2007). Mortality distributions in time and space during tsunamis: Implications for the performance of tsunami warning systems and the importance of Education for Self Warning and Voluntary Evacuation (ESWAVE). American Geophysical Union, Fall Meeting 2007, abstract #S53A-1045. http://adsabs.harvard.edu/abs/2007AGUFM.S53A1045R. Accessed 12 June 2011.

Scheffers, A., & Kelletat, D. (2003). Sedimentologic and geomorphologic tsunami imprints worldwide: A review. *Earth-Science Reviews, 63*(1–2), 83–92.

Schiermeier, Q. (2005). On the trail of destruction. *Nature, 433*(7024), 350–354.

Strunz, G., Post, J., Zosseder, K., Wegscheider, S., Mück, M., Riedlinger, T., Mehl, H., Dech, S., Birkmann, J., Gebert, N., & Harjono, H. (2011). Tsunami risk assessment in Indonesia. *Natural Hazards and Earth System Sciences, 11*(1), 67–82.

Tappin, D. R. (2017). Tsunamis from submarine landslides. *Geology Today, 33*(5), 190–200.

Thomas, C., Burbidge, D., & Cummins, P. (2007). A preliminary study into the tsunami hazard faced by southwest Pacific nations. Risk and Impact Analysis Group, Geoscience Australia.

United Nations (UN). (2006). *Global survey of early warning systems. Gaps and opportunities towards building a comprehensive global early warning system for all natural hazards.* Report prepared at the request of the Secretary-General of the United Nations.

US Indian Ocean Tsunami Warning System Program. (2007). *Tsunami warning center reference guide.* Bangkok: US Agency for International Development. https://nctr.pmel.noaa.gov/education/IOTWS/program_reports/TsunamiWarningCenterGuide.pdf

Verbeek, R. D. M. (1884). The Krakatoa eruption. *Nature, 30*(757), 10–15.

Wang, J. F., & Li, L. F. (2008). Improving tsunami warning systems with remote sensing and geographical information system input. *Risk Analysis: An International Journal, 28*(6), 1653–1668.

Ward, S. N., & Day, S. (2001). Cumbre Vieja volcano: Potential collapse and tsunami at La Palma, Canary Islands. *Geophysical Research Letters, 28*(17), 3397–3400.

Whelan, F., & Kelletat, D. (2003). Submarine slides on volcanic islands – A source for mega-tsunamis in the Quaternary. *Progress in Physical Geography, 27*(2), 198–216.

Witze, A. (2022). Why the Tongan eruption will go down in the history of volcanology. *Nature, 602*(7897), 376–378.

FURTHER READING

Bryant, E. (2001). *Tsunami: The underrated hazard.* Cambridge: Cambridge University Press.

Kelman, I. (2006). Warning for the 26 December 2004 Tsunamis. *Disaster Prevention and Management, 15*(1), 178–189.

O'loughlin, K. F., & Lander, J. F. (2003). *Caribbean Tsunamis: A 500-year history from 1498–1998.* Dordrecht: Springer.

WEB LINK.S

Caribbean Tsunamis. https://uwiseismic.com/tsunamis/caribbean-tsunamis

International Oceanographic Commission Tsunami Programme. www.ioc-tsunami.org

NOAA Center for Tsunami Research. www.nctr.pmel noaa.gov

The Tsunami Society. http://tsunamisociety.org

US Tsunami Warning System. www.tsunami.gov

CHAPTER NINE

Mass Movements

9

OVERVIEW

Mass movements encompass a wide range of hazards, commonly called landslides and avalanches. In this chapter the various types and mechanisms of mass movements are explored with a particular focus on falls, flows and slides. The causes and triggers are diverse and vary from geological, geomorphological, meteorological and environmental sources, although increasingly the activities of humans are becoming the most significant driver of mass movements. The impact of these events are typically localised, but they can cause significant deaths and destruction, along with secondary hazards such as flooding. Whilst avalanche hazards typically affect skiers and climbers, many rural and urban areas are at risk, and considerable work is being done to enhance the monitoring of mass movements to integrate these into effective warnings to aid community awareness and preparedness. Engineering solutions are commonly devised and integrated to help prevent mass movements and help preserve key infrastructure such as roads and rivers that often dam following a mass movement. Whilst many of these approaches are costly, for some the integration of better land use zoning and planning is vital to mitigate against the risks. Biological solutions to maintaining slope safety is a cheap, effective and environmentally sound method in contrast to building engineering structures. With climate change and the increasing vulnerabilities of large populations living on unsafe lands, mass movements will be a hazard seen far more frequently in the future.

A. THE NATURE OF MASS MOVEMENTS

Mass movement is a term used for any type of downslope movement of earth materials such as soil and rock debris in response to the pull of gravity, or the rapid or gradual sinking of the Earth's ground surface in a predominantly vertical direction. Formerly, the term mass wasting was used, but it only

DOI: 10.4324/9781351261647-11

referred to processes by which large masses of crustal materials are moved by gravity from one place to another. Mass movement includes the sinking of confined areas of the Earth's ground surface. Mass movement is a common hazard in mountainous areas and is often aided by water generating significant alteration of landforms. The movements vary greatly in size (ranging from a few cubic metres to over 100 cubic kilometres) and in speed (ranging from millimetres per year to hundreds of metres per second).

Rapid mass movements generally cause the greatest loss of life, whilst slower downslope movements create significant economic costs. It is convenient to classify mass movements according to the material forming most of their mass. *Landslides* consist mostly of rock and/or soil; *snow avalanches* are formed predominantly from snow and/or ice. Mass movements are usually triggered by natural processes, notably earthquakes, weathering or erosional debris (often from intense and/or prolonged rainfall and snowmelt); the character and structure of the underlying bedrocks; the removal of the vegetation cover; and artificial or natural increases in the slope's steepness. Some damaging landslides occur in materials deposited by humans that include mining waste, landfill and garbage. People play an increasing role in the causation of these mass movements. Figure 9.1 illustrates the different types of mass movement, indicating their materials, nature of motion and velocity.

Slopes are some of the most common landforms on Earth and are dynamic evolving systems. Materials are constantly moving down the slope at varying speeds, and the slopes can be convex, straight or concave in shape. The forces that influence slope stability are a balance between the driving forces that move earth materials down a slope (gravity and the weight of the materials, e.g. vegetation, fill materials and buildings) and the resisting forces that oppose this movement (the shear strength of the slope materials). Slope stability is determined by a safety factor (SF). This is the ratio between the resisting forces to the driving forces. If the factor is greater than 1, then the driving force exceeds the resisting forces, and subsequently slope failure is expected. All of these forces are dynamic, with local conditions driving changes that include type of earth materials, slope angle and topography, climate, vegetation, water and time.

1. Landslides

The term 'landslide' is a loosely defined term but is shorthand to describe a wide range of downslope movements of soil and/or rock (dry earth materials) under the influence of gravity. Although many landslides occur through the process of rock or soil sliding on a distinct surface, this is not always so. In practice, there is a wide variety of types of movement, including falling (free falling), sliding (movement of material as a coherent body over a shear plane) and flowing (turbulent motion of material with a water content of less than 21%).

Landslides are concentrated in five major types of terrain (Jones, 1995):

- *Upland areas subject to seismic shaking*: Earthquakes in hilly or mountainous terrain often trigger large numbers of landslides. For example, in the 1999 Chi-Chi earthquake in Taiwan over 9,200 large landslides were triggered in 35 seconds (Hung, 2000). In the years after a large earthquake further landslides may occur, as slope materials have been destabilised by the shaking.
- *Mountainous environments with high relative relief*: Mountain areas are subject to high levels of rock falls and landslides, due to the steep terrain, deformed rock

MECHANISM		MATERIAL			VELOCITY
		Rock	Fine-grained soil	Coarse-grained soil	
Slide	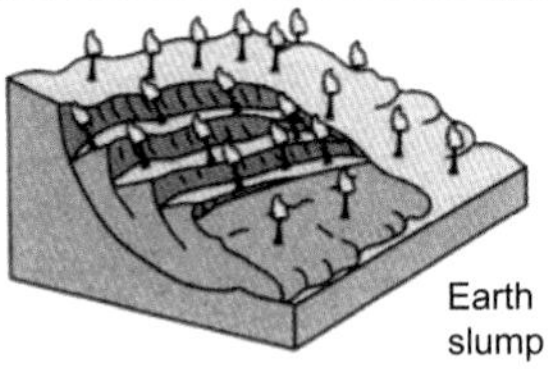 Earth slump	Slump	Earth slump	Debris slump	Slow (days to weeks)
			Stair-stepped slices with surfaces that slope downward in an uphill direction. Trees and shrubs within a slice have a similar orientation.		
	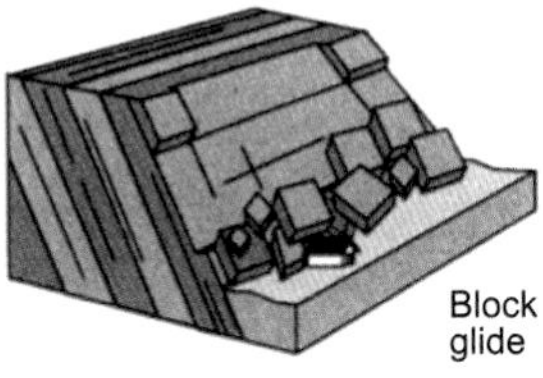 Block glide	Block glide *Down the plane of inclined rock layers where inclination is parallel to slope.*	Earth slide 	Debris slide	Variable (Secondsfor rock, minutes to days for debris/earth)
			Chaotic jumble of soil and debris. Mixed orientations of trees and shrubs.		
Flow	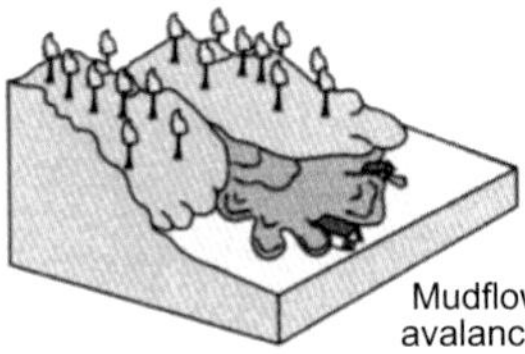 Mudflow, avalanche	Rock avalanche	Mudflow, avalanche	Debrisflow, avalanche	Very rapid (minutes)
		Behaves asa viscous fluid. Follows valleys.			
	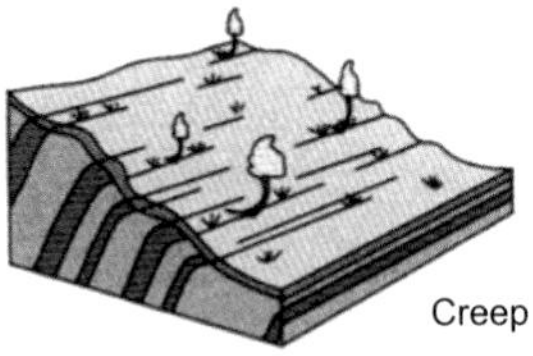 Creep	Creep	Creep	Creep	Extremely slow (months)
		Marked by thin, weak layers of rock dragged downward by surface flow. Trees bowed as the result of tilting downslope followed by upward growth.			
Fall	 Rockfall	Rockfall	Earthfall	Debrisfall	Extremely rapid (seconds)
			Undercut rock and debris fall vertically for lack of support.		

Figure 9.1 A classification of landslides, in accordance with Varnes (1978), based on mechanism, material and velocity of movement (Freeman, 2004). Landslides are classified according to the materials involved and the mechanism of movement.

Source: Varnes (1978) and Freeman (2004)

masses and high rainfall totals. Rock avalanches are a particular hazard. They involve massive quantities of rock debris, with volumes over 100×10^6 m^3, able to travel very large distances. In New Zealand, for example, some rock avalanches with volumes in excess of 100 km^3 have travelled more than 15 km. On average, about one massive rock avalanche occurs worldwide each year, generally in high, tectonically active mountain chains such as the Himalayas, the Rockies and the Andes.

- *Areas of moderate relief suffering severe land degradation*: Human activity and land degradation is the cause of many events. North Korea is prone to landslides because the population has been forced to remove almost all of the forests on the hills to burn as firewood. South Korea, on the other hand, has similar terrain, but a proactive policy of slope management and afforestation has resulted in a much lower incidence of landslides.
- *Areas with high rainfall*: Intense or prolonged rainfall is the most common trigger of slope instability, and areas subject to very high rainfall totals are inevitably susceptible to landslides. This process is especially active in humid tropical areas, where rock weathering can penetrate tens of metres below the ground surface. In Malaysia the weathered material can extend to a depth of 30 m or more, and landslides are common in the intense rainfall during the passage of a tropical cyclone. Areas affected by monsoonal rainfall, such as the Indian subcontinent, are also vulnerable to landslides. In temperate regions landslides tend to result from the disturbance of more shallow surface layers, brought about by seasonal rainfall in the form of either high winter totals or convectional storms in summer.
- *Areas with thick deposits of fine-grained materials*: Fine-grained deposits, such as loess and tephra, are weak and vulnerable to the effects of saturation. As a result, they are prone to landslides. Notable areas at risk include the loess plateaux of Gansu in northern China. In the 1920 earthquake, flowslides in the loess triggered by ground shaking are estimated to have killed over 100,000 people, the largest landslide disaster in recorded history. Areas mantled in tephra on and around volcanic sites are also at high risk.

2. Falls

Rock falls involve the movement of material through the air and occur on very steep rock faces as free fall. The blocks that fall often detach from the cliff face along an existing weakness, such as a joint, bedding or exfoliation surface. The fall starts as an initial slip along a joint or bedding plane which then transitions into falling, due to the steepness of the cliff. The scale of rock falls varies from individual blocks through to *rock avalanches* that are hundreds of millions of cubic metres in size. Whilst the largest rock falls clearly have the greatest hazard potential, an individual block the size of an egg cup can be fatal if it hits a person on the head. Once the blocks of bedrock fall and accumulate at the base of the cliff, it is called a 'talus' slope.

The triggering of rock falls is complex. Earthquakes are an important factor because seismic waves can literally shake blocks off a cliff. For example, in the 1999 Chi-Chi earthquake in Taiwan, the road system was severely damaged by rock-fall activity in the epicentral zone, which greatly hindered the government response. Rock falls are also triggered by the presence of water in joints and fissures, which can loosen material. During winter periods, the process of freeze–thaw, in which water repeatedly freezes and expands in cracks in a rock face, is particularly important. Rock-fall activity often increases during heavy rain and in periods when the temperature passes though many freeze – thaw cycles in springtime in high-altitude areas. In some high mountain areas, permanent ice serves to hold fractured blocks on the slope, although recent atmospheric warming has led to some thawing of the ice and more rock-fall activity (Sass, 2005). In some cases, there is no obvious trigger. In May 1999 an unexpected rock fall occurred at Sacred Falls State Park in Oahu, Hawaii. A mass about 50 m^3 detached from the walls of a steep gorge and fell 160 m into the valley below, striking a group of hikers and killing eight people (Jibson & Baum, 1999).

3. Slides

Many landslides involve the sliding of a mass of soil and/or rock, usually along a slip surface which either forms as the slide develops or results from the activation of an existing weakness, such as a bedding plane, joint or fault. There are many ways in which the strength of the material comprising the slope is exceeded by downslope stress to create instability. Sometimes the sliding occurs because of the deformation of a weak layer within the rock or soil, in which case a shear zone is formed (see Figure 9.2).

In a slope, the forces that cause movement are called *shear forces* and are due to

Figure 9.2 The Ferguson rockslide above Highway 140 in the Merced River canyon, California. This slide became active in 2006 and is closely monitored because it is near a main entrance to Yosemite National Park. There is a threat to all road traffic between mileposts 103 and 104 as well as recreational users of the river in the run-out path below.

Source: Photo: Mark Reid, USGS

gravity trying to pull the mass down the slope. There are two main forces that resist the movement of the landslide. These are:

- *Cohesion*. This is the resistance arising from the 'stickiness' of particles or from interlocking. So, for instance, sandstone gets some of its strength from the cement that glues the particles of sand together. As a result of cohesion, sandstone cliffs are often vertical, to substantial heights. Beach sand, however, has no bonds between the particles and rests at a much lower angle. In a sandcastle, the initial wetness of the sand generates a suction force between the particles that holds them together – this also is a form of cohesion.
- *Friction*. This arises from the resistance of particles to sliding across each other. In a pile of dry sand the gradient of the slope is sustained by the friction between the sand grains. The magnitude of the friction force depends upon the weight of the material above the surface, just as it is more difficult to move a chair with someone sitting in it than when it is empty. Friction and cohesion together provide the resistive forces that maintain stability in a slope. Movement of the landslide occurs when the shear forces exceed the resistive forces. The crucial relationship between the resistive forces and the shear forces varies continually in most slopes for the following reasons:
- *Weathering*. Through time, weathering of the rock or soil mass can reduce its strength. In particular, weathering can attack the bonds between particles that provide cohesion and so weakens the rock or soil. As this occurs the slope becomes progressively less stable.
- *Water*. When the rock or soil forming the slope has water in its pore space, the overall weight of the slope increases, which slightly increases the shear forces. More importantly, this water provides a buoyancy force to the landslide mass that acts to reduce the friction, in much the same ways as a car skids once it starts to aquaplane. Thus, as the slope gets wetter it usually becomes less stable.
- *Increased slope angle*. In some situations the slope angle may increase, perhaps due to erosion of the toe of the slope by a river or through cutting of the slope by humans, perhaps for road construction or other developments. This serves to increase the shear forces.
- *Earthquake shaking*. During earthquakes, the magnitude of the forces varies as shaking of the slope occurs. This can both increase the shear forces and reduce the resistive forces, thereby triggering failure.

Quite often the surface on which movement occurs has a planar form. In this case, a *translational landslide* will result. This occurs because the landslide has activated an existing plane of weakness, such as a bedding plane. Commonly, down-cutting by a river causes an inclined bedding plane to be exposed in the riverbank (Figure 9.3). The block on the slope is then free to move when the resisting forces become sufficiently weak. This is a significant hazard during road construction in mountain areas, when cutting off the slope to create a bench for the road often exposes inclined bedding surfaces or joints (Petley, 2009). Translational landslides are often rapid. Once sliding starts, the materials in the shear zone lose their cohesion as inter-particle bonds are broken, and their frictional strength is reduced as the shear surface becomes smooth or even polished. This allows the

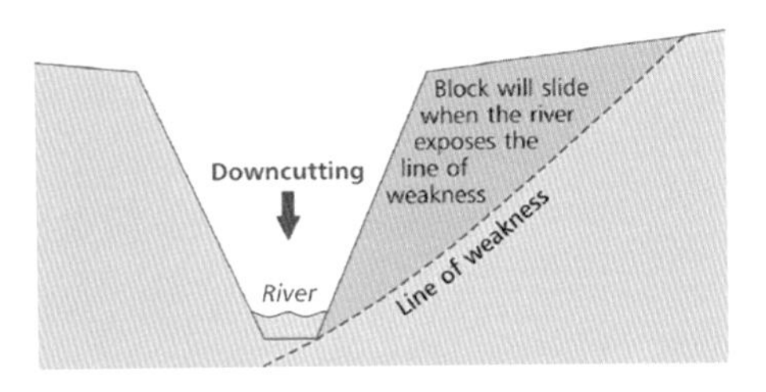

Figure 9.3 Down-cutting by rivers, construction activity or erosion by glaciers can cause landslides by exposing a weak layer of rock that permits sliding to occur.

Source: After Petley (2009)

landslide to accelerate rapidly, sometimes reaching very high speeds.

A specific hazard is the *landslide dam*, caused by large quantities of rock and other debris blocking river flow in a narrow valley or gorge to create a lake. A dual problem ensues. Valley bottomland is flooded upstream and may damage agriculture and other infrastructure, whilst the unregulated release of the stored water in a flood wave may create even greater losses downstream. A typical example is the Tsatichhu dam, Bhutan, when the Kurichuu River was blocked by $7–12 \times 10^6$ m^3 of material in September 2003 (Dunning et al., 2006). The lake was released by the collapse of the dam almost a year later when a flood peak of 5,900 $m^3 s^{-1}$ was recorded 35 km downstream.

Another more recent example is the deadly landslide that occurred at the Sunkoshi River, in the Sindhupalchok District, Nepal, on 2 August 2014. Heavy rainfall led to a landslide that blocked the Sunkoshi River to form an artificial lake (47 m deep and 400 m long), killing 156 people and blocking over 5 km of the Araniko Highway, the main (and only) artery of goods and people flow to China, causing a huge traffic jam. The landslide had a volume of 5.5 million cubic metres, swept away 24 houses and threatened to flood hundreds of downstream villages. It took the Nepali Army 45 days to dig a canal through the blockage to allow water in the lake to drain, causing damage to houses downstream and threatening to damage the Lamusanghu Hydropower Dam.

In other instances, sliding occurs on a surface with a curved form and will produce a *rotational landslide*. This type of landslide is most commonly seen in comparatively homogeneous materials, such as clays, and in horizontally bedded rock masses. The mobile block rotates as movement occurs, leading to a characteristic set of landforms (Figure 9.4). These landslides tend to be less rapid than translational slides, even though the same processes of loss of cohesion and friction apply. This is because, as long as the block stays intact, the geometry of the movement usually prevents rapid forward acceleration. In some cases, especially where the landslide is formed of weak materials, such as strongly weathered bedrock, the mobile block breaks up to form a flow. This commonly occurs during very intense rainfalls in New Zealand, where it has left a landscape covered in a pattern of scars and depositional features. Intact rotational landslides tend to cause substantial amounts of property damage but few fatalities. About 400 houses in the town of Ventnor on the Isle of Wight, England, are built on an active rotational landslide. Fortunately, the rate of movement is low and loss of life is unlikely, even though the estimated annual cost of damage caused by ground movement exceeds £2 million.

Scientists tend to differentiate between the *causes* of a landslide, which are factors that render a slope susceptible to instability, and the *trigger* of a landslide, which is the final event causing failure. Causes and triggers serve either to decrease

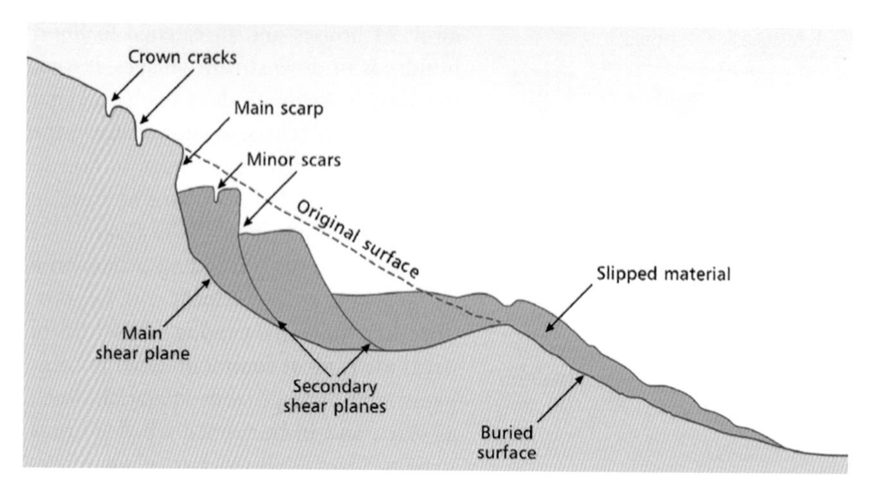

Figure 9.4 The characteristic profile of a rotational landslide. Surface changes, such as the prior opening of crown cracks, can be used as warning signs and lead to evacuation of people downslope.

the strength of the slope materials or to increase the shear forces.

Causes of landslides tend to be long term and include the following:

- *Weathering* of the slope materials may serve to reduce their strength through time until they are no longer strong enough to support the slope during periods of high pore water pressure. Weathering often occurs as a front that moves down through the rock or soil from the surface, but it can also occur preferentially along joints and fractures in the slope or even deep in the slope, due to the circulation of hydrothermal fluids.
- *An increase in slope angle and removal of lateral support.* Landslides in undeveloped terrain are often caused by river erosion at the base of the slope. This increases the angle of the slope and decreases support to the upper layers. Of course, human activity can have the same effect. For example, Jones et al. (1989) described how cutting of a road into the base of a slope in Turkey left 25 metre-high faces in colluvium standing at an angle of 55° but supported only by a 3 metre-high masonry wall. Eventual collapse of this slope led to the 1988 Catak landslide disaster in which 66 people died.
- *Head loading* is a common cause of human-induced slope failure. This occurs when additional weight is placed on a slope, often through the dumping of waste material or the emplacement of fill for house or road construction. This increases the forces driving the landslide and may also increase the slope angle. Head loading can also occur naturally, for instance when a small slope failure flows onto material further downslope, thereby increasing its weight and rendering it more likely to fail.
- *Changes to the water table* can destabilise a slope. Sometimes climate or weather conditions increase the level of the water table, rendering a slope more vulnerable to intense rainfall events. Human activity can produce similar effects. For example, in the case of the Vaiont landslide (Box 9.1 and Figure 9.5) a rise in the water table as the lake

filled led to the destabilisation of the slope. Leaking pipes are a problem in urban areas when small movements along a slope crack drains, water supply pipes or main sewers.

- *Removal of vegetation* either by wildfires or by human activities like logging, overgrazing or construction is important. Trees are particularly good at limiting landslides, partly because of the retaining strength of the roots and partly because of the role that trees play in controlling water movement on slopes. Clear-felling often leads to an increase in landslide activity, although the effects may be delayed for a few years until the roots rot.

Triggers of landslides are necessary for most events and include:

- *Increase in pore water pressure* in the slope materials to the point where shear stress exceeds shear resistance. In most cases, this occurs because of intense storms or prolonged periods of rain. Important relationships between rainfall and landslides over different timescales are illustrated in Figure 9.6. Figure 9.6 (a) shows the average monthly total of rainfall-induced fatal landslides during the years 2003–2009 across South Asia, together with the monthly precipitation. A strong seasonal pattern emerges, reflecting the dominant role of summer monsoon rainfall in the creation of landslides in this region. Figure 9.6 (b) illustrates hourly rainfall totals recorded during a tropical storm on 2 July 2002 that triggered fatal landslides in Chuuk, Federated States of Micronesia. Once again, the timing of the rainfall is closely associated with the triggering of landslides and the resulting fatalities.

Box 9.1 The Vaiont Landslide

The Vaiont disaster was triggered in October 1963 by the filling of a reservoir constructed for hydroelectric power generation. The landslide, which had a volume of approximately 270 million m^3, slipped into the reservoir at a velocity of about 30 m s^{-1} (approx. 110 km h^{-1}), displacing about 30 million m^3 of water (Figure 9.5). This water wave swept over the dam and crashed onto the village below, killing about 2,500 people.

The dam site managers were aware of the landslide and had been monitoring the movement of the slope since 1960. During 1962 and 1963, they deliberately induced movement of the landslide by raising and lowering the lake level with the intention of causing the mass to slide slowly into the lake. Although this would have led to a blockage of that section of the reservoir by the landslide mass, it was believed that the volume of water in the unblocked section would be sufficient to allow the generation of electricity. A bypass tunnel was constructed on the opposite bank so that, when the reservoir was divided into two sections, the level of the lake could still be controlled. Unfortunately, the catastrophic nature of the landslide was not anticipated and these plans never came to fruition.

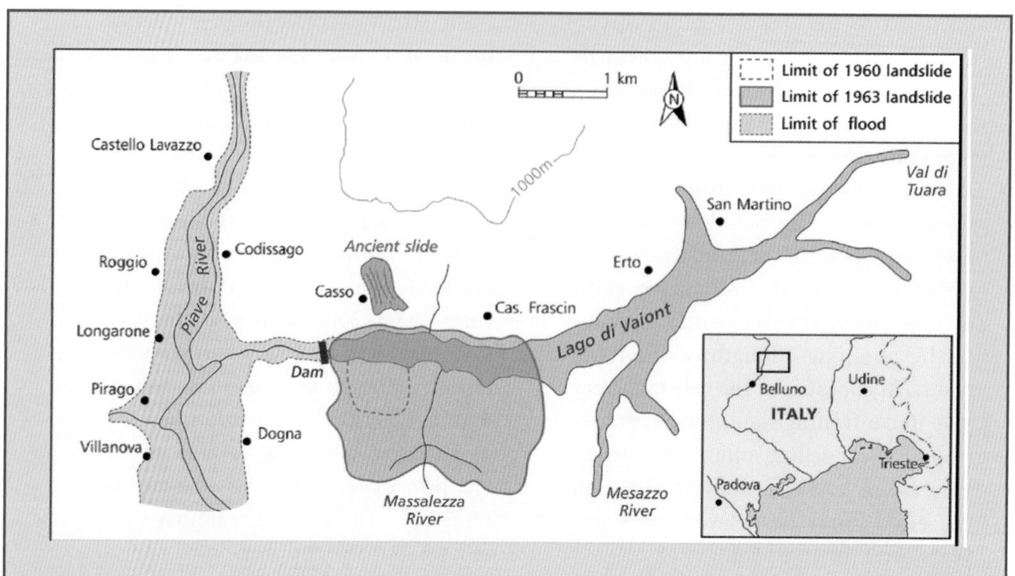

Figure 9.5 A map showing the area of land disturbed in the major Vaiont landslide of 1963 and the downstream area inundated by the subsequent flood wave.

Source: After Petley (2009)

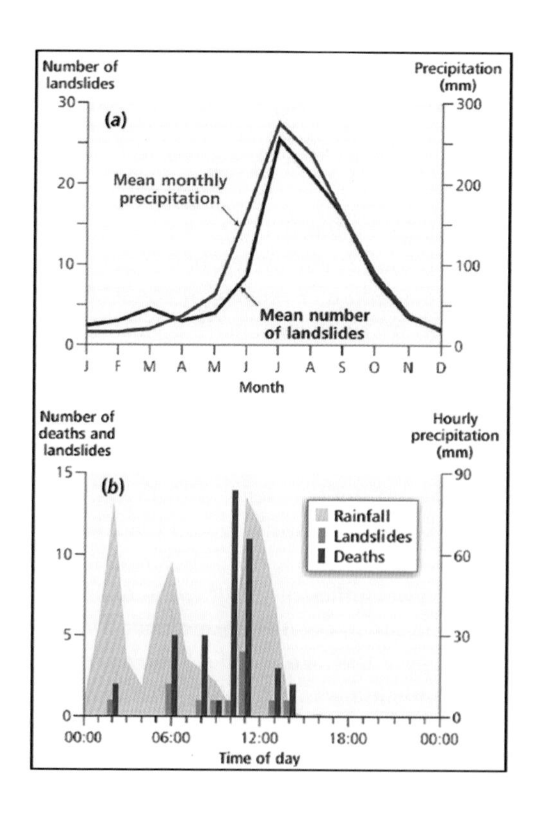

Figure 9.6 Landslide activity in relation to rainfall in the tropics. (a) Average monthly precipitation and the number of rainfall-induced fatal landslides in South Asia 2003–2009. (b) Total hourly rainfall, number of fatalities and the timing of associated landslides, State of Chuuk, 2 July 2002.

Source: (a) After Petley (2010). On the impact of climate change and population growth on the occurrence of fatal landslides in South, East and SE Asia. *Quarterly Journal of Engineering Geology and Hydrogeology* 43(4) (2010): 487–96. Reproduced with permission of the Geological Society. (b) After C. Sanchez et al., Risk factors for mortality during the 2002 landslides in Chuuk, Federated States of Micronesia. *Disasters* 33 (2009): 705–20. Copyright John Wiley and Sons 2009. Reproduced with permission.

- *Earthquake shaking*. During powerful earthquakes, such as the 2005 event in Kashmir, vertical ground accelerations can exceed 1g. This means that landslide material instantaneously becomes weightless, reducing friction on the base to zero. According to an analysis by Keefer (1984), earthquakes of magnitude 4.0 and greater are able to trigger slope failure, and earthquakes with a magnitude greater than about $M_L = 7.0$ are able to generate thousands of slope failures in hilly areas (see Figure 9.7 for the landslide in Colinas, El Salvador, triggered an Mw 7.6 earthquake.
- *Human activity*. In some cases, human activity is the trigger. This is common in quarries, where excavation and blasting can destabilise a slope to the point of failure. Human-induced failures are common in mountain areas where road construction and slope cutting has taken place. This is a cause of mortality amongst road maintenance teams and road users in the Himalayas. For example, in December 2007 in Hubei Province, China, a rock fall triggered by the construction of a road tunnel buried a bus; 33 passengers and two construction workers were killed.

Figure 9.7 The Las Colinas landslide was one of thousands of slope failures triggered by the Mw = 7.6 earthquake that hit El Salvador in January 2001. Ground shaking at this location was amplified by ridge topography at the edge of a steep escarpment. Weak deposits, mainly of volcanic origin, quickly lost strength and a rapid flow slide swept into Santa Tecla city, a residential suburb of San Salvador. Many houses were buried and about 580 lives were lost.

Source: Photo: Ed Harp, USGS

For a small group of cases, it is impossible to determine the final landslide trigger. Investigations of the 1991 Mount Cook landslide in New Zealand, in which the country's highest mountain lost some 10 m from its peak, failed to reveal any trigger event (McSaveney, 2002). The failure may have occurred as a result of a time-dependent process or a triggering mechanism not yet understood. The 1979 Abbotsford landslide at Dunedin in New Zealand is a good example of how physical and human factors may combine to increase landslide risk and thus prevent attribution to a single cause. This urban landslide, which destroyed 69 houses across an area of 18 ha at a cost of NZ$10–13 million, resulted from at least three contributory factors (Hancox, 2008):

- Unfavourable geology – the site was on a 7° dip slope composed of Tertiary sediments with very weak clay layers.
- A rise in the water table – this was due to increased rainfall over the preceding 10 years and leakage from a Dunedin City Council water main at the top of the slope.
- Quarry activity – the excavation of some 300,000 m^3 of sand from a quarry at the toe of the slope.

4. Debris Flows

Debris flows are movements of fluidised soil and rock fragments acting as a viscous mass that can range from a thick soup texture to wet concrete. They occur when loose materials become saturated and start to behave as a fluid rather than a solid. These flows can be slow or rapid depending on the conditions and can vary from small to moderate scales (transporting 100–1000 m^3 (down a valley) or large flows that can transport several km^3 of materials, typically from failures of a flank of a mountain, volcanic mudflows (lahars) and debris flows. Debris flows have volumetric sediment concentrations exceeding about 40% to 50%, and the remainder of a flow's volume consists of water. Debris includes sediment grains with diverse shapes and sizes, commonly ranging from microscopic clay particles to great boulders.

Debris flows can be triggered by intense rainfall or snowmelt, by dam-break or glacial outburst floods, or by landsliding (which may be associated with intense rain or earthquakes). The conditions required for debris flow initiation include:

- slopes steeper than 25 degrees;
- abundant loose sediment, soil, or weathered rock; and
- sufficient water to bring this loose material to a state of almost complete saturation.

Debris flows can be more frequent following forest and brush fires, and they pose a significant hazard in many steep, mountainous area. In Japan a large debris flow or landslide is called yamatsunami (山津波), literally a mountain tsunami.

Flows most commonly occur in very heavy rain but, in most cases, the flow actually starts as a different type of landslide. For example, in tropical environments, the heavy precipitation associated with the passage of tropical cyclones, when rainfall totals often exceed 600 mm day^{-1} and intensities reach 100 mm hr^{-1} (Thomas, 1994), can trigger large numbers of small, shallow translational landslides in the soil that mantles the hill slopes. In some cases, the initial movement of the landslide allows the saturated soil mass to break up, changing the movement into a debris flow. Debris flows often accelerate rapidly down the slope, disrupting and entraining soil and other material. In this way, landslides of just a few cubic metres can turn into debris flows with a volume of tens of thousands of cubic metres and cause high levels of loss.

Debris flows follow existing stream channels. Consequently, the areas likely to be affected may be predictable. However, steep rock gullies provide little resistance to the flow and allow the mass to reach high rates of movement. Debris flows are accelerated downhill by gravity and tend to follow steep mountain channels that fall onto alluvial fans or floodplains. The front or 'head' of a debris-flow surge often contains an abundance of coarse material such as boulders and logs that cause significant friction. Problems can arise when the flow starts to spread out; when this occurs in inhabited areas, the amount of damage can be large because the density of a debris flow is greater than that of a flood (typically by a factor of 1.5 to 2.0) and the rate of movement is rapid. As a result, debris flows claim the majority of lives lost in landslides.

Many tropical cities, such as Rio de Janeiro and Hong Kong, are at risk from both landslides and debris flows. Jones (1973) documented the effects of exceptionally heavy rainfall, often linked to stationary cold fronts, around Rio de Janeiro, Brazil. In 1966, landslides produced over 300,000 m^3 of debris in the streets of Rio and more than 1,000 people died when many slopes, over-steepened for building construction, failed. One year later, further storms hit Brazil and mudflows caused a further 1,700 deaths and some disruption of the power supply for Rio. In February 1988 further debris flows took at least 200 lives and made 20,000 people homeless. Most victims in this area live in unplanned squatter settlements on deforested hillsides (Smyth & Royle, 2000). On 15 February 2022, following intense rainfall in Petrópolis (Rio de Janeiro), mudslides and flooding destroyed parts of the city, killing 231 people. Rural environments are not immune; most of the 30,000 fatalities in the 1999 Venezuela landslides resulted from debris flows down the river valleys (IFRC, 2002).

Other geological flows that can be described as debris flows are typically given more specific names such as lahars and jökulhlaups (as discussed in Chapter 7).

5. Snow Avalanches

Just like landslides, an avalanche risk exists when the shear stress exceeds the shear strength of the material, in this case a mass of snow located on a slope. There are three key conditions that drive unstable conditions for snow avalanches: slope steepness, stability of the snowpack and the weather. The strength of a snowpack is related to its density and temperature. When compared to other solids, snow layers have the ability to undergo large changes in density. Thus, a layer deposited with an original density of 100 kg m^{-3} can densify to 400 kg m^{-3} during a winter, largely due to compression by the weight of overlying snow, pressure melting and the re-crystallisation of the ice. This densification increases the strength of the snow. However, the shear strength decreases as the temperature warms towards 0°C. As the temperature rises further, such that liquid meltwater is present in the pack, the risk of movement within the snow blanket increases.

Most snow loading on slopes occurs slowly. This gives the snowpack an opportunity to adjust by internal deformation, because of its plastic nature, without any damaging failure. The most important triggers of pack failure tend to be heavy snowfall, rain, thaw or some artificial increase in dynamic loading, such as skiers traversing the surface (Box 9.2). For failure to occur in a hazardous snowpack, the slope must also be sufficiently steep to allow the snow to slide. Therefore, avalanche frequency is related to slope angle, with most events occurring on intermediate slope gradients of between 30° and 45°. Angles below 20° are generally too low for sliding to occur and most slopes above 60° rarely accumulate sufficient snow to pose a major threat. The steepest angle at which snow is stable is its angle of repose.

Most avalanches start at fracture points in the snow blanket where there is high tensile stress, such as a break of ground slope, at an overhanging cornice or where the snow fails to bond to another surface, such as a rock outcrop. Avalanches are most likely to run during, or immediately following, a heavy snowstorm. This is because the weight of new snow places extra stress on the snowpack, especially if it does not bond well to the pre-existing surface. Snow is a good thermal insulator, and diurnal temperature changes may have little influence on pack stability. However, a warm front bringing in milder weather can raise temperatures sufficiently to induce melting in the surface layers.

Three distinct sections of an avalanche track can usually be identified. These are the *starting zone*, where the snow initially breaks away, the *track* or path followed and the *run-out zone*, where the flowing snow decelerates and stops. Because avalanches tend to recur at the same sites, the risk can often be detected from the identification of previous avalanche paths in the landscape. Clues in the terrain include breaks of slope, eroded channels on the hillsides and damaged vegetation. In heavily forested mountains, avalanche paths can be identified by the age and species of trees and by sharp 'trim-lines' that separate the mature, undisturbed forest from the cleared slope. Once the hazard location is determined, a range of potential adjustments is available, some of which are shared with landslide hazard mitigation.

Two types of snow avalanches can be classified. Loose-snow avalanches typically start at a point and widen as they move downslope. Slab avalanches are coherent blocks of snow and ice that move downslope, which is more dangerous, resulting in millions of tons of snow that can travel up to 100 km per hour (see Box 9.2 and Figure 9.7). Avalanches tend to move down chutes that have been created by prior avalanches.

Avalanche movement depends on the type of snow and the terrain. Most avalanches start with a gliding motion but then rapidly accelerate on slopes greater than 30°. It is common to recognise three types of avalanche motion:

- Powder avalanches are the most hazardous and are formed of an aerosol of fine, diffused snow behaving like a body of dense gas. They flow in deep channels but are not influenced by obstacles in their path. The speed of a powder avalanche is approximately equal to the prevailing wind speed but, being of much greater density than air, the avalanche is more destructive than windstorms. At the leading edge its typical speed is 20–70 m s^{-1} and victims often die by inhaling snow particles.
- Dry flowing avalanches are formed of dry snow travelling over steep or irregular terrain with particles ranging in size from powder grains to blocks of up to 0.2 m diameter. These avalanches follow well-defined surface channels, such as gullies, but are not greatly influenced by terrain irregularities. Typical speeds at the leading edge range from 15–60 m

Box 9.2 How Snow Avalanches Start

Snow avalanches result from two different types of snowpack failure:

- Loose-snow avalanches occur in snow where inter-granular bonding is very weak, thus producing behaviour rather like dry sand (Figure 9.8a). Failure begins near the snow surface when a small amount of snow, usually less than 1 m^3, slips out of place and starts to move down the slope. The sliding snow spreads to produce an elongated, inverted V-shaped scar.
- Slab avalanches occur where a strongly cohesive layer of snow breaks away from a weaker underlying layer, to leave a sharp fracture line or crown (Figure 9.8b). Rain or high temperatures, followed by re-freezing, create ice-crusts which may provide a source of instability when buried by subsequent snowfalls. The fracture often takes place where the underlying topography produces some upward deformation of the snow surface, leading to high tensile stress, and the associated surface cracking of the slab layer. The initial slab which breaks away may be up to 10,000 m^2 in area and up to 10 m in thickness. Such large slabs are very dangerous because when a slab breaks loose, it can bring down 100 times the initial volume of snow.

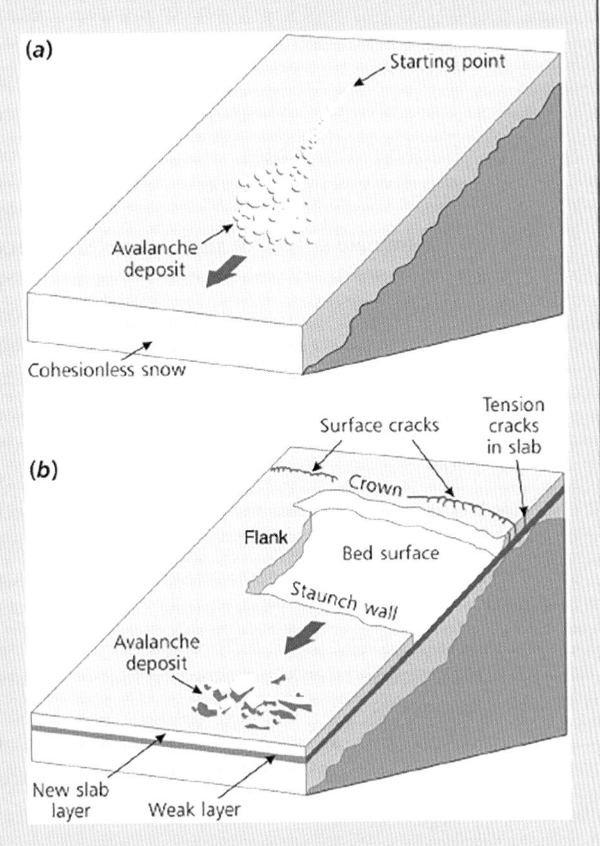

Figure 9.8 The two most common types of snow-slope failure: (a) loose-snow avalanche; (b) slab avalanche. Slab avalanches normally create a greater hazard because of the larger volume of snow released.

s^{-1} but can reach speeds up to 120 m s^{-1} whilst descending through free air.

- Wet flowing avalanches occur mainly in the spring season and are composed of wet snow formed of rounded particles (0.1 to several metres in diameter) or a mass of sludge. Wet snow tends to flow in stream channels and is easily deflected by small terrain irregularities. Flowing wet snow has a high mean density (300–400 kg m^{-3} compared to 50–150 kg m^{-3} for dry flows) and can achieve considerable erosion of its track despite reaching speeds of only 5–30 m s^{-1}.

B. IMPACTS AND CONSEQUENCES

In the past, the losses associated with mass movements have probably been underestimated. This is partly because many events occur in rural mountainous environments remote from population centres. Also, a substantial proportion of the cumulative loss results from numerous small mass movements, rather than large incidents. The process is often attributed to the trigger event, such as an earthquake or a rainstorm, rather than to the mass movement itself (Jones, 1992). In recent years, mass movements have attracted more attention. Typically, mass-movement investigations are conducted either on an intensive site-specific basis for risk assessment on individual slopes or as a regional-scale evaluation to identify trends and hazardous zones relevant to wider land use planning.

Estimates of loss vary widely. According to the EM-DAT database, about 4,517 people died as a result of mass movements. However, the Durham Fatal Landslide Database shows that in total 55,997 people were killed in 4,862 distinct landslide events between January 2004 and December 2016 (Froude & Petley, 2018). Most of the deaths were in geographically distinct regions: the Pacific Rim, Central America and the Caribbean, mainland China, Southeast Asia and along the southern edge of the Himalayan Arc. These areas have hilly or mountainous terrain, active tectonic processes (including uplift and earthquakes), intense rainfall events (associated with tropical cyclones, El Niño/La Niña events or monsoon weather patterns) and comparatively large populations of poor people who live in unsafe locations. The Durham Fatal Landslide Database also establishes that landslide occurrence triggered by human activity

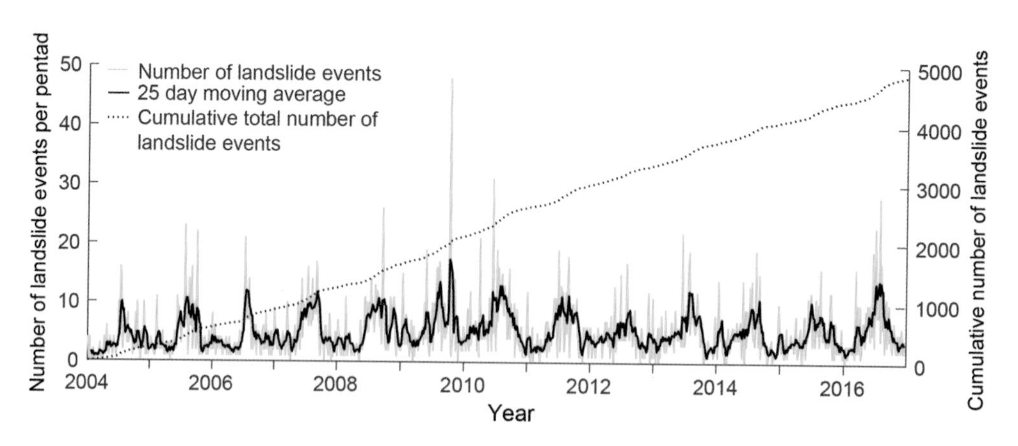

Figure 9.9 The occurrence of non-seismically triggered landslide events from 2004 to 2016, and cumulative total of recorded events. The data are arranged by pentads (5-day bins), starting on 1 January each year; thus the first pentad includes records for 1–5 January, and there are a total 73 pentads. A simple 25-day moving average is shown.

Source: Froude & Petley (2018, p. 2166)

is increasing, mostly due to construction, illegal mining and hill cutting. Therefore, whilst there are highly active years of landslides that coincide with regional rainfall driven by climate anomalies, human disturbance appears to be more detrimental to future landslide incidence than climate.

Mass movements kill in two chief ways. Some victims die through fatal trauma (collisions with rocks, trees or ice slabs), but the main cause of death is suffocation. Following a tropical storm over the volcanic islands of Chuuk State, Micronesia, when over 250 landslides and debris flows killed 43 people, 90% of the recorded deaths were attributed to suffocation. This is similar to the pattern of mortality associated with snow avalanches. Humans die quickly if buried beneath snow, soil or rock because they are unable to breathe under a depth of such material greater than 30 cm. Even shallow burial can be fatal. In addition, victims are at risk from hypothermia in the case of avalanches and from drowning in the case of water-rich debris slides. Most of the survivors from mass movement disasters are rescued quickly and are often saved by some small-scale physical protection, perhaps the shelter offered by a building or a vehicle.

Almost 90% of people who are trapped by an earthquake, and survive, are rescued within the first 24 hours. Information from the USA suggests that the survival rate for avalanche victims buried in the snow declines even more rapidly with time (Figure 9.10a). After only 15 minutes, almost 90% of avalanche victims are still alive, but this falls to about 30% after 60 minutes and to fewer than 5% after three hours. The time taken in locating people buried deeply in the snow, and then removing the snow cover, also explains why survival rates for avalanche victims also decline with the depth of snow (Figure 9.10b). At depths in excess of 1.2 m, the survival rate drops to 20% or less. *Avalanche beacons*, sometimes called transceivers, are standard equipment for some ski-slope patrollers because they represent the best means of finding a buried victim, as long as a device is also carried by the victim. Collapsible ski-pole probes are used to locate buried bodies.

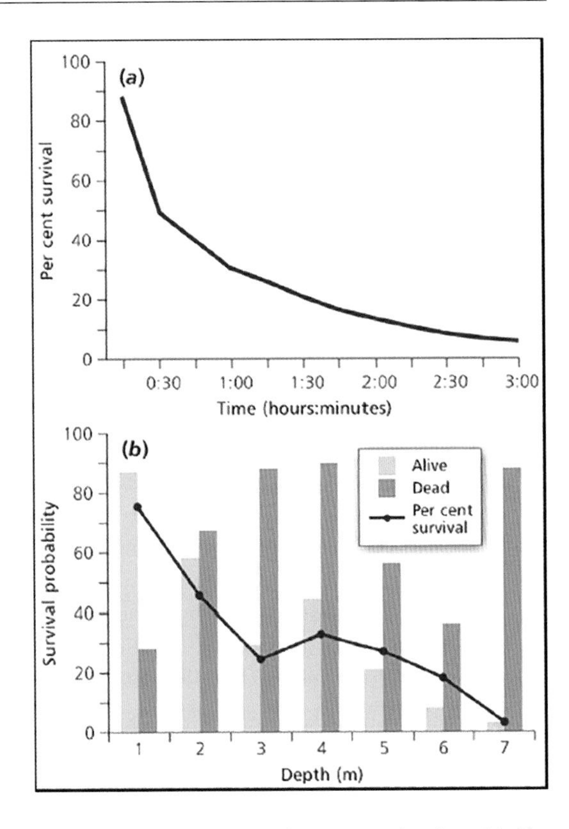

Figure 9.10 Survival after an avalanche. (a) The percentage chance of survival against time for US avalanche victims buried in the snow. After 15 minutes the survival rate is almost 90%, but it falls to less than 30% after 1 hour. (b) The probability of survival in US avalanches 1950–2006 in relation to depth of burial.

Source: After Colorado Avalanche Information Center at https://avalanche.state.co.us/ (accessed 31 May 2011). Reproduced with permission of Colorado Avalanche Information Center.

The need for a locally based rapid-response search-and-rescue capability is crucial for all mass movement hazards.

In Canada, local offices of Parks Canada regularly monitor the stability of relevant snowpacks and issue avalanche risk warnings in collaboration with the British Columbia Ministry of Highways. The provincial government coordinates most local search and rescue in the Canadian Rockies, and the Royal Canadian Mounted Police have people and dogs trained for avalanche rescue work. Specialised weather forecasts are available from the Atmospheric Environment Service, and avalanche awareness is promoted through bodies like the Avalanche Canada (now known as Avalanche Canada) with technical courses, films and videos. Although these methods may increase the general knowledge of avalanche threats, Butler (1997) found that local residents were often unaware of the danger at a given time. A full avalanche search is a complex operation, but the increasing use of avalanche airbags and digital transceivers by winter sports enthusiasts will reduce the chances of burial and increase the chances of being found, respectively. Such advanced arrangements are rare for landslides.

Mortality from mass movements is comparatively low in more developed countries. For example, Italy has the highest fatality rate from slope failures in Europe. Deaths average 60 per year and around 80% occur in fast-moving events (Guzzetti, 2000). However, the economic losses are high. In the USA, Canada and India the estimated costs of landslides exceed US$1 billion per year (Schuster & Highland, 2001), whilst the direct damage caused by landslides between 1945 and 1990 exceeded $15 billion in Italy. Indirect losses are poorly quantified but include damage to transport links, electricity transmission systems and gas and water pipelines, flooding due to landslide dams across rivers, impaired agricultural and industrial production, loss of trade and a reduction in property values.

In 2014 the Oso landslide in Washington, USA, occurred on 22 March killing 43 people and destroying 49 homes and other structures, making it the deadliest landslide in American history (Wartman et al., 2016). The event is described as a complex slide with a single large slump block that blocked the Stillaguamish River for ten days, causing flooding upstream and blocking State Route 530 (see Figure 9.11). It was determined that the forestry practices over the last 50 years had contributed to the instability of the area, and a $60 million settlement was paid in 2016 from the State of Washington and the forestry company responsible for the logging to the Oso Landslide survivors and victims funds.

It is generally accepted that mass movement risks and losses have increased over time. The reasons for this trend are unclear, mostly because landslide disasters are complex events with physical triggers, such as earthquake activity and high rainfall, combined with societal causes such as rapid economic development, population growth and urbanisation.

In China, about 80% of major landslides occur on the tectonically active flank of the Tibet Plateau. Since 1980, there has been a general increase in landslide activity, mainly attributed to more construction work and a shift in climatic conditions, causing about 1,000 fatalities per year (Runqiu, 2009). The expansion of many cities in developing nations has forced more people to live in unregulated *barrio* settlements located on steep slopes. Hong Kong suffered a major increase in landslide-related fatalities during the 1970s, primarily due to the growth of illegal communities of immigrants on unstable slopes, although government action

Figure 9.11 Oblique aerial photograph of the 2014 landslide in northwest Washington. This image shows the entire extent of the landslide source area and path. This event is commonly named the 'Oso Landslide' in many official reports. It is also referred to as the 'SR530 Landslide,' as named by Snohomish County and Washington State.

Source: Mark Reid, USGS

to move these people onto safer terrain reduced the death toll. Petley et al. (2007) found considerable variation in landslide fatalities from year to year in Nepal. A strong control was exercised by heavy rains linked to the strength of the summer monsoon, but an upward trend outside the monsoon cycle was tentatively attributed to land use disturbances arising from rural road construction.

Snow avalanches tend to occur in Arctic and temperate regions whenever snow is deposited on slopes steeper than about 20°. Each year, avalanches claim more than 150 lives worldwide. The USA annually suffers up to 10,000 potentially damaging avalanches, although only about 1% affect life or property. The worst avalanche disaster in the USA occurred in 1910 in the Cascade Range, Washington, when three snowbound trains were swept into a canyon with the loss of 118 lives. In February 1999 two avalanches struck the towns of Galtür and Valzur in Austria, leading to the loss of 38 lives. The risk posed by avalanches is more severe in Europe than in North America because the population density is higher in the Alps than in the Rockies.

These examples highlight the ways in which the development of mountain areas for winter recreation over the last 50 years has increased avalanche risk. About 70% of avalanche fatalities in highly developed countries are associated with the voluntary activities of ski touring and mountain climbing. Many countries have recorded an increase in the number of fatalities since the early 1950s, as illustrated for the US in Figure 9.12. Most victims are male skiers or backpackers around 30 years of age and fatalities cluster in the late winter months, when snowfall accumulations are at a maximum.

Apart from the growth of winter recreation, the increase of transportation routes through mountain areas has led to more avalanche-related mortality. The construction of roads and railways leads to the removal of mature timber that, if left intact, would help to stabilise the snow cover and protect people and infrastructure in the valley bottom. For example, the Trans-Canada Highway runs under nearly 100 avalanche tracks in a 145 km section near Rogers Pass, and at least one vehicle is under an avalanche path at any given time. Much less is known about avalanche hazards in mountainous areas such as the Himalayas, but the threats are serious. In the Kaghan Valley, Pakistan, avalanches pose a threat to local residents over-wintering on the valley floor, and 29 people were killed in single event in the winter of 1991–92 (De Scally & Gardner, 1994), whilst in December 2005, 24 people were killed by a single avalanche in the Northwest Frontier Province of Pakistan.

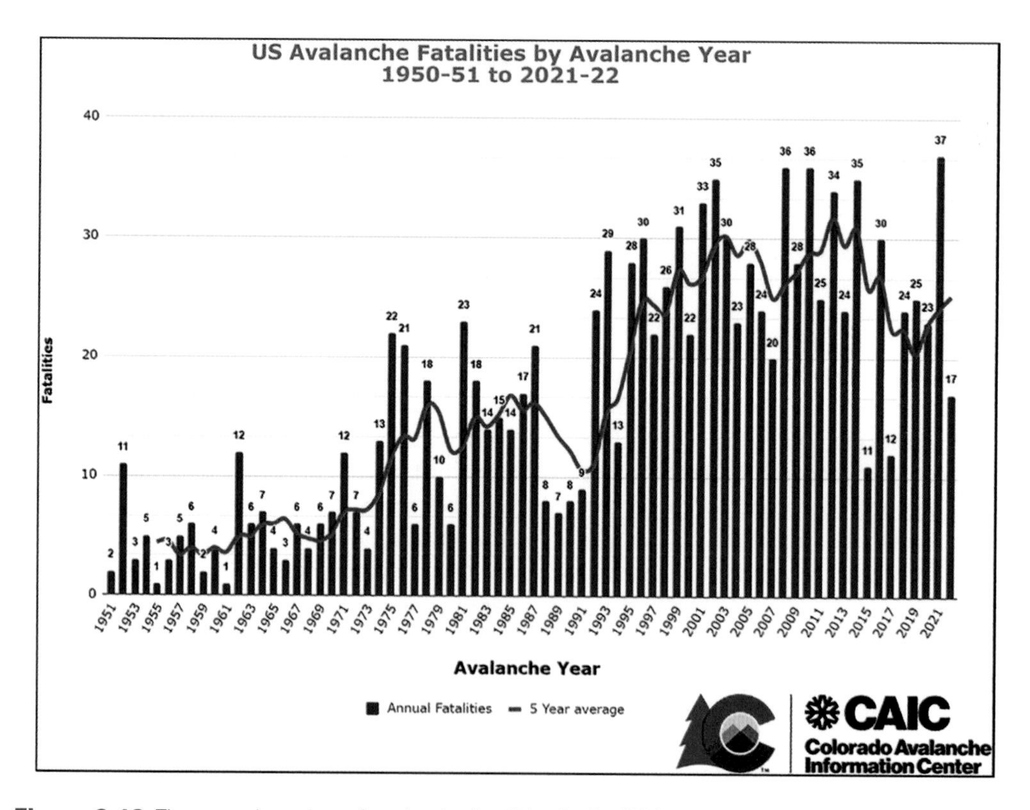

Figure 9.12 The annual number of avalanche fatalities in the USA over the 1950/51 to 2020/21 winter seasons. The rising deaths reflect growing participation in winter sports activities. Over half the victims were mountain climbers or backcountry skiers.

Source: After Colorado Avalanche Information Center at https://avalanche.state.co.us/accidents/statistics-and-reporting (accessed 8 August 2022). With permission of the Colorado Avalanche Information Center.

Snow avalanches can exert high external loadings on built structures. Using reasonable estimates for speed and density, it can be shown that maximum direct impact pressures should be in the range of 5–50 t/m^2, although some pressures have exceeded 100 t/m^2 (Perla & Martinelli, 1976). Table 9.1 provides a guide to avalanche impact pressures and the likely damage to human-made structures. The Galtür disaster in Austria, which occurred in February 1999, was the worst in the European Alps for 30 years. In this event, 31 people were killed and seven modern buildings were demolished in a winter sports village previously thought to be located in a low-hazard zone. A series of storms earlier in the winter deposited nearly 4 m of snow in the starting zone, a previously unrecorded depth. By the time the highest level of avalanche warnings was issued, the snow mass in the starting zone had grown to approximately 170,000 tonnes. During its track down the mountain, at an estimated speed in excess of 80 m s^{-1}, the avalanche picked up sufficient snow to double the original mass. By the time it reached the village the leading powder wave was over 100 m high, with sufficient energy to cross the valley floor and reach the village.

Table 9.1 Relationships between impact pressure and the potential damage from snow avalanches

Impact pressure (tonnes m^2)	Potential damage
0.1	Break windows
0.5	Push in doors
3.0	Destroy wood-frame houses
10.0	Uproot mature trees
100.0	Move reinforced concrete structures

Source: After Perla and Martinelli (1976)

C. RISK ASSESSMENT AND MANAGEMENT

There are a number of precursory signs that can be detected to indicate a potential landslide that will vary depending on the geological, geomorphological, climatic and human contexts. Signs such as ground cracks, slow movement, the behaviours of seepages and springs following rainfall and the tilting of trees and other vegetation are all signs of subsurface movements. In built environments cracking in walls or tilting infrastructure or window frames with gaps are all potential precursors. Monitoring mass movements is vital to help detect movement and monitor their changes over time to provide warnings to vulnerable populations and infrastructure. Once monitoring is in place it is possible to develop warning systems to give people as much notice as possible, but with rapid mass movements this time may not be enough to get to safety. Therefore, engineering solutions are vital to help protect as best as possible against the forces of the mass movements. Community awareness and action is key, particularly in developing nations where expensive engineering solutions are not viable, alongside more effective land use planning, to try to prevent any damage.

1. Warnings

Considerable effort has gone into the development of warning systems for both avalanches and landslides as rapid mass movements (RMM) as they pose a substantial risk to people and infrastructure. This has been done by developing various monitoring tools using a range of technologies as summarised in Table 9.2. Numerous types of instrumentation and monitoring techniques are available, but many are not widely used for early warning systems

Table 9.2 Technologies typically used in current EWSs and proposed for future EWSs

Observed parameter		**Technology**	**Type of EWS**	**References**
Precipitation	Sum, intensity	Rain gauge Precipitation radar	All Forecasting systems	Panziera et al. (2011)
Snow cover	Depth Wetness		Forecasting systems Forecasting systems	
Soil moisture	Water content Water suction/ pressure Groundwater table	TDR Tensiometer Piezometer	Forecasting systems Forecasting systems Forecasting systems	
Rock/soil surface	Precursor of failure Displacement		Warning systems Alarm systems Warning systems	Michlmayr et al. (2013) Caduff et al. (2015) Wegmuller et al. (2013)
Triggered mass movement	Vibration Flow surface height Flow characteristics	Geophone Seismometer Radar Video	Alarm systems Alarm systems Alarm systems Alarm and warning systems	

Source: Stähli et al. (2015, p. 908)

(EWSs) in low- and middle-income countries due to their complexity and high cost. Consequently, new technologies such as Community Slope SAFE (Sensors for Acoustic Failure Early-warning) using acoustic emissions approaches can be used that are reliable and more cost-efficient to enable communities to monitor and self-warn (Dixon et al., 2018).

Site-Specific Warnings Based Upon Surface Movement

Most rainfall-induced landslides are preceded by a period of slow movement, called *creep*. This phenomenon has been employed in some warning schemes, and Hungr et al. (2005) described how site-specific changes in earth movement have been successfully used to predict the time of final failure. Slope movement is monitored with field instruments such as inclinometers, tiltmeters, theodolites and electronic distance recorders, supplemented by Global Positioning Systems (GPS) and radar satellite (InSAR) techniques (Casagli et al., 2010, and Box 9.3). Data is sent in real time to software that can compare the observed movement against predetermined trigger factors, often based on the rate of movement or the acceleration, so that a warning can be issued.

In the past, traffic routes were protected from avalanches by avalanche sheds, but now avalanche warning systems are available for key transportation corridors. These systems detect the movement of an avalanche high on a slope before operating a set of barriers or traffic controls that close the road or railway. Surface movement of the snow is detected using tripwires, radar, geophones or wire-mounted tiltmeters. Although these systems are expensive, they have been deployed in both North America and Europe. Evidence suggests that the site-specific nature of avalanche paths, plus the economic importance of the traffic route, are crucial factors in optimising such schemes (Rheinberger et al., 2009). Figure 9.13 depicts an avalanche management scheme operated along a 14 km corridor of Idaho State Highway 21 that crosses 56 avalanche tracks (Rice et al., 2002). Automatic avalanche detectors, using tilt switches, are suspended from a cableway near to the road over the most active avalanche track. When these switches exceed a preset threshold, the system can initiate a call by radio telemetry to alert the highway authority of the avalanche and can advise road users of the blockage immediately, either by activating flashing warning signs or by automatically closing snow gates at each end of the corridor.

General Warnings Based on Weather Conditions

For landslides, historical precipitation and landslide records are combined to identify the rainfall level (intensity and duration) at which past landslides have been initiated over the area in question. Such regional alerts are mainly relevant to the recognition of precipitation thresholds that trigger shallow landslides (Brunetti et al., 2010). For example, in 1986 the United States Geological Survey trialled an early landslide warning system for the San Francisco Bay region, based upon six-hour forecasts of rainfall duration and intensity (Keefer

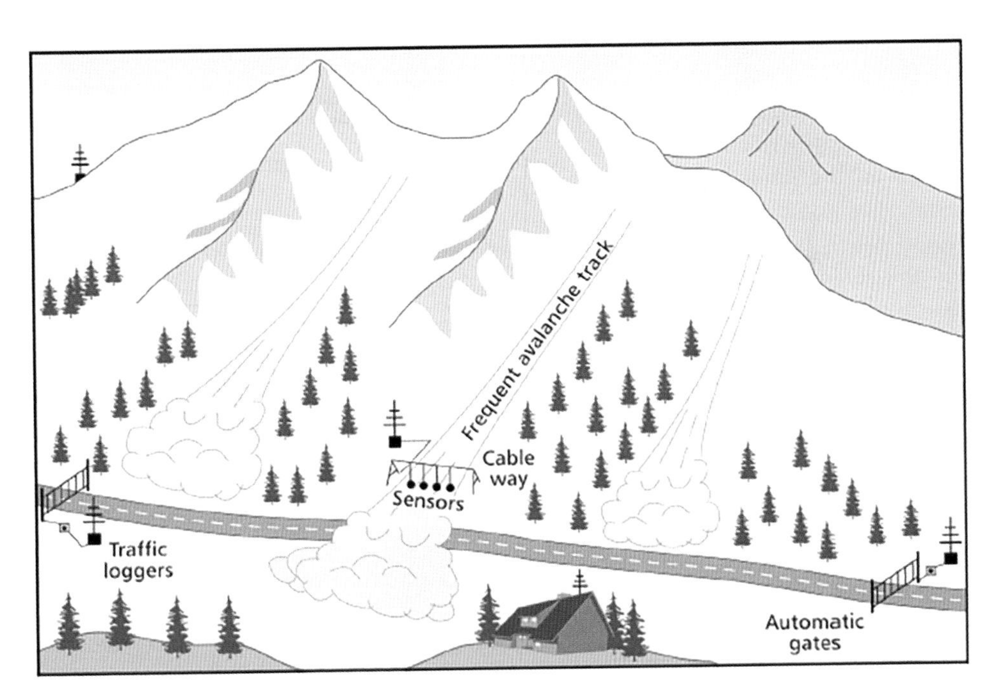

Figure 9.13 Avalanche hazard management on mountain roads in the western USA. Avalanches reaching the highway are detected by sensors suspended from a cableway above the track. The relevant stretch of road can then be closed for snow clearance by the highway agency alerted by telemetry.

Source: Adapted from Rice et al. (2002). Reprinted from *Cold Regions Science and Technology* 34, R. Rice Jr et al., Avalanche hazard reduction for transportation corridors using real-time detection and alarms, 31–42. Copyright (2002), with permission from Elsevier.

et al., 1987), but the maintenance costs proved to be prohibitive.

One of the most effective systems is that operated in Hong Kong. This uses a network of 110 rain gauges scattered across the territory, together with analysis of Doppler radar data, to issue warnings of the occurrence of landslides. The warning threshold is based on the accumulated rainfall over the previous 21 hours plus the forecast rainfall for the next three hours. Using GIS technology, the forecast rainfall is used to calculate an estimated number of landslides across the territory. This number is the basis for issuing a warning, which is released to the media, along with advice to the public to maximise safety. But the physical relationships are not entirely straightforward. Antecedent soil moisture conditions are an important

Box 9.3 The Tessina Landslide Warning System

The Tessina landslide is a 3 km long earth flow located in the Dolomite Mountains of northern Italy (Figure 9.14). It has been active since 1960 but the rate of movement, and the volume of material involved, increased in 1992 (Petley, 2009). Given that the village of Funes was located on the margin of the landslide, there was a risk that a substantial movement would cause the landslide to overrun the settlement, creating casualties and large economic losses. In response, the Italian government research agency CNT-IRPI designed and implemented a landslide warning system (Menorri, 1994).

The warning system consisted of two elements:

- In the source area of the landslide, 13 survey prisms (used to reflect an electronic distance measurement (EDM) beam to measure changes in distances) were installed on the landslide. On the margin of the active movement, in stable ground, a robot theodolite was installed, powered

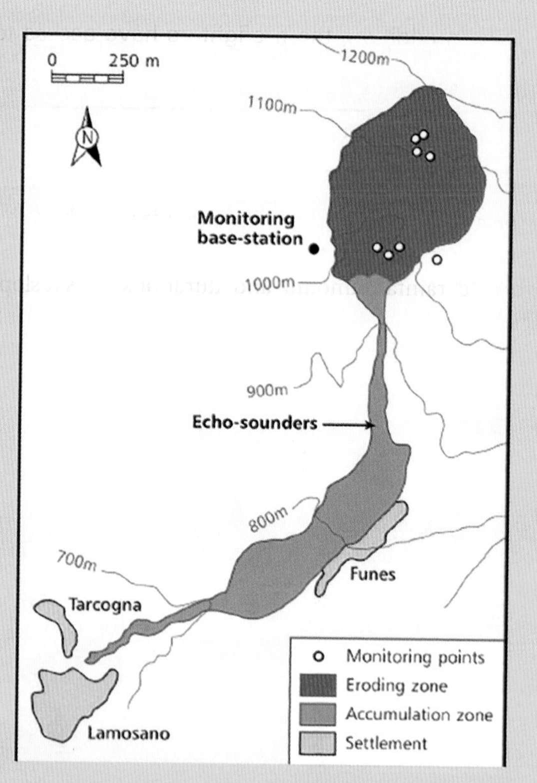

Figure 9.14 A map showing the area of the Tessina landslide threat in the Dolomites of northern Italy. The warning system has protected the town of Funes for many years.

Source: After Petley (2009)

by solar cells. Every 30 minutes this instrument measures the location of each of the prisms. The data are then stored on a computer, which also determines the level of movement. Located alongside the prisms are two wire extenometers that also measure the displacement of the landslide, again feeding data to the central computer unit.

- About 100 m above the village, two tiltmeters were installed approximately 2 m above the landslide. These consist of 2 m long steel bars containing a device to measure the angle at which the bar is hanging. If a flow were to come down the slope, the bars would tilt. The instruments are designed such that if the bar is tilted at more than 20° for over 20 seconds, the central computer would be alerted. A backup echometer located on one of the wires provides an indication of any rapid height change on the surface of the flow that might indicate a slide event.

The central computer is programmed to compare all information against preset thresholds. If these thresholds are exceeded, an alarm is sounded in the local fire station. In addition, the firefighters have access to three video cameras situated at key locations that offer an additional opportunity to verify the indicated movements.

factor, and threshold values vary considerably with the size of the area. For example, threshold rainfall amount and durations vary by three orders of magnitude over the USA and by over one order of magnitude across smaller areas such as a county (Baum & Godt, 2010).

Avalanche warning systems using forecasts and predictions have existed for many years. Forecasts are used in the day-to-day management of winter sports facilities, whilst predictions aid long-term land zoning. Avalanche forecasting involves the testing of the stability of snow, with an emphasis on the detection of weak layers. The results are then evaluated in conjunction with weather forecast information. Regional avalanche schemes are often computer aided. The method introduced in Switzerland in 1996 relies on model calculations of the snowpack, inputs from about 60 weather stations and a GIS-based mapping system of avalanche tracks to provide daily forecasts for areas of about 3,000 km^2 (Brabec et al., 2001). In conditions of severe risk, it is normal practice to clear ski slopes and to restrict traffic on dangerous sections of highway or railway track. Today the working group of the European Avalanche Warning Services consists of 29 avalanche warning services from 16 countries. The aim is to develop common standards for the avalanche warning services in Europe and support the avalanche warning services in place.

2. Mitigative and Engineering Solutions

Landslides

The design and construction of measures to prevent slope failure is a routine task within geotechnical engineering. For example, within the 1,100 km^2 area of Hong Kong, over 57,000 slopes have been engineered to prevent failure. Similarly, the railway agency in the UK has to maintain over 16,000 km of earthworks designed to prevent slope failures (Petley, 2009). Methods of slope protection are well developed and include the following (and as illustrated in Figure 9.15):

- *Drainage*. As slope failures are generally linked to the presence of high water pressures in a slope, drainage is a key tool for improving stability. The aim is either to prevent water from entering a critical area of slope by installing gravel-filled trench drains around that area or to remove water from within a slope by installing horizontal drains. Simple methods can work well. Holcombe and Anderson (2010) reported the success of a community-based, low-cost scheme to improve slope stability in the Eastern Caribbean. Following the construction of a network of open drains and a 1:100-year rainfall event, no landslides were recorded on previously unstable slopes occupied by densely populated urban communities. Problems can arise through a lack of maintenance. Drains can become blocked with fine particles or even by animals using them as burrows. In addition, small amounts of movement in a slope can cause drains to become cracked or broken and so leak water into a slope at critical locations.
- *Regrading*. In many cases, the landslide threat can be minimised by reducing the overall slope angle. This can be achieved by excavating the upper parts of the slope or by placing material at the toe, an approach used during road construction in upland areas. In some cases, good results can be achieved by removing the natural slope soil or rock and replacing it with a lighter material.
- *Supporting structures*. Supporting structures such as piles, buttresses and

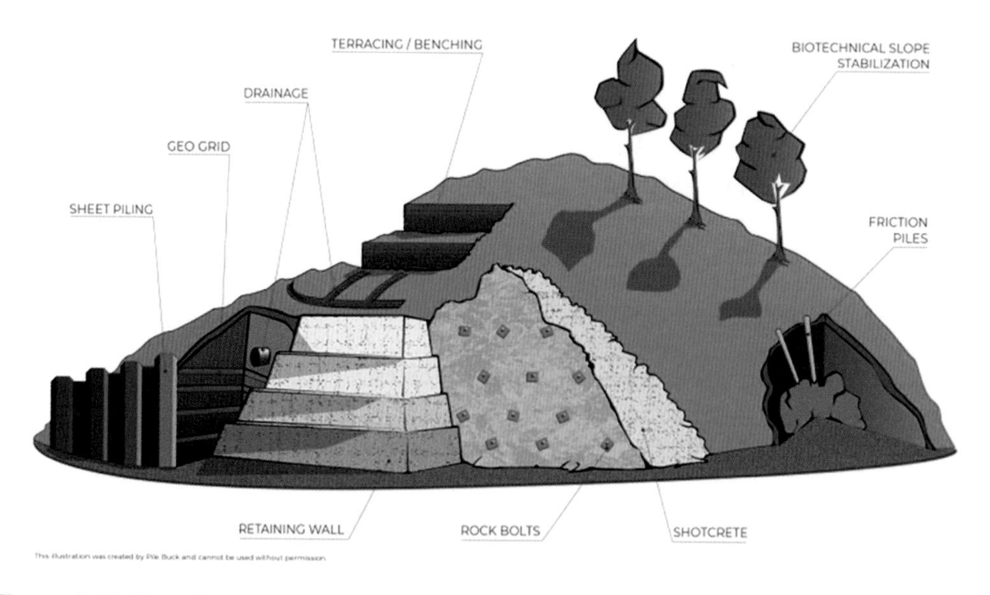

Figure 9.15 Slope failure repair options.

Source: https://pilebuck.com/engineering/landslide-mitigation-techniques/

retaining walls are widely used for slopes adjacent to buildings and transportation routes. For example, the UK rail network has over 7,000 slopes supported by retaining walls. Although effective, this is an expensive and visually intrusive approach and there is a move towards measures that sit within the soil or rock rather than on the surface. Examples include soil nails and rock bolts. In addition, structures can be designed to deflect small landslides around vulnerable facilities. For example, diversion walls are often constructed around electricity pylons in mountain areas to control localised debris flows.

- *Vegetation*. Vegetation on slopes performs several functions. Tree and plant roots help to bind soil particles together and provide stability. The vegetation canopy protects the soil surface from rain splash impact, whilst transpiration processes reduce the water content of the slope. In recent years, bioengineers have developed new ways of controlling shallow landslides in soil and for preventing soil erosion. It is important to ensure that the vegetation can thrive in the chosen location, and preference is normally given to local species. This approach is considered to be more environmentally aware than traditional engineering methods. The capital costs of bioengineering are usually less than those for conventional structures, although maintenance costs tend to be higher.
- *Other methods* include the chemical stabilisation of slopes and the use of grouting to reduce soil permeability and increase its strength. On some construction sites, moving soil has been frozen temporarily while soil-retaining structures were completed, but this is a very expensive option. In many tropical countries, shallow and localised slope failures are covered in plastic sheets to reduce the impacts of rainfall until full stabilisation can be achieved.

Whilst engineering remains predominant in landslide protection, it is increasingly pursued alongside planning legislation to limit new development on dangerous slopes. In the USA, urban planning is managed by the *Uniform Building Code*. This specifies a maximum slope angle for safe development of 2:1, which approximates to a 27° angle, as well as minimum standards for soil compaction and surface drainage. Similarly, in New Zealand, there is a legal requirement that proposed new buildings must have resource *consent* for which slope stability is a major component. The success of these schemes depends on the availability of technically trained inspectors to enforce the regulations. This is often a problem, especially where local corruption also exists, but long-term success can be achieved. The city of Los Angeles introduced a grading ordinance as early as 1952. Before this date more than 10% of all building lots were damaged by slope failure, but more recent losses at new construction sites have been estimated at fewer than 2%.

3. Avalanches

Two main physical techniques are used to protect against the hazard posed by snowpacks.

Artificial Release

Artificial release is accomplished through the use of small explosive charges to trigger controlled avalanches. This technique is used surprisingly often; in the USA about 10,000 avalanches are artificially released each year. The main advantages are:

- The snow release occurs at predetermined times when the downslope areas – perhaps containing recreational facilities and roads – can be closed.
- Measures to allow snow clearance can be put in place before the avalanche occurs, thereby minimising inconvenience.
- The snowpack can be released safely in several small avalanches, rather than allowing the build-up of a major threat.

Explosive charges are most effective when placed in the initiation zone, or near the centre of a potential slab avalanche, when the relationship between stress and strength within the snowpack is delicately balanced. These requirements can be met only through close liaison with a snow stability monitoring and avalanche forecasting service. In some cases, dedicated teams are dropped by helicopter into the initiation zone in order to place the charges. Needless to say, this is a hazardous task with respect to both the handling of explosives and the possibility of the team's triggering an avalanche with unexpected consequences. Alternatively, it is possible to use military field guns to fire explosives onto the slope from a safe zone, in order to protect key facilities. For example, the Rogers Pass in the Canadian Rockies funnels both the Canadian Pacific rail route and the Trans-Canada highway through the Selkirk Mountains of British Columbia. At this location, Parks Canada and the Canadian Armed Forces work together to trigger avalanches with field artillery.

Defence structures are a common adjustment to avalanches throughout the world. In Switzerland alone, the total amount spent on avalanche defence structures in the period 1950–2000 was approximately €1 billion (Fuchs & McAlpin, 2005). There are four main types of avalanche defence structure (Figure 9.16):

- *Retention structures* are designed to trap and retain snow on a slope and thus to prevent the initiation of an avalanche or to stop a small avalanche before it can develop fully. Above the starting zone, *snow fences* and *snow nets* are used to hold back snow. On ridges and gentle slopes, large volumes of snow can be intercepted and retained in this way. In the starting zone *snow rakes* or *arresters* are used to provide external support for the snowpack, thus reducing internal stresses. They may also stop small avalanches before they gain momentum. The earliest structures were massive walls and terraces made of rocks and earth. Today they are made of combinations of wood, steel, aluminium and/or pre-stressed concrete. Such structures are effective, but they do have a negative impact on the visual appearance of the landscape, an important consideration in areas dependent on tourism. Their use is also less effective in regions with large seasonal accumulations of snow, like the northwest USA, where snowfall may bury the structures and so limit their effectiveness.
- *Redistribution structures* are designed to prevent snow accumulation by drifting. In particular, they are used to prevent the build-up of cornices that often break off steep slopes and initiate an avalanche.
- *Deflectors and retarding devices.* The easiest way to control an advancing avalanche is to guide it along a gently curving path, so deflectors built of earth, rock or concrete are placed in the avalanche track and run-out zone. However, the scope for lateral diversion is limited, and changes of direction no greater than 15°–20° from the original avalanche path have been proved to be most successful. In addition, wedges pointing upslope can be used to split an

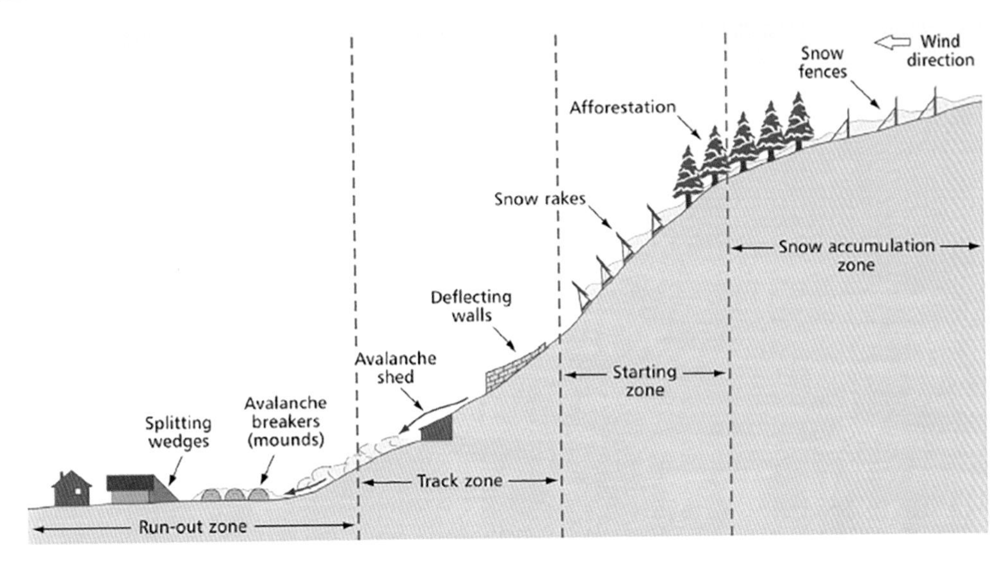

Figure 9.16 Idealised slope section showing avalanche hazard-reduction measures. Snow retention is achieved by defensive structures in the snow-accumulation and starting zones, often complemented by a mature forest cover. Deflection of the avalanche away from people and infrastructure, and lowering the destructive potential of the flow, is achieved by structures in the track and run-out zones.

avalanche and divert the flow around isolated facilities like electricity transmission towers or isolated buildings. Towards the run-out zone, on ground slopes of less than 20°, other structures – earth mounds and small dams – are useful to obstruct an avalanche as it loses energy. Braking mounds are widely used against dense, wet snow avalanches.

- *Direct-protection structures* such as avalanche sheds and galleries provide the most complete avalanche defence. They are designed to allow the flow to pass over key built facilities, and avalanche sheds typically act as protective roofs over roads or railways. They are expensive to construct and need careful design to ensure that they are correctly located and can bear the maximum snow loading on the roof.

Some techniques for managing avalanche hazards mimic the natural protection offered by mature forests. Therefore, where possible, it is desirable to plant forests on avalanche-prone slopes and avoid the need for unsightly, maintenance-intensive structures. As shown in Figure 9.17, forest cover is often used in combination with structures, with structures retaining snow on the upper slope and trees providing further safety above transport routes or settlements. Allowances have to be made for differences in forest structure and terrain roughness, but while large avalanches released far above the forest are unlikely to be completely stopped by tree cover, small avalanches triggered close to the tree line are expected to experience significant deceleration.

A major difficulty is that avalanche tracks offer poor prospects for successful tree planting and growth. Gully erosion by previous events means that avalanche-prone slopes are often characterised by thin soils with limited water retention. Furthermore, young trees may be destroyed by

Figure 9.17 A combination of snow fences and mature forest cover protect valley development and surrounding infrastructure from avalanches in the Austrian Alps.

Source: Photo: ALMIDI.NET/J.W. Alker, © SuperStock 1848–450211 by imagebroker.net

avalanches before they can provide stability to the snowpack. Therefore, expensive site preparation, coupled with soil fertilisation, is frequently required, and it may be necessary to stabilise the snow in the starting zone while the tree cover establishes itself. Sometimes it can take over 75 years before slower-growing species have reached the point where they are strong enough to resist avalanche forces. In some cases, the natural forest may already have been removed to allow the introduction of economic activities, such as skiing, and attempts to re-establish tree cover may not be wholly welcome.

4. Community Preparedness

There are numerous programs that exist within regions and nations focused on preparing for a landslide. The National Disaster Management Authority of India publishes example preparedness handouts on what to do before, during and after a landslide. For some communities there are low-cost monitoring approaches that could be used as part of a community-driven warning system, such as those currently under test in Malaysia (Dixon et al., 2018). Even without any monitoring equipment, some communities can monitor the landscape using rough rules of thumbs. For example in Manizales, Colombia, the 'slope guardians' are a group of young mothers that work to prevent and mitigate the effects of landslides via training received from the government. In the Philippines, slope angles are effectively measured using folded papers to approximate the angles and aid the observations and warnings of potential mass movements.

5. Disaster Aid

Mass movement disasters rarely attract substantial disaster aid, due to the relatively small scale of the losses, although the occurrence of multiple mass movement events has become more common. Large-scale relief operations were required in the aftermath of the 1997 hurricane 'Mitch' disaster in Nicaragua and Honduras, the 1999 Vargas landslide disaster in Venezuela (Box 9.4), the 2005 Kashmir earthquake (which induced thousands of landslides), the 2006 Leyte landslide disaster in the Philippines and the 2007 landslide disaster in North Korea. In the case of the Leyte landslide, a large rock slope collapsed onto the town of St Bernard, burying a town of 1,400 people, including 268 pupils in a school. Search-and-rescue teams were dispatched from the Philippines, Taiwan, the UK and the USA, although few people, if any, were rescued by them. The low success rate achieved by international emergency responders re-emphasises the need for the prior establishment of local capability.

6. Insurance

In many countries, including the UK, private insurance against mass movement hazards is unavailable because the industry fears the prospect of many high-cost claims. Unavailability of insurance can discourage development in hazardous areas but, because information about landslide hazards is not widely disseminated, many people remain unaware of the risk. Limited insurance is provided by some government schemes. For example, in the USA some cover is provided through the National Flood Insurance Program, which requires areas subject to 'mudslide' hazards associated with river flooding to have insurance in order to qualify for federal aid. Unfortunately, technical difficulties in mapping 'mudslide' hazard areas have led to limited use of this provision. A more successful example exists in New Zealand, where the government-backed Earthquake Commission (EQC) provides limited coverage for houses and land, in cooperation with private insurers. The EQC pays out for landslide damage to residential property on a regular basis. Interestingly, in the period 2000–2007, the EQC paid out more to cover damage from landslides than it did for earthquakes, due to a marked upward trend in landslide claims.

Generally speaking, legal liability forms a basis for financial compensation after landslide losses. American jurisprudence recognises civil liability for death, bodily injury and a wide range of economic losses associated with landslides. In many countries the legal defence of 'act of God' carries decreasing credibility because of assumptions that landslide risks are reasonably well understood, that access to relevant information for particular sites is available and that a wide range of landslide mitigation measures can be applied. As a result, court judgments have tended to identify developers, and their consultants, as responsible for damage related to mass movements. In some areas, local planning agencies have shared the liability because it has been successfully argued that the issue of a permit for residential development implied the warranty of safe habitation. Litigation can be costly. For example, legal claims arising from the landslide-induced failure of the Ok Tedi tailings dam in Papua New Guinea in 1984 included a lawsuit for over US$1 billion for costs for direct damage and US$4 billion for compensation arising

Box 9.4 The Vargas Landslides

On 14–16 December 1999, a huge rainstorm struck Vargas state in Venezuela. It deposited approximately 900 mm of rainfall over a three-day period and triggered many landslides in the hills. The landslides transitioned into a series of debris slides and mudflows which struck urban areas located on alluvial fans beside the coast (see Figure 9.18). Whilst the precise death toll of these landslides will never be known, the best estimate is that about 30,000 people lost their lives and the economic losses reached US$1.8 billion (Wieczorek et al., 2001).

More than 8,000 homes were destroyed, displacing up to 75,000 people. Over 40 km of coastline was significantly altered. In the aftermath of the disaster, the national government attempted to evacuate about 130,000 people from the northern coastal strip (IFRCRCS, 2002). The government used its own resources, and an unexpected opportunity, to attempt a permanent relocation of people from the coast, where living conditions were poor and population densities exceeded 200 km^2, to less crowded parts of the country. By August 2000, some 5,000 families were resettled in new, if sometimes unfinished, houses; 33,000 remained in temporary accommodation. Many evacuees opposed the resettlement programme and drifted back to their original location. In 2006 the population of the area had reached the pre-disaster level, leaving the area highly vulnerable to a repeat of the disaster.

Figure 9.18 The large fan of the Quebrada San Julián at Caraballeda was one of the most heavily damaged areas in the December 1999 event USGS.

Source: https://pubs.usgs.gov/of/2001/ofr-01-0144/

from pollution of the Fly River (Griffiths et al., 2004). Both cases were settled out of court, showing that litigation is an inadequate substitute for proper hazard-reduction strategies.

7. Land Use Planning

The recurrence of mass movements at the same topographic site means that mapping is important for hazard mitigation, if only through the qualitative identification, and avoidance, of unsafe sites (Parise, 2001). Most nations possess sufficiently accurate topographic and geological databases to create slope maps and digital elevation models that, together with field surveys, produce fairly reliable risk assessments. These methods are appropriate for small vulnerable nations. For example, Figure 9.19 illustrates the relatively large areal extent of landslides mapped on Tonoas island, Chuuk State, Micronesia, following tropical storm Chata'an in 2002 (Harp

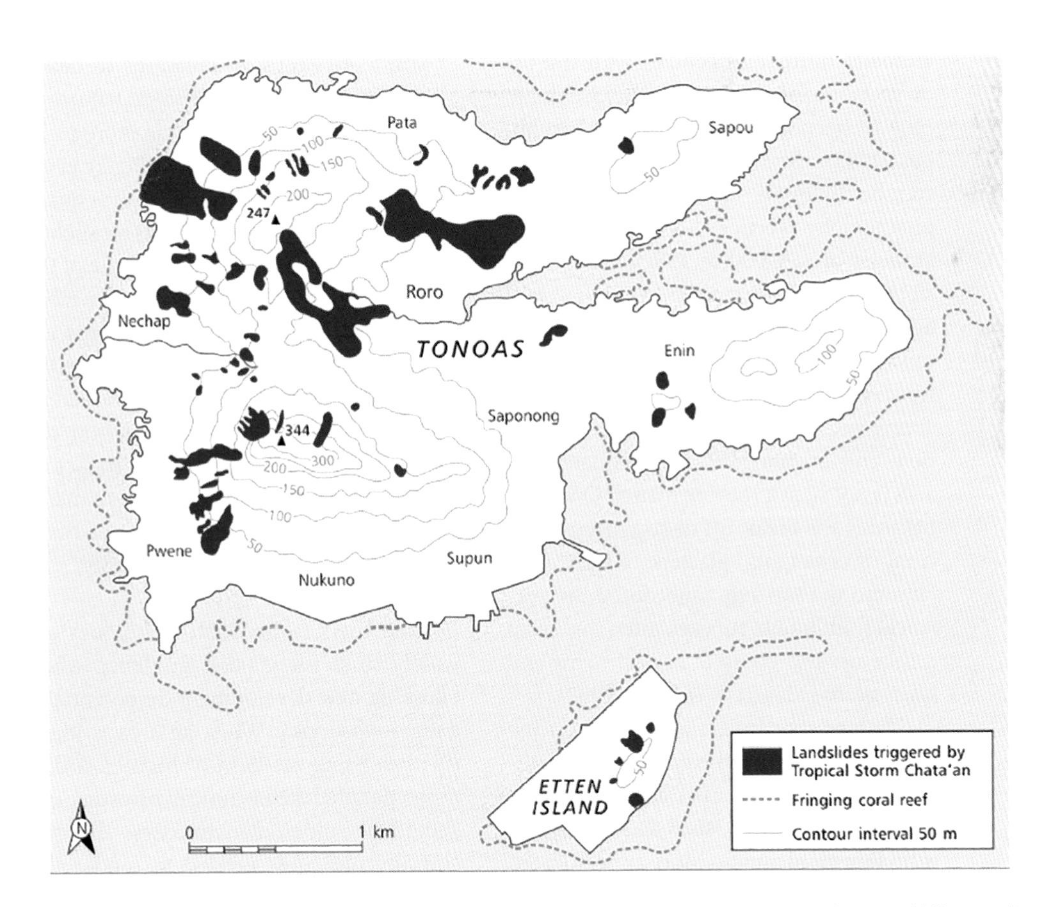

Figure 9.19 The widespread distribution of landslides on Tonoas Island, Federated States of Micronesia, in 2002. These debris flows and mudslides were triggered by intense tropical rainfall during the passage of typhoon Chata'an and disrupted several small coastal communities.

Source: After Harp et al. (2009). Reprinted from *Engineering Geology* 104, E.L. Harp et al., Mapping of hazard from rainfall-triggered landslides in developing countries: examples from Honduras and Micronesia, 295–311. Copyright (2009), with permission from Elsevier.

et al., 2009). The largest debris flows travelled several hundred metres and crossed coastal flatlands to reach populated areas. Such landslide-inventory maps can be used to identify safer zones suitable either for the temporary evacuation of people during future storms or for the permanent relocation of residents.

Remote sensing has been used for many years to produce preliminary maps of both landslide and avalanche tracks (Sauchyn & Trench, 1978; Singhroy, 1995). More comprehensive post-disaster assessments have become available (Tsai et al., 2010), although difficulties remain with the use of space-borne sensors for hazard mapping in steep mountain areas (Buchroithner, 1995). Reconnaissance information can be followed up with low-level air photography. Vertical aerial photographs at scales of 1:20,000 to 1:30,000 are often suitable, especially if taken when tree foliage and other vegetation cover is at a minimum. For example, many avalanche tracks function as landslide gullies during the spring and summer seasons. The recognition, and mapping, of less frequent hazards is not such a routine matter. One such hazard is the break-off of large ice masses from overhanging glaciers, which then fall onto the starting zones for snow avalanches, although surveys after the event can be used to compile hazard maps and safety plans (Margreth & Funk, 1999).

The pressure for building land on the edge of many cities means that the application of development restrictions based upon simple criteria, such as slope angle, is no longer adequate. More sophisticated land use planning approaches based upon the assessment of susceptibility and hazard are required, and two basic approaches have been employed:

- *Geological techniques* involve producing a very detailed risk map of landslides in a defined study area. Apart from mapping previous slides, information is collected on causal factors such as rock types, slope angle, the presence or absence of vegetation and the rainfall distribution. Attempts are then made, often using GIS, to correlate the location of the landslides with the possible causal factors. However, such approaches have had mixed success, sometimes because they overestimate the area likely to be affected by slope failure (van Asch et al., 2007).
- *Geotechnical techniques* use mathematical slope-stability equations to determine the likelihood of slope failure. In recent years, it has become more usual to attempt this using GIS (Petley et al., 2005). These techniques require quantitative estimates of parameters such as the strength of the soil, the angle of the slope and the depth of the water table. Geological and topographic maps are then used to determine the spatial distribution of the key parameters. Unfortunately, general figures from the literature have to be assigned to factors such as the soil strength. This is a major weakness because the values of these parameters can vary considerably.

Despite their faults, such techniques are widely used for general guidance when planning new development on potentially dangerous slopes. When an area is identified as being medium or high hazard, a more detailed geotechnical investigation should be undertaken to assess the risk and any measures needed to render the site safe. Some results have been impressive. For example, in 1958 the Japanese government enacted the 'Sabo' legislation to mitigate landslides and debris flows triggered by typhoon rainfall. In 1938 nearly 130,000 Japanese homes were destroyed and more than 500 lives were lost in landslides. In 1976, the worst year for landslides in that country for two decades, only 2,000 homes were lost and fewer than 125 people died. Similar evidence exists for Hong Kong, where hillside development was not properly regulated until the 1970s (Morton, 2016). A slope safety scheme was introduced in the 1990s and the rolling average annual fatality rate, which peaked at about 20 during the 1970s, fell sharply (Figure 9.20).

However, systematic approaches to landslide hazard reduction remain the exception rather than the rule. A major problem arises when potential slope failures are identified in areas already used for housing. There is often a demand for mitigation to be undertaken at national or local government expense, although most administrations fund emergency works only. This is because the cost of permanent mitigation is considered to be the responsibility of householders, even though building insurance rarely covers landslides.

As with landslides, the most effective mitigation of avalanche hazards is through land use planning based on the identification of site-specific risk. In Switzerland, avalanche zoning laws were mandated by the government as early as 1951. Zoning begins with the collection of historical avalanche data, supplemented with terrain models and an understanding of avalanche dynamics, to determine detailed degrees of risk. Where sites are near existing settlements, avalanche frequency will be a matter of local knowledge. At more remote locations, other methods are necessary, for example the use of satellite imagery and a digital elevation model. The long-term pattern of avalanche activity can be compiled from dendrochronological information. For example, Muntán et al. (2009) reconstructed avalanche paths in the Pyrenees over a 40-year period. The scarring of tree rings in tracts of forest damaged by previous events, but remaining standing in the avalanche track, can produce reliable

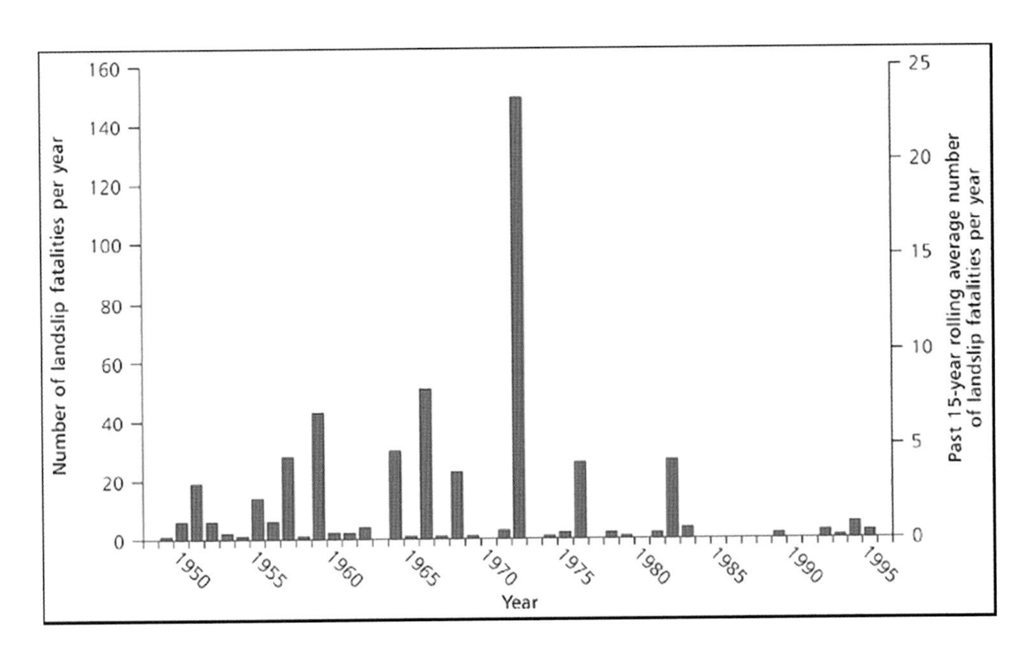

Figure 9.20 The reduction of landslide incidence by hazard management in Hong Kong, 1948–1996.

Source: After Malon (1997, p. 17)

Table 9.3 Vegetation characteristics in avalanche tracks as a rough indicator of avalanche frequency

Minimum frequency (years)	Vegetation clues
1–2	Bare patches, willows and shrubs; no trees higher than 1–2 m; broken timber
2–10	Few trees higher than 1–2 m; immature trees or pioneer species; broken timber
10–25	Mainly pioneer species; young trees of local climax species; increment core data
25–100	Mature trees of pioneer species; young trees of local climax species; increment core data
>100	Increment core data needed

Source: After Perla and Martinelli (1976)

frequency estimates over the past 200 years or so (Hupp et al., 1987). Where trees have been destroyed, close inspection of the residual damaged vegetation, including height and species, can be a useful guide. Table 9.3 shows how this evidence can be used when initial mapping is undertaken at a scale of about 1:50,000.

In British Columbia, snow avalanche atlases are used as operational guides for highway maintenance personnel. The maps are accompanied with a detailed description of the terrain and vegetation for each avalanche site, together with an assessment of the hazard impact. Where avalanches threaten settlements, larger-scale maps (from 1:25,000 to 1:5,000) are needed. Two important parameters are always difficult to determine in avalanche hazard studies. These are the length of the run-out zone, which determines whether a particular site will be reached by flowing snow, and the impact pressure at any given point, which determines the likely level of damage. Computer modelling of avalanche dynamics is increasingly sophisticated, but the precision required for the model to be reliable remains high (Barbolini & Savi, 2001). As the scale resolution of terrain data and the models improve, such methods grow in importance.

The avalanche hazard maps found in most countries normally adopt a three-zone, colour-coded system (Table 9.4). These schemes need regular updating. For example, following the 1999 avalanche disaster at Galtür, Austria, the exclusion zone for buildings, previously drawn up for a 1-in-150-year event, was extended, and revised regulations required all new buildings to be reinforced against

Table 9.4 The Swiss avalanche zoning system

High-hazard (red) zone

- Any avalanche with a return interval <30 years
- Avalanches with impact pressures of 3 t/m^2 or more and with a return interval up to 300 years
- No buildings or winter parking lots allowed. Special bunkers needed for equipment

Moderate-hazard (blue) zone

- Avalanches with impact pressures <3 t/m^2 and with return intervals of 30–300 years
- Public buildings that encourage gatherings of people should not be erected
- Private houses may be erected if they are strengthened to withstand impact forces
- The area may be closed during periods of hazard

Low-hazard (yellow) zone

- Powder avalanches with impact pressures 0.3 t/m^2 or less with return intervals >30 years
- Extremely rare flowing avalanches with return periods >300 years.

No-hazard (white) zone

- Very rarely may be affected by small air blast pressures up to 0.1 t/m^2
- No building restrictions

Source: After Perla and Martinelli (1976)

specified avalanche pressures. In addition, snow rakes were installed for the first time in the starting zone and an avalanche dam was constructed across part of the run-out zone on the valley floor.

In 2019 the WSL Institute for Snow and Avalanche Research SLF has developed maps covering the territory of the Swiss Alps that illustrate where threats of typical skier-triggered avalanches exist, and they also are able to automatically and efficiently pinpoint avalanche starting zones.

Landslides are a diverse, complex hazard that with anthropogenically forced global warming are likely to become more common, larger and more destructive. There are many costly engineering solutions or biological solutions to help manage the land, and technology is helping to provide more up-to-date maps of risk to be used to minimise risk exposure.

KEY TAKE AWAY POINTS

- Mass-movement hazards vary significantly in type, scale and duration, although most impacts are confined to the localised area affected.
- Increasingly, it is the actions of humans driving mass movement hazards and consequently it is expected these types of hazard events will happen more frequently in the future, and potentially on bigger scales.

Figure 9.21 The new maps at map.geo.admin.ch and in the White Risk app. www.slf.ch/en/news/2018/12/new-maps-illuminate-avalanche-terrain.html

Source: Harvey, S.; Schmudlach, G.; Bühler, Y.; Dürr, L.; Stoffel, A.; Christen, M., 2018: Avalanche terrain maps for backcountry skiing in Switzerland. In: 2018: International snow science workshop proceedings 2018. International snow science workshop, ISSW 2018, Innsbruck. 1625–1631. www.whiterisk.ch

- Engineering solutions are frequently adopted to act as a preventative measure to mass movements, but more natural biological-based solutions can provide cheaper and longer-lasting solutions.
- Community preparedness and swift response can save lives, alongside the integration of new GIS and remote sensing technologies that can help identify areas of high risk. However, the data needs to be acted on, otherwise it cannot prevent or reduce the disaster risk.

BIBLIOGRAPHY

Barbolini, M., & Savi, F. (2001). Estimate of uncertainties in avalanche hazard mapping. *Annals of Glaciology, 32*, 299–305.

Baum, R. L., & Godt, J. W. (2010). Early warning of rainfall-induced shallow landslides and debris flows in the USA. *Landslides, 7*(3), 259–272.

Brabec, B., Meister, R., Stöckli, U., Stoffel, A., & Stucki, T. (2001). RAIFoS: Regional avalanche information and forecasting system. *Cold Regions Science and Technology, 33*(2–3), 303–311.

Brunetti, M. T., Peruccacci, S., Rossi, M., Luciani, S., Valigi, D., & Guzzetti, F. (2010). Rainfall thresholds for the possible occurrence of landslides in Italy. *Natural Hazards and Earth System Sciences, 10*(3), 447–458.

Buchroithner, M. F. (1995). Problems of mountain hazard mapping using spaceborne remote sensing techniques. *Advances in Space Research, 15*(11), 57–66.

Butler, D. R. (1997). *A Major Snow-Avalanche Episode in Northwest Montana, February, 1996.* San Marco: Natural Hazards Research and Applications Information Center.

Casagli, N., Catani, F., Del Ventisette, C., & Luzi, G. (2010). Monitoring, prediction, and early warning using ground-based radar interferometry. *Landslides, 7*(3), 291–301.

De Scally, F. A., & Gardner, J. S. (1994). Characteristics and mitigation of the snow avalanche hazard in Kaghan Valley, Pakistan Himalaya. In *Recent studies in geophysical hazards* (pp. 197–213). Dordrecht: Springer.

Dixon, N., Smith, A., Flint, J. A., Khanna, R., Clark, B., & Andjelkovic, M. (2018). An acoustic emission landslide early warning system for communities in low-income and middle-income countries. *Landslides, 15*(8), 1631–1644.

Dunning, S. A., Rosser, N. J., Petley, D. N., & Massey, C. R. (2006). Formation and failure of the Tsatichhu landslide dam, Bhutan. *Landslides, 3*(2), 107–113.

Freeman, T. (2004). *Geoscience laboratory* (3rd ed., p. 310). New York: John Wiley & Sons.

Froude, M. J., & Petley, D. N. (2018). Global fatal landslide occurrence from 2004 to 2016. *Natural Hazards and Earth System Sciences, 18*(8), 2161–2181.

Fuchs, S., & McAlpin, M. C. (2005). The net benefit of public expenditures on avalanche defence structures in the municipality of Davos, Switzerland. *Natural Hazards and Earth System Sciences, 5*(3), 319–330.

Griffiths, J. S., Hutchinson, J. N., Brunsden, D., Petley, D. J., & Fookes, P. G. (2004). The reactivation of a landslide during the construction of the Ok Ma tailings dam, Papua New Guinea. *Quarterly Journal of Engineering Geology and Hydrogeology, 37*(3), 173–186.

Guzzetti, F. (2000). Landslide fatalities and the evaluation of landslide risk in Italy. *Engineering Geology, 58*(2), 89–107.

Hancox, G. T. (2008). The 1979 Abbotsford Landslide, Dunedin, New Zealand: A retrospective look at its nature and causes. *Landslides, 5*(2), 177–188.

Harp, E. L., Reid, M. E., McKenna, J. P., & Michael, J. A. (2009). Mapping of hazard from rainfall-triggered landslides in developing countries: Examples from Honduras and Micronesia. *Engineering Geology, 104*(3–4), 295–311.

Holcombe, E., & Anderson, M. (2010). Tackling landslide risk: Helping land use policy to reflect unplanned housing realities in the Eastern Caribbean. *Land Use Policy, 27*(3), 798–800.

Hung, J. J. (2000). Chi-Chi earthquake induced landslides in Taiwan. *Earthquake Engineering and Engineering Seismology, 2*(2), 25–33.

Hungr, O., Corominas, J., & Eberhardt, E. (2005). Estimating landslide motion mechanism, travel distance and velocity. In *Landslide risk management* (pp. 109–138). London: CRC Press.

Hupp, C. R., Osterkamp, W. R., & Thornton, J. L. (1987). *Dendrogeomorphic evidence and dating of recent debris flows on Mount Shasta, Northern California* (No. 1396). USA: Department of the Interior, US Geological Survey.

IFRC. (2002). *The pain remains for Venezuela's displaced flood victims*. https://reliefweb.int/report/venezuela-bolivarian-republic/pain-remains-venezuelas-displaced-flood-victims. Accessed 9 August 2022.

IFRCRCS. (2002). *World disasters report*. International Federation of Red Cross and Red Crescent Societies.

Jibson, R. W., & Baum, R. L. (1999). *Assessment of landslide hazards in Kaluanui and Maakua gulches, Oahu, Hawaii, following the 9 May 1999 Sacred Falls landslide* (No. 99–364). USA: US Geological Survey.

Jones, D. K. C. (1992). Landslide hazard assessment in the context of development. In *Geohazards* (pp. 117–141). Dordrecht: Springer.

Jones, D. K. C. (1995). The relevance of landslide hazard to the International Decade for Natural Disaster Reduction. Proceedings of Conference on Landslide Hazard Mitigation with Particular Reference to Developing Countries (pp. 19–33).

Jones, D. K. C., Lee, E. M., Hearn, G. J., & Genc, S. (1989). The Catak landslide disaster, Trabzon Province, Turkey. *Terra Nova, 1*(1), 84–90.

Jones, F. O. (1973). *Landslides of Rio de Janeiro and the Serra das Araras escarpment, Brazil* (No. 697). USA: USGPO.

Keefer, D. K. (1984). Landslides caused by earthquakes. *Geological Society of America Bulletin, 95*(4), 406–421.

Keefer, D. K., Wilson, R. C., Mark, R. K., Brabb, E. E., Brown III, W. M., Ellen, S. D., Harp, E. L., Wieczorek, G. F., Alger, C. S., & Zatkin, R. S. (1987). Real-time landslide warning during heavy rainfall. *Science, 238*(4829), 921–925.

Malon, D. A. W. (1997). Risk management and slope safety in Hong Kong. *HKIE Transactions, 4*(2–3), 12–21.

Margreth, S., & Funk, M. (1999). Hazard mapping for ice and combined snow/ice avalanches—Two case studies from the Swiss and Italian Alps. *Cold Regions Science and Technology, 30*(1–3), 159–173.

McSaveney, M. J. (2002). *Recent rockfalls and rock avalanches in Mount Cook national park, New Zealand*.

Menorri, R. (1994). A system of monitoring and warning in a complex landslide in Northeastern Italy. *Landslide News, 94*(8).

Morton, B. (2016). A 62-year analysis of historical aerial images showing human impacts on beach topography in Hoi Ha Wan Marine Park, Hong Kong: Implications for marine park management and sustainability with a changing global climate. *Journal of Coastal Conservation, 20*(5), 375–396.

Muntán, E., García, C., Oller, P., Martí, G., García, A., & Gutiérrez, E. (2009). Reconstructing snow avalanches in the Southeastern Pyrenees. *Natural Hazards and Earth System Sciences, 9*(5), 1599–1612.

Parise, M. (2001). Landslide mapping techniques and their use in the assessment of the landslide hazard. *Physics and Chemistry of the Earth, Part C: Solar, Terrestrial & Planetary Science, 26*(9), 697–703.

Perla, R. I., & Martinelli, M. (1976). *Avalanche handbook* (No. 489). USA: US Department of Agriculture, Forest Service.

Petley, D. N. (2009). On the impact of urban landslides. *Geological Society, London, Engineering Geology Special Publications, 22*(1), 83–99.

Petley, D. N. (2010). On the impact of climate change and population growth on the occurrence of fatal landslides in South, East and SE Asia. *Quarterly Journal of Engineering Geology and Hydrogeology, 43*(4), 487–496.

Petley, D. N., Hearn, G. J., Hart, A., Rosser, N. J., Dunning, S. A., Oven, K., & Mitchell, W. A. (2007). Trends in landslide occurrence in Nepal. *Natural Hazards, 43*(1), 23–44.

Petley, D. N., Mantovani, F., Bulmer, M. H., & Zannoni, A. (2005). The use of surface monitoring data for the interpretation of landslide movement patterns. *Geomorphology, 66*(1–4), 133–147.

Rheinberger, C. M., Bründl, M., & Rhyner, J. (2009). Dealing with the white death: Avalanche risk management for traffic routes. *Risk Analysis: An International Journal, 29*(1), 76–94.

Rice Jr, R., Decker, R., Jensen, N., Patterson, R., Singer, S., Sullivan, C., & Wells, L. (2002). Avalanche hazard reduction for transportation corridors using real-time detection and alarms. *Cold Regions Science and Technology, 34*(1), 31–42.

Runqiu, H. (2009). Some catastrophic landslides since the twentieth century in the southwest of China. *Landslides, 6*(1), 69–81.

Sass, O. (2005). Rock moisture measurements: Techniques, results, and implications for weathering. *Earth Surface Processes and Landforms: The Journal of the British Geomorphological Research Group, 30*(3), 359–374.

Sauchyn, D. J., & Trench, N. R. (1978). Landsat applied to landslide mapping. *Photogrammetric Engineering and Remote Sensing, 44*(6), 735–741.

Schuster, R. L., & Highland, L. (2001). *Socioeconomic and environmental impacts of landslides in the Western Hemisphere* (pp. 1–50). Denver, CO: US Department of the Interior, US Geological Survey.

Singhroy, V. (1995). SAR integrated techniques for geohazard assessment. *Advances in Space Research, 15*(11), 67–78.

Singhroy, V. (2005). Remote sensing of landslides. *Landslide Hazard and Risk*, 469–492.

Smyth, C. G., & Royle, S. A. (2000). Urban landslide hazards: Incidence and causative factors in Niterói, Rio de Janeiro State, Brazil. *Applied Geography, 20*(2), 95–118.

Stähli, M., Sättele, M., Huggel, C., McArdell, B. W., Lehmann, P., Van Herwijnen, A., Berne, A., Schleiss, M., Ferrari, A., Kos, A., Or, D., & Springman, S. M. (2015). Monitoring and prediction in early warning systems for rapid mass movements. *Nat. Hazards Earth Syst. Sci., 15*, 905–917. https://doi.org/10.5194/nhess-15-905-2015.

Thomas, M. F. (1994). The Quaternary legacy in the tropics: A fundamental property of the land resource. *Soil Science and Sustainable Land Management in the Tropics*, 73–87.

Tsai, F., Hwang, J. H., Chen, L. C., & Lin, T. H. (2010). Post-disaster assessment of landslides in southern Taiwan after 2009 Typhoon Morakot using remote sensing and spatial analysis. *Natural Hazards and Earth System Sciences, 10*(10), 2179–2190.

van Asch, T. W., Malet, J. P., van Beek, L. P., & Amitrano, D. (2007). Techniques, issues and advances in numerical modelling of landslide hazard. *Bulletin de la Société géologique de France, 178*(2), 65–88.

Varnes, D. J. (1978). Slope movement types and processes. *Special Report, 176*, 11–33.

Wartman, J., Montgomery, D. R., Anderson, S. A., Keaton, J. R., Benoît, J., dela Chapelle, J., & Gilbert, R. (2016). The 22 March 2014 Oso landslide, Washington, USA. *Geomorphology, 253*, 275–288.

Wieczorek, G. F., Larsen, M. C., Eaton, L. S., Morgan, B. A., & Blair, J. L. (2001). *Debris-flow and flooding hazards associated with the December 1999 storm in coastal Venezuela and strategies for mitigation* (No. 2001–144). USGS Open File Report 01-0144, Reston, Virginia, USA.

FURTHER READING

Ahmed, B. (2021). The root causes of landslide vulnerability in Bangladesh. *Landslides, 18*, 1707–1720.

Hewitt, K. (1992). Mountain hazards. *Geojournal, 27*, 47–60. A good general account of problems in high terrain.

McClung, D. M. (2016). Avalanche character and fatalities in the high mountains of Asia. *Annals of Glaciology, 57*(71), 114–118.

WEB LINKS

American Avalanche Association (A3) and the US Forest Service National Avalanche Center (NAC). www.avalanche.org

Avalanche Canada. www.avalanche.ca

International Programme on Landslides. http://iplhq.org

National Disaster Management Authority of India: Landslide Awareness Handout. https://ndma.gov.in/Resources/awareness/landslide/visuals

New Zealand Landslide Database. http://data.gns.cri.nz/landslides

WSL Institute for Snow and Avalanche Research SLF. www.slf.ch

CHAPTER TEN

Storm Hazards

10

OVERVIEW

This chapter covers storm hazards through precipitation, wind, lightning, air temperature and humidity. These can compound for winter storms, thunderstorms, hailstorms, cyclonic storms, tornadoes and others. Storm hazards impact people through multiple causes of deaths, injuries and mental health consequences as well as the results of damage to infrastructure, agriculture and systems for water, energy and health. This chapter describes these connected categories of storm hazards, and how they compound, and then examines two specific categories: cyclonic storms and tornadic storms. Risk assessment and management include warnings, building codes, planning regulations and giving people resources and opportunities to get information and to act on it. Challenges emerge in determining the appropriate actions when dealing with multiple storm hazards.

A. THE NATURE OF STORMS

Storm hazards comprise three main phenomena: precipitation, wind and lightning, all of which are enfolded within air temperature and humidity. As shown in Table 10.1, these phenomena can compound to form different kinds of storms.

1. Precipitation

Precipitation can be liquid water, mainly rain (Figure 10.1), or solid water, such as snow, ice, sleet, freezing rain and graupel. When precipitation falls as rain, a common hazard consequence is floods (Chapter 11). Some storms involving snow are termed blizzards (Figure 10.2), although the differentiation is arbitrary. For example, a blizzard is defined by Australia's Bureau of Meteorology as strong winds in conjunction with blowing or falling snow with an expected reduction in horizontal visibility

DOI: 10.4324/9781351261647-12

Table 10.1 Storms as compound hazards

Tropical cyclones	Extratropical cyclones	Tornadoes	Hailstorms	Snowstorms and ice storms
Wind	Wind	Wind	Hail	Snow
Pressure drop	Pressure drop	Pressure drop	Wind	Ice
Rain	Rain	Rain	Lightning	Glaze
Storm surge	Storm surge	Lightning	Cold	Cold
Waves	Waves	Updraughts		Wind
Flooding	Flooding	Flooding		Flooding
Landslides and rockslides	Landslides and rockslides	Landslides and rockslides		Landslides, rockslides and avalanches
Saline intrusion	Saline intrusion			Slush
				Frost

Figure 10.1 An intense rainstorm in Victoria, Seychelles.

Source: Photo: Ilan Kelman

to less than 200 metres. By comparison, the USA's National Weather Service requires more than three hours of sustained wind or frequent gusts at or over 56 km h^{-1} and snow impeding visibility over 400 m. Neither definition mentions temperature.

Dozens of blizzards can easily occur every winter from the mid-latitudes and

Figure 10.2 A blizzard in Boulder, Colorado.

Source: Photo: Ilan Kelman

northwards in countries such as Canada, Denmark including Kalaallit Nunaat (Greenland), Finland, Germany, Iceland, Norway, Poland, Russia, Sweden, Switzerland and the USA. The Baltic, Balkan and central and eastern Asian countries, too, experience intense snowfall and blizzards. Five such events occurring January–February 2008 in China's Yungui Plateau are summarised in Table 10.2. The table evidences how environmental hazards do not necessarily occur in isolation. Anyone dealing with them, such as cold weather, must be ready for sequential and simultaneous hazards, sometimes being the same such as in Table 10.2 and sometimes being different.

Many other countries experience snowfall, ice and frost, certainly at high elevations from Tanzania to Chile, but also in cities from Delhi to Sydney. Blizzards and ice storms can occur at relatively low latitudes and elevations, such as in Texas in February 2011 and February 2021. In 1993 a storm affected much of eastern North America and was referred to as the '1993 Superstorm' and 'the Storm of the Century'. Dramatic epithets are not uncommon in the news media. The February 2010 North American blizzard is referred to as Snowmageddon, while Snowzilla and Snowpocalypse have also been used in the media. The 1993 North American storm was initiated over the warm waters of the Gulf of Mexico and moved north along the Atlantic seaboard as a rapidly intensifying surface low. In comparison, the centre of 2010's Snowmageddon came from Baja California and moved across the USA to reach the Atlantic.

The Northeast Snowfall Impact Scale (NESIS) was developed by Kocin and Uccellini (2004) as a relative measure of winter storms in the 13-state Northeast region of the USA. NESIS accounts for the total snowfall, its geographic distribution and the density of the population affected. The 5-point scale was calibrated on the severity of 30 snowstorms recorded from 1956 to 2000 and

Table 10.2 The five sub-weather events affecting China's Yungui Plateau in early 2008

Sub event	Date	Snow areas	Sleet areas
1	January 10–16	Central part of Shaanxi, south part of Shanxi, Henan, central and north part of Anhui, north part of Jiangsu, Hubei, Hunan and northeast part of Jiangxi	Central and south part of Hunan, western and southern parts of Guizhou
2	January 18–22	Eastern part of Hubei, southern part of Henan, middle and northern parts of Anhui, northern part of Jiangsu, and northern part of Hunan	Southern part of Anhui, majority areas in Hunan, all around Guizhou, northeastern part of Guangxi
3	January 25–29	Snow appeared in southern part of Henan, eastern part of Hubei, Anhui, Jiangsu and the northern part of Zhejiang. The depth of snow on Jan. 28 reached 20–45 cm.	Majority areas in Jiangxi, majority areas in Guizhou, some areas in Hunan
4	January 31–February 2	Snow appeared in the middle part of Hunan, northern part of Jiangxi, southern part of Anhui, southern part of Jiangsu and northern part of Zhejiang etc. The depth of snow in local areas on Feb. 2 reached 20–35 cm.	Guizhou, Hunan, Jiangxi, Zhejiang, Yunnan etc.
5	February 4–6	Hunan, northern part of Jiangxi, western part of Guizhou, eastern part of Yunnan	Eastern part of Yunnan, western part of Guizhou

Source: Provided by Ye et al. (2012) from the Committee of Experts of State Disaster Relief Commission (2008)

describes storms as Category 1 notable (NESIS value 1 to < 2.5), Category 2 significant (2.5 to < 4), Category 3 major (NESIS value 4 to < 6), Category 4 crippling (NESIS value 6 to < 10), and Category 5 extreme (≥ 10). Table 10.3 lists the top five recorded snowstorms in the northeast USA with their NESIS score and description.

Ice storms or glaze storms lead to thick accretions of ice on structures and vegetation whenever there is liquid precipitation, or cloud droplets, and the temperature of both the air and the object surfaces is below freezing point. Electric lines and trees are at high risk because the weight of ice can bring them down. Freezing rain also slickens surfaces, as

Table 10.3 The top five highest impact snowstorms in the northeast USA

Rank	Date	NESIS value	Category	Description
1	March 12–14, 1993	13.20	5	Extreme
2	January 6–8, 1996	11.78	5	Extreme
3	March 2–5, 1960	8.77	4	Crippling
4	January 22–24, 2016	7.66	4	Crippling
5	February 15–18, 2003	7.50	4	Crippling

Source: www.ncdc.noaa.gov/snow-and-ice/rsi/nesis

it falls as water into below-zero temperatures and so freezes on contact with roads, vehicles and structures. Sometimes, patches of road (especially on bridges or in shaded areas) become coated with a thin layer of transparent ice. Drivers cannot see it, so they drive over the ice and are unable to stop or steer, and so vehicle crashes result.

Hail consists of ice particles falling from clouds to reach the ground. The damage potential depends on the number of particles, their sizes and the surface wind speed driving them. Most hail is produced when strong vertical motions are present, giving rise to cumulonimbus clouds which also produce lightning and possible thunderstorms. Hailstorms result from strong surface heating and tend to be features of the warm season. Graupel is seen as being distinct from hail and ice pellets, even though graupel can be referred to as soft hail or snow pellets. It forms during snow when water droplets below freezing point (called supercooled) freeze on snowflakes' surface leading to balls of rime of diameter from around 2–5 mm.

2. Wind

Wind is simply fast-moving air: its steady impact and pressure differential as well as localised extremes in the form of gusts can cause hazard and risk. The physical impact, pressure differential and debris lead to casualties and damage to property and infrastructure.

In October 1987, for example, a small depression rapidly deepened in the Bay of Biscay and then moved over Western Europe, tracking directly over southwest England to Norfolk with gusts of up to 190 km h^{-1} and then moving across the North Sea and up Norway's coast. Aside from dozens of deaths and extensive property damage, in England alone more than 15 million trees were downed. Forestry losses are common due to wind, and branches and trees fall on power lines and across transportation routes. Switzerland exemplifies estimated increasing trends in woodland losses, perhaps because warming winters make soils less frozen and wetter, making it easier for wind to fell trees (Usbeck et al., 2010).

Also in Europe, between January and March 1990, four severe storms ('Daria', 'Herta', 'Vivian' and 'Wiebke') caused extensive damage (Figure 10.3). Over 200 people died, with a significant proportion of deaths due to trees falling onto vehicles, and the insurance loss was over €8 billion (Munich Re, 2002). In December 1999, three separate storms – 'Anatol', 2–4 December; 'Lothar', 24–27 December; and 'Martin',

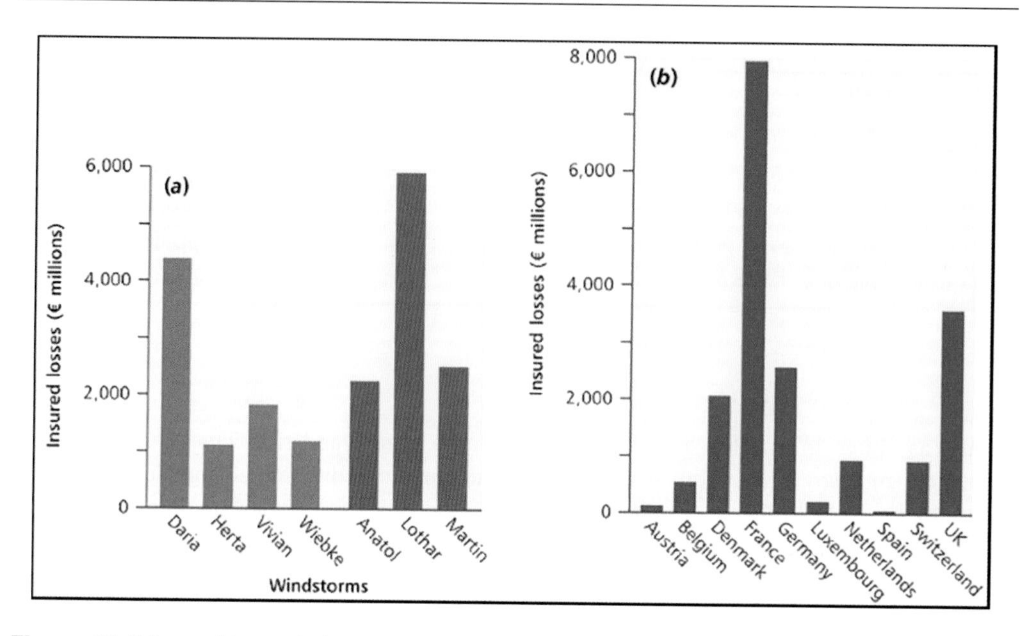

Figure 10.3 Insured losses in four named European windstorms in 1990 ('Daria', 'Herta', 'Vivian' and 'Wiebke') compared to the losses in three 1999 storms ('Anatol', 'Lothar' and 'Martin'): (a) the losses created by individual storms; (b) the aggregate losses in individual countries.

Source: Compiled from data in Munich Re (2002)

25–28 December – set new wind speed records and killed more than 130 people. In France, 'Lothar' and 'Martin' caused many electricity pylons to collapse and 25% of all transmission lines were lost, causing one of Europe's greatest recorded failures of an electrical supply network (Abraham et al., 2000). Total insured losses in 1999 were estimated at almost €11 billion, but, after adjustment for price inflation, the real cost of insured losses was greater in 1990. Most insured losses arise from minor damage to buildings sustained over large areas.

3. Lightning

Lightning occurs when a large positive electrical charge builds up in the upper, often frozen, layers of a cloud and a large negative charge, together with a smaller positive force, forms in the lower cloud. Since the cloud base is negatively charged, there is attraction towards the normally positive earth and the first (leader) stage of the flash brings down negative charge towards the ground. The return stroke is a positive discharge from the ground to the cloud and is seen as lightning. The extreme heating and expansion of air immediately around the lightning path sets up sound waves heard as thunder. The dangers from lightning (including volcanic lightning) are the electricity and the heat, the latter of which can also start fires (Chapter 14).

4. Cyclonic Storms

Every year brings dozens of cyclonic storms from the low to high latitudes, with variability in numbers and parameters being high from year to year and within individual ocean basins. Tropical cyclones are defined as non-frontal, low-pressure, synoptic-scale systems with strongly organised convection that form over warm oceans. The terms 'cyclone', 'hurricane' and 'typhoon' all refer to tropical cyclones, depending on the specific location where the system forms. Aside from being geographically arbitrary, the thresholds of moving between systems such as tropical depressions, tropical waves, tropical storms and tropical cyclones are also arbitrary, although there is a clear distinction between a rotating storm and a non-rotating storm. For instance, wind speeds of about 65 km h^{-1} or higher represent tropical storms while the threshold for transition up to a tropical cyclone is around 119 km h^{-1}.

The winds blow around low-pressure centres with strong isobaric gradients. Severity tends to be classified according to central pressure, wind speed or ocean surge on the Saffir/Simpson scale (Table 10.4). Concerns have long been raised that this approach to classifying the hazard does not correlate well with potential disaster severity, in particular given how damaging and lethal freshwater flooding can be during cyclonic storms.

Three main hazards are associated with tropical cyclones:

- *Strong winds* are given categories, with the highest, Category 5, starting at wind speeds of around 252 km h^{-1}, although wind gusts over 300 km h^{-1} have been observed. The wind's force experienced by a structure perpendicular to the moving air mass is proportional to the square of the wind speed, and so the damage potential increases rapidly with storm severity.
- *Rainfall* leads to freshwater flooding, landslides and other slides. Hurricane Harvey in 2017 led to significant inundation in Houston and the surrounding area, while Hurricane Mitch in 1998 caused devastating landslides and mudflows in Central America. At any one location, the total rainfall during the passage of a tropical cyclone can exceed 250 mm, all of which may fall in a period as short as 12 hours. Higher falls of precipitation are likely if there are mountains near the coast. For instance, intense rains have been recorded on La Réunion in the Indian Ocean, with a 12-hour fall of 1,144 mm and a 24-hour fall of 1,825 mm in January 1966. Rainfall and wind associated with a decaying cyclone can continue inland a long way from the coast, as Toronto, Ontario, experienced with Hurricane Hazel in 1954, Hurricane Isabel in 2003 and Hurricane Sandy in 2012.
- *Storm surge* (Figure 10.4) is a raised dome of seawater perhaps 60–80 km across. In addition to deaths, injuries, infrastructure damage and transport interruption, the surge contaminates agricultural land and freshwater supplies with salt. Swell waves move outward from the storm

Table 10.4 The Saffir–Simpson hurricane scale

Category	Central pressure (mb)	Wind speed (m/s)	Surge height (m)
1	>980	33–42	1.2–1.6
2	965–979	43–49	1.7–2.5
3	945–964	50–58	2.6–3.8
4	920–944	59–69	3.9–5.5
5	< 920	> 69	> 5.5

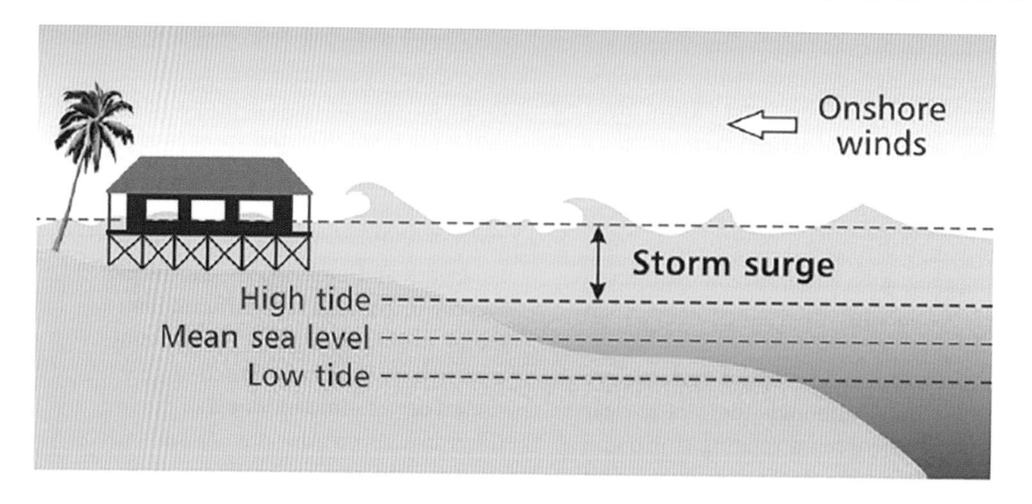

Figure 10.4 The storm surge hazard. Strong onshore winds drive water inland well above high tide levels.

centre, perhaps three to four times faster than the storm itself, and can act as a warning of its approach to coastlines over 1,000 km distant. The wind-driven waves pile water up along shallow coasts to a maximum height that depends on the intensity of the tropical cyclone, its forward speed of movement, the angle of approach to the coast, the submarine contours of the coast and the phase of the tide. The farther the wind blows over water, called 'fetch' (or 'fetch length' or 'wind fetch'), the higher the waves. Storm surge heights can easily reach 5 m above the expected high tide level. Confined bays with low-lying coasts, such as the Gulf of Mexico and the Bay of Bengal, have high risk. A further increase in sea level is due to low atmospheric pressure, at a rate of 260 mm for every 30 mb fall in air pressure.

Tropical cyclones form through heat and moisture and so tend to have specific areas of origin (Figure 10.5). Most of these storm systems decay rapidly over land areas, although some then gain strength if they leave the land to track over open water again. Their meteorological evolution typically begins with a small, low-pressure disturbance, perhaps a vortex near the inter-tropical convergence zone (ITCZ), which is a region around the equator where key winds meet. If surface pressure continues to fall, perhaps by 25–30 mb, this creates a circular area with a typical radius of only 30 km with strong in-blowing winds.

This disturbance can then develop into a tropical cyclone if certain environmental conditions are satisfied:

- The rising air, convected over a wide area, must be warmer than the surrounding air masses up to 10–12 km above sea level. This warmth comes from latent heat taken up by water when it evaporates from the ocean and/or when it is liberated by condensation in bands of cloud spiralling around the low-pressure centre. There must also be high atmospheric humidity up to about 6 km. If the rising air has insufficient moisture for the release of latent heat as the water condenses, or if it is too cool

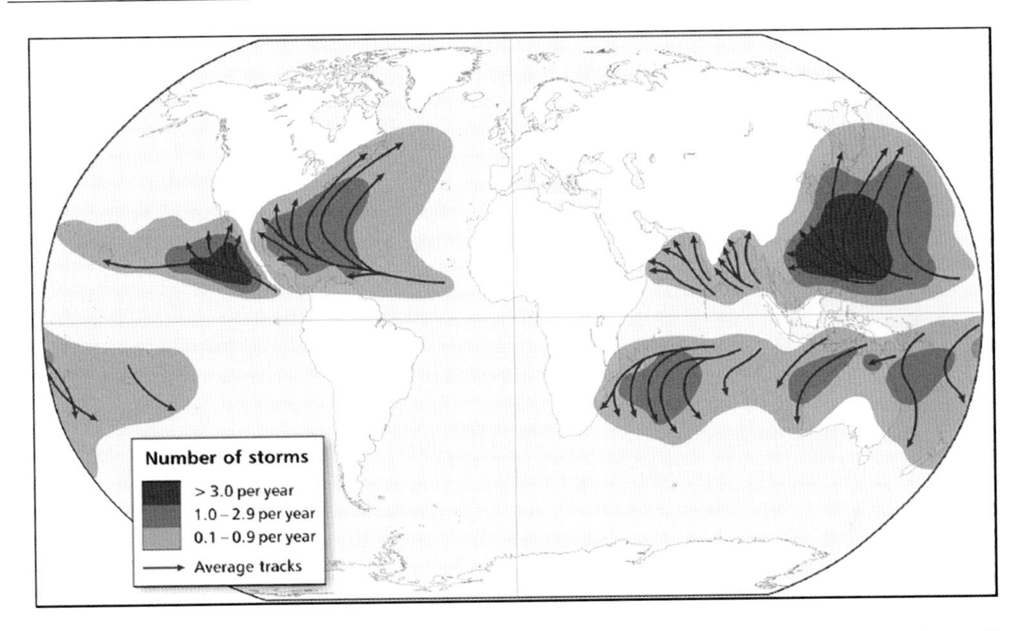

Figure 10.5 World map showing the location and average annual frequency of tropical cyclones. The way in which storm tracks curve polewards is evident.

in the first place, the chain reaction will not start. This is why cyclones form only over tropical oceans with surface temperatures of around 26°C or more.

- Vorticity must be present to give the low-pressure system an initial rotation. Tropical cyclones tend not to develop within about 5° latitude of the equator, where vorticity is minimal (zero at the equator) and inflowing air will quickly fill a strong surface low. Between 5° and 12° north and south of the equator, the air flow converging on a low is deflected to produce a favourable spiral structure.
- The broad air current in which the cyclone is formed should have weak vertical wind shear because wind shear inhibits vortex development. Vertical shear of the horizontal wind of less than 8 m s^{-1} allows the main area of convection to remain over the centre of lowest pressure in the cyclone. Although this is not a difficult condition to satisfy in the tropics, it does explain why no cyclones develop in the strong, vertically sheared current of the Asian summer monsoon. Most cyclones occur there after the monsoon season, in late summer and autumn, when sea-surface temperatures are at their highest level.
- In combination with the developing surface low, an area of relatively high pressure should exist above the growing storm. As this happens rarely, few tropical disturbances develop into cyclones. If high pressure exists aloft, a strong divergence or outflow of air is maintained in the upper troposphere. Crudely stated, the system acts like a suction pump, drawing away rising air and strengthening the sea-level convergence.

Mature tropical cyclones can be regarded as thermodynamic heat engines where the energy derived from evaporation at the ocean surface is lost partly by thermal radiation (where the moist air rises and diverges) and partly by surface friction as it moves over the sea. As the wind speed increases and the storm intensifies, these energy losses grow relative to the energy gain and a theoretical upper limit to storm development is set. The wind velocity increases towards the eye, and the lowest central pressures produce the highest-velocity winds.

A ring-like wall of towering cumulus cloud rises to 10–12 km around the eye of the storm. Most of the rising air flows outward near the top of the troposphere (Figure 10.6) and acts as the main 'exhaust area' for the storm. The release of rain and latent heat when water vapour condenses into water droplets encourages even more air to rise, and violent spiralling produces strong winds and heavy rain. A small proportion of the air sinks towards the centre, to be compressed and warmed in the eye of the storm. The warm core helps to maintain the system because it exerts less surface pressure, thus maintaining the low-pressure heart of the storm. Although storm systems tend to move westward at about 4–8 m s^{-1}, driven by the upper-air easterlies, they generally recurve erratically towards the pole (Figure 10.5).

Away from the tropics, extra-tropical cyclones (ETCs) form. In the northern hemisphere, they are found near the Aleutian and Icelandic low-pressure areas. Polar lows form in the Barents Sea and other Arctic areas. Some mid-latitude cyclones develop very quickly and become rapidly deepening depressions. Figure 10.7 illustrates the tracks of seven major European ETCs from 1999 to 2010 (Kafali, 2011).

5. Tornadoes

A tornado is difficult to define as there is no agreed description, although it usually refers to a narrow, violently rotating column of air, averaging about 100 m in diameter, that extends towards the ground. Most tornadoes are associated with 'parent' cumulonimbus clouds and are recognised by a funnel-shaped cloud that appears to hang from the cloud base above. They are highly localised storms, sometimes associated with thunder and hail, and tend to form in warm, moist air ahead of a strong cold front. This is because the contrast in air masses produces latent heating, as water releases heat when changing from gas to liquid, and creates a low-pressure area near the surface. Violent tornadoes can also be found in weakly unstable conditions in the lower atmosphere. The greatest hazard exists when the funnel cloud touches the ground and creates some of the strongest horizontal pressure gradients and highest wind speeds seen in nature.

A scale of tornado intensity was devised by Fujita (1973) and is shown in Table 10.5, although many updates have occurred since, to refine and expand it. It is believed that about one-third of all tornadoes exceed the Fujita scale of F-2 and attain wind speeds greater than 50 m/s. The forward speed of a tornado is perhaps only 5–15 m/s and most are of short duration with a destructive path rarely more than 0.5 km wide and 25 km long. Nonetheless, some tornadoes have shown destruction along a kilometre-wide swathe and for hundreds of kilometres. The difficulty in identifying a tornado compared to similar phenomena, as well as differentiating individual tornadoes within a swarm, leads to uncertainties in how extensive the impacts can be from a single funnel, even knowing that most structures could be completely destroyed by a strong tornado.

Tornadoes are widespread globally with Antarctica being the only continent

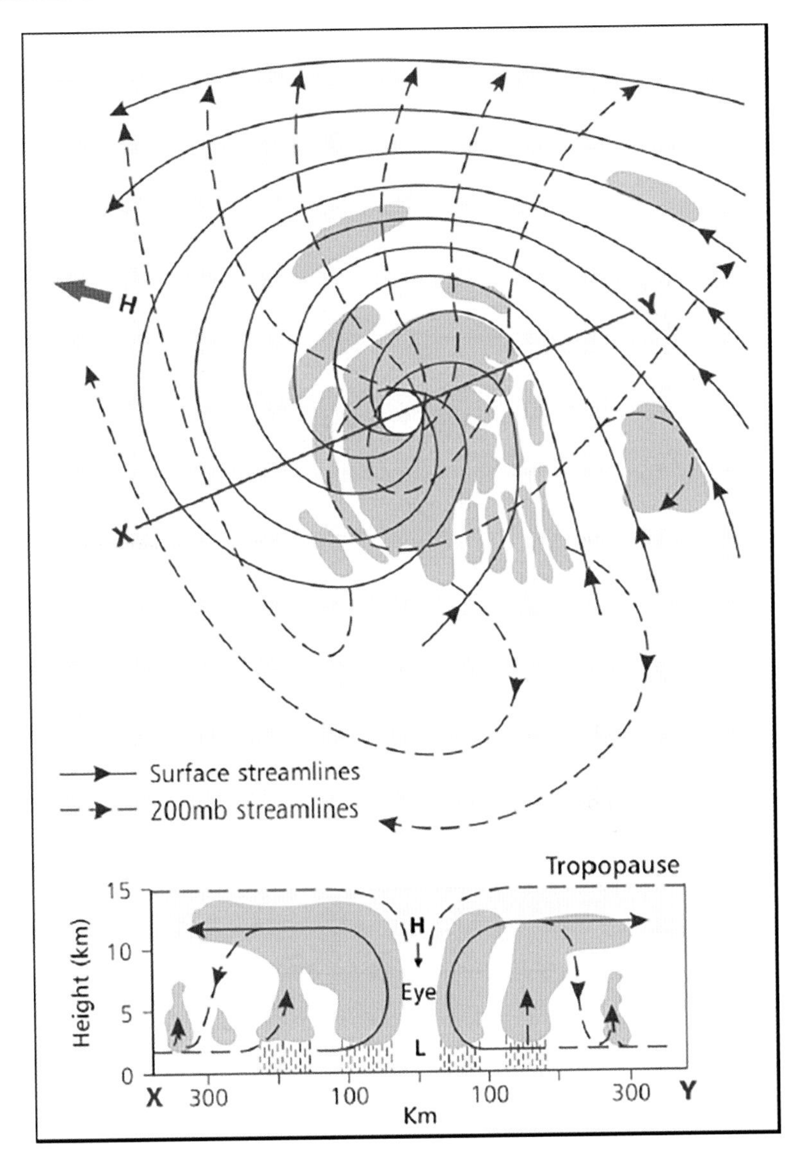

Figure 10.6 A model of the areal (above) and vertical (below) structure of a tropical cyclone. Spiral bands of cloud and rainfall are indicated in the vertical section X–Y across the system. The streamline symbols refer to the upper diagram.

Source: After Barry and Chorley (1987). Reproduced with permission.

not reporting any so far. A tornado in Bangladesh in May 1989 killed between 800 and 1,300 people, whilst in 1996 another tornado in the country killed about 700 people and destroyed approximately 17,000 homes (Paul, 1997). Mexico developed a database listing 378 tornado and 99 waterspout reports since 2000 (León-Cruz et al., 2022). Tornadoes happen in New Zealand year round (Kreft & Crouch,

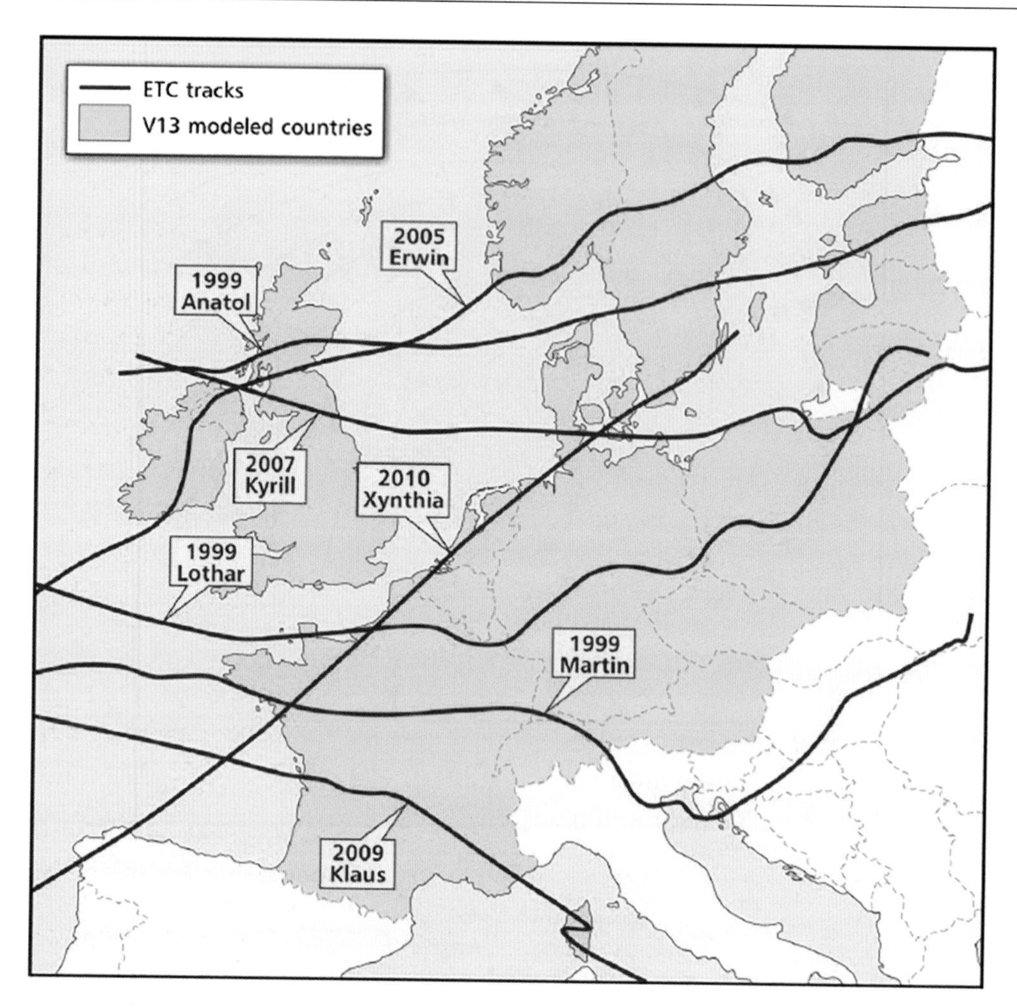

Figure 10.7 The generalised tracks of seven severe windstorms crossing Europe from 1999 to 2010.

Source: After Kafali (2011). Reproduced with permission from AIR Worldwide.

2011). In the southeast USA from 1954 to 2015, tornadoes peaked in number in May with another peak of about half the yearly maximum in late November and early December (Long et al., 2018). American emergency managers might say that the tornado season lasts from January through to December while the time of day to be most concerned about tornadoes runs from just after midnight until just before midnight. That is, tornado season is every second of every minute of every hour of every day of the year.

B. IMPACTS AND CONSEQUENCES OF STORMS

Storm hazards impact people through deaths and injuries. In floods, people can drown, experience physical trauma from infrastructure collapse or debris in the water, be affected by chemical contamination and microbes in the floodwater and encounter displaced animals such as snakes and spiders. All these effects, and many more, are addressed in Chapter 11. In

Table 10.5 The Fujita scale of tornado intensity.

Category	Damage	Wind speed (m s^{-1})	Typical impact
F-1	Light	18–32	Damage to trees, free-standing signs and some chimneys
F-2	Moderate	33–50	Roofs damaged, mobile homes dislodged, cars overturned
F-3	Severe	51–70	Large trees uprooted, roofs removed, mobile homes demolished, damage from flying debris
F-4	Devastating	71–92	Masonry buildings damaged, cars become airborne, extensive damage from large missiles
F-5	Disastrous	93–142	Wood-frame buildings lifted from foundations and disintegrated, cars airborne for more than 100 m

Source: Originally from Fujita (1973), with many updates since

storms, people also die or are injured from windborne debris, trees falling or structural collapse, all of which are common occurrences in tropical cyclones and tornadoes. Sometimes, exact hazard-related death tolls are muddied when they are attributed to a particular storm without indicating which specific phenomena killed each person.

Physical trauma results from hail hitting people, with single hailstorms reported to have killed dozens or hundreds of people. According to Shi et al. (2022), hail is one of the hardest storm-related hazards for insurers to work out damage from, often costing them the most of storm phenomena within a single storm. Reasons include hail-related damage not always being immediately obvious to claimants while historical hail-related damage claim data are not always readily available. A single storm in the USA, Europe or Australia can produce losses exceeding $1 billion.

Lightning produces casualties through burns, heat and electricity with results including scarring, hair singeing, eardrum ruptures, skin lesions, eye injuries from the flash, skin burns and cardiac arrest. Examples of lightning death rates from some Asian countries are in Table 10.6. In compiling statistics, Holle and Cooper (2019) indicate that for a single lightning-related event, so far, the highest known number of deaths was 21 in Zimbabwe (then Rhodesia) in 1975 and the highest known number of casualties was seven killed and 67 injured in South Africa in 2010. All the given tolls do not factor in casualties from lightning-related fires, particularly vegetation fires.

Thunderstorms can also bring on asthma in certain people with sensitivities. Asthma is commonly linked with spring when increased levels of grass pollen are in the air. If a thunderstorm happens at this time, very high concentrations of pollen can be pushed down to near ground levels, and if this occurs across populated areas, those with predisposed health conditions may experience increased severity of asthmatic symptoms. These events are known as epidemic thunderstorm asthma (ETSA) with 26 known events recorded worldwide (Price et al., 2021). Of this, seven of the 11 ETSA events in Australia occurred in Melbourne with the most hazardous impacting approximately 10,000 individuals in 2016 and resulting in ten related deaths.

Table 10.6 Lightning fatalities from some Asian countries

Country	Timeframe	Annual fatality rate per million	Average fatalities per year
Bangladesh	1990–2016	1.6	251
China	1997–2009	0.3	360
India	1967–2012	2.0	1755
Japan	1990–1997	>0	2
Malaysia	2008–2011	0.8	22
Mongolia	2004–2013	1.5	5
Singapore	1970–1979	1.5	3
Sri Lanka	2003	2.6	49

Source: Holle & Cooper (2019)

Deaths and injuries are not limited to people meeting the hazard directly. For all storm hazards, downed power lines can electrocute people. Vehicle fires and crashes are common during evacuations, although the rates are not necessarily different from day-to-day life.

Power outages, as with most hazards, can make evacuating more dangerous due to lack of lighting or can induce more caution in people. Power outages in storms add to casualties by making it more difficult to continue indoor temperature control, by leading to carbon monoxide poisoning when people use local generators and by resulting in fires when people use candles or fireplaces for heat and light. During hurricanes in the US, disproportionate numbers of ethnic minorities and children die from carbon monoxide poisoning (Alfaro et al., 2022). Medical services can be interrupted. While formal health centres and hospitals should have local, backup power, they do not always, and people might not be able to get to hospital due to transport interruptions and transport route blockages. Many people at home rely on electricity-powered medical and assistance devices, from stair lifts for people with mobility restrictions to continuous positive airway pressure (CPAP) machines for obstructive sleep apnoea (OSA).

In addition to physical health impacts, mental health impacts are always prominent, as with all other hazards. However, they are typically not tabulated in reports. One aspect of storm hazards which matches some but differs from other hazards is speed of onset and decay. Tornado warnings can be hours in advance of a touchdown with many other storms giving days. The immediate encounter with them all is typically over in minutes or hours, with some provisos. Freshwater flooding from tropical cyclones can continue for days after the storm's centre has passed while storm recurrence, especially for tornadoes and lightning, could occur at any point after one strike. It can be mentally anguishing to be picking up the pieces from one weather phenomenon when another one comes roaring through – or even anticipating what might happen next time.

Animals and plants, too, can experience injuries and deaths from similar mechanisms to people (Figure 10.8). Certainly, plants cannot take shelter themselves. How much damage occurs, such as for cotton (Yue et al., 2019) and corn (Battaglia et al., 2019), depends on hailstone size and fall density along with the timing of the hail compared to plant maturity and rates of crop defoliation, branch breaking and fruit fall. Following

Figure 10.8 A tree in Apia, Samoa, said to have been felled by Cyclone Heta in 2004

Source: Photo: Ilan Kelman

storm surge, salinisation of groundwater and soil impedes the growth and survival of many plants, also affecting animals and people when they cannot find uncontaminated food and water.

Overall death tolls from storms can be hard to determine, particularly in terms of hypothermia (freezing), hyperthermia (overheating) and carbon monoxide poisoning. These situations are not rare. Tropical cyclones occur around the world year round while cold-weather storms impact every winter. As an example of ice storms from Canada, Etkin and Maarouf (1995) detail regular incidents:

- In March 1958, 43 hours of continuous freezing rain hit St John's, Newfoundland.
- In February 1961, power was out for a week in parts of Montréal, Québec, after wires broke during an ice storm.
- In January 1968, southern Ontario had three days of snow and ice falling leading to power outages.
- In April 1984, St John's, Newfoundland, lost electricity for several days after an ice storm.
- In December 1986, one-quarter of Ottawa's homes lost power during an ice storm.

The deadliest winter storm or deadliest blizzard is generally said to have happened in Iran in 1972, killing 4,000 people. The deadliest known tropical cyclones are listed in Table 10.7.

Aside from numbers being hard to verify from storms throughout history, not every death was necessarily recorded, particularly among people subjected to slavery or considered to be of a low class. Additionally, cold weather and storms sometimes merge into a single season's toll, with many

Table 10.7 The ten deadliest known tropical cyclones

Modern-day country	Year	Minimum death toll
Bangladesh	1970	300,000
India and Bangladesh	1737	300,000
Vietnam	1881	300,000
India	1839	300,000
Bangladesh	1584	200,000
Bangladesh	1876	200,000
Bangladesh	1897	175,000
China	1975	171,000
Bangladesh	1991	138,000
Myanmar	2008	138,000

Source: Adapted from www.wunderground.com/hurricane/articles/deadliest-tropical-cyclones

armies in human history being destroyed in a winter through combat and ill preparedness for the weather. In the UK every year, winter deaths due to being unable to afford heating – referred to as *fuel poverty* – occur. These numbers are difficult to specify, as the aggregate of 'excess winter deaths' recorded by the UK's Office of National Statistics does not differentiate this cause.

For tornadoes, casualties and damage are mainly a function of people getting to shelter and the robustness of infrastructure, rather than the tornado wind speed, width or path length. In Canada and the USA, occupants of mobile homes and road vehicles have been most likely to die in tornadoes. These patterns shift at times when flimsy structures such large warehouses or small sheds (Figure 10.9) are hit rather than mobile homes. Many vehicle-related tornado deaths occur when people try to shelter in underpasses. Meanwhile, hospitals can have difficulties treating the large number of survivors with soft tissue injuries and fractures, again being more a result of being in the path of debris than of the tornado's specific characteristics.

Tropical cyclones illustrate further. When Hurricane Fifi struck Honduras as a Category 2 storm in 1974, it produced landslides on the steep hills where many poorer people had relocated after being forced off more fertile valley land due to conflict and land grabs. Over 8,000 people died. Then, in 1998 Hurricane Mitch peaked at Category 5 but made landfall in Honduras as Category 1. As it slowly swept over Central America dumping its rain, over 11,000 people died, perhaps the same number were injured, and 80,000 were left homeless with aid helping 2.5 million people. In the mountains of Honduras, 100–150 cm of rain fell within 48 hours and created over 1 million landslides and mudflows. About 60% of bridges, 25% of schools and 50% of the agricultural base, mainly in the cash-crop sector of bananas and coffee, were destroyed. This latter impact demonstrates vulnerability: decades of oppression of Hondurans and outside demands on them created the cash cropping which reduced diversity of available food and increased dependence on outside markets. No one would buy wrecked crops and the farmers knew few alternatives for getting through the disaster, meaning that humanitarian assistance was essential.

Another storm to hit the area was Hurricane Stan in 2005, peaking at Category 1 and killing over 1,500 people. Then in 2020, back-to-back Category 4 hurricanes made landfall on Central America: Hurricane Eta and Hurricane Iota. The total death toll from both was around 500 people. Across Fifi, Mitch, Stan, Eta and Iota, few patterns between peak or landfall wind speed and hurricane deaths are evident in Central America. The reason is that the people's vulnerability was different for the times and places hit, showing how hazard characteristics do not necessarily match with disaster characteristics.

Similarly, Hurricane Katrina in 2005 was Category 5 as it traversed Florida causing damage without a significant calamity. The storm was borderline Category 3/4 upon landfall in Louisiana and Mississippi, and then over 1,800 people died with thousands more lives upended as New Orleans' levees failed and parts of the city flooded. The disaster's cause was not the hurricane but rather building a city below sea level and engraining huge social inequalities and inequities, while not being prepared for a hurricane – not having options to prepare due to poverty and marginalisation – despite being in the hurricane belt during hurricane season (Box 10.1). Corruption and ineptitude from local to national levels provided a baseline for a disaster to happen irrespective of the hazard's characteristics.

Katrina's storyline is repeated in different forms across other tropical cyclones and storm hazards more generally. Similar stories are documented for Cyclone Nargis in Burma in 2008 and Typhoon Haiyan (Yolanda) in the Philippines in 2013, among many others. Impacts and consequences – that is, the disaster – do not come from the environmental hazard but from vulnerability.

Figure 10.9 A small shed as potentially lethal tornado debris after the 24 April 2007 tornadoes in Rosita Valley, Texas.

Source: Photo: Ilan Kelman

Box 10.1 The Hurricane Katrina Disaster

Much of New Orleans, Louisiana, lies mostly below sea level, up to 3.0 m below on sinking alluvial and peat soils (Waltham, 2005). Rainwater drainage is routinely pumped from low-lying areas into Lake Pontchartrain, but the city relies on the surrounding wetlands and barrier islands, plus a complex system of artificial floodwalls and levees, to prevent inundation from hurricane storm surge. In the years before Hurricane Katrina, these defences weakened due to lack of maintenance and upgrades while levee construction and dredging limited sediment supply for delta renewal. Wetlands lost up to 75 km^2 each year as the barrier islands along the Louisiana coast eroded at rates of up to 20 m each year. As a result, the entire delta is subsiding, New Orleans is sinking further below sea level and the natural coastal buffer has been largely destroyed.

Human vulnerability compounds this picture. Unemployment, underemployment and racism have been chronic concerns for Louisiana and Mississippi. Perhaps 25% of all families in New Orleans lived in poverty in 2005. Much infrastructure needs repair and maintenance, especially people's homes in poorer areas.

Soon after 6:00 a.m. local time on 29 August 2005, Hurricane Katrina made landfall in southeast Louisiana. The Mississippi coastal towns of Biloxi and Gulfport also suffered major damage, even to engineered structures, from a storm surge at least 7.5 m above sea level, before the storm swept inland. Driven by strong northerly winds, the waters of Lake Pontchartrain were pushed against the flood defences. Many failed and floodwater flowed into the northern areas of New Orleans below sea level, reaching depths of 1.5–2.0 m over about 80% of the city (Figure 10.10). Slightly higher areas – formed from fossil beaches and fluvial deposits related to a previous course of the Mississippi River – were wet from rainfall, but did not flood. The depth of flooding was directly linked to land elevation, so the lowest parts of the city, which were largely residential, suffered most. Almost 80% of all direct property damage was in the residential sector.

About two-thirds of the flooding was due to breaks in the levee system and one-third was due to overtopping, with rainfall adding to the internal water levels. Altogether, 50 levees were damaged, 46 of them due to breaching and overtopping caused by a mix of underseepage, scour erosion behind the structures and erosion along the tops of levees. For example, the Industrial Canal levee was undermined by seepage through the underlying silt and sand, whilst the 17th Street canal failure was due to overtopping and collapse. Although the severity of Hurricane Katrina technically exceeded the design criteria for the New Orleans flood defences, the system should have performed better than it did (Interagency Performance Evaluation Taskforce, 2009).

Engineering Problems Included

- Over-reliance on a series of hurricane design models dating back to 1965.
- The piecemeal development and maintenance of the levees giving inconsistent levels of protection.

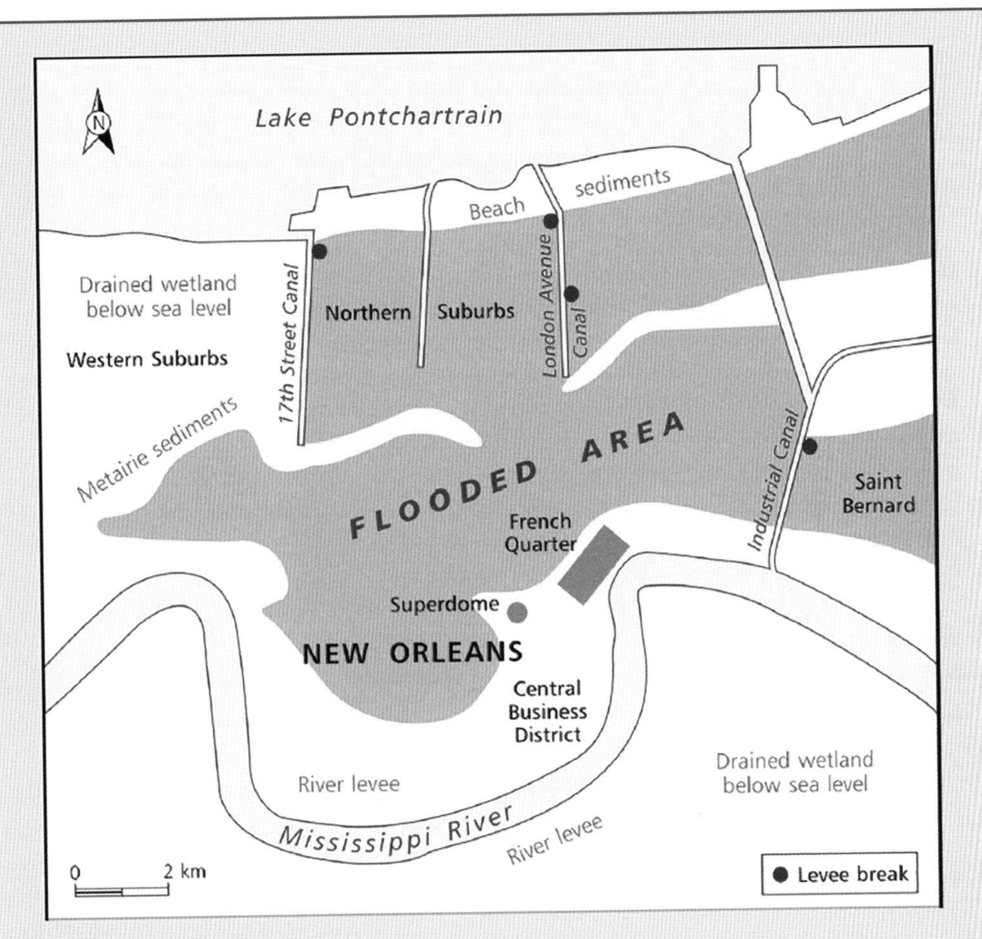

Figure 10.10 Some effects of Hurricane Katrina on New Orleans in August 2005. The location of the main levee breaks is indicated, along with the extent of flooding in the central urban area. Many residents found temporary shelter in the Superdome.

Source: After T. Waltham (2005), The Flooding of New Orleans. *Geology Today* 21 (2005): 225–31. Copyright John Wiley and Sons 2005. Reproduced with permission.

- Failure to take into account varying rates of ground subsidence across the area and to build structures, like flood outfalls, above the appropriate datum.
- Inability of the pumping stations to cope with the demand, so that only 16% of the total capacity operated during the storm.
- Underestimation of the dynamic forces acting on the flood defences and the erodible soils.

Chronic Social Problems Included

- Inadequate policy and preparedness, including protocols for evacuation (especially catering to households without private vehicles), post-disaster shelter and humanitarian aid.

- Insufficient social services and health care over the long-term meaning that the poor, the elderly, people with disabilities and ethnic minorities tended to suffer most.
- A mayor who was later convicted for fraud, bribery and money laundering.
- The head of the US Federal Emergency Management Agency (FEMA) had almost no prior disaster-related experience.

Despite these known long-term technical and societal issues, and despite New Orleans having a prior history of hurricane disasters such as Hurricane Betsy in 1965 and Hurricane Camille in 1969, the city's governance was underprepared. In July 2004, a training exercise called Hurricane Pam involved emergency officials responding to a scenario of a direct Category 3 hurricane strike. Search and rescue, hospital evacuation, temporary shelters and debris removal were lacking. Yet as Hurricane Katrina bore down just over a year later, the mandatory evacuation order was not issued until the day before landfall. Eighty per cent of the population managed to get out, leaving around 100,000 people, generally the least able to prepare or evacuate. An estimated 1,833 people died during the hurricane and subsequent floods.

These reasons explain why a bipartisan US government report on Hurricane Katrina was called 'A Failure of Initiative' (US House of Representatives, 2006).

C RISK ASSESSMENT AND MANAGEMENT

Risk assessment and management for storms covers a huge range of time and space scales. Cyclonic storms and winter storms can affect places for hours or days, tornadoes typically pass by in minutes and lightning strikes last around 1/5 of a second on average even while the threat might be present for hours at a time and subsequent fires can continue for longer. Flash floods resulting from storms are addressed in Chapter 11. Warnings covering a large area can be issued days in advance, with spatial precision being refined as the weather's behaviour becomes clearer.

As Typhoon Haiyan (Yolanda) approached the Philippines in November 2013, maps detailing the probabilities of its path were released illustrating ranges of 10% bands of likelihood that the storm's centre would pass within 120 km over the next ten days. On 4 November, it was clear that it would make a direct pass over the central Philippines with a greater than 5% chance of it reaching anywhere from Burma's capital Naypyidaw and Thailand's capital Bangkok to Vietnam's capital Hanoi and Nanning in southern China. As Haiyan made landfall on the Philippines, eventually killing over 6,000 people mainly from storm surge, the greater than 5% chance no longer included Naypyidaw or Bangkok, although there was a slight enlargement in area covered within southern China.

Tornado warnings can learn from and offer lessons covering the need to plan and to be aware of warnings and response actions long before a hazard appears. And actions do show success. Despite over 40 recorded instances around the world throughout human history of disasters involving tornadoes killing over 100 people each, only one confirmed one (so far) happened after 2000. This was on 22 May 2011 with Joplin, Missouri, particularly hard hit, leading to a total death toll of at least 158. On 10–11 December 2021, especially in Kentucky, fatalities from a tornado outbreak came close to 100, demonstrating that the potential remains for disasters involving tornadoes to reach several hundred fatalities.

In fact, since the historical catastrophes involving tornadoes – from multiple capsized ships in Valletta, Malta's harbour in the 16th century to hundreds of bodies each in the USSR and Bangladesh in the 1980s – population numbers have increased and are not always living and working in tornado-safer infrastructure and conditions. Even while knowledge, awareness, planning regulations, building codes, shelters and desire to act have improved in many places – for tornadoes, tropical cyclones, lightning (Figure 10.11), wind, blizzards and other storm hazards including floods (Chapter 11) – monitoring and enforcement of regulations and giving people resources and opportunities to act have not necessarily improved. A workplace tornado shelter might exist for everyone and warnings might be issued which are received and understood, but not all employers are amenable to their employees losing 15–60 minutes of work for every single tornado warning.

At home, only a minority around the world can afford a well-stocked family or community shelter, and not everyone has mobile devices for alerts, whether a wind-up handheld radio or the latest smartphone. If messages are issued in only a few languages, not everyone in an area will be covered, so wording should ensure that it remains clear when put through automatic translators. People with hearing impairments will not necessarily pick up on local sirens, especially when asleep – which also goes for people with full hearing who are inebriated, using cognition-inhibiting medications or substances, using earplugs, or in buildings with noise-reduction glazing.

Even when being aware of storm-related sirens sounding, should the response be to go below ground for tornadoes or to higher storeys in case the storm brings a flood (Chapter 11)? Both can occur simultaneously. How do people balance the need to stay indoors for lightning and hail with the need to evacuate to safety for tornadoes, and so being outdoors for a short time to get into an external shelter. For people who own private vehicles, how should they balance getting away from a tropical cyclone while avoiding vehicles for tornadoes and never driving through floodwater (Chapter 11)? If in a shelter for tornadoes, could debris block the exits meaning that supplies are needed for days rather than hours? Electrocutions and fires are commonly documented after a tropical cyclone, suggesting the importance of these considerations for tornadoes.

Safer buildings help prevent casualties and disruptions during and after severe storms. Tropical Cyclone Tracy devastated Darwin, Australia, on 25 December 1974, resulting in the deaths of 71 people and leaving only 6% of the city's housing habitable, which led to a mass post-storm evacuation of almost 35,000 people (Haynes et al., 2011). Because of Darwin's remoteness, people were evacuated to other cities around Australia, hundreds of kilometres from Darwin rather than near the Darwin area, with some families split up due to priority-led protocols where fathers were separated from their wives and children – noting the gendered assumptions and constructions regarding

Figure 10.11 Lightning protection test link to inspect a lightning conductor and installation for earth grounding for a building in Singapore.

Source: Photo: Ilan Kelman

family structure and who should stay with whom. These decisions forced on disaster-affected people caused further stress associated with displacement, separation and isolation. To prevent this situation from recurring, significant changes were made to Australian building codes to help buildings withstand the impacts of severe winds. These codes are designed to ensure that buildings remain stable and do not collapse during severe winds, thereby minimising loss and damage (Henderson & Ginger, 2008). Continuous improvements to building design in the Building Codes of Australia have contributed to newly built structures remaining structurally sound during tropical cyclones, noting that wind is not the only or necessarily most severe hazard from tropical cyclones.

Our abilities to live with storm hazards have demonstrated significant advances in risk assessment and management, helping to avoid casualties and mitigate damage. Numerous aspects remain ready for further, deeper investigation and action.

KEY TAKE AWAY POINTS

- Storm hazards include precipitation, wind, lightning, air temperature and humidity with specific storms often referred to as winter storms, thunderstorms, hailstorms, cyclonic storms, tornadoes and others.
- Due to measures taken or not taken to deal with storms before they happen, there is poor correlation between the hazards and the impacts and consequences.
- Some of the measures can be different depending on the storm hazard, meaning that careful assessment and planning is needed to make the most effective decisions.
- A combination of actions is needed of warnings, building codes, planning regulations and giving people resources and opportunities to get information and to act on it.

BIBLIOGRAPHY

Abraham, J., Abraham, J., Bendimerad, F., Berger, A., Boissonnade, A., Collignon, O., Couchmann, E., Grandjean, F., McKay, S., Miller, C., Mortgat, C., Muir-Wood, R., Page, B., Shah, T., Smith, S., Wiart, P., & Xien, Y. (2000). *Windstorms Lothar and Martin, December 26–28, 1999.* Newark, CA: Risk Management Solutions.

Alfaro, S., Sen-Crowe, B., Autrey, C., & Elkbuli, A. (2022). Trends in carbon monoxide poisoning deaths in high frequency hurricane states from 2014–19: The need for prevention intervention strategies. *Journal of Public Health*, article fdac053. https://doi.org/10.1093/pubmed/fdac053

Barry, R. G., & Chorley, R. J. (1987). *Atmosphere, weather and climate.* Methuen, MA and London: Routledge.

Battaglia, M., Lee, C., Thomason, W., Fike, J., & Sadeghpour, A. (2019). Hail damage impacts on corn productivity: A review. *Crop Science, 59*(1), 1–14.

Committee of Experts of State Disaster Relief Commission. (2008). *Comprehensive evaluation of the sleet and snow disaster in South China.* Beijing: Committee of Experts of State Disaster Relief Commission.

Etkin, D., & Maarouf, A. (1995, February 11–14). An overview of atmospheric natural hazards in Canada. In D. Etkin (Ed.), *Proceedings of a tri-lateral workshop on natural hazards* (pp. 1–63 to 1–92). Merrickville, Canada: Sam Jakes Inn.

Fujita, T. T. (1973). Tornadoes around the world. *Weatherwise, 26*, 56–62, 79–83.

Haynes, K., Bird, D. K., Carson, D., Larkin, S., & Mason, M. (2011). Institutional response and Indigenous experiences of Cyclone Tracy. Report to National Climate Change Adaptation Research Facility, Gold Coast, Australia.

Henderson, D., & Ginger, J. (2008). Role of building codes and construction standards in windstorm disaster mitigation. *The Australian Journal of Emergency Management, 23*, 40–46.

Holle, R. L., & Cooper, M. A. (2019, June 12–14). Overview of lightning injuries around the world. Paper from The 11th Asia-Pacific International Conference on Lightning, Hong Kong.

Interagency Performance Evaluation Taskforce. (2009). Performance evaluation of the New Orleans and Southeast Louisiana Hurricane Protection System: Final report of the interagency performance evaluation task force. US Army Corps of Engineers (USACE), New Orleans, Louisiana.

Kafali, C. (2011). Regional wind vulnerability in Europe. Report 04.2011, AIR Currents, AIR World-wide, Boston.

Kocin, P. J., & Uccellini, L. W. (2004). A snowfall impact scale derived from North-East storm snow fall distributions. *Bulletin of the American Meteorological Society, 85*, 177–194.

Kreft, P., & Crouch, J. (2011). Albany Tornado, Tuesday 03 May 2011. *Weather and Climate, 31*, 67–80.

León-Cruz, J. F., Pineda-Martínez, L. P., & Carbajal, N. (2022). Tornado climatology and potentially severe convective environments in Mexico. *Climate Research, 87*, 147–165.

Long, J. A., Stoy, P. C., & Gerken, T. (2018). Tornado seasonality in the southeastern United States. *Weather and Climate Extremes, 20*, 81–91.

Munich Re. (2002). Winter storms in Europe (II): Analysis of 1999 losses and loss potentials. Geo Risks Research Department, Munich Re Group, Munich.

Paul, B. K. (1997). Survival mechanisms to cope with the 1996 tornado in Tangail, Bangladesh: A case study. Quick Response Report 92, Natural Hazards Research and Applications Information Center, Boulder, CO.

Price, D., Hughes, K. M., Thien, F., & Suphioglu, C. (2021). Epidemic thunderstorm asthma: Lessons learned from the storm downunder. *The Journal of Allergy and Clinical Immunology: In Practice, 9*(4), 1510–1515.

Shi, P., Fung, G. M., & Dickinson, D. (2022). Assessing hail risk for property insurers with a dependent marked point process. *Journal of the Royal Statistical Society: Statistics in Society*, Series A, *185*(1), 302–328.

Usbeck, T., Wohlgemuth, T., Dobbertin, M., Pfister, C., Bürgi, A., & Rebetez, M. (2010). Increasing storm damage to forests in Switzerland from 1858 to 2007. *Agricultural and Forest Meteorology, 150*, 47–55.

US House of Representatives. (2006). A failure of initiative: Final report of the select bipartisan committee to investigate the preparation for and response to hurricane Katrina. 109th Congress, 2nd Session, Report 109–377, United States Government, Washington, DC.

Waltham, T. (2005). The flooding of New Orleans. *Geology Today, 21*, 225–231.

Ye, Q., Wang, M., & Han, J. (2012). Integrated risk governance in the Yungui Plateau, China: The 2008 ice-snow storm disaster. *Revue de géographie alpine/Journal of Alpine Research*, 100–101. www.doi.org/10.4000/rga.1731

Yue, Y., Zhou, L., Zhu, A.-X., & Ye, X. (2019). Vulnerability of cotton subjected to hail damage. *PLoS One, 14*(1), e0210787.

FURTHER READING

Cerveny, R. S., Bessemoulin, P., Burt, C. C., Cooper, M. A., Cunjie, Z., Dewan, A., Finch, J., Holle, R. L., Kalkstein, L., Kruger, A., Lee, T.-C., Martínez, R., Mohapatra, M., Pattanaik, D. R., Peterson, T. C., Sheridan, S., Trewin, B., Tait, A., & Wahab, M. M. A. (2017). WMO assessment of weather and climate mortality extremes: Lightning, tropical cyclones, tornadoes, and hail. *Weather, Climate, and Society, 9*(3), 487–497.

Di Gangi, E., Lapierre, J., Stock, M., Hoekzema, M., & Cunha, B. (2022). Analyzing lightning characteristics in central and southern South America. *Electric Power Systems Research, 213*, article 108704.

Doswell, C. A. (Ed.). (2015). *Severe convective storms.* Boston, MA: American Meteorological Society.

WEB LINKS

National Lightning Safety Council. www.lightningsafetycouncil.org

NOAA National Severe Storms Laboratory. www.nssl.noaa.gov

Philippine Atmospheric, Geophysical and Astronomical Services Administration. www.pagasa.dost.gov.ph

Tornado and Storm Research Organisation. www.torro.org.uk

World Meteorological Organization. www.wmo.int

CHAPTER ELEVEN

Flood Hazards

11

OVERVIEW

This chapter covers flood hazards, explaining the term's ambiguity and the parameters chosen as descriptors. Floods are generally characterised through physical properties (depth, velocity, duration, temperature, capillary rise, suction, scour, waves, frozen water, water vapour affecting humidity, non-water substances including sediment), chemical properties including salt from the ocean, energy properties including electricity, nuclear properties from radioactive substances and biological properties from animals, plants and microbes. Impacts and consequences are seen through multiple causes of deaths, injuries and mental health influences as well as the results of damage to infrastructure, agriculture and systems for water, energy and health. Risk assessment and management should use the advantages of floods, such as creating fertile soil and supporting livelihoods based on coastal ecosystems, while enacting physical and social changes to protect infrastructure and support people for a wide variety of preparedness, warning and response techniques to cover the gamut of flood hazards. Often, floodwalls such as dikes and levees are relied on, even though they can increase flood risk over the long term.

A. THE NATURE OF FLOODS

1. Defining Floods

A flood hazard is, in effect, too much water, sometimes articulated as water where it is not wanted. What either conceptualisation and variation means is subjective, since some people and infrastructure can function safely when inundated and some cannot. Nonetheless, areas which are for the most part dry can sometimes end up wet. This situation is generally termed a flood, and the flood can be hazardous if the assumption is that the area is and should be dry all the time.

DOI: 10.4324/9781351261647-13

Flood hazard classification also has subjectivities. Flood types at the first order are generally and arbitrarily categorised by the type of water and the speed of its appearance.

Slow freshwater floods emerge from rainfall or snow melt filling up a waterway or water body so that the water spreads out beyond the banks. An ice jam, landslide, felled vegetation or other blockage in a watercourse can lead to the water steadily flowing beyond its usual channel. For ice jams, floating ice generally from the usual spring break-up can temporarily block a waterway and can lodge at bridges, within channel constrictions or at shallows, sometimes freezing the entire channel in the vicinity so that water behind it backs up and spreads out. Near shorelines of lakes, pressure ridges in the ice can dislodge houses from their foundations.

Fast freshwater floods refer to a lot of water from a storm or sudden ice or snow melt rapidly covering land, through what is called flash flooding or ponding (sometimes referred to as surface water flooding, although this term has various meanings). Ponding is not necessarily associated with a waterway or body of water, but is when water cannot drain fast enough and accumulates. It is often small scale and can simply be because drainage is not designed or maintained for the location (Figures 11.1 and 11.2). Flash flooding can occur in steep canyons where the river accumulates run-off and suddenly rises, as well as in a wadi (or wād) which

Figure 11.1 Shallow flooding in Malé, Maldives, after rainfall because the water does not drain fast enough.

Source: Photo: Ilan Kelman

Figure 11.2 Water ponding in Vilnius, Lithuania, during rainfall because drains are not maintained.

Source: Photo: Ilan Kelman

is a generally dry streambed in the Middle East or North Africa which becomes a flowing stream with little warning during the rainy season.

Glacial lake outburst floods (GLOFs), jökulhlaups (Chapter 7) and other forms of sudden melt of frozen water lead to fast freshwater floods.

Another source of fast freshwater floods is the failure of structures such as dams, levees and dikes. Around the world, many of these structures are old, unmonitored or unmaintained, with unmapped inundation zones and with unclear ownership or responsibility for monitoring, maintenance and repair. An earthquake or heavy rainfall could lead to a sudden break. Wars across centuries have targeted enemy dams, an act that is now illegal under international humanitarian law through Article 56 of the Protocol I of 8 June 1977 Additional to the Geneva Conventions of 12 August 1949. Structurally sound dams can also lead to fast freshwater floods. On 9 October 1963, a landslide into the reservoir behind Italy's Vaiont Dam led to a massive wave overtopping the structure, surging downstream and killing around 2,000 people. The dam did not break. Few countries have prepared inundation maps or made emergency plans for any such events.

Slow saltwater floods come from seas and oceans, where most of the world's water lies, and it contains mineral ions that we taste, smell, feel and interpret as 'salt'. Around 3.5% of ocean water's weight is from these dissolved minerals. Tidal cycles occur over several hours with annual peaks and troughs, meaning that some land is mainly dry but then gets wet at times. In Richmond in west London,

UK, the Thames River is tidal, and the highest tides flood parked cars left there by unwary or uncaring drivers (Figure 11.3). Over decades and centuries, sea level changes because the oceans rise and because coasts subside or erode, leading to inland areas being flooded with saltwater (Figure 11.4).

Fast saltwater floods have different sources. Storm surge flooding can inundate vast swathes with mere hours of warning – and only seconds if sea defences breach and the water rushes in. Storm surges occur when a weather system lowers the air pressure so that the sea level rises while strong winds push water to the shore. When these phenomena coincide with high tide or exceed the usual high tide level, then storm surge flooding occurs, with Figure 11.5 showing a map of high-risk areas around England and Wales. Storm surges can propagate up rivers, leading to saltwater flooding inland. Other rapid saltwater floods include seiching, which is when the shaking from an earthquake (Chapter 6) leads to a body of water (which could also be freshwater) sloshing around so that the water ends up on land, tsunamis (Chapter 8) and tidal bores, all of which can also propagate inland by going upstream along waterways.

Subjectivities emerge in that the binaries of slow–fast and fresh–salty (or saline) are continua, not dichotomies. Speed of the flood can range from seconds (a canyon experiencing a cloudburst) to millennia (ice ages ending due to changes in the Earth's orbit around the sun, so the glacial melt leads to sea level rising dozens of metres), with all timescales in between represented.

Figure 11.3 Tidal flooding from the River Thames in Richmond, west London, the UK, a regular occurrence.

Source: Photo: Ilan Kelman

Figure 11.4 Water not draining on the landward side of the sea wall in western Tongatapu, Tonga, expected to be more common as land erodes and sea level rises.

Source: Photo: Ilan Kelman

Today, sea-level rise is occurring due to anthropogenic climate change (IPCC, 2021–2022). Global warming means that snow, ice and permafrost melt, with the water running into the oceans and increasing its volume. The ocean water absorbs excess heat from the atmosphere, warming up and expanding in volume (called 'thermal expansion'). These two phenomena together might lead to oceans up to 2 m higher than pre-industrial periods by 2100. If the Greenland or Antarctic ice sheets melt, then sea level could rise dozens of metres over coming centuries, leading to severe slow-rise coastal flooding. In some places, coastal flooding is exacerbated by land subsidence, coming from natural factors as long-term land shifts for geological reasons and from short-term human-caused factors as extraction of groundwater and petroleum. The weight of infrastructure and construction techniques can compact land over years, lowering it with respect to the ocean.

Consequently, various timescales for floods are enmeshed. What is fast and what is slow depends on definitions, as demonstrated in the circularity of the preceding definitions, for instance a 'fast flood' occurs 'rapidly'. Similarly, the freshwater–saltwater axis includes all possibilities from almost no mineral ions at some river headwaters to seas so salty that people float in them. Estuaries, deltas and tidal waterways are brackish while sea salinity varies around the world, according to the depth at which the water's saltiness is measured and over time (Durack & Wijffels, 2010).

Other parameters describe floods further, with the most prominent being depth (deep or shallow) and duration (short or long). These terms are also continuous with the decision of where to separate the descriptors being subjective. Water holds impurities other than mineral ions, such as sediment and human-made chemicals, yet fresh–salty is often the principal divider of floodwater compositions. A location can be subject to many different types of floods, as with Bangladesh in Figure 11.6, demonstrating again that indisputable flood categories are difficult to justify.

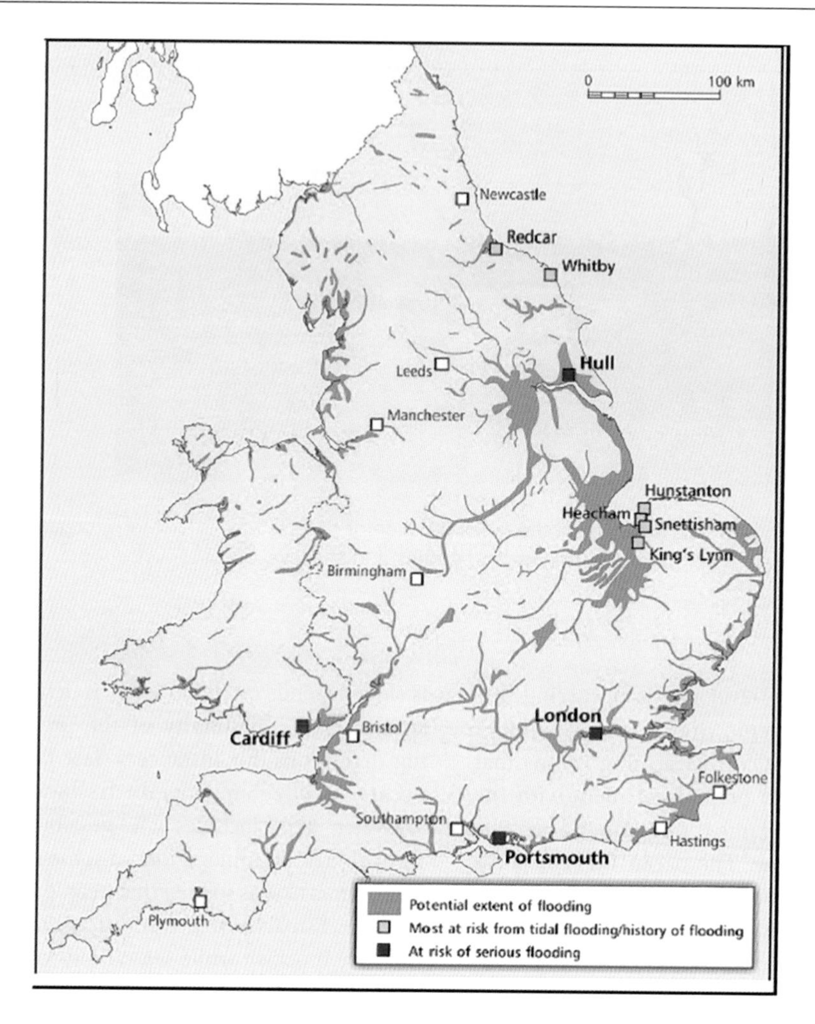

Figure 11.5 Areas of potential coastal flooding in England and Wales.

Source: After a map provided by the Environment Agency of England and Wales accessed on 19 September 2011 but no longer available

Similarly, what is a flood and what is a flood hazard is not clearly delineated. River floodplains, lake shorelines, coastal tidal zones, and marshes, bogs, swamps and other wetlands are examples of areas perpetually flooding with thriving ecosystems based on and requiring the flooding. These water–land interfaces are not hazardous per se unless people, infrastructure and their needs are interfered with. Placing infrastructure upon, living in or setting up transportation routes across locations known to get wet, under the assumption that they will remain dry, manufactures a flood hazard from a flood. That is, being in a floodable area under the assumption that it will not flood means that a flood becomes a flood hazard. The hazardous nature is subjective and not consistently characterisable due to human decisions determining the level of potential and actual hazard.

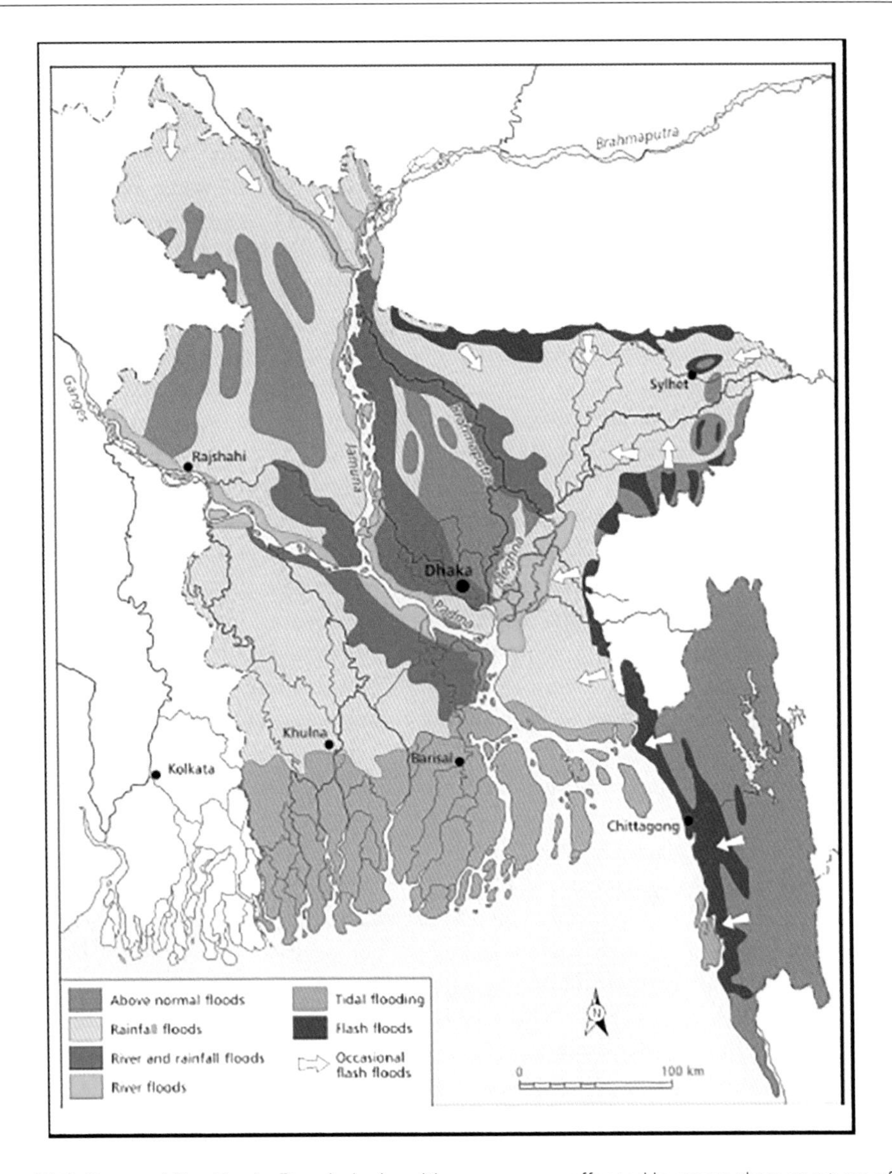

Figure 11.6 Types of flooding in Bangladesh, with some areas affected by more than one type of flood. The statements 'above normal' and 'occasional' are subjective in terms of how 'normal' is calculated and how much 'above' and 'occasional' must be for these classifications.

Source: After Brammer (2000) and reproduced with permission

Conversely, many locations are not frequently wet, so inundation is unusual. A storm surge or tsunami several metres high reaching hundreds of metres inland could flood areas outside of a floodplain by most measures. If rainfall continues with the standing water not draining, then ponding can cover a fairly large area. In such circumstances, water has expanded beyond the water–land interface, so the flood by definition is a flood hazard. As with fast–slow and saline–fresh, hazardous–unhazardous is a continuum and can depend on the specific circumstances and adopted definitions.

2. Flood Parameters

As alluded to in the previous section, floods are characterised by a variety of parameters. The description of water as different types of floods is often inseparable from its interactions with different human environments.

Floodwater depth indicates how high the water level is, either above the water surface or on land that is usually dry. Again, what is a 'usual' water surface or what 'usually dry' means is subjective, as rain falling on land and being absorbed could mean that the land should not be called 'dry'.

Where water meets infrastructure, its weight imparts a hydrostatic (non-moving water) pressure laterally (horizontally). Another hydrostatic component of flood depth is capillary rise. Capillary rise refers to water flowing through narrow spaces, which could be pores in soil or tiny gaps between a toothbrush's hairs. This type of flow can occur upwards irrespective of gravity, saturating soil above the level where soil saturation typically stops (called the 'water table'). Definitionally, wet soil above the water table is technically dry land getting wet, although terming it a 'flood' is debatable. Capillary rise can impart pressures on and can moisten structures it encounters, leading to water or flood damage.

Flood velocity describes the speed and direction of water. The three-dimensional velocity field of a flood is complicated due to large speed variations over short times and distances. Shorelines and infrastructure produce drag or present a barrier, substantially slowing the nearby water. Eddies, vortices, pools and surface choppiness can produce rapid direction changes for the water. Waves (Figure 11.7) generate complex three-dimensional multi-directional speeds, and some waves break imparting significant and swiftly changing forces. Water, including in floods, has varying flow regimes, the two most common ones called 'laminar' (comparatively smooth) and 'turbulent' (comparatively complicated), with transition states between those two. Velocities change quickly and substantively over small times and distances in all dimensions, leading to dangerous forces and pressures, often experienced as water currents. During storm surges, the wind can shift suddenly, meaning that the water drains swiftly producing an ebb surge, with a similar phenomenon observed after a tsunami wave has come ashore. Most calculations for velocity-related forces and pressures must make numerous assumptions. They may include the properties of the floodwater's viscosity, that the floodwater cannot be compressed, that it lacks heat and other energy form transfers internally and externally, that it behaves as a continuum rather than comprising discrete particles and that the functions of its properties obey some rules of calculus.

As water flows around an object, further forces can be experienced with suction and scour being common. These are a consequence of flood velocity encountering solids – whether soil, vegetation or infrastructure – rather than being flood velocity per se. Through suction and scour, materials can be entrained in water leading to erosion, as can occur with water seeping through materials. Through such forces, floods can

Figure 11.7 The complex forces of breaking waves are good for surfing, as in Huntington Beach, California, but not good for infrastructure.

Source: Photo: Ilan Kelman

destabilise slopes leading to mass movements (Chapter 9) with the slipping material covering the full range from mainly solid such as rocks to mainly liquid such as water with soil. In between, mudflows/lahars (Chapters 7 and 9), slush avalanches and many others are floodlike even if not within a standard vision of what a flood is.

These instances demonstrate materials in floods other than mineral ions. Rocks, soil, sand, grains, vegetation, infrastructure, vehicles, animals and people are often referred to as 'debris' within floodwater. All affect the flood's characteristics, changing depth, velocity, duration and subsequent forces and pressures. Substances in floods can be chemical, including petroleum, vehicle or cooking oil, paints, household chemicals, pesticides and fertilisers. Sewage produces a foul-smelling slurry that floods infrastructure. Energy-related flood hazards are electrocution and water temperature, while the presence of a flood can humidify the air. Nuclear contamination is possible with radioactive substances, natural or anthropogenic, such as from a nuclear power plant. Animals and plants, alive or dead, in floodwater can bring dangers. Snakes and spiders, venomous or not, commonly bite people wading through or swimming in floodwater, as do alligators and crocodiles, while casualties during floods from fire ants are documented. Venomous fish and shark bites are a more theoretical flood possibility, although they could pose dangers as with jellyfish, man o' war and other such creatures, if stranded inland or if people in floods encounter them. Few such instances have been verified.

More common biological flood hazards are microbes. Fungi and moulds thrive in flooded locales. Without careful management, endemic diseases including cholera, leptospirosis including Weil's disease, Legionnaires' disease, typhoid fever and hepatitis A are among the ailments that can increase after a flood. Unless measures are taken, insects may breed post-flood increasing rates of already present vector-borne diseases, notably dengue, malaria and various viral fevers, among others. Diarrhoeal and gastro-intestinal diseases including cryptosporidiosis become prevalent in the absence of clean water and appropriate sanitation and hygiene. If people are evacuated through crowded transport or to crowded shelters without precautions, then human-to-human transmission can occur of numerous diseases such as colds, flu, tuberculosis and COVID-19.

In summary, flood parameters are:

- Physical: depth, velocity, duration, temperature, capillary rise, suction, scour, waves, frozen water, water vapour affecting humidity, non-water substances.
- Chemical: non-water substances.
- Energy: electricity.
- Nuclear: radioactive substances.
- Biological: animals, plants, microbes.

Other possible parameters appear only in the context of impacts and consequences (Section 11.B), as do many combinations and interactions of flood parameters.

B. IMPACTS AND CONSEQUENCES OF FLOODS

Irrespective of how a flood or flood hazard is defined, water can have lethal and destructive consequences. The depth–speed combination can sweep people and vehicles away, while bringing down infrastructure, without reaching excessive values (Figure 11.8). Cold water rapidly reduces body temperature leading to quick death. The danger to life from floodwater leads to the mantra of never entering a flooded area without appropriate training and equipment, including by driving, walking, cycling, skateboarding, scootering, snowmobiling or any other transport mode. Possible causes of death are:

- Drowning, when a person cannot get enough oxygen due to being underwater.
- Physical trauma from being hit by debris or from debris due to collapsing infrastructure such as buildings and bridges.
- Hypothermia, which can drain the strength of the fittest people leading to drowning or death from cold.
- Electrocution.
- Power outages so that people die from heat or cold or they improperly use fires or generators, burning down their dwelling or perishing from carbon monoxide (CO) poisoning.
- Vehicle fires and crashes while evacuating or sheltering in vehicles.
- Exposure to chemical or nuclear contamination in the floodwater.
- Animals and plants in the floodwater.
- Disease.

Examples of pathways to becoming a flood casualty are:

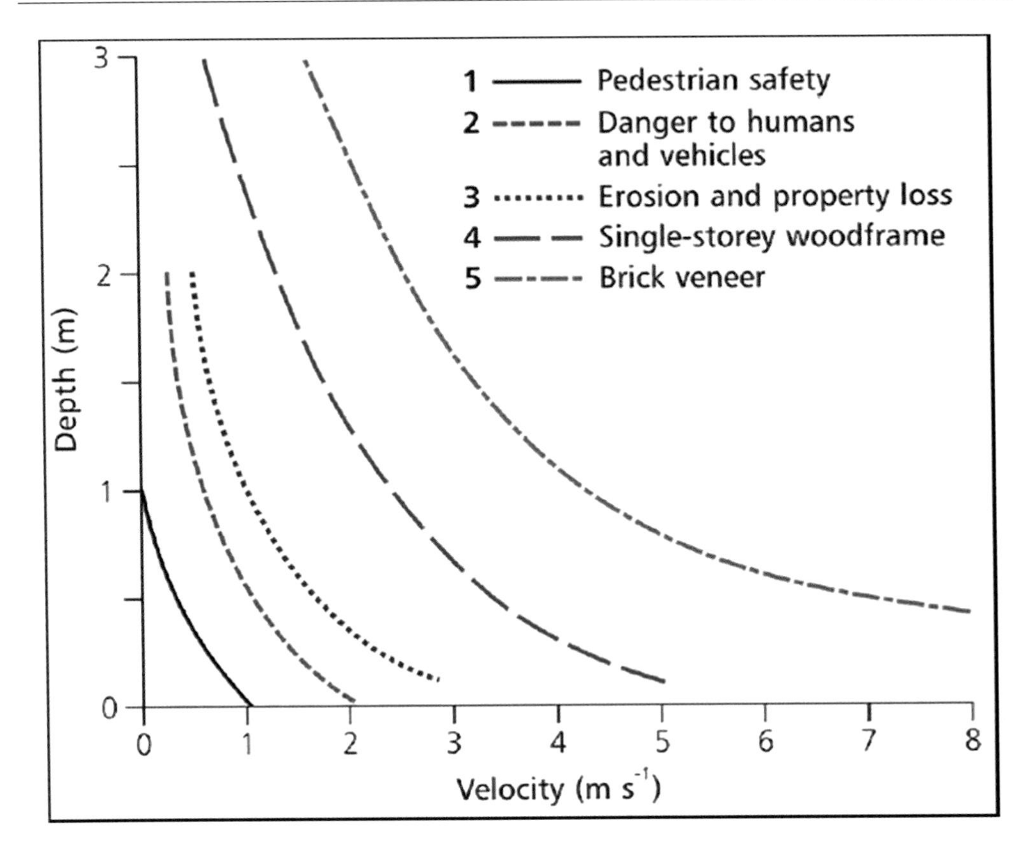

Figure 11.8 Flood hazard thresholds as a function of depth and velocity of water flow.

Source: Adapted from Smith (2000) and other sources

- A sewer cover pops off in the flood, so someone walking through the water drops through the remaining hole, drowning or succumbing to physical trauma.
- Physical trauma or drowning when the building they are in collapses due to physical force, scouring or undermining.
- Hypothermia, heat or vector-borne disease while atop a roof of a flooded building.
- A vehicle floats and tumbles, leading to physical trauma.
- Respiratory difficulties from post-flood mould.

As a specific example of hazard combinations, in November 1994 over 100 people were killed in Durunqa, Egypt, when floods destroyed a petroleum storage facility and carried burning oil into the town.

Again, the ambiguity of floods and flood hazards emerges in terms of what would be categorised as a flood death compared to a non-floodwater death. Drownings occur while on calm beaches, in swimming pools, when boats sink, in buckets, in puddles and in toilets, with the toll estimated at over 26 people per hour (WHO, 2021). Many fluids other than water have led to deaths, including beer and molasses, with both being referred to as 'floods' on occasion. On 17 October1811, in central London, UK, a vat burst sending the equivalent of over 2.5 million pub servings of beer smashing through the brewery's wall and coursing through the streets of St Giles at heights of up to 4.5 m. At least eight people died while drunkenness and alcohol poisoning were reported afterward. On 15 January1919 in Boston, Massachusetts, a storage tank burst and molasses surged through the streets at over 50 km/h and up to 7 m high, killing 21 people and injuring 150.

Most analyses of the impacts and consequences of flood hazards becoming flood disasters rely on EM-DAT with all the problems described in Chapter 1. Due to limited and inconsistent data within EM-DAT, analyses rightly supplement the data with other sources and limit the timeframe of their analysis, leading to overview statements such as:

- 'In the period 1975–2001 a total of 1816 worldwide freshwater flood events killed over 175,000 persons and affected more than 2.2 billion persons' (Jonkman, 2005, p. 171).
- From 1970–2019, 'Of the 22,326 disasters, 11,072 have been attributed to weather, climate and water hazards. These disasters resulted in 2.06 million deaths and US$3.64 trillion in losses. Thus, over the last 50 years, 50% of all recorded disasters, 45% of related deaths and 74% of related economic losses were due to weather, climate and water hazards' (WMO, 2021).
- 'During the period 1985–2003, the world experienced between 1700 and 2500 (major) flood events. . . . On a global scale the frequency of floods is increasing, although the reliability of flood data tends to overestimate this trend. . . . Between 1985 and 2003 approximately 300 000 people lost their lives as a result of flooding' (Douben, 2006).

Table 11.1 provides details of some major, recent flood disasters, but numbers must always be viewed with caution and the list is not meant to be comprehensive or representative. Similarly, a continuous sequence of flood disasters continued after 2013, although the death tolls tended to be in the dozens or hundreds rather than the thousands, leading to the lack of entries in Table 11.1. Death toll estimates for the 2016 floods in India range from dozens to thousands, with the disparity perhaps partly explained by people aggregating or not aggregating numerous floods covering a wide spatial and temporal range.

Aside from morbidity due to disease and physical injury, mental health impacts following floods are commonplace, and these have long been known. Eighteen months after the 1972 flood disaster at Buffalo Creek, West Virginia, over 90% of survivors had mental health impacts attributed to their flood experience (Newman, 1976). Even earlier, after the 1968 floods in Bristol, England, deaths and physical health issues increased substantially among the people directly affected compared to those who were not, with a particularly poignant statement being (Bennet, 1970, p. 457):

> A number of patients (not necessarily from the survey population) were referred for psychiatric care whose symptoms dated from the floods. Some ascribed their troubles to the flooding, others did not. All of them had been having difficulties in their lives before July 1968 – some only just managing to keep going – and the floods came as the last straw. In the same way those who are referred to psychiatrists after

a bereavement are often found to have had difficulties beforehand with interpersonal relationships or in their general ability to cope and to run their lives.

Attribution of a health impact to a flood hazard is challenging even when the flood hazard likely had a major influence.

Other such issues arise, mirroring the challenges raised about data in previous chapters. As alluded to in the discussion on Table 11.1, many flood health impacts, including fatalities, happen in relatively small numbers during small, highly localised events. Compiling consistent and comprehensive statistics is not easy. For instance, during 2010 floods in Ghana's Central Gonja District, roads were washed out so people were travelling to hospital via boat when it capsized and two died (Bempah & Øyhus, 2017). Responding to a particular hazard may mean that other forms of risk emerge that can cause fatalities not officially listed as due to flooding. Poor boat handling skills might be the cause of death more than floods.

Additionally, as noted in other chapters with respect to all hazards, livelihood interruption, stress, access cut off to education, uncertainties and many other fundamental and long-term impacts emerge

Table 11.1 Examples of major, recent flood disasters

Year	Place	Environmental hazards	Deaths
1970	Bangladesh	Cyclone	300,000
1974	Bangladesh	Floods	28,700
1974	Central America	Hurricane	8,000
1978	India	Floods	3,800
1980	China	Floods	6,200
1988	Afghanistan	Floods	6,345
1991	Bangladesh	Cyclone	139,000
1998	Central America	Hurricane	11,000
1998	China	River floods, storms and landslides	3,656
1999	Venezuela	Flash floods and landslides	30,000
2004	India, Bangladesh, Nepal	Floods	2,200
2004	Haiti, Dominican Republic	Floods	2,000
2004	Caribbean, USA	Hurricane Jeanne, floods	2,000
2005	Central America	Hurricane	1,800
2005	USA	Hurricane	1,800
2007	Bangladesh	Cyclone Sidr	3,295
2007	Bangladesh, India, Nepal	Floods	2,030
2008	Myanmar	Cyclone Nargis	140,000
2010	Pakistan (Figure 11.9)	Floods	1,760
2013	Philippines	Typhoon Haiyan	6,334
2013	India	Flash floods	5,500

Source: Reported by Jonkman (2005), Kron (2015) and WMO (2021), among others

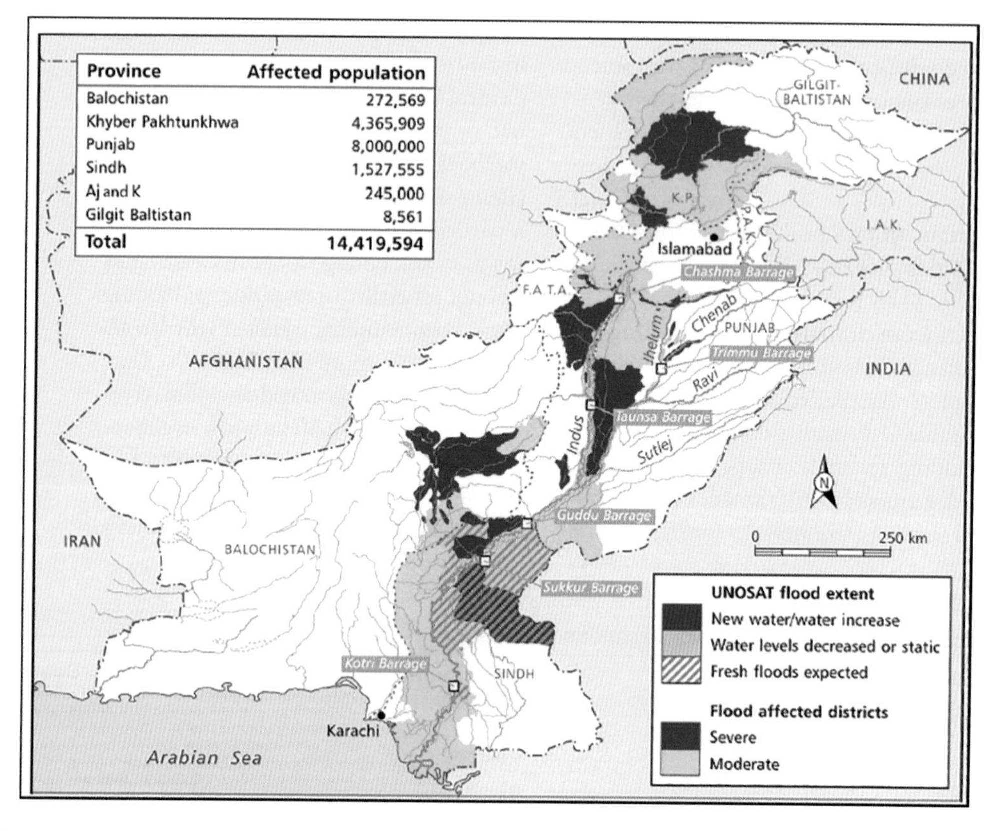

Figure 11.9 Map showing extensive flooding over Pakistan in August 2010 at which point over 14 million people were affected. By the end, more than 20 million people were affected.

Source: After OCHA (2010)

from floods. As always, these are hard to quantify and compare, so they rarely make it into tables listing data on flood hazards or flood disasters.

Impacts and consequences also relate to infrastructure and vehicle damage. When water rises around or impacts a structure, then various pressure functions could emerge:

- Water covers the entire structure from the outside, yielding a linear pressure.
- Water rises part way up the structure from the outside, yielding a linear pressure up until the water level and then no pressure.
- Water rises part way up the structure inside and outside, but to different levels, yielding a linear pressure pushing in or out across part of the structure.
- Water rises to the same level inside and outside, yielding no pressure.

Deposited sediment or debris would change any pressure differential, as would localised currents and waves. An inundated house would nominally have no pressure on the walls since the inside and outside levels are equal, yet waves, currents and debris would be more variable outside than inside, imposing some level of transient pressure differential. Peaks of waves can add a maximum force approximately equivalent to the hydrostatic force of the inside–outside depth difference, while troughs can decrease the total force by up to 40% – beyond these limits, the wave would break; the exact change in total force depends on the ratio of wave height to water depth which is less than about 0.70 for non-breaking waves (USACE, 1984).

Pressure differentials and impact forces can destroy infrastructure, vehicles and other objects. Windows, doors, walls and roofs of buildings break, ruining the structure or bringing it down entirely. Bridge supports can be taken out, leading to the bridge's collapse. Bridges over rivers can collect debris on the upstream side until the pressure is enough to break them. Floating vehicles and other objects – as diverse as cargo containers, trees and machinery – collide with each other, infrastructure and, as noted earlier, people. Any chemicals burning on the floodwater or explosive substances could lead to further structure damage through fire and blasts.

Another potentially damaging force from floods which exists when a structure or object is inundated is buoyancy. The buoyancy force is a function of the submerged volume of the object and is an uplift force, ripping buildings from their foundations and floating other objects. Moving structures can break pipes, leading to sewage and gas leaks.

The discussion of inundated properties and vehicles indicates another set of impacts and consequences: damage from contact with the water and any salinity or contaminants, including through capillary rise. Even if the structure remains intact despite the forces and pressures – including through scour and undermining – many structures and objects are ruined by water contact because water resistance measures have not been taken. Few electronic devices or paper documents are usable after flooding, meaning that people without backups lose their phone contacts, passports, birth certificates, property and land deeds, and – whether electronic or hard copy – photos, videos, work material and books. Family heirlooms, antiques, furniture, clothes and medical devices are typically unsalvageable. Any structure's systems also need thorough cleaning or replacement, including electrics, plumbing, HVAC (heating, ventilation and air conditioning) and appliances such as refrigerators, stoves and washing machines (Figure 11.10). At times, the structure remains intact post-flood yet must be gutted and refurbished especially to ensure thorough drying and cleaning (Figure 11.11).

The stress, time, effort and resources required for all these actions tend to be underestimated. Re-occupation might not be feasible for several months. Insurance companies and repair contractors can be overwhelmed with demand, even if they are not trying to deny claims or inflate prices. Both materials and labour are stretched as everyone in the flooded location tries to repair and reinhabit their properties, even while some essential services might not be restored. Once people have moved back into their home and started working again, the feeling is rarely the same, because the refurbishment and new possessions are a reminder of their traumatic experience. Figure 11.12 typifies an affected community site.

Yet moving away is not straightforward if property values decline due to the flood, flood risk, people moving away or uninsurability. Property owners can be trapped in decaying and continually flooding properties because they cannot afford, or do

Figure 11.10 Contents ruined by muddy floods in Bevendean, England.

Source: Photo: Ilan Kelman

not have access to, options for repairs, insurance or moving.

By ripping up vegetation and putting water onto land that is expected to be dry, environmental changes are a flood impact and consequence. Plants and animals can drown or be harmed by other flood characteristics such as fire, electrocution, debris impact, salinity and floodwater contaminants. Floods that realign coasts and banks change those ecosystems, as with erosion and accretion.

C. RISK ASSESSMENT AND MANAGEMENT

Fundamentally, many flood hazards could avoid becoming flood disasters if people did not depend on floodable areas for living, working or transportation. This approach is not much of a solution, because it could force people into locations with other hazards and because using floodplains has many advantages.

The regular, sometimes seasonal or annual, 'flood pulse' is a vital element in bringing water and nutrients to ecosystems, maintaining a diverse range of habitats and in some freshwater environments maintaining the soil's fertility. Many cultures around the world use 'flood farming', deliberately awaiting the flood pulse or other regular floods to support their agriculture. Freshwater floods can be used for drinking water for livestock, irrigation for crops and fisheries,

Figure 11.11 House drying out before repairs in Scott's Ferry, New Zealand.

Source: Photo: Ilan Kelman

Figure 11.12 A flooded cemetery in Barbados.

Source: Photo: Ilan Kelman)

in addition to capturing and storing the water for human use, keeping in mind the need for proper water treatment for many human uses. Periodic inundation in freshwater and saltwater environments helps to maintain wetland ecosystems which can help society without direct and tangible livelihood benefits.

Not everyone can live amphibiously, with regular flooding. Coastal mangroves, salt marshes and swamps, if managed well, can contribute to reducing flood damage by diminishing the force of waves and currents while absorbing and redirecting water. This principle applies to the development or conservation and maintenance of green spaces such as ravines, parks, fields and beaches, especially in or near urban areas. Creating and restoring such ecosystems provides an outlet or storage space for water during times of floods, preventing the water from affecting people and properties. Thus, the flood does not become a flood hazard. The premise is that floodplains are permitted to behave as floodplains, with added benefits of recreational amenities for people living nearby.

How else could flood risk be assessed, managed and balanced to avoid the potential of floods as hazards from becoming flood disasters? Aside from reducing vulnerability (Part 1), knowing more about the hazard, namely the parameters of concern (Section 11.A) in a specific place, can contribute. Any flood parameter, and several in tandem, can be subject to the techniques described in Chapter 4 followed by the options described there for risk communication.

Flood warnings, especially, must vary substantially, given the vast time and space scales of possible flood hazards. Flash floods might give hours to minutes for a narrow canyon while we have had and still have decades of warning that sea-level rise is affecting most of the world's coastlines (Section 11.A.1). A flood warning can be as complicated as a digital elevation model of a shore's topography coupled with real-time satellite monitoring of a cyclonic storm to run a bathymetric model that determines storm surge height as it overlaps with the tides. It can be as simple as stretching a chain across a river so that people downstream hear the water rattling it, giving them seconds to scramble up the slopes out of harm's way.

While centuries and millennia of efforts for flood risk assessment and management are not always formalised, including for warnings, more contemporary examples demonstrate possible actions. Gruntfest et al. (1978) reconstructed choices made by the 140 people who died in the 1976 Big Thompson, Colorado, flash flood. Their analysis determined what information might have been available to people in the inundation zone and what information they might have needed in order to survive. This work produced an education and information programme with simple advice signs dotted throughout the region's canyons. (Figure 11.13). People growing up in the area would see these signs while visitors would view their wording and diagram. In 2013, significant flooding happened across the region, including in Big Thompson Canyon. Although population numbers had increased substantially in the intervening years, fewer than a dozen people died across Colorado. As described in Chapter 5, warnings do not necessarily lead to successful action (Box 11.1).

Numerous techniques exist to reduce flood damage for individual properties. Most notably, deployable temporary barriers tend to be the cheapest response-related measure, as long as sufficient flood warning is provided and someone is able to set up the barrier without putting

Figure 11.13 Warning and advice sign for flash flooding along Boulder Creek, Colorado.

Source: Photo: Ilan Kelman

Box 11.1 Flash Floods in Europe

In July 2021, storms produced floods in at least a dozen European countries. Over 200 people died, the majority in flash floods in Germany. Five days before some of the worst inundations started, the European Flood Awareness System (EFAS) issued a forecast that major flooding would be expected with a high probability. Hours before deluges killed and damaged, warning messages explained that imminent and dangerous inundation was expected. The Netherlands, however, reported no confirmed deaths, part of which was disaster risk reduction and part of which might have been extra time to be ready for the water as well as lower flow rates. For instance, rather than flash floods, the Netherlands experienced more slow-rise floods with levees and dikes giving time to evacuate before inhabited areas flooded.

This situation exemplifies how fluvial monitoring and forecasting can work. Yet, the warnings failed to avert a disaster, partly because those affected could not imagine the suddenness, intensity and danger of the floods. This despite many of the damaged places having long histories of flooding. Survivors could identify and explain how to close any gaps they encountered. Gruntfest et al. (1978) used and explained how to reconstruct aspects of the decision-making of those who did not survive.

themselves in danger. A more permanent approach is to raise the property with a foundation preventing scour and undermining. Alternatively, the property owner could assume that a basement, cellar and ground floor will be flooded, so use it for non-essential purposes to minimise disruption during post-flood reinstatement. These areas would not have power mains, would not be inhabited and would not store valuable or irreplaceable items. To use all parts of a building for any purpose, floor and wall materials, including insulation and doors, could be made from quick-drying material and finishes that manage with quick cleaning and disinfecting rather than long-term drying or full replacement. They would also need to be strong enough to withstand impacts from a flood wave and debris. Plumbing and electrics could be raised and the building constructed so that water easily drains out. No approach is entirely safe from floods, as the depth, velocity, temperature, contaminants or debris could always cause damage. A suite of options ranging across a wide variety of costs and inconveniences exist from which the property owner would select, knowing the advantages and limitations.

Bridges can be designed to live with floods. Rather than letting debris in floodwater pile up and wash away a bridge, it can be designed with the connection to one riverbank or a mid-river support being weaker than the other end. At a certain pressure, it breaks (or is released deliberately) and the bridge swings round while remaining connected to the other end. The free end can be latched temporarily to the bank to avoid it swinging out across the river again, provided no one is placed in danger to do so. These 'breakaway bridges' then sit parallel to the river as the debris and flood flow past, with straightforward reattachment after the water subsides.

Rather than using a combination of techniques to live with flooding, often separation of people and water dominates by using structural approaches, focusing on levees, dams, walls and dikes. Aside from the huge social and environmental costs of improperly implemented structures, notably with large-scale initiatives (Box 11.2), the structural approach alters

Box 11.2 Flood Hazards on the Yangtze River, China

China has a long record of flood disasters. Attempts to reduce flood damage date back over 4,000 years (Wu, 1989). Most human settlements and cultivable land are on the alluvial plains of seven great rivers. The largest of these is the Ch'ang Chiang (Yangtze) – the third-longest river in the world – which flows for 6,300 km toward the Pacific Ocean through an area populated with dozens of millions of people and which has produced floods that have killed more than 300,000 people over the last century and a half. In the middle reaches, two huge connected depressions – the Dongting Lake and the smaller Poyang Lake – provide flood storage and help to protect downstream areas. In an average year, the Yangtze carries 500 million tonnes of sediment, mostly in the flood season. Under natural conditions, most rivers in the Yangtze basin would change course frequently in response to the progressive rise of the land, due to silt deposition.

The catchment area is subject to summer monsoon rains and tropical cyclones. Records over the last 2,000 years show that damaging floods occur, on average, once in every ten years, and the variability of floods and droughts can be linked to ENSO episodes (Tong et al., 2006). In 1998, 320,000 km^2 of land were flooded, over 3,600 people were killed (Table 11.1) and more than 200 million people were affected with the direct property damage estimated at $20 billion.

Flood management for the Yangtze includes:

- *Levees.* The earliest flood levees date back to AD 345 and are designed to control floods with a 10- to 20-year return period (Zhang, 2004). There are now about 3,600 km of main river levees and 30,000 km of tributary levees for farmland, oilfields and cities. Due to silt deposition in the constricted river channel, the level of the Yangtze during high floods can attain 10 m above the land behind the levees, which themselves reach 16 m in height in places. Many levees are old and weak and are subject to breaching.
- *Lakes.* Many surface depressions are used to regulate large floods that overtop the levees. Lake Dongting provides detention storage but starts to break its banks when full and threatens 6,670 km^2 of densely populated farmland and over 10 million people, many in cities such as Yueyang and Wuhan. Major lake failures are more frequent, due to silt deposition and land reclamation for economic development. The capacity of the lake has been reduced by nearly 80% since 1950. In August 2002, the Dongting lake reached a record high level of 35.9 m, a state of emergency was declared and more than 80,000 people were mobilised to strengthen the levees.
- *Dams.* China has built a major proportion of all the world's large dams. This tradition ensured the construction of the Three Gorges Dam (TGD) near Chingquing, designed to produce hydro-electricity and control floods in the middle and lower reaches of the Yangtze. The TGD stores up to 39 billion m^3 of water, held in a reservoir 600 km long behind a dam 175 m high and nearly 2 km long. At this time, it is the largest hydropower station and dam in the world. Water impounded by the dam flooded 13 cities, 140 towns and over 1,300 villages. At an estimated cost of $22.5 billion, it is expensive and controversial, especially in terms of the adverse social impacts and consequences. Hwang et al. (2007) recorded high levels of stress and depression in residents subjected to involuntary relocation, especially for poor rural people who had never previously moved in their lives.

The reservoir began to fill on 1 June 2003. The TGD project began operating immediately but was not declared complete until 2009. During the dry season between December and March, the dam discharges water for agriculture and industry down river. Before the onset of the flood season in June, storage is reduced to capture flood flows equivalent to 22 billion m^3. This should reduce the peak flow of the 100-year flood in the downstream Jinjiang section from 86,000 m^3 s^{-1} to less than 60,000 m^3 s^{-1}, a discharge within the safe capacity of this part of the river. As an example, inflows to the dam during July 2010 peaked at 70,000 m^3 s^{-1}. This exceeded the maximum discharge recorded during the 1998 floods, but the regulated outflow remained at about 40,000 m^3 s–1.

Future consequences of the TGD project remain uncertain. Sediment deposition behind the dam is likely to limit its efficiency, and many alleged negative aspects include landslides, loss of biodiversity and wildlife and the destruction of over 300 heritage sites. Legislation to encourage afforestation, combined with controls on illegal logging, now exists but is unlikely to be fully implemented or to have much influence on peak flows. Yet immediate impacts indicated that the TGD had swift and significant effects on the flow of the Yangtze river below the dam, although the effects decay downstream, due to inflows from downstream tributaries (Guo et al., 2012). Complex interactions between the river and Poyang Lake have been found, including the partial mitigation of flood risk in the lake basin during the July–September wet season. Ongoing monitoring of erosion and sediment deposition and efficient flow management will help in optimising the flood-control benefits of the TGD (Fang et al., 2012). Emergency flood evacuation could involve up to 1 million people with their livestock and possessions. Flood management along the Yangtze River remains daunting.

flood hazards such that low-level hazards do not encounter people and infrastructure, but higher-level hazards can remain problematic. Over the long term, relying on the structures can prove counterproductive if changes to perception and behaviour, along with other flood risk reduction measures, are not implemented simultaneously.

In what is called 'the levee effect', assuming that structural measures give full protection from flood hazards means that floodable areas are erroneously perceived to be completely safe for investment and development without other measures. As the demand for building on floodplains without additional measures increases, prices can rise and more property and people are placed at risk. Overall, relying on only structural approaches to eliminate low-level flood hazards makes people vulnerable to the inevitable high-level flood hazards and to flood hazards that the structures are not designed to prevent, such as ponding or flood velocity from a breach. Overall flood risk increases because of the structures, while acknowledging the gains from settling and using the land (Etkin, 1999; Fordham, 1999; Montz & Gruntfest, 1986; Tobin, 1995).

This analysis does not conclude that structural approaches must be avoided. It presents the adverse consequences which can be addressed to some extent with appropriate actions to reap the benefits. Multiple flood risk reduction measures need to be implemented in tandem to avert a false sense of security from the structural approaches. Then, any structural approaches can be balanced and work together with wider disaster risk reduction, including dealing with perceptions, behavioural changes and non-structural actions.

Canada has shown the possible successes of a multi-pronged approach which uses but does not rely on structural measures. Faced with mounting flood losses in insurance claims and disaster relief, Canada introduced a comprehensive Flood Disaster Reduction Program in 1971 (Shrubsole, 2008). The aim was to decrease reliance on structural schemes in favour of a strategy based on floodplain mapping and public education, which yielded demonstrably significant cost savings (Brown et al.,

1997), although changes to government priorities in 1999 then reduced its implementation (de Loë & Wojtanowski, 2001). Canada nonetheless continues to use a balance of structural and non-structural approaches for living with flood hazards. Winnipeg, Manitoba, opened the Red River Floodway in 1968 and finished an expansion in 2010 to divert floodwaters around the city, a structural approach with plans in place for when the floodway's capacity is exceeded. Toronto, Ontario, accepts that a major highway and railway line near the city centre will on occasion flood from the adjacent river, disrupting transport. The city aims to improve warnings to prevent trains and vehicles ending up in floodwater. In both cities, awareness of the flood-related measures and their limitations varies, suggesting the need for more risk communication, awareness and education (Chapter 4).

Numerous flood risk assessment and management pathways are particularly

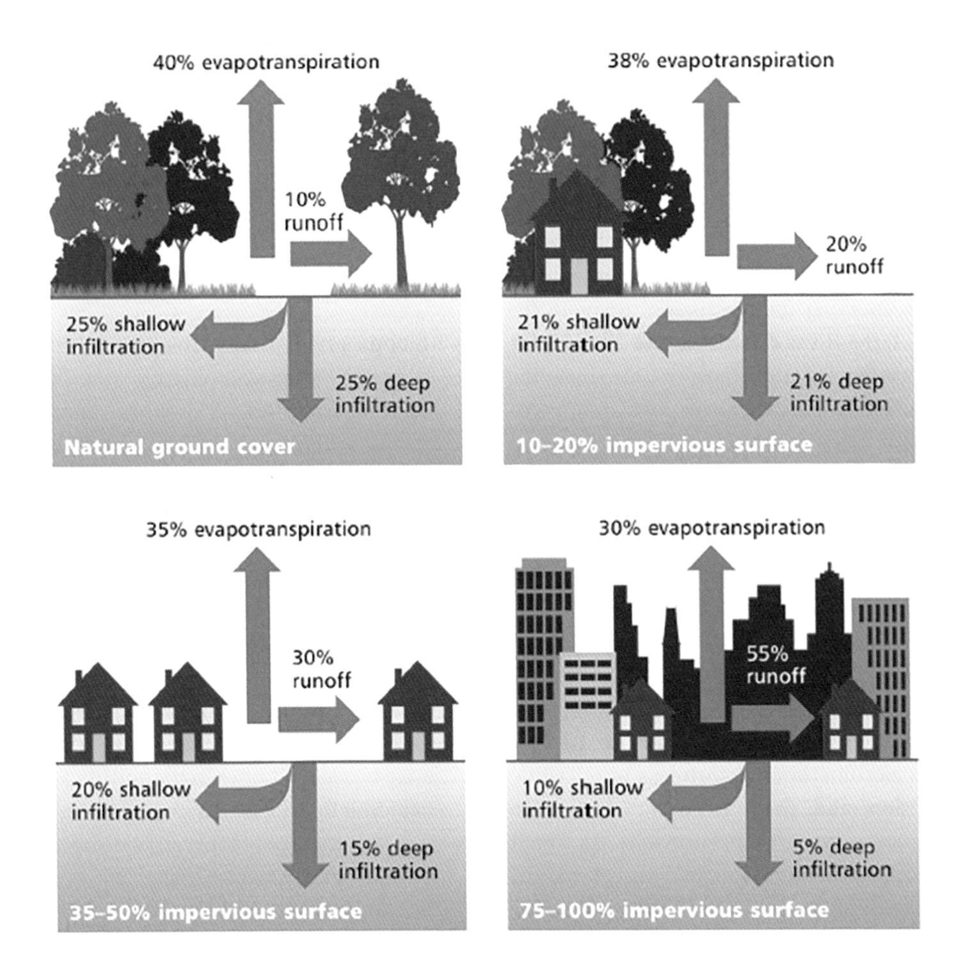

Figure 11.14 Schematic illustration of the influence of urbanisation on hydrology, showing a progressive increase in surface run-off unless other measures are taken.

Source: After Federal Interagency Stream Restoration Working Group (2001)

important as the flood hazard changes from many factors simultaneously. In Canada and China, and many other places around the world, a principal factor augmenting flood hazards is urban development (Figure 11.14), particularly when green areas are paved over without permeable surfaces or places to store excess water (discussed earlier in this section). Then, a higher proportion of rainfall is translated into surface run-off. Hydraulically smooth urban surfaces, serviced with dense networks of drains and underground sewers, deliver water more rapidly to the river channel than before. This affects flood onset, reducing the lag period between rainfall and peak flow. Higher flood peaks occur and arrive more quickly. Meanwhile, waterway channels can be constricted by bridge supports or riverside facilities, reducing the carrying capacity so that high flows overtop the banks more frequently. This phenomenon is seen along the Mississippi River in the USA, the River Thames in the UK and the Singapore River in Singapore, also meaning that return period calculations are inaccurate because the baseline is ever changing with river engineering. Meanwhile, stormwater drainage capacity does not always keep up with building development, so old, poorly maintained sewage systems are overwhelmed during rainstorms.

Other land use changes contribute to increased run-off. Deforestation especially has some (not complete) correlation with increased run-off, because treeless slopes sometimes absorb less water while erosion and then sediment deposition reduce water channel capacity. In small drainage basins, fourfold increases in peak flows have been recorded, together with suspended sediment concentrations as much as 100 times greater than in rivers draining forested land. The 1966 flood that claimed 33 lives and damaged 1,400 works of art and 300,000 rare books in the city of Florence, Italy, was partially attributed to long-term deforestation in the upper Arno River basin.

Meanwhile, changing weather – precipitation, heat and wind patterns – affects flood hazards. Many natural climate variabilities affect the amount of water in floods while anthropogenic climate change augments rainfall intensity in many places (IPCC, 2021–2022), so more water falls in one place, including during some severe storms (Chapters 10, 16 and 18). The danger from flood hazards and the devastation of flood disasters should never be underestimated. A variety of techniques for reducing vulnerability can mean that risk assessment and management reduces the problems and increases the benefits.

KEY TAKE AWAY POINTS

- Floods are typically characterised by physical, chemical, energy and biological properties, all of which impact people and infrastructure.
- Plenty of measures can be adopted to try to avoid adverse flood impacts, but many measures often used – such as dams and levees – end up making vulnerability worse.
- Floods have many positive impacts on society, including fertile soil alongside rivers and supporting livelihoods based on coastal ecosystems, so a balance is needed between permitting and avoiding flood hazards.
- Caution is needed in attributing all flood hazard characteristics to weather, or in assuming that weather is the dominating influence on floods, since many other factors – such as land use and river or coastal engineering – are important.

BIBLIOGRAPHY

Bempah, S. A., & Øyhus, A. O. (2017). The role of social perception in disaster risk reduction: Beliefs, perception, and attitudes regarding flood disasters in communities along the Volta River, Ghana. *International Journal of Disaster Risk Reduction, 23*, 104–108.

Bennet, G. (1970). Bristol floods 1968: Controlled survey of effects on health of local community disaster. *British Medical Journal, 3*, 454–458.

Brammer, H. (2000). Flood hazard vulnerability and flood disasters in Bangladesh. In D. J. Parker (Ed.), *Floods* (Vol. 1, pp. 100–115). London: Routledge.

Brown, D. W., Moin, S. M. A., & Nicolson, M. L. (1997). A comparison of flooding in Michigan and Ontario: 'Soft' data to support 'soft' water management approaches. *Canadian Water Resources Journal, 22*(2), 125–139.

de Loë, R., & Wojtanowski, D. (2001). Associated benefits and costs of the Canadian. *Flood Damage Reduction Program Applied Geography, 21*, 1–21.

Douben, K.-J. (2006). Characteristics of river floods and flooding: A global overview, 1985–2003. *Irrigation and Drainage, 55*, S9–S21.

Durack, P. J., & Wijffels, S. E. (2010). Fifty-year trends in global ocean salinities and their relationship to broad-scale warming. *Journal of Climate, 23*(16), 4342–4362.

Etkin, D. (1999). Risk transference and related trends: Driving forces towards more mega-disasters. *Environmental Hazards, 1*, 69–75.

Fang, H., Han, D., He, G., & Chen, M. (2012). Flood management selections for the Yangtze River midstream after the Three Gorges Project operation. *Journal of Hydrology, 432–433*, 1–11.

Federal Interagency Stream Restoration Working Group. (2001). Stream corridor restoration: Principles, processes, and practices. GPO Item No. 0120-A; SuDocs No. A 57.6/2:EN3/PT.653. ISBN-0-934213-59-3. Federal Interagency Stream Restoration Working Group, Washington, DC.

Fordham, M. (1999). Participatory planning for flood mitigation: Models and approaches. *The Australian Journal of Emergency Management, 13*(4), 27–34.

Gruntfest, E. C., Downing, T. E., & White, G. F. (1978, February). Big Thompson flood exposes need for better flood reaction system to save lives. *Civil Engineering, 48*, 72–73.

Guo, H., Hu, W., Zhang, Q., & Feng, S. (2012). Effects of the Three Gorges Dam on Yangtze River flow and river interaction with Poyang Lake, China: 2003–2008. *Journal of Hydrology, 416–417*, 19–27.

Hwang, S. S., Xi, J., Cao, Y., Feng, X., & Qiao, X. (2007). Anticipation of migration and psychological stress and the Three Gorges Dam project, China. *Social Science & Medicine, 65*(5), 1012–1024.

IPCC. (2021–2022). Sixth assessment report. IPCC (Intergovernmental Panel on Climate Change), Geneva, Switzerland.

Jonkman, S. N. (2005). Global perspectives on loss of human life caused by floods. *Natural Hazards, 34*, 151–175.

Kron, W. (2015). Flood disasters: A global perspective. *Water Policy, 17*, 6–24.

Montz, B., & Gruntfest, E. C. (1986). Changes in American urban floodplain occupancy since 1958: The experiences of nine cities. *Applied Geography, 6*(4), 325–338.

Newman, C. J. (1976). Children of disaster: Clinical observations at Buffalo Creek. *American Journal of Psychiatry, 133*, 306–312.

OCHA. (2010). Pakistan: Initial floods emergency response plan. OCHA (UN Office for the Coordination of Humanitarian Affairs), New York and Geneva.

Shrubsole, D. (2008). From structures to sustainability: A history of flood management strategies in Canada. *International Journal of Emergency Management, 4*, 183–196.

Smith, D. I. (2000). Floodplain management: Problems, issues and opportunities. In D. J. Parker (Ed.), *Floods* (Vol. 1, pp. 254–267). London: Routledge.

Tobin, G. A. (1995). The levee love affair: A Stormy relationship. *Water Resources Bulletin, 31*(3), 359–367.

Tong, J., Qiang, Z., Deming, Z., & Yijin, W. (2006). Yangtze floods and droughts (China) and teleconnections with ENSO activities (1470–2003). *Quaternary International, 144*(1), 29–37.

USACE. (1984). *Shore protection manual* (4th ed., 2 vols.). Washington, DC, USA: USACE (United States Army Corps of Engineers) Coastal Engineering Research Centre.

WHO. (2021). *Drowning*. World Health Organization, Geneva, Switzerland. www.who.int/news-room/fact-sheets/detail/drowning

WMO. (2021). WMO atlas of mortality and economic losses from weather, climate and water extremes (1970–2019). World Meteorological Organization, Geneva.

Wu, Q. (1989). The protection of China's ancient cities from flood damage. *Disasters, 13*, 193–227.

Zhang, H. L. (2004). China: Flood management. Case Study in Integrated Flood Management. WMO/GWP Associated Programme on Flood Management, Geneva.

FURTHER READING

Criss, R. E., & Shock, E. L. (2001). Flood enhancement through flood control. *Geology, 29*(10), 875–878.

Green, C. H., Parker, D. J., & Penning-Rowsell, E. C. (1993). Designing for failure. Chapter 6 in P. A. Merriman & C. W. A. Browitt (Eds.), *Natural disasters: Protecting vulnerable communities, proceedings of the conference held in London, 13–15 October 1993* (pp. 78–91). London: Thomas Telford.

WCD. (2000). Dams and development: A new framework for decision-making. The Report of the United Nations World Commission on Dams, Earthscan, London.

White, G. F. (1942/1945). Human adjustment to floods: A geographical approach to the flood problem in the United States. Doctoral Dissertation, University of Chicago (1942) and republished as Research Paper No. 29 (1945), University of Chicago, IL, Department of Geography.

WEBSITES

Floodlist. https://floodlist.com

Turn around, Don't Drown. www.weather.gov/safety/flood-turn-around-dont-drown

World Drowning Prevention Day. www.who.int/campaigns/world-drowning-prevention-day

World Meteorological Organization. www.wmo.int

CHAPTER TWELVE

Drought Hazards

12

OVERVIEW

Drought is often called a 'creeping' or slow-onset hazard because it develops slowly. Droughts have a prolonged duration, impacting regions for several months to years. Single events can extend over regions that are subcontinental in scale and cover multiple countries at a time. Droughts occur when there is a shortage of water in the landscape across a distinct period. However, droughts are not confined to areas of low rainfall or other precipitation. The disaster consequences of drought increase with their severity and duration. This chapter recognises four different categories of drought: meteorological, hydrological, agricultural and famine. Meteorological drought is defined on statistical criteria relating to shortfalls of precipitation alone. Hydrological drought occurs when streamflows and/or groundwater levels are sufficiently reduced to threaten water resources. Agricultural drought has implications for food production, with the main consequences being reduced crop and animal production. Famine droughts are complex, where hydrometeorological conditions are exacerbated by the social relations organising resource use and by economic insecurity. Risk assessments are improving our understanding of drought, resulting from our increased understanding of climate variability and its implications for agriculture, food supplies and the welfare of rural communities. Management practices require proactive approaches that integrate environmental data, alongside monitoring and forecasting of significant shortfalls in seasonal rainfall or snow/ice melt with efficient and just food and nutrition management.

A. THE NATURE OF DROUGHTS

1. Definitions

Put simply, drought is an exceptional dry period, but such periods are hard to recognise, especially in the early stages. A simple definition is 'any unusual dry period which results in a shortage of water'. This indicates

DOI: 10.4324/9781351261647-14

that precipitation deficiency is the 'trigger' but that the impacts are the most important characteristics. It is the shortage of *useful water* – in the soil, in rivers or in reservoirs – that creates the hazard, but it should not be confused with aridity, water scarcity or desertification. These are common characteristics of arid and semi-arid regions around the world. While drought and aridity both relate to water deficiency, drought occurs across distinct periods, whereas aridity is a constant and relatively permanent part of particular landscapes (Haile et al., 2020).

Drought is dependent not only on weather, but also on the hydrological processes that determine what happens to the precipitation and on the societal consequences of water shortages. That is, drought disasters are better understood in terms of the impacts on natural resources and human activities, such as water supplies, agricultural production and food availability, rather than on the basis of rainfall statistics alone.

Specifically, drought hazards are difficult to define because:

- Many indices have been developed for identifying and measuring aspects of drought such as onset, severity, duration and geographical scale (Heim, 2002; Keyantash & Dracup, 2002). One outcome is that an overall consensus on drought events can be hard to find. Similar issues apply to famine assessment.
- Vulnerability to drought depends on both social and physical factors, including geology and soil types, water storage facilities, crop types and access to warning information. For example, areas dependent on rivers for irrigation or urban water supplies may be far downstream from the headwaters where precipitation has failed.
- Drought impacts might not always be immediately or obviously visible. It can be difficult to translate agricultural losses of crops and livestock due to water shortage into measures that adequately describe a rise in poverty or decrease in livelihoods and opportunities.

In summary, drought as an environmental hazard should be regarded as a *process* which develops over time, with increasingly severe effects, rather than an *event*. Above all, the link between drought and famine is unclear.

2. Spatial and Temporal Distribution

Droughts are not confined to areas of low rainfall or other precipitation, any more than floods are confined to areas of high rainfall. Drought is an integral part of natural climatic variations – which are increasingly being transformed by anthropogenic climate change (Chapter 18) – and influences both rural and urban areas. It is important to view water shortages in terms of *resource need* rather than absolute rainfall amounts. In other words, drought and aridity are not the same. This is because people normally adapt their activities to the expected moisture environment: a yearly rainfall of 200 mm might be tolerable for a semiarid sheep farmer but disastrous for a wheat farmer accustomed to an average 500 mm of rain per year. What might be termed *drought hazards* create major socio-economic impacts. They are unlikely to result from a single dry year within the normal range of precipitation variability, but rather from a period with consecutive years of below-average rainfall (Figure 12.1). Rainfall patterns create uncertainty in several ways:

- *variability* – from season to season and year to year
- *trends* – towards wetter or drier conditions over several years

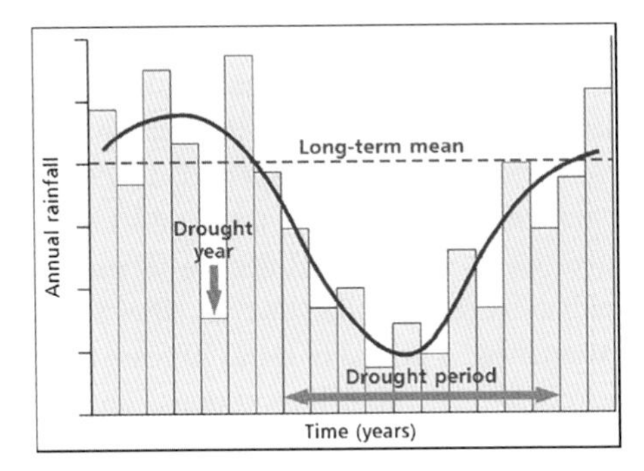

Figure 12.1 The development of a drought regime due to a persistent lack of rainfall over several years. Most drought hazards build up during a period with below-average precipitation.

- *persistence* – wetness or dryness grouped over a period of years.

Failure to account for rainfall variability, rather than low absolute amounts, is often a major cause of drought. Humanity's history is replete with examples of over-optimistic agricultural expansion into marginal areas during wetter climatic phases, followed by retreat in the drier years later.

3. Environmental Causes

Meteorological drought originates from anomalies within the general atmospheric circulation, but the climate processes that cause drought are not completely understood. More research is required into *climate dynamics*, a broad field of science that examines the entire coupled system of atmosphere, oceans, cryosphere, biomass and land surface as interacting components that produce the global climate. Much of this work is geared to the analytical and numerical modelling necessary to explain – and ultimately predict – aspects of climatic variability on all scales of space and time (Goosse et al., 2008). Some variations depend on external forcing factors, such as solar activity or volcanic eruptions, whilst others, like the El Niño Southern Oscillation (ENSO) and the North Atlantic Oscillation (NAO) occur due to internal actions and reactions within the system (see also Part 3). Feedbacks within the system and lag effects, due to inertia in the oceans and ice sheets, apply, and small changes in the forcing factors can lead to important consequences for the climate if a significant threshold is crossed.

Climate dynamics stresses the role of *teleconnections*. These are links between climatic anomalies occurring at long distances apart, especially interactions between the atmosphere and the oceans. For example, it is known that sea-surface temperature anomalies (SSTAs) played a role in the late 20th-century Sahel drought. Direct sea–air interactions influence the flux of both sensible heat and moisture at the ocean–atmosphere interface. One example would be the initiation of drought by negative (relatively cold) SSTAs leading to descending air and anti-cyclonic weather. It is possible that the starting-point for the 1975–1976 drought over northwest Europe was very low sea-surface temperatures over the Atlantic Ocean north of 40°N which caused near-surface stability in the atmosphere and led to blocking anticyclones. The case for oceanic forcing of drought is best established for ENSO events (see Part 3). According to Dilley and Heyman (1995), worldwide meteorological drought hazards double during the second year of an El Niño episode, as compared with all other years. This happened in 1982–1983, when droughts in Africa, Australia, India, northeast Brazil and the United States coincided with a major El Niño phase. Over smaller areas, other climate drivers, like the NAO, may not fully capture the detailed aspects of drought, due to weather types and synoptic-scale processes (Vicente-Serrano & López-Moreno, 2006).

The African Sahel is drought sensitive because it occupies a transitional climatic zone between the Sahara to the north and tropical rainforests to the south. Mean annual rainfall typically varies from about 100 mm (hyper-arid) on the Saharan edge to 800 mm (dry sub-humid) along the southern margins. But mean values are highly misleading. These rainfall values are characterised by high variability on all climatic timescales – seasonal, year to year and decadal. More than 80% of the annual rainfall is expected in the summer months of July, August and September. The year-to-year variability, expressed by the coefficient of variation, is high, ranging from approximately 20% to 40%, and leads to low reliability of the rainy season (Figure 12.2). This pattern has been disrupted by prolonged rainfall anomalies such as the relatively wet periods in 1905–1909 and 1950–1969. From the mid-1960s, there was a pronounced decline in annual rainfall (Figure 12.3). Overall, the Sahel experienced a period of some 30 years with below-average rainfall conditions, broken only by widespread rains in the 1994 wet season.

The agricultural impact of the drought was made worse by good rains in the 1950s and 1960s, which encouraged rain-fed cropping into marginal lands and larger herd sizes. The drought was driven mainly by a rise in sea-surface temperatures in the southern hemisphere and the Indian

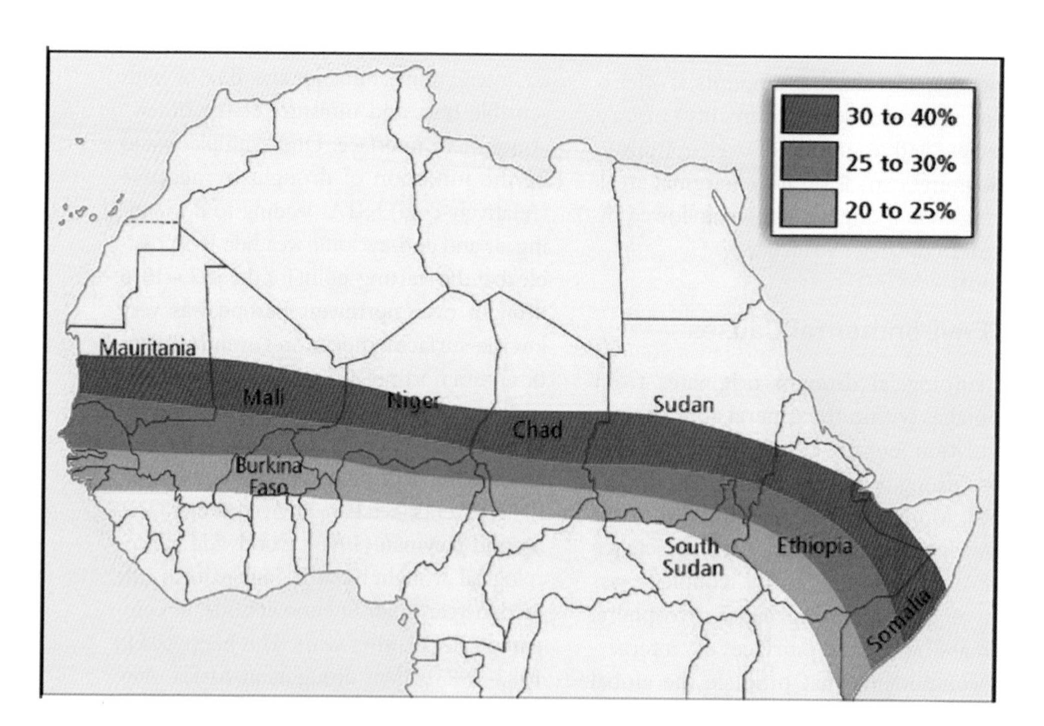

Figure 12.2 Countries of the Sahel region prone to drought. The coloured areas show the average annual departures from normal rainfall. When rainfall totals are low and variability is high, drought is likely to be a recurrent feature.

Source: USAID. US government material not subject to copyright protection in the United States.

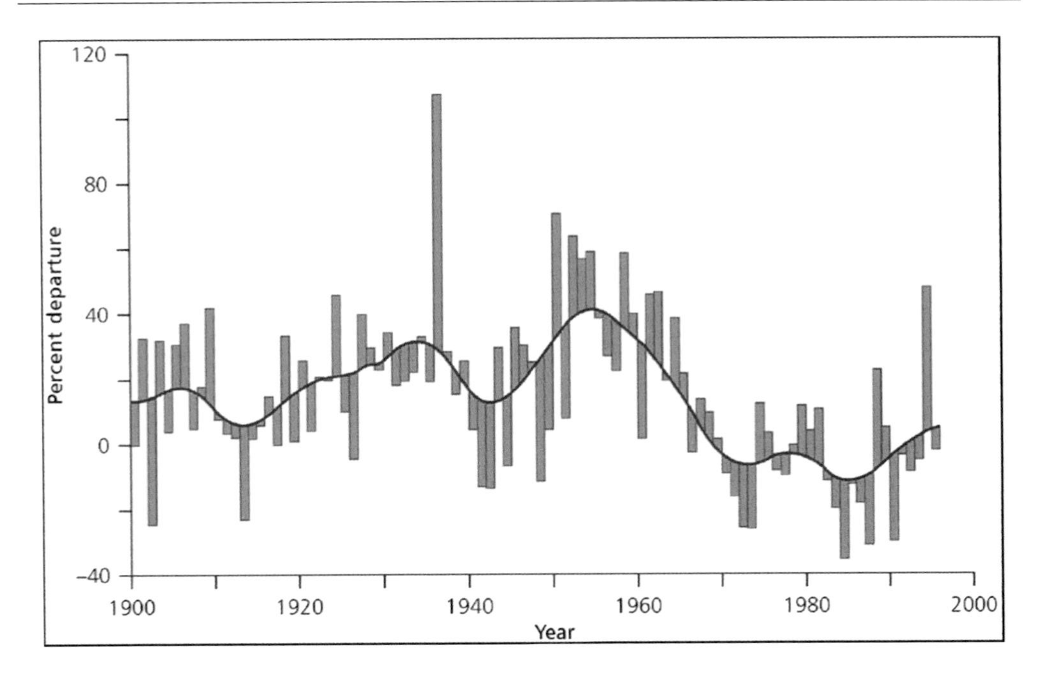

Figure 12.3 Sahelian rainfall during the rainy season (June–September) expressed as a percentage of the 1961–1990 mean. The downturn in the late 1960s was a major factor in the famine disasters of the later 20th century.

Source: After M. Hulme at https://www.uea.ac.uk/groups-and-centres/climatic-research-unit (accessed on 31 May 2023), used with permission

Ocean, which produced changes in the atmospheric circulation. Broadly speaking, it was believed that the tropical rain belts were attracted to the relative warmth of the southern hemisphere, so that the Sahel, on the northern extension of these rain belts during the northern summer, effectively lost its rainy season. Further work confirmed that the Sahel region is highly sensitive to the variability of SSTAs in all tropical ocean basins, including the remote Pacific. This is because unusually warm low-latitude waters favour deep convection over the oceans that, in turn, weakens the summer monsoon convergence and creates drought from Senegal to Ethiopia (Giannini et al., 2003). Building on this, Sheen et al. (2017) concluded that prolonged droughts in the Sahel region are linked to cooler temperatures in the north Atlantic and Mediterranean Sea, which cause the migration of the tropical rainbelt toward the equator, weakening of southerly monsoon winds and reduced low-level moisture content.

Following a wide geographical survey, Peel et al. (2002) concluded that the variability of annual precipitation was appreciably higher in ENSO-affected climatic zones than in other areas. Teleconnections involving SSTAs and rainfall have been established for several parts of the world. Across eastern Australia (Queensland, New South Wales, Victoria and Tasmania), annual spring and summer rainfall is strongly linked to the Southern Oscillation Index (SOI), with a correlation coefficient of 0.66 (Nicholls, 2011). Figure 12.4 shows a clear relationship between low and negative SOI values and individual drought years such as 1940, 1982 and 1997.

Whilst ENSO-based models can reproduce year-to-year variability quite well, they have been less successful in handling multi-decadal drying trends. McCabe et al. (2004), for example, demonstrated that SSTAs over the North Atlantic Ocean, as well as the Pacific, contribute to drought incidence in the United States. Together, these influences accounted for more than half of the spatial and temporal variation in multi-decadal drought in this region.

Another complex climatic link exists between ocean surfaces and the strength of the Indian summer monsoon system. Whilst ENSO alone accounts for about 30% of the inter-annual variability in monsoon rainfall over India, unexpected drought years – like 2002 – have occurred. The cause has been attributed to interactions between ENSO and weather conditions over the more local Indian Ocean, where deep convection in the atmosphere over the equatorial marine surface varies annually between the eastern and western parts of the ocean, to create the Indian Ocean Oscillation (Gadgil et al., 2003). This feature, termed the *Indian Ocean Dipole* (IOD), modulates the ENSO–Indian monsoon teleconnection on decadal timescales. Strong ENSO events are associated with a weak monsoon airflow and potential drought hazards over India. Conversely, in the absence of any ENSO tendency, the Indian Ocean effect is strong, and positive IOD years (pIOD) produce above-average monsoon rainfall. When El Niño events coincide with pIOD conditions, the drought-forming subsidence of ENSO is counteracted so that a near-average monsoon season results (Ummenhofer et al., 2011). Figure 12.5

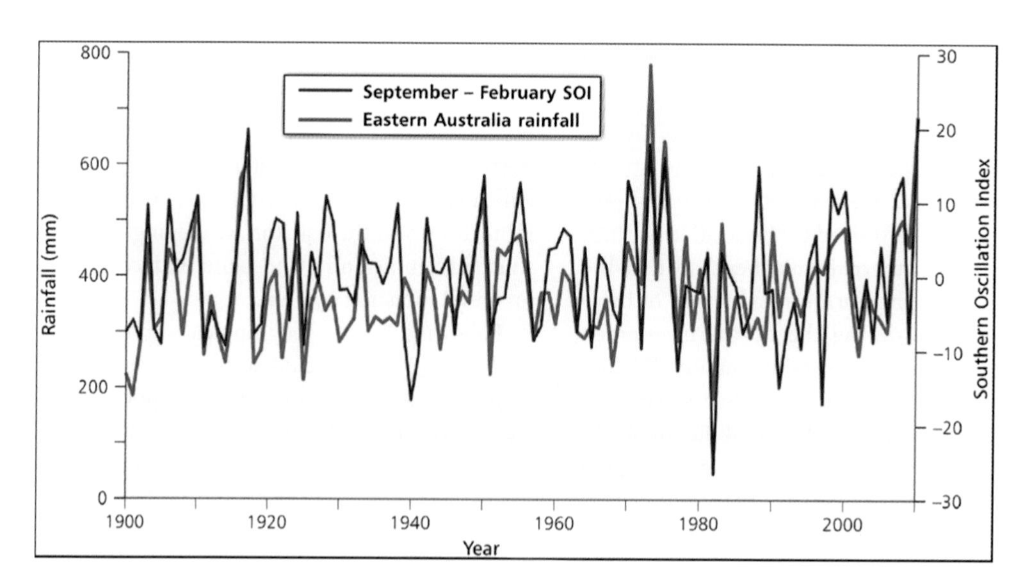

Figure 12.4 Rainfall in eastern Australia during the September–February period in relation to the Southern Oscillation Index (SOI). The SOI measures the strength of La Niña and El Niño events, a normal part of climate, which are strongly correlated with summer rainfall in eastern Australia and other parts of the world.

Source: After Nicholls (2011). Used with permission.

shows annual rainfall anomalies over the core monsoon region of India for the summer monsoon season (June, July, August, September . . . (JJAS)) from 1877 to 2006 in relation to ENSO, pIOD and joint ENSO–pIOD events. More than 85% of ENSO years experienced anomalously low rainfall, some deficits exceeding 800 mm per month; pIOD years had above-average rainfall, whilst co-occurring ENSO/pIOD events led to near-average conditions.

Droughts are initiated by large-scale atmospheric processes, but the dryness may be prolonged by *regional forcing*. This is due to localised feedback mechanisms between the land surface and the lower atmosphere. Excessively dry ground helps to maintain the rainless atmospheric state because a higher proportion of the incoming solar radiation is used to heat the ground and the air, compared to normal conditions when more energy would be used in evaporation. Where prolonged dryness has reduced the vegetation cover, changed the surface albedo and created greater dustiness, it is possible that drought may become almost self-perpetuating. The theory is attractive for the Sahel, where a lack of rain, combined with pressure on land resources, has produced environmental deterioration and even desertification. Other regional feedbacks include the increase in aerosols due to large-scale biomass burning, such as occurred in Indonesia (Field et al., 2009).

Similar mechanisms have been found elsewhere. The 1988 North American drought was linked to an ENSO episode which led to a northward displacement of the inter-tropical convergence zone southeast of Hawaii and the eventual appearance

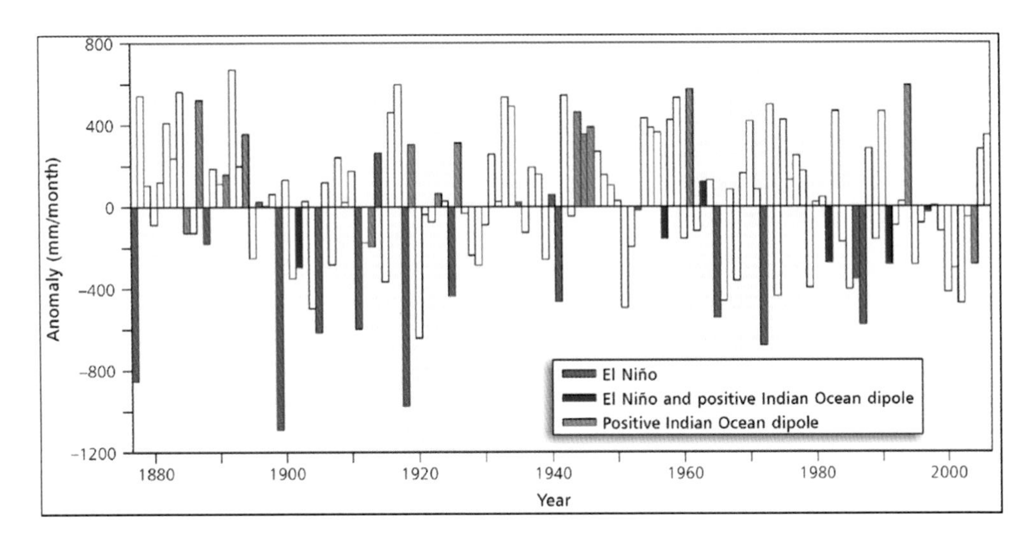

Figure 12.5 Time-series of rainfall anomalies (mm per month) during June–September 1877–2006 for three interacting categories of climatic conditions over the core Indian monsoon region. Anomalies are shown for El Niño events (blue), El Niño co-occurring with positive Indian Ocean dipole (pIOD) events (purple) and positive Indian Ocean dipole events (orange).

Source: After Ummenhofer et al. (2011). Reproduced from C.M. Ummenhofer et al. (2011) Multi-decadal modulation of the El Niño–Indian monsoon relationship by Indian Ocean variability. *Environmental Research Letters* 6: 1–8. With permission from IOP Publishing Ltd.

of a strong anticyclone at upper levels over the American Midwest (Trenberth et al., 1988). In fact, most droughts in the USA can be explained by primary forcing and anomalous SSTAs, but the exceptional 'Dust Bowl' drought of the 1930s was probably amplified by regional factors. Once dry conditions are established, the subsiding air, combined with the drying out of open water surfaces and the upper soil layers, reduces the relative humidity in the lower atmosphere.

Potentially, this limits the ability of local weather disturbances to create rainfall. Cook et al. (2007) and Cook et al. (2011) simulated climatic conditions at the time using general circulation models (GCMs) and found that anomalous sea-surface temperatures alone were insufficient to account for the severity of the 1930s drought. However, when land-surface factors – devegetation and dust – were incorporated into the model, together with feedbacks between the atmosphere and the land, the rainfall anomalies and other drought features were represented more accurately.

B. IMPACTS AND CONSEQUENCES OF DROUGHTS

It is common to recognise four categories of drought types (Figure 12.6) based on defining components and hazard impacts. The potential disaster consequences increase with the severity and duration of these types of drought. Each drought type has different criteria for definition and assessment, although most drought indices use rainfall values, either singly or in combination with other hydrometeorological factors (Mishra & Singh, 2010).

1. Meteorological Drought

Meteorological drought is defined on statistical criteria relating to shortfalls of precipitation alone. As stated, precipitation deficiency is not necessarily hazardous because the links between precipitation and the useful water necessary to meet demands are indirect. Rainfall itself does not supply water to plants: the soil does this. Equally, rainfall does not supply water for irrigation or domestic use: rivers and groundwater do this.

The concept of meteorological drought has led to many rainfall-based definitions. The simplest method is to define drought on the minimum duration of a rain-free period, the length of which has differed from six days (Bali), to 30 days (southern Canada) and up to two years (Libya). As early as 1887 the UK Meteorological Office distinguished between an *absolute drought*, a period of at least 15 consecutive days with less than 0.2 mm of rain on any one day, and a *partial drought*, defined as at least 29 consecutive days when the mean daily rainfall failed to exceed 0.2 mm.

Other definitions depend on the rainfall totals that fall within a stated percentile value below the long-term average, usually during the main crop growing season or a calendar year. These definitions are of limited value unless they recognise that the impact of any rainfall deficiency is likely to vary through the period in question. The Australian Bureau of Meteorology, with its interest in agricultural drought, has used such a period-specific rainfall system, declaring a drought if the rainfall in a given area fails to exceed 10% of all previous totals for the same period of the year and if the situation persists for at least three months.

The *Standardized Precipitation Index* (SPI) is a better rainfall-based measure. It uses the long-term rainfall record for a given location and period and estimates the probability of an observed precipitation deficit occurring over a given time

Types of drought and their characteristics

Drought duration and severity (increasing downward)

	Major features	Major impacts
Meteorological drought	**Rainfall deficit** Low precipitation High temperatures Strong winds Increased solar radiation Reduced snow cover	Loss of soil moisture Supply of irrigation water declines
Hydrological drought	**Streamflow deficit** Reduced infiltration Low soil moisture Little percolation and ground-water recharge	Reduced storage in lakes and reservoirs Less water for urban supply and power generation – restrictions Poorer water quality Threats to wetlands and wildlife habitats
Agricultural drought	**Soil moisture deficit** Low evapotranspiration Plant water stress Reduced biomass Fall in ground-water levels	Poor yields from rain-fed crops Irrigation systems start to fail Pasture and livestock productivity declines Rural industries affected Some government aid required
Farming drought	**Food deficit** Loss of natural vegetation Increased risk of wildfires Wind-blown soil erosion Desertification	Widespread failure of agricultural systems Food shortages on seasonal scale Rural economy collapses Rural–urban migration Increased malnutrition and related mortality Humanitarian crisis International aid required

Figure 12.6 A classification of drought types based on defining components and hazard impacts. Disaster potential increases with the severity and duration of drought. Rainfall deficit alone may not produce a hazard.

period. The advantage is that, because the SPI can be calculated for a variety of timescales - commonly 1 to 36 months - droughts of different lengths and severities can be assessed. It can also be related to a range of hydrological parameters, such as streamflow and groundwater level, applied to various agricultural needs and can also be adapted for spatial–temporal analysis.

2. Hydrological Drought

Hydrological drought occurs when streamflows and/or groundwater levels are sufficiently reduced to threaten water resources. Hydrological drought tends to be measured by relating the reduced availability of water to the demands necessary for various uses. These demands vary greatly. For example, some approaches estimate the moisture deficit within the soil, in an attempt to assess the availability of water for plants and crops (Hunt et al., 2009).

The *Palmer Drought Severity Index* (PDSI) is a widely used index and is based on a soil-moisture budgeting system that analyses precipitation and temperature for a given area over a period of months or years (Palmer, 1965). Under the PDSI, drought is defined in terms of available moisture relative to the norm. The severity is assessed as a function of the length of period of abnormal moisture deficiency, as well as the magnitude of this deficiency. The PDSI produces a single hydrological measure for drought effects on soil moisture, groundwater and streamflow and, like the SPI, can be used to rank the severity of drought episodes (Table 12.1). In some ways the PDSI is superior to the SPI, but the numerical values cannot be related directly to highly specific hazard impacts, for example reduced yields of different crop types (Alley, 1984; Guttman, 1997).

More recently, advances in paleoclimatology have led to the reconstruction of PDSI data across hundreds of years using dendrochronology, which is the study of tree-ring width to determine present and reconstruct past climate variability. Tree rings depict the radial growth of

Table 12.1 Drought severity ranked by return period according to the Standardized Precipitation Index, the Palmer Drought Severity Index and typical impacts

Drought severity	Return period (years)	Typical impacts	SPI	PDSI
Minor	3–4	Short-term dryness; slower than normal crop and pasture growth; some lingering water deficits after the event.	−0.5 to −0.7	−1.0 to −1.9
Moderate	5 to 9	Reduced yields for crops and pastures; streams, wells and reservoirs at low levels; high fire risk, voluntary restrictions on water use publicised.	−0.8 to −1.2	−2.0 to −2.9
Severe	10 to 17	Appreciable crop and pasture losses; stock levels fall; fire risk very high; isolated urban and rural water shortages common; mandatory water restrictions imposed.	−1.3 to −1.5	−3.0 to −3.9
Extreme	18 to 43	Major crop and pasture losses; death and emergency sales of stock; extreme risk of wildfires; widespread water shortages combined with severe limitations on use.	−1.6 to −1.9	−4.0 to −4.9
Exceptional	>44	Complete failure of arable and pastoral production; collapse of rural economy; major wildfire outbreaks; severe water shortages with emergency arrangements in action for rural and urban supply systems; government aid required.	< −2.0	−5.0 or less

Source: Modified from Department of Space and Climate Physics and AON Benfield Hazard Research Centre, University College London, at http://drought.mssl.ucl.ac.uk/class.html (accessed 20 August 2011). Courtesy of the UCL Global Drought Monitor.

trees which are sensitive to seasonal precipitation rates or lack thereof, in times of drought. They are therefore useful for determining past climates across millennia (Luckman, 2013). For example, Wang et al. (2021) used tree-ring chronologies to reconstruct historical PDSI for the Tianshan Mountains in Central Asia. The reconstructed model identified extreme droughts in 1912 and 1914 along with widespread and persistent drought in the 1910s, more regional persistent drought through the 1880s and extremely dry climate episodes in the 1640s and 1650s.

Hydrological drought is mainly linked to water supplies. Therefore, direct hydrological measurements, such as streamflow and reservoir storage, can also help to determine drought conditions relevant to water resources. For example, in the rural areas of northeastern Brazil there are no permanent rivers and water supplies depend on seasonal rainfall stored in shallow reservoirs and ponds prone to high rates of evaporation. After two or three years with below-average rains, these storages dry up. Drought here gives rural areas less access than usual to clean water supplies; isolated communities have to rely on the distribution of water by road tankers, with negative consequences for community health.

3. Agricultural Drought

Agricultural drought is important because of the implications for food production. For example, during the drought years of 1992 and 1994 in Malawi, agricultural sector output fell 25% and 30%, respectively, below normal. All farmers, whether arable or pastoral, rely on the water available for plant growth in the soil, and an agricultural drought exists when soil moisture is insufficient to maintain average crop and grass yields.

Ideally, the severity of agricultural drought should be based on direct soil-moisture measurements, but indirect assessments are usually made through water-balance calculations like the PDSI. Regional water stress on plants can be monitored from space through the *Vegetation Condition Index* (VCI), described by Liu and Kogan (1996). Unfortunately, indices like the VCI and the PDSI cannot be directly linked to drought impacts on farm production because each crop responds differently to heat and moisture stress. Consequently, attempts have been made to derive a *Crop-Specific Drought Index* (CSDI), as pioneered by Meyer (1993) for corn.

Drought accounted for over 40% of the estimated US$349 billion cost of all weather-related disasters in the USA between 1980 and 2003 (Ross & Lott, 2003). Droughts are a long-term feature of North America and occur on the Great Plains about every 20 years. During droughts in the 1890s and 1910s there were deaths due to malnutrition. A turning point was reached in the 'Dust Bowl' years of the 1930s, with spells when almost two-thirds of the USA was in drought (Figure 12.7). The impact of this drought was amplified by poor farming techniques and prompted massive injections of aid, greater control of soil erosion and improved irrigation practices. Together with more sustainable farm management and crop insurance, these measures ensured that the 1950s drought had less severe impacts. Even so, problems of US drought management remain (Pulwarty et al., 2007).

The main consequence of agricultural drought is reduced crop and animal production. When fodder is inadequate, mass slaughter of livestock follows. It may take up to five years for animal

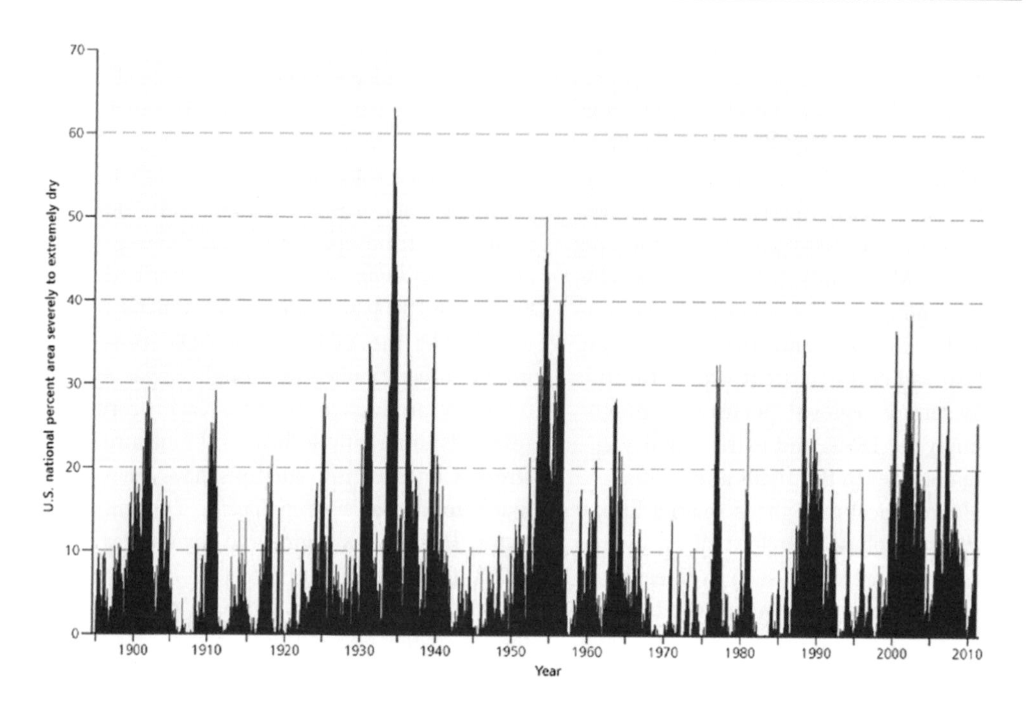

Figure 12.7 Monthly percentage area of the United States in severe and extreme drought from January 1895 to August 2009. This indicates that drought is a normal part of climate, with more severe episodes recorded in the 1930s, 1950s and more recently in 1988 and around 2000.

Source: After National Climatic Data Center, NOAA at www.ncdc.noaa.gov/ (accessed 6 February 2012). US government material not subject to copyright protection in the United States.

stocking levels to recover. In 1988 the USA experienced a costly agricultural drought over the Midwest. The 1988 corn yield was 31% below the progressive upward trend that is driven by improved technology, the largest drop since the mid-1930s (Figure 12.8). More than one-third of the American corn crop was destroyed, at a loss of US$4.7 billion (Donald, 1988). Agricultural drought on this scale disrupts international trade in food, and world grain stocks fell to a 63-day supply, the lowest since the mid-1970s. At the farm level, severe drought disrupts normal activities and causes a diversion of capital from farm development to drought-reducing strategies, a fall in cash liquidity and a rise in debt.

Drought can disrupt food supplies leading to hunger and famine, yet action can avert the worst human consequences. This happened during the 1990–1992 drought in southern Africa. In general, the harvest yield was 30%–80% below normal and 86 million people were affected over an area of almost 7×10^6 km^2. Although this drought did not result in famine, there was severe hardship, but the countries of the region came together to avoid a major disaster despite the drought (Holloway, 2000). In Zimbabwe, the volume of agricultural production fell by one-third. By November 1992, half the population had registered for drought relief. In some administrative districts in Zambia, yields of maize were down by 40%–100% and 2 million rural people were

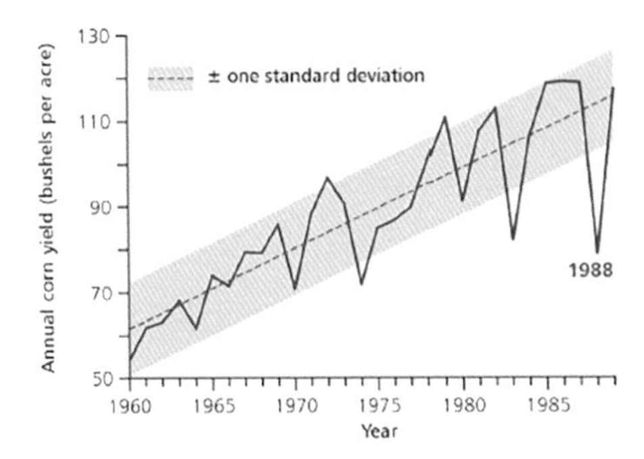

Figure 12.8 Annual corn yields in the USA 1960–1989 showing the effect of the 1988 drought. Yields in 1988 were more than 30% below trend, the largest annual drop recorded since the 1930s Dust Bowl years.

Source: After Donald (1988). US government material not subject to copyright protection in the United States.

affected (International Federation of Red Cross and Red Crescent Societies, 1994). According to Kajoba (1992), part of the grain shortfall was due to the cultivation of hybrid maize under imported fertiliser regimes, rather than a reliance on more traditional drought-resistant crops like sorghum, millets and cassava. Communities in remote areas suffered badly. The drought closed schools and led to a decline in tourism as wildlife camps were deserted. Due to low water levels, the Kariba, Kafue and Victoria Falls hydropower stations worked at 30% capacity and the government imposed daily power breaks. However, as outlined by Eldridge (2002) and Holloway (2000), and shown in Table 12.2, responses of drought-affected communities, particularly among women, ensured that a famine did not occur.

4. Famine Drought

The interrelationships between drought and famine are complex because drought can be hydrometeorological, whereas famine is a disaster that emerges from complex social relations, typically when those with access to power and resources cause problems for those without. In fact, many long-standing theories based on extensive evidence explain how there is usually plenty of food available during famines, but the people who need it most cannot access it, so they go hungry (Sen, 1983, 1999). Box 12.1 – which summarises some of the drought and famine challenges faced by communities in Ethiopia on the Horn of Africa – illustrates the role of both environmental and social factors in causing widespread distress. Both droughts and famines create problems of definition and measurement and represent complex and intersecting processes occurring over a range of timescales rather than events. For example, the early stages of a famine disaster are difficult to distinguish from chronic, lesser states of hunger, such as malnutrition, and other forms of food scarcity, like seasonal shortfalls in harvest yields.

C. RISK ASSESSMENT AND MANAGEMENT

Various drought monitoring and forecasting agencies exist to assess the temporal and spatial risk of drought. The best prospects for forecasting meteorological droughts lie with climate-based models that couple the atmosphere and the oceans in order to predict events like ENSO which affect food production in many countries.

One such agency is the *Intergovernmental Climate Prediction and Applications Centre* (ICPAC) set up in Nairobi in 2003 with responsibility for 11 countries in the Greater Horn of Africa. ICPAC provides climate monitoring and prediction services for early warning and mitigation

Table 12.2 Overview of the impacts and consequences of the 1992 drought in southern Africa, including household responses that averted an ensuing famine and lessons for policy.

The importance of household responses to food crises:

- Those most affected by a crisis, especially poor people, often contribute a large and underestimated proportion of the overall responses to it.
- Famine following the 1992 southern Africa drought was averted largely by the activities of those it affected most, particularly women.
- Some activities helped poor households survive the 1992 drought in the short term at the cost of compromising their medium-term livelihood security.

Cereal purchases and their consequences:

- Purchases comprised the largest source of cereals for most villagers in 1992; for many households purchases provided 2–3 times as much staple food as relief programmes.
- Financing these purchases had adverse longer-term consequences, mainly due to severe deteriorations in the rural terms of trade.
- Poverty forced the poorest drought-affected households to variously cut expenditure on education and agricultural inputs (as well as on other less essential items), work off their own land for longer than usual; and, in some cases, sell assets.

Poverty, drought and inequity:

- Poverty amplified the harmful effects of the 1992 drought on livelihoods. Correspondingly, the drought deepened the poverty of the most under-resourced households.
- Markets for both food and rural commodities worked against most households. Earnings from many income sources (particularly water-dependent sources) decreased, especially among poor smallholders.
- Economic inequities were worsened by the drought. These inequities also reduced access by some of the poorest people to relief and rehabilitation programmes.

The relatively small role of relief and rehabilitation programmes:

- Food aid received by the poorest drought-affected people accounted for just 15–25 per cent of a household's cereal requirements. In addition to relatively small receipts, villagers also complained of late or infrequent deliveries, and poor targeting: richer villagers sometimes received more drought relief food than poor villagers.
- Villagers received little external support for their livestock, very little help to reduce non-survival expenditure, and no assistance to maintain their purchasing power.
- The 1992 southern Africa relief and rehabilitation programme was a success, in conventional terms. But it illustrated a double deficiency of most food aid programmes of the last two decades: a failure to adequately involve those people most affected by food crises in these programmes, and a failure to address the adverse impacts of changes in market processes on household livelihood security during those crises.

Policy implications for food crises:

- Interventions before and during food crises should complement and support the activities of the poorest and most vulnerable households.
- The people most affected by crises should participate in the planning and implementation of emergency aid programmes, to maximise their effectiveness.
- Particular emphasis should be placed on interventions that minimise reductions in the purchasing power of the poorest households during crises.

Source: Taken from Eldridge, C. (2002), Why Was There No Famine Following the 1992 Southern African Drought? *IDS Bulletin*, 33: 79–87. https://doi.org/10.1111/j.1759-5436.2002.tb00047.x, p. 84.

of extreme climate events by issuing regular regional advisories along with weekly, monthly and seasonal bulletins. Climate Outlook Forums are held three times a year where experts come together to provide consensus outlooks and identify drought management strategies. Other bodies are attached to academic institutions, like the

Box 12.1 Drought and Famine in the Horn of Africa: The Case of Ethiopia

The Horn of Africa, a peninsula in the northeast of the continent, contains the nations of Somalia, Djibouti, Eritrea and Ethiopia, along with adjacent parts of Kenya and Sudan. It has been an arid region for at least 4,000 years and is regularly affected by drought, as well as floods and violent conflict.

In Ethiopia, 5%–10% of the population has typically experienced chronic food shortages. In the 1984–1985 drought, more than one-fifth were at risk of starvation after widespread rainfall deficits over several years. Excess mortality was estimated at 700,000, significantly exacerbated by power brokers manipulating the situation for political gain. Following three consecutive low rainfall years, a similar crisis emerged in 1999–2000, characterised by market failure, a ban imposed on livestock trading and a border conflict. Of the 66 million population, 85% live in the countryside, and 90% of exports depend on agriculture. Yet only 12% of the land is arable and is mostly dependent on rain-fed production. Less than 2% is in permanent crops, often grown for export and exposed to fluctuations in world commodity prices. Cattle herding has been the main traditional activity, occupying over 60% of the land area, but drought and mismanagement have degraded the rangeland vegetation to favour of camels and small ruminants, with an attendant rise in poverty (Kassahun et al., 2008).

It was previously believed that rainfall deficiencies in Ethiopia were driven by the moist summer monsoonal flow of air over Africa. It is now known that the key rainy period from June through to September (the *Kiremt* season) is largely dependent on the El Niño Southern Oscillation (ENSO) cycle, with sea-surface temperature anomalies in the Atlantic and Indian oceans of lesser significance (Korecha & Barnston, 2007). *Kiremt* rains result principally from localised convective activity over the Ethiopian highlands of western and west-central Ethiopia. The northeastern area and the southeastern pastoral lowlands are normally relatively dry at this time of year (Figure 12.9a). Figure 12.9b shows that this season provides about half the country with over 60% of its annual rainfall; the resulting harvest supplies over 90% of the nation's food. Unfortunately, these rains are unreliable. Ethiopia's farmers are well aware of this and adopt what strategies are available to them in unfavourable conditions. For example, agriculturalists in the north of the country have shifted towards more drought-resistant crops and a shorter growing season (Meze-Hausken, 2004), combined with more supplementary irrigation during September, taken from surface run-off water captured in farm ponds in July and August, in an attempt to prolong the growing period (Araya & Stroosnijder, 2011).

In a year with good rains, Ethiopia can produce enough food for its needs, although households still require the financial means to access supplies. Food emergencies are often complex, as in the 1999–2000 drought, when meagre *belg* rains during spring 1999 led to crop losses in the northeast highlands and cattle-herd reductions of up to 80% in the south and southeast were coincident with a border dispute with Eritrea. By January 2000, an estimated 7.7 million people were affected, and by July 2000 a humanitarian crisis existed over much of the country.

In 2022, while recovering from drought in 2017, Ethiopia has experienced failed rainy seasons for three consecutive years, with 8.2 million people affected by this current prolonged drought (WHO, 2022). As with the nature of drought, Ethiopia's neighbouring countries of Somalia, Kenya, South Sudan, Sudan and Djibouti were also experiencing food insecurity at stressed, crisis or emergency levels (Figure 12.10). There is a very high risk that this prolonged drought will lead to high levels of excess mortality and morbidity in relation to malnutrition and acute watery diarrhoea including cholera (WHO, 2022). Other public health risks include measles, malaria, sexual and reproductive health impacts, COVID-19, chronic infectious disease, trauma and mental health impacts.

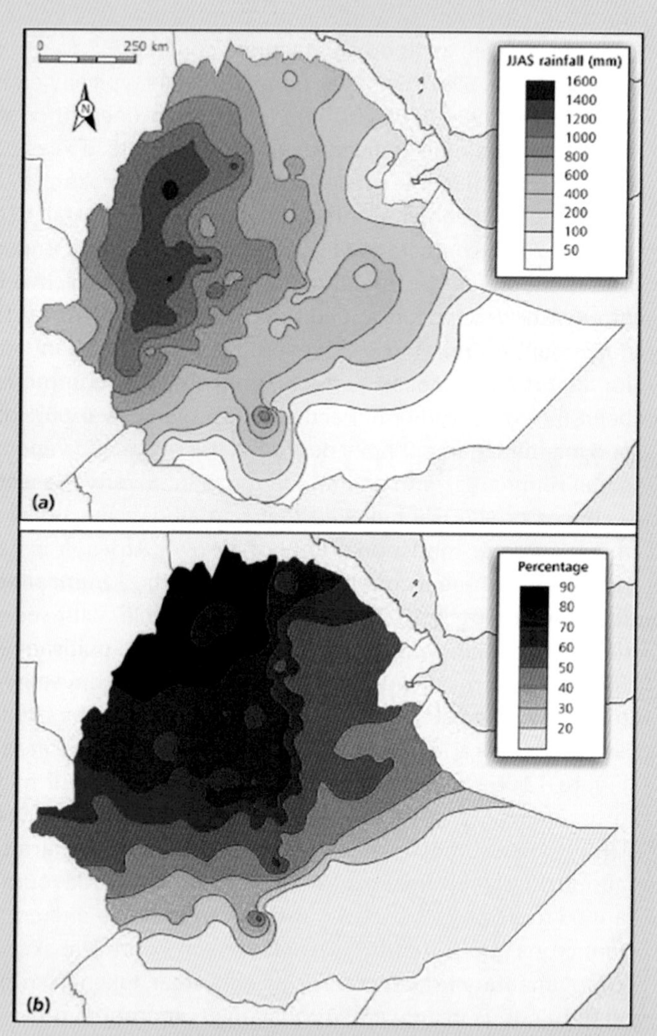

Figure 12.9 Rainfall patterns over Ethiopia: (a) total summer (JJAS) amounts over Ethiopia 1971–2000; (b) percentage of 1971–2000 mean annual rainfall contributed by JJAS rains. Summer rainfall over Ethiopia varies greatly in amount and timing. Both factors are crucial for food production and water management.

Source: After Korecha and Barnston (2007). Reproduced from D. Korecha and A.G. Barnston, Predictability of June–September rainfall in Ethiopia. *Monthly Weather Review* 135 (2007): 628–50. Copyright American Meteorological Society, reproduced with permission.

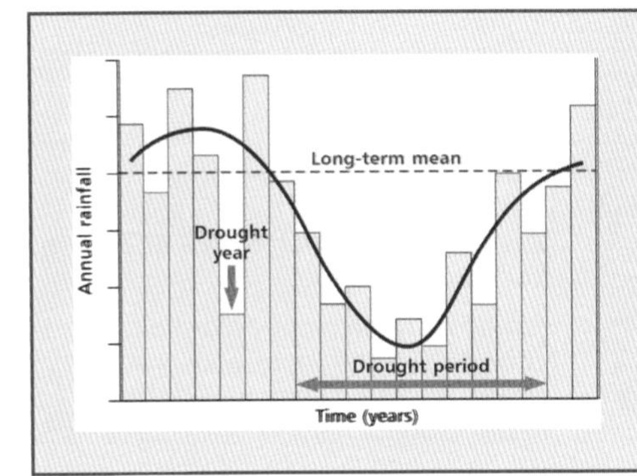

Figure 12.10 Food security outcome predictions for June–September 2022 across the Horn of Africa.

Source: https://cdn.who.int/media/docs/default-source/documents/emergencies/phsa_horn-of-africa_2022-06-09.pdf?sfvrsn=d4306a36_2&download=true. Accessed 18 August 2022.

US National Drought Mitigation Center at the University of Nebraska.

Crop-yield models can be incorporated into climate forecasts. Developments in satellite remote sensing have led to much more accurate agricultural monitoring of cropped areas, crop development and estimates of production. For example, the *Normalized Difference Vegetation Index* (NDVI), described by Gouveia et al. (2009), is used as an early indicator of both crop and pasture failure. Paleoclimatic studies are also important because research has shown that risk assessments based on instrumental hydrological data alone can underestimate the risk of persistent or extreme droughts in some regions (Ho et al., 2016, O'Donnell et al., 2021).

Management of drought therefore requires a proactive approach that integrates environmental data, alongside monitoring and forecasting of significant shortfalls in seasonal rainfall or snow/ice melt with efficient food and nutrition management, including the distribution of food aid. This process has become increasingly sophisticated, due to improved understanding of climate variability and its implications for agriculture, food supplies and the welfare of rural communities.

Two examples of global warning systems are the UN-sponsored *Global Information and Early Warning System* (GIEWS) and the USA-sponsored *Famine Early Warning System* (FEWS NET). They assess the risk of crop failure and food shortage (Table 12.3). These systems were established after severe food emergencies in the 1970s and 1980s. They rely on multi-agency support and focus on large-scale monitoring and forecasting in preparation for intervention at a more local level. For example, GIEWS operations rely on NDVI and rainfall estimates, along with an Agricultural Stress Index that provides early detection of agricultural areas most likely affected by dry spells and drought. Rainfall and vegetation estimates are then processed into maps of current and forecast conditions for staple food crops and pastureland. Regular reports on rainfall, food production and famine vulnerability are published and, when danger threatens, local offices undertake rapid-assessment surveys to clarify the situation on the ground.

In summary, the agricultural monitoring process integrates field observations, crop models and remotely sensed data to assess harvest production in a particular year and the likelihood that food shortages requiring humanitarian intervention will occur. Haile (2005) argued that if this process was streamlined, the needs assessment window could be brought

Table 12.3 Global monitoring and warning for drought and food shortages

Organisations	Global information and early warning systems (GIEWS)	Famine early warning system (FEWS NET)
Origins	Started 1975 by the Food and Agriculture Organization (FAO) of the United Nations; HQ Rome, Italy	Started 1985 as the FEW program by USAID, an agency of the US federal government; renamed FEWS NET in 2000; HQ Washington, DC
Objectives	Continuously monitor food supply and demand and other key indicators for assessing the overall food security situation internationally	Provide early warning and evidence-based analysis on food security across approx. 30 countries to help decision-makers plan for humanitarian crises
Cooperating agencies	Mainly other UN and FAO bodies – including WFP, UNDP, EU, OCHA	NASA, NOAA, USDA, USGS and CHC-UCSB, Chemonics International Inc., Kimetrica
Routine operations	Regular reports on global and regional crop production, demand for staple foodstuffs, reserve stock levels, agricultural trade	Monthly reports including maps showing current and projected food insecurities and alerts on emerging crises
Emergency operations	FAO HQ issues Special Alerts for areas where crops or food supplies are threatened to activate decision-makers and aid donors	FEWS/Chemionics head office issues warnings to decision-makers based on advice from field-based staff

forward to extend the lead-time for decision-making in the vulnerable areas. Figure 12.11a shows the situation for sub-Saharan countries as depicted by Haile (2005), with the agro-meteorological monitoring phase extending from February to October or November. If required, emergency appeals are launched in January under this model, which allows only a limited time for raising disaster funds before the onset of seasonal hunger. Haile (2005) postulated that it is desirable to link the monitoring and assessment process more closely to humanitarian decision-making. For example, if the appeal process began as early as October, more time would be available for farmers and others to respond to the oncoming drought (Figure 12.11b). Other key improvements would include taking greater account of the needs of pastoralists and along with a two-way flow of information contributing to early warnings of drought so that indigenous knowledge and experience is easily incorporated into the system.

After the famine droughts of the mid-1980s, several sub-Saharan countries – notably Chad and Mali – set up comprehensive food and nutrition monitoring systems (Autier et al., 1989). Regional expertise and indigenous support are vital, as shown by the establishment of the *Southern African Development Community (SADC) Remote Sensing Unit* in Harare, Zimbabwe and the *AGRHYMET Regional Centre* in Niamey, Niger. Created in 1974, this is a specialised institute sponsored by nine sub-Saharan states for improving food supplies and natural resource management in the Sahel. It

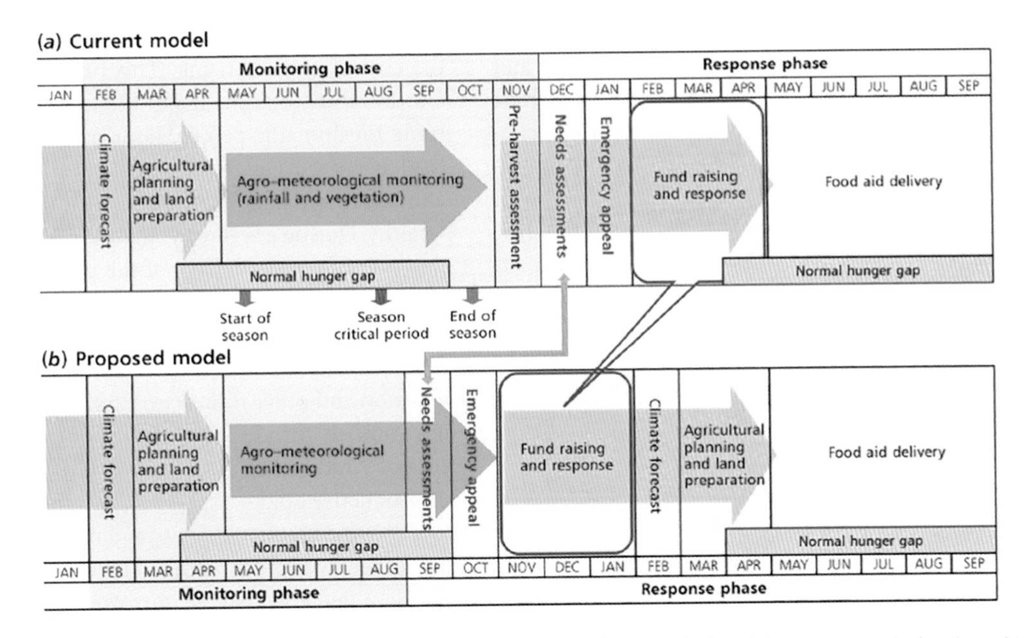

Figure 12.11 Drought response and food security in sub-Saharan Africa: (a) arrangements for drought monitoring, fundraising and food-aid delivery; (b) proposed model for earlier response and preparedness. Decision-making will be improved by bringing the whole humanitarian process forward by about four months.

Source: After Haile (2005). Reproduced from M. Haile, Weather patterns, food security and humanitarian response in sub-Saharan Africa. *Philosophical Transactions of the Royal Society B – Biological Sciences* 360 (2005): 2169–2182. With permission from the Royal Society.

provides food security assessments and NDVI product enhancement but also trains local staff in agro-meteorological and hydrological monitoring, statistics and data compilation and dissemination designed to predict food shortages.

The difference between regional food availability and household-level access to food means that an early indication of the downward spiral into famine is dependent on local nutritional field surveys. These surveys measure body conditions such as height for age, weight for age and weight for height, to identify those, like preschool-age children, with the greatest needs. Other reasonably reliable famine precursors are rising grain prices combined with falling livestock prices and wages as the economic balance shifts from assets and services, like jewellery and labour, to food, which rises in value both absolutely and relatively. However, the range of risk assessment methods used can produce discrepancies in the assessment of need and the distribution of food aid. Despite these uncertainties, it is most unfortunate that an early humanitarian response to famine drought seems to be increasingly dependent on clear evidence of excess mortality, without which many potential donors may be unwilling to act.

Conversely, many people evolved strategies for dealing with drought in order to avoid famine. For instance, nomadic communities in the Sahel adopted the practice of herd diversification, involving camels, cattle, sheep and goats, all with different grazing habits, water requirements and breeding cycles, which helped to spread any risk of pasture failure. During years with abundant rainfall, people increased their herds for food storage and as an insurance against drought. When drought occurred, people migrated to find good pasture and, in the most severe episodes, could either eat or sell off surplus livestock. Informal systems of communal loss sharing allowed the transfer of gifts or loans of any spare animals available to those in greatest distress and various fallback activities, such as gazelle hunting or caravan trading, were intensified as temporary measures. Similarly, villagers in rural Mali adjusted to lower harvests by diversifying their income sources from non-agricultural activities (Cekan, 1992).

Under severe drought conditions, people have to do more. They often start by eating less, in an attempt to conserve food stocks. Agricultural adjustments include crop replacement (drought-resistant crops preferred at the normal planting time), gap-filling in fields (where germination of an earlier crop has been poor) and resowing or irrigating crops. When food stocks are exhausted, they turn to a range of wild 'famine' foods that are not part of the normal diet because of their low nutritional value. In Zambia, for example, this includes eating honey mixed with soil, wild fruits and wild roots, some of which are poisonous unless boiled for several hours before eating. The selling of livestock is usually underway by this stage, although a case has been made for external intervention to ensure that destocking takes place early in a drought to prevent ecological damage to the grazing land (Morton & Barton, 2002). Unfortunately, destocking accelerates the fall in the value of livestock at a time when the cost of grain is rising. This produces a disadvantageous change in the terms of trade for nomadic people. Poor pastoralists have to sell a larger proportion of their animals in order to buy food than do the wealthy. During a severe drought, many of the poor are squeezed out of the pastoral economy and forced to settle in towns to live on famine relief or from wages paid to herders or labourers (Haug, 2002). Without food and other resources, rural dwellers routinely turn to local wage labour for support, rather than work on their own unproductive land.

For some households, outstanding debts and other favours can be called in. Cash or food entitlements may be borrowed from more prosperous relatives, neighbours or other support groups in a non-agricultural response strategy. Without access to external help, there is little option but to resort to the trading of valuables, such as jewellery, or other capital assets, such as radios, bicycles or firearms, which can be sold to buy grain. As incomes decline, health conditions also deteriorate. This deterioration is exacerbated by poor nutrition and growing competition for declining, and increasingly polluted, water supplies. Wherever possible, villagers poach wild game in order to survive. Table 12.4 shows that during a severe drought in 1994–1995 that affected over 10% of Bangladesh, households adopted a variety of non-agricultural adjustments. Over half of those questioned sold livestock and over 70% of the respondents either sold or mortgaged land (Paul, 1995).

In drought conditions families can start to break up. Some children may be sent to distant relatives out of the famine zone and male members may seek work in the towns. This can lead to large-scale migration that may be permanent if families

Table 12.4 Adoption of non-agricultural adjustments to drought by households in Bangladesh

Adjustment	Number of households	%
Sold livestock	166	55
Sold land	112	37
Mortgaged land	106	35
Mortgaged livestock	2	1
Sold possessions	26	9
Family members migrated	1	0

Source: After Paul (1995)

Note: 265 households were surveyed. Multiple responses are possible.

lose their land rights because of moving. In many cases, those with more resources in a location can choose to stay or leave. Some choose to leave before conditions become too bad, meaning that the earliest migrants tend to have some level of personal resources. When conditions are atrocious, those left behind are usually those who had the fewest resources to begin with. As with other hazards, prior experience sometimes enhances survival possibilities, but some traditional responses are now less available than in the past due to rapid social and environmental changes.

More research on staple grains and better dryland farming techniques, such as terracing, strip cropping and soil erosion control, is needed. It may even be more productive to support nomadic pastoralism rather than irrigation schemes. New attitudes are required in order to change funding priorities. Improving the physical infrastructure in areas at risk by the provision of better roads will not only reduce short-term vulnerability by helping the distribution of emergency food aid but will also allow the optimum location of new facilities, such as well-equipped health clinics, which will support longer-term benefits.

Longer-term strategies that move away from reactive, short-term drought management solutions can also include the coupling of drought risk insurance with sustainable land management techniques. Traditional drought insurance has been used as a short-term solution to recover from drought losses but it attracts high associated costs and can lead to a moral hazard, whereby farmers take more risks because they are covered by insurance (Choudhury et al., 2016). The latter can occur particularly when that insurance is subsidised by the government. That is, the government is bearing the costs of those risks. Subsidised crop insurance programs have often failed because they became too expensive for the governments that implemented them. Index-based insurance, such as that based on rainfall indices, are becoming more popular particularly among small-scale farmers in places like Ghana because of their lower insurability threshold making them more serviceable for the insurer (Choudhury et al., 2016). Agricultural index insurance, based on rainfall indices, links insurance payouts to historical rainfall data in relation to crop and livestock losses when rainfall amounts are below the historical threshold. Agricultural index insurance can also be based on NDVI. Tsegai and Kaushik (2019) argue that with current advancements in drought monitoring, forecasting and early warning systems, alongside drought risk assessments and sustainable land management practices, insurance tools like the agricultural index insurance have the ability to revolutionise small-scale farmers' access to formal insurance, particularly when applied to the community level.

The longer-term management solutions for increasing the supply of water to meet anticipated demands is the use of dams and pipelines for the artificial storage and transfer of water supplies. The emphasis on these engineering solutions is symbolised by the global spread of large dams. Reservoirs have been used extensively to maintain urban water supplies. The greatest buffering against drought exists for those areas with a sufficient margin between the supply capacity of the system and the maximum use. Figure 12.12 illustrates how water-supply reservoirs smooth out seasonal variations in river flow to manage supplies in order to meet demand. The annual surplus available for supply will be eroded if storage is not increased to cope with the progressive rise in demand, due either to population increase or greater water use by individual households, typical of most areas. During a drought period, water shortages, with attendant restrictions on use, are likely. A drought exists when the managed supply available for distribution falls below the average annual runoff, leading to an interrupted service and the possible need to import extra supplies from elsewhere.

Further management solutions involve improving efficiency in water use and

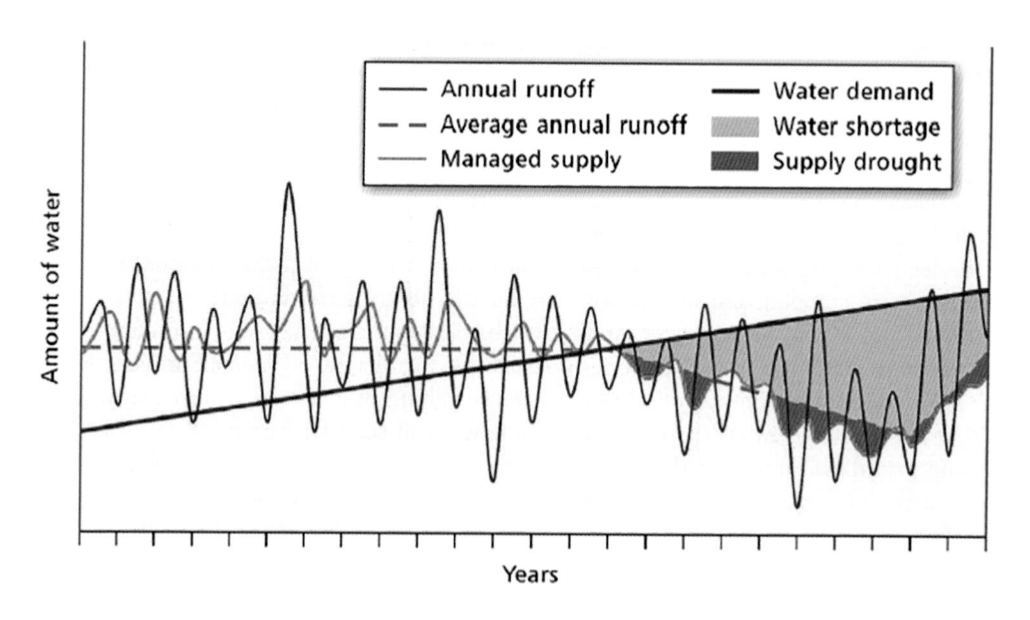

Figure 12.12 Idealised emergence of a water supply drought. Water demand rises progressively over time. Without any increase in supply, relatively minor reductions in run-off create a water shortage. When the managed supply falls below average annual or seasonal run-off, an urban water-supply drought exists.

managing the demand for water as well as the supply. The short-term adjustments used by water authorities during hydrological droughts are aimed mainly at the domestic consumer. They include both supply-management and demand-management practices. Supply-management methods tend to concentrate on the more flexible use of available supplies and storage. This can be achieved by switching water abstraction between surface and ground sources and by water transfers between different supply authority areas to ease the greatest shortages. Temporary engineering, such as the laying of emergency pipelines, may allow importation of water from more distant sources. Other technical measures include reducing the water pressure in the main supply pipes and repairing as many leaking pipes as possible in the distribution system. When all else fails, water can be rationed in the worst-hit areas by rota cuts. Attempts to manage (reduce) consumer demand normally include a mix of legal measures and public appeals to conserve water. Special legislation may be introduced, such as the Drought Act rushed through the British Parliament in August 1976 to prohibit non-essential domestic uses of water like washing cars or watering gardens. Combined with 'save water' publicity campaigns, these measures can significantly reduce residential demand for short periods.

But crisis management is no substitute for longer-term planning and water conservation in urban areas. Where hydrological drought is more common, as in Adelaide, South Australia, the management of water demand has been a central plank of policy for many years. During the summer months, when rainfall is almost entirely absent, as much as 80% of the water consumed within the metropolitan area has been used to irrigate gardens. As part of an overall conservation strategy, domestic use can be reduced through a combination of financial measures (seasonal peak pricing), technical measures (curbs on inefficient water-using appliances and advice on watering methods) and social measures (persuading people to grow native plants in their gardens rather than more water-demanding European varieties).

The response to water shortages depends on the severity of the conditions. For example, the responsible bodies in England and Wales – Environment Agency (EA), water companies, local authorities and government – cooperate along the following lines:

- *Stage One*. EA and water companies conserve water by encouraging all consumers to reduce demands and by in-house operational measures, such as lowering the water pressure in the distribution system.
- *Stage Two*. Water companies introduce restrictions on hosepipes for domestic garden use and car washing. The EA uses powers to ban agricultural spray irrigation. Water conservation publicity is stepped up by the EA, water companies and local authorities.
- *Stage Three*. Legally enforceable *Drought Orders* or *Drought Permits* are granted, either by the EA or by central government, to restrict or ban water abstraction for agriculture and other specified uses, whilst allowing water companies to abstract increased amounts of surface and groundwater
- *Stage Four*. Water companies cut off domestic supplies at pre-advertised times of day, standpipes are installed in urban areas and tankers are used to import supplies into critical areas.

Many reservoir-based urban water supply systems are designed to provide a predetermined minimum supply during roughly 98% of the time (2% probability of failure), although minor shortages may be accepted more frequently. With an element of over-design and careful crisis management, it is possible for these systems to perform well during droughts of a magnitude beyond the 1:100-year drought event. Figure 12.13 shows the operation of domestic water restrictions on the river Tone, Japan, which helped to maintain supplies during the summer drought of 1994 (Omachi, 1997). Without these actions, the content of the reservoirs would have declined more quickly and storage would have been exhausted by 12 August 1994.

Drought increases pressure on land resources if land and water are not properly managed. Overgrazing, poor cropping methods, deforestation and improper soil conservation techniques may not create drought, but they can amplify drought-related disasters.

Small-scale farming in marginal drylands is so dependent on seasonal rains that it is desirable to spread risk by engaging in forms of mixed agriculture that include livestock and tree crops as well as annual arable produce. In parts of Asia, an integrated pattern of sugar cane and livestock farming is found. This is complemented with cattle rearing combined with vegetable and maize cultivation in the uplands. Where irrigation is possible, the development of improved crop varieties and fertiliser application

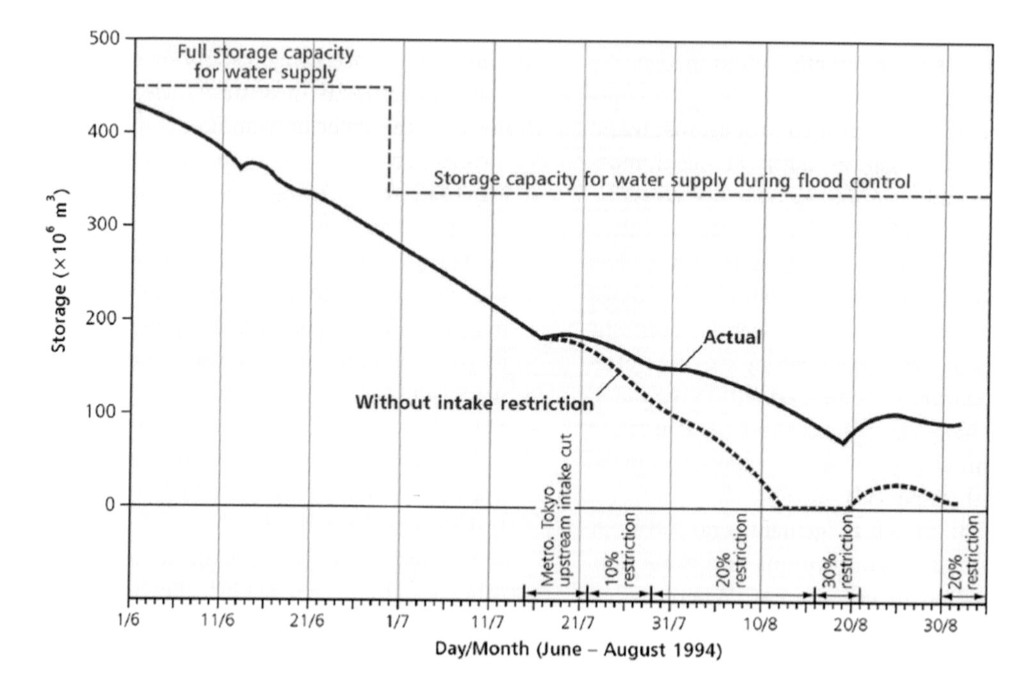

Figure 12.13 Changes in water storage in reservoirs along the upper river Tone, Japan, showing the effect of intake restrictions during the summer drought of 1994.

Source: After Dr Toshikatsu Omachi (1997). Originally published in IDI Water Series No. 1, *Drought Conciliation and Water Rights: Japanese Experience* (Infrastructure Development Institute, Japan, March 1997). Reproduced with permission from the Infrastructure Development Institute, Japan.

can increase output, but these options are less relevant for rain-fed subsistence farming. Short-term responses are frequently required, not least in the regions dependent on monsoon rains. For example, if the seasonal rains are weaker than expected at the time of planting, possible mid-season adaptations include thinning of field crops and the spreading of green material as a soil to reduce water loss. When the monsoon is late but produces adequate rainfall, the selection of short-duration crops such as pulses and oilseed compensate for the reduced length of the growing season. In years when rainfall is inadequate overall, leguminous crops and oilseeds tend to provide the best yields. At all times, it is necessary to control weeds. These compete with crops for moisture and nutrients, thereby reducing yields, especially in dry periods.

Sustained dry-land farming is dependent on soil conservation measures against water and wind erosion. A grass or legume cover is an effective control against water erosion, as are strip cropping and contour cultivation, which retard the flow of water downhill. Wind erosion can be reduced by maintaining a trash cover at the soil surface, plus the use of crop rotations and shelterbelts – where possible – to lower the wind velocity at the soil surface. Local watersheds provide the most suitable land use unit for managing land and water in drought-prone regions where the top priority is to conserve rainwater for crops, livestock and people. In many places, maintaining the quantity and quality of water in the village's reservoir is critical because this acts as a common-property, multipurpose resource.

The basis of successful water management is the retention of surplus resources in the wet season for later use in the growing season. An obvious method is to divert intermittent surface flows into farm and village reservoirs, but high evaporation rates, combined with sedimentation and pollution, limit the survival time of these resources in the dry season. The artificial recharge of groundwater is often a better method. This is achieved by delaying and spreading the downslope movement of water to promote greater infiltration into the soil and shallow aquifers. Several methods exist. Contour ridging and ploughing is extensively used, sometimes combined with earthen bunds at the terrace edge, to retain the maximum amount of water on a slope. These techniques also minimise the threat of soil erosion. In limited areas with intermittent concentrated flows of water, like small ravines or vegetated ditches, *check dams* are useful. The lowest-cost structures are made of rock, vegetation or sandbags and form barriers that reduce water velocities and pond-back temporary high flows to increase infiltration into adjacent aquifers and wells (Figure 12.14). They cannot be used for permanent streams and are rarely employed for channels draining more than 0.05 km^2, unless constructed from concrete.

More permanent check dams have been installed in parts of India. The state

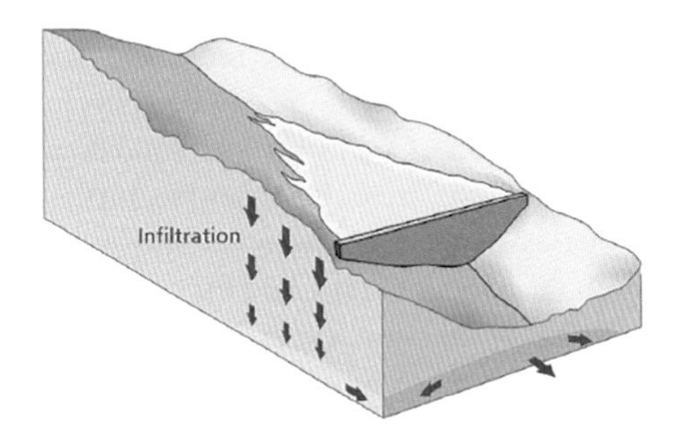

Figure 12.14 The use of check dams across intermittent water courses. Small quantities of water can be stored for domestic supply or plot irrigation, whilst infiltration aids local soil moisture conditions around the dam.

of Uttar Pradesh is estimated to have lost about 15,000 × 10^4 km^2 of productive land, due to erosion of light alluvial soil in ravines. Small concrete structures can arrest up to 80% of the seasonal run-off in some ravines and have raised the local water table by several metres.

Even in the wetter places, land use policies to combat drought are necessary because rural areas rarely have the large water storages and the options for reducing consumer demand that are available in the cities (Campbell et al., 2011). Therefore, drought strategies should prepare the agricultural sector to withstand shortfalls of precipitation. This involves the careful management of surpluses in good years, adoption of appropriate stocking rates so that the pasture is not easily exhausted, the build-up of a reserve of fodder and the improvement of on-farm water supplies. The installation of an irrigation system may offer some security, but the amount of storage may not provide complete drought proofing. Pigram (1986) cited heavy losses sustained by irrigators of rice and cotton in New South Wales, Australia, during the latter stages of the 1979–1983 drought, when water allocations were suspended in the middle of the irrigation season. Flexible decision-making is always needed, and disaster risk reduction will be strengthened by a greater diversity of cropping patterns and income sources in drought-prone areas. For example, further scope exists for the development of more drought-resistant crops, and for crops with varying production cycles, that make it easier for rural communities to exist when the rains fail.

KEY TAKE AWAY POINTS

- Drought as an environmental hazard should be regarded as a process which develops over time, with increasingly severe effects, rather than an event.
- Drought disasters are better understood in terms of the impacts on natural resources and human activities, such as water supplies, agricultural production and food availability, rather than on the basis of rainfall statistics alone. Droughts can therefore be categorised as hydrological, agricultural and famine as well as meteorological.
- Climate-based models that link atmospheric and oceanic processes, such as ENSO, are used to forecast meteorological droughts. These can be combined with agricultural monitoring processes that integrate field observations, crop models and remotely sensed data to determine the likelihood of food shortages occurring in affected regions and the need for humanitarian intervention.
- Management solutions include improving water efficiency, reducing water demand, careful management of water surplus in good years, improving agricultural practices and ensuring more equitable household-level access to food.

BIBLIOGRAPHY

Alley, W. M. (1984). The Palmer drought severity index: Limitations and assumptions. *Journal of Climatology and Applied Meteorology, 23*, 1100–1109.

Araya, A., & Stroosnijder, L. (2011). Assessing drought risk and irrigation need in northern Ethiopia. *Agricultural and Forest Meteorology, 151*, 425–436.

Autier, P., D'Altilia, J.-P., Delamalle, J.-P., & Vercruysse, V. (1989). The food and nutrition surveillance systems of Chad and Mali: The 'SAP' after two years. *Disasters, 13*, 9–32.

Campbell, D., Barker, D., & McGregor, D. (2011). Dealing with drought: Small farmers and environmental hazards in

southern St Elizabeth, Jamaica. *Applied Geography, 31*, 146–158.
Cekan, J. (1992). Seasonal coping strategies in Gental Mali: Five villages during the 'Soudiere'. *Disasters, 16*, 66–73.
Choudhury, A., Jones, J., Okine, A., & Choudhury, R. (Lena). (2016). Drought-triggered index insurance using cluster analysis of rainfall affected by climate change. *Journal of Insurance Issues, 39*(2), 169–186. www.jstor.org/stable/43921956
Cook, B. I., Seager, R., & Miller, R. L. (2011). Atmospheric circulation anomalies during two persistent North American droughts; 1932–1939 and 1948–1957. *Climate Dynamics, 36*, 2339–2355.
Cook, E. R., Seager, R., Cane, M. A., & Stahle, D. W. (2007). North American drought: Reconstructions, causes and consequences. *Earth Sciences Review, 81*, 93–134.
Dilley, M., & Heyman, B. N. (1995). ENSO and disaster: Droughts, floods and E1 Niño-Southern Oscillation warm events. *Disasters, 19*, 181–193.
Donald, J. R. (1988, October 18). Drought effects on crop production and the US economy: The drought of 1988 and beyond. Proceedings of a Strategic Planning Seminar, National Climate Program Office, Rockville, MD, 143–162.
Eldridge, C. (2002). Why was there no famine following the 1992 Southern African drought? *IDS Bulletin, 33*, 79–87. https://doi.org/10.1111/j.1759-5436.2002.tb00047.x
Eldridge, K. A., & Christensen, A. (2002). Demand-withdraw communication during couple conflict: A review and analysis. In *Understanding marriage: Developments in the study of couple interaction* (pp. 289–322).
Field, R. D., van der Werf, G. R., & Shen, S. S. P. (2009). Human amplification of drought-induced biomass burning in Indonesia since 1960. *Nature Geoscience, 2*, 185–188.
Gadgil, S., Vinayachandran, P. N., & Francis, P. A. (2003). Droughts of the Indian summer monsoon: Role of clouds over the Indian Ocean. *Current Science, 85*, 1713–1719.
Giannini, A., Saravanan, R., & Chang, P. (2003). Oceanic forcing of Sahel rainfall on inter-annual to inter-decadal time scales. Science, 302, 1027–1030.
Goosse, H., Barriat, P. Y., Lefebvre, W., Loutre, M. F., & Zunz, V. (2008). Introduction to climate dynamics and climate modelling [Online textbook]. www.climate.be/textbook Université Catholique de Louvain, Belgium. Accessed 4 September 2011.
Gouveia, C., Trigo, R. M., & Dacamara, C. C. (2009). Drought and vegetation stress monitoring using satellite data. Natural Hazards and Earth System Sciences, 9, 185–195.
Guttman, N. B. (1997). Comparing the Palmer drought index and the standardized precipitation index. Journal of the American Water Resources Association, 34, 113–121.
Haile, G. G., Tang, Q., Li, W., Liu, X., & Zhang, X. (2020). Drought: Progress in broadening its understanding. WIREs Water, 7, e1407. https://doi.org/10.1002/wat2.1407
Haile, M. (2005). Weather patterns, food security and humanitarian response in sub-Saharan Africa. Philosophical Transactions of the Royal Society B: Biological Sciences, 360, 2169–2182.
Haug, R. (2002). Forced migration, processes of return and livelihood construction among pastoralists in northern Sudan. Disasters, 26, 70–84.
Heim, R. (2002). A review of twentieth-century drought indices used in the United States. Bulletin of the American Meteorological Society, 83, 1149–1165.
Ho, M., Lall, U., & Cook, E. R. (2016). Can a paleodrought record be used to reconstruct streamflow?: A case study for the Missouri River Basin. Water Resources Research, 52, 5195–5212.
Holloway, A. (2000). Drought emergency, yes . . . drought disaster, no: Southern Africa 1991–93. Cambridge Review of International Affairs, 14(1), 254–276.
Hunt, E. D., Hubbard, K. G., Wilhite, D. A., Arkebauer, T. J., & Dutcher, A. L. (2009). The development and evaluation of a soil moisture index. International Journal of Climatology, 29, 747–759.
Kajoba, G. M. (1992, October). Food security and the impact of the 1991–92 drought in Zambia. Unpublished text of lecture delivered at the University of Stirling.

Kassahun, A., Snyman, H. A., & Smit, G. N. (2008). The impact of rangeland degradation on the pastoral production systems, livelihoods and perceptions of the Somali pastoralists in Eastern Ethiopia. Journal of Arid Environments, 72, 1265–1281.

Keyantash, J., & Dracup, J. A. (2002). The quantification of drought: An evaluation of drought indices. Bulletin of the American Meteorological Society, 83, 1167–1180.

Korecha, D., & Barnston, A. G. (2007). Predictability of June-September rainfall in Ethiopia. Monthly Weather Review, 135, 628–650.

Liu, W. T., & Kogan, F. N. (1996). Monitoring regional drought using the Vegetation Condition Index. International Journal of Remote Sensing, 17, 2761–2782.

Luckman, B. H. (2013). Dendrochronology. In S. A. Elias & C. J. Mock (Eds.), Encyclopedia of quaternary science (2nd ed., pp. 459–470). Amsterdam: Elsevier.

McCabe, G. J., Palecki, M. A., & Betancourt, J. L. (2004). Pacific and Atlantic Ocean influences on multidecadal drought frequency in the United States. Proceedings of the National Academy of Sciences, 101, 4136–4141.

Meyer, S. J. (1993). A crop specific drought index for corn: Application in drought monitoring and assessment. Agronomy Journal, 85, 396–399.

Meze-Hausken, E. (2004). Contrasting climate variability and meteorological drought with perceived drought and climate change in northern Ethiopia. Climate Research, 27, 19–31.

Mishra, A. K., & Singh, V. P. (2010). A review of drought concepts. Journal of Hydrology, 391, 202–216.

Morton, J., & Barton, D. (2002). Destocking as a drought-mitigation strategy: Clarifying rationales and answering critiques. Disasters, 26, 213–228.

Nicholls, N. (2011, March 11 Friday). What caused the eastern Australian heavy rains and floods of 2010/11? Bulletin of the Australian Meteorological and Oceanographic Society 24: President's Column.

International Federation of Red Cross and Red Crescent Societies (IFRCRCS) (1994) World Disasters Report 1994. Dordrecht: Martinus Nijhoff.

O'Donnell, A. J., McCaw, W. L., Cook, E. R., & Grierson, P. F. (2021). Megadroughts and pluvials in southwest Australia: 1350–2017 CE. Climate Dynamics, 57(7), 1817–1831. DOI: 10.1007/s00382-021-05782-0

Omachi, T. (1997). Drought conciliation and water rights: Japanese experience. Water Series 1, Infrastructure Development Institute, Tokyo.

Palmer, W. C. (1965). Meteorological drought. Research Paper 45, US Weather Bureau, Department of Commerce, Washington, DC.

Paul, B. K. (1995). Farmers' and public responses to the 1994–95 drought in Bangladesh: A case study. Quick Response Report 76, Natural Hazards Research and Applications Information Center, Boulder, CO.

Peel, M. C., McMahon, T. A., & Finlayson, B. L. (2002). Variability of annual precipitation and its relationship to the El Niño-Southern Oscillation. Journal of Climate, 15, 545–551.

Pigram, J. J. (1986). Issues in the management of Australia's water resources. Melbourne: Longman Cheshire.

Pulwarty, R. S., Wilhite, D. A., Diodato, D. M., & Nelson, D. I. (2007). Drought in changing environments: Creating a roadmap, vehicles and drivers. Natural Hazards Observer, 31(5), 10–12.

Ross, T., & Lott, N. (2003). A climatology of 1980–2003 extreme weather and climate events. Technical Report No. 2003-01, NOAA/NESDIS, National Climate Data Center, Ashville, NC.

Sen, A. K. (1983). Poverty and famines: An essay on entitlement and deprivation. Oxford: Oxford University Press.

Sen, A. K. (1999). Development as freedom. New York: Anchor Books.

Sheen, K., Smith, D., Dunstone, N., et al. (2017). Skilful prediction of Sahel summer rainfall on inter-annual and multi-year timescales. Nature Communications, 8, 14966. https://doi.org/10.1038/ncomms14966

Trenberth, K., Branstator, G. W., & Arkin, P. A. (1988). Origins of the 1988 North American drought. Science, 242, 1640–1645.

Tsegai, D., & Kaushik, I. (2019). Chapter 13: Drought risk insurance and sustainable land management: What are the options for integration? In E. Mapedza, D. Tsegai, M. Bruntrup, & R. McLeman (Eds.), Current directions in water scarcity research (Vol. 2, pp. 195–210). Elsevier.

Ummenhofer, C. M., Sen Gupta, A., Li, Y., Taschetto, A. S., & England, M. H. (2011). Multi-decadal modulation of the El Niño-Indian monsoon relationship by Indian Ocean variability. Environmental Research Letters, 6, 1–8.

Vicente-Serrano, S. M., & López-Moreno, J. I. (2006). The influence on atmospheric circulation at different spatial scales on winter drought variability through a semi-arid climatic gradient in north-east Spain. International Journal of Climatology, 26, 1427–1453.

Wang, T., Bao, A., Xu, W., Yu, R., Zhang, Q., Jiang, L., & Nzabarinda, V. (2021). Tree-ring-based assessments of drought variability during the past 400 years in the Tianshan mountains, arid Central Asia. Ecological Indicators, 126, 107702. https://doi.org/10.1016/j.ecolind.2021.107702

WHO. (2022). Greater Horn of Africa (Food insecurity and drought). Horn of Africa Crisis: Public Health Situation Analysis. https://cdn.who.int/media/docs/default-source/documents/emergencies/phsa_horn-of-africa_2022-06-09.pdf. Accessed 30 July 2022.

FURTHER READING

Devereux, S. (1993). Theories of famine. London: Harvester Wheatsheaf.

Fleurett, A. (1986). Indigenous responses to drought in sub-Saharan Africa. Disasters, 10(3), 224–229.

Glantz, M. H., & Katz, R. W. (1977). When is a drought a drought? Nature, 267, 192–193.

Wilhite, D. A. (2002). Combating drought through preparedness. Natural Resources Forum, 26, 275–285.

Wilhite, D. A., & Easterling, W. E. (Eds.). (1987). Planning for drought: Toward a reduction of societal vulnerability. Boulder, CO and London: Westview Press.

Wilhite, D. A., & Glantz, M. H. (1985). Understanding the drought phenomenon: The role of definitions. Water International, 10(3), 111–120.

WEB LINKS

FEWS NET, the Famine Early Warning Systems Network. https://fews.net

GIEWS: Global Information and Early Warning System on Food and Agriculture. www.fao.org/giews/en

The North American Drought Atlas. www.ncdc.noaa.gov/paleo/pdsi.html

CHAPTER THIRTEEN

Extreme Temperature Hazards

13

OVERVIEW

Extreme temperatures considered in this chapter cover cold waves and heatwaves, both of which are defined by periods of extremely cold or extremely hot temperatures over consecutive days measured in comparison to the usual temperatures experienced at that location for that time of year. Both result in the occurrence of excess deaths, based on the expected rate of fatalities for that time period and in that location, as well as other health impacts such as hypothermia and frostbite during periods of extreme cold and heat stress and heat stroke during periods of extreme heat. Infrastructure and critical services can be impacted during extreme temperature events, with both cold waves and heatwaves often leading to a loss of electricity, which causes conditions to deteriorate further. Heatwaves are becoming more problematic due to the urban heat island effect where the infrastructure of urbanised areas absorbs and transfers heat back into the surrounding environment, causing temperatures to be hotter than in bordering semi-urban and rural areas. Anthropogenic climate change is also leading to increases in the intensity, frequency and duration of heatwaves across the globe. While risk assessment and management of extreme temperatures is lacking in comparison to many other environmental hazards, extreme heat and heatwaves are the leading cause of natural hazard-related fatality in many countries.

A. THE NATURE OF EXTREME TEMPERATURES

Extreme temperature hazards are characterised by consecutive days of extreme hot or cold temperatures, outside the normal temperature range for that region and at that time of year. They are not one-off days of extreme temperatures. Such extreme temperature events are known as cold waves or heatwaves.

DOI: 10.4324/9781351261647-15

1. Cold Waves

Cold waves are described by the American Meteorological Society as a rapid decrease in temperature and are associated with the minimum that they reach, which is considered in relation to the location in which they occur and the time of year. That is, cold waves that occur early in the cooler season can be more dangerous than those that occur later in the cooler season (Barnett et al., 2012).

Cold waves, also referred to as cold spells, do not have a standard definition due to this spatial and temporal variability. Wang et al. (2016) acknowledged that using a standard temperature threshold to define cold waves is not ideal for large spatial areas, such as in countries like China. Instead, they determined a cold wave based on weather fluctuations where the mean daily temperature fell below the fifth percentile of a particular time period at a specific location for at least two consecutive days. Wang et al. (2016) also characterised cold waves by their duration, intensity and timing within the season. In terms of duration, two days is considered short, three to five days is moderate and six or more days is long. Timing is measured by the number of days between the start of the cold wave and the beginning of the cold season.

The Intergovernmental Panel on Climate Change (IPCC) reports that cold temperature extremes have decreased globally since 1950 (Seneviratne et al., 2021). As the climate continues to warm due to human-caused climate change (Chapters 16 and 18), the IPCC projects that cold extremes will continue to decrease in intensity and frequency and cold waves will be shorter in length (Hoegh-Guldberg et al., 2018), even if global warming stabilises at 1.5°C (Seneviratne et al., 2021). Nevertheless, Wang et al. (2016) and others argue that because cold waves continue to significantly increase mortality rates in places like China, more needs to be done to prevent cold temperature-related fatalities.

2. Heatwaves

The Australian Bureau of Meteorology describes heatwaves as unusually hot maximum and minimum temperatures occurring over a three-day period at a given location, in consideration of that location's local climate and past weather (BOM, 2022). This takes into account the high overnight temperatures, where people have little respite, making it harder for the body to recover from the previous day's heat.

As datasets improve, it has become clear that heatwaves impact almost all regions across the globe and that this trend is increasing. Using the HadGHCND (a gridded daily temperature dataset based upon near-surface maximum (TX) and minimum (TN) temperature observations). Perkins-Kirkpatrick and Lewis (2020) examined the Berkeley Earth observational dataset to assess changes in global heatwave trends in relation to frequency, duration, intensity and cumulative heat over the period from 1950 to 2014 (Figure 13.1). Their research shows that heatwave frequency is increasing significantly and rapidly across almost all regions (Figure 13.1b). Heatwave duration is increasing particularly across South America, Africa, the Middle East and Southwest Asia (Figure 13.1d). Southern Australia and small areas of Africa and South America are the only regions showing increases in heatwave intensity (Figure 13.1f). Linked to heatwave frequency, cumulative heat trends show significant increases almost everywhere, with the

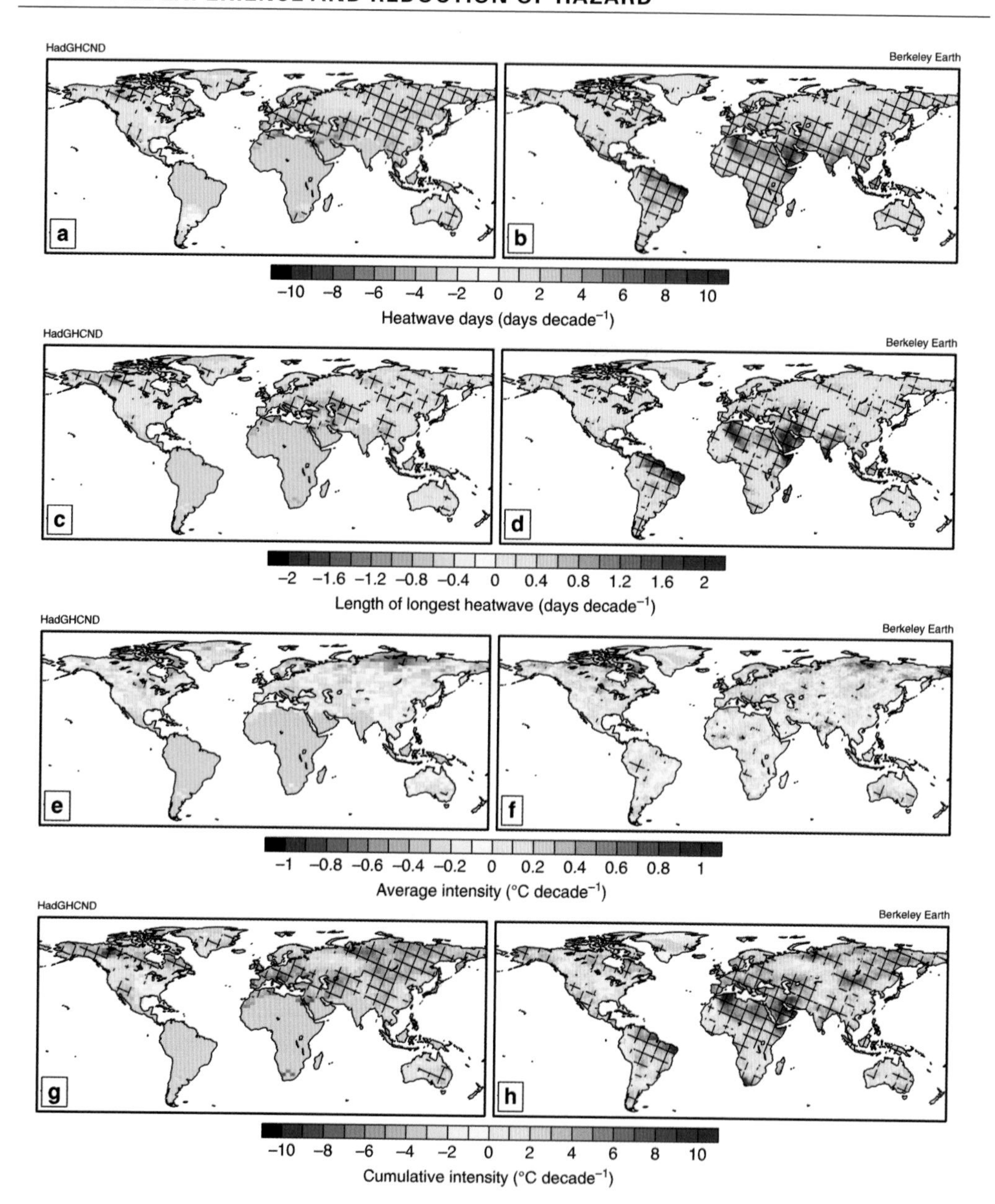

Figure 13.1 Examination of the Berkeley Earth observational dataset alongside the HadGHCND gridded daily temperatures dataset to determine changes in global heatwave trends in relation to frequency, duration, intensity and cumulative heat over the period from 1950 to 2014.

Source: Perkins-Kirkpatrick, S.E., Lewis, S.C. Increasing trends in regional heatwaves. *Nat Commun* 11, 3357 (2020). https://doi.org/10.1038/s41467-020-16970-7.

greatest changes seen in the Middle East and parts of Africa and South America (Figure 13.1h). Perkins-Kirkpatrick and Lewis (2020) conclude that changes in global heatwave trends are accelerating due to the effects of anthropogenic climate change.

These increases were evident during the northern hemisphere summer in 2022, with unprecedented heatwaves experienced in many parts of the world. Tokyo recorded nine consecutive days above 35°C, temperatures in the United Kingdom reached above 40°C for the first time since records first began there, China was subjected to multiple heatwaves with one impacting over 400 cities and wildfires sparked by extreme heat burnt across parts of France, Spain, Greece and Germany (Witze, 2022).

In addition to a changing climate, rapidly increasing urbanisation is also contributing to heatwave intensity. When urbanisation takes over the natural land cover, replacing it with asphalt/bitumen roads, concrete footpaths, brick buildings and other heat-absorbing materials, 'urban heat islands' (UHI) occur. This urban infrastructure absorbs and retains the heat of the day, causing day and nighttime temperatures to be higher than what they would naturally be in areas with vegetation and natural land cover (Figure 13.2). Adding to the UHI is reduced evapotranspiration and shade due to the lack of vegetation, impermeable concrete areas reducing surface moisture, dark surfaces of car parks, roads and rooftops retaining further heat and a concentration of heat-producing

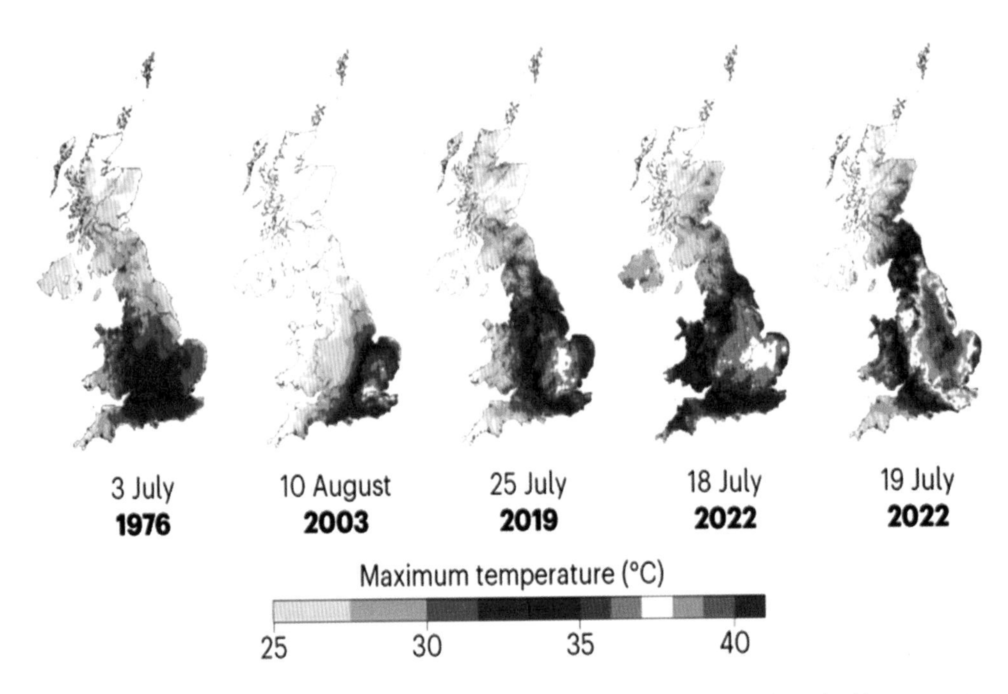

Figure 13.2 The July 2022 heatwave that impacted the United Kingdom and resulted in temperatures reaching more than 40°C for the first time in recorded history in the country.

Source: www.nature.com/articles/d41586-022-02114-y. Accessed: 26 September 2022.

air-conditioning units, buildings, factories and vehicles. The UHI temperatures therefore vary spatially and temporally across the urban area in relation to local microclimates, vegetation cover and urban infrastructure (Figure 13.3).

The UHI is typically measured by the difference in near-surface air temperature between highly urbanised areas and surrounding rural locations. The intensity of a UHI reaches its maximum at night when absorbed heat is released into the atmosphere, particularly when the weather is clear and still. In countries like Australia, Canada, China, Europe, India and the USA, where dense and often rapid urbanisation continues to occur, the impacts of UHI are increasingly known and studied. However, in places like Africa and in tropical mountain cities, such as Baguio City in the Philippines, research into UHI effects and impacts is lacking (Estoque & Murayama, 2017; Simwanda et al., 2019). UHI is becoming a growing concern, however, particularly in African cities where urbanisation is increasing at a very rapid rate, and urbanisation is often unplanned and unregulated and marked by overcrowded informal settlements.

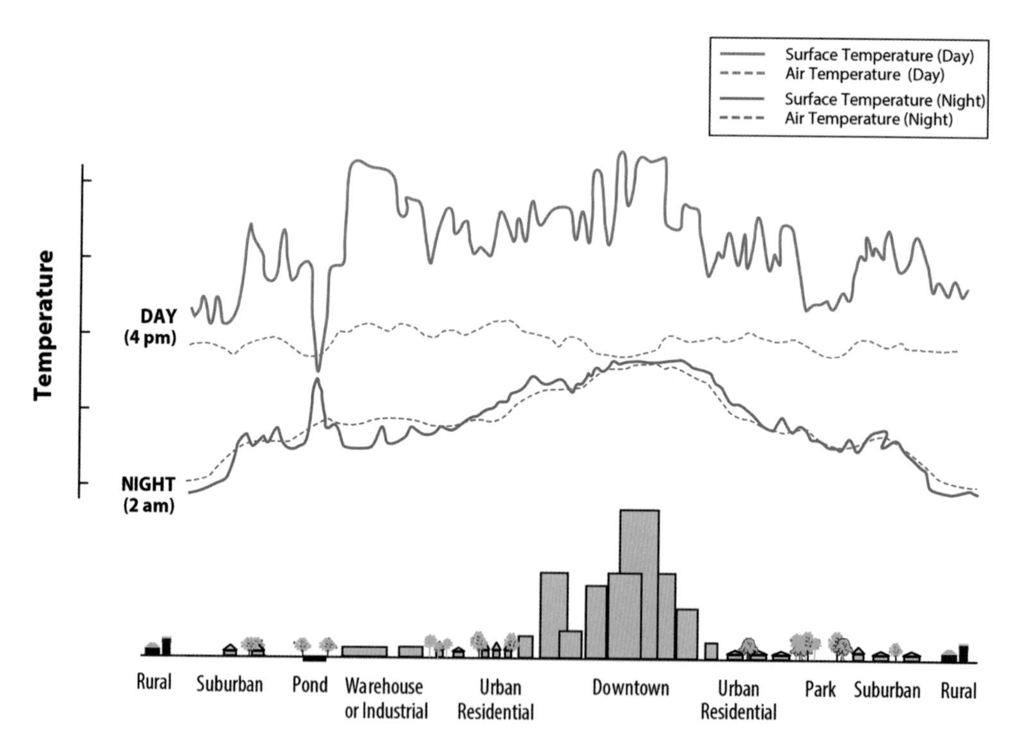

Figure 13.3 Urban heat island effect occurs where there is a concentration of urban infrastructure and a lack of vegetation cover. There are greater differences between the daytime surface and atmospheric air temperatures than the nighttime. The figure shows the near constant day and nighttime temperature of the pond water and cooler temperatures associated with parks and residential areas, as compared to the inner city area of downtown.

Source: www.epa.gov/heatislands/learn-about-heat-islands. Accessed: 21 August 2022.

B. IMPACTS AND CONSEQUENCES OF EXTREME TEMPERATURES

Extreme temperature events have significant impacts on human health, particularly among the most vulnerable such as the elderly, young children and those suffering from chronic disease. Extreme temperature events also have significant impacts on the agriculture, energy and transport sectors along with their associated infrastructure. Workplace productivity may decline, and health services may become over-stretched as more people succumb to the ongoing extreme temperature. Such events can also lead to or are associated with an increase in other hazards, such as wildfires, snowstorms and extreme winds.

The average human body is most efficient at a core temperature of 37°C. Compared with the natural variations of ambient air temperature, physiological comfort and safety can be maintained within only a relatively narrow thermal range. Irreversible deterioration

(a)

Figure 13.4 Thermal images taken by the City of Melbourne during a heatwave event in January 2017. (a) Heat camera image taken outside a market area showing reduced temperatures where trees are shading the road. (b) Heat camera image of a major inner city road where temperatures across large parts of the surface area reach up to 60°C.

Sources: City of Melbourne; www.theguardian.com/sustainable-business/2017/feb/21/urban-heat-islands-cooling-things-down-with-trees-green-roads-and-fewer-cars. Accessed: 20 August 2022.

(b)

Figure 13.4 (Continued)

and death frequently occur if the internal body temperature falls below 26°C or rises above 40°C. However, the epidemiology of temperature-related death is complex. In the US, for example, the National Climatic Data Centre's *Storm Data* showed that heat-related fatalities outweighed those associated with cold stress, whereas the US's National Centre for Health Statistics Compressed Mortality Database showed that excessive cold was responsible for four times more deaths than excessive heat in any given year (Dixon et al., 2005). The authors related this discrepancy to a number of factors, including the incompleteness of source information, limited quality control and subjectiveness of the medical examiner or coroner when determining whether extreme temperatures were the cause of death as a primary, secondary or major contributing factor. Authorities therefore often rely on daily excess death rates or 'premature deaths', determined by recorded daily deaths against the typical rates for that time period in that location.

Barnett et al. (2010) concluded that whilst relatively hot and cold temperatures do increase mortality around the world, it was difficult to identify the best temperature measure, due to wide variations in test results between age groups, seasons and cities. In Montréal,

Canada, Goldberg et al. (2011) examined whether higher or colder than optimal temperatures increased risk of dying. The authors found a clear link between summer mortality and daily maximum temperatures above 27°C, although no equivalent relationship existed for colder temperatures in winter.

In comparison, Gasparrini et al. (2015) examined temperature-related fatalities across 384 mostly urban locations in Australia, Brazil, Canada, China, Italy, Japan, South Korea, Spain, Sweden, Taiwan, Thailand, UK and USA. They found that most fatalities were attributable to cold stress rather than heat. Expanding on this work, Zhao et al. (2021) collected data across 750 locations and 43 countries and found over 5 million non-optimal temperature-related fatalities from 2000 to 2019 (Zhao et al., 2021). The greatest number occurred in Asia, but the largest number of heat-related deaths occurred in Eastern Europe while the largest number of cold-related fatalities occurred in sub-Saharan Africa. Examination of data from 2000–2003 against 2016–2019 showed a global reduction in cold-related fatalities but an increase in heat-related fatalities, albeit it slightly less resulting in a net reduction overall (Zhao et al., 2021).

Gasparrini et al. (2015) also showed that extreme temperatures associated with heatwaves and cold waves caused substantially fewer fatalities than that attributable to milder non-optimum weather. In many cases, thermal effects are obscured by other factors such as air pollution, timelags in response and socio-economic conditions such as housing conditions, ill health in the population and other demographic indices. However, when heatwaves and cold waves do occur, they cause significant spikes in fatality rates which cause significant issues to health and emergency response services.

1. Cold Stress

Cold stress can cause physiological damage in the form of hypothermia or frostbite. The effects of low temperature alone are compounded by wind and moisture, so that *windchill*, for example, is caused by the combination of low temperature and high wind speed. Cold temperature-related mortality is often associated with outbursts of cold, Arctic air into the mid-latitudes during spells of severe winter weather.

Generally, the highest cold-related death rates occur during the most severe weather, which is often associated with winter anticyclones. These events also favour the trapping of high concentrations of air pollution in the cold air of the lower atmosphere, a contributory factor to excess deaths resulting from hypothermia and cold-aggravated illnesses. Older people (over 65 years) living in poorly heated houses with low energy efficiency are at greatest risk. This has been attributed to the adverse effect of low temperatures on existing underlying illnesses, such as influenza, coronary thrombosis and respiratory disease, but it has proved difficult to establish strong links between winter mortality and deprivation.

During such weather, critical infrastructure can be impacted. Water pipes can freeze over, preventing people from accessing drinking water. Widespread power outages further exacerbate the disaster as people are left in their homes without adequate heating. For example, a winter storm that impacted Texas, USA, in February 2021 caused 246 fatalities among people ranging in age from 102 years to less than one year old (DSHS, 2021). On average, people were without power for almost two days, bringing their indoor temperature down to below freezing (Watson et al., 2021), but some remained without power for up to five days.

The main cause of death was due to extreme cold exposure-related injuries (65.5%) resulting in hypothermia (64.2%) and frostbite (1.2%). The risk factors associated with hypothermia included undertaking repair work outside during the storm, inadequate clothing, the consumption of alcohol and/or illicit drugs and homelessness. Other fatalities occurred where people were unable to receive treatment for pre-existing health conditions. In some cases, people's medical equipment or medication had frozen, or they were reliant on electricity-dependent equipment for life support, or they were unable to reach dialysis or oxygen treatment facilities due to the extreme weather. The roads claimed 22 lives, as driving conditions were exceptionally hazardous. A further 19 people died of carbon monoxide poisoning when attempting to heat their homes during power outages with items unsuitable for indoor use such as generators, grills and vehicles running in enclosed spaces (DSHS, 2021).

Indirect winter deaths also occur due to heart attacks from snow shovelling and house fires due to the use of emergency heaters. Once again, the elderly and the poor, including the homeless, suffer the most. Children are also highly vulnerable. Liu et al. (2021) found a prevalence of asthma among school-aged children exposed to cold spells in China with effects increasing as the cold spell duration increases.

Within Europe, the highest excess winter death rates are in Portugal, Spain and Ireland (rather than Scandinavia) and have been linked to poverty and poor housing conditions (Healy, 2003). The UK also fares badly in comparison to countries with colder climates. Wilkinson et al. (2001) found that deaths in England from heart attacks and strokes were 23% higher during December and March than in other months. Mortality rose by 2% with every 1°C fall in outdoor temperature below 19°C. For England and Wales, total excess winter mortality has been estimated at 30,000 per year. In Scotland, Gemmell et al. (2000) showed a difference of 30% between the summer trough of weekly death rates and the winter peak. In Spain, Montero et al. (2010) found that mortality was linked to the duration of cold spells and high relative humidity. Such periods towards the end of winter created the greatest mortality.

2. Heat Stress

Heat stress is greatest when both atmospheric temperature and humidity are high and physical discomfort leads to mortality. The amount by which the temperature exceeds the local seasonal mean is more important than the absolute value of temperature. The threat is also high in the first heatwave of the season, before acclimatisation occurs. After several days of excessive heat, the typical mortality rate may rise to two or three times the normal seasonal rate. For example, during one of Australia's worst heatwaves that impacted southern Australia in late January and early February 2009, Victoria recorded 374 excess deaths over the period from 26 January to 1 February (Figure 13.5). This equated to a 62% rise in total all-cause mortality with 980 deaths recorded compared to a mean of 606 for the same period from 2004–2008 (DHHS, 2009).

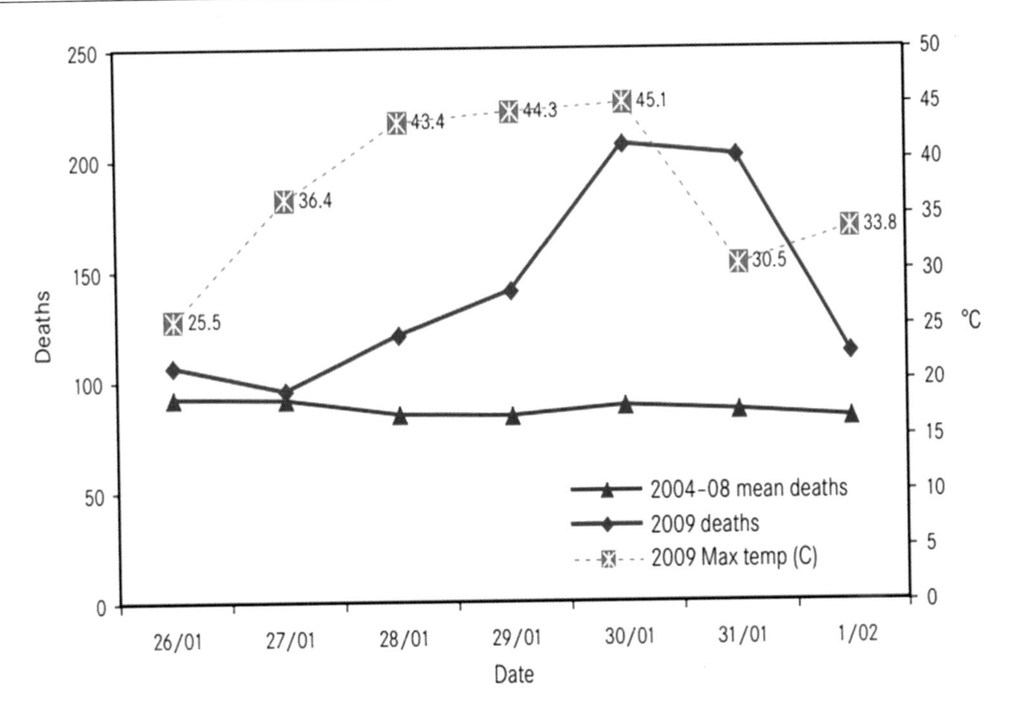

Figure 13.5 The 2009 heatwave in Victoria, Australia, from 26 January to 1 February, shown by the number of deaths alongside maximum recorded temperatures and in comparison with the mean death rate between 2004 and 2008 over that same time period. Mortality rates were taken from the Victorian Registry of Births, Deaths and Marriages and the State Coroner's Office.

Source: DHHS (2009) January 2009 Heatwave in Victoria: An Assessment of Health Impacts. Published by the Victorian Government Department of Human Services Melbourne, Victoria. jan-2009-victoria-heatwave-assessment-of-health-impacts.pdf (aidr.org.au). Accessed: 31 August 2022.

Many parts of southern Australia recorded temperatures 12 to 15 degrees above normal, which was exacerbated by high nighttime temperatures. The maximum temperature in Victoria was 45.1°C on 30 January 2009, which immediately followed two consecutive days of temperatures over 43°C, including overnight highs giving people no respite. In Melbourne, ambulance call outs were 46% higher than that recorded in the same three-day period 27–30 January the previous year. In particular, heat-related illnesses, including heat stress, heat stroke and dehydration recorded a 34-fold increase from the previous year, with a total of 514 cases (see Figure 13.6) compared to 499 in 2008. The majority of these were among people aged 75 years and older. Another significant increase occurred with the number of people recorded as deceased on arrival (DOA) at hospital emergency departments. The week of 26 January to 1 February recorded 126 DOAs compared to an expected number of 44 (DHHS, 2009).

High temperatures remained after the three-day January peak period of over 43°C and then peaked again over the three-day period of 6–8 February, culminating in the highest ever temperature, of

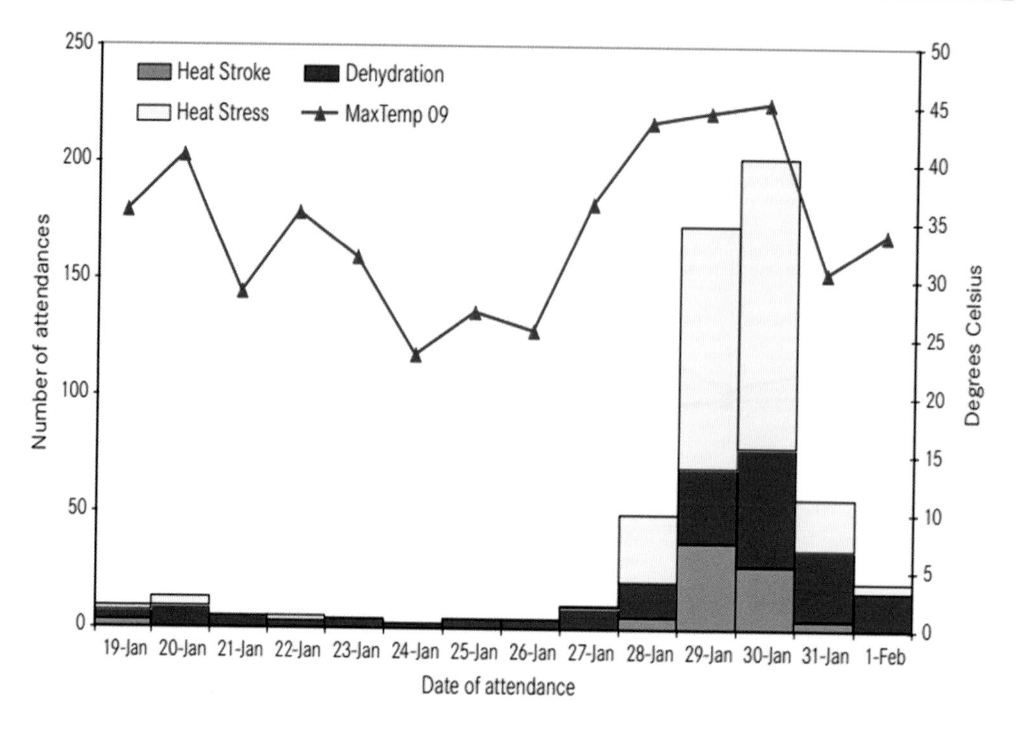

Figure 13.6 Ambulance call outs for heat-related illnesses in Melbourne from 19 January to 1 February 2009 alongside maximum recorded temperatures.

Source: DHHS (2009) January 2009 Heatwave in Victoria: An Assessment of Health Impacts. Published by the Victorian Government Department of Human Services Melbourne, Victoria. jan-2009-victoria-heatwave-assessment-of-health-impacts.pdf (aidr.org.au) Accessed: 31 August 2022.

46.4°C, recorded in Victoria on 7 February, the day of the Black Saturday wildfires (see Chapter 14). However, the heat was not confined to just Victoria; many parts of South Australia, including Adelaide, also recorded extreme temperatures with a maximum of 45.7°C, overnight temperatures around 30°C and an estimated excess mortality rate of 50–150 people. It also recorded eight consecutive days where maximum temperatures were above 40°C. Across both cities, there was widespread failure of critical infrastructure and disruption to services, including power outages and train and tram cancellations, with estimated financial losses at $AU800 million (Reeves et al., 2010).

Although large urban areas pose the greatest risks, Loughnan et al. (2010) found evidence of increased heat-stress mortality in persons aged 65 and older living in small towns in rural parts of Victoria, Australia. Metropolitan centres are especially problematic, due to intensification of heat stress by the UHI effect and socio-economic disadvantage. There is growing awareness of the need to reduce the heat-related health inequalities in inner cities (Harlan et al., 2006). This problem was highlighted for a heatwave in the New York–New Jersey metropolitan area in 1966, when up to 200 extra deaths occurred in the city core, over the number expected in the suburbs (Buechley et al., 1972).

Based on intensity, the Australian Bureau of Meteorology (BOM, 2022) recognises three types of heatwaves:

- *Low-intensity heatwaves* – are more common but most people can cope with the conditions.
- *Severe heatwaves* – are less common but cause greater impacts to vulnerable people such as the elderly and those with underlying health issues.
- *Extreme heatwaves* – are rare but they can impact anyone who does not take heed and adapt their behaviour accordingly. People who undertake outdoor activities, such as for work or exercise, are at greater risk of being affected.

Southern Australia's 2009 heatwave was in the extreme category, and it is clear that due to anthropogenically induced climate change, heatwave intensity, frequency and duration is increasing in Australia (Perkins-Kirkpatrick et al., 2016). When combined with ageing populations and continued urbanisation exacerbating the UHI effect, Australia can expect increasing rates of morbidity and mortality resulting from heat-related illnesses. This has already been witnessed by the spikes in fatality rates during the 2009 southern Australia heatwave and again during a 2014 event (Coates et al., 2022). These researchers predominantly found fatalities were highest among men with risk increasing with age, socio-economic disadvantage, social isolation and geographical remoteness, among people with a mental or physical disability and a lack of air-conditioning, where with some cases it was present but not used.

In China, heatwave-related mortality is also rising, with an estimated increase of an additional 1,000 deaths every 1.2 years over the past decade (Cai et al., 2021). Yang et al. (2019) showed that women, the elderly, people with lower educational attainment, along with those who live in cities of high latitudes, higher levels of air pollution and fewer hospital beds per capita were at greatest risk. Yang et al. surmised that people with lower educational attainment were more likely to be vulnerable to heatwaves due to poorer living and working conditions and more limited access to health care.

In 2019, heatwaves were responsible for approximately 26,800 deaths in China and in that same year, people aged over 65 years experienced 13 more heatwave days compared to baseline records between 1986 and 2005. Cai et al. (2021) also showed that heat-related productivity losses among outdoor workers was 0.5% of total national work hours and cost China 1% of its gross domestic product. While many studies have focused on the impact of heatwaves in cities across China, heatwave-related mortality extends beyond the city limits, with high-risk areas found in rural settings where the population is ageing (He et al., 2019) and vulnerability exists based on the aforementioned factors.

Heatwaves are also experienced frequently in the United States, where they are the number one weather-related cause of death. For example, a 36% increase in deaths was recorded over a five-day period in 1966 (Bridger & Helfand, 1968); in 1955, 946 excess deaths resulted in Los Angeles (more than twice the mortality recorded in the 1906 San Francisco earthquake and fire) (Oechsli & Buechly, 1970); and over 700 people died in Chicago in 1995 (Klinenberg, 2002). Even so, Ostro et al. (2009) claimed that existing statistics on heatwave-related deaths, as reported in California, probably under-represent the real extent of the hazard, due to a lack of clear case definition and the multifactorial nature of such deaths. The elderly and those with existing heart disease are at greatest risk. During the month of July 1993, 213 heat-related deaths were recorded in Philadelphia, mostly at the place of residence (Johnson & Wilson, 2009). Figure 13.7 illustrates the frequency of heat-related death by age group. The mean age of the deceased was 65.7 years, with maximum values in the 60–90 years band. Other high-risk groups are the urban poor, especially those who lack domestic air conditioning or are dependent on alcohol or drugs.

Records continue to break. The Canadian town of Lytton broke the country's highest recorded temperature three times in three consecutive days, with temperatures reaching 49.6°C on the third day (WMO, 2021). Stott et al. (2004) claimed that anthropogenic climate change has already more than doubled the risk of European summers as hot as that of 2003. The 2003 summer in Europe was the warmest since 1500, with temperatures around 40°C and with an initial estimate of over 30,000 excess heat-related deaths reported (Haines et al., 2006). Robine et al.

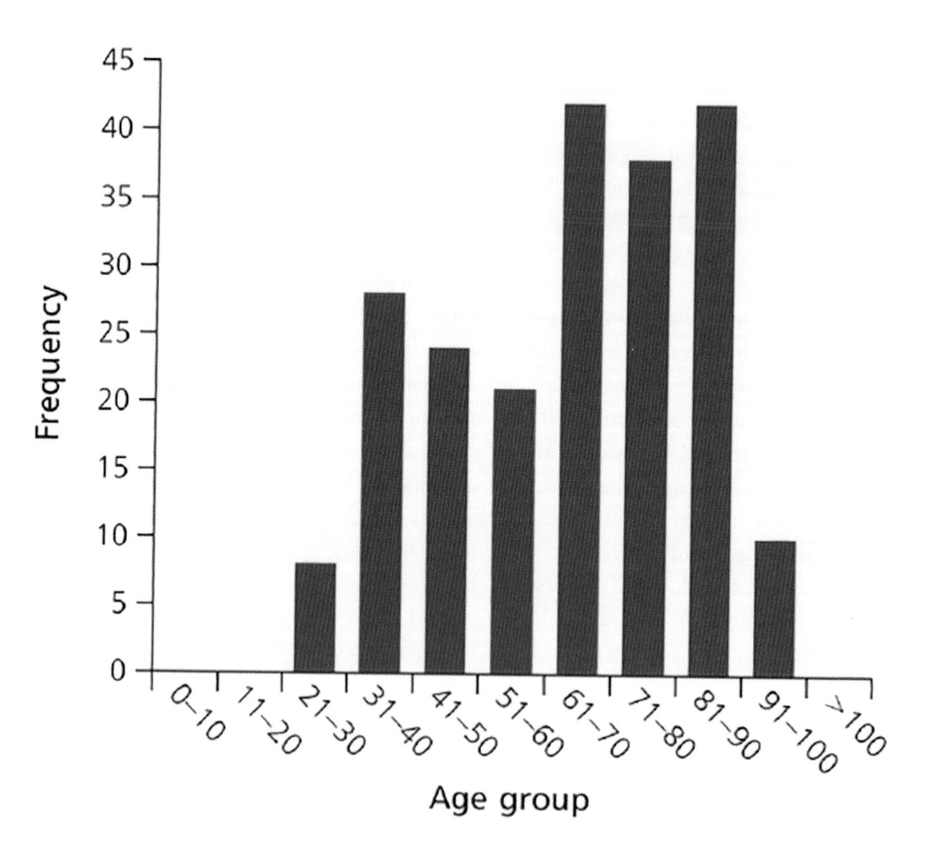

Figure 13.7 Frequency of heat-related deaths by age group in Philadelphia, USA, during July 1993. Mortality was concentrated in the 60–90 age band.

Source: After Johnson and Wilson (2009). Reproduced from *Applied Geography 29*, D.P. Johnson and J.S. Wilson, The socio-spatial dynamics of extreme urban heat events: the case of heat-related deaths in Philadelphia, 419–434. Copyright (2009), with permission from Elsevier.

(2008) revised this figure to an estimated 70,000 excess deaths during the 2003 summer. Half of these were in France during August (see Figure 13.8), when overall excess mortality averaged 60%; parts of Paris experienced rates over 150% (Poumadère et al., 2005). Apart from temperature, many socio-economic factors were implicated, including age, gender, dehydration, medication, urban residence, poverty and social isolation. Most at risk were elderly people living alone in cities when many medical staff were absent for the traditional summer holiday. According to Lagadec (2004), this disaster exposed a lack of preparedness, and France has subsequently developed a national heatwave plan, which helped to reduce mortality when a larger-scale, but less intense, event took place in 2006 (Fouillet et al., 2008).

C. RISK ASSESSMENT AND MANAGEMENT

Extreme temperature hazards have attracted much less attention when compared to other environmental hazards, mostly due to the difficulties in determining direct impacts in terms of morbidity and mortality. Between the two, heatwaves have received far greater focus although strategies devised to reduce the risk are still lacking when compared to other hazards. In previous times, there has been a perception that it is just cold or just hot and society must deal with it. However, with increasing trends in the intensity, frequency and duration of heatwaves due to anthropogenically induced climate change and greater international reach of media drawing greater attention to the impacts of

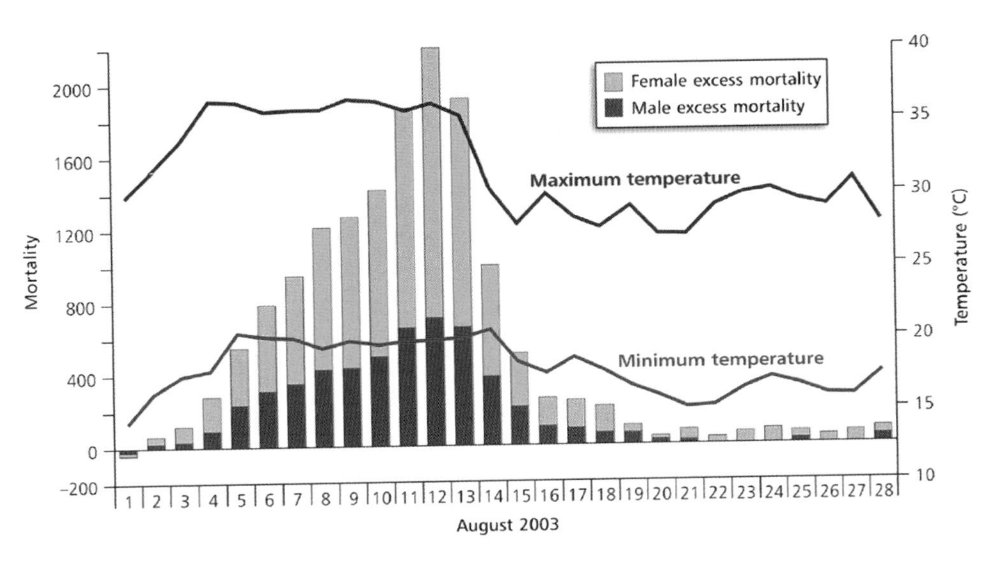

Figure 13.8 Number of excess deaths recorded in France each day during the 2003 heatwave in relation to maximum and minimum temperatures. Between 4 and 15 August, almost 15,000 deaths were directly attributed to heat stress caused by dehydration, hyperthermia and heat stroke.

Source: After Poumadère et al. (2005). Reproduced from M. Poumadère et al., The 2003 heat wave in France: dangerous climate change here and now. *Risk Analysis* 25 (2005): 1483–1494. Copyright John Wiley and Sons 2005. Reproduced with permission.

cold waves and heatwaves, more is being done to assess and manage the risk.

It must be acknowledged, however, that in some regions where extreme temperatures are the norm, people do successfully live and work. In the Arctic, for example, where the monthly average temperature is below 10°C, more than 4 million people were recorded as living there in 2013 (Larsen & Fondahl, 2014). This also applies to regions of extreme heat. In Coober Pedy in outback Australia, for example, where average temperatures are well above 30°C for five months of the year and extreme days reach almost 48°C, people also successfully live and work. In both these environments, people have learned to adapt their behaviours and homes to suit the environmental conditions in which they live.

1. Cold Waves

As a weather phenomenon, national, regional and local weather services provide forecasts for cold waves and associated extreme cold events. In India, for example, the Cold Wave Warning Services, operated by the India Meteorological Department (IMD), generates and disseminates impact-based forecasts and risk-based warnings for cold wave events (IMD, n.d.). These warnings are disseminated to various government ministries, national, state and district disaster management authorities and critical transport networks. Public dissemination occurs via social media (e.g. Facebook and Twitter), the national IMD website and electronic and print media services. IMD also disseminates warnings via specifically designed mobile applications such as Meghdoot, which is a joint initiative between IMD, Indian Institute of Tropical Meteorology and Indian Council of Agricultural Research developed to provide farmers with critical meteorological information in an easy-to-use mobile phone application.

For a more broadscale, global perspective, ESRI developed a global cold wave model which forecasts the potential for cold waves to occur four days from the current day (ESRI, 2022). These forecasts can be used by government officials, including public safety and natural resource managers, to assess potential risks of a forecasted cold wave. The global cold wave model classifies extreme cold events as:

- Cold wave – low intensity cold wave, frequent during winter months, most people can cope with this level.
- Significant cold wave – moderate intensity cold wave, less frequent during winter months, challenging for people to cope with this level.
- Extreme cold wave – high intensity cold wave, rare during winter months, more challenging for people to cope with this level.
- Extreme cold wave – very high intensity cold wave, exceptionally rare, high impact to daily life and supply chain.

In the United States, cold waves are defined by National Weather Services' local weather forecast office and forecasts are disseminated accordingly. This information feeds into the Cold Wave Risk Index, which is part of the National Risk Index, where the score and rating represent a community's relative risk for cold waves compared to all other areas across the US (FEMA, 2021).

Communities at risk of cold waves can assess their own risk by informing themselves of their local weather patterns and keeping tuned into communication channels for early warnings. This is particularly pertinent for farmers as livestock and crops can die in freezing weather conditions, but precautions can be taken to ensure survival, such as providing adequate shelter for livestock and maintaining drinkable freshwater and food supply.

Individually, households can prepare themselves for extreme cold and cold waves. Regularly checking and knowing what to expect in terms of weather enables people to dress accordingly and adjust their schedules if needed to avoid being out during the worst of the weather. The National Weather Service in the US advises people to ensure they have a full tank of fuel in case they are stranded while out driving and need to keep the car running for warmth. They also advise keeping a winter car survival kit that includes jumper cables, flashlights, a first aid kit, food, water, a basic tool kit, a battery or hand-cranked radio, cat litter or sand for tyre traction, shovel, ice-scraper, warm clothes and blankets and a charged mobile phone.

To avoid hypothermia during cold waves, people should dress in adequate clothing and layers, particularly if they need to venture outside, avoid breezes and drafts indoors, eat nutritious, hot and healthy food and consume warm drinks regularly throughout the day. If people are taking any form of medication, they should ask their doctor whether it increases their risk of hypothermia. After the cold wave passes, the National Weather Service in the US advises residents to check for any frozen or burst water pipes, remove any built-up snow from footpaths and driveways by shovelling or salting the area and look out for stairs covered in ice and salt if needed to avoid the risk of slipping. Residents should also check on neighbours, friends and family in case anyone is in need of assistance. Social outreach programs are critical for people who do not have others that can check on them, along with reaching people that are suffering from homelessness.

2. Heatwaves

In Australia, the Bureau of Meteorology runs a heatwave forecast service (www.bom.gov.au/australia/heatwave/) from October to March each year, providing two different products:

- Heatwave *assessments* identify heatwaves and their intensity (described as severity) in the preceding three-day periods (Figure 13.8).
- Heatwave *forecasts* predict heatwaves and their intensity (described as severity) in upcoming three-day periods (Figure 13.9).

To assess potential risks, the BOM examines the forecast maximum and minimum temperatures across upcoming three-day periods against average temperatures and temperatures over the last 30 days at that location. This means that if temperatures have been lower than average for the previous 30-day period, or part thereof, and there is a predicted sustained spike in temperature across a three-day period for that location, a heatwave will be forecast. Based on this, conditions signifying a heatwave in a specific location will be different at the start of summer compared to the end of summer. It also recognises that extreme heat experienced in one location is not necessarily extreme if it were to occur in another location.

As with cold waves, local and state governments use the information produced by weather services to trigger heat health warnings. In Victoria, Australia, the Department of Health monitors the BOM forecasts for heatwaves and issues heat health alerts to Victorian communities in

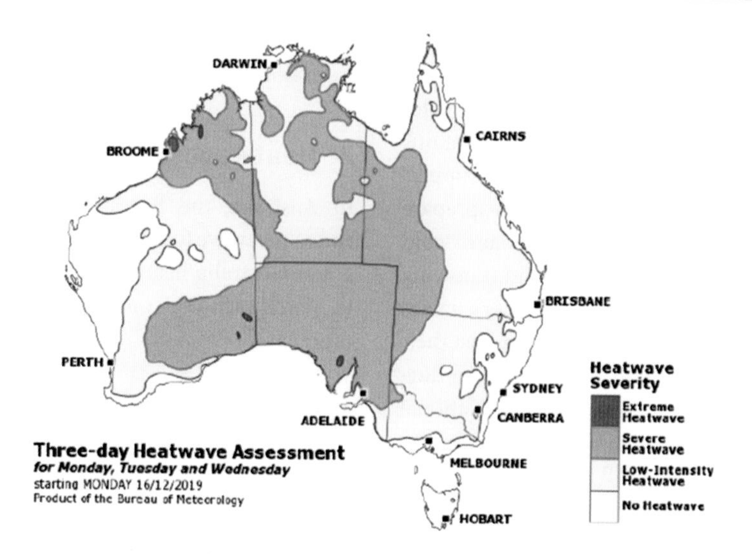

Figure 13.9 An example of a heatwave assessment map, as produced by Australia's Bureau of Meteorology.

Source: Bureau of Meteorology, Australia, www.bom.gov.au/australia/heatwave/knowledge-centre/heatwave-service.shtml. Accessed: 6 August 2022.

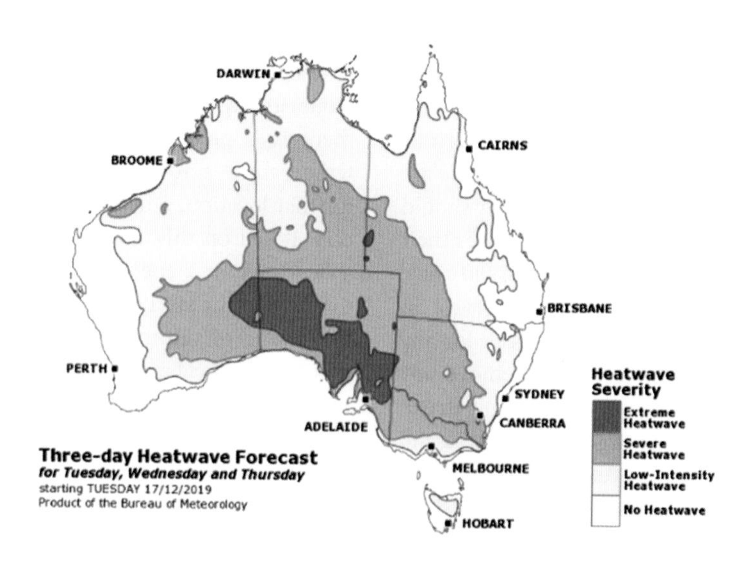

Figure 13.10 An example of a heatwave forecast map, as produced by Australia's Bureau of Meteorology.

Source: Bureau of Meteorology, Australia, www.bom.gov.au/australia/heatwave/knowledge-centre/heatwave-service.shtml. Accessed: 6 August 2022.

a specific weather forecast district when thresholds are met.

In the United States, NOAA recently released (July 2022) an interagency National Integrated Heat Health Information System – www.heat.gov/. This resource provides information on current conditions and future outlooks alongside heat forecast maps, a heat and health tracker that details historic and recent local heat and health information to help inform community preparedness and response, and other planning, education and action guides. One such education tool is a community science mapping campaign where more than 40 communities across the US have mapped their urban heat island (www.heat.gov/pages/mapping-campaigns). This campaign aims to foster local partnerships and engage local residents to build understanding of how heat is spatially distributed across their urban environment with the information used to develop sustainability plans and improve public health practices and urban forestry.

Microclimates in urban areas can be modified through the shading of houses by trees against direct solar radiation and by the use of high albedo materials to reflect more of the incident radiation on individual properties (Solecki et al., 2005). A study in Lisbon, Portugal, found that a small green space, 0.24 ha in extent, lowered the temperature in the surrounding densely urbanised area significantly, an effect that was most apparent on the hottest days (Oliveira et al., 2011). The planting of trees can help to reduce urban air pollution, often a contributory factor in heat stress, although inner cities rarely have space for large tracts of vegetation. More domestic air conditioning seems a likely response, but not everyone can afford it and it uses a lot of energy.

In addition to planting more trees and other vegetation, UHI can be reduced by installing green roofs that consist of vegetative layers such as grasses or shrubs that cover the rooftop or by growing a rooftop garden that consists of grasses, shrubs and trees but is also accessible to people. Cooler roofing material and cooler footpath surfaces can also be used to reduce the absorption of solar radiation.

To reduce the risk of suffering health impacts during a heatwave, the Heat Health initiative (http://heathealth.cvga.org.au/heatproof-your-home/) funded by the Victorian government, Australia, recommends households implement the following measures:

- Seal off living areas where people spend the most time so that only one room needs to be cooled. It is also wise to seal external doors and windows with a rolled-up towel where needed to prevent draughts of hot air penetrating the home.
- Ensure all windows and doors are kept closed. However, as the evening air cools, it is wise to open these again to reduce the internal temperature of the house.
- Avoid using ovens and other cooking appliances that produce heat.
- Use a fan if no air conditioning unit is available. The air movement cools the body, rather than the room temperature, but can be just as effective.
- Protect external windows from direct sunlight where possible with shades or awnings. Internally, use curtains, sheets or blankets to minimise heat transference from warm window glass to internal rooms.

As with cold waves, it is pertinent to regularly check and know what to expect in terms of weather and to adjust schedules accordingly to avoid being out during the hottest parts of the heatwave. During the hottest periods, people are advised to stay indoors where it is cool and avoid physical activity. If they are unable to cool their home, respite from the heat might be available in a local shopping centre, library or other government building. However, it is advisable to plan ahead and retreat to such locations early to avoid being out during the hottest part of the day. Lightweight clothing is recommended along with drinking plenty of cool water throughout the day and reducing consumption of caffeinated drinks and alcohol. Similarly with cold waves, certain medications can increase a person's risk of heat stress and these should be checked with medical professionals. Social outreach is also critical as neighbours, friends and family living alone may need assistance along with those suffering from homelessness. Special care is also needed with pets and livestock. To ensure their safety during extreme heat and heatwaves, they must have access to a plentiful supply of cool, clean water and shade with adequate ventilation.

KEY TAKE AWAY POINTS

- Extreme temperature events are characterised by consecutive days of extreme hot or cold temperatures, outside the normal temperature range for that region and at that time of year.
- Extreme temperature events that occur early in the season are more dangerous to human health than are those that occur later in the season, as people have not yet had a chance to acclimatise to the cold winter or hot summer weather.
- Heatwaves impact almost all regions across the globe, and this trend is increasing due to anthropogenic climate change.
- On the other hand, cold waves are expected to decrease in intensity and frequency as the climate warms due to anthropogenic climate change.
- Extreme heat and heatwaves are the leading cause of natural-hazard related fatalities in many countries.
- The increase in heatwave intensity around urbanised areas is known as the urban heat island (UHI) effect. The UHI is caused by urban infrastructure (asphalt/bitumen roads, concrete footpaths, brick buildings, etc.) absorbing and retaining the heat of the day causing day and nighttime temperatures to be higher than what they would naturally be in areas with vegetation and natural land cover.
- Extreme temperature hazards have attracted much less attention when compared to other environmental hazards, mostly due to the difficulties in determining direct impacts in terms of morbidity and mortality.
- Individually, people should monitor weather warnings for extreme temperature events. When such an event is forecasted, people should dress accordingly and avoid being out during the worst of the weather, if possible.

BIBLIOGRAPHY

Barnett, A., Hajat, S., Gasparrini, A., & Rocklov, J. (2012). Cold and heat waves in the United States. *Environmental Research, 112*, 218–224.

Barnett, A. G., Tong, S., & Clements, A. C. A. (2010). What measure of temperature is the best predictor of mortality? *Environmental Research, 110*, 604–611.

BOM. (2022). *Understanding heatwaves.* www.bom.gov.au/australia/heatwave/knowledge-centre/understanding.shtml. Accessed 6 August 2022.

Bridger, C. A., & Helfand, L. A. (1968). Mortality from heat during July 1966 in Illinois. *International Journal of Biometeorology, 12*, 51–70.

Buechley, R. W., van Bruggen, J., & Truppi, L. E. (1972). Heat island = death island? *Environmental Research, 5*, 85–92.

Cai, W., Zhang, C., Suen, H. P., Ai, S., Bai, Y., Bao, J., Chen, B., Cheng, L., Cui, X., Dai, H., Di, Q., Dong, W., Dou, D., Fan, W., Fan, X., Gao, T., Geng, Y., Guan, D., Guo, Y., Hu, Y., Hua, J., Huang, C., Huang, H., Huang, J., Jiang, T., Jiao, K., Kiesewetter, G., Klimont, Z., Lampard, P., Li, C., Li, Q., Li, R., Li, T., Lin, B., Lin, H., Liu, H., Liu, Q., Liu, X., Liu, Y., Liu, Z., Liu, Z., Liu, Z., Lou, S., Lu, C., Luo, Y., Ma, W., McGushin, A., Niu, Y., Ren, C., Ren, Z., Ruan, Z., Schöpp, W., Su, J., Tu, Y., Wang, J., Wang, Q., Wang, Y., Wang, Y., Watts, N., Xiao, C., Xie, Y., Xiong, H., Xu, M., Xu, B., Xu, L., Yang, J., Yang, L., Yu, L., Yue, Y., Zhang, S., Zhang, Z., Zhao, J., Zhao, L., Zhao, M., Zhao, Z., Zhou, J., & Gong, P. (2021). The 2020 China report of the Lancet Countdown on health and climate change. *The Lancet Public Health, 6*(1), e64–e81.

Coates, L., van Leeuwen, J., Browning, S., Gissing, A., Bratchell, J., & Avci, A. (2022). Heatwave fatalities in Australia, 2001–2018: An analysis of coronial records. *International Journal of Disaster Risk Reduction, 67*, 102671.

DHHS. (2009). *January 2009 heatwave in Victoria: An assessment of health impacts.* Published by the Victorian Government Department of Human Services Melbourne, Victoria. jan-2009-victoria-heat-wave-assessment-of-health-impacts.pdf (aidr.org.au). Accessed 31 August 2022.

Dixon, P. G., Brommer, D. M., Hedquist, B. C., Kalkstein, A. J., Goodrich, G. B., Walter, J. C., Dickerson, IV, C., Penny, S. J., & Cerveny, R. (2005). Heat mortality versus cold mortality: A study of conflicting databases in the United States. *Bulletin of the American Meteorological Society, 86*(7), 937–943. https://doi.org/10.1175/BAMS-86-7-937

DSHS. (2021). *February 2021 winter storm-related deaths – Texas. Dec. 31, 2021.* Texas Department of State Health Service. www.dshs.texas.gov/news/updates/SMOC_ebWinterStorm_MortalitySurvReport_12-30-21.pdf. Accessed 30 August 2022.

ESRI. (2022). *Overview: Global cold wave model.* www.esri.com/en-us/arcgis-marketplace/listing/products/c438967f20504862a83cb99d6e0c0536. Accessed 11 September 2022.

Estoque, R. C., & Murayama, Y. (2017). Monitoring surface urban heat Island formation in a tropical mountain city using Landsat data (1987–2015). *ISPRS Journal of Photogrammetry and Remote Sensing, 133*, 18–29. https://doi.org/10.1016/j.isprsjprs.2017.09.008

FEMA. (2021). *National risk index, technical documentation, November 2021.* www.fema.gov/sites/default/files/documents/fema_national-risk-index_technical-documentation.pdf. Accessed 20 September 2022.

Fouillet, A. R., et al. (2008). Has the impact of heat waves on mortality changed in France since the European heat wave of summer 2003? A study of the 2006 heat wave. *International Journal of Epidemiology, 37*, 309–317.

Gasparrini, A., Guo, Y., Hashizume, M., Lavigne, E., Zanobetti, A., Schwartz, J., Tobias, A., Tong, S., Rocklöv, J., Forsberg, B., Leone, M., De Sario, M., Bell, M. L., Guo, Y.-L. L., Wu, C.-f., Kan, H., Yi, S.-M., de Sousa Zanotti Stagliorio Coelho, M., Saldiva, P. H. N., Honda, Y., Kim, H., & Armstrong, B. (2015). Mortality risk attributable to high and low ambient temperature: A multicountry observational study. *The Lancet, 386*(9991), 369–375.

Gemmell, I., McLoone, P., Boddy, F. A., Dickinson, G. J., & Watt, G. C. M. (2000). Seasonal variation in mortality in Scotland. *International Journal of Epidemiology, 29*, 274–279.

Goldberg, M. S., Gasparrini, A., Armstrong, B., & Valois, M.-F. (2011). The short-term influence of temperature on daily mortality in the temperate climate of Montreal, Canada. *Environmental Research, 111*, 853–860.

Haines, A., Kovacs, R. S., Cambell-Lendrum, D., & Corvalan, C. (2006). Climate change and human health: Impacts, vulnerability and mitigation. *Lancet, 367*, 2101–2109.

Harlan, S. L., Brazel, A. J., Prashad, L., Stefanov, W. L., & Larsen, L. (2006). Neighbourhood microclimates and vulnerability to heat stress. *Social Science and Medicine, 63*, 2847–2863.

He, C., Ma, L., Zhou, L., Kan, H., Zhang, Y., Ma, W., & Chen, B. (2019). Exploring the mechanisms of heat wave vulnerability at the urban scale based on the application of big data and artificial societies. *Environment International, 127*, 573–583.

Healy, J. D. (2003). Excess winter mortality in Europe: A cross-country analysis identifying key risk factors. *Journal of Epidemiology and Community Health, 57*, 784–789.

Hoegh-Guldberg, O., Jacob, D., Taylor, M., Bindi, M., Brown, S., Camilloni, I., Diedhiou, A., Djalante, R., Ebi, K. L., Engelbrecht, F., Guiot, J., Hijioka, Y., Mehrotra, S., Payne, A., Seneviratne, S. I., Thomas, A., Warren, R., & Zhou, G. (2018). Impacts of 1.5°C global warming on natural and human systems. In V. Masson-Delmotte, P. Zhai, H.-O. Pörtner, D. Roberts, J. Skea, P. R. Shukla, A. Pirani, W. Moufouma-Okia, C. Péan, R. Pidcock, S. Connors, J. B. R. Matthews, Y. Chen, X. Zhou, M. I. Gomis, E. Lonnoy, T. Maycock, M. Tignor, & T. Waterfield (Eds.), *Global warming of 1.5°C: An IPCC special report on the impacts of global warming of 1.5°C above pre-industrial levels and related global greenhouse gas emission pathways, in the context of strengthening the global response to the threat of climate change, sustainable development, and efforts to eradicate poverty* (pp. 175–312). Cambridge, UK and New York, NY, USA: Cambridge University Press. DOI: 10.1017/9781009157940.005

IMD. (n.d.). *Cold wave warning services*. India Meteorological Department. https://mausam.imd.gov.in/imd_latest/contents/pdf/pubbrochures/Cold%20Wave%20Warning%20Services.pdf. Accessed 20 September 2022.

Johnson, D. P., & Wilson, J. S. (2009). The socio-spatial dynamics of extreme urban heat events: The case of heat-related deaths in Philadelphia. *Applied Geography, 29*, 419–434.

Klinenberg, E. (2002). *Heat wave: A social autopsy of disaster in Chicago*. Chicago and London: University of Chicago Press.

Lagadec, P. (2004). Understanding the French 2003 heat wave experience: Beyond the heat, a multilayered challenge. *Journal of Contingencies and Crisis Management, 12*, 160–169.

Larsen, J. N., & Fondahl, G. (2014). Arctic human development report. Regional Processes and Global Linkages, Norden, TemaNord, 567, 504.

Liu, X., He, Y., Tang, C., Wei, Q., Xu, Z., Yi, W., Pan, R., Gao, J., Duan, J., & Su, H. (2021). Association between cold spells and childhood asthma in Hefei, an analysis based on different definitions and characteristics. *Environmental Research, 195*, 110738.

Loughnan, M., Nicholls, N., & Tapper, N. (2010). Mortality-temperature thresholds for ten major populations centres in rural Victoria, Australia. *Health and Place, 16*, 1287–1290.

Montero, J. C., Mirón, I. J., Criado-Álvarez, J. J., Linares, C., & Diaz, J. (2010). Mortality from cold waves in Castile-La Mancha, Spain. *Science of the Total Environment, 408*, 5768–5774.

Oechsli, F. W., & Buechly, R. W. (1970). Excess mortality associated with three Los Angeles September hot spells. *Environmental Research, 3*, 277–284.

Oliveira, S., Andrade, H., & Vaz, T. (2011). The cooling effect of green spaces as a contribution to the mitigation of urban heat: A case study in Lisbon. *Building and Environment, 46*(11), 2186–2194.

Ostro, B. D., Roth, L. A., Green, R. S., & Basu, R. (2009). Estimating the mortality effect of the July 2006 California heat wave. *Environmental Research, 109*, 614–619.

Perkins-Kirkpatrick, S. E., & Lewis, S. C. (2020). Increasing trends in regional heatwaves. *Nat Commun, 11*, 3357. https://doi.org/10.1038/s41467-020-16970-7

Perkins-Kirkpatrick, S. E., White, C. J., Alexander, L. V., Argüeso, D., Boschat, G., Cowan, T., Evans, J. P., Ekström, M., Oliver, E. C. J., Phatak, A., & Purich, A. (2016). Natural hazards in Australia: Heatwaves. *Climatic Change, 139*(1), 101–114.

Poumadère, M., Mays, C., Le Mer, S., & Blong, R. (2005). The 2003 heat wave in France: Dangerous climate change here and now. *Risk Analysis, 25*, 1483–1494.

Reeves, J., Foelz, C., Grace, P., Best, P., Marcussen, T., Mushtaq, S., Stone, R., Loughnan, M., McEvoy, D., & Ahmed, I. (2010). *Impacts and adaptation response of infrastructure and communities to heatwaves: The southern Australian experience of 2009*. National Climate Change Adaptation Research Facility.

Robine, J.-M., Cheung, S. L. K., Le Roy, S., Van Oyen, H., Griffiths, C., Michel, J.-P., & Herrmann, F. R. (2008). Death toll exceeded 70,000 in Europe during the summer of 2003. *Comptes Rendus Biologies, 331*(2), 171–178.

Seneviratne, S. I., Zhang, X., Adnan, M., Badi, W., Dereczynski, C., Di Luca, A., Ghosh, S., Iskandar, I., Kossin, J., Lewis, S., Otto, F., Pinto, I., Satoh, M., Vicente-Serrano, S. M., Wehner, M., & Zhou, B. (2021). Weather and climate extreme events in a changing climate. In V. Masson-Delmotte, P. Zhai, A. Pirani, S. L. Connors, C. Péan, S. Berger, N. Caud, Y. Chen, L. Goldfarb, M. I. Gomis, M. Huang, K. Leitzell, E. Lonnoy, J. B. R. Matthews, T. K. Maycock, T. Waterfield, O. Yelekçi, R. Yu, & B. Zhou (Eds.), *Climate change 2021: The physical science basis: Contribution of working group I to the sixth assessment report of the intergovernmental panel on climate change* (pp. 1513–1766). Cambridge, United Kingdom and New York, NY, USA: Cambridge University Press. DOI: 10.1017/9781009157896.013

Simwanda, M., Ranagalage, M., Estoque, R. C., & Murayama, Y. (2019). Spatial analysis of surface urban heat Islands in four rapidly growing African cities. *Remote Sens, 11*, 1645. https://doi.org/10.3390/rs11141645

Solecki, W. D., Rosenzweig, C., Parshall, L., Pope, G., Clark, M., Cox, J., & Wiencke, M. (2005). Mitigation of the heat island effect in urban New Jersey. *Global Environmental Change Part B: Environmental Hazards, 6*(1), 39–49.

Stott, P. A., Stone, D. A., & Allen, M. R. (2004). Human contribution to the European heatwave of 2003. *Nature, 432*(2 December), 610–614.

Wang, L., Liu, T., Hu, M., Zeng, W., Zhang, Y., Rutherford, S., Lin, H., Xiao, J., Yin, P., Liu, J., Chu, C., Tong, S., Ma, W., & Zhou, M. (2016). The impact of cold spells on mortality and effect modification by cold spell characteristics. *Scientific Reports, 6*, 38380. DOI: 10.1038/srep38380

Watson, K. P., Cross, R., & Jones, M. P. (2021). *The winter storm of 2021*. https://uh.edu/hobby/winter2021/storm.pdf. Accessed 30 August 2022.

Wilkinson, P., Armstrong, B., & Landon, M. (2001). *Cold comfort: The social and environmental determinants of excess winter deaths in England 1986–1996*. Bristol: The Policy Press.

Witze, A. (2022). Extreme heatwaves: Surprising lessons from the record warmth. *Nature, 608*(7923), 464–465. DOI: 10.1038/d41586-022-02114-y

WMO. (2021). *June ends with exceptional heat*. https://public.wmo.int/en/media/news/june-ends-exceptional-heat. Accessed 23 August 2022.

Yang, J., Yin, P., Sun, J., Wang, B., Zhou, M., Li, M., Tong, S., Meng, B., Guo, Y., & Liu, Q. (2019). Heatwave and mortality in 31 major Chinese cities: Definition, vulnerability and implications. *Science of the Total Environment, 649*, 695–702.

Zhao, Q., Guo, Y., Ye, T., Gasparrini, A., Tong, S., Overcenco, A., Urban, A., Schneider, A., Entezari, A., Vicedo-Cabrera, A. M., Zanobetti, A., Analitis, A., Zeka, A., Tobias, A., Nunes, B., Alahmad, B., Armstrong, B., Forsberg, B., Pan, S.-C., Íñiguez, C., Ameling, C., De la Cruz Valencia, C., Åström, C., Houthuijs, D., Dung, D. V., Royé, D., Indermitte, E., Lavigne, E., Mayvaneh, F., Acquaotta, F., de'Donato, F., Di Ruscio, F., Sera, F., Carrasco-Escobar, G., Kan, H., Orru, H., Kim, H., Holobaca, I.-H., Kyselý, J., Madureira, J., Schwartz, J., Jaakkola, J. J. K., Katsouyanni, K., Hurtado Diaz, M., Ragettli, M. S., Hashizume, M., Pascal, M., de Sousa Zanotti Stagliorio Coélho, M., Valdés Ortega, N., Ryti, N., Scovronick, N., Michelozzi, P., Matus Correa, P., Goodman, P., Nascimento Saldiva, P. H., Abrutzky, R., Osorio, S., Rao, S., Fratianni, S., Dang, T. N., Colistro, V., Huber, V., Lee, W., Seposo, X., Honda, Y., Guo, Y. L., Bell, M. L., & Li, S. (2021). Global, regional, and national burden of mortality associated with non-optimal ambient temperatures from 2000 to 2019: A three-stage modelling study. *The Lancet Planetary Health, 5*(7), e415–e425.

FURTHER READING

Herring, S. C., Hoell, A., Christidis, N., & Stott, P. (Eds.). (2021). Explaining extreme events in 2020 from a climate perspective. *Bulletin of the American Meteorological Society, 103*, S1–S17.

Son, J.-Y., Liu, J. C., & Bell, M. L. (2021). Temperature-related mortality: A systematic review and investigation of effect modifiers. *Environmental Research Letters, 14(7)*, paper 073004.

Stillman, J. H. (2019). Heat waves, the new normal: Summertime temperature extremes will impact animals, ecosystems, and human communities. *Physiology, 34*(2), 86–100.

WEB LINKS

Mayo Clinic for Hypothermia. www.mayoclinic.org/diseases-conditions/hypothermia/diagnosis-treatment/drc-20352688 and heatstroke www.mayoclinic.org/diseases-conditions/heat-stroke/symptoms-causes/syc-20353581

Red Cross and Red Crescent Societies Heat Action Day. www.ifrc.org/heat-action-day and cold wave preparedness and key messages www.ifrc.org/cold-waves

UK Fuel Poverty Statistics. www.gov.uk/government/collections/fuel-poverty-statistics

Victorian Government: Planning for Extreme Heat and Heatwaves: Resources and Guides. www.health.vic.gov.au/environmental-health/planning-for-extreme-heat-and-heatwaves

World Meteorological Organization. www.wmo.int

CHAPTER FOURTEEN

Wildfire Hazards

14

OVERVIEW

Wildfires present complex problems arising from the interaction of physical, biological and social factors in different landscape settings. Fire ignition can occur naturally (e.g. by lightning strikes) or by people inadvertently during agricultural, recreational and sometimes everyday activities or deliberately (e.g. arson or when a controlled land management burn becomes out of control). Wildfires cause significant impacts and consequences in many countries around the world, with death and injury to people caused by heat exposure, smoke inhalation and falling debris along with mental health issues associated with psychological stress. This chapter foregrounds wildfire risk to lives and property on the rural–urban fringe – often termed the wildland–urban interface (WUI). As wildfire risk increases due to anthropogenically induced climate change and because more people are living in the WUI, wildfire management – including risk maps, fire ban legislation and volunteer firefighting – is becoming increasingly important.

A. THE NATURE OF WILDFIRES

Wildfire is a generic term for uncontrolled fires fuelled by natural vegetation. Locally, wildfires are described by the dominant vegetation type fuelling the fire: bushfire, brushfire, coastal fire, veldt fire, forest fire and grassfire. The latter is the burning of finer fuels such as native grasslands, agricultural crops and pastures as well as open grassy spaces in and around suburban areas.

Apart from Antarctica, no continent is free from the various combinations of ignition source, fuel and weather conditions necessary for wildfire hazard. A seasonal fire pattern is found in most areas with a Mediterranean-type climate, where most rain falls in the winter and vegetation is dry during the annual summer drought. The dry season in tropical climates has rather similar seasonal biophysical conditions,

DOI: 10.4324/9781351261647-16

whilst large continental interiors – like those of the USA or Eurasia – experience dry air for much of the year, with the potential for a long fire season.

Australia is one of the most fire-prone countries in the world. Fires caused by lightning strikes have been characteristic for at least 100 million years, and the native vegetation is adapted to regular burning as practised by Aboriginal and Torres Strait Islander people living across Australia for over 60,000 years. This burning, undertaken on certain land, in certain ways in the right seasons, is a land management practice to ensure animals and plants remain abundant and diverse for food consumption, as well as making the land more accessible (Woodward et al., 2020). In more recent times, however, about 2,000 wildfires occur each year, many now started illegally, with some extending to over 100,000 ha. The most hazardous feature is the speed with which they spread. According to Mercer (1971), Australian bushfires can engulf up to 4 km^2 of forest in 30 minutes, as compared with as little as 0.005 km^2 over the same period in the slower-burning coniferous forests of the northern hemisphere.

1. Fire Ignition

Naturally occurring wildfires are commonly ignited by lightning strikes. However, human actions are the prime cause in current times, especially when high temperatures and drought follow a period of active vegetation growth. Figure 14.1 compares the causes of wildfire ignition on public land in the state of Victoria, Australia (a high-risk area averaging 600 fires per year), and in Bages County, Catalonia, Spain (a typical western Mediterranean area of rural depopulation averaging 15 fires per year). In both cases, natural causes (lightning) are small; arson may be higher than indicated in Bages County, due to the high percentage of unknown causes.

Inadvertent sources are a mix of agricultural and recreational activities. In dry conditions, grassfires can ignite when using machinery such as chainsaws, lawnmowers and welding equipment or when a vehicle's hot exhaust pipe comes in contact with fine fuel. Mechanical issues with farm machinery can also spark a grassfire. In countries like Australia, France and Greece, there is an added risk of wildfire in and around popular tourist areas where hikers, campers and people fishing unintentionally start wildfires when campfires become out of control or are not properly extinguished, or when cigarettes are carelessly discarded. The latter is a widespread common occurrence and is not isolated to wildfire-prone countries. Iceland's west was impacted by wildfire in March 2006 when a carelessly discarded cigarette set the winter grass alight.

Deliberate fire-raising is a widespread problem that has attracted the attention of criminologists (Willis, 2005). In California one-quarter of all wildfires are believed to be due to arson but only 10% of police investigations lead to an arrest. In Australia, about 13% of all bushfires for which there is a recorded cause are logged as deliberate but a further one-third are suspicious, making it possible that around half the total may be deliberately lit.

2. Availability of Fuel

The amount and moisture status of vegetation influences both the intensity of a wildfire (heat-energy output) and the rate of spread. Thus, grassfires tend to be less intense than bushfires, brushfires, forest fires and scrub fires. Apart from its quantity, the moisture content of fuel is

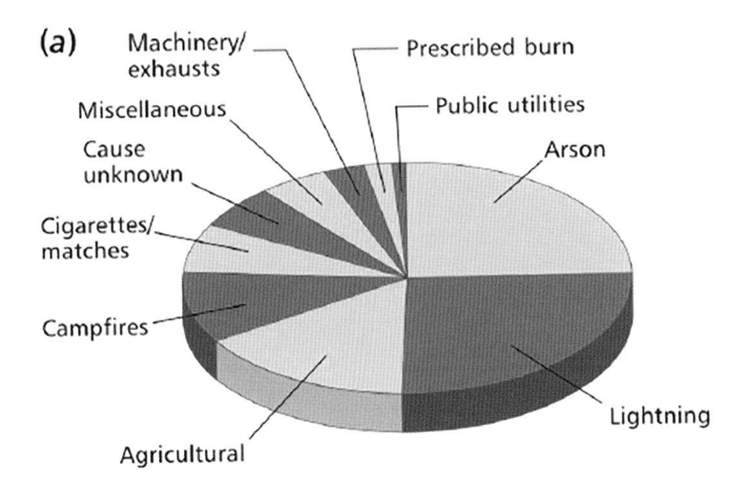

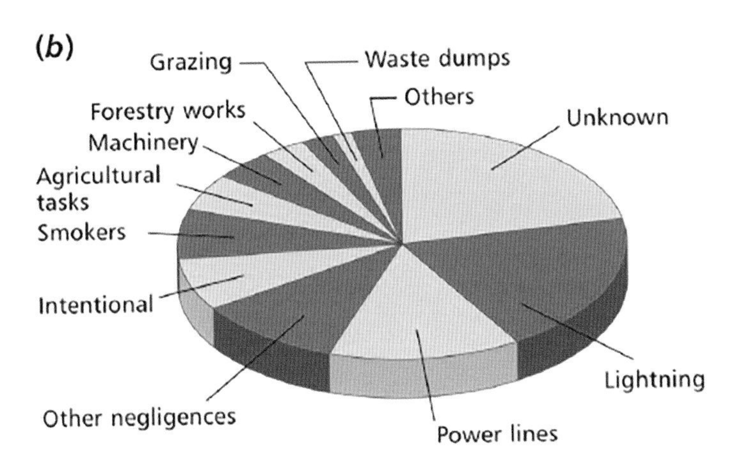

Figure 14.1 Sources of wildfire ignition in two different regions. (a) The State of Victoria, Australia. (b) Bages County, Catalonia, Spain. Adapted from State of Victoria at http:// www.nre.vic.gov (accessed 29 January 2003) and after Badia et al. (2002). Reprinted from Global Environmental Change 4, Badia et al., Causality and management of forest fires in Mediterranean environments, 23–32. Copyright (2002), with permission from Elsevier.

important. This depends largely on the weather and climate. Climatic conditions ensure that there is a marked seasonal procession of risk in most countries. Figure 14.2 illustrates the seasonal variation of fire danger across Australia, where the risk is driven by weather conditions.

According to Cunningham (1984), southeast Australia is the most hazardous wildfire region on Earth. This is because many forests are dominated by the genus *Eucalyptus*. Most Australian forests accumulate a great deal of litter on the forest floor, mainly from bark shedding, after a number of fire-free years. Apart from creating a source of fuel, bark shedding creates a special problem of rapid fire-spread known as 'spotting'. This occurs when ignited fuel or embers are blown ahead of an advancing fire front by strong winds to create 'spot'

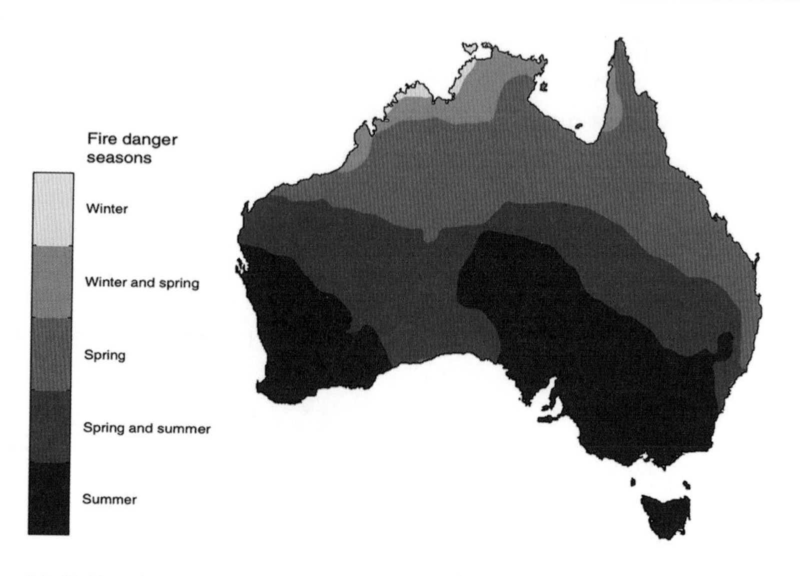

Figure 14.2 Fire danger seasons across Australia, with each area showing the highest average value of the Forest Fire Danger Index (FFDI, based on daily data from 1950 to 2016 (Dowdy, 2018)) with a 100 km smoothing applied to highlight broad-scale regional features. The seasons are averaged for the following months: winter – June, July and August; winter and spring – half of July, August, September and half of October; spring – September, October and November; spring and summer – half of October, November, December and half of January; and summer – December, January and February. No notable regions were evident across summer and autumn; autumn; autumn and winter.

Source: Dowdy, Andrew J. (2020). Seamless climate change projections and seasonal predictions for bushfires in Australia. *Journal of Southern Hemisphere Earth Systems Science* 70, 120–138.

fires. Australian eucalypts have the longest ember spotting distances in the world. The reason is bark shedding by, for example, the *stringybark* and *candlebark* species that produce loose, fibrous tapers easily torn loose by strong winds and carried as embers in convection currents. Spotting distances of 30 km or more have been authenticated, at least twice the distance recorded in the deciduous hardwood and coniferous forest fires of North America. In addition, eucalyptus trees contain volatile waxes and oils in their leaves that release a high heat output when burned and greatly increase the flammability of the vegetation (Chapman, 1999). At ambient fire temperatures of around 2,000°C, these oils can create a spontaneous gas explosion.

3. Prevailing Fire Weather

Weather conditions are crucial for the ignition and development of wildfires. Drought periods provide a drying effect on vegetation and may also have atmospheric conditions suitable for 'dry lightning' storms, when no appreciable rain falls. Such storms are most active during unstable weather conditions in the summer months and ignite 60% of all fires on public land in the western states of the USA. In the summer of 2000, 122,000 wildfires were started in this area and burned 32,000 km^2 (Rorig & Ferguson, 2002). There is evidence that increased spring and summer temperatures in the western USA since the mid-1980s have led to a longer wildfire season and, in turn, to more frequent large wildfires (events >400 ha) that burn for longer durations (Westerling et al., 2006).

Brotak (1980) compared extreme fire hazard situations in the eastern USA and southeast Australia and found that most fire outbreaks occur near weather fronts, particularly in warm, dry conditions ahead of a well-developed cold front with unstable temperature lapse rates and strong winds at low levels. In California, easterly *Santa Ana winds*, which occur mainly in September and October – the driest and warmest months in the San Francisco Bay Area – create an extreme hazard during autumn. Strong northeasterly Santa Ana-type winds developed in late July 1977 and led to a disastrous wildfire which began in the hills and advanced to within a mile of the downtown area of the city of Santa Barbara. Over 230 homes were destroyed (Graham, 1977). In October and November 1993, 21 major wildfires developed in six Southern Californian counties, fanned by hot, dry Santa Ana winds. Three people were killed, 1,171 structures were destroyed and some 800 km^2 were burned. The combined property loss was estimated at US$1 billion.

Once ignited, the rate of fire spread is closely related to the surface wind strength and direction. This is because the burning fire front advances by first heating and then igniting vegetation in its path through a combined process of convection and radiation. Most wildfire damage, including loss of life, occurs during a relatively short period of time – usually a few hours – compared with the total duration of the fire. These high-loss episodes are associated with extreme fire-risk weather, often involving high winds that shift in direction and cause the fire to accelerate in unexpected directions. Fire acceleration is greatly aided by the topography. For a fire driven upslope, wind and slope acting together increase the propagating heat flux by exposing the vegetation ahead of the fire to additional convective and radiant heat (Figure 14.3).

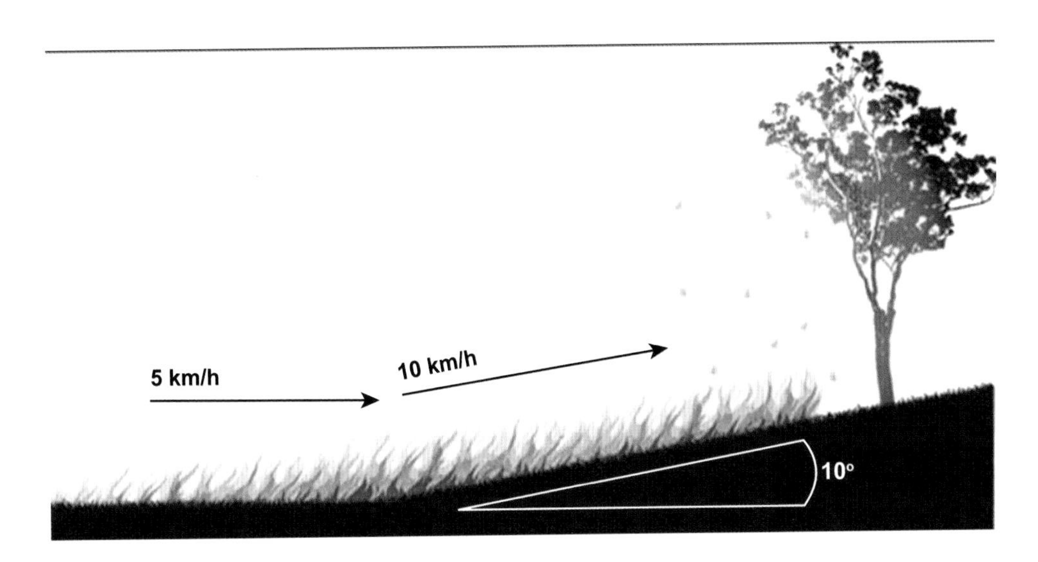

Figure 14.3 A fire will increase in speed as it travels uphill.

Source: www.cfa.vic.gov.au/plan-prepare/am-i-at-risk/how-fire-behaves. Accessed: 24 July 2022.

The combined effect of wind and slope positions the advancing flames at an acute angle so that once the slope exceeds 15°–20° the flame front is effectively a sheet moving parallel to the slope. Data from experimental fires in eucalypt and grassland areas in Australia have shown that the rate of forward progress of a fire on level ground doubles on a 10° slope and increases nearly four times when travelling up a 20° slope (Luke & McArthur, 1978). Such conditions pose a serious risk to firefighters as well as residents. The 1994 South Canyon fire in Colorado, USA, killed 14 firefighters when a change in wind conditions caused flames to leap across the canyon floor and ignite a steep slope of Gambel oak trees immediately below the men. Within seconds, flames up to 90 m high spread up the slope at a speed impossible to outrun.

The combined effects of fuel and weather conditions were evident in the Ash Wednesday bushfires that raged across southeast Australia in February 1983 (Figure 14.4a). These fires included the largest conflagration experienced in the Forest Reserves of South Australia (Keeves & Douglas, 1983). The area had been in drought for the previous six months and the fires were all ignited between 11.00 and 16.30 h, when air temperatures and solar radiation were high, relative humidity was low and the winds were strong and gusty. The first outbreak (the Narraweena fire) started at about 12.10 h in grassland and, within four hours, travelled 65 km southeast through intensively managed agricultural land before veering with a change in wind direction. The parallel Clay Wells fire began at 13:30 local time in roadside grassland and quickly spread to native forest and adjacent pine plantations where large quantities of fuel created crown fires and allowed the development of spot fires downwind (Figure 14.4b). By 16.00 h the wind had changed from northwest to west-southwest and increased in speed from 30–60 km h^{-1} to 50–80 km h^{-1}, with gusts over 100 km h^{-1}, before dying down several hours later. In total, fire damaged about 30% of the area planted with conifers in the forests of South Australia. The 'Ash Wednesday' fires in Victoria and South Australia during February 1983 caused 76 deaths; 8,000 people were made homeless and economic losses were put at AU$200 million (Bardsley et al., 1983).

Unfortunately, fire conditions are worsening. Harris and Lucas (2019) examined the variability of fire weather in Australia and found a long-term upward trend, particularly prevalent in southern Australia, and concluded that this is most likely due to anthropogenic climate change.

B. IMPACTS AND CONSEQUENCES OF WILDFIRES

Wildfires cause significant impacts and consequences in many countries around the world, with death and injury to people caused by heat exposure, smoke inhalation and falling debris (e.g. burnt infrastructure and trees) along with mental health issues associated with psychological stress. One of the world's greatest wildfire disasters occurred in October 1871 when about 17,000 km² of land in Wisconsin and Michigan, USA, was burned. About 1,500 lives were lost. The fires began on the night that an urban fire in Chicago killed 250 people and were preceded by a 14-week drought across the Midwest. Many small fire outbreaks were not considered a threat until strong winds whipped up the flames to create a widespread, uncontrollable blaze.

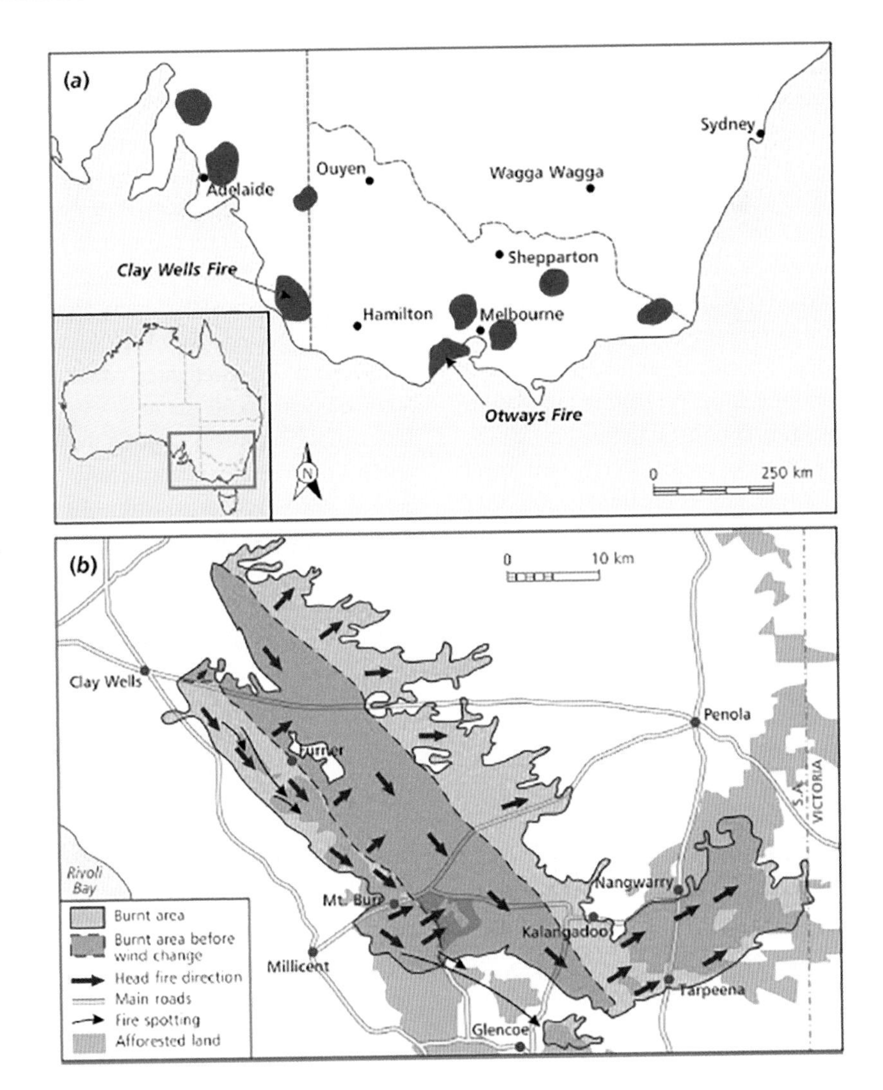

Figure 14.4 The 'Ash Wednesday' bushfires of 16 February 1983 in southeastern Australia: (a) the location of major fires; (b) the progress of the Clay Wells fire, South Australia. This fire originated with a long, narrow shape created by strong, dry, northwesterly winds ahead of a cold front. The change in shape was due to the later onset of a southwesterly wind. Various spot fires can also be seen.

Source: After Keeves and Douglas (1983) Reprinted from A. Keeves and D.R. Douglas, Forest fires in South Australia on 16 February1983 and consequent future forest management aims. *Australian Forestry* 46(3) (1983): 148–64. With permission from the Institute of Foresters of Australia.

Today, wildfire risk to lives and property can be highest on the rural–urban fringe – often termed the *wildland–urban interface* (WUI) – where housing blocks are intermixed with natural vegetation. The lifestyle attractions of a semi-rural environment have encouraged the expansion of low-density suburbs in places like

Sydney, Melbourne and Adelaide in Australia, and the Los Angeles and San Francisco Bay communities in the USA, into the surrounding bushland. Highlighting this risk, Handmer (1999) described a wildfire that affected Sydney in January 1994, when four deaths occurred and 200 houses were destroyed, despite the efforts of over 20,000 firefighters mobilised from all over Australia. In Canberra, the Australian capital, a series of semi-natural ridges, used for open-space recreation and nature conservation, run through the city, which, in many suburbs, backs directly onto rural areas without any transitional land uses (Lucas-Smith & McRae, 1993). During January 2003, bushfires in the Canberra suburbs killed four people, injured almost 500 others, destroyed 488 homes and forced more than 5,000 residents to evacuate their homes. More recently, the 'Black Saturday' fires of February 2009 claimed 173 lives and destroyed over 2,000 homes in the state of Victoria. Over the summer of 2019–2020, wildfires impacted all of Australia's states and territories. Known as the 'Black Summer' fires, they destroyed over 3,000 homes and claimed 33 lives.

Blanchi et al. (2014) examined bushfire-related deaths across Australia from 1901 to 2011 and found that of the known 674 civilian and 151 firefighter fatalities, 78% occurred within 30 m and 85% within 100 m of the bushland edge. Of those fatalities that occurred within structures (e.g. the home), 76% were located within 10 m, 88% within 30 m and 95% within 50 m of the bush. Expanding on Blanchi et al.'s research, Haynes et al. (2020a) examined wildfire-related fatalities among civilians and firefighters at the WUI across Australia, Greece, Portugal, Sardinia and Spain (Figure 14.5).

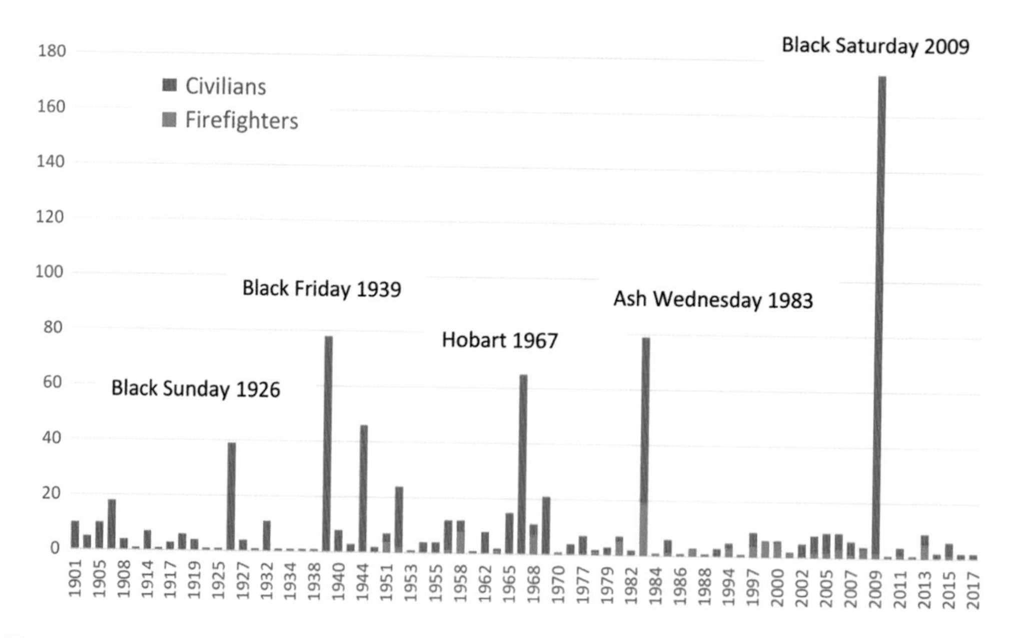

Figure 14.5 Distribution of fatalities among firefighters and civilians between the time period 1901 and 2017.

Source: Haynes, K., Short, K., Xanthopoulos, G., Viegas, D. X., Ribeiro, L. M., & Blanchi, R. (2020a). Wildfires and WUI Fire Fatalities. In Encyclopedia of Wildfires and Wildland-Urban Interface (WUI) Fires (pp. 1073–1088). doi:10.1007/978-3-319-52090-2_92.

The Australian dataset, covering 1901–2017 with a total of 846 fatalities and an average rate of 7.3 civilian deaths per year, showed a marked increase in the number of female deaths over this period (Figure 14.6). The Greece, Portugal, Sardinia and Spain dataset covered the period of 1945–2016 with a total of 865 deaths, noting that it is not a complete set across all four jurisdictions due to a lack of records. Also, in the following years, significant fires occurred in Portugal in June and October 2017 and in Greece in July 2018. The 2017 fires in Portugal took the lives of a further 66 civilians and one firefighter in June and 51 civilians in October, while the fires in Greece in July 2018 took the lives of 102 people.

Wildfires in natural bushland and forests damage ecosystems, kill wildlife and create indirect losses. While it is difficult to quantify, wildfires such as the Australian 'Black Summer' fires of 2019–2020 can devastate ecosystems, flora and fauna as fire regimes quickly change from those that are essential ecosystem processes (Dickman, 2021). After a major event, timber and forage resources may be destroyed alongside animal habitats, soil nutrient stores depleted and amenity value reduced for many years. If the burned-out areas consist of steep canyons, debris flows, rill erosion and floods are likely to follow. Such fires also adversely influence timber production, outdoor recreation, water supplies and other natural assets, and land restoration is expensive. A major threat exists in the dry, inland part of the western USA, where over 150,000 km^2 of forests are at risk. In 1988 nearly

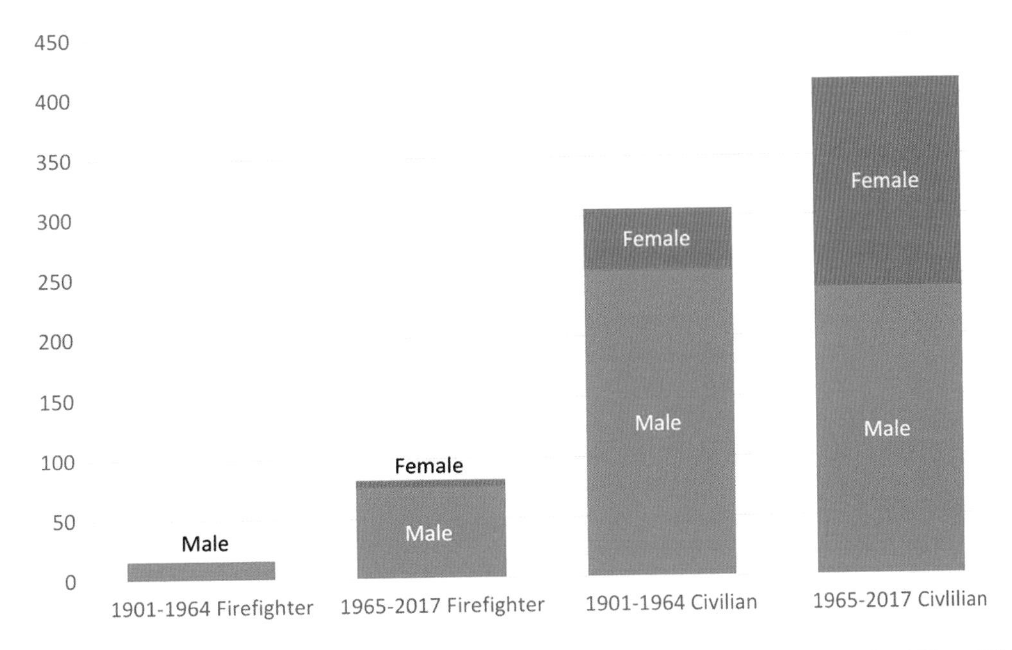

Figure 14.6 Number of civilian and firefighter fatalities by gender from the time periods 1901 to 1964 and 1965 to 2017.

Source: Haynes, K., Short, K., Xanthopoulos, G., Viegas, D. X., Ribeiro, L. M., & Blanchi, R. (2020a). Wildfires and WUI Fire Fatalities. In Encyclopedia of Wildfires and Wildland-Urban Interface (WUI) Fires (pp. 1073–1088). doi:10.1007/978-3-319-52090-2_92.

3,000 km^2 of Yellowstone National Park was burned out, despite the efforts of more than 9,000 firefighters, raising important policy issues about fire management in areas with an important heritage status (Romme & Despain, 1989).

In recent decades the large-scale use of fire to clear forested land for development has increased risk. This has become an international issue in parts of Southeast Asia, often because of the resulting smoke pollution. For example, smoke haze in 2015 from fires ignited to clear logging waste and agricultural waste and to secure land tenure in Indonesia blanketed the neighbouring countries of Malaysia, Singapore and Thailand. Field et al. (2016) showed that the 2015 fires ranked as one of the worst events on record for the region, alongside those that occurred in 1991 and 1994 but behind a 1997 event. The 2015 event was significant in regard to the number of fires that were burning uncontrolled for months. As a result, air pollution measurements across Sumatra and Borneo – measuring particulate matter, sulphur dioxide, nitrogen dioxide, carbon monoxide and ozone – recorded levels above 2,000 on Indonesia's Pollutant Standards Index, where anything above 350 is considered hazardous to human health (Voiland, 2015). Voiland (2015) reported that government records recorded 19 pollution-related fatalities, 500,000 people suffering from respiratory problems and more than 43 million people exposed to the smoke haze. Fire risk from land clearing is not confined to Southeast Asia. For example, biomass burning across Greece extended to over 12% of the forested area of the country during 2007; 67 people died and dense smoke plumes travelled thousands of kilometres to the south (Kaskaoutis et al., 2011).

In some instances, wildfires can lead to other disasters in a cascading effect. For example, two bushfires were burning out of control near the town of Morwell, Victoria, in February 2014. Adjacent to Morwell is the Hazelwood open cut coal mine. Embers from these bushfires set the northern, eastern and southeastern batters and floor of the mine alight. Over 7,000 firefighters from across Australia were called in to fight the fire, which proved extremely difficult to extinguish due to the highly combustible nature of the brown coal. Exacerbating the situation was the fact that the mine operator was not prepared for such an event, which meant that there was either poorly maintained or no firefighting equipment present in the areas of the mine that were on fire. After 45 days, the fire was officially out but not before it had blanketed Morwell in a considerable amount of ash, causing significant physical issues – including respiratory symptoms, skin and eye irritations, nosebleeds and headaches – and psychological distress among the local population (Bird & Taylor, 2021).

C. RISK ASSESSMENT AND MANAGEMENT

Wildfires are multifactorial events. The most obvious risk factor is increased human exposure, due to the spread of people into areas of predominantly natural vegetation. In the USA during the early 1990s, over one-quarter of the fires attended by public fire departments were in timber, brush and grass areas housing rural communities of less than 2,500 people (Rose, 1994). It was estimated that these rural residents were almost twice as likely to die in a fire as people living in larger communities of 10,000–100,000 people (Karter, 1992). The chief risk is in

California, where 80,000 km^2 of highly flammable brushland has been developed to create an urban/wildland 'intermix' (Hazard Mitigation Team, 1994). At this time (1994), five fires creating the greatest loss to buildings in California had occurred within the previous five years. In 1991 a wildfire in the East Bay Hills area of San Francisco killed 25 people, injured more than 150 and made over 5,000 homeless (Platt, 1999). With estimated losses of US$1.5 billion, it was the third-most costly urban fire in US history. The fire started under classic conditions of high temperatures, low air humidity and strong winds and spread rapidly, aided by a dry vegetation cover. Firefighters were hampered by congested access roads, plus a loss of water pressure, and some 60 years of urban development in this area was destroyed, with only the building foundations remaining.

With the increasing risk of fire weather due to anthropogenically induced climate change (Harris & Lucas, 2019) and an increase of people living in the WUI, wildfire management is becoming more and more important.

1. Risk Maps

Geospatial technology is increasingly used to assess and map wildfire risk for management purposes. Wildfire risk maps typically combine the three key variables of *fuel* (type, moisture level, size, shape), *topography* (elevation, slope, aspect) and *weather* (temperature, precipitation, relative humidity, wind speed), as shown by Lein and Stump (2009). For example, in Victoria, Australia, hazard and risk assessments are undertaken to determine bushfire prone areas (BPA) based on weather, topography and vegetation (VBA, 2022). A bushfire attack level (BAL) assessment, based on the Fire Danger Index – an annual accumulated value of daily measurements of vegetation dryness, air temperature, wind speed and humidity (CSIRO, n.d.) – slope of the land, surrounding vegetation type and proximity of other buildings, must be carried out to determine the type of construction required before building in a BPA. These rules ensure that all new homes are constructed to a minimum standard to withstand ember attack. In areas of extreme bushfire risk, further planning and building permits are required under a Bushfire Management Overlay (BMO).

Wildfire risk maps are becoming more accessible and interactive (Figure 14.7). For example, the US Department of Agriculture encourages communities at risk to stay informed by exploring their wildfire risk (https://wildfirerisk.org). Designed for use by community leaders, planners and fire managers, the Wildfire Risk to Communities maps are interactive and nationwide, built from nationally consistent data that consists of vegetation and fuels, weather and community data. Based on the information provided by these maps, community leaders, planners and fire managers might consider the following land use planning and more broad-scale landscaping options:

- cluster development – individual homes or apartments built in small groups (saving land for community open space);
- low overall housing density – individual residential lots at least 0.5 ha;
- minimum spaces between buildings (approx. 10 m) and clusters;
- access lanes wide enough for firefighting equipment;
- all properties with a minimum setback from the natural bush (approx. 30 m);

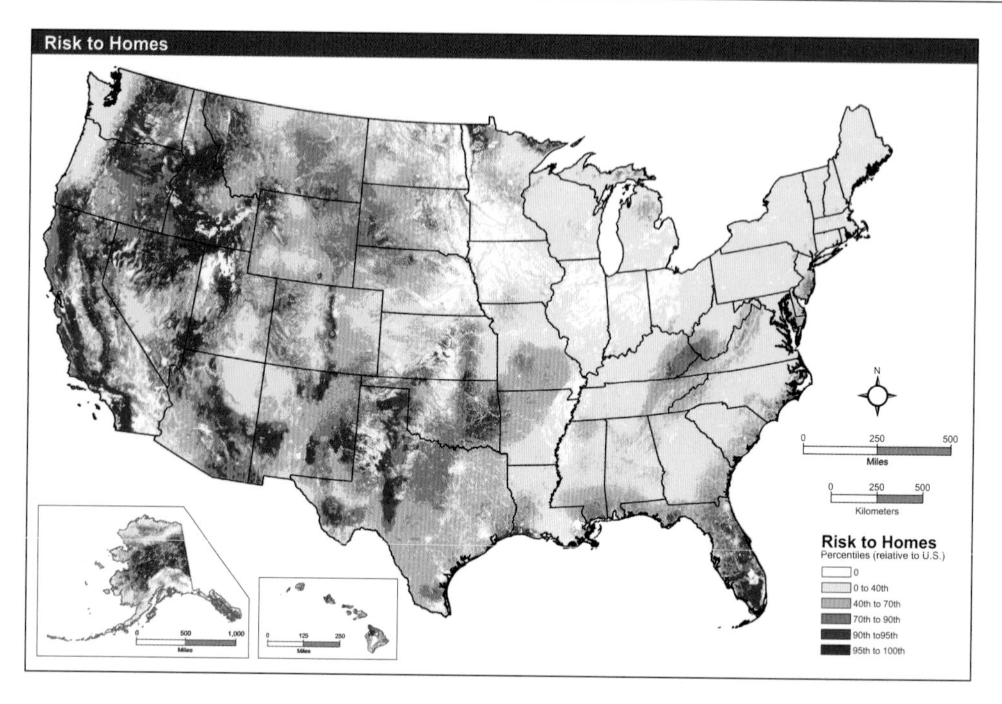

Figure 14.7 An example of a wildfire risk map highlighting the State of California, USA.

Source: https://wildfirerisk.org/explore/0/06/. Accessed: 26 September 2022.

- mature trees and large shrubs on edge of bush pruned to avoid direct spread of fire from one tree to another; and
- intermediate area managed by clearing all dead vegetation and planting grass or small, low fuel-volume vegetation.

More thought should also be given to limiting occupancy rates, and to raising the capability of road networks for emergency evacuation, when fire-prone areas are approved for new residential development (Cova, 2005). It is important that neighbouring local authorities work together on land and vegetation management and develop a common database for land planning in the WUI. When areas have been burnt out, consideration should be given to the government acquisition of land for public open space and the rebuilding of properties at lower densities on larger plots.

2. Fire Ban, and Controlled Fire, Legislation

Another management solution is the implementation of strict fire ban legislation. Such measures are difficult to enforce, although *total fire bans* on days of extreme fire danger are needed and are used. They usually apply to a particular weather forecast district and last for 24 hours, during which period no fire may be lit in the open. Authorities can also potentially control where fires might be lit with the provision of barbecue places set up in safe clearings alongside roads to discourage indiscriminate fire lighting.

A lack of burning can increase the risk of a future wildfire, due to the growing supply of fuel that has been allowed to build up over time. The recognition of this relationship has led to the increasing use of low-intensity fires referred to as controlled, planned or prescribed burns – a practice that has been undertaken for more than 60,000 years by Aboriginal and Torres Strait Islander people in Australia and is often referred to as cultural burns among these First Nations people.

The purpose of controlled burns is to consume existing fuel in relatively small, supervised fires that reduce the risk of major outbreaks later. This may be a cost-effective policy for genuine wildland areas but is difficult to implement in the mixed landscape of the rural–urban fringe where farmland, forest plantations and suburban gardens coexist. Controlled burning is labour intensive and can lead to uncontrolled fires, cause air pollution and have controversial effects on local ecosystems, such as a reduction in the diversity of flora. A simulation study of prescribed burning around Sydney, southeast Australia, showed that very frequent levels of burning were needed to improve fire safety significantly (Bradstock et al., 1998). Such levels are difficult to achieve because of the high costs in steep terrain and the need for mild and stable weather conditions outside of the fire season. Due to these difficulties, excess fuel can be removed by mechanical means – including commercial timber harvesting and through the creation of firebreaks – as well as controlled burning.

3. Firebreaks

Firebreaks are a strip of land, up to 40 m wide, that is devoid of vegetation (i.e. fuel) and are used to stop or reduce the spread and intensity of fire. They are typically created on the WUI and around farmlands and reserves with heavy machinery such as a bulldozer, grader or tractor. This fuel-free area also provides easy access to firefighting vehicles, if the need arises. Firebreaks may be constructed in preparation for an impending fire season or during a wildfire event with the aim of keeping the fire contained within a certain area.

On a smaller scale, residents can create their own firebreaks by landscaping their properties for fire. This includes ensuring the area around the house and other structures is clear of easily flammable vegetation. Australia's Country Fire Authority in Victoria outlines the following principles for landscape design for the reduction of fire risk: create a defendable space and remove flammable objects from around the home, break up fuel continuity and carefully select, locate and maintain trees (Figure 14.8).

4. Volunteer Firefighting and Community Awareness

When wildfire occurs, local firefighting groups are often the first line of defence. These bodies are composed of volunteers and are often taken for granted by state and federal governments. For example, in the USA the value of rural firefighting services to the nation was estimated to exceed US$36 billion each year, but the firefighters feel they neither influence policy nor obtain the resources needed to work effectively (Rural Fire Protection in America, 1994). In Australia there are over 200,000 volunteers but, as in North America, the

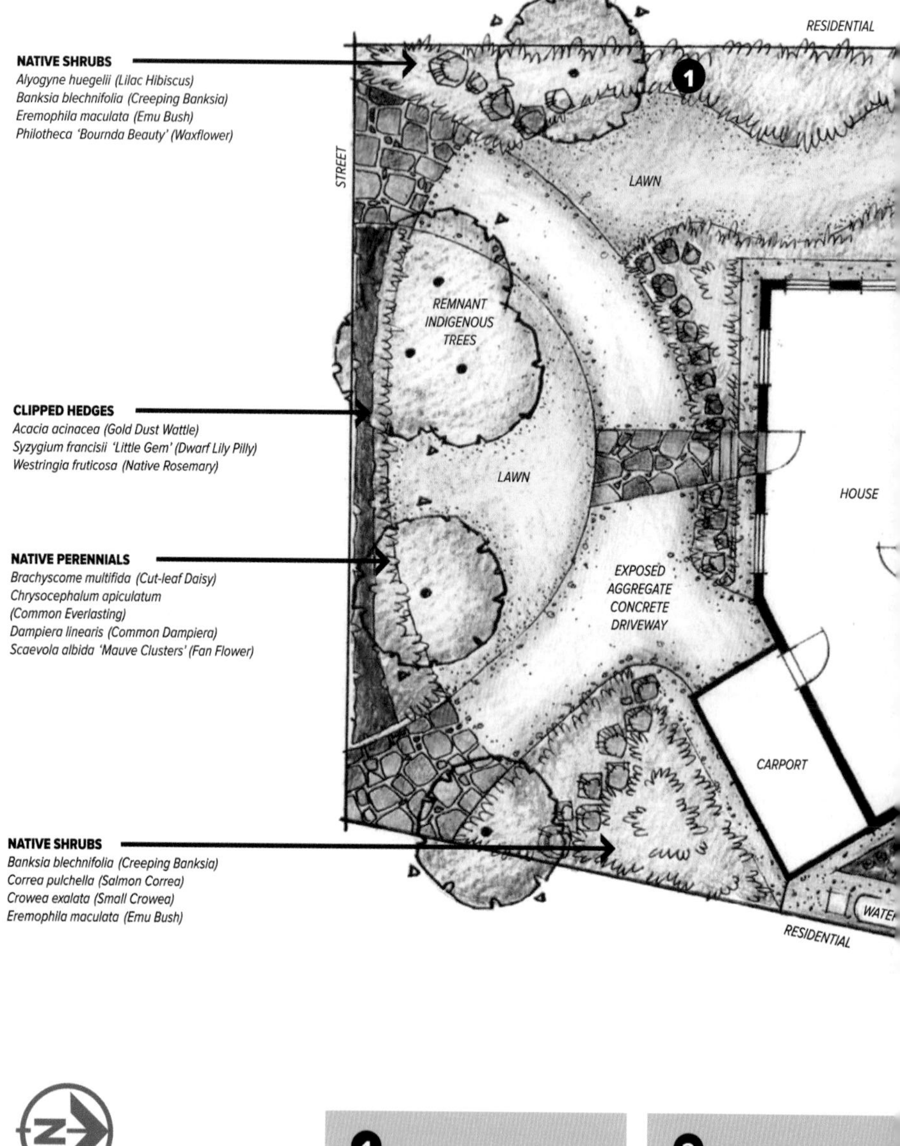

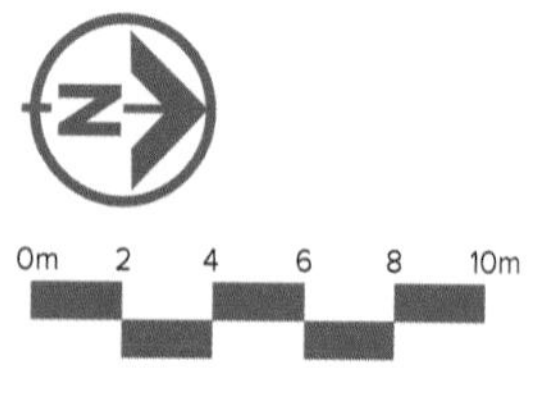

1 Avoid planting shrubs under trees. Instead use groundcovers with low flammability and low-growing succulents under trees to provide maximum separation between fuel at the ground level and the tree canopy.

2 Use decorative paving and gravel around the house to keep plants away from windows, doors and other vulnerable parts of the building.

Figure 14.8 A mock-up plan provided by Australia's Country Fire Authority, showing residents how to best landscape their property to reduce its potential flammability.

Source: CFA (2021) Suburban Garden. Landscaping for Bushfire: Garden Design and Planting Selection. https://www.cfa.vic.gov.au/plan-prepare/how-to-prepare-your-property/landscaping. Accessed: 12 August 2022.

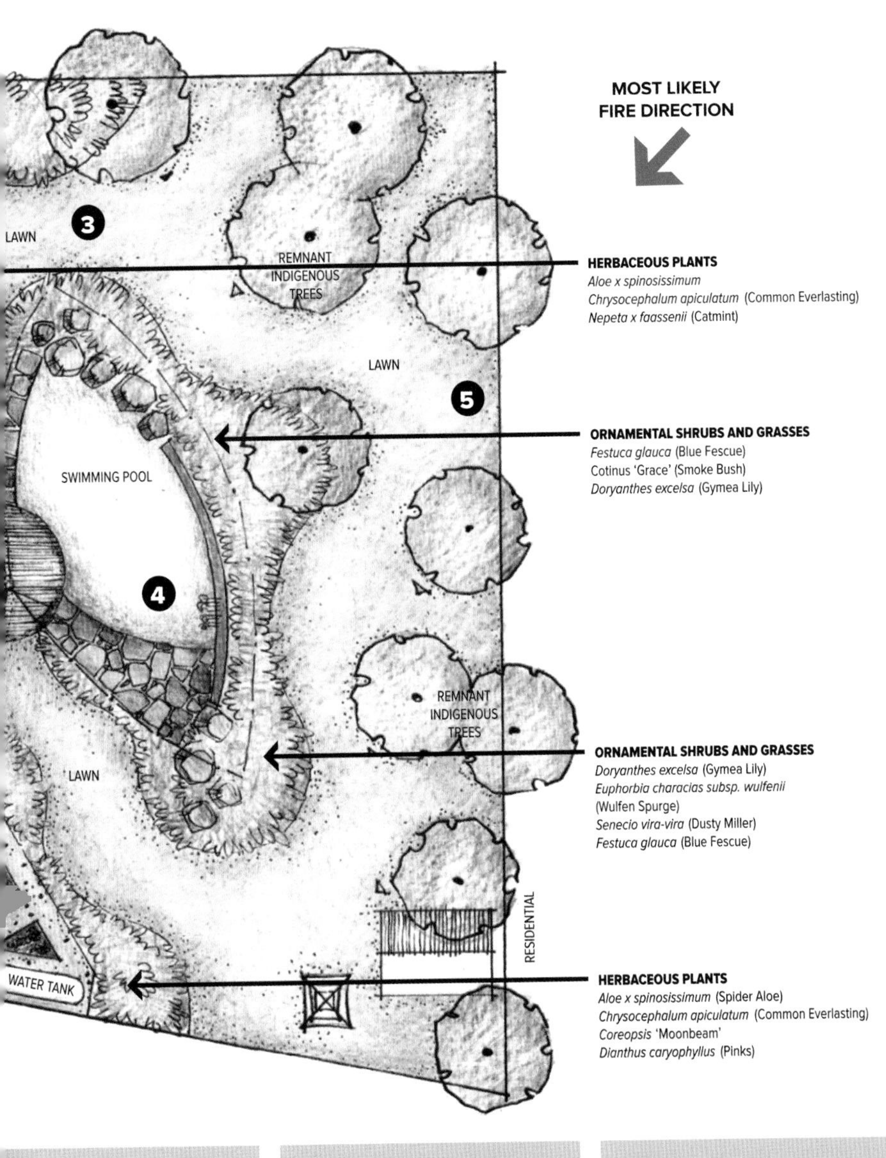

se lawn areas to provide
paration between garden
ds. This breaks up the
ntinuity of vegetation in
e garden to prevent fuel
rridors.

Locate landscaping features with low flammability such as pools, tennis courts or maintained lawn between the house and the most likely direction of a bushfire.

Clump trees to create at least two metres distance between tree canopies. This helps prevent fire spreading from tree to tree.

number is declining rapidly due to socio-economic factors, including demographic trends that show a reduction of the population between 25 and 45 years of age in rural areas (McLennan & Birch, 2005). Rural fire services require costly training and specialist equipment. Because piped water supplies are not always available, fire teams need methods to deliver and use water more efficiently. This implies dedicated items such as tankers for transporting water or access to aircraft. There is also a need for more general tools such as earth-moving plant equipment to construct access tracks and firebreaks.

In the United States, major wildfires cross local government boundaries and affect land managed by private landowners and state and federal agencies. A comprehensive fuel modification plan should be agreed to reduce fire intensity, including prescribed burns and vegetation thinning. It is also necessary to have an overall view of firefighting requirements, including water supply and equipment. This approach was tried in California after the Oakland–Berkeley Hills firestorm of 1991, when the cities of Oakland and Berkeley formed a consortium with other 'intermix' landowners to develop a coordinated hazard-reduction plan. Similar bushfire management committees, representative of local interest groups, exist in Australia. Other more localised efforts include the Community Fire Units in New South Wales and Canberra, the Community Fireguard programme in Victoria and the Bushfire Ready Neighbourhoods programme in Tasmania. Essentially, these programs educate and train residents living in the WUI on how to plan and prepare themselves for wildfire and respond with informed decision-making, if the need occurs. Similar programs elsewhere include Canada's FireSmart-ForestWise program and the USA's Firewise Communities. These programs not only help residents reduce their risk, but residents also report feeling empowered to help themselves, their neighbours and official fire services, which is instrumental in creating connections and building trust, solidarity and cooperation between each of these groups (Haynes et al., 2020b). Furthermore, these localised programs can provide significant economic savings. Gibbs et al. (2015) showed that over a ten-year period, Victoria's Community Fireguard program saved each group AU$732,747 from reduced property losses and AU$1.4 million in reduced fatalities in the event of a major fire, while only costing the group AU$10,884 over that same time period.

When considering community preparedness, there are many reasons why it is often low for wildfires. In Edmonton, Canada, households hold a variety of views on the effectiveness of fire reduction and rarely implement the full range of measures available (McGee, 2005). From a study in California, Collins (2005) concluded that residents were reluctant to remove vegetation from around their property because they attached a high amenity value to their semi-natural environment. Others, such as those living in areas lacking basic community services (e.g. roads, piped water) and those who did not own their properties, lacked the incentives and the financial means to make hazard adjustments. In parts of Victoria, Australia, residents recognised the risk but relied on firefighters for protection and made few preparations of their own (Beringer, 2000). Awareness of fire hazard tends to grow with residence time in the area. Figure 14.9 shows that the deployment of self-reliant protection measures can increase fourfold with residence periods of 25 years or longer. While these protection measures are moderately priced, retrofitting homes to comply with

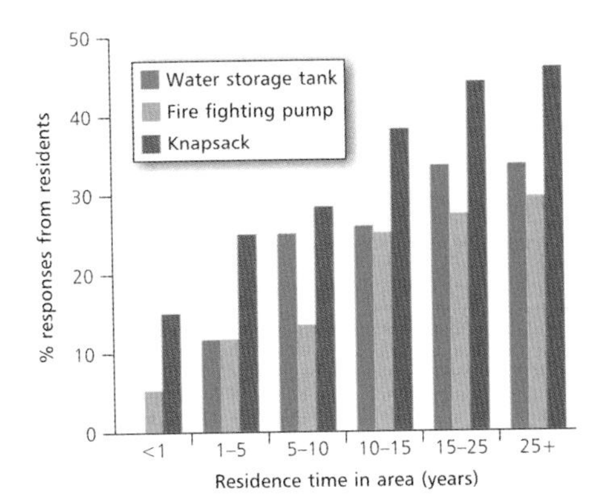

Figure 14.9 Relationship between length of residence in the North Warrandyte area, Victoria, Australia, and the ownership of firefighting equipment. This area is exposed to high bushfire risk.

Source: After Beringer (2000). Reprinted from *Fire Safety Journal* 35, J. Beringer, Community fire safety at the urban/rural interface, 1–23. Copyright (2000), with permission from Elsevier.

requirements outlined under Victoria's BPA or BMO is unaffordable for many residents.

Residents living in wildfire areas are alerted to potential threats through forecasts and warnings. In Australia, the Fire Danger Rating is a forecast produced by the Bureau of Meteorology, providing information up to four days in advance. Australia's Fire Danger Rating system is currently being simplified from six levels to four and contains clear information on the public action that should be undertaken under each level (Figure 14.10).

5. Fire Detection

In some areas, lookout towers are used for early fire detection. In more remote regions, regular surveys by aircraft or other remote sensing means may be necessary. During dry weather, plants reduce the amount of evapotranspiration from their leaves, with a consequent increase in the surface temperature of large vegetation stands, such as forests. These temperature changes can be detected on satellite images, and the derivation of an appropriate 'vegetation stress index' can be used as an indication of where wildfire outbreaks are most likely to occur

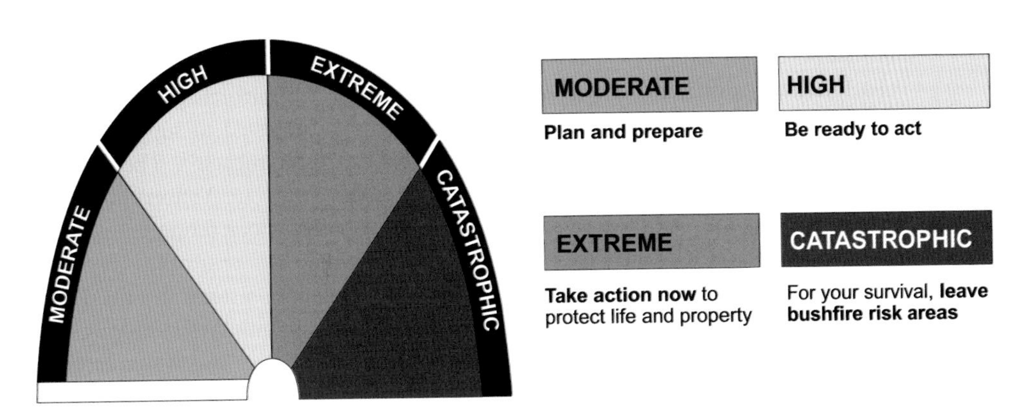

Figure 14.10 Australia's new Fire Danger Rating System was applied across the whole country from September 2022. It includes four levels of alert and will be used to forecast fire-risk days. A clear set of messages will be attached to each level and will include actions that the public can adopt to reduce their risk.

Source: www.cfa.vic.gov.au/warnings-restrictions/total-fire-bans-and-ratings/about-fire-danger-ratings. Accessed: 27 July 2022.

(Patel, 1995). It is then possible to intensify ground surveillance in these areas and to exclude the public until the fire risk starts to fall. In the western USA, most fires are caused by summer lightning strikes, so investment in automated lightning detection systems, with options for follow-up aerial survey, is prudent.

With the advancement in technologies, Barmpoutis et al. (2020) compared terrestrial, uncrewed aerial vehicles (UAVs) and satellite-based systems for the early detection of fire and smoke related to wildfires. These researchers found that terrestrial systems were more efficient in terms of accuracy and response times but have limited coverage. UAVs provide a more accurate broader view of the fire under dangerous conditions and in rough, inaccessible terrain but are affected by weather conditions and have limited flight time. Satellite-based systems provide the largest scale of coverage but, depending on the type, they have either high temporal resolution with low spatial resolution or vice versa. Overall, fire agencies ultimately use, when available, a combination of all three, combined with lookout towers, aerial surveys, Fire Danger Rating forecasts alongside crowdsourcing information and monitoring social media.

Early fire detection is important because risk reduction options decline rapidly once an outbreak occurs. In Australia, Handmer and Tibbits (2005) showed that late evacuation from homes, as opposed to staying inside, resulted in high fatality rates. More recent data reinforced the dangers of late evacuation and illustrated a gendered difference in bushfire deaths; men are mostly killed outside, protecting assets, whilst women and children died sheltering in the house or attempting to flee (Haynes et al., 2010). Even when fatalities do not occur, the impacts of late and/or a lack of preparedness for evacuation can cause significant impacts, resulting in psychosocial trauma and post-traumatic stress. In May 2016, a wildfire threatened the Fort McMurray area, prompting the evacuation of 88,000 people. The fire destroyed 2,400 buildings and resulted in over $3.5 billion in insured losses (Mamuji & Rozdilsky, 2019). While no fatalities occurred as a direct result of the fire, two people lost their lives in a car crash during the evacuation. As a result, evacuees considered it to be a frightening, unpredictable and traumatic experience, with impacts still being felt three years after the event (Thériault et al., 2021). Should these risk-reduction efforts be considered a success because there were no direct fatalities or a failure because of the massive financial costs?

Attention has turned to self-help measures, as residents are encouraged to develop a household strategy for voluntary evacuation well before a fire outbreak. The newly developed Fire Danger Rating system in Australia is a testament to this with the highest category, Catastrophic, instructing people for their survival – they must leave bushfire risk areas when this forecast is issued, regardless of whether there is fire in the landscape.

6. Disaster Aid

As with other hazard-related disasters, many people are compelled to support residents that have been impacted. The 1983 'Ash Wednesday' fires in Victoria and South Australia raised a total fund of some AU$12 million, which was channelled through an appeal fund administered by the Department of Community Welfare (Healy, 1985). About three-quarters of this sum originated within Australia itself, including federal funds released under the National Disaster Relief Arrangements. A large part of the federal assistance was in the form of interest-free repayable loans, rather than direct grants.

To ensure funds are distributed equitably and consistently and that they comply with charitable law requirements, donations are often held in separate trust accounts. For example, the Victorian Bushfire Appeal Fund (VBAF) was established following the 2009 Victorian bushfires by the Victorian Government and in partnership with the Red Cross and Commonwealth Government. The fund totalled AU$402 million, with interest, and was disseminated through Community Foundations and Trusts within the communities impacted by the 2009 fires and via a small grants program. In doing so, the VBAF aimed to ensure that long-term recovery was driven and delivered by the community. The VBAF also supported individuals and families with psychological services to help them deal with the trauma of living through such an event and housing assistance for those who had lost their homes during the fires. The VBAF (2016) Progress Report indicates that almost all of the money had been distributed after eight years.

The Australian 2019–2020 Black Summer fires also attracted significant public support, with AU$640 million raised. However, getting money out to where it is most needed, and to where donors expect their money is going, is a difficult process. The charities receiving and managing the funds work in an environment of uncertainty, not knowing how much and when money would be received or requested (ACNC, 2020). They are also restricted by governing rules on how that money can be used. For example, significant money was donated to the New South Wales Rural Fire Service (NSW RFS) Trust, but funded activities under that trust can only include support for NSW RFS, including purchasing and maintaining equipment and providing training and resources. Donations could not be used in other impacted jurisdictions elsewhere around the country, nor could they go directly to impacted residents (ACNC, 2020). Donors to this fund, however, were geographically dispersed around Australia and internationally, and were not aware that their donations would have been restricted in this way.

Disaster appeals also raise questions about compensation for insured and uninsured residents with some members of the public questioning whether those who fail to insure should be eligible to receive disaster aid. However, wildfire insurance can be prohibitively expensive and unaffordable for some, particularly in high-risk areas as insurance premiums rapidly increase following each wildfire disaster. These high prices may be seen as a mechanism to discourage homeownership in high-risk areas. However, those already living in this environment may have no choice but to relinquish their insurance cover due to escalating prices.

Insurance enables wildfire-impacted residents to replace their home and its contents along with other infrastructure that has been destroyed or damaged by smoke and fire. While insurance coverage helps to alleviate some post-disaster stress and trauma when claims and payouts are dealt with swiftly, impacted residents still suffer from long-term disruption to their lives and are susceptible to the physical and mental toll of recovery (Eriksen & de Vet, 2021). There is also the risk of underinsurance, which occurs when residents have not insured their home and contents for what they are really worth, due to financial hardship or increasing premiums or because they are simply unaware of the true value of home and contents replacement costs. In this case, underinsured residents may need to find alternative solutions to make payouts stretch across their needs, compounding existing stress levels (Eriksen & de Vet, 2021).

KEY TAKE AWAY POINTS

- Wildfires are more commonly known by the dominant vegetation type fuelling them. For example, as bushfire, brushfire, coastal fire, veldt fire, forest fire and grassfire.
- No continent, apart from Antarctica, is free from the various combinations of ignition source, fuel and weather conditions necessary for wildfire hazard.
- Wildfire risk to lives and property can be highest on the rural–urban fringe – often termed the *wildland–urban interface* (WUI) – where residential areas are intermixed with natural vegetation.
- Wildfire risk maps, typically combining fuel (type, moisture level, size, shape), topography (elevation, slope, aspect) and weather (temperature, precipitation, relative humidity, wind speed), are used to inform land use planning and building development policies and recommendations.
- Strict fire ban legislation is also implemented to help reduce possible ignition sources on days of extreme fire weather conditions.
- At the community level, volunteers are trained to prepare for and respond to wildfires threatening homes and lives. Volunteer firefighters and associated personnel also play a crucial role in raising awareness of fire risk in their communities.

BIBLIOGRAPHY

ACNC. (2020). Bushfire response 2019–20: Reviews of three Australian charities. *Australian Charities and Not-for-Profits Commission*. www.acnc.gov.au/tools/reports/bushfire-response-2019-20-reviews-three-australian-charities. Accessed 13 August 2022.

Anna Badia, David Saurí, Rufí Cerdan & Joan-Carles Llurdés (2002) Causality and management of forest fires in Mediterranean environments: an example from Catalonia, Global Environmental Change Part B: Environmental Hazards, 4:1, 23-32, DOI: 10.3763/ehaz.2002.0403

Bardsley, K. L., Fraser, A. S., & Heathcote, R. L. (1983). The second Ash Wednesday: 16 February 1983. *Australian Geographical Studies, 21*, 129–141.

Barmpoutis, P., Papaioannou, P., Dimitropoulos, K., & Grammalidis, N. (2020). A review on early forest fire detection systems using optical remote sensing. *Sensors, 20*, 6442. https://doi.org/10.3390/s20226442

Beringer, J. (2000). Community fire safety at the urban/rural interface: The bushfire risk. *Fire Safety Journal, 35*, 1–23.

Bird, D., & Taylor, A. (2021). Disasters and demographic change of 'single-industry' towns: Decline and resilience in Morwell, Australia. In D. Karácsonyi, A. Taylor, & D. Bird (Eds.), *The demography of disasters: Impacts for population and place* (pp. 125–151). Cham: Springer International Publishing.

Blanchi, R., Leonard, J., Haynes, K., Opie, K., James, M., & Oliveira, F. D. (2014). Environ mental circumstances surrounding bushfire fatalities in Australia 1901–2011. *Environmental Science & Policy, 37*, 192–203.

Bradstock, R. A., Gill, A. M., Kenny, B. J., & Scott, J. (1998). Bushfire risk at the urban interface estimated from historical weather records: Consequences for the use of prescribed fire in the Sydney region of South-Eastern Australia. *Journal of Environmental Management, 52*, 259–271.

Brotak, E. A. (1980). Comparison of the meteorological conditions associated with a major wild land fire in the United States and a major bushfire in Australia. *Journal of Applied Meteorology, 19*, 474–476.

Chapman, D. (1999). *Natural hazards*. Melbourne: Oxford University Press.

Collins, T. W. (2005). Households, forests and fire hazard vulnerability in the American West: A case study of a California community. *Environmental Hazards, 6*, 23–37.

Cova, T. J. (2005). Public safety in the urban-wildland interface: Should fire-prone communities have a maximum occupancy? *Natural Hazards Review, 6*, 99–107.

CSIRO. (n.d.). *Forest fire danger index*. https://research.csiro.au/bushfire/assessing-bushfire-hazards/hazard-identification/fire-danger-index/. Accessed 26 July 2022.

Cunningham, C. J. (1984). Recurring natural fire hazards: A case study of the Blue Mountains, New South Wales, Australia. *Applied Geography, 4*, 5–57.

Dickman, C. R. (2021). Ecological consequences of Australia's 'black summer' bushfires: Managing for recovery. *Integr Environ Assess Manag, 17*, 1162–1167. https://doi.org/10.1002/ieam.4496

Dowdy, A. J. (2018). Climatological variability of fire weather in Australia. *Journal of Applied Meteorology and Climatology, 57*(2), 221–234.

Dowdy, A. J. (2020). Seamless climate change projections and seasonal predictions for bushfires in Australia. *Journal of Southern Hemisphere Earth Systems Science, 70*(1), 120–138.

Eriksen, C., & de Vet, E. (2021). Untangling insurance, rebuilding, and wellbeing in bushfire recovery. *Geographical Research, 59*, 228–241. https://doi.org/10.1111/1745-5871.12451

Field, R. D., van der Werf, G. R., Fanin, T., Fetzer, E. J., Fuller, R., Jethva, H., Levy, R., Livesey, N. J., Luo, M., Torres, O., & Worden, H. M. (2016). Indonesian fire activity and smoke pollution in 2015 show persistent nonlinear sensitivity to El Niño-induced drought. *Proceedings of the National Academy of Sciences, 113*(33), 9204–9209.

Gibbs, L., Sia, K.-L., Block, K., Baker, E., Nelsson, C., Gilbert, J., Cook, A., & MacDougall, C. (2015). Cost and outcomes associated with participating in the community fireguard program: Experiences from the Black Saturday bushfires in Victoria, Australia. *International Journal of Disaster Risk Reduction, 13*, 375–380.

Graham, N. E. (1977). Weather surrounding the Santa Barbara fire: 26 July 1977. *Weatherwise, 30*(4), 158–159.

Handmer, J. W. (1999). Natural and anthropogenic hazards in the Sydney sprawl: Is the city sustainable? In J. K. Mitchell (Ed.), *Crucibles of hazard* (pp. 138–185). Tokyo: United Nations University Press.

Handmer, J. W., & Tibbits, A. (2005). Is staying at home the safest option during bushfires? Historical evidence for an Australian approach. *Environmental Hazards, 6*, 81–91.

Harris, S., & Lucas, C. (2019). Understanding the variability of Australian fire weather between 1973 and 2017. *PLoS One, 14*(9), e0222328. https://doi.org/10.1371/journal.pone.0222328

Haynes, K., Bird, D. K., & Whittaker, J. (2020b). Working outside 'the rules': Opportunities and challenges of community participation in risk reduction. *International Journal of Disaster Risk Reduction, 44*, 101396.

Haynes, K., Handmer, J., McAneney, J., Tibbits, A., & Coates, L. (2010). Australian bushfire fatalities 1900–2008: Exploring trends in relation to the 'prepare, stay and defend or leave early' policy. *Environmental Science and Policy, 13*, 185–194.

Haynes, K., Short, K., Xanthopoulos, G., Viegas, D. X., Ribeiro, L. M., & Blanchi, R. (2020a). Wildfires and WUI fire fatalities. In *Encyclopedia of wildfires and Wildland-Urban Interface (WUI) fires* (pp. 1073–1088). DOI: 10.1007/978-3-319-52090-2_92

Hazard Mitigation Team. (1994). Southern California firestorms. FEMA-1005-DR-CA Report, Federal Emergency Management Agency, San Francisco, CA.

Healy, P. M. (1985). The effect of bonus schemes on accounting decisions. *Journal of Accounting and Economics, 7*(1–3), 85–107.

Karter, M. J. (1992). Fire Loss in the United States during 1991. Fire Analysis and Research Division, National Fire Protection Association, Quincy, MA.

Kaskaoutis, D. G., et al. (2011). Satellite monitoring of the biomass-burning aerosols during the wildfires of August 2007 in Greece: Climate implications. *Atmospheric Environment, 45*, 716–726.

Keeves, A., & Douglas, D. R. (1983). Forest fires in South Australia on 16 February 1983 and consequent future forest management aims. *Australian Forestry, 46*, 148–164.

Lein, J. K., & Stump, N. I. (2009). Assessing wildfire potential within the wildland-urban interface: A south-eastern Ohio example. *Applied Geography, 29*, 21–34.

Lucas-Smith, P., & McRae, R. (1993). Fire risk problems in Australia. *STOP Disasters, 11*, 3–4.

Luke, R. H., & McArthur, A. G. (1978). *Bushfires in Australia.* Canberra: Australian Government Publishing Service.

Mamuji, A. A., & Rozdilsky, J. L. (2019). Wildfire as an increasingly common natural disaster facing Canada: Understanding the 2016 Fort McMurray wildfire. *Natural Hazards, 98,* 163–180. https://doi.org/10.1007/s11069-018-3488-4

McGee, T. K. (2005). Completion of recommended WUI fire mitigation measures within urban households in Edmonton, Canada. *Environmental Hazards, 6,* 147–157.

McLennan, J., & Birch, A. (2005). A potential crisis in wildfire emergency response capability? Australia's volunteer firefighters. *Environmental Hazards, 6,* 101–107.

Mercer, D. (1971). Scourge of an arid continent. *Geographical Magazine, 45,* 563–567.

Patel, T. (1995, March 11). Satellite senses risk of forest fires. *New Scientist,* 12.

Platt, R. H. (1999). Natural hazards of the San Francisco Bay mega-city: Trial by earthquake, wind and fire. In J. K. Mitchell (Ed.), *Crucibles of hazard* (pp. 335–374). Tokyo: United Nations University Press.

Romme, W. H., & Despain, D. G. (1989). The Yellowstone fires. *Scientific American, 261,* 37–46.

Rorig, M. L., & Ferguson, S. A. (2002). The 2000 fire season: Lightning-caused fires. *Journal of Applied Meteorology, 41,* 786–791.

Rose, G. A. (Ed.). (1994). Fire protection in rural America: A challenge for the future. Rural Fire Protection in America Steering Committee, Report to Congress sponsored by the National Association of State Foresters, Washington, DC.

Rural Fire Protection in America. (1994). Fire protection in rural America: A challenge for the future. Rural Fire Protection in America, Washington, DC.

Thériault, L., Belleville, G., Ouellet, M. C., & Morin, C. M. (2021, November 11). The experience and perceived consequences of the 2016 fort McMurray fires and evacuation. *Front Public Health, 9,* 641151. DOI: 10.3389/fpubh.2021.641151. PMID: 34858911; PMCID: PMC8632018.

VBA. (2022). Bushfire areas and overlays. www.vba.vic.gov.au/consumers/bushfire/areas-overlays. Accessed 25 July 2022.

VBAF. (2016). Victorian bushfire appeal fund 2016 progress report. Report prepared by the VBAF Independent Advisory Panel. www.dhhs.vic.gov.au/publications/victorian-bushfire-appeal-fund-vbaf-reports. Accessed 13 August 2022.

Voiland. (2015). *Seeing through the Smoky Pall: Observations from a Grim Indonesian Fire Season.* www.giss.nasa.gov/research/features/201512_smoke/. Accessed 24 July 2022.

Westerling, A. L., Hidalgo, H. G., Cayan, D. R., & Swetnam, T. W. (2006). Warming and earlier spring increase in Western US forest wildfire activity. *Science, 313*(18 August), 940–943.

Willis, M. (2005). Bushfires: How can we avoid the unavoidable? *Environmental Hazards, 6,* 93–99.

Woodward, E., Hill, R., Harkness, P., & Archer, R. (Eds.). (2020). Our knowledge our way in caring for country: Indigenous-led approaches to strengthening and sharing our knowledge for land and sea management. Best Practice Guidelines from Australian Experiences. NAILSMA and CSIRO.

FURTHER READING

Carroll, M. S., Higgins, L. L., Cohn, P. J., & Burchfield, J. (2006). Community wildfire events as a source of social conflict. *Rural Sociology, 71*(2), 261–280.

Finney, M. A., McAllister, S. S., Grumstrup, T. P., & Forthofer, J. M. (2021). Wildland fire behaviour: Dynamics, principles and processes. CSIRO, Clayton South, Australia.

Kadry, M. (1997). Co-operation in the Mediterranean area in fire disasters. *Annals of Burns and Fire Disasters, 10*(2), 67–75.

Paton, D. (2015). Wildfire hazards, risks, and disasters. Elsevier, Amsterdam, Nether lands.

WEB LINKS

Bushfire Help. www.bushfirehelp.org

Global Fire Monitoring Center. https://gfmc.online

Global Observations of Forest and Land Cover Dynamics (GOFC/GOLD) Fire Imple mentationTeam. https://gofcgold.org/gofc gold-fire-implementation-team

Global Wildland Fire Network. https://gfmc.online/GlobalNetworks/globalNet.html

International Association of Wildland Fire. www.iawfonline.org

Wildfire Partners. https://wildfirepartners.org

CHAPTER FIFTEEN

Epidemic Disease Hazards

15

OVERVIEW

This chapter introduces the general nature of infectious diseases and their profound impact on communities over time in the form of epidemics and pandemics. Understandings of the nature of disease – and hence how to effectively manage it – have varied over time and space. The chapter discusses some of these ideas concerning disease and its management, and the significant ways in which these ideas have shaped modern-day politics, such as the formation of monitored borders and the emergence of both a 'global health infrastructure' and 'public health'. A key theme of this discussion is the way in which a modern-day understanding of and approaches to disease management have been shaped by colonialism. A more in-depth discussion of specific infectious diseases – influenza, malaria and cholera – emphasises the complex ways in which environmental and social conditions intersect to transform a hazard into a disaster, but also highlights the efficacy of disease prevention measures. The chapter goes on to list key approaches to disease hazard reduction – from immunisation to animal and vector control – as well as the recent emergence of a 'One World, One Health' approach.

A. THE NATURE OF DISEASE

Disease is a disorder of the structure and functioning capacity of people and other organisms that produces specific symptoms, and as such is different to a direct physical injury. Of the four main classifications of disease – deficiency diseases, physiological diseases, hereditary diseases and infectious diseases – it is the last of these that has preoccupied academics and policy-makers working on environmental risk. Infectious diseases are those that are communicable between people via different modes of transmission (such as through contaminated food or water or through insect bites) and includes those that are 'zoonotic', meaning they

DOI: 10.4324/9781351261647-17

are communicable between animals and people.

Infectious diseases have had a profound impact on communities over time. The most impactful diseases now and in the past are caused by microparasites (such as viruses, bacteria and protozoa) that multiply within their hosts, causing an immediate threat to health. Diseases such as smallpox, cholera, leprosy, typhoid, typhus, plague, tuberculosis, measles, yellow fever, malaria and now coronavirus have exerted tremendous changes on the history of the world as high fatalities, but also state responses to these, have guided and even transformed the ways in which societies work.

Infectious diseases are *endemic* when they are consistently present but limited to a particular region. Malaria, for example, is considered endemic in certain countries and regions. This pattern means that disease spread and rates can be predicted to an extent, if historical and geographical records are available. The World Health Organization (WHO) defines a *disease outbreak* as:

> the occurrence of cases of disease in excess of what would normally be expected in a defined community, geographical area or season. Outbreaks are maintained by infectious agents that spread directly from person to person, from exposure to an animal reservoir or other environmental source, or via an insect or animal vector. Human behaviours nearly always contribute to such spread.
>
> (WHO website, www.emro.who.int/health-topics/disease-outbreaks/index.html)

Outbreaks can be noted when cases of a disease are higher than predicted, and they are especially noticeable when the disease is newly identified in a population. *Epidemics* are unusually large and/or unexpected outbreaks for a given place and time, but are also an indication of a lack of containment of the disease. A variety of statistics-based thresholds are used to identify epidemics by entities such as the WHO, as well as governments. A *pandemic* is a disease outbreak that occurs over a very wide region or even worldwide, crossing international boundaries and affecting a large number of people. It is also a highly charged political term, as Fischer (2020) explains with regard to the WHO's declaration of COVID-19 as pandemic:

> The formal declaration of . . . infectious disease as pandemic tells governments, agencies and aid organizations worldwide to shift efforts from containment to mitigation. It has economic, political and societal impacts on a global scale, and the WHO takes extreme care when making this determination.

The impacts of coronavirus have been many and are deeply felt (Chapter 1, Box 1.1). Though its effects are still very much underway, we can begin to see where this disease sits amidst other pandemics that have emerged over time. In Figure 15.1 the deadliest pandemics – recorded in various ways, and thus an approximation of relative deadliness – is presented, and indicates the immense scope of the 'Black Death' (a form of bubonic plague); of smallpox, which was transmitted into populations in the 'New World' via colonialism with devastating effect; of 'Spanish flu' (which is influenza), so-called because it was initially reported on by the Spanish press during World War I; and more recently the HIV/AIDS pandemic.

Bacterial, viral and parasitic infections are all capable of causing health disasters

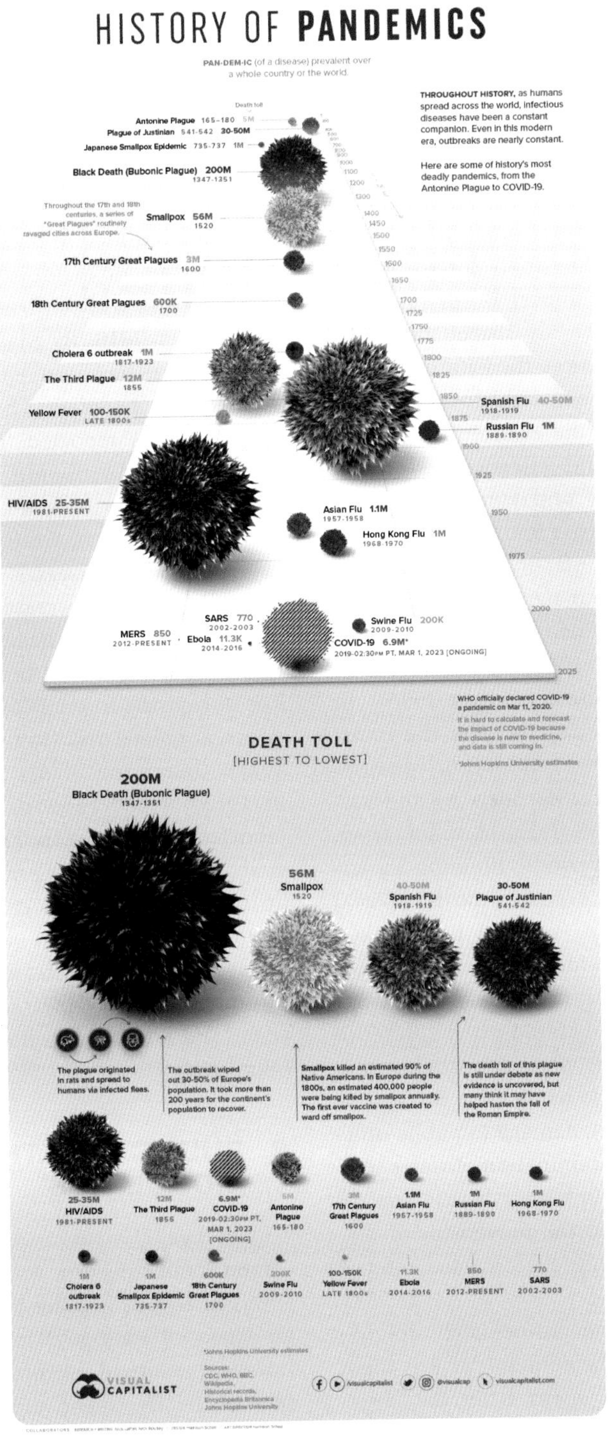

HISTORY OF PANDEMICS

PAN-DEM-IC (of a disease) prevalent over a whole country or the world.

THROUGHOUT HISTORY, as humans spread across the world, infectious diseases have been a constant companion. Even in this modern era, outbreaks are nearly constant.

Here are some of history's most deadly pandemics, from the Antonine Plague to COVID-19.

DEATH TOLL [HIGHEST TO LOWEST]

Figure 15.1 Visualising the history of pandemics.

Source: Visual Capitalist, https://www.visualcapitalist.com/history-of-pandemics-deadliest

through the transmission to people of pathogens via insects, rodents or other *vector organisms*. Some vectors (lice, bugs, fleas and certain mosquitoes) benefit from human-aided transport, such as troop movements. Mosquitoes, biting midges and flies are sensitive to weather conditions like temperature, humidity and rainfall (which provides more surface water). The scope and impact of all infectious diseases, however, are shaped by factors such as malnutrition, the nature of public health management (which includes the management of the environment as well as regulations and guidance designed to reduce exposure and vulnerability), the local level of disease immunity and previous exposure to infection.

Epidemics and pandemics are disasters in and of themselves. But they are also associated with the aftermath of other forms of disaster (Ligon, 2006). For example, floods can increase the transmission of both *water-borne* diseases (typhoid, cholera, hepatitis A) and *vector-borne* diseases (malaria, yellow fever, dengue fever). Periodic flooding associated with ENSO events in the dry coastal area of northern Peru has been linked with malaria epidemics in the region. In practice, disease outbreaks arise from many causes, including the presence of disease in the population before the event, ecological change (e.g. the spread of surface water), damage to public utilities (e.g. the contamination of water supplies), the interruption of disease-control programmes and the movement of displaced people into overcrowded refugee camps.

In the period immediately following a disaster, the epidemic risk is high. A major threat arises from the large-scale displacement of people (Watson et al., 2007). Aid agencies give priority to basic post-disaster precautions, such as disease-vector control, the management of human-waste disposal, good personal hygiene and the safe preparation of food. Cholera is a major concern because of its capacity to spread rapidly within a population. As outlined in Box 1.3 of Chapter 1, following an earthquake in January 2010, cholera was confirmed in Haiti on 21 October 2010. By March 2011, over 250,000 cases had been reported and over 4,600 people had died. The source of the epidemic was reported to be drinking water from the Artibonite River contaminated by sewage from a United Nations camp with deficient sanitation.

Vector-borne diseases – like malaria and yellow fever – increase after floods in tropical areas, due to the increase in mosquito and other insect breeding sites. A loss of housing forces people to live outdoors, at greater risk from biting insects. When hurricane 'Flora' struck Haiti in 1963, 75,000 cases of malaria were reported in the next six months (Mason et al., 1964). Interacting factors included an incomplete malaria eradication programme, the washing of insecticide from houses by heavy rain, an increase of mosquito breeding in areas of standing water and a lack of shelter for the local population.

B. PUBLIC HEALTH AND BIOPOLITICS

'Public health' – as a combination of sanitary and infection control measures – was introduced in the 19th century as an aspect of state governance. There is, however, a broad history and geography to the treatment and management of what is now regarded as infectious disease. For centuries disease formed part of a 'pestilential' world informed by religious texts and classical texts. Pestilence was a capacious term that included famine, crop failures, swarming animals and 'corrupted air'. This air could be created by noxious effluvia exhaled by the sick, which might be contagious, but also by 'miasmas' generated by the Earth through decomposing filth and putrefying marshes and stagnant water. In Classical science, these airs impacted some more than others depending on the internal 'humours' of the body – its balance of blood, phlegm, bile – which in turn was impacted by diet, exercise and moral behaviour.

Local quarantines were established in many communities to try to restrict the movement of disease, particularly at ports. From at least the 14th century quarantine (from the Italian 'quaranta,' meaning 40) was adopted as a means of separating persons, animals and goods that may have been exposed to pestilence. Quarantine camps tended to be sited on islands where ship passengers and goods could be isolated. In the 19th century, as epidemics of cholera became pandemic several times with tremendous loss of life across the globe, the passage of quarantine laws became a key part of state government. These became a feature of international borders – particularly in settler colonial countries such as Australia, New Zealand, the US and Canada – monitoring and regulating the movement of people and goods. Largely focused on non-English-speaking, and non-White-presenting people, this racialised surveillance included checks for vaccination records, including smallpox inoculation. Internally, countries such as the UK introduced Contagious Disease Acts that targeted women in port towns as sources of sexually transmitted disease and enacted forced medical examinations as well as custodial sentences for what was found to be prostitution. Moving into the 20th and even 21st centuries this focus on physical quarantine, the monitoring of the health status of people and organisms, but also the stigmatisation of certain groups, is still very much in evidence.

It is in the early 20th century that a *global health infrastructure* takes shape, as colonial powers increasingly monitored and regulated the international flows of people and things – including the unintended transportation of disease pathogens – but also undertook experiments in public health outside, as well as inside, of their borders. In the 1920s UK authorities, for example, singled out various colonies for public health and medical grants, through either the Empire Marketing Board or the Colonial Development Fund (CDF), to support long-term field studies. These experiments were concerned with retaining the health of troops stationed in colonies, who had long been impacted by epidemic disease, but also the health of local communities and their livestock and crops that, it was hoped, would become increasingly commercialised and thus produce higher tax returns. New Schools of Tropical Medicine were accordingly established in the UK, as well as hospitals and field stations in the colonies themselves (Rottenburg, 2009; Sunil et al., 2006; Tilley, 2004).

The precursor to the WHO, the League of Nations Health Organization (LNHO), was funded to deal with infectious disease and epidemics, as well as malnutrition and drug abuse, and the compilation of epidemiological records. In doing so it worked extensively with the philanthropic Rockefeller Foundation, setting a pattern for the funding of public health initiatives that is still very much in evidence today. The Rockefeller Foundation had begun to practice its own public health interventions at 'home' in the south of the US but rapidly expanded operations into colonies as well as countries under the domination of the US, working directly with local officials in Africa, Central and South America and Asia (Farmer et al., 2013).

In the process, a *disease ecology* approach was developed that hinged on examining how a disease such as malaria emerged from the interactions between microbes, vectors, human hosts and landscapes. Nevertheless, because this research was undertaken in colonial territories, the subjects of experimentation were more often than not people subject to colonisation, who were often dehumanised and afforded very little voice in how infectious disease and its control were investigated. What is more, there is evidence that in some of this research infectious disease was deliberately introduced into vulnerable groups to produce experimental data (Box 15.1).

These developments point to a much more expansive history of public health than that usually offered, which tends to centre new technologies, such as new forms of inoculation and water purification, and administrative reforms such as health boards and sanitary boards. Indeed, an awareness of how such reforms were shaped by prevailing (though not uncontested) actions and ideas around race and gender might be better termed *biopolitics*, which analyses the 'how', 'why' and 'with what effect' of (1) the construction of statistical populations and subpopulations along dimensions such as gender and race, as well as age and occupation, which factors then become 'baked in' to policy formulation; and (2) the simultaneous hollowing out of social explanations of epidemic disease and the encouragement to individuals to monitor and change their behaviours so as to become responsible citizens. In reviewing the challenges of managing epidemic disease today, and in the future, funding for public health services is vital for *disease prevention* by means of the greater use of clean water, sanitation, safe food, prophylactic drugs and immunisation, health education and mass screening. But, it is important to be aware of this broader biopolitical history and the problems that have been created.

C. INFECTIOUS DISEASE AND WEATHER

As the preceding sections indicate, there is a complex relationship between the pathogens that cause infection and the vectors that transmit these; the environments within which parasites and vectors live; and the conditions that exacerbate the vulnerability of specific individuals and groups to exposure and morbidity/mortality from disease. When the pathogen develops in the environment – or in an intermediate host or vector – rather than in the human body, it comes under the influence of the prevailing weather conditions. Most viruses, bacteria and parasites are unable to complete their development if the ambient temperature is below a certain threshold, for example 18°C for the malaria parasite. Conversely, temperatures slightly above a given threshold will shorten the time

Box 15.1 Creating a Disease Outbreak; Introducing Sexually Transmitted Diseases (STDs)

The US Public Health Service Sexually Transmitted Disease Inoculation Study of 1946–1948 hinged on the deliberate creation of a disease outbreak. In 1947 staff members from the US Public Health Service – concerned with reducing STD infections among US troops stationed abroad – intentionally exposed over 1,600 sex workers, soldiers, prisoners and psychiatric patients in Guatemala to STDs to test the effectiveness of their prophylactic intervention. As Spector-Bagdady and Paul A. Lombardo explain, 'There are no records indicating that consent was obtained from the participants, and there is evidence that some were, in fact, deceived. In addition, 83 subjects died during the experiments, although the connection between the deaths and involvement in the experiments is unclear' (Spector-Bagdady & Lombardo, 2019, pp. 30–31). The lead investigator, John C. Cutler, was also responsible for leading the 1932–1972 public health study in Tuskegee, Alabama, that allowed syphilis in nearly 400 poor African-American men to progress without treatment. In Guatemala, Cutler

> also began gonorrhoea experiments in the Guatemalan military in February 1947 and, over a year and a half, exposed almost 600 soldiers to disease. Methods of deliberate exposure included having the soldiers have intercourse with infected sex workers and using a needle to insert gonorrheal pus taken from one man into the penis of another . . . Guatemalan physicians actively assisted with chancroid transmission, scratching soldiers' arms and rubbing infection into their wounds.
>
> (Spector-Bagdady & Lombardo, 2019, p. 31)

In 2010, after details of the study were researched and released by historian Professor Susan Reverby, an apology was offered by then president Barack Obama, then secretary of state Hillary Clinton and then secretary of health and human services Kathleen Sebelius to their Guatemalan counterparts.

Further Reading

Gutmann, Amy, et al. 'Presidential commission for the study of bioethical issues.' Moral science: protecting participants in human subjects research. Washington DC (2011).

Spector-Bagdady, Kayte, and Paul A. Lombardo. 'US Public health service STD experiments in Guatemala (1946–1948) and their aftermath.' Ethics & Human Research 41.2 (2019): 29–34.

needed for development and increase reproduction rates and the risk of a disease outbreak.

This section focuses more directly on the linkages between infectious disease and weather. Table 15.1 summarises some of these links between weather and disease, while also noting the role of other factors that help make a disease hazard into a disaster. It can be seen that cholera and malaria have *important* links with weather whilst the seven other diseases

Table 15.1 Selected infectious diseases with epidemic potential, in which climate plays at least a moderate role, ranked according to global significance

Disease	Mode of transmission	Distribution	Weather–epidemic link	Strength of weather link and comments
Influenza	Air borne	Worldwide	Decrease in temperature (winter)	Moderate A range of human-related factors more significant
Diarrhoeal diseases	Food and water borne	Worldwide	Increases in temperature and decreases in rainfall	Moderate Sanitation and human behaviour more important
Cholera	Food and water borne	Africa, Asia, Russian Federation, South America	Increases in sea and air temperature, plus El Niño events	Important Sanitation and human behaviour also important
Malaria	Bite of female *Anopheles* mosquitoes	Epidemic in over 100 countries	Changes in temperature and rainfall	Important Local factors also relevant
Meningococcal meningitis	Air borne	Worldwide	Increases in temperature and decreases in humidity	Significant
Lymphatic filariasis	Bite of female *Culex*, *Anopheles* and *Mansonia* mosquitoes	Africa, India, South America, South Asia and Pacific Islands	Temperature and rainfall affect distribution	Moderate
Leishmaniasis	Bite of female phlebotomine sandflies	Africa, Central Asia, Europe, India, South America	Increases in temperature and rainfall	Significant
African trycomopariasis	Bite of male and female tsetse flies	Sub-Saharan Africa	Changes in temperature and rainfall	Moderate Cattle density and vegetation patterns relevant
Dengue	Bite of female *Aedes* mosquitoes	Africa, Europe, South America, Southeast Asia, Western Pacific	High temperature, humidity and rainfall	Significant Other factors also important
Japanese encephalitis	Bite of female *Culex* and *Aedes* mosquitoes	Southeast Asia	High temperature and heavy rains	Significant Animal factors important

St Louis encephalitis	Bite of female *Culex* and *Aedes* mosquitoes	North and South America	High temperature and heavy rain	Moderate Animal factors important
Rift Valley fever	Bite of female culicine mosquitoes	Sub-Saharan Africa at end of epidemic	Heavy rains, cold weather	Significant Animal factors important
West Nile virus	Bite of female culicine mosquitoes	Africa, Central Asia, Southwest Asia, Europe	High temperatures and heavy precipitation at onset	Moderate Non-climatic factors may be more important
Ross River virus	Bite of female culicine mosquitoes	Australia and Pacific Islands	High temperatures and heavy precipitation at onset	Significant Host immune factors and animals important
Murray Valley fever	Bite of female *Culex* mosquitoes	Australia	Heavy rains and below-average atmospheric pressure	Significant
Yellow fever	Bite of female *Aedes* and *Haemagogus* mosquitoes	Africa, South and Central America	High temperature and heavy rain	Moderate Population factors important

Source: After Kuhn et al. (2005). Reproduced from K. Kuhn et al., Using Climate to Predict Infectious Disease Epidemics (World Health Organization, 2005). With permission of the World Health Organization.

Notes: Global significance assessed on the calculation of disability-adjusted life years (DALYs). Strength of climate link assessed on 5-point scale: weak, moderate, significant, important, primary.

listed have a *significant* link. Of course, weather conditions are never the sole driver of disease epidemics. In almost all cases temperature conditions are relevant; increases in temperature associated with high rainfall or humidity are the most common combinations.

Box 15.2 provides a more in-depth look at how a specific group – or 'family' – of diseases are linked with weather conditions.

D. SPECIFIC DISEASE HAZARDS

In this section we focus on specific infectious diseases that have become pandemic and have had a tremendous impact on societies over time.

1. Influenza

Influenza is one of the world's most common and most deadly diseases. 'Flu' has been known for over 2,000 years; the first well-described pandemic was in 1580. A respiratory illness, there are actually four influenza 'types' reflecting small differences in the virus's antigenic proteins; only two of these, A and B, are of clinical relevance to people. While influenza B infections occur only between people, influenza A viruses also circulate in animal populations, such as birds, pigs and horses. All of these animal viruses are distinct from human influenza viruses, however, and so while some may occasionally infect humans – causing symptoms ranging from mild conjunctivitis to severe pneumonia and death – this is relatively rare. Human infections of zoonotic influenza are acquired through direct contact with infected animals or contaminated environments and do not spread quickly between people. Over the past decades there have been instances of sporadic transmission of influenza viruses between animals and humans, such as the 2009 swine flu outbreak.

Pandemics result from the introduction into human populations of influenza A subtypes that many people do not have a 'learned' immune response to, such as the 'Asian influenza' of 1957 and the 'Hong Kong flu' of 1968. During the 1918–1919 'Spanish flu' pandemic many more died of influenza than in World War I, though as Figure 15.2 shows estimates vary greatly on the total number. Lung damage was the major cause of these deaths, at a time when antibiotics were unavailable.

Most people recover from the infection, but it can cause fatal complications like pneumonia. Seasonal influenza epidemics occur regularly both in the northern and the southern hemispheres each winter, causing approximately 500,000 deaths per year worldwide (Fauci, 2006). In 1984, the World Health Organization (WHO) established the Global Influenza Surveillance Network. Tasked with conducting year-round surveillance of influenza, assessing the risk of pandemic influenza and assisting in preparedness measures, the network currently comprises institutions in 124 WHO member states, from national influenza centres (NICs) that collect virus specimens in their country to reference and regulatory laboratories.

2. Malaria

This is the world's major vector-borne parasitic disease. Single-celled parasites – *Plasmodium* – are pathogenic and are transmitted from person to person by infected female mosquitoes. Their bite draws blood into their bodies and nurtures their eggs; it also allows the parasite to enter an individual's blood system, causing

Box 15.2 The Flaviviruses

One of the most significant groupings of diseases impacting people are the Flavivirus family. 'Flavus' is Latin for yellow, and the diseases are associated with jaundice and the yellowing of a patient's skin. More than 70 flaviviruses have been identified, of which about half cause disease in humans. The natural viral host is local wildlife and generally the disease is carried to humans by arthropod (insect) vectors – mosquitoes in the tropics and ticks in the higher latitudes. Tick-borne flaviviruses are less important for human disease than mosquito-borne viruses because tick species feed on animals in the wild and tick-borne diseases tend to be more restricted geographically. Many mosquito-borne diseases have been known for centuries but have recently upsurged, due to combinations of climate change, the rolling back of public health measures, increased socio-economic insecurities under austerity and viral factors.

Dengue fever, for example, has been widespread in the tropics for over 200 years, with intermittent pandemics emerging at roughly 10- to 40-year intervals. It is now the most widely distributed mosquito-borne disease of humans. Together with dengue haemorrhagic fever (DHF), it is caused by one of four related virus serotypes of the genus *Flavivirus*. Infection with one serotype provides no immunity against any of the other three. It is also unusual in that humans are the natural hosts for the virus. Humans are infected by *Aedes aegypti*, a domestic, daytime-biting mosquito. The population density of this insect is highly dependent on human habitation. Water-storage facilities and the availability of breeding sites around residential buildings are key factors in promoting the disease.

Emergence during the late 20th century is attributed to poor vector control, overcrowding of refugee and urban populations and more frequent international travel. For a time, dichlorodiphenyltrichloroethane (DDT) sprays eliminated *Aedes aegypti* from many countries, but dengue fever epidemics have increased, partly because effective mosquito control has been relaxed in areas where it is endemic. In the Pacific region, dengue viruses were reintroduced in the 1970s after an absence of over 25 years, and there has been a re-emergence in Central and South America, where the geographical distribution is believed to be wider than it was before the eradication programme. For the patient, dengue fever produces a range of viral symptoms capable of developing into severe and fatal haemorrhagic disease. There are approximately 100 million cases of infection per year, with 2.5 billion people at risk (Ligon, 2004). No dengue vaccine is available.

Yellow fever has been an important tropical disease for nearly 500 years. As early as 1908 the deliberate suppression of mosquito breeding sites had eliminated yellow fever from many urban centres. However, in 1932 the disease was found to have an independent *zoonotic* (animal-borne) transmission cycle involving monkeys. There are three types of transfer:

- *Sylvatic (jungle) yellow fever* occurs in tropical rainforests when monkeys become infected by wild mosquitoes, which pass on the virus when bitten by other mosquitoes. The infected wild mosquitoes then bite humans in the forest, such as timber workers. Disease incidence is low, due to the sparse population, but the virus can be transferred to unvaccinated inhabitants of nearby towns.

- *Intermediate yellow fever* occurs mainly in the savannahs of Africa, where semi-domestic mosquitoes infect both monkey and human hosts to create small epidemics. Infected mosquito eggs can survive several months of drought before hatching in the rainy season, so the virus is well suited to this climatic environment. Increased contact between humans and infected mosquitoes in the wet-and-dry tropics, where water projects and other developmental changes increase mosquito density, is a major cause of African outbreaks.
- *Urban yellow fever* of epidemic proportions typically occurs when migrants introduce the virus into crowded townships, where the disease spreads by domestic mosquitoes directly from person to person. In the savannah areas of Africa, water is commonly stored in large earthen pots, and the consequent high rates of household breeding for *Aedes eygypti* have been implicated in several yellow fever epidemics in Senegal, Ghana, the Gambia, Côte d'Ivoire, Nigeria and Mauritania during the 1965–1987 period.

An estimated population of over 500 million people, living in Africa between latitude 15°N and 10°S of the equator, is at risk of yellow fever infection, whilst the disease is endemic in nine South American countries and some Caribbean islands. There are an estimated 200,000 cases of yellow fever per year, with 30,000 deaths, but the disease is significantly under-reported. There is no recognised treatment for yellow fever. The most important preventive measure is a highly effective vaccine. For adequate protection 80% of the population should be vaccinated, but the immunisation cover is below 40% in most countries where yellow fever is endemic.

West Nile virus was not recognised until 1937, when it was clinically isolated in the West Nile district of northern Uganda (Campbell et al., 2002). It is endemic in Africa, Asia, Europe and Australia and was introduced into the USA in 1999 via an outbreak in New York City. There have been several West Nile fever epidemics, notably that of 1973–1974 in South Africa. The virus is maintained in endemic disease areas through a mosquito–bird–mosquito transmission cycle. The transfer of the disease to new areas is mainly by migratory birds. The incubation period for West Nile fever is typically from two to six days and, in the worst 15% of cases, the development of encephalitis leads to coma. In some areas of Africa, immunity to West Nile virus is thought to reach 90% in adults. But in Europe and North America, where the disease is likely to become more prevalent, such background immunity is almost non-existent.

illness and at worst coma and death. Until the 20th century malaria was common in Europe, the eastern seaboard of North America, northern Australia, parts of the Caribbean and much of South America. This pattern was changed, however, as lands were cleared and drained, insecticides such as DDT became prevalent and housing was adapted via, for example, the use of window screens, and mosquito

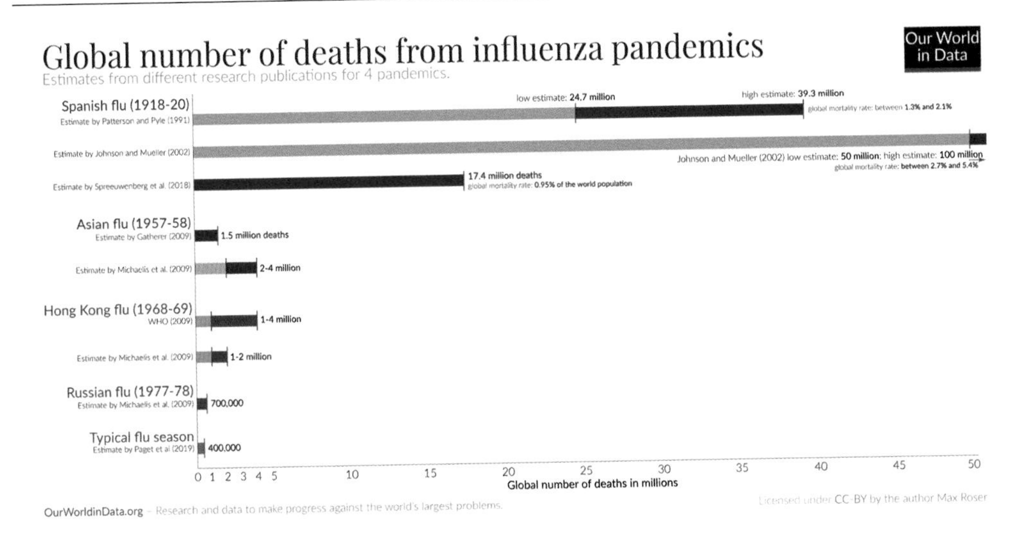

Figure 15.2 Global number of deaths from influenza pandemics.

Source: https://ourworldindata.org/spanish-flu-largest-influenza-pandemic-in-history

Figure 15.3 Changing prevalence of malaria over time and space.

Source: https://ourworldindata.org/malaria

habitats and habits were changed (Figure 15.3). In the 21st century malaria remains a pressing problem in central Africa and the Amazon region, the Indian subcontinent and Southeast Asia. In these areas it is one of the leading causes of child mortality, especially among those under five years old (Figure 15.4).

Epidemics of malaria are major disasters and are caused by many factors, including an influx of non-immune people into endemic areas, a break in normal

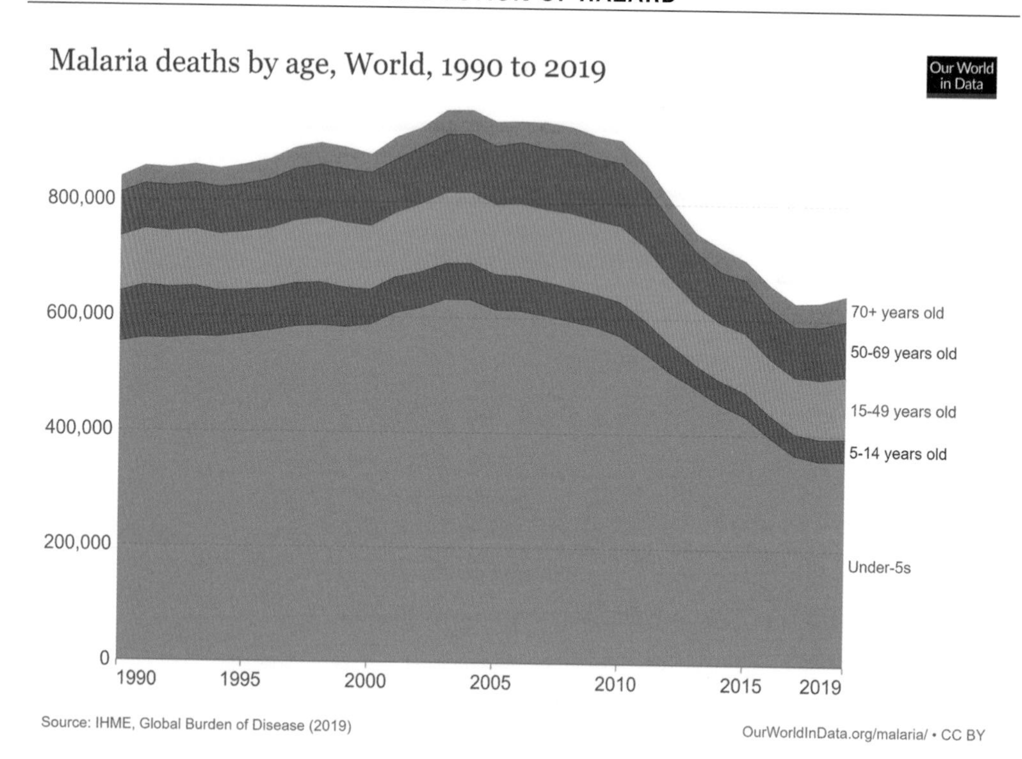

Figure 15.4 Deaths from malaria by age, 1900 to 2019.
Source: https://ourworldindata.org/malaria

malaria control measures and an increase in associated diseases or malnutrition. Environmental changes, like deforestation, irrigation and flooding, also play a part (Guintran et al., 2006). Cities with unplanned and expanding peripheries, which encroach upon the surrounding malaria-infested rural areas, are at high risk. Key weather influences are:

- *Temperature.* At ambient temperatures below 20°C the malaria parasite has a limited life cycle, although it can exist in small, warmer locations such as houses. As the temperature rises from 21°C to 27°C it thrives because the time required for development progressively decreases.
- *Rainfall. Anopheline* mosquitoes require bodies of surface water for breeding. Adequate rainfall is essential, but survival rates depend on the season and on rainfall intensity, e.g. heavy rainfall can wash away the breeding larvae.
- *Relative humidity.* Atmospheric humidity is normally high in the wet season and during spells of rainfall. If the relative humidity falls below 60%, the life of the mosquito is shortened and transmission of the parasites to humans is limited.

In Africa, weather-related epidemics of malaria afflict mostly highland or semi-arid areas and occur in two- to seven-year cycles linked to climate variability

(Pascual et al., 2008). Similar links have been established in other regions, such as Sri Lanka (Briët et al., 2008) and Thailand, where outbreaks of malaria are closely associated with seasonal rains, although farming practices and population movements are involved too (Wiwanitkit, 2006; Childs et al., 2006). In Venezuela and elsewhere, outbreaks of malaria have been linked to El Niño Southern Oscillation (ENSO) fluctuations.

While efforts to control malaria have been in evidence for a long time – primarily involving insecticides and the use of bednets – a key shift came in 2021. The WHO endorsed a malaria vaccine for 'broad use' in children, which had been trialled in 2019 in Malawi, Kenya and Ghana as part of regular inoculation programmes.

3. Cholera

Cholera is an acute intestinal infection caused by the bacterium *Vibrio cholerae*. This bacterium is part of the flora found in brackish water and estuaries in tropical areas and is spread to humans via contaminated food and water. The disease is endemic in tropical latitudes (Figure 15.5). During the 19th century cholera followed troop movements from its original source in the Ganges delta of India to abroad, and around the world. The current pandemic began in South Asia in 1961, reached Africa in 1971 and the Americas in 1991. Up to 6 million cases of infection are reported annually. About 10% of cases lead to severe dehydration that, without treatment, can lead to death. Each year there are 1.3 to 4.0 million cases of cholera, and 21,000 to 143,000 deaths worldwide due to cholera. Most fatalities occur in people with low immunity and with no access to oral rehydration solutions, intravenous fluids or antibiotics.

Cholera has an impressive ability to spread rapidly into new areas and to become epidemic. This is because *Vibrio cholerae* is so common in the aquatic environments of the tropics and has a very short incubation period of between two hours and five days. Consequently, there is a rapid onset of the early symptoms of diarrhoea and vomiting. Like most infectious diseases, it is particularly dangerous when newly introduced into an unprepared area, where the fatality rates can rise to 50% of infections. For example, in 1970 and 1991 the disease struck West Africa and Latin America respectively, although both regions had not suffered major outbreaks for around 100 years. It is now endemic in both continents.

In 2014 the WHO's Global Task Force on Cholera Control (GTFCC) was reactivated. The GTFCC is a network of more than 50 partners – including academic institutions, nongovernmental organisations and United Nations agencies – working on the design and implementation of strategies to contribute to capacity development for cholera prevention and control globally.

E. RISK ASSESSMENT AND REDUCTION

As with other environmental hazards, reducing the risks from infectious diseases depends on long-term strategies of disaster risk reduction, focusing on prevention but including control.

1. Protection

Where an effective vaccine is available, immunisation is the best approach to infectious disease. For example, a good

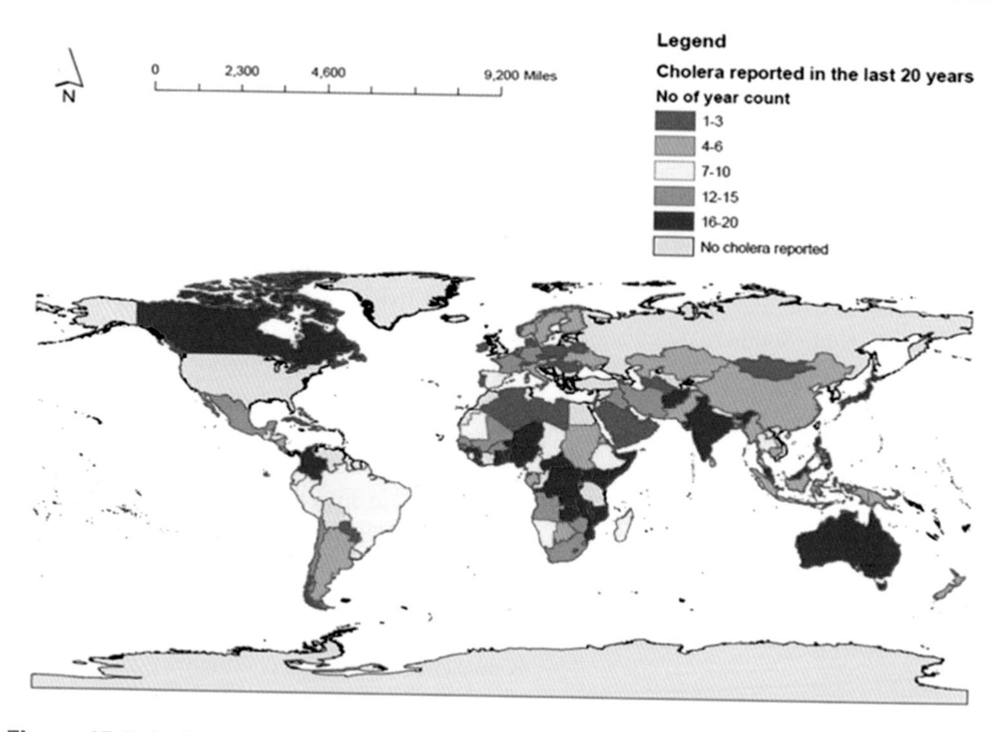

Figure 15.5 Cholera map showing the number of years of reported cholera patients in WHO database.
Source: Jutla et al. (2017).

vaccine exists for yellow fever. This provides immunity within one week for 95% of people vaccinated. The immunity lasts at least ten years and has few side effects. Following the success of global immunisation programmes to eradicate smallpox in the 1960s and 1970s, the WHO reorganised its emergency division in 1993 with a view to providing more effective responses to epidemics. Mass vaccinations, covering 80% of the population, would save millions of lives each year from common diseases like tuberculosis, measles, whooping cough, tetanus and diphtheria.

Vector control has been the other key element in disease prevention. For example, special attempts are made to suppress mosquitoes by pesticide applications at the beginning of a disease outbreak, before vaccination can take effect. Such responses could be improved by more detailed maps showing mosquito breeding sites and preparedness campaigns designed to increase local awareness of such sites. Although liquid pesticides can be effective in controlling diseases like malaria, the longer-term ecological effects of large-scale applications of agents such as DDT are worrying. According to the US Centers for Disease Control and Prevention, DDT and its related chemicals persist for a long time and have multiple health impacts:

> DDT and its related chemicals persist for a long time in the environment and in animal tissues. . . . Following

Figure 15.6 A man uses a 'fogging' technique to spray a village in northern Thailand with insecticide against mosquitoes. Such aerial spraying has had only limited success in the control of malaria.

Source: (Photo: Panos/William Daniels WDA00040THA

exposure to high doses, human symptoms can include vomiting, tremors or shakiness, and seizures. Laboratory animal studies show DDT exposure can affect the liver and reproduction. DDT is a possible human carcinogen.

(CDC National Biomonitoring Program Website, www.cdc.gov/biomonitoring/DDT_FactSheet.html#:~:text=Following%20exposure%20to%20high%20doses,to%20U.S.%20and%20International%20authorities)

2. Mitigation

Surveillance is critical for the prompt recognition and control of an emerging epidemic, especially in areas where there is poor vaccination cover. The longer an epidemic remains undetected and uncontrolled, the higher the morbidity and mortality. Cholera was the first disease for which surveillance and reporting was employed on a large scale. If a disease is common – like cholera –it will be easily diagnosed and appropriate responses will quickly follow. Other diseases, like dengue fever, are more difficult to identify. They can spread quickly into areas with high concentrations of *Aedes aegypti* mosquitoes and produce new virus strains and serotypes against which local resistance is limited. The use of remotely sensed data in association with terrain analysis is now providing more timely information on vector breeding habitats for malaria and other diseases.

Once a disease outbreak is suspected, there is a need for rapid access to laboratory testing and diagnostic facilities. Guintran et al. (2006) described the requirements for early detection of malaria epidemics in Africa, which include improved data collection, the use of early indications in the population and a better definition of epidemic thresholds. Here the overall aim is to progress from surveillance to early warning.

Treatment can be effective in dealing with the clinical symptoms of some infectious diseases. In the case of malaria, it has been suggested that infants and children in Africa should be treated even before they show any symptoms (Vogel, 2005). Simple treatments can be effective. For example, the dehydration and fever associated with yellow fever can be treated with oral rehydration salts and paracetamol, although any bacterial infection will need antibiotics. Unfortunately, many patients die before reaching hospital – often through a lack of local transport.

Reliable hospital therapies are not available for all diseases. For example, malaria parasites quickly become resistant to drugs. Resistance to chloroquine is high, particularly in Southeast Asia, and there is a need for new antimalarial drugs. In other instances, such as cholera, an oral vaccine may be available, but in such small quantities that it is used for individual travellers rather than the public as a whole. Because public health facilities are under threat in so many countries, there is a need for international partnerships capable of taking a longer-term view.

3. Forecasts and Warning

Early warning systems exist for many health hazards. For example, following the heatwaves in summer 2003, a number of European countries introduced heat-health warning systems (Matzarakis et al., 2011). In 2000 the WHO initiated the *Global Outbreak and Response Network* to identify unusual agents and pathogens, in order to improve international rapid responses. The prospect of effective early warning systems (EWS) for disease epidemics based on seasonal climate forecasts, or other environmental indices, provides one of the most hopeful ways forward (World Health Organization, 2005). This is because forecasts of an epidemic many months in advance would – in theory – justify increased surveillance and preparedness in the areas at risk. Thus, cholera epidemics may be predictable by monitoring 'environmental signatures' such as the seasonal abundance of zooplankton and chlorophyll and water temperatures using remotely sensed vegetation images (Colwell, 1996; Constantin de Magny et al., 2008).

4. The WHO and One Health

Since its establishment in 1948 the WHO has relied on contributions from member states (both assessed and voluntary) and private donors. According to Kelland, writing for Reuters,

> In all, 93 percent of the money given to the WHO for its health programmes is now 'specified' – earmarked – by donors for particular projects. The largest of the non-government donors is the Bill & Melinda Gates Foundation . . . contributing between $250 million and $300 million a year. In one year – 2013 – it was the largest donor bar none, overtaking even total contributions from the U.S. government.
>
> (Kelland, 2016)

One of the Gates Foundation's priorities is the eradication of polio, Kelland continues, and this is by far the best-resourced WHO programme, 'accounting for 23.5 percent of the WHO's current programme budget' (ibid, no page).

A current priority for the WHO is the realisation of a 'One World, One Health' approach to epidemic disease in people that links this with plant and animal health, food security and antibiotic resistance strategies, as well as surveillance strategies and pandemic preparedness. This effort to cut across diverse domains of expertise ensues in large part from the perceived rise in emerging and re-emerging infectious diseases, a majority of which are zoonotic, and many of which are either food borne or vector borne, and are thus entangled with the health of animal, plants and ecologies. But also there is awareness of the drawbacks of 'silo thinking' with regard to environmental risks, as different agencies focus on specific aspects of what are interconnected events (Hinchliffe, 2017).

The 'One Health' approach advocated by the WHO has been visualised in a number of ways, but perhaps the most effective is that provided by *The Lancet* (Figure 15.7), which has its own One Health commission.

The One World, One Health approach promises much in regard to the 'joined up' thinking that infectious disease, but also all environmental hazards, requires in order to prevent, mitigate and respond to disasters. Nevertheless, such efforts are difficult to realise when hazard-centred organisations have their own particular remit and criteria for success, political administrations and parties have their manifestos and constituencies, funders have their specific priorities, and experts have their own particular foci and skills. It is not too surprising to find, then, that by now well-entrenched approaches to disease prevention – animal culling, containment via quarantines, and the monitoring of sanitation and hygiene measures – remain at the heart of One World, One Health. What is more, the COVID-19 pandemic – as well as the resurgence of diseases thought to be contained and on the wane – has very much hit home the fact that global infrastructures of disease management need considerable work.

KEY TAKE AWAY POINTS

- Infectious disease has been, and continues to be, a world-shaping environmental hazard, impacting societies across time and space. Epidemics and pandemics have produced a profound loss of life, the scale of which can be gleaned from a range of historical records.
- 'Public Health' – comprising infection control and sanitation measures – is a relatively recent set of policies and practices that builds on centuries of disease management. Though public health policies and practices have had a significant impact on disease management, these have also been shaped by ideologies that stigmatise particular groups because of racism, sexism, ableism and classism.
- There is a direct link between the prevalence of specific infectious diseases and weather, though this relationship is complicated by social factors that impact exposure and vulnera bility.
- Disease hazard reduction comprises a number of approaches, including immunisation, early detection and forecasting, and containment and culling

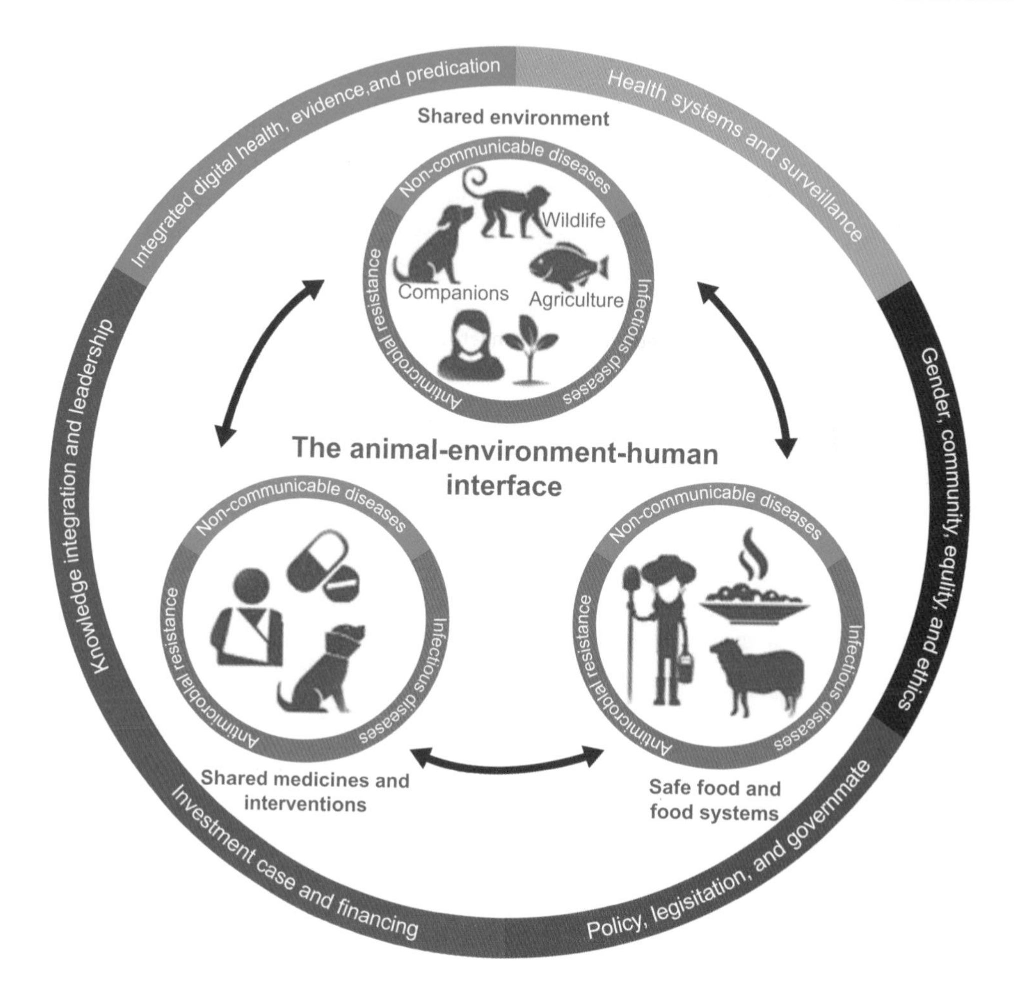

Figure 15.7 Approach of the One Health Lancet Commission.

Source: Amuasi et al. (2020)

(of animals), each managed by different organisations and networks and undertaken by experts trained in particular specialities. Though efforts are underway to 'join up' organisations and experts, there remains much work to be undertaken.

BIBLIOGRAPHY

Amuasi, J. H., et al. (2020). Reconnecting for our future: The lancet one health commission. *The Lancet, 395*(10235), 1469–1471.

Briët, O. J., Vounatsou, P., Gunawardena, D. M., Galappaththy, G. N., & Amerasinghe, P. H. (2008). Models for short term malaria prediction in Sri Lanka. *Malaria Journal, 7*(1), 1–11.

Campbell, G. L., et al. (2002). West Nile virus. *The Lancet Infectious Diseases, 2*(9), 519–529.

Childs, D. Z., Cattadori, I. M., Suwonkerd, W., Prajakwong, S., & Boots, M. (2006). Spatiotemporal patterns of malaria incidence in northern Thailand. *Transactions of the Royal Society of Tropical Medicine and Hygiene, 100*(7), 623–631.

Colwell, R. R. (1996). Global climate and infectious disease: The cholera paradigm. *Science, 274*(5295), 2025–2031.

Constantin de Magny, G., Murtugudde, R., Sapiano, M. R., Nizam, A., Brown, C. W., Busalacchi, A. J., Yunus, M., Nair, G. B., Gil, A. I., Lanata, C. F., & Calkins, J. (2008). Environmental signatures associated with cholera epidemics. *Proceedings of the National Academy of Sciences, 105*(46), 17676–17681.

Farmer, P., et al. (2013). *Reimagining global health: An introduction* (Vol. 26). USA: University of California Press.

Fauci, A. S. (2006). Seasonal and pandemic influenza preparedness: Science and countermeasures. *The Journal of Infectious Diseases, 194*(suppl 2), S73–S76.

Fischer, R. B. (2020, March 9). What's the difference between pandemic, epidemic and outbreak? *The Conversation*. https://theconversation.com/whats-the-difference-between-pandemic-epidemic-and-outbreak-133048

Guintran, J.-O., et al. (2006). Systems for the early detection of malaria epidemics in Africa: An analysis of current practices and future priorities. WHO Report WHO/HTM/MAL/2006.1115. https://apps.who.int/iris/bitstream/handle/10665/43584/9789241594882_eng.pdf

Hinchliffe, S. (2017). More than one world, more than one health: Re-configuring inter-species health. In *Global health and geographical imaginaries* (pp. 159–175). Abingdon, UK: Routledge.

Jutla, A., Khan, R., & Colwell, R. (2017). Natural disasters and cholera outbreaks: Current understanding and future outlook. *Current Environmental Health Reports, 4*(1), 99–107.

Kelland, K. (2016, February 8). The World Health Organization's critical challenge: Healing itself. *Reuters Investigates*. www.reuters.com/investigates/special-report/health-who-future/

Ligon, B. L. (2004, July). Emerging and re-emerging infectious diseases: Review of general contributing factors and of West Nile virus. In *Seminars in pediatric infectious diseases* (Vol. 15, No. 3, p. 199). Elsevier.

Ligon, B. L. (2006, April). Reemergence of an unusual disease: The chikungunya epidemic. In *Seminars in pediatric infectious diseases* (Vol. 17, No. 2, pp. 99–104). WB Saunders.

Mason, J., Cavalie, P., & World Health Organization. (1964). *Malaria epidemic in Haiti following a hurricane* (No. WHO/Mal/483.64). World Health Organization.

Matzarakis, A., Muthers, S., & Koch, E. (2011). Human biometeorological evaluation of heat-related mortality in Vienna. *Theoretical and Applied Climatology, 105*, 1–10.

Pascual, M., Cazelles, B., Bouma, M. J., Chaves, L. F., & Koelle, K. (2008). Shifting patterns: Malaria dynamics and rainfall variability in an African highland. *Proceedings of the Royal Society B: Biological Sciences, 275*(1631), 123–132.

Rottenburg, R. (2009). Social and public experiments and new figurations of science and politics in postcolonial Africa. *Postcolonial Studies, 12*(4), 423–440.

Spector-Bagdady, K., & Lombardo, P. A. (2019). US Public health service STD experiments in Guatemala (1946–1948) and their aftermath. *Ethics & Human Research, 41*(2), 29–34.

Sunil, T. S., Rajaram, S., & Zottarelli, L. K. (2006). Do individual and program factors matter in the utilization of maternal care services in rural India? A theoretical approach. *Social Science & Medicine, 62*(8), 1943–1957.

Tilley, H. (2004). Ecologies of complexity: Tropical environments, African trypanosomiasis, and the science of disease control in British colonial Africa, 1900–1940. *Osiris, 19*, 21–38.

Vogel, G. (2005). Will a Preemptive Strike Against Malaria Pay Off? *Science, 310*(5754), 1606–1607.

Watson, J. T., Gayer, M., & Connolly, M. A. (2007). Epidemics after natural disasters. *Emerging Infectious Diseases, 13*(1), 1.

Wiwanitkit, V. (2006). An observation on correlation between rainfall and the prevalence of clinical cases of dengue in Thailand. *Journal of Vector Borne Diseases, 43*(2), 73.

World Health Organization. (2005). *Using climate to predict infectious disease epidemics* (pp. 21–24). Geneva: WHO

Document Production Services. https://apps.who.int/iris/bitstream/handle/10665/43379/9241593865.pdf

WEB LINKS

Centers for Disease Control and Prevention, USA. www.cdc.gov

Health and Disaster Management Handbook. https://knowledge.aidr.org.au/resources/health-and-disaster-management-handbook

Health in Emergency and Disaster Research Center. https://rcedh-en.uswr.ac.ir

Pan American Health Organization. www.paho.org/disasters

Visual Capitalist, Visualising the History of Pandemics. www.visualcapitalist.com/history-of-pandemics-deadliest/

World Health Organization. www.who.int

Part Three

ENVIRONMENTAL HAZARD AND RISK IN AN ANTHROPOCENE

A new version of the image known as 'the Pale Blue Dot'. The image was processed by JPL engineer and image processing enthusiast Kevin M. Gill with input from two of the image's original planners, Candy Hansen and William Kosmann.

Source: NASA/JPL-Caltech

DOI: 10.4324/9781351261647-18

There is perhaps no better demonstration of the folly of human conceits than this distant image of our tiny world. To me, it underscores our responsibility to deal more kindly with one another, and to preserve and cherish the pale blue dot, the only home we've ever known.

– Carl Sagan, Pale Blue Dot:
A Vision of the Human Future in Space

CHAPTER SIXTEEN

Anthropocene Challenges

16

OVERVIEW

This chapter explores the multi-scale, wide-ranging changes that human activity is causing locally and planet-wide. We begin by introducing the term 'Anthropocene' along with some of its critics and a selection of alternatives. We outline the emergence of this term in scientific debates on Earth system science, including climate change science. We then discuss the ways in which communities across the globe are faced with both slow- and fast-moving hazards and the assessment of multiple risks that need to be communicated and managed, as well as the many uncertainties. We emphasise the geographic diversity with regard to who is vulnerable and how and the importance of identifying which social processes are responsible for producing disaster conditions. Finally, we note that as the hazards and risks associated with an Anthropocene become the subject of inquiry across academia – and well beyond the hazard and risk subfield – the key terms are all under assessment even as we strive to articulate their importance to policy-makers, communities and other groups.

A. INTRODUCTION

Environmental hazards can become centre stage due to high-profile disasters, from those involving floods and fires to air pollution and chemical exposures. In Part 2 of this book the scope of some of these hazards – those which have traditionally been considered as 'natural' because of the triggering role of physical events – is outlined, as well as some of the ways in which they are responded to. Though much research into environmental hazard and risk still uses traditional classification schemes – such as 'natural' and 'technological', in Part 1 of this book we emphasised the difficulties in sustaining this distinction. In Part 3, we return to this theme. We do so by introducing – though not uncritically – the concept of the Anthropocene.

DOI: 10.4324/9781351261647-19

While this is a contested term, one of the key features of Anthropocene debate is a consideration of the extent to which global-scale physical processes – climatic, but also terrestrial, maritime and cryogenic, and associated regional- and local-scale events – are shaped by historical and present-day social actions. We see the role of social activity not only in creating the disaster, but also in actively shaping many of the physical phenomena that produce hazards. What is more, these hybrid social/natural processes and events create new kinds of vulnerabilities and exposures, such that a significant area of interest is now the scope and impact of 'Anthropocene challenges'.

B. WELCOME TO THE ANTHROPOCENE?

The Anthropocene is a difficult term to pin down. Initially proposed by work in the natural sciences, the Anthropocene has been used to describe the increasingly dominant role of people in shaping global-scale physical processes and in degrading the ecosystems that these processes sustain. For many scientists, this role has become so profoundly impactful, and with such long-lasting, disastrous consequences for people, that we are entering a new period of Earth history. Here, the '-cene' part of the word refers to a geological period, and 'Anthro-' refers to *Anthropos*: that is, human beings. One foundational commentary (Crutzen & Stoermer, 2000) reviews some major changes of humanity on the environment and then using the term 'mankind', suggests:

> Considering these and many other major and still growing impacts of human activities on earth and atmosphere, and at all, including global, scales, it seems to us more than appropriate to emphasize the central role of mankind in geology and ecology by proposing to use the term 'anthropocene' for the current geological epoch.
>
> (p. 17)

Should the Anthropocene be dated back to the domestication of plants and animals? Or to a later date when we have evidence for the global-scale anthropogenic modification of soils? More and more scholars are looking to the colonisation of South America, which resulted in pandemics, massive depopulation of indigenous societies, forest regrowth on abandoned agricultural lands that impacted global climate systems and the large-scale enslavement of people to produce an entrenched plantation-based economic system. Here, it is a sustained social action – colonisation – that leads to a series of impacts that are disastrous in the short and long term and that create and exacerbate new kinds of vulnerability.

For some, it is the global presence of radioisotopes released during atmospheric nuclear weapons testing from 1945 onwards, transported in the atmosphere and deposited that is a key indicator of anthropogenic activity. For others, however, it is the cumulative impact of a series of transformations that – taken together – define the nature of the Earth's Anthropocene condition. This approach is exemplified by the identification of a 'Great Acceleration' in the mid-20th century comprising both 'Earth system' and 'socio-economic' trends (Figure 16.1).

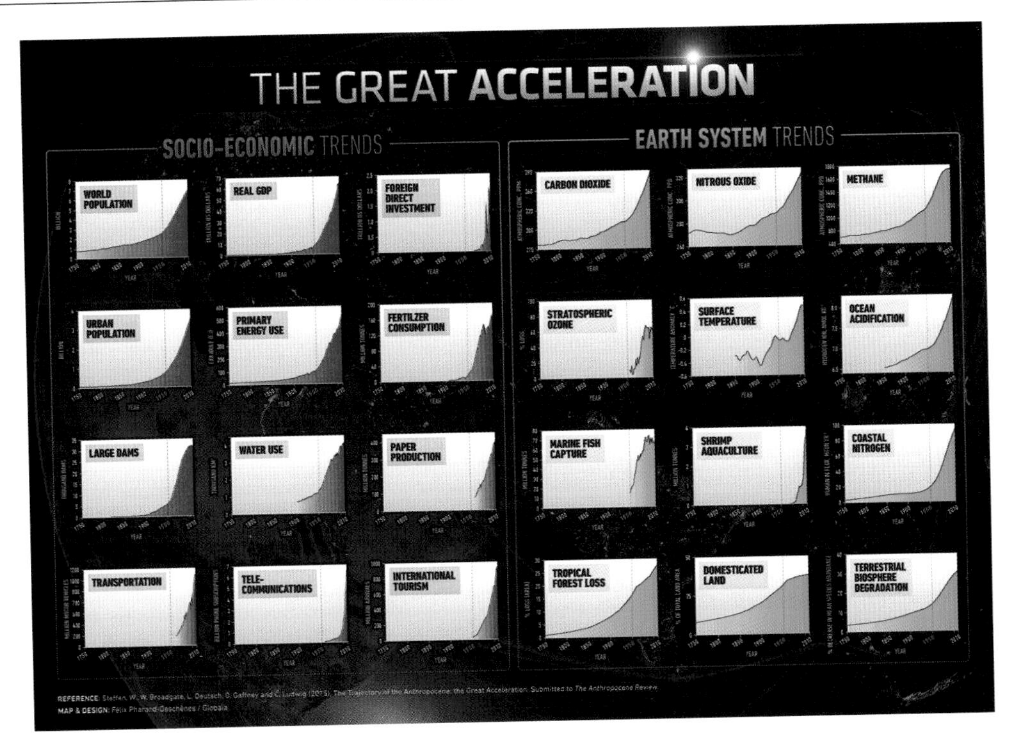

Figure 16.1 The Great Acceleration graphs.

Source: IGBP (2015) www.igbp.net/globalchange/greatacceleration.4.1b8ae20512db692f2a680001630.html

The Great Acceleration graphs – which identify a period of extensive technological, demographic, economic and resource use expansion from 1945 onward – were originally designed as part of the synthesising project of the International Geosphere-Biosphere Programme (IGBP), undertaken 1999–2003. This synthesis aimed to pull together a decade of research in the IGBP's core projects and, importantly, generate a better understanding of the structure and functioning of the Earth system as a whole.

An important insight from this work is the range of new hazardous conditions identified. These conditions include anthropogenic global warming, but also biodiversity loss leading to mass extinction and the ubiquity of microplastics in terrestrial and marine ecosystems. As shown by the IGBP and reiterated by the Intergovernmental Panel on Climate Change (IPCC), proposed evidence for these conditions includes increasing global average temperatures and carbon dioxide concentrations, rising sea levels and ocean acidification. On the basis of such evidence, in 2016 the Anthropocene Working Group of the International Commission on Stratigraphy provisionally recommended that the Anthropocene be formally recognised as a distinct unit of geological time. As of January 2022, the working group's members now largely favour the Great Acceleration as the origin point for this proposed epoch and as cause for the devastation to ecologies and habitats.

The Anthropocene challenges are multiple, and often interconnected; cut across borders; and work to a range of timescales, from the immediate to the millennial. These include (e.g. IPBES, 2019; IPCC, 2021–2022):

- Extinction rates of animals and plants rising far above the long-term average. The Earth is on course to see 75% of species become extinct in the next few centuries if current trends continue.
- Rising levels of human-produced greenhouse gases in the atmosphere leading to global warming.
- The proliferation of plastics in waterways and oceans, with microplastic particles now virtually ubiquitous. Plastics will leave identifiable fossil records for future generations to discover.
- A doubling of the nitrogen and phosphorus in our soils in the past century with fertiliser use. This is likely to be the largest impact on the nitrogen cycle in 2.5 billion years.
- The laying down of a layer of airborne particulates such as black carbon from fossil fuel burning in sediment and glacial ice.

C. CRITIQUING THE ANTHROPOCENE

The Anthropocene has become a pervasive term in academia but also in policy circles and the media. As the term makes its way into the social sciences, arts and humanities, it has been challenged for its overly simplistic (and even misleading) approach to the 'Anthropos' and the implication that all people are equally responsible for the socio-economic actions that have been noted in the Great Acceleration graphs. In efforts to distinguish contributions, sometimes three broad categories of countries are suggested: the OECD (Organisation for Economic Co-operation and Development) countries such as the UK, US, France and Germany, the BRICS (Brazil, Russia, India, China and South Africa) countries which are labelled as rapidly developing economies, and 'all other' countries. This categorisation has numerous flaws, such as the huge diversity of countries within each group, the huge diversity of characteristics within countries, the focus on macroeconomic indicators for classification and the rapid changes which many countries in all groups are undergoing.

Moreover, because the focus of attention in the IGBP graphs is on large-scale datasets that indicate global conditions, the more specific, day-to-day impacts on particular communities are overlooked. Irrespective of how well or how poor national-level indicators might appear – and the choice of indicators can provide vastly different pictures, such as the misleading characteristics of GDP – people's typical experiences can diverge substantially from the aggregated metrics. For instance, the UK might appear to have a high GDP and large trade numbers with the world, while thousands die every winter because they cannot afford to insulate or heat their homes; raw sewage is being discharged into the sea; and individual debt levels climb for people trying to feed their families each day. Drawing on critical social theory, science studies, feminist and indigenous studies, as well as history, philosophy and performance, many researchers examine how the Anthropocene as a series of ideas and claims about how the world works is constructed, contested and strategically deployed for political rather than scientific purposes.

Certainly, the IGBP research programme was very much dominated by the earth sciences. It struggled to integrate social science, humanities and arts approaches to data collection, analysis, interpretation and communication. Rather than unpack a history and geography of colonialism, political ideologies and the detrimental environmental and social transformations, the IGBP relied on a species-scale approach to humanity (Nixon, 2017, p. 24). It is unsurprising to find, then, that population growth was identified as a causal factor in regard to increasing environmental degradation. It has remained prominent in several reiterations of the graphs, including the animation called 'Welcome to the Anthropocene' (Figure 16.2; IGBP, 2012) shown at the *Planet Under Pressure* conference in London, UK, from 26–29 March 2012. As described by IGBP (2012), 'the graph's y-axis title rotates like an airline departure board providing latest information on key parameters: population, energy use, GDP, urbanization, fertilizer use, deforestation, biodiversity loss, and others. The graph overlay is an artist's interpretation of the great acceleration in human activity'.

Critiques of this material are not about denying the specific impact of people in particular places – topics which are the subjects of lengthy books by themselves. They are about recognising that it is not about numbers of bodies but about many factors, including consumption and impact per person and unequal land distributions. One example among many, certainly with its own critiques and long-term debates and disputes, is the simple mnemonic IPAT equation $I = P \times A \times T$ (Ehrlich & Holdren, 1972) for which:

I – Impacts, which needs to be described specifically.
P – Population numbers.

Figure 16.2 A screen capture from Welcome to the Anthropocene.

Source: IGBP (2012) www.igbp.net/multimedia/welcometotheanthropocene.4.1081640c135c7c04eb480001382.html

A – Affluence, which might be a monetary term or another form.
T – Technology, which for a calculation would need to be quantified.

The importance of IPAT is its age. Anthropocenic ideas are not new. They have been proposed and evaluated for a long time, providing a rich body of work to draw upon.

Fundamentally, there is no accepted or definitive description or approach to the current state of humanity and the Earth, whether that be IPAT, the Anthropocene or other concepts. The key is that this science and its applicability for policy and practice are not fixed but are ever evolving. Terminology and definitions should not be accepted at face value. Critiques and alternatives should be searched for, examined and critiqued themselves. Advantages and limitations should be admitted openly, meaning that certain ideas might be useful and usable in some contexts but not in others. The Anthropocene is by no means a necessary, unique, focused or appropriate term to capture the diverse, globe-spanning social activities that create and exacerbate hazard and risk for people and ecosystems.

Another concept that addresses widespread environmental hazards is the Plantationocene (e.g. Haraway, 2015). This situates environmental degradation and associated hazards alongside exploitation and suffering within the plantation agriculture of colonialism. It also foregrounds how plantation logics – that is, the global circulation of people and plants, the simplification of agricultural landscapes into controlled plantations and the role of long-distance capital investments in the homogenisation of agriculture – continue to organise modern economies and environments. It also has its critics, mainly about it downplaying the racial elements within the development of plantation economies alongside assumptions underlying the formation, structure and contribution of plantations to economies (Davis et al., 2019).

The Capitalocene (e.g. Haraway, 2015) highlights capitalism as a cause of the problems, thereby downplaying other exploitive ideologies, some of which put as much faith into capital-centric approaches as capitalism and which exploit people and the environment as much as capitalism. Further alternatives to the Anthropocene include Anthrobscene, Chthulucene, Homogenocene, Manthropocene, Northropocene, Phagocene, Plasticene, Poubellocène, Thanatocene and Thermocene. More terms and phrases are generated all the time, producing continuing discussions and debates.

Arguably, what the Anthropocene does is provide an umbrella term for the analyses that have been undertaken by natural scientists into the 'forcing' of physical processes. These analyses comprise, for example, the creation of an evidence base that disaggregates environmental patterns (of global temperature, for example) that are natural from those that are anthropogenic. It is important to consider the evidence base of what would have happened to the climate and other global characteristics without humanity and what particular social activities have done to the environment. We might also add that in such an approach those suffering most from 'forced' environmental change must not be assumed to be passive victims. Rather, the communities in which they live and work are sites of care and ingenuity, and each person and group has plenty to offer regarding abilities and knowledges.

Aspects of these ideas – just some aspects, as being comprehensive would not be possible – are explored in Part 3 to close this book. Chapter 17 covers industrial and waste hazards, noting how infrastructure and pollution contribute to environmental hazards. Chapter 18 focuses on one large-scale human contribution to the planet melded with natural processes, that of climate change. Climate change is not the only planetary environmental hazard influencer, so it is placed within the contexts of wider global environmental hazards as well as their interaction with climate. Chapter 19 concludes Part 3 and the book through a succinct call to action on what could and should be done with the knowledge you have gained by reading and learning about environmental hazards through this book.

KEY TAKE AWAY POINTS

- Science has provided insight into the diverse ways in which physical processes have been impacted by social actions. Significant geographic diversity remains regarding which groups and individuals are primarily associated with these actions and which groups and individuals are most impacted by the changes positively and negatively.
- The 'Anthropocene' as a term has been used to describe the current scale of environmental changes associated with human activities, but it is vague and contested and might not be useful for communication and action.
- The 'Anthropocene' is not the only term that has been used to describe and assess the impact of human activities on physical processes at wide scales, with better differentiation sometimes made among events, processes, concepts and labels for communication and education.

BIBLIOGRAPHY

Crutzen, P. J., & Stoermer, E. F. (2000). The 'Anthropocene'. *IGBP Newsletter*, *41*, 17–18.

Davis, J., Moulton, A. A., Van Sant, L., & Williams, B. (2019). Anthropocene, Capitalocene, . . . Plantationocene?: A manifesto for ecological justice in an age of global crises. *Geography Compass*, *13*(5), article e12438.

Ehrlich, P. R., & Holdren, J. P. (1972, May). One-dimensional ecology. *Bulletin of the Atomic Scientists*, *16*, 18–27.

Haraway, D. (2015). Anthropocene, Capitalocene, Plantationocene, Chthulucene: Making kin. *Environmental Humanities*, *6*(1), 159–165.

IGBP. (2012). *Welcome to the Anthropocene*. www.igbp.net/multimedia/welcometotheanthropocene.4.1081640c135c7c04eb480001382.html

IGBP. (2015). *Great acceleration*. www.igbp.net/globalchange/greatacceleration.4.1b8ae20512db692f2a680001630.html

IPBES. (2019). Global assessment report on biodiversity and ecosystem services of the Intergovernmental Science-Policy Platform on Biodiversity and Ecosystem Services. IPBES (Intergovernmental Science-Policy Platform on Biodiversity and Ecosystem Services), Bonn, Germany.

IPCC. (2021–2022). Sixth assessment report. IPCC (Intergovernmental Panel on Climate Change), Geneva, Switzerland.

Nixon, R. (2017). The Anthropocene and environmental justice. In J. Newell, L. Robin, & K. Wehner (Eds.), *Curating the future: Museums, communities and climate change* (pp. 23–31). Abingdon: Routledge.

FURTHER READING

Demos, T. J. (2017). *Against the Anthropocene: Visual culture and environment today*. Boston, MA: MIT Press.

Ehrlich, P. R., & Holdren, J. P. (1971). Impact of population growth. *Science*, *171*(26), 1212–1217.

Holdren, J. P. (2018). A brief history of 'IPAT'. *The Journal of Population and Sustainability*, *2*(2), 66–74.

Horton, J. L. (2017). Indigenous artists against the Anthropocene. *Art Journal*, *76*(2), 48–69.

Lewin, J., & Macklin, M. G. (2014). Marking time in geomorphology: Should we try to formalise an Anthropocene definition? *Earth Surface Processes and Landforms*, *39*(1), 133–137.

WEBLINKS

Intergovernmental Panel on Climate Change (IPCC). www.ipcc.ch

Intergovernmental Science-Policy Platform on Biodiversity and Ecosystem Services (IPBES). www.ipbes.net

International Commission on Stratigraphy. www.stratigraphy.org

International Geosphere-Biosphere Programme (IGBP). www.igbp.net (closed in 2015 without updates, so the website is archival).

CHAPTER SEVENTEEN

Industrial and Waste Hazards

17

OVERVIEW

This chapter addresses hazards that are generated by industrial and waste hazards. Whilst historically these hazards would be classed under 'technological' or 'human-made' hazards, this distinction has become far less clear over time. This chapter first reviews the hazardous environmental conditions that endure from a period of extensive development of resources and technologies. Differing frameworks on how accidents form are explored before sections summarise issues relating to *industrial processes*, with a particular focus on chemicals/petrochemicals and nuclear energy; *large-scale structures* that produce risks such as the probability of structural failure during the lifetime of the structure; and *transport systems* that produce risks through the method of travel but also transport networks and infrastructures at scale. The chapter then reviews the global impact of environmental hazards that have emerged from the cumulative waste products of industrialisation, focusing on air pollution, plastics and microplastics, and radioactive materials.

A. INTRODUCTION

In Part 1 of this book we noted that despite the difficulty in separating out 'natural' from 'technological' and 'socionatural' hazards, these categories are still very much used as a way of describing particular hazardous conditions and events and their consequences for health and safety. A significant number of these hazardous conditions emerge from technological trajectories that have been planned for and engineered as a means of pursuing economic progress and comprise built structures, industrial-scale processes and the extensive transformation of ecosystems. Some of the hazards associated with industrial technologies and infrastructures are rapid-onset events, such as crashes and wrecks and toxic substance releases. More difficult to identify and track are the slow-onset hazards – often described as 'socionatural' – and how these intersect

DOI: 10.4324/9781351261647-20

with vulnerability to take a toll on life and well-being. These include the long-term exposure of individuals to pollutants associated with industrialisation and the truncated lives that ensue from a combination of these with economic insecurity and lack of access to health care.

What the Anthropocene debates outlined in Chapter 16 alert us to are (1) the hazardous environmental conditions that endure from a period of extensive technological and resource use expansion from 1945 onward (Section B of this chapter); and (2) the planetary scope of environmental hazards that have emerged from the cumulative waste products of industrialisation, including air pollution, plastics and microplastics, and radioactive materials (Section C of this chapter).

Many accounts of industrial (technological) accidents dwell on case-study detail but, as with natural hazards, there are at least two opposing schools of thought about the root causes of such disasters. These views can be summarised as the following.

The High Reliability School

This view admits the potential for human error in complex, dangerous technologies and the possibility of accidents, but argues that properly designed and managed organisations can compensate for such errors. The argument is that:

- High-risk industries always seek failure-free performance and give top priority to reliability and safety. All personnel are subject to constant on-the-job training and are well informed about the potential dangers.
- Complex organisations have in-built redundancy. Therefore, duplication and overlap of components and procedures provides a backup system and a fail-safe environment if problems arise.
- In large organisations there is a culture of local decision-making so that delegated authority can lead to swift accident-preventing decisions.

The Normal Accidents School

Conversely, this view believes that serious technological accidents are 'normal', much as natural disasters are a normal part of Earth processes (Perrow, 1999). In particular, the failure to control minor breakdowns or recurrent disruptions suggests that more serious accidents are likely to follow. Key features of this school are:

- Safety and reliability are not undisputed priorities. They compete with other objectives, such as increasing demands on performance and the need for profitability. Built-in redundancy simply increases the complexity of the technology and may encourage complacency within the organisation just when the system becomes more difficult to understand.
- Competitive pressures for innovation can produce design faults in new equipment. At the same time, routine maintenance of older components may be overlooked and lead to eventual failure. These risks are increased whenever dangerous technologies spread into remote or hostile geographical environments.
- Constant training and local decision-making cannot eliminate operational failure. Operators often work unsocial shift patterns in relative isolation, a pattern associated with boredom and the temptation for substance abuse. The substitution of computer control is not a guaranteed response, due to

the potential for hardware failure and defects in software.

There are plausible elements to both viewpoints, and a more thorough overview can be found in *Foundations of Safety Science: A Century of Understanding Accidents and Disasters* (Dekker, 2019). It is true that – so far – the nightmare scenario of an accidental nuclear war has not yet materialised. It is also possible that the rigorous formal checks in place continue to make this unlikely. On the other hand, defective design and inadequate management have already created near-accidents with nuclear weapons systems. Since 1950, there have been 32 nuclear weapon accidents that involved nuclear weapons that accidentally launched, fired, detonated or were stolen. Worryingly to date, six nuclear weapons have been lost and never recovered.

Industrial/technological disasters can occur anywhere so long as the responsible bodies fail to prioritise safety measures and governments do not enforce compliance with health and safety regulations. The balance between industrial/technological risk and health and safety is constantly changing. Technical and legislative developments that foster occupational and public safety have to be set against factors and trends that increase risks. Glickman (1992) found that major industrial accidents tended to occur at refineries and manufacturing plants or during the transportation of dangerous materials. These accidents were linked to the nature and scale of industrial activity; harmful energy was released in either mechanical impact form (dam burst, waste tip slippage, vehicle deceleration) or chemical impact form (explosion, fire). The most hazardous materials were seen as high-level radioactive materials, explosives and a limited number of gases and liquids that are poisonous when inhaled or ingested. Chemicals are most hazardous if they are flammable, explosive, corrosive or toxic in low concentrations. In order to constitute a community-scale risk, such substances must be present in large quantities and be stored or transported in an insecure manner. Toxic materials are hazardous if they are transferred to the affected population by severe air pollution in the form of a 'toxic cloud'. An important feature of severe pollution episodes is that the adverse effects, both on the human body and on the environment, can often outlast the impacts associated with 'natural' hazards.

B. INDUSTRIAL ENVIRONMENTAL HAZARDS AND RISKS

The rise of the modern chemical and petrochemical industry created a suite of entirely new technologies and associated risks. This industry, like some other processing operations, has tended to group on large sites near to significant concentrations of population. Over 30 years ago, for example, a study of Canvey Island – a major chemical and oil-refining complex on the north shore of the river Thames about 40 km downstream from London, England – revealed that the quantities of flammable and toxic materials either in process, store or transport represented a serious threat to public safety (Great Britain, Health and Safety Executive, 1978). The hazards included fire, explosion, missiles and the spread of toxic gases. It was concluded that the existing industrial installations possessed a quantifiable risk of killing up to 18,000 people.

Environmental hazards that emerge from industrial development are by no means new to the 20th century. These include:

- *Industrial processes* (including extractive mining, manufacturing, power production, storage and transport of hazardous materials) that produce risks such as the probability of death or injury per person per number of hours exposed to the materials.
- *Large-scale structures* (such as public buildings, bridges and dams) that produce risks such as the probability of structural failure during the lifetime of the structure.
- *Transport systems* (including road, air, sea and rail travel) that produce risks such as the probability of death or injury per distance travelled.

Specific disasters associated with industry, structures and transport tend to be reported on a case-by-case basis, and such reporting can foreground the particular impacts on people from these. As outlined in Chapter 16, however, these are accruing at a fast rate, and their accumulated impact has global consequence. What is more, many of these disasters are associated with the extraction and use of carbon fuels (coal, oil and gas), as noted in Table 17.1.

As Table 17.1 demonstrates, science and technology can also be used to help mitigate against industrial hazards, as exemplified by the management of mining tailings, oil spills and water management for sinkholes. There is work being undertaken, for example, on how slag heaps are able to absorb carbon dioxide from the atmosphere through a process known as sequestration, reducing the impact of greenhouse gases. In this case, the outputs of one type of environmental hazard – slag wastes – can potentially be used to help address another type of hazard, global warming. Such measures, however, in and of themselves are by no means the solution to climate change and its attendant challenges; what is required for this is a move away from carbon fuels.

Industrial Processes

There are a number of key hazards relating to chemical and petrochemical industries and the nuclear energy industry. Driven by the development of science and technology, chemical/petrochemical and nuclear hazards both create significant and long-lasting devastation.

1. Chemicals and Petrochemicals

The rise of the modern chemical and petrochemical industry created a suite of entirely new technologies. This industry, like some other processing operations, has tended to accumulate on sites near to significant concentrations of population. Major industrial disasters tend to occur at refineries and manufacturing plants or during the transportation of dangerous materials. These disasters are linked to the nature and scale of industrial activity; harmful energy is released in either *mechanical impact form* (dam burst, waste tip slippage, vehicle deceleration) or *chemical impact form* (explosion, fire).

The most hazardous materials are high-level radioactive materials, explosives and a limited number of gases and liquids that are poisonous when inhaled or ingested. Chemicals are most hazardous if they are flammable, explosive, corrosive or toxic in low concentrations. In order to constitute a community-scale risk, such substances must be present in large quantities and be stored or transported in an insecure manner. Toxic

Table 17.1 The environmental hazards of carbon fuel extraction

Hazard	Illustrative event	Impact	Past and potential responses
Mining waste (tailings)	Aberfan disaster, 1966, Wales UK	Mine tailings result in environmental contamination, erosion and are difficult to rehabilitate. They can also result in loss of life as seen by heavy rain that led to a build-up of water within coal tailings in Aberfan. On October 21 the tailings slid downhill as a slurry, killing 116 children and 28 adults. Although the National Coal Board (NCB) in Wales were blamed, they were never prosecuted.	Microorganisms can now be used to obtain metals from the tailings, and soil and vegetation can be used to rehabilitate the tailings using a 'capping' technique. This provides a safer, more stable structure that also limits contamination.
Oil spills	Deepwater Horizon, 2010, USA (see Figure 17.1)	Technological disasters can be the source of major environmental pollution such as that following the Deepwater Horizon drilling accident in the Gulf of Mexico. On April 20 an oil rig operated by British Petroleum (BP) exploded and subsequently sank, with the loss of 11 crew members. The sea-floor well was not successfully capped for almost three months and created the largest offshore oil spill in US history.	BP was obliged to set aside a US$20 billion fund to compensate coastal residents and the many businesses serving the seafood industry. 1.8 million gallons of dispersants (substances that emulsify the oil, thus allowing for easier break down by bacteria) were pumped directly into the leak and applied aerially to the slick, creating further impacts on the ecosystem.
Natural gas flares	Shell Haven refinery, 1994, England UK	Flare gas is a by-product of numerous industrial processes including oil and gas recovery, petrochemical process, landfill gas production and wastewater treatment. It is a popular but controversial way of eliminating unwanted gas but introduces toxic pollutants such as sulphur dioxide into the atmosphere, contributing to acid rain and greenhouse gases.	By developing smaller-scale uses of gas at/near the source or by capturing it and transporting via a pipeline, the gas can be used. The cost-effectiveness of these depends on the economic and political state of the energy markets.
Occupational diseases e.g. black lung	Appalachian mining communities, USA, 2005 onwards	Mining at the coal face produces clouds of dust. Larger particles settle in the main air passages of miners, while smaller particles travel deeper into the lung tissue. Prolonged exposure can lead to chronic bronchitis, characterised by large, dense masses of fibrous tissue in the lungs. In the US, cases declined after the 1969 Federal Coal Mine Health and Safety Act, which set limits of dust exposure.	Black lung is an entirely preventable environmental hazard. Though policy and surveillance programs can be introduced, the efficacy of these needs to be monitored. A key means of ensuring this is strong labour protection laws and unionisation.

(Continued)

Table 17.1 (Continued)

Hazard	Illustrative event	Impact	Past and potential responses
		In 2005, however, the National Institute for Occupational Safety and Health (NIOSH) identified regional clusters of rapidly progressing severe black lung cases, especially in Appalachia. These indicate a lack of compliance by coal mine operators, longer working hours and increased mechanisation.	
Sinkholes and subsidence	The Guatemala City Sink Hole, 2010, Guatemala	Two types of sinkholes exist. A natural one where the roof of a cave collapses and exposes the underground cavern, or another type when water dissolves the rock underneath soil and creates an underground chasm. Without rock to support it, the soil layer collapses and creates a hole on the surface. Human sinkholes are created when city development compromises the structural integrity of the underlying rock. Roads, buildings and other types of construction can result in water collecting and washing away the supporting rock layer (especially at low sea levels and after a heavy storm). The Guatemala City sinkhole was ~ 20 m by 90 m and swallowed a three-story factory, killing 15 people on May 30. The sinkhole occurred due to Tropical Storm Agatha, the Pacaya Volcano eruption and leakage from sewer pipes.	By proactively replacing structurally failing infrastructures such as water pipes and sewage systems, maintaining infrastructure, it is possible to prevent sinkholes. Underwater remotely operated vehicles (ROVs), pipe crawlers and utility crawlers can be used to monitor and solve underwater problems to resolve sinkhole issues.

materials are hazardous if they are transferred to the affected population by severe air pollution in the form of a 'toxic cloud'. This is exemplified by the 1984 Union Carbide plant in Bhopal, India, event. What is also evident, as summarised in Box 17.1, is how the disastrous nature of such an incident can be profoundly exacerbated by efforts of corporation management to head off compensation claims and by geopolitics.

Although progress has been made since the 1984 Union Carbide disaster to make safety an integral function for all process industries, disastrous events still happen. Qi et al. (2012) summarised present-day challenges under three categories:

- failure to learn from past incidents in order to improve process design, procedures, site training etc.;
- insufficient attention paid to leading indicators of risk; and
- growing complexity of operations, combined with inadequate communication.

Figure 17.1 Supply boats attempt to extinguish fires on the Deepwater Horizon oil rig located off the southeast tip of Louisiana in the Gulf of Mexico on 21 April 2010. The fires were started the previous day by an explosion when 126 personnel were on board, and the rig sank into the sea on 22 April.

Source: Photo: AP Photo/Gerald Herbert 100421038603, © Alamy Images/Geopix

2. The Nuclear Industry

As of 2022 there are 439 nuclear reactors in operation in 33 countries around the world, with major clusters in Europe, Japan and the US. The US has the largest number of nuclear power reactors – 92 – in operation (Figure 17.3). In 2022, 55 power reactors were currently being constructed, primarily in China, India, South Korea, Taiwan and Pakistan.

Large nuclear power stations have the capability to cause many deaths, extreme societal disruption and long-lasting pollution. Because of this, nuclear plants are rarely sited near urban areas and the industry is highly regulated. For example, the *Nuclear Regulatory Commission* (NRC) supervises the use of nuclear power for civilian purposes in the USA; in the UK the *Office for Nuclear Regulation* (ONR) is responsible for authorizing use in both the civilian and the defence sectors. On a world scale, these responsibilities fall to the *International Atomic Energy Agency* (IAEA), which in 1992 formalised a nuclear event scale that defines various severities of incidents leading up to major disasters (Figure 17.3).

A major nuclear power plant disaster occurred during the night of 25–26 April 1986, at Chernobyl, which is about 130 km north of the city of Kiev, situated in the Vyshhorod Raion of northern Kyiv Oblast, Ukraine (Box 17.2). There are many European nuclear reactors located either on national boundaries, such as rivers, or within

Box 17.1 The 1984 Gas Disaster at the Union Carbide Plant in Bhopal, India

Methyl isocyanate (MIC) is a fairly common industrial chemical used in the production of pesticides but has qualities that make it hazardous (Lewis, 1990). First, it is extremely volatile and vaporises easily. Since MIC can boil at a temperature as low as 38°C, it has to be kept cool. Second, MIC is active chemically and reacts violently with water. Third, MIC is highly toxic, perhaps 100 times more lethal than cyanide gas and more dangerous than phosgene, a poison gas used in World War I. Fourth, MIC is heavier than air and, when released, stays near ground level.

During the early morning of 3 December 1984, over 40 metric tones of MIC gas leaked from a pesticide factory in the industrial town of Bhopal, India, within a two-hour period and created the world's worst industrial disaster in a city of over 1 million people (Hazarika, 1988). The chemical was stored in an underground tank that became contaminated with water and produced a chemical reaction, followed by a rise in gas pressure and subsequent emissions into the atmosphere. The Bhopal factory was built within 5 km of the city centre by Union Carbide, a multinational company based in the USA. A dense cloud of gas drifted over an area with a radius of some 7 km. It is believed that over 3,000 people may have been killed by cyanide-related poisoning, with a further 300,000 injured or harmed by genetic defects passed on to following generations. In fact, a total of 600,000 injury claims and 15,000 death claims were ultimately filed with the Indian government.

Most fatalities were in the poor neighbourhoods located in low-lying parts of the city, including a shantytown of some 12,000 people near the gates of the factory. Most of the victims were the very young and the very old, although pregnant women suffered badly too. The disaster was severe because of the large numbers of people inhaling the gas and the lack of any emergency planning. There was no local knowledge of the nature of the chemicals in the factory, no adequate warning and only limited means of evacuation. The company provided no information about the medical treatment required by the victims, and key resources such as oxygen (needed to treat respiratory problems) were in short supply.

An investigation revealed that safety devices failed through a combination of faulty engineering and poor maintenance practice, although the company claimed that the cause was sabotage. A contributory factor was that the air-conditioning system, normally in use to keep the MIC cool, was shut down at the time of the disaster. Safety procedures were clearly inadequate. For example, the plant lacked the computerised warning and fail-safe system used by the company in the USA.

The plant was unprofitable at the time of the disaster and, because cutbacks had been made in maintenance, blame was attached to the local management team. Over the following two years, the parent company slimmed down, partly by distributing assets to shareholders and creditors, who were mainly banks. This strategy was deemed necessary in order to fend off a hostile take-over bid, but it also served to off-load assets that were not then exposed to compensation claims. At the same time, the US legal system overturned precedent and opposed compensation claims for such an overseas liability, on the grounds that it would unfairly tax the US courts. So the responsibility was passed back to the Indian government.

The Indian government made itself the sole representative of the victims and filed compensation claims against the company both in the USA and in India. In 1989 Union Carbide made a final out-of-court compensation payment of US$470 million. This compares unfavourably with the US$5 billion awarded in the USA after the *Exxon Valdez* oil spill. Special compensation courts typically awarded sums of £500 for injury and £2,000 for death. In the meantime, the Indian government distributed relief at about £4 per month for each family affected. But the Indian government failed to organise efficient legal aid or medical care for the victims. As a result, victims found it difficult to have their cases brought to court without resorting to bribes or paying private lawyers. Many medicines that should have been supplied free to patients were obtainable only on the black market. Families sometimes had to spend double their monthly government allowance on medicines. Ten years after the event, it was estimated that fewer than one-quarter of the total claims had been settled and that less than 10% of the damages paid by Union Carbide had reached the victims.

This event did promote safety actions in the US. The main Union Carbide plant in West Virginia was quickly closed and about US$5 million was spent on technical

Figure 17.2 The Atal Ayub colony at Bhopal, Madhya Pradesh, India, in 1992. People continue to live in the shadow of the now defunct and abandoned Union Carbide pesticide plant, responsible for the worst industrial disaster on record.

Source: Photo: Panos/Rod Johnson RJH00053IND

improvements at other Union Carbide sites in the USA (Cutter, 1996). The Bhopal disaster left a legacy of improved regulation in the chemical industry, although safety innovation is more apparent at plants in some countries but not others (Lacoursiere, 2005). In general, there is greater attention to safety issues, better regulatory consultation between the chemical process industry and governments and more preparedness for emergencies.

As an example, the term 'process-safety' was used around 500 times per year as a keyword in science and engineering journals at the time of the event, but usage rose to an annual frequency of over 2,500 by 2004 (Mannan et al., 2016).

Despite these advances, the toxic release from the Union Carbide plant in Bhopal is a reminder of how an inability to learn from comparatively minor events can lead to disaster. According to Gupta (2002), at least six serious incidents occurred at the Bhopal plant in the four years before the tragedy. More significantly, it is claimed that there had been nearly 60 leaks of MIC at Union Carbide's West Virginia plant between 1980 and 1984, most of which were unreported. Clearly, in 1984 a Bhopal-style disaster could have happened anywhere in the world. The question today is – how far has that situation changed?

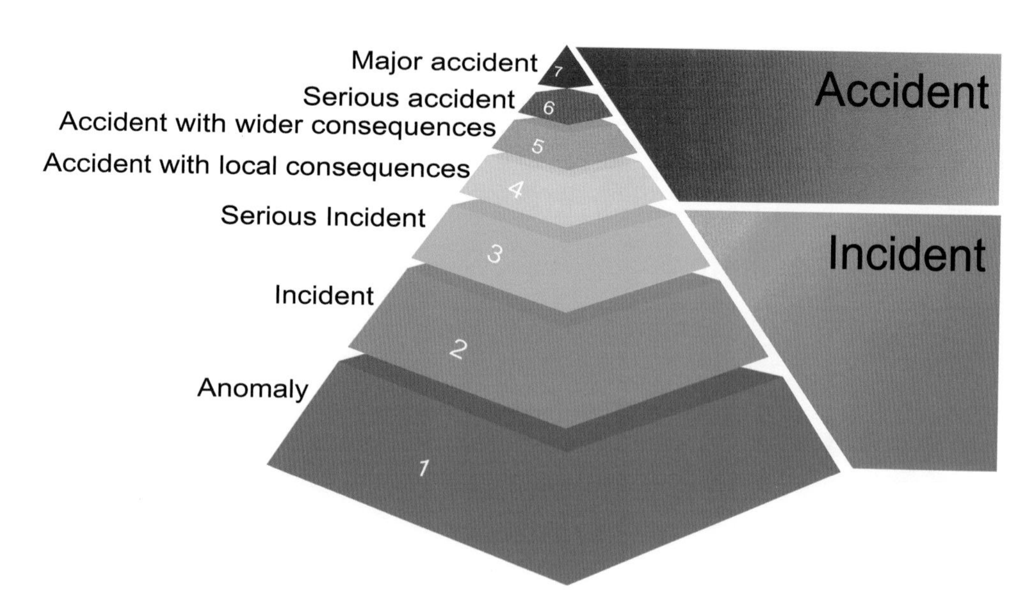

Figure 17.3 The International Nuclear Event Scale.

Source: After International Atomic Energy Authority (personal communication)

25 km of a national border, so that the risk of another transboundary incident remains.

Large-Scale Structures

Many changes have taken place that help to reduce these kinds of risks. For example, in the case of fire hazard, whole urban areas rarely burn down as in the past because of improved fire regulations and more efficient firefighting services. During the 20th century, improvements in engineering design and a growing awareness of health and safety issues, reinforced by government legislation, have made large structures much safer than in the past. However, with increased numbers of large-scale structures there is an increased number of low-risk, high-impact potential hazards such as dams (Box 17.3; see also Chapter 11).

High-hazard installations will almost always be unwanted by most of the local population. The least acceptable facilities tend to be nuclear waste and toxic chemical disposal sites, together with chemical processing plants, nuclear power plants and fuel storage depots. Major industrial accidents tend to result from a planning decision which locates a dangerous technology in an inappropriate place, often combined with a failure to control subsequent intensive land uses from invading the area around the site. It is rarely possible to place other industrial activities, housing and people sufficiently far away from high-hazard sites to guarantee zero risk. At the simplest level, land use planning aims to separate densely populated areas from very high-risk facilities and their transport routes and to reduce any

Box 17.2 The 1986 Nuclear Disaster at Chernobyl

The Chernobyl disaster in April 1986 occurred at a nuclear reactor constructed to a flawed design that was operated by poorly trained personnel. The immediate cause of the disaster was an unauthorised experiment conducted by workers at the plant to determine the length of time that mechanical inertia would keep a steam turbine free-wheeling, and the amount of electricity it would produce, before the diesel generators needed to be switched on. During the experiment, the routine supply of steam from the reactor was turned off and the power level was allowed to drop below 20%, well within the unstable zone for this type of water-cooled, graphite-moderated reactor design.

During the experiment, the reactor was not shut down and a number of the built-in safety devices were deliberately overridden. In this situation vast quantities of steam and chemical reactions built up sufficient pressure to create an explosion, which blew the heavy protective slab off the top of the reactor vessel (see Figure 17.4a). The resulting steam explosions and fires ejected at least 5% of the radioactive reactor core into the atmosphere, where it was carried downwind. Lumps of radioactive material were ejected from the reactor and deposited within 1 km of the plant, where they started other fires. The main plume of radioactive dust and gas sent into the atmosphere was rich in fission products and contained iodine-131 and caesium-137, both of which can be readily absorbed by living tissue.

Immediate efforts were made to control the release of radioactive material. A major limitation was that water could not be used on the burning graphite reactor

Figure 17.4 (a) This photo was taken from a helicopter several months after the Chernobyl explosion of the destroyed Chernobyl reactor #4 building in 1986.

Source: Photo: IAEA Imagebank – https://en.wikipedia.org/wiki/Chernobyl_disaster#/media/File:IAEA_02790015_(5613115146).jpg.

core because this would have created further clouds of radioactive steam. Instead the fire had to be starved of oxygen by the dumping of many tonnes of material (lead, boron, dolomite, clay and sand) from over-flying helicopters. Two plant workers died on the night of the disaster and a further 28 were killed during the next few weeks by acute radiation poisoning. A further 200 people were exposed to over 2,000 times the

Figure 17.4 (b) The Chernobyl reactor #4 building as of 2006, including the later-built sarcophagus and elements of the maximum-security perimeter.

Source: Photo: Carl Montgomery

normal annual dose from background levels of radiation. Eventually some 135,000 people were evacuated from within a 30 km radius exclusion zone around the plant, and the nearby town of Pripyat was abandoned.

In the two weeks following the disaster, the radioactive plume circulated over much of northwestern Europe. Away from Chernobyl itself, the greatest depositions of radioactive material occurred in areas affected by rain, which flushed much of the particulate material out of the atmosphere. These areas included Scandinavia, Austria, Germany, Poland, the UK and Ireland. Some of the heaviest fallout was experienced in the Lapland province of Sweden, where it affected the grazing land of reindeer, contaminated the meat and dealt the Lapp culture a great blow. More widely, the immediate consequence was a general contamination of the food chain, and restrictions on the sale of vegetables, milk and meat were imposed. Some countries also issued a ban on grazing cattle out of doors and warnings to avoid contact with rainwater.

It has proved difficult to assess the long-term health consequences from potentially fatal cancers attributable to the Chernobyl disaster. The 50,000 soldiers who fought to control the fire on the reactor roof suffered huge exposure to radiation, as did the 500,000 workers who subsequently cleaned up the site. Others subjected to

Figure 17.4 (c) The New Safe Confinement in final position over reactor 4 at Chernobyl Nuclear Power Plant.

Source: Photo: Tim Porter

high doses include some of the total 400,000 people who were relocated. Over 15 years later, it was estimated that 2 million people were affected with various health disorders, including a marked drop in the human birth rate. Rahu (2003) was more cautious and claimed that the only direct public health evidence of radiation exposure was 1,800 cases of childhood thyroid cancer recorded in the 1990–1998 period, but did acknowledge many cases of psychological illness attributable to factors like fear of radiation, relocation and economic hardship.

The Russian government hastily built a Shelter Sarcophagus, starting just 24 days after the disaster, to limit radioactive contamination of the environment (see Figure 17.4b). Taking 206 days to build, Chernobyl Unit 4 is enclosed within a large concrete shield, which had a 30-year life expectancy. In 2016 the New Safe Confinement (NSC or New Shelter) mega structure was put in place enclosing the temporary Shelter Sarcophagus, and funded by Chernobyl Shelter Fund (see Figure 17.4c). The New Safe Confinement is designed to prevent the release of radioactive contaminants, protect the reactor from external influence and ultimately facilitate the disassembly and decommissioning of the reactor and prevent water intrusion; however, plans to return the land to productive agriculture and forestry have been profoundly disrupted by the Russian invasion of Ukraine from 2022 onwards.

Box 17.3 Dam Failures

Dam failure is an example of a low-risk, high-impact hazard; it does not occur often but the consequences can be catastrophic. In August 1975, the Banqiao Dam on the upper Ruhe River in Henan province, China, failed after heavy rainfall and contributed to floods that inundated over 1×10^6 ha of land and killed some 20,000 people. In 1993 the Gouhou Dam in Qinghai province suffered structural failure and a further 1,200 lives were lost in floods. Within Europe, the 1959 failure of the high gravity-arch Maupassant Dam in southern France led to more than 450 deaths in the town of Fréjus. This dam collapsed only five years after completion and illustrates the fact that about 70% of all dam failures occur within ten years of construction.

The rate of dam failure is much lower than in the past. More recently, the average failure rate has fallen below 0.5%. Despite the lower failure rates, there are significantly more dams globally, and Figure 17.5 illustrates catastrophic dam failures over the past century, highlighting the increasing volumes of material released and loss of human life. Most dams worldwide are small, and most failures can be linked to the type of dam involved. The most common dams are older *fill-type* structures, built of compacted earth or rock. They are vulnerable both to overtopping by floods and to inadequate foundations that fail to prevent subsurface erosion. Most of the world's highest and largest-capacity dams were built within the last 25 years and

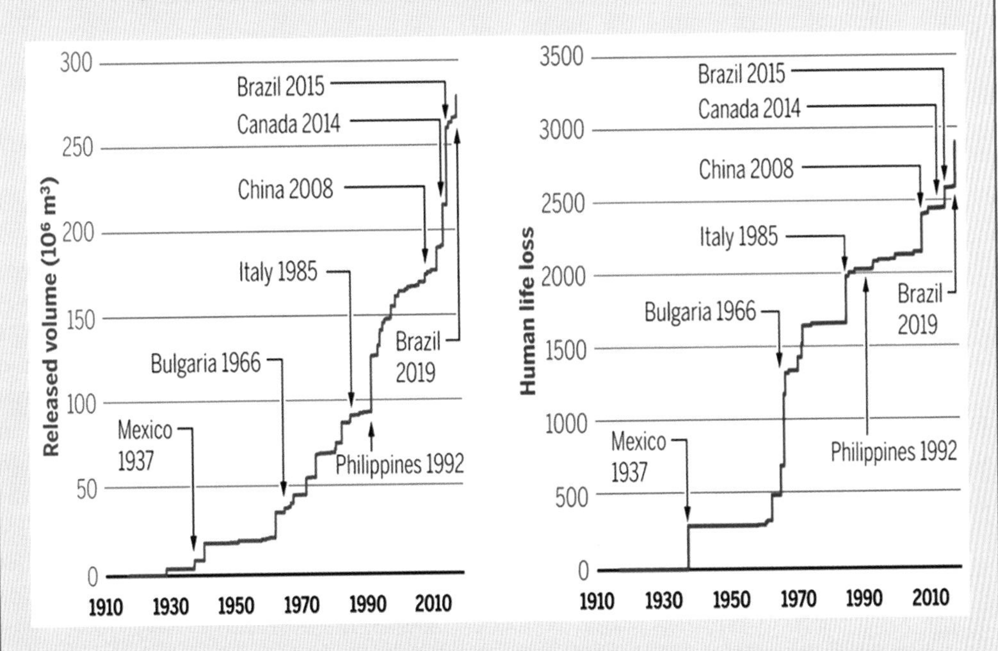

Figure 17.5 Catastrophic dam failures over the past century, tailings dam and ash pond failures and the resulting fast-moving mudflows which have led to a cumulative loss of almost 3,000 lives.

Source: Graphic is taken from: Santamarina et al. (2019). Drawn by C. Bickel, p. 527. Data from: State of World Mine Tailings Portfolio (2020)

have proved controversial on ecological and socio-economic grounds rather than on account of safety standards. No major failures of *concrete-gravity* dams have been recorded in recent times, although doubts have been expressed about the Three Gorges Dam on the upper Yangtze River, China. This is built on granite bedrock and designed so that its weight resists the pressure from the stored water, but concerns have been expressed about the possibility of induced earthquake activity in the area, due to the great weight of the stored water, and the possible risk from large landslides.

residual threat through the use of buffer zones. Other actions include limiting the risks, for example by restricting the amount and nature of substances stored on industrial sites or requiring tanker delivery rather than onsite storage. This was one of the key issues in the Buncefield accident in the UK (see Box 17.4).

This disaster raised many concerns about high-hazard sites. Apart from technical issues around the design and operation of these facilities, many recommendations were made for improving emergency preparedness and land use planning (Buncefield Major Incident Investigation Board, 2008). The area surrounding Buncefield had been subject to incremental development during recent decades, placing more and more people and property at risk, and the main recommendation from the Investigation Board was for a thorough review of the existing land planning system for high-hazard sites. Specific matters suggested for review included moving away from expressing harm in terms of 'dangerous doses' towards a risk of fatality, better integration of land use planning with the UK 1999 Act Control of Major Accident Hazards Regulations (COMAH), taking account of societal risk by factoring in the size and distribution of the population around the site, proper cost-benefit analysis of restricting development near hazardous sites and making the entire process more transparent and accessible to the general public (Buncefield Major Incident Investigation Board, 2008). It seems essential that, in a crowded island like Britain, more attention should be given to the population that is at risk from high-hazard facilities, especially those – like Buncefield – where off-site development already exists.

Transport Systems

The risks of individual transport, such as bicycles and cars, have become more manageable with the development of increased safety measures. Risk reduction has been achieved at relatively low cost through improvements in car design, more use of motorways and legislation such as the compulsory wearing of seat belts and stricter enforcement of drink-driving laws that help protect an individual during an accident. In the case of public transport, ships, trains and aircraft, alongside individual cars, are all built to higher safety standards. Nevertheless, traffic crashes claimed over 30 million lives worldwide during the 20th century. Today, every year, more than 1.35 million people die on the world's roads and millions more are seriously injured. Some interesting statistics from Brake, the road safety charity (www.brake.org.uk), states:

Box 17.4 The 2005 Explosions and Fires at the Buncefield Fuel Site, UK

On 11 December 2005, serious explosions and fires occurred at the Buncefield oil storage and transfer depot on the outskirts of Hemel Hempstead, Hertfordshire (Buncefield Major Incident Investigation Board, 2005). This facility began operating in 1968, when few other activities or buildings existed nearby. All the evidence suggests that the main explosion occurred due to the ignition of a vapour cloud originating from Tank 912 in Bund A (see Figure 17.6). In turn, this was due to an escape from the tank of about 300 t of unleaded petrol, due to over-filling caused by the failure of safety systems designed to prevent such an accident. In total, over 20 large fuel storage tanks were involved. The resulting fire, described as the largest in peacetime Europe, burned for 32 hours and closed sections of the M1 motorway, despite the efforts of up to 1,000 firefighters.

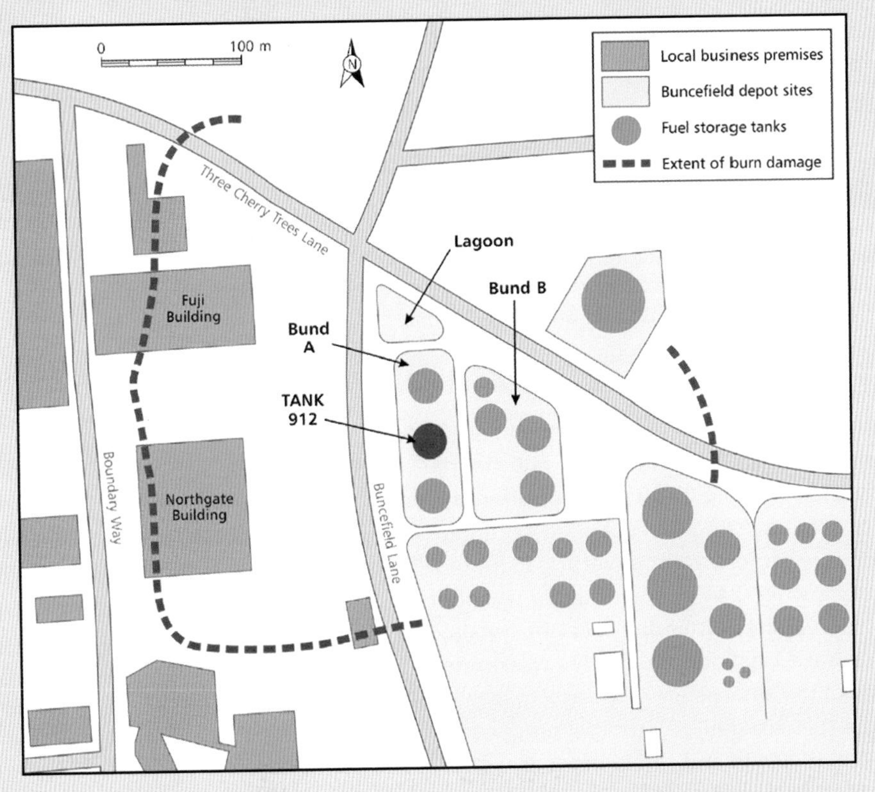

Figure 17.6 Map of the pre-incident layout of the Buncefield fuel depot site showing the extent of burn damage.

Source: BUNCEFIELD MAJOR INCIDENT INVESTIGATION; Initial Report to the Health and Safety Commission and the Environment Agency of the investigation into the explosions and fires at the Buncefield oil storage and transfer depot, Hemel Hempstead, on 11 December 2005. Buncefield Major Incident Investigation Board, 2006. Initial Report to the Health and Safety Commission and the Environment Agency of the investigation into the explosions and fires at the Buncefield oil storage and transfer depot, Hemel Hempstead, on 11 December 2005 (icheme.org)

Figure 17.7 A view of the fire at the Buncefield oil storage terminal near Hemel Hempstead, England, on 11 December 2005. Buncefield, the fifth-largest oil products storage depot in the UK, was destroyed following a series of explosions.

Source: Photo: Courtesy of Chiltern Air Support Unit and Buncefield Investigation at www.buncefieldinvestigation.gov.uk/images/index.htm, © Alamy Images/PA Images

No one died, but 43 people were injured and 2,000 persons were evacuated. Twenty business premises, employing 500 people, were destroyed and a further 60 businesses, employing 3,500 people, suffered damaged. At least 300 residential properties were damaged, and fuel supplies to London and parts of southeast England, including Heathrow airport, were severely disrupted. There was a significant impact on health services when 244 people, mainly members of the emergency services with respiratory complaints, attended local hospitals (Hoek et al., 2007). Economic losses were placed at £1 billion, mainly due to compensation claims against site operators, plus costs of the emergency response, losses in the aviation sector and the cost of the subsequent investigation. Criminal proceedings were brought against five defendants for neglect of health and safety procedures. After a complex corporate criminal trial, guilty verdicts were delivered against the defendants in June 2010.

- Road deaths are the eighth highest cause of death for people of all ages.
- Road deaths are the number one killer of those between the ages of 5 and 29.
- Pedestrians, cyclists and motorcyclists make up more than half of all road deaths (these road users are collectively known as vulnerable road users).

- Road deaths affect the poorest hardest. Of all road deaths, 93% are in low- or middle-income countries, and there has been no reduction in the number of road deaths in any low-income country since 2013.
- Africa has, by far, the highest rate of road deaths of any WHO region.

The implementation and access to transportation safety measures is subject to geography (except for international airline companies). What is not so obvious is the environmental consequence of transportation systems at scale. The air pollution from airplanes and road traffic on greenhouse gases is significant, and global. The movement of invasive species due to sea transportation is devastating for some species resulting in extinction. The transport of oil, gas and nuclear waste can result in severe accidents (such as Deepwater Horizon). Significant hazard events can occur via the infrastructure established to support the transportation of goods, as seen during the explosion in Beirut in 2020 (see Box 17.5).

C. SHORT- AND LONG-TERM WASTE HAZARDS

The cumulative waste products of industrialisation (e.g. air pollution, plastics and microplastics, and radioactive materials) requires us to consider the interconnected nature of our planet and society. This section examines the role waste hazards create in the context of Anthropocene hazards.

1. Ambient Air Pollution

Air pollution is the contamination of air due from substances in the atmosphere that are harmful to the health of people and other organisms, or that cause damage to the climate or to materials. 'Ambient' refers to uncontaminated air, and so measures of ambient air pollution register the impacts of social activities on the atmosphere.

Box 17.5 Port of Beirut Explosion

On 4 August 2020, a large amount of ammonium nitrate stored at the Port of Beirut in the capital city of Lebanon exploded, killing at least 218 people, causing 7,000 injuries and resulting in over US$15 billion in property damage. Three hundred thousand people were left homeless. A cargo of 2,750 tonnes of the substance (equivalent to around 1.1 kilotons of TNT) had been confiscated by the Lebanese authorities from the abandoned ship *MV Rhosus*. It was stored in a warehouse for over six years without suitable safety measures, therefore when a fire broke out in the warehouse, the explosion occurred, also damaging the adjacent grain silos (see Figure 17.8). The blast was so powerful that it physically shook the whole country of Lebanon and was felt as far as parts of Europe and was heard in Cyprus, more than 240 km away. The event happened during an economic crisis of Lebanon, COVID-19 and rising levels of poverty, placing enormous strain on the nation that declared a two-week state of emergency.

Figure 17.8 Damage after the Beirut explosion. Here you can see the 140 m crater created by the explosion.

Source: Photo: Mehr News Agency, © Alamy Images/James Chehab

In July and August 2022, part of the silos collapsed following a weeks-long fire in the remaining grain. As of 2022, the investigation by the Lebanese government is ongoing.

Regional-scale ambient air pollution was detected in the mid-1960s when acid deposition was recorded over rural parts of Europe and North America, due to the transport of oxides of sulphur and nitrogen from industrial sources hundreds of kilometres away. Since the 1980s, rapid urbanisation and industrialisation has led to an increase in air pollutant emissions termed 'ambient' because they occur in outdoor environments. Currently, fossil fuel emissions from industry, transport systems and power stations are the chief sources of ambient air pollution, alongside biomass burning of forests and agricultural wastes. In 2019, the World Health Organization reported:

- 99% of the world population was living in places where the WHO air quality guidelines levels were not met.
- Ambient air pollution in both cities and rural areas was estimated to cause 4.2 million premature deaths worldwide in 2016.
- Some 91% of those premature deaths occurred in low- and middle-income countries, with the greatest number in the WHO Southeast Asia and Western Pacific regions.

- Policies and investments supporting cleaner transport, energy-efficient homes, power generation, industry and better municipal waste management would reduce key sources of outdoor air pollution (from www.who.int/news-room/fact-sheets/detail/ambient-(outdoor)-air-quality-and-health).

According to the Organisation for Economic Co-operation and Development (OECD), nearly 1.5 million people die each year from exposure to particulate matter, more than from malaria or water pollution (www.oecd.org/environment/indicators-modelling-outlooks/49928853.pdf). Furthermore, 'If no new policies are implemented, the OECD Environmental Outlook Baseline scenario projects that urban air quality will continue to deteriorate globally. By 2050, outdoor air pollution (particulate matter and ground-level ozone) is projected to become the top cause of environmentally related deaths worldwide' (Figure 17.9). Several mega-cities (Beijing, Delhi, Jakarta, Calcutta and Mumbai) exceed WHO standards for suspended particulate matter and sulphur dioxide in the atmosphere, and increases in these pollutants will create additional risk. The OECD projects that without profound reduction measures in place 3.5 million deaths per year will occur by 2050, with most of these located in rapidly industrialised countries such as India and China.

There is immense concern about the effects of ambient air pollution on health hazards. Air pollution mortality is due to exposure to fine particulate matter of 2.5 microns or less in diameter (PM2.5), which contributes to strokes, heart disease, lung cancer and both chronic and acute respiratory diseases, including asthma. Reviewing the medical evidence available, the WHO estimated that in 2016 58% of ambient air pollution-related premature deaths were

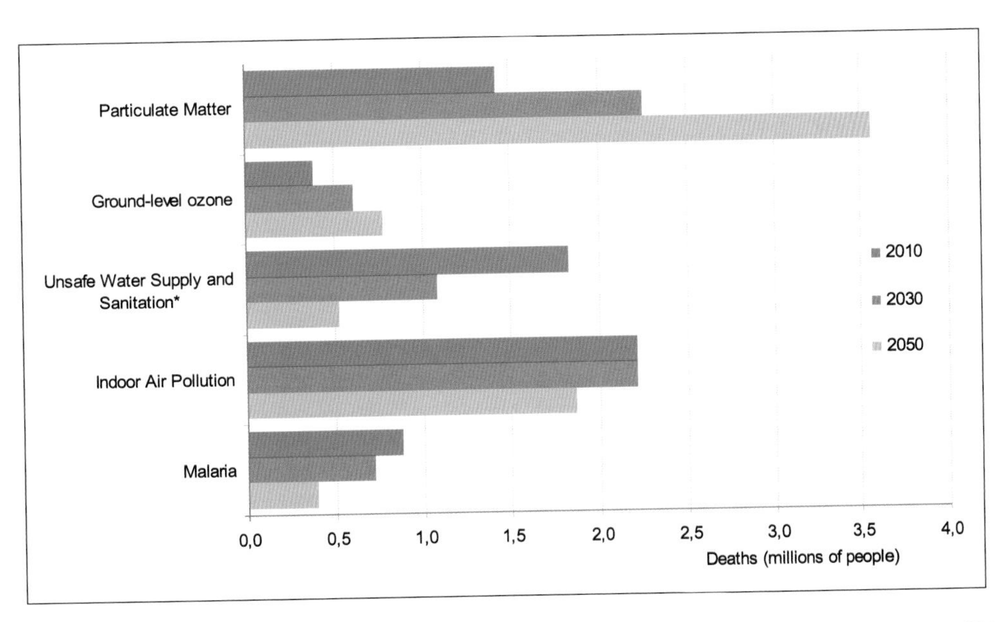

Figure 17.9 Global premature deaths from selected environmental risks, baseline from 2010 to 2050.

Source: OEDC (2012), www.oecd.org/environment/indicators-modelling-outlooks/49928853.pdf.

due to ischaemic heart disease and stroke; 18% of deaths were due to chronic obstructive pulmonary disease and acute lower respiratory infections; and 6% of deaths were due to lung cancer.

Ambient air pollution has knock-on effects on climate and weather. Individual particles remain in the atmosphere for only a few days, but the resulting haze clouds absorb incoming solar radiation and exert a cooling effect directly opposed to the warming trend associated with greenhouse gas emissions. The aerosols also reduce precipitation by weakening the hydrological cycle (Ramanathan et al., 2001). This problem is most marked in Asia, where an intermittent haze layer, known as the *Asian brown cloud*, overlies the region from Pakistan to China. It is estimated that anthropogenic sources are responsible for three-quarters of the Asian brown cloud. This haze layer, up to 3 km deep, reduces the amount of sunlight and solar energy received at the Earth's surface by 10%–15% during the winter monsoon season (December to April). There is also likely to be a reduction in evaporation, especially over the ocean surfaces, with effects on the regional climate and hydrology on a scale equivalent to that arising from global warming. Several scientists (Fu & Zheng, 1998; Xu, 2001; Bollasina et al., 2011) have observed a downward trend in summer precipitation over south Asia linked to migration and weakening of the summer monsoon rain belt. The effects of aerosol pollution may extend beyond Asia and disturb precipitation patterns in Australia (Rotstayn et al., 2007). Ramanathan (2007) further suggested that warming in the Himalayas associated with the Asian brown cloud might be responsible for glacier retreat in the high mountains.

Satellite remote sensing has been a valuable tool to better understand the scope and composition of ambient air pollution. The variety of satellites monitoring environmental conditions means that estimates can be made of nitrogen dioxide (NO_2), sulfur dioxide (SO_2), ammonia (NH_3), carbon monoxide (CO), some volatile organic compounds (VOCs) and surface particulate matter (PM2.5).

In terms of response, WHO member states adopted a resolution (World Health Organization, 2015) and a road map (World Health Organization, 2016) for an enhanced global response to the adverse health effects of air pollution. There are also three air pollution-related Sustainable Development Goals indicators:

- 3.9.1 Mortality from air pollution;
- 7.1.2 Access to clean fuels and technologies; and
- 11.6.2 Air quality in cities.

2. Plastics

Plastics are materials of polymers, which are repeating chains of molecules that can be processed and shaped. These molecules can be derived from crude oil and natural gas through distillation, polymerisation and processing. As Zalasiewicz et al. note,

> They are fundamental to contemporary hygiene, as wrapping for foodstuffs and other materials, as disposable gloves, coats and medicine encapsulations used in hospitals, and in providing inexpensive clean water systems via water bottles and pipelines.
>
> (Zalasiewicz et al., 2016, p. 5)

A by-product of the fossil field sector – and made popular and profitable through planned obsolescence and advertising (Liboiron, 2013) – the industrial

Figure 17.10 Firefighters attempt to control a forest fire deliberately started in 1997 to clear land for development in East Kalimantan, Indonesia. Thousands of fires raged for several months and created a dense cloud of air pollution across much of Southeast Asia.

Source: Photo: Panos/Dermot Tatlow DTA00167INN

manufacturing of plastics involves the burning of large quantities of fossil fuels, thus adding to the problem of ambient air pollution. These plastics are such a large and rapidly proliferating source of pollution in the environment that they are one of the suggested key indicators that we are indeed entering a new geological period. Plastic pollution is also one of the major challenges facing communities, with the release of toxic chemicals and plastic ingestion being core concerns.

Most plastics used in manufactured goods are petroleum-derived thermoplastics, which can be melted, moulded and resolidified repeatedly. Plastics also have additives, however, such as plasticizers, pigments, stabilizers and lubricants that increase structural and impact strength, improve scratch resistance and provide brightness and colour (Corcoran et al., 2018). Organic compounds are often used as additives, and the release of these – including phthalates, organotins and bisphenol A (BPA) – means that living organisms become exposed to potentially harmful chemicals.

On land, plastic waste is abandoned as litter, as well as being deposited in managed or unmanaged landfill sites. Waste plastics are exposed to solvents that leach away additives – some more easily than others. This process transports potentially harmful chemicals into ecosystems and is very much more rapid if no plastic liner is present under the landfill. These then flow alongside plastics and microplastic to the world's waterways and oceans.

Jambeck et al. (2015) estimate that 4.8–12.7 million metric tonnes of mismanaged plastic waste entered the world's oceans in 2010. According to the World Economic Forum (2019),

- Synthetic textiles are the biggest source of microplastics in the world's oceans.
- Car tyres are the second biggest source at 28% – a consequence of their erosion while driving (from https://www.weforum.org/agenda/2019/12/microplastics-ocean-plastic-pollution-research-salps?fbclid=IwAR1q91lhTD-AvcS6Qq99zR2UxXWwNS--6Sm9r3x5Yoq7WlbvIBGb3jrSNFI).

Figure 17.12 is a model of floating debris dispersal – with higher quantities in red – based on sampling and correcting for wind-driven vertical mixing by Eriksen et al. (2014). It is estimated that there is a minimum of 5.25 trillion particles weighing 268,940 tons in the oceans.

In a marine environment plastic debris can transport alien invasive species and microbial pathogens, while larger plastics can fragment into microplastics (<5 mm). Because of their size, microplastics are susceptible to the absorption of contaminants – such as polychlorinated biphenyls (PCBs) – that can be harmful to organisms ingesting the particles, including people (Figure 17.13).

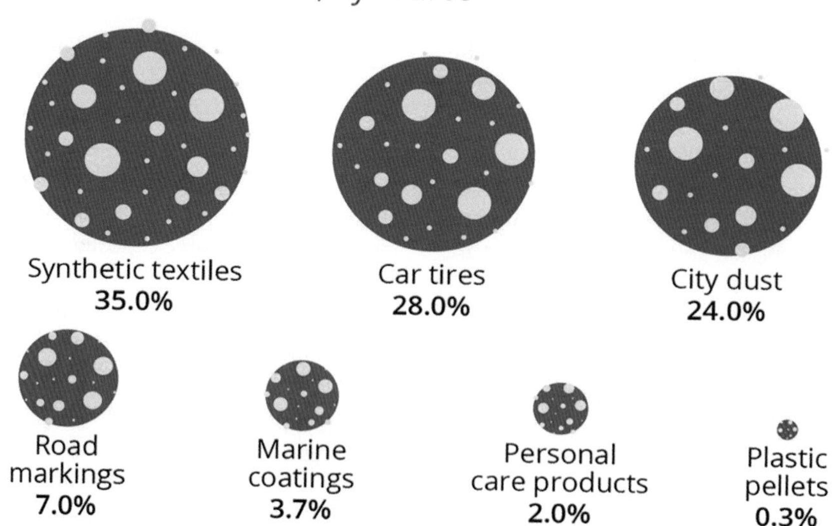

Figure 17.11 Where the ocean's microplastics come from.

Source: International Union for Conservation of Nature, 2022, This is where the ocean's microplastics come from | World Economic Forum (weforum.org)

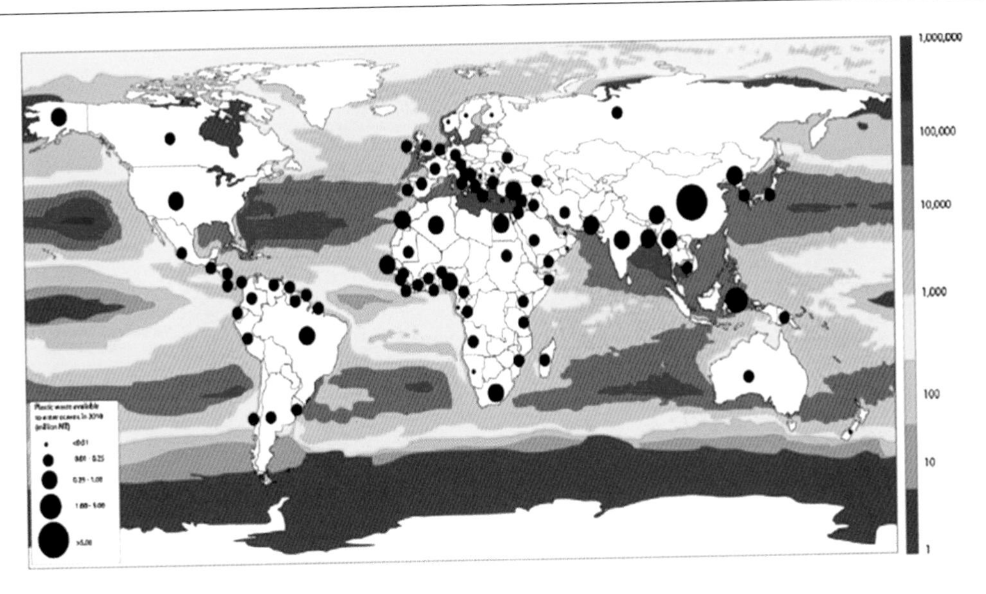

Figure 17.12 Floating debris dispersal.

Source: Eriksen et al. (2014)

Though plastic pollution creates an environmental hazard that clearly crosses borders, it is only belatedly that a sustained effort has been made to manage and mitigate this environmental hazard. One hundred eighty-seven nations – not including the United States – have agreed to restrict international trade in plastic scrap and waste. In February 2022, the UN Environmental Assembly met in Nairobi, Kenya, to debate an endorsement of the beginning of official negotiations over a plastics treaty. Two distinct approaches have been put forward. On the one hand, a *waste management approach* led by Japan argues for subsidies to waste management and chemical recycling; a strategy that would benefit the waste-to-energy sector and hinges on building up waste management facilities. On the other hand, a *life cycle approach* led by Rwanda and Peru, with support from the EU, argues for the governance of plastics from production to their use and disposal or recycling. This would entail the limiting of virgin plastic production and controlling the use of toxic chemicals that are added to plastic products.

3. Forever Chemicals and Chemical Mixing

'Forever chemicals' is the name given to chemicals that play a profound role in the manufacturing of all kinds of products, including smart technologies and consumer goods, which are not easily degradable and which are transported widely across the globe. Key examples include:

- Insecticides such as aldrin, chlordane, DDT, dieldrin, endrin, heptachlor, hexachlorobenzene, mirex and toxaphene.
- Industrial chemicals such as hexachlorobenzene and polychlorinated biphenyls (PCBs).

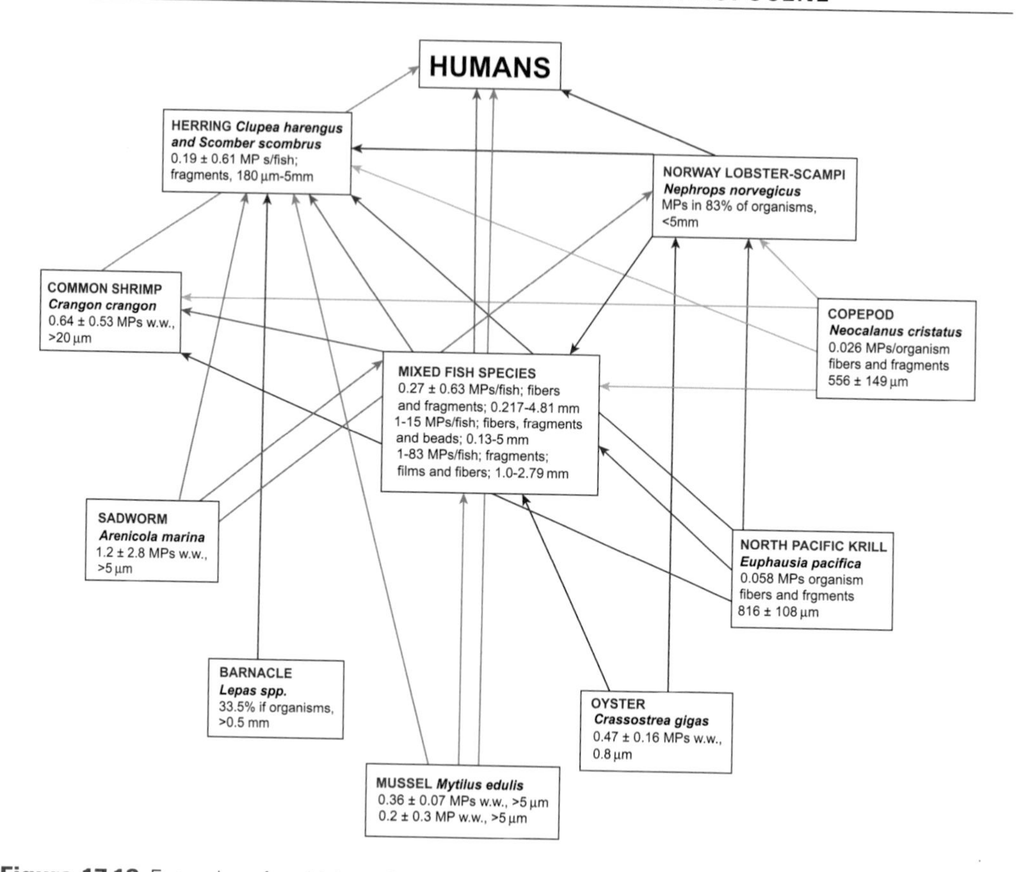

Figure 17.13 Examples of multiple pathways for human exposure to microplastics through seafood.

Source: Center for International Environmental Law (CIEL), 2019, www.ciel.org/wp-content/uploads/2019/03/ciel-rpt-HE-Fig8-path.png

- Industrial by-products such as hexachlorobenzene, polychlorinated dibenzo-*p*-dioxins and polychlorinated dibenzofurans (PCDD/PCDF), and PCBs.

According to the UN Stockholm Convention on Persistent Organic Pollutants (POPs) – which is an international treaty tasked with identifying and restricting the production and use of the most toxic chemicals of global concern – these carbon-based chemicals:

- remain intact for exceptionally long periods of time;
- become widely distributed through water, soil and air via physical processes;
- accumulate in living organisms including people, and are found at higher concentrations at higher levels in the food chain; and
- are toxic to people and wildlife (from http://chm.pops.int/TheConvention/ThePOPs/The12InitialPOPs/tabid/296/Default.aspx).

The extensive contamination of the environment from POPs means that many species, including humans, will be exposed to acute and toxic effects of the chemicals for generations to come, including cancer, allergies and hypersensitivity, damage to the central and peripheral nervous systems, reproductive disorders and disruption of the immune system. Some POPs are also considered to be endocrine disrupters: this means that they alter the hormonal system, damaging the reproductive and immune systems of exposed individuals as well as their offspring.

PFAs (per- and polyfluorinated alkyl substances) are a large chemical family of over 9,000 highly persistent chemicals that have become global pollutants. PFAs are used in a wide range of consumer goods to help repel grease and water, and can be found in:

- paper and cardboard food packaging (such as takeaway containers, pizza boxes, ready-made cakes etc.);
- textiles (such as carpets, mattresses and waterproof outdoor clothing and equipment);
- electronics (such as smartphones); and
- cosmetics and sunscreens.

PFAS can be released during the manufacturing process as well as while the product is in use or abandoned as waste. They are very mobile in water, meaning that they can be transported over long distances, entering ecosystems across the globe. Currently, two subgroups of PFAs are listed in the UN Stockholm Convention.

Assessing the scope and more detailed nature of the hazard posed by POPs is very challenging because these chemicals do not persist in the environment in isolation. People and other living organisms are daily exposed to a wide mix of chemicals originating from various sources. The safety of chemicals is usually assessed through the evaluation of single substances – as seen in Figure 17.14, which summarises one of the POPs that the Stockholm Convention is assessing – or in some cases of mixtures intentionally added for particular uses. Very little is known about the dangers of chemical mixing as an environmental hazard or the impacts from combined exposure to multiple chemicals from different sources and over time.

4. Nuclear Waste Materials

Nuclear Weapons Waste

For many, a key global-scale signal for the onset of the Anthropocene is the presence of radioisotopes produced during nuclear weapons testing, and in particular the long-lived plutonium isotopes (239Pu and 240Pu). Since 1945, seven either countries have undertaken nuclear weapons testing (Figure 17.15), releasing radioisotopes into the environment. Though some countries such as the (former) USSR carried out testing within their domestic territory, most – including the UK, France and the US – carried out tests in overseas territories (Figure 17.16).

5. Nuclear Industry Waste

High-level nuclear wastes are products from reactors, including spent or used fuel, with a radioactive half-life (the period taken for half the atoms to disintegrate) of more than 1,000 years. Intermediate-level waste has a shorter half-life but exists in larger quantities. The usual solution to the disposal of nuclear waste has been to store it for years in pools of water near the power plant so that the temperature falls and some radioactivity decays. The material is then taken to a permanent storage site elsewhere, via public highways.

By the year 2000, the USA had some 40,000 tonnes of spent nuclear fuel stored at about 70 sites awaiting disposal. Sharp differences between public opinion and the technical community were highlighted

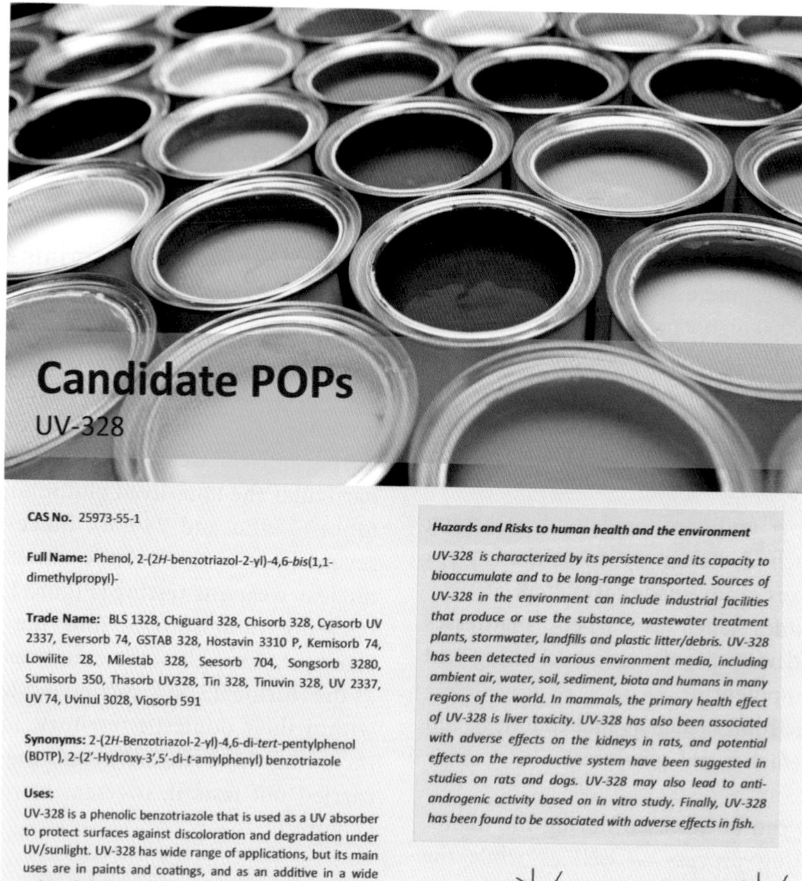

CAS No. 25973-55-1

Full Name: Phenol, 2-(2*H*-benzotriazol-2-yl)-4,6-*bis*(1,1-dimethylpropyl)-

Trade Name: BLS 1328, Chiguard 328, Chisorb 328, Cyasorb UV 2337, Eversorb 74, GSTAB 328, Hostavin 3310 P, Kemisorb 74, Lowilite 28, Milestab 328, Seesorb 704, Songsorb 3280, Sumisorb 350, Thasorb UV328, Tin 328, Tinuvin 328, UV 2337, UV 74, Uvinul 3028, Viosorb 591

Synonyms: 2-(2*H*-Benzotriazol-2-yl)-4,6-di-*tert*-pentylphenol (BDTP), 2-(2'-Hydroxy-3',5'-di-*t*-amylphenyl) benzotriazole

Uses:

UV-328 is a phenolic benzotriazole that is used as a UV absorber to protect surfaces against discoloration and degradation under UV/sunlight. UV-328 has wide range of applications, but its main uses are in paints and coatings, and as an additive in a wide variety of plastics, including in the non-food contact layer of food packaging. In the automobile sector, UV-328 is used in paints, coatings and sealants, as well as in liquid crystal panels and meters mounted on vehicles, and resin for interior and exterior parts of vehicles. In food packaging, it is used as an additive in plastics, printing ink and adhesives.

Hazards and Risks to human health and the environment

UV-328 is characterized by its persistence and its capacity to bioaccumulate and to be long-range transported. Sources of UV-328 in the environment can include industrial facilities that produce or use the substance, wastewater treatment plants, stormwater, landfills and plastic litter/debris. UV-328 has been detected in various environment media, including ambient air, water, soil, sediment, biota and humans in many regions of the world. In mammals, the primary health effect of UV-328 is liver toxicity. UV-328 has also been associated with adverse effects on the kidneys in rats, and potential effects on the reproductive system have been suggested in studies on rats and dogs. UV-328 may also lead to anti-androgenic activity based on in vitro study. Finally, UV-328 has been found to be associated with adverse effects in fish.

Open form   Closed form

Reference

1. Risk profile for UV-328. UNEP/POPS/POPRC.17/13/Add.3.
2. Proposal to list UV-328 in Annex A to the Stockholm Convention on Persistent Organic Pollutants. Risk profile for UV-328. UNEP/POPS/POPRC.16/4.

Secretariat of the Basel, Rotterdam and Stockholm Conventions
11-13, Chemin des Anémones
1219 Châtelaine, Switzerland
Tel: +41 22 917 8271
Email: brs@brsmeas.org
Website: www.pops.int

UN environment programme

Figure 17.14 According to the UN Stockholm Convention, 'Any Party may submit proposal for listing a new chemical in Annex A, B, or C of the Convention. The POPs Review Committee evaluates the proposals and makes recommendations to the Conference of the Parties on such listing in accordance with Article 8 of the Convention'.

Source: © Secretariat of the Stockholm Convention, from https://hlpf.un.org/inputs/stockholm-convention-on-persistent-organic-pollutants-0

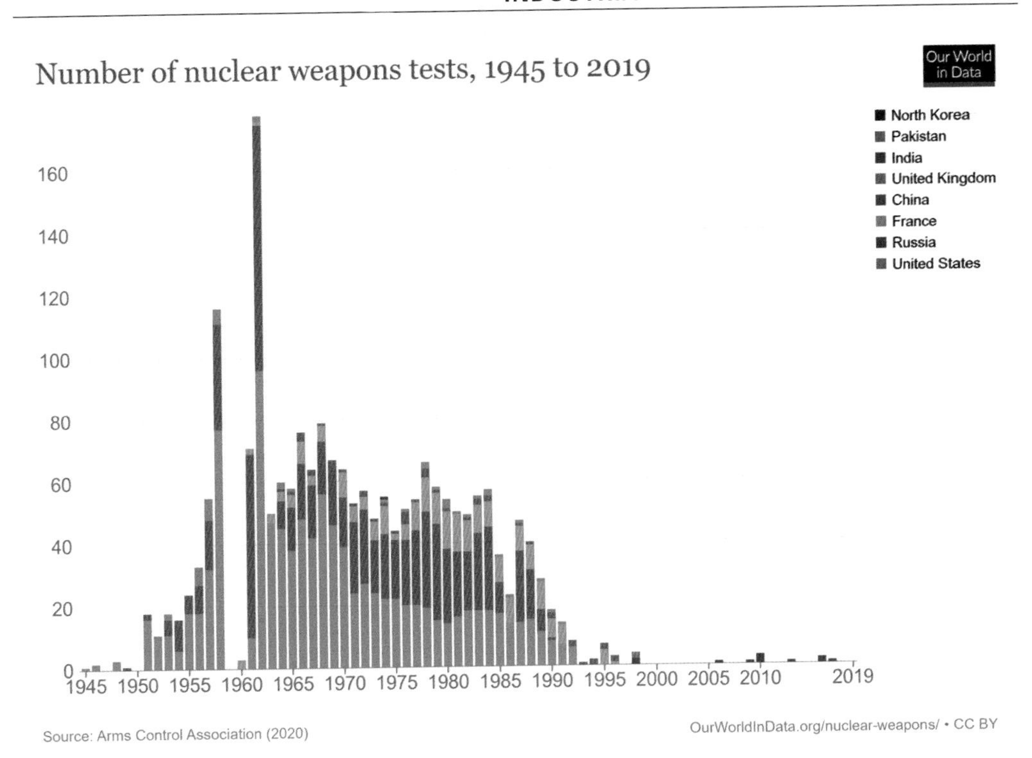

Figure 17.15 Numbers of nuclear weapon tests, 1945–1998.

Source: Our World in Data, 2022

when, in 2002, the US Congress designated Yucca Mountain, Nevada, as the sole repository site for the nation's high-level nuclear waste. Yucca Mountain, a long ridge of volcanic ash 1,500 m high, was scheduled to open in 2017 to receive deliveries of waste from across the USA for many years. Once inside the storage tunnels, it was planned that up to 70,000 t of spent fuel would be placed in titanium-covered tubes and monitored for 300 years before the mountain was sealed. Faced with charges of environmental racism due to the impacts on neighbouring communities (and with land ownership contested by the Shoshone), as well as legal setbacks due to the 'arbitrary' assumption that the depository should safely hold radioactive materials for 10,000 years, the site was abandoned in 2010.

A repository – named Onkalo, which is the Finnish word for 'cave' or 'deep hole' – is currently being constructed in Eurajoki by Posiva, which is owned by two Finnish power companies. The final facility is intended to be operational starting from 2025 and is sited next to two of the country's four existing nuclear power plants. Spent nuclear fuel rods will be packed into copper and cast-iron canisters and will be sealed by copper lids welded to the canisters. The sealed canisters will then be transported by a lift to the underground repository for final disposal. Here, the canisters will be installed in 6–8 m

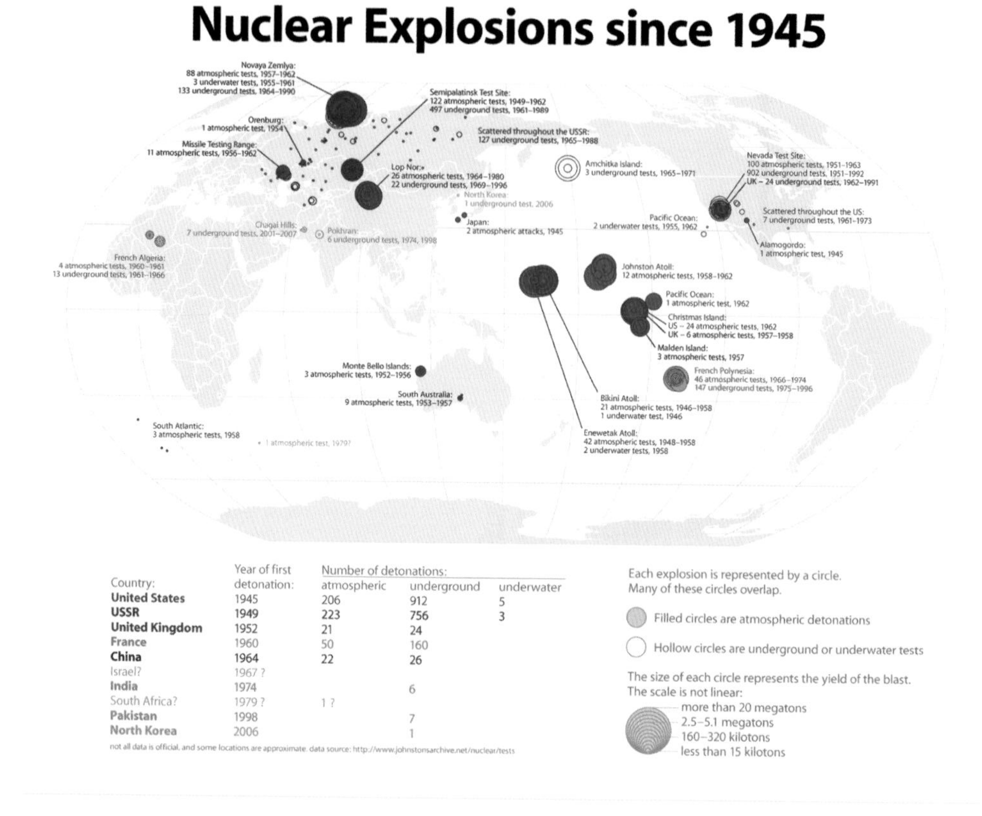

Country:	Year of first detonation:	Number of detonations: atmospheric	underground	underwater
United States	1945	206	912	5
USSR	1949	223	756	3
United Kingdom	1952	21	24	
France	1960	50	160	
China	1964	22	26	
Israel?	1967 ?			
India	1974		6	
South Africa?	1979 ?	1 ?		
Pakistan	1998		7	
North Korea	2006		1	

not all data is official, and some locations are approximate. data source: http://www.johnstonsarchive.net/nuclear/tests

Figure 17.16 Nuclear explosions since 1945.

Source: Bill Rankin (2007), www.radicalcartography.net/index.html?nuclear

Box 17.6 (Mis)managing Nuclear Weapons Waste in the US

On 16 July1945, scientists from the US's Manhattan Project gathered in a New Mexico desert for the Trinity atomic bomb test. The plutonium bomb – named Gadget – instantly vaporised the tower from which it was launched and pulled up the desert sand into the heart of the explosion, where high temperatures liquefied it. The 'stuff' later rained down, cooled and turned solid. Trinitite, a glasslike rock, was created from the sand and other materials at the site: most is green, due to the iron present in the sand, but other samples contain iron from the tower and are black. Other trinitites are red from the miles of copper electrical wire used in the test. The radiation created by the bomb had a lasting and damaging effect on the surrounding communities known as 'Downwinders'. Radioactive fallout landed on vegetables and the grasses eaten by livestock, accumulating radionuclides.

The plutonium for the bomb, and its successors, was produced at the Hanford site, on the Columbia River, Washington State. By 1960 it comprised nine nuclear reactors and five large plutonium processing complexes. From 1944 to 1987 the site produced liquid and solid radioactive waste which was buried on site. Though leaks released radioactive isotopes into the air and contaminated groundwater, much of this was undocumented. When decommissioned, the site held 53,000,000 US gallons of high-level radioactive waste stored in 177 underground tanks. By 1998, about a third of these tanks were leaking. Clean-up has been privatised and involves 'vitrifying' the liquid into glass and reburying most of the solid waste.

A salt mine in New Mexico was selected as a geological container for the toxic detritus of waste nuclear weapons production. This waste comprises chemical sludge, clothing, tools, debris, soil and other items, all laced with radioactive plutonium (https://wipp.energy.gov/) Several hundred metres underground, waste drums are placed in the crystalline salt caverns. Computer projections predict that within 1,000 years, the ceilings and walls will collapse, sealing the plutonium in place (Pillar, 2006). In 2014, however, a burst drum released small quantities of plutonium and americium to the surface, spreading 900 metres from the repository's exhaust shaft. The repository is to be sealed in 2033.

Plutonium-239 has a half-life of 24,100 years, forming a hazard that will far outlast the society that created it. According to Tracy et al.:

> Once the repository is closed, its contents cannot be monitored or problems fixed. We cannot be certain that future inhabitants of the area will even know WIPP is there. To put the timescales in perspective, agriculture was developed just over 10,000 years ago. . . . Reliance on the geological barrier is so great that the form and composition of the waste is assumed to be unimportant; it need not even be treated. Human intrusion could release radioactivity to the environment. Salt deposits, layered as sediments or as salt domes, are often associated with mineral and energy resources, such as potash and hydrocarbons – oil and gas. . . . The probability of a borehole piercing the repository in the next 10,000 years is significant.
>
> (Tracy et al., 2016, p. 149)

Further Reading: Tracy, C. L., Dustin, M. K., & Ewing, R. C. (2016). Policy: Reassess New Mexico's nuclear-waste repository. *Nature, 529*(7585), 149–151.

deep shafts within the bedrock, surrounded by bentonite clay that will act as a protective barrier from nuclear waste leakage. After installation, the tunnels will be backfilled with clay blocks and pellets. The shafts used for access will also be backfilled to block human entry as well as the downward movement of water (Ikonen et al., 2020).

Onkalo was the subject of *Into Eternity*, a 2010 film documentary by Michael Madsen, which provides insight into the engineering of the site, but also of the aims and concerns of those tasked with constructing it. According to one film critic,

> The film is framed as a message to the future, to those of us who might have blundered into this place We Were Never Meant to Go. Mr. Madsen himself appears in the darkness, illuminated by a burning match just long enough to drop rhetorical bombs, like the idea that we are encountering the last remnant of the fires that once warmed our civilization. I found myself wondering just who, after another ice age, he might be talking to: Computers? Cockroaches? Ourselves, reduced to Stone Age lifestyles after the collapse of civilization under the weight of ice or nuclear or biological apocalypse? Citizens of the galaxy on a sentimental tour of the old home world? Does future history go up or down or sideways? . . . As a species, we are good at forgetting. So maybe the best, ultimate, defense against people messing with Onkalo would be simply to forget that it is there. The best way to keep a secret is not to let on that there is a secret at all. But what about the ethical duty to warn those future generations with some kind of marker that would survive the scouring of Finland by glaciers and evolution of language? If, in fact, the canisters are rediscovered a few hundred years or a few thousand years from now, we can imagine our descendants' reaction at having been left such a nasty surprise.
>
> Dennis Overbye, writing for *The New York Times*, May 10, 2010, 'Finland's 100,000-Year Plan to Banish Its Nuclear Waste'.

Radioactivity can exist on a timeline of millennia. This means that we now face the problem of protecting future generations, as well as communities across the globe, from the effects of nuclear waste and, consequently, effectively communicating these efforts to a society at least 10,000 years in the future. This issue, dubbed *nuclear semiotics*, confronts the problem that language changes much more quickly than some forms of radioactivity. While radioactive materials being created now will survive as environmental hazards to potentially harm future generations tens of thousands of years in the future, our languages will not survive that same length of time.

KEY TAKE AWAY POINTS

Though environmental hazards emerging from industry, structures and transport have a long-term history, there has generally been an uptick in these as carbon fuel extraction, and the production, movement and consumption of commodities, has continued to grow.

Chemical and petrochemical industrial activities, products and wastes have produced a proliferating range of hazards that are increasingly cross-border and that have outstripped international governance mechanisms.

Policies and procedures to reduce and mitigate low-risk but high-impact accidents associated with large-scale structures such as dams and nuclear plants are increasingly evident. But these require sustained investment from design to the warning alert systems.

Many of the risks emerging from industry, structures and transport come from the proliferation of pollutants and wastes, including air contamination, toxic radiation materials, nuclear waste, plastics and microplastics. Though scientists can track the origin of these materials it remains challenging to pinpoint their specific – and combined – impacts on the health of people, other organisms and ecologies.

Many of the environmental hazards noted in Anthropocene debates – such as forever chemicals – have both short-term and long-term impacts on health and proliferate across borders, meaning that their reduction and mitigation need to be planned for at the international as well as national scale over several timescales.

BIBLIOGRAPHY

Bollasina, M. A., Ming, Y., & Ramaswamy, V. (2011). Anthropogenic aerosols and the weakening of the South Asian summer monsoon. *Science, 334*(6055), 502–505.

Buncefield Major Incident Investigation Board. (2005). *Buncefield major incident investigation*. Initial report to the Health and Safety Commission and the Environment Agency of the investigation into the explosions and fires at the Buncefield oil storage and transfer depot, Hemel Hempstead, on 11th December.

Buncefield Major Incident Investigation Board. (2008). *Recommendations on land use planning and the control of societal risk around major hazard sites*. Buncefield Investigation.

Corcoran, P. L., Jazvac, K., & Ballent, A. (2018). Plastics and the Anthropocene. In *Encyclopedia of the Anthropocene* (pp. 163–170). Oxford/Waltham, MA: Elsevier. DOI: 10.1016/B978-0-12-809665-9.10000-X

Cutter, S. L. (1996). Vulnerability to environmental hazards. *Progress in Human Geography, 20*(4), 529–539.

Dekker, S. (2019). *Foundations of safety science: A century of understanding accidents and disasters*. Abingdon, UK: Routledge.

Eriksen, M., Lebreton, L. C. M., Carson, H. S., Thiel, M., Moore, C. J., Borerro, J. C., et al. (2014). Plastic pollution in the world's oceans: More than 5 trillion plastic pieces weighing over 250,000 tons afloat at Sea. *PLoS One, 9*(12), e111913. https://doi.org/10.1371/journal.pone.0111913

Fu, C., & Zheng, Z. (1998). Monsoon regions: The highest rate of precipitation changes observed from global data. *Chinese Science Bulletin, 43*, 662–666.

Glickman, T. S. (1992). *Acts of god and acts of man: Recent trends in natural disasters and major industrial accidents*. Washington, DC: Resources for the Future.

Great Britain, Health and Safety Executive. (1978). *Canvey: An investigation of potential hazards from operations in the Canvey Island/Thurrock area*. HM Stationery Office.

Gupta, J. P. (2002). The Bhopal gas tragedy: could it have happened in a developed country?. *Journal of Loss Prevention in the process Industries, 15*(1), 1–4.

Hazarika, S. (1988). *Bhopal, the lessons of a tragedy*. Penguin Books India.

Hoek, M. R., Bracebridge, S., & Oliver, I. (2007). Health impact of the Buncefield oil depot fire, December 2005: Study of accident and emergency case records. *Journal of public health, 29*(3), 298–302.

Ikonen, A., Engelhardt, J., Fischer, T., Gardemeister, A., Karvonen, S., Keto, P., . . . Wanne, T. (2020). Concept description for Norwegian national disposal facility for radioactive waste. *Espoo, Finland, 88*.

Jambeck, J. R., Geyer, R., Wilcox, C., Siegler, T. R., Perryman, M., Andrady, A., . . . Law, K. L. (2015). Plastic waste inputs from land into the ocean. *Science, 347*(6223), 768–771.

Lacoursiere, P. J. P. (2005). Bhopal and its effects on the Canadian regulatory framework. *Journal of Loss Prevention in the Process Industries, 18*(4–6), 353–359.

Lewis, J. R. (1990). Amaryllidaceae alkaloids. *Natural Product Reports, 7*(6), 549–556.

Liboiron, M. (2013). Modern waste as strategy. *Lo Squaderno: Explorations in Space and Society, 29*, 9–12.

Mannan, M. S., Reyes-Valdes, O., Jain, P., Tamim, N., & Ahammad, M. (2016). The evolution of process safety: Current status and future direction. *Annual Review of Chemical and Biomolecular Engineering, 7*, 135–162.

Perrow, C. (1999). *Normal accidents: Living with high-risk technologies* (2nd ed.). Princeton, NJ: Princeton University Press.

Pillar, C. (2006, May 3). An alert unlike any other. *Los Angeles Times*. www.latimes.com/archives/la-xpm-2006-may-03-fi-forever3-story.html

Qi, R., Prem, K. P., Ng, D., Rana, M. A., Yun, G., & Mannan, M. S. (2012). Challenges and needs for process safety in the new millennium. *Process Safety and Environmental Protection, 90*(2), 91–100.

Rahu, M. (2003). Health effects of the Chernobyl accident: fears, rumours and the truth. *European Journal of Cancer, 39*(3), 295–299.

Ramanathan, V. (2007). Global dimming by air pollution and global warming by greenhouse gases: Global and regional perspectives. In *Nucleation and Atmospheric Aerosols: 17th International Conference, Galway, Ireland, 2007* (pp. 473–483). Dordrecht: Springer Netherlands.

Ramanathan, V. C. P. J., Crutzen, P. J., Kiehl, J. T., & Rosenfeld, D. (2001). Aerosols, climate, and the hydrological cycle. *Science, 294*(5549), 2119–2124.

Rankin, B. (2007). *Nuclear explosions since 1945*. www.radicalcartography.net/index.html?nuclear. Accessed 28 March 2022.

Rotstayn, L. D., Cai, W., Dix, M. R., Farquhar, G. D., Feng, Y., Ginoux, P., Herzog, M., Ito, A., Penner, J. E., Roderick, M. L. and Wang, M., (2007). Have Australian rainfall and cloudiness increased due to the remote effects of Asian anthropogenic aerosols?. *Journal of Geophysical Research: Atmospheres, 112*(D9).

Santamarina, J. C., Torres-Cruz, L. A., & Bachus, R. C. (2019). Why coal ash and tailings dam disasters occur. *Science, 364*(6440), 526–528. https://www.science.org/doi/full/10.1126/science.aax1927?casa_token=OExNKC4-5bSYAAAAA%3AKMnm9v45O8v5qHGWyA7tTjJwD05TAMSkKUTfAlNglF3gYnFx6l2jx2p9PNqEiPEgO22dC80E0KnH

State of World Mine Tailings Portfolio. (2020). https://worldminetailingsfailures.org. Accessed: 9 October 2022.

Toon, O. B., Tabazadeh, A., Browell, E. V., & Jordan, J. (2000). Analysis of lidar observations of Arctic polar stratospheric clouds during January 1989. *Journal of Geophysical Research: Atmospheres, 105*(D16), 20589-20615.

Tracy, C. L., Dustin, M. K., & Ewing, R. C. (2016). Policy: Reassess New Mexico's nuclear-waste repository. Nature, 529(7585), 149-151.

World Economic Forum. (2019). *The ocean is teeming with microplastic – a million times more than we thought, suggests new research*, 13 December 2019. https://www.weforum.org/agenda/2019/12/microplastics-ocean-plastic-pollution-research-salps?fbclid=IwAR1q91IhTD-AvcS6Qq99zR2UxXWwNS--6Sm9r3x5Yoq7WlbvIBGb3jrSNFI

World Health Organization. (2015). *Economic cost of the health impact of air pollution in Europe: clean air, health and wealth* (No. WHO/EURO: 2015-4102-43861-61759). Regional Office for Europe, World Health Organization.

World Health Organization. (2016). *Health and the environment: Draft road map for an enhanced global response to the adverse health effects of air pollution: Report by the Secretariat* (Executive Board, 138). World Health Organization. https://apps.who.int/iris/handle/10665/250653

Xu, J. (2001). An analysis of the climatic changes in eastern Asia using the potential evaporation. *Journal of Japan Society of Hydrology and Water Resources, 14*(2), 151–170

Zalasiewicz, J., Waters, C.N., Summerhayes, C.P., Wolfe, A.P., Barnosky, A.D., Cearreta, A., Crutzen, P., Ellis, E., Fairchild, I.J., Gałuszka, A. and Haff, P., (2017). The Working Group on the Anthropocene: Summary of evidence and interim recommendations. Anthropocene, 19, pp.55-60.

FURTHER READING

Carrigan, A. (2012). 'Justice is on our side'? Animal's people, generic hybridity, and eco-crime. *The Journal of Commonwealth Literature, 47*(2), 159–174.

Chapman, J. (2005). Predicting technological disasters: Mission impossible? *Disaster Prevention and Management, 14*, 343–352.

Mannan, M. S., et al. (2005). The legacy of Bhopal: The impact over the last 20 years and future direction. *Journal of Loss Prevention in the Process Industries, 18*, 218–224.

Parks, V., Ramchand, R., & Edelman, A. F. (2019). *Database of literature on oil spills and public health*. Santa Monica, CA: RAND Corporation.

Roe, E., & Schulman, P. (2008). *High reliability management: Operating on the edge*. Stanford, CA: Stanford Business Books.

WEB LINKS

Bhopal Disaster Information Centre. www.bhopal.com

Environmental Emergencies Centre. www.eecentre.org

International Atomic Energy Authority. www.iaea.org

UNEP (UN Environment Programme) Data. www.unep.org/publications-data

CHAPTER EIGHTEEN

Climate and Beyond

18

OVERVIEW

Environmental hazards have many influences, including longer-term processes such as climate. Climate is weather statistics averaged over decades and changes naturally in many ways over decades, centuries, millennia, and epochs. Today, human activity releases gases into the atmosphere and destroys ecosystems absorbing these gases. As the concentration of these greenhouse gases rises, the sun's heat is trapped, raising the Earth's temperature and changing the climate with impacts on weather. This chapter begins by describing what is meant by climate, climate change, climate variability and anthropogenic global warming. It then focuses on how weather has changed, and the consequences for environmental hazards, with some parameters becoming more extreme, some becoming less extreme and others changing their characteristics and distributions away from the extremes. Plenty can and should be done to stop the changing environmental hazards from becoming disasters, just as disaster risk reduction is needed even without human-caused and natural climate change. The chapter goes on to consider some of the many other large-scale hazards affecting the Earth, from collisions with space objects to flood basalts, some of which are incredibly rare yet could threaten the extinction of humanity.

A. INTRODUCTION

Climate is defined as average weather statistics (IPCC, 2022), usually examined over 30 years, although there is no fixed definition and climate is sometimes measured over 20 to 50 years. *Climate change* is defined as a sustained shift in the average value of climatic parameters (e.g. temperature, sunshine, precipitation, wind), either singly or in combination. The shift must be maintained for sufficient time, which, as with climate, is typically calculated over 30-year periods. In addition to human influences on the climate – which the bulk of this chapter describes – climate

DOI: 10.4324/9781351261647-21

change happens naturally, and for the Earth's entire history has happened naturally. Natural climate change might be local, such as from forests growing or dying. It might be regional or continental, such as from mountain building through tectonic forces or mountain erosion from wind and rain. Natural climate change also happens globally, such as from the Earth's orbit changing over millennia which affects how much radiation from the sun the planet receives or from a volcanic eruption or meteorite strike throwing dust into the atmosphere which blocks incoming solar radiation.

In contrast to climate change, *climate variability* is expressed by differences in climatic elements that might vary over shorter time periods or might occur over cycles or quasi-cycles, so shifting up and down rather than a particular trend. Climate variabilities can group over a period of years and may be temporarily mistaken for part of a trend, even though it happens that their oscillations happen to coincide and so appear to have changed the climate for some years. These climate variabilities are regional and global, are often poorly understood and can be defined arbitrarily with sometimes changing definitions (Table 18.1).

Today, climate change due to *anthropogenic global warming* is a major influence on climate. Global warming has been detected by observed increases in the average surface and atmospheric temperatures, linked to the growth in worldwide emissions of greenhouse gases (GHGs) including, but not limited to, carbon dioxide. GHGs change the climate by absorbing outgoing long-wave infrared radiation from the Earth's surface. The concentration of CO_2 and other atmospheric greenhouse gases is higher than at any other time in the last 800,000 years, with continuing emissions pushing it to the likely

Table 18.1 Examples of climate variabilities

Acronym	Name	Definition	Timeframe
ENSO	El Niño Southern Oscillation	Five consecutive three-month periods of sea-surface temperature in one specific Pacific region being 0.5°C different from expectations, reported as the Oceanic Niño Index (ONI).	The cycle lasts about 2–7 years.
IOD	Indian Ocean Dipole	The difference in sea-surface temperature between one location in the western Indian Ocean and one location in the eastern Indian Ocean.	Extremes happen about every 7–15 years.
MJO	Madden Julian Oscillation	An eastward moving band of cloud and rainfall near the equator.	Occurs typically every 30–60 days.
NAO	North Atlantic Oscillation	The sea's surface pressure difference between two locations in the north and south of the North Atlantic Ocean.	Might have different phases of 1–3 decades each.
PDO	Pacific Decadal Oscillation	Defined by changes in sea-surface temperatures and sea level pressures in different locations of the Pacific Ocean.	The oscillation occurs over 20–30 years.

maximum during the Pliocene epoch, which was from 5.3 until 2.6 million years ago (IPCC, 2022). The increase of these gases in the atmosphere over the last half century seems to be about 100 times the rate of natural increases before, although there is much that we do not know about our planet's history, and knowledge continues to evolve. Concentrations of greenhouse gases 'have continued to increase in the atmosphere, reaching annual averages of 410 parts per million (ppm) for carbon dioxide (CO_2), 1866 parts per billion (ppb) for methane (CH_4), and 332 ppb for nitrous oxide (N_2O) in 2019' (IPCC, 2021, p. 4).

The human signal over the long-term temperature record is clear, especially when considering the development of modern humanity since the last Ice Age ended (Figures 18.1 and 18.2).

The anthropogenic forcing of average temperatures is made clearer when considering how much of an observed change in temperature can be attributed to natural processes alone and how much is excess beyond this. As can be seen in Figure 18.3, observations of surface temperature 'fit' simulations of the same when both social and natural factors are included in the model, indicating the excess attributed to social activities.

Some of the CO_2 released by human activities into the atmosphere is absorbed by the oceans. Combining CO_2 and water (H_2O) leads to carbonic acid (H_2CO_3), so seawater is increasing in acidity at the same time as its temperature is rising. All these climate change impacts – including interactions among land, water and air – are affecting, most typically adversely, human societies, ecosystems and their

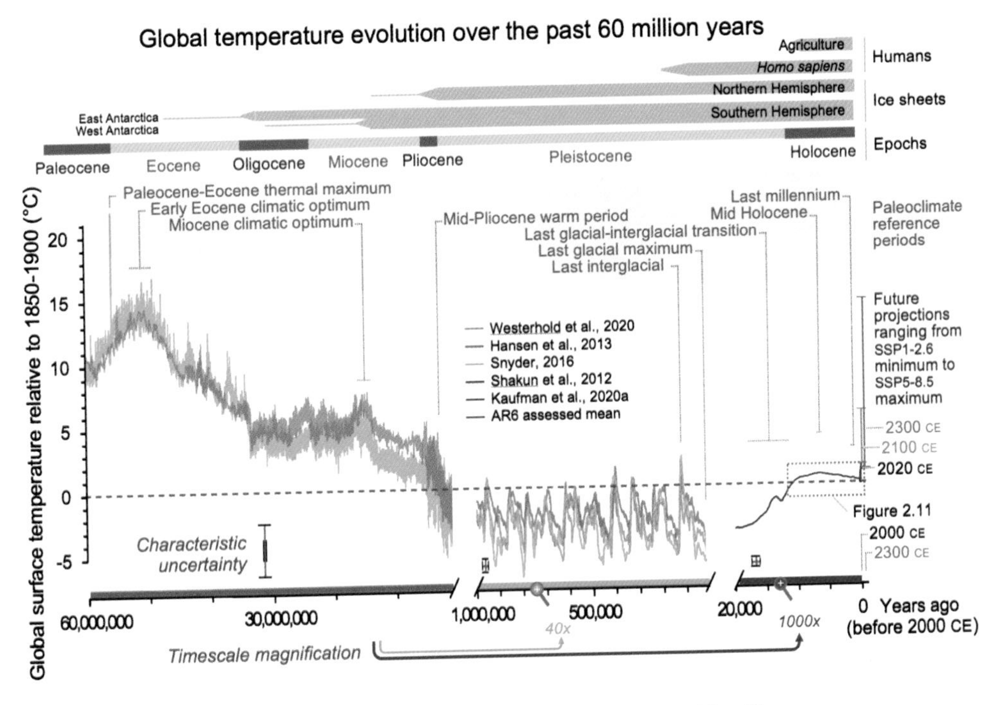

Figure 18.1 Global mean surface temperature (GMST) over the past 60 million years.

Source: IPCC (2022), Working Group 1. Cross-Chapter Box 2.1, Figure 1.

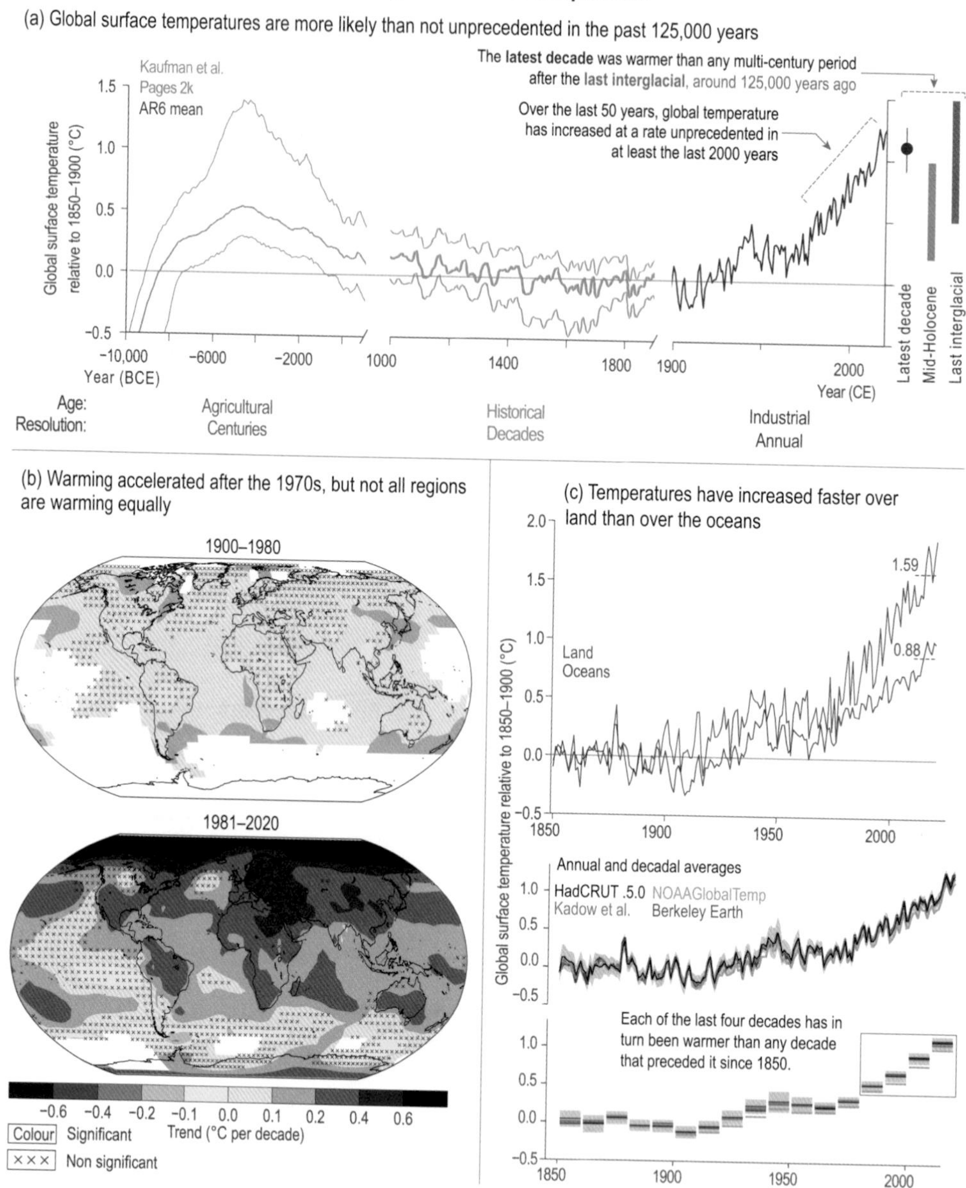

Figure 18.2 Earth's surface temperature history (global mean surface temperature) over the Holocene divided into three timescales: (a) 12 kyr–1 kyr in 100-year time steps; (b) 1000–1900 CE, 10-year smooth; and (c) 1900–2020 CE.

Source: IPCC (2022), Working Group 1. Figure 2.11(a).

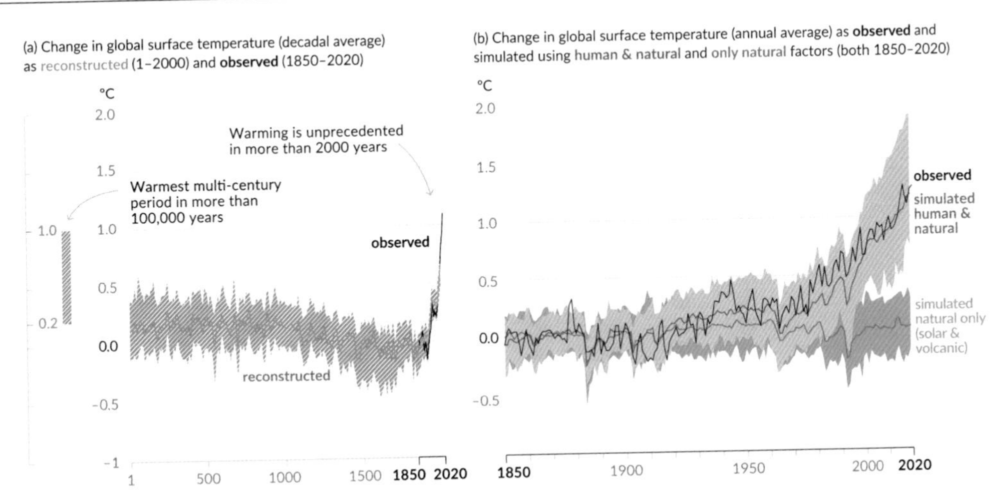

Figure 18.3 Changes in global surface temperature over the past 170 years (black line) relative to 1850–1900 and annually averaged, compared to climate model simulations of the temperature response to both human and natural drivers (brown) and to only natural drivers (solar and volcanic activity, green). Solid coloured lines show the multi-model average, and coloured shades show the very likely range of simulations.

Source: IPCC (2022), Figure SPM.1(b).

inseparable interactions (IPCC, 2022; Figure 18.4).

Yet natural climate change cannot be forgotten. Climate change science looks to other global environmental hazards and connects beyond climate. Extraterrestrial objects such as comets and asteroids (Section D) as well as large volcanic eruptions (Chapter 7), sometimes called supervolcanoes, are part of the Earth's geological history and future. The non-linear, mathematically chaotic nature of the climate system can lead to abrupt climate change, as in previous ice ages, while possibly changing the climate variabilities and having knock-on impacts on the large-scale global oceanic currents (thermohaline circulation) that balances our current climate. All these factors cascade to affecting the day-to-day weather and so environmental hazards too.

B. ANTHROPOGENIC GLOBAL WARMING

Outcomes from human-caused climate change, often referred to as anthropogenic global warming, are highly relevant for understanding environmental hazards and hence risks and disasters. Climate change scenarios carry some uncertainties, and the balance between negative and positive impacts varies according to location and context (Glantz, 1995). Most changes from AGW require major adjustments to how societies live and work, within the context of social changes as well.

The Intergovernmental Panel on Climate Change (IPCC), established in 1988 and currently representing 195 member states, is the United Nations body for providing a political approval of the

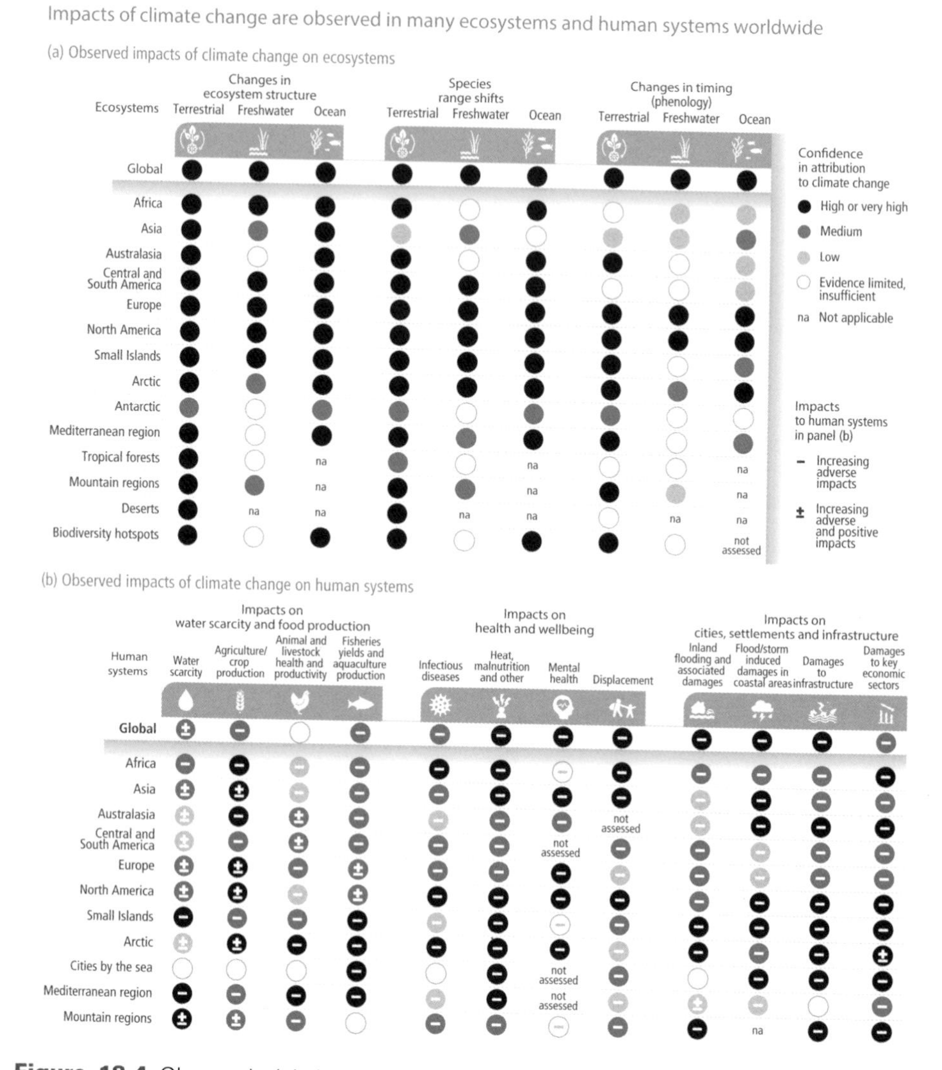

Figure 18.4 Observed global and regional impacts on ecosystems and human systems attributed to climate change. Confidence levels reflect uncertainty in attribution of the observed impact to climate change

Source: IPCC (2022), Working Group 2, Figure SPM.2.

scientific assessment of climate change, its impacts and future risks, and options for acting, termed adaptation and mitigation. Climate change adaptation is defined as adjusting to climate change's adverse impacts and accepting its benefits. Climate change mitigation is defined as reducing greenhouse gas emissions and increasing their 'uptake'. Despite evident overlaps of adaptation and mitigation, as well as the influences they have on each other, the IPCC initially separated the two but has slowly accepted their interconnection in its reports and recommendations.

Through its review of thousands of scientific papers – followed by member states having the opportunity to comment on and request changes in reports – the IPCC identifies the strength of scientific agreement in different areas. Specifically:

- IPCC Working Group 1 'combines observations, palaeoclimate, process studies, theory and modelling into a complete picture of the climate system and how it is changing, including the attribution (or causes) of change'. This comprises reviewing work on 'greenhouse gases and aerosols in the atmosphere; temperature changes in the air, land and ocean; the hydrological cycle and changing precipitation (rain and snow) patterns; extreme weather; glaciers and ice sheets; oceans and sea level; biogeochemistry and the carbon cycle; and climate sensitivity.'
- IPCC Working Group II assesses 'vulnerabilities and the capacities and limits of these natural and human systems to adapt to climate change and thereby reduce climate-associated risks together with options for creating a sustainable future for all through an equitable and integrated approach to mitigation and adaptation efforts at all scales'.
- IPCC Working Group III 'focuses on climate change mitigation, assessing methods for reducing greenhouse gas emissions, and removing greenhouse gases from the atmosphere', looking at 'energy, transport, buildings, industry, waste management, agriculture, forestry, and other forms of land management' (from www.ipcc.ch/working-groups/).

Each Working Group produces Assessment Reports and Special Reports that contribute to the IPCC Synthesis Report that integrates the findings of all three groups.

The IPCC finished its Sixth Assessment cycle in 2023, when the Synthesis Report was released in time to inform the 2023 Global Stocktake by the UN Framework Convention on Climate Change. In addition to the Working Group Assessment Reports, there were three Special Reports: on *Global Warming of 1.5°C* (2018), on *Climate Change and Land* (2019) and on the *Ocean and the Cryosphere in a Changing Climate* (2019).

The first of these Special Reports looks at the potential for, and the impacts of, a rise in global warming of 1.5°C – and potentially of 2°C – above pre-industrial levels 'in the context of strengthening the global response to the threat of climate change, sustainable development, and efforts to eradicate poverty'. The key findings are (IPCC, 2018, pp. 4–5):

- Human activities are estimated to have caused approximately 1.0°C of global warming above pre-industrial levels, with a *likely* range of 0.8°C to 1.2°C. Global warming is *likely* to reach 1.5°C between 2030 and 2052 if it continues to increase at the current rate. (*high confidence*).
- Estimated anthropogenic global warming is currently increasing at 0.2°C (*likely* between 0.1°C and 0.3°C) per decade due to past and ongoing emissions (*high confidence*).
- Warming greater than the global annual average is being experienced in many land regions and seasons, including two to three times higher in the Arctic. Warming is generally higher over land than over the ocean (*high confidence*).
- Trends in intensity and frequency of some climate and weather extremes have been detected over time spans during which about 0.5°C of global

warming occurred (*medium confidence*). This assessment is based on several lines of evidence, including attribution studies for changes in extremes since 1950.

- Reaching and sustaining net zero global anthropogenic CO_2 emissions and declining net non-CO_2 radiative forcing would halt anthropogenic global warming on multi-decadal timescales (*high confidence*). The maximum temperature reached is then determined by cumulative net global anthropogenic CO_2 emissions up to the time of net zero CO_2 emissions (*high confidence*) and the level of non-CO_2 radiative forcing in the decades prior to the time that maximum temperatures are reached (*medium confidence*).
- On longer timescales, sustained net negative global anthropogenic CO_2 emissions and/or further reductions in non-CO_2 radiative forcing may still be required to prevent further warming due to Earth system feedbacks and to reverse ocean acidification (*medium confidence*) and will be required to minimise sea-level rise (*high confidence*).
- Climate-related risks for natural and human systems are higher for global warming of 1.5°C than at present, but lower than at 2°C (*high confidence*). These risks depend on the magnitude and rate of warming, geographic location, levels of development and vulnerability, and on the choices and implementation of adaptation and mitigation options (*high confidence*).
- Impacts on natural and human systems from global warming have already been observed (*high confidence*). Many land and ocean ecosystems and some of the services they provide have already changed due to global warming (*high confidence*).
- Future climate-related risks depend on the rate, peak and duration of warming. In the aggregate, they are larger if global warming exceeds 1.5°C before returning to that level by 2100 than if global warming gradually stabilises at 1.5°C, especially if the peak temperature is high (e.g. about 2°C) (*high confidence*). Some impacts may be long lasting or irreversible, such as the loss of some ecosystems (*high confidence*).
- Adaptation and mitigation are already occurring (*high confidence*). Future climate-related risks would be reduced by the upscaling and acceleration of far-reaching, multilevel and cross-sectoral climate mitigation and by both incremental and transformational adaptation (*high confidence*) (from www.ipcc.ch/site/assets/uploads/sites/2/2019/06/SR15_Headline-statements.pdf).

The first bullet point was then updated in IPCC (2022) to:

> Emissions of greenhouse gases from human activities are responsible for approximately 1.1°C of warming since 1850–1900, and finds that averaged over the next 20 years, global temperature is expected to reach or exceed 1.5°C of warming.

C. CLIMATE CHANGE AND ENVIRONMENTAL HAZARDS

AGW is substantially changing environmental hazards. The IPCC documents observed increases in weather parameters – such as frequency, intensity, spatial extent and duration – across timescales. For example, heatwaves are becoming more common, leading to more physiological heat stress and more land evaporation, which contributes to droughts. More atmospheric moisture contributes to intense precipitation and more atmospheric energy drives hurricanes, tornadoes and other storms. Higher sea levels and storm surges impact coastal flooding and freshwater. Yet the relationships are not straightforward. Hurricanes, and many other cyclonic storms, are decreasing in frequency while increasing in intensity due to AGW.

1. Observed Changes to Extremes

Understanding how AGW affects weather involves the statistical interpretation of values that fall close to the upper or lower end of the observed or actual ranges of variation. Confidence in the analysis, including the detection of trends, depends on the *quality of the data* (e.g. consistency of observation, reliability of reporting) and on the *quantity of the data* (e.g. sample size and areal coverage). By definition, extremes are rare events, so probability statistics are not well established. The nature of the data changes over the years, between seasons, across regions and for individual extremes. What is more, observational techniques, baselines and qualities change over time, sometimes leading to a step change in data series.

Some evidence can be localised in space, whilst short datasets invite confusion between variations and trends. For example, tropical cyclones – and their disaster impacts – are highly variable from year to year, largely depending on where and when a storm makes landfall. The level of scientific understanding across different hazards inevitably differs, as it does for disasters; and, as noted in Part 1, there can be poor correlation between hazard parameters including extremes and disaster impacts. The more extreme a hazard parameter, the less frequent (by definition) it occurs and so the more difficult to assess in terms of significance and impacts. Scientists also lack knowledge regarding aspects of the physical processes of many hazards. AGW is one (extremely important) factor among many within these wider contexts of environmental and social changes.

2. Projected Changes to Extremes

Many different climate simulation models are used to explain expectations and projections regarding weather and climate. These models can be grouped into *global circulation models* (GCMs) and the *regional climate models* (RCMs) used to downscale the GCM outputs. The resulting simulations depend on the specific inputs to the models, such as physical quantities, vertical atmospheric levels and the thresholds set. In addition, model confidence varies according to the spatial and temporal scales of resolution and the type of extreme phenomenon involved, as well as the quality of the data and the programming.

Unsurprisingly, differing results emerge between models. Consequently, the IPCC assigns confidence levels to climate change projections. Much of this projected future, though, depends on how we act. Will we reduce greenhouse gas emissions and increase their uptake? Are we able to implement DRR in order to deal with weather and climate changes? How could we adjust to AGW impacts that modern humanity has never experienced? These are ongoing areas of research, policy and practice regarding the interaction between AGW, environmental hazards and disasters.

3. Climate Change and Heat

There is much interest in the effects of climate change on human health, notably the impact of heatwaves (Chapter 13), especially in conjunction with humidity and when nights do not cool down sufficiently for people to recover from hot days. Despite statistical complications arising from changes to temperature variability as well as mean values, there is agreement that the main health hazard comes from multi-day heatwaves

Actual outcomes are modified by societal factors, and 'heat response plans' (HRPs) have been introduced by some cities. These address initiatives such as the installation of (1) 'cool' roofs and pavements using highly reflective and lighter coloured materials to increase the reflectance (albedo) of these surfaces, and (2) green roofs and walls, which use rooftop plantings to cool buildings via evapotranspiration. A common theme in HRPs is artificial indoor cooling to help people avoid the worst impacts of heat and humidity. Many cities open public, cooled spaces during heatwaves, such as for people who are homeless or who do not have their own cooling. Caution is needed. First, artificial cooling is energy intensive and, in using electricity, might be contributing to AGW. Second, fans are not a good approach since, when it is hot enough, the fan simply blows hot air onto people, dehydrating them within cooling. Third, as per one of the rationales of cooling centres, many people do not have access to artificial cooling or might not be able to afford. Even if people know about cooling centres, they might fear crime in getting there or while there – or might not be able to get to and from the cooling centre due to affordability or mobility restrictions. Fourth, many workers would lose their livelihood if they spent time in cooler areas. Examples are construction workers, agricultural workers and garment workers. Their only choice for retaining their work is to be outdoors or in hot spaces.

Without appropriate action to stop AGW swathes of land are becoming too hot for working and too hot for producing food. According to Vicedo-Cabrera et al. (2021) – who note empirical data from 732 locations in 43 countries from 1991 to 2018 – 37.0% of warm-season heat-related deaths can be attributed to anthropogenic climate change (Figure 18.5).

Meanwhile, in higher latitudes, growing seasons are extending and agriculture is more feasible in areas where it was previously limited, so people need to adjust their livelihoods and skills to take advantage of it. It is not clear that increases in food production in some areas can offset the heat-related losses in other places.

4. Climate Change and Disease

Rising temperatures encourage the poleward spread of some important vectors carrying diseases. Many other factors influence disease spread, most notably human responses to vectors. For instance, warmer temperatures not only support the migration of vectors to higher latitudes and elevations, but also mean that many insects and parasites breed more frequently and mature earlier. The number of life cycles per year they go through can increase along with biting rates. Some viruses and parasites use intermediary hosts such as birds or rats to reach humans, meaning that climate change's impacts on those hosts affects disease spread. Meanwhile, higher temperatures over certain thresholds inhibit some insects from breeding as can saltwater encroachment (due to sea-level rise and increased storminess) into

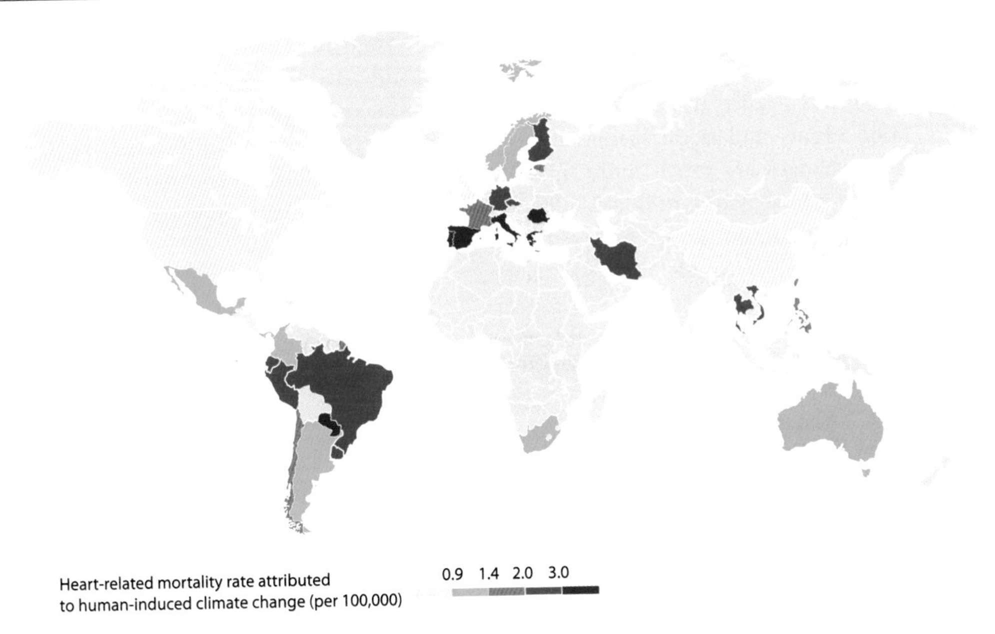

Figure 18.5 Heat-related mortality rate attributable to human-induced climate change, 1991–2018.

Source: Vicedo-Cabrera et al. (2021)

previously freshwater areas. Overall, relationships between climate change's specific parameters such as air temperature and specific diseases need to be individualised to determine the impact.

Human decisions can nonetheless break any links augmenting disease due to climate change. Malaria used to be endemic in places such as Colorado and southern England. Measures including draining wetlands in England and using window netting in Colorado helped to eradicate it. Pesticides certainly contributed, but then poisoned people and the environment. New malaria cases continue in these locations, not just from returning travellers but also from mosquitoes hitching rides on intercontinental flights. Yet malaria has not returned as a major concern. Even where the environment in Colorado and England is becoming more amenable to some vectors, human actions related to health, hygiene, sanitation, health systems and insect control can avoid AGW increasing disease rates.

5. Precipitation and Floods

There is evidence of increased precipitation totals and more intense rainfall trends over certain regions due to AGW, which is understandable given that warmer air holds more water, so it can rain more. Some regional patterns are not uniform. A study of extreme rainfall trends in southwest India found large intra-regional differences (Pal & Al-Tabbaa, 2009). It is not a straightforward task to translate precipitation trends into higher flood risks, although the incidence of floods has been related to past changes in the climate system. According to Knox (2000), geological evidence suggests that the magnitude and frequency of floods shows a sensitivity to past climate changes smaller than those expected in the 21st century.

Various studies anticipate increases in flood events and flood losses, although these changes are much more linked to placing infrastructure in flood-vulnerable locations without flood risk reduction measures (Chapter 11) than to changes in climate and weather. There is concern about future flood risk in the deltaic areas of the Netherlands and Bangladesh, due to a combination of population growth, infrastructure development, land subsidence and sea-level rise. Additionally, any projected increases in disaster-related impacts are often partly explained by continuing growth in the asset values at risk, even as wealth inequities also expand.

6. Climate Change and Storms

Tropical cyclones (Chapter 10) are being affected by AGW. There are regional variations, but globally the projections and the trends seem to be decreasing numbers and increasing intensity (Chand et al., 2022; Knutson et al., 2020). That is, fewer storms seem to be appearing, yet those that form are tending to be stronger, producing an increase in the numbers of intense storms. Storms might also be moving more slowly along their tracks, leaving more time to dump rain on a single location while holding more rain to dump. Other factors are being examined including formation latitude, the range over which a storm can track, how close to coastlines formation occurs and the time between two storm formations or landfalls in nearby locations.

Cyclonic storms happen away from the tropics or near tropics. Extra-tropics storms are also being affected by AGW, including polar lows that happen in the Barents Sea and North Atlantic Ocean as well as medicanes that are cyclonic storms in the Mediterranean. The known influences on these storms vary seasonally and geographically, with some areas seeing declining frequency and others with unclear trends (Landgren et al., 2019; Romero & Emanuel, 2017).

Not all storms are cyclonic (Chapter 10), with AGW affecting rain and wind. Separating AGW impacts from other climate variabilities is not straightforward, although attribution is improving (Herring et al., 2021). Yet long-term views are needed, with comparative datasets not always available. Wang et al. (2009) found substantial decadal-scale fluctuations of storms between 1874 and 2007 in the North Sea region, with a maximum of winter storm activity in the early 1990s. Conversely, tropical cyclone records are not known to be complete so far back, meaning that calculating trends must account for changing baselines in observations as well as the possibility of missing data.

7. Climate Change and Droughts

Trends in droughts are difficult to identify, because so much about drought is related to water (over)use and (mis)management. Dai (2011) drew attention to increased global aridity since the 1970s. He concluded that recent warming has increased evaporation demands and probably altered atmospheric circulation patterns to produce an overall drying effect, although much is seasonal and can be highly spatially contextual.

Winter droughts in the Mediterranean region have become more common since the 1970s (Hoerling et al., 2012), although water use has also increased during this time period. Climate models suggest increased aridity during the 21st century over much of Africa, southern Europe, the Middle East, most of the Americas, Southeast Asia and Australia (Dai, 2011). Simulations for the southwestern USA show dry events persisting for periods of 12 years or more, a situation exacerbated by globally warmer temperatures that reduce spring snowpacks and soil-moisture levels later in the year (Cayan et al., 2010). Climate change may also disturb conventional drought links such as the NAO (Table 18.1).

8. Climate Change and the Cryosphere (Glacial, Periglacial and Permafrost Hazards)

A glacier is an ice mass, typically forming over decades and moving slowly. A major hazard occurs when a glacier advances into human infrastructure or melts and retreats with the water becoming floods. Large-scale glacial advances have occurred over thousands of years when the Earth's orbital path around the sun changes, reducing the amount of heat to our planet and resulting in ice ages. Glaciers kilometres high moved toward lower latitudes, which if it happened today would raze cities. Then, as the glaciers retreated, the meltwaters raised sea levels dozens of metres, inundating settlements in coastal areas.

Today, glacial retreat is accelerated by AGW. Meanwhile, regional factors affect glaciers' changes. Ramanathan et al. (2007) suggested that warming in the Himalayas associated with air pollution, referred to as the Asian brown cloud, might be responsible for exacerbating glacier retreat in the high mountains.

Changing glaciers lead to sudden hazards. A serac is an irregular glacial or ice formation, often present as a pinnacle or outcrop. As wind and temperature changes affect the ice, seracs can collapse, striking people or infrastructure below. Single chunks of ice, multiple blocks in icefalls and wider ice mass movements can similarly cause harm with little warning. Sometimes, environmental shifts lead to glaciers accelerating or moving several times their usual rate for short bursts, which are called glacial surges.

Another large-scale, sudden hazard is a glacial lake outburst flood (GLOF), sometimes labelled by its Icelandic name jökulhlaup. The latter tends to occur when an under-ice volcano erupts, quickly melting large amounts of water that surge down the slopes. Icelanders have recorded jökulhlaups since the island was first permanently settled and one to two large ones typically occur each generation. The 1996 Vatnajökull jökulhlaup cut Iceland's main road (the ring road) and communications in the southeast.

Another form of glacial flood can occur when an earthen barrier forming a glacial lake fails, perhaps in heavy rain or during an earthquake, so the lakewater drains downslope. Sudden glacial melts or GLOFs can release large amounts of rocks and sediments, leading to the environmental hazards addressed in Chapter 9. Carrivick and Tweed (2016) compiled 1,348 glacier outburst floods from 20 countries during the past millennium killing over 12,000 people.

Periglacial and permafrost hazards are prevalent around frozen environments. Periglacial areas are adjacent to or near glaciers, with the land often experiencing frequent motion due to successive freezing and thawing of the soil along with continual meltwater which freezes and thaws. Waterways can end up blocked, with sediment accumulating behind it and then the dam fails when it warms, leading to an outburst flood. Periglacial environments are known for solifluction, referring to the creeping movement of loose material (such as gravel, soil, pebbles and rock), mainly due to gravity. When the material encounters a steep slope or sudden warming, it can become a rapid mass movement (Chapter 9).

Permafrost is continually frozen ground. Under natural and human-caused climate change, much permafrost is melting, destabilising structures built on it. A significant hazard from permafrost melting could be the release of methane hydrates, yielding potent greenhouse gas emissions and hence accelerated human-caused climate change.

9. Climate Change and Sea-Level Rise

Sea-level rise is an ongoing outcome from AGW (Figure 18.6). Mean global sea level has been rising since the late 19th century, and the process is accelerating beyond the average rate 1.8±0.5 mm yr^{-1} from later in the 20th century. Current sea-level rise is from two main sources. First, snow, ice and permafrost are melting, leading to water run-off into the oceans and raising sea levels by roughly 1.5 mm yr^{-1}. Second, and leading to greater rises than this melt, is thermal expansion. The ocean waters absorb the extra heat from the atmosphere, so the seas are warming. Water above 4°C becomes less dense as the temperature rises, meaning that ocean volume is expanding, hence the term 'thermal expansion', which we see as sea-level rise of about 1.7 mm yr^{-1} and getting faster.

These two factors – melt and thermal expansion – contribute to absolute sea-level rise, meaning that the ocean's level is rising. Relative sea-level rise means apparently rising seas – 'apparently' because it is actually the land that is sinking. Local land subsidence can occur due to groundwater or fossil fuel extraction as well as the weight of buildings or other construction on a soft substrate. Regional land subsidence in the northern hemisphere is continuing from the end of the last Ice Age. The weight of the glaciers pushed down land, which meant that land to the south was pushed up. The glaciers melted, releasing that weight, so the northern lands are rising again, in a process called 'isostatic uplift', which pushes down places to the south. This process is evident in the UK where much of Scotland's coastline is rising while eastern England is sinking, with the latter experiencing relative sea-level rise.

Similarly, the community of Isle de Jean Charles in Louisiana needs to move as the sea encroaches onto their land. Many state that this forced relocation is due to AGW leading to rising seas, but data on local sea level thus far tends to indicate that the land is subsiding due to resource extraction and harm to coastal wetlands. Absolute sea level might be a factor now and will likely be a factor in the future. The statement of AGW's influence nonetheless requires evidence, and there is clear evidence of relative sea-level rise due to land subsidence.

Even larger concerns about absolute sea level are expected in the future. Huge ice sheets cover Greenland and Antarctica, both of which show high melt rates, notably the West Antarctic Ice Sheet. Possibilities exist for runaway melting which would be hard to stop once it starts, a process sometimes referred to as ice sheet collapse. Our knowledge on ice sheet behaviour is expanding swiftly, although for now the expectation is that we will be certain toward the year 2100 about the extent to which Antarctic ice will add to sea-level rise. Depending on

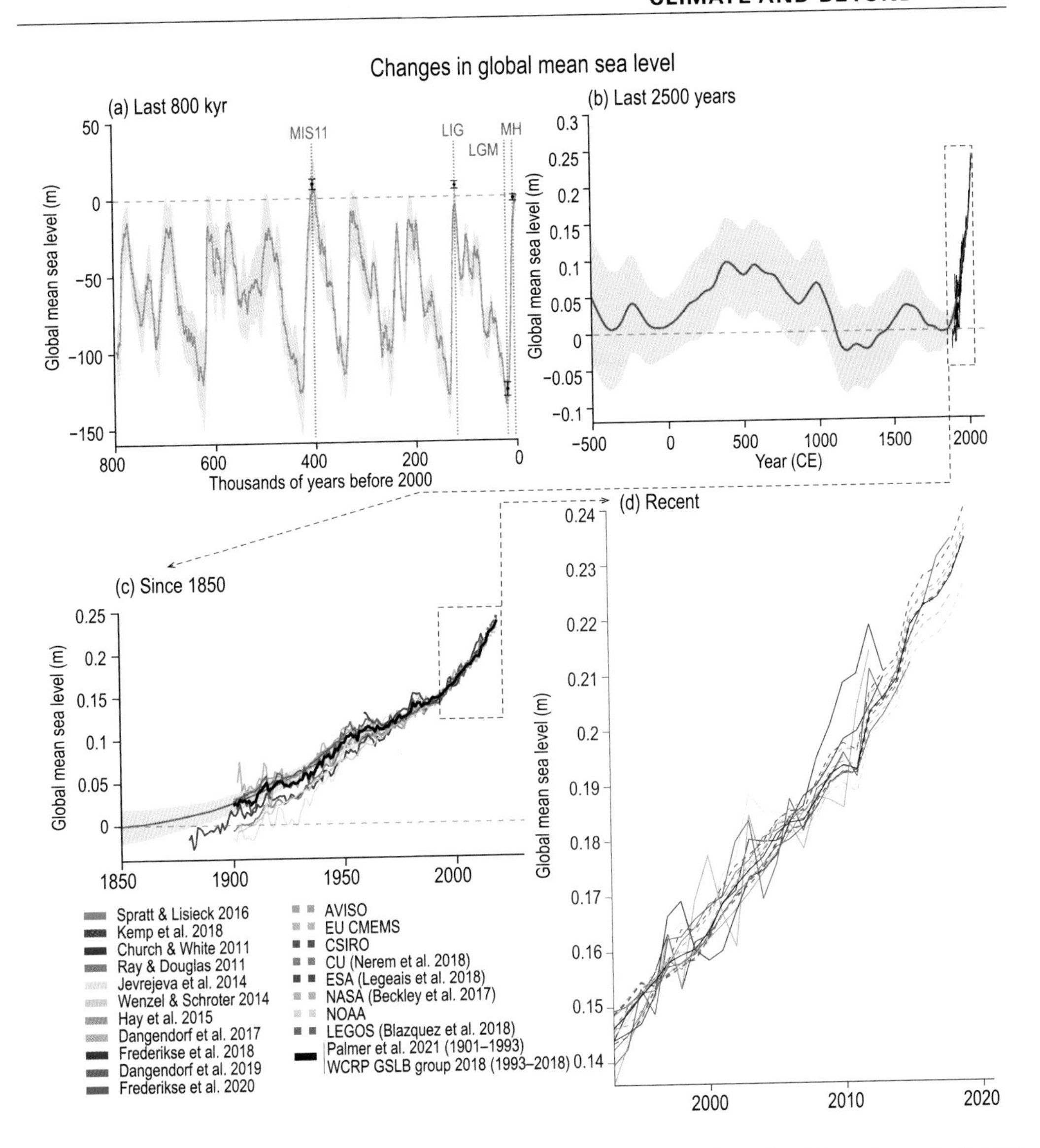

Figure 18.6 Changes in global mean sea level. (a) Reconstruction of sea level from ice core oxygen isotope analysis for the last 800 kyr. Note the much broader axis range (200 m) than for later panels (tenths of metres). (b) Reconstructions for the last 2,500 years based upon a range of proxy sources with direct instrumental records superposed since the late 19th century. (c) Tide-gauge and, more latterly, altimeter-based estimates since 1850. The consensus estimate used in various calculations in Chapters 7 and 9 is shown in black. (d) The most recent period of record from tide-gauge and altimeter-based records.

Source: IPCC (2022), Working Group 1. Figure 2.28.

specific scenarios, this might be 1–3 m by 2300 expanding to over 15 m by 2500 if we do not stop AGW. If all ice covering Antarctica melts, then its contribution to sea-level rise could reach 60 m, which means that the world's coastlines, several megacities and some countries would need to be reconfigured.

Many complexities require further research. We still have a rudimentary understanding about how ice sheets melt and move, with factors including the ice's composition, the stones or soil on which it sits, and what happens to floating ice sheets that block land-based glaciers from flowing into the sea. Different factors intersect. When the base of an ice sheet thaws, the ice sheet more readily flows, so the speed of melt increases. Meanwhile, when the top of an ice sheet melts, less weight presses down on the land, so it rises in isostatic uplift. Air temperature decreases with elevation, so as the ice sheet rises, the top is colder and melting slows. Some scenarios lead to the sea-based ice shelves not melting extensively, blocking the land-based ice sheets from flowing into the ocean and then raising sea levels.

D. SUPERHAZARDS

The geological record provides evidence of global catastrophes leading to 'mass extinctions' of life on Earth. The most extreme of these occurred 251 million years ago at the boundary between two of the great geological periods, the Permian and the Triassic. Fossil evidence suggests that this event killed 96% of all marine species and 70% of all land species (including plants, insects and vertebrate animals). A major example is the *K/T extinction*, which occurred 65 million years ago, at the boundary between the Cretaceous and Tertiary periods, and destroyed more than half the species on Earth, including the child-inspiring dinosaurs (hopefully also inspiring adults). Science continues to debate the exact cause of this mass extinction, since a major flood basalt (Chapter 7) erupted around the same time, plus there appears to have been two large meteorite strikes. Were all three independent environmental hazards? Did the meteorite strike(s) trigger the flood basalt and were they part of a swarm? Did one of these two or three environmental hazards dominate the mass extinction or was it all of them combined? How much did tsunamis from the meteorites affect the extinctions compared to climate change or other factors? Did all events overlap, pushing the Earth's climate over various tipping points?

Although the Earth is constantly threatened by showers of debris from space, most approaching material burns up in the atmosphere so that only the very largest masses survive to reach the surface. These tend to be either *asteroids* (a range of solid objects varying in size from less than 1 km to about 1,000 km in diameter) or *comets* (diffuse bodies of gas and solid particles that orbit the sun). Until recently, the chance of a large extraterrestrial object striking Earth was deemed highly unlikely. This attitude was fostered by the limited surface evidence of previous impacts. Global geography decrees that any impactor from space has the greatest chance of hitting a marine surface, where it will leave little, if any, surface trace, apart from possible tsunami deposits around coastlines. Even if the object reached a land mass, the probability was – at least in the past – that it would strike an uninhabited region and the event would pass unnoticed.

Risk assessments have changed, following the increased ability of telescopes to search space for *near-Earth objects* (NEOs), coupled with the recognition of additional fossil crater sites. Table 18.2 lists a sample of these sites. The consequences for society resulting from a NEO strike, or one just skimming through the atmosphere and producing a heat and shock blast, depend only partially on size and mass. Other factors include the velocity of the body on impact (noting that mass times velocity equals momentum, which is typically a parameter indicating possible impact), whether the strike is on land or sea, and – most notably, given that vulnerability causes disasters – people and infrastructure affected by the resulting hazards, which could be local or global.

In order to create a global-scale catastrophe – defined by Chapman and Morrison (1994) arbitrarily as the death of more than one-quarter of the world's population (over two billion people at the moment) – an impactor would need to be between 0.5 and 5 km in diameter. At the upper end of this range, Toon et al. (1997) identified other processes related to an NEO strike which alter the composition of the atmosphere and bring about climate change. Examples are blast waves injecting dust and water into the atmosphere, soot production from burning forests, acid rain and ozone depletion.

Obviously, such concerns and analyses are not new. For instance, according to a report back in 2000 by the UK Task Force on Potentially Hazardous Near Earth Objects (NEOs) (Toon et al., 2000) at least 150 m in diameter is on an orbit that will bring it within 7.5 million kilometres of the Earth. The risk of impact from comets is assessed at 10%–30% of that for asteroids. These hazards can be accommodated into conventional DRR strategies. Some forecasting and warning is possible. For example, a lead-time of 250–500 days between detection and impact has been estimated for long-period comets (Marsden & Steel, 1994), while the period for asteroids might extend to decades or more. NEOs could be detected earlier by the deployment of larger, wide-angled search telescopes dedicated to whole-sky observation. The

Table 18.2 Some known impact craters ranked by age – millions of years before the present (Ma)

Crater name	Country	Diameter (km)	Age (ma)
Barringer	United States	1.1	0.049
Zhamanshin	Kazakhstan	13.5	0.9
Ries	Germany	24	15
Popiga	Russia	100	35.7
Chicxulub	Mexico	170	64.98
Gosses Bluff	Australia	22	214
Manicouagan	Canada	100	290
West Clearwater	Canada	36	290
Acraman	Australia	90	>450
Kelly West	Australia	10	>550
Sudbury	Canada	250	1,850
Vredefort	South Africa	300	2,023

Source: After Cintala and Grieve (1998)

NASA Spaceguard Survey, which has been operational since 1996, is concerned with NEOs greater than 1 km in size.

What to do when a threatening NEO is discovered continues to evolve in a field termed 'planetary defence' (Schmidt, 2019). Reducing worldwide vulnerability to a major object is not feasible since everyone could be in trouble. Evacuating huge swathes of land to permit a hit and then dealing with the climate-related aftermath would hardly seem viable given the disruption and destruction. Instead, deflecting an object is much cheaper and more effective, for instance by landing a rocket on it or nudging it with a tug spacecraft, with destruction (often depicted with nuclear weapons) as a last resort. Tradeoffs work in our favour.

The greater an NEO's momentum (mass times velocity), the greater its potential cataclysm and the greater its detectability. The earlier we detect an object, the more time we have to respond and the easier it is to deflect it. Consequently, the key to avoid an NEO-related disaster is continual, systematic, comprehensive monitoring of outer space. Some work is ongoing, although it is not enough. Here, DRR including for this type of climate change and for AGW truly illustrates the maxim that 'Prevention is better than cure'.

NEOs, flood basalts, an explosive supervolcano and heat-humidity plus sea-level rise are not the only possible environmental hazards threatening the planet and humanity. The field of 'existential risks' was alluded to in Chapters 4 and 5 (see also Ord et al., 2021; Ord, 2020). Other possible environmental hazards in this realm include unknown consequences of significant ocean acidification, the Earth's magnetic field flipping as it does frequently throughout geological history, a nearby supernova (the explosive death of certain star types), a nearby radiation burst from stars that have so far been observed too far away to cause us harm, or a small and hence difficult-to-detect black hole colliding with the Earth.

KEY TAKE AWAY POINTS

- Human activities are now changing the global climate quickly and substantially. AGW is a major concern affecting us now with huge consequences and even more serious difficulties in the near and distant futures.
- Climate change's impacts on diverse environmental hazards vary widely, with many disasters avoidable by reducing vulnerability.
- Climate change is not just about AGW, and climate-related hazards are not exclusively related to climate change. For example many other global-scale environmental hazards can impact the climate, such as volcanic eruptions, or even asteroids, that can create mass/human extinction due to secondary impacts on the climate.

BIBLIOGRAPHY

Atkinson, H., Tickell, C., & Williams, D. A. (2000). Report of the task force on potentially hazardous near Earth objects.

Carrivick, J. L., & Tweed, F. S. (2016). A global assessment of the societal impacts of glacier outburst floods. *Global and Planetary Change, 144*, 1–16.

Cayan, D. R., Das, T., Pierce, D. W., Barnett, T. P., Tyree, M., & Gershunov, A. (2010). Future dryness in the southwest US and the hydrology of the early 21st century drought. *Proceedings of the National Academy of Sciences, 107*(50), 21271–21276.

Chand, S. S., Walsh, K. J. E., Camargo, S. J., Kossin, J. P., Tory, K. J., Wehner, M. F., Chan, J. C. L., Klotzbach, P. J., Dowdy, A. J., Bell, S. S., Ramsay, H. A., &

Murakami, H. (2022). Declining tropical cyclone frequency under global warming. *Nature Climate Change, 12*, 655–661.

Chapman, C. R., & Morrison, D. (1994). Impacts on the Earth by asteroids and comets: Assessing the hazard. *Nature, 367*(6458), 33–40.

Cintala, M. J., & Grieve, R. A. (1998). Scaling impact melting and crater dimensions: Implications for the lunar cratering record. *Meteoritics & Planetary Science, 33*(4), 889–912.

Dai, A. (2011). Drought under global warming: A review. *Wiley Interdisciplinary Reviews: Climate Change, 2*(1), 45–65.

Glantz, M. H. (1995). Assessing the impacts of climate: The issue of winners and losers in a global climate change. *Studies in Environmental Science, 65*, 41–54.

Herring, S. C., Christidis, N., Hoell, A., Hoerling, M. P., & Stott, P. A. (Eds.). (2021). Observations of the rate and acceleration of global mean sea level change: Explaining extreme events in 2019 from a climate perspective. *Bulletin of the American Meteorological Society, 102*, S1–S112.

Hoerling, M., Eischeid, J., Perlwitz, J., Quan, X., Zhang, T., & Pegion, P. (2012). On the increased frequency of Mediterranean drought. *Journal of Climate, 25*(6), 2146–2161.

IPCC. (2018). Summary for policymakers. In V. Masson-Delmotte, P. Zhai, H.-O. Pörtner, D. Roberts, J. Skea, P. R. Shukla, A. Pirani, W. Moufouma-Okia, C. Péan, R. Pidcock, S. Connors, J. B. R. Matthews, Y. Chen, X. Zhou, M. I. Gomis, E. Lonnoy, T. Maycock, M. Tignor, & T. Waterfield (Eds.), *Global Warming of 1.5°C. An IPCC Special Report on the impacts of global warming of 1.5°C above pre-industrial levels and related global greenhouse gas emission pathways, in the context of strengthening the global response to the threat of climate change, sustainable development, and efforts to eradicate poverty* (pp. 3–24). Cambridge and New York, NY: Cambridge University Press. https://doi.org/10.1017/9781009157940.001

IPCC. (2021). Summary for policymakers. In V. Masson-Delmotte, P. Zhai, A. Pirani, S. L. Connors, C. Péan, S. Berger, N. Caud, Y. Chen, L. Goldfarb, M. I. Gomis, M. Huang, K. Leitzell, E. Lonnoy, J. B. R. Matthews, T. K. Maycock, T. Waterfield, O. Yelekçi, R. Yu, & B. Zhou (Eds.), *Climate change 2021: The physical science basis*. Contribution of Working Group I to the Sixth Assessment Report of the Intergovernmental Panel on Climate Change. https://www.ipcc.ch/report/ar6/wg1/downloads/report/IPCC_AR6_WGI_SPM_final.pdf

IPCC. (2022). Sixth Assessment Report. IPCC (Intergovernmental Panel on Climate Change), Geneva.

Knox, J. C. (2000). Sensitivity of modern and Holocene floods to climate change. *Quaternary Science Reviews, 19*(1–5), 439–457.

Knutson, T., Camargo, S. J., Chan, J. C. L., Emanuel, K., Ho, C.-H., Kossin, J., Mohapatra, M., Satoh, M., Sugi, M., Walsh, K., & Wu, L. (2020). Tropical cyclones and climate change assessment: Part II: Projected response to anthropogenic warming. *Bulletin of the American Meteorological Society, 101*(3), E303–E322.

Landgren, O. A., Batrak, Y., Haugen, J. E., Støylen, E., & Iversen, T. (2019). Polar low variability and future projections for the Nordic and Barents Seas. *Quarterly Journal of the Royal Meteorological Society, 145*(724), 3116–3128.

Marsden, B. G., & Steel, D. I. (1994). Warning times and impact probabilities for long-period comets. In *Hazards due to comets and asteroids* (pp. 221–239).

Ord, T. (2020). *The precipice: Existential risk and the future of humanity*. Hachette Books.

Ord, T., Mercer, A., & Dannreuther, S. (2021). *Future proof: The opportunity to transform the UK's resilience to extreme risks*. London, UK: The Centre for Long-Term Resilience.

Pal, I., & Al-Tabbaa, A. (2009). Trends in seasonal precipitation extremes–An indicator of 'climate change'in Kerala, India. *Journal of Hydrology, 367*(1–2), 62–69.

Ramanathan, V., Ramana, M. V., Roberts, G., Kim, D., Corrigan, C., Chung, C., & Winker, D. (2007). Warming trends in Asia amplified by brown cloud solar absorption. *Nature, 448*, 575–578.

Romero, R., & Emanuel, K. (2017). Climate change and hurricane-like extratropical cyclones: Projections for North Atlantic polar lows and medicanes based on CMIP5 models. *Journal of Climate, 30*(1), 279–299.

Schmidt, N. (Ed.). (2019). *Planetary defense: Global collaboration for defending Earth from asteroids and comets.* Switzerland: Springer.

Toon, O. B., Zahnle, K., Morrison, D., Turco, R. P., & Covey, C. (1997). Environmental perturbations caused by the impacts of asteroids and comets. *Reviews of Geophysics, 35*(1), 41–78.

Vicedo-Cabrera, A. M., Scovronick, N., Sera, F., Royé, D., Schneider, R., Tobias, A., . . . Gasparrini, A. (2021). The burden of heat-related mortality attributable to recent human-induced climate change. *Nature Climate Change, 11*(6), 492–500.

Wang, X. L., Zwiers, F. W., Swail, V. R., & Feng, Y. (2009). Trends and variability of storminess in the Northeast Atlantic region, 1874–2007. *Climate Dynamics, 33,* article 1179.

FURTHER READING

Bankoff, G., & Hilhorst, D. (Eds.). (2021). *Vulnerability and the politics of disaster risk creation.* Abingdon, UK: Routledge.

Glantz, M. H. (Ed.). (1999). *Creeping environmental problems and sustainable development in the Aral Sea Basin.* Cambridge: Cambridge University Press.

Glantz, M. H. (2003). *Climate affairs: A primer.* Covelo: Island Press.

WEB LINKS

Climate Change and Health Indicators. https://lancetcountdown.org

Earth Impact Database. www.passc.net/EarthImpactDatabase/New%20website_05-2018/Index.html

Intergovernmental Panel on Climate Change. www.ipcc.org

UN Climate Change. https://unfccc.int

World Meteorological Organization. www.wmo.ch

CHAPTER NINETEEN

19

Using Environmental Hazards Knowledge for Action

This book is not just about knowledge. It is also about actions – actions to stop people being harmed when we generally recognise why disasters happen and how disasters could be avoided. As has been evident throughout this book, and through continuing research, we are far from knowing everything about environmental hazards. In fact, it is unlikely that we could ever know everything about environmental hazards. Indeed, the more we discover about hazards, the more complex we realise they are; moreover, we become more aware of what we do not know. The world is full of surprises, uncertainties and unknowns. And, of course, change is constant, so even near-complete data about the past (if that would be possible) is not necessarily a suitable guide for the future.

None of these uncertainties should stop us from acting to help people by implementing disaster risk reduction. This is because we do know why particular people and groups are vulnerable, and we know how to reduce vulnerability. Certainly, policy and practice can be informed by science. As this book has demonstrated, whilst there is scope to improve our scientific knowledge and reduce uncertainties, many disasters remain the result of a lack of communication and action based on all 'expert' knowledges, a designation that is not confined to scientists. Rather than policy and practice being informed by science only, it should be integrated and co-produced, enabling those most vulnerable to be fully involved.

While much of the content of the book – particularly Part 3 – presents a disturbing picture of threats and risks based on current science, it is important not to fall into a simple 'doom' narrative that makes all of us passive victims incapable of responding with creativity and care to these challenges. To be sure, environmental hazards abound while, for the first time in this planet's history, the cumulative impacts of humanity are being felt everywhere. Although many of humanity's physical impacts are 'baked into' future conditions, the scope and impact of environmental

DOI: 10.4324/9781351261647-22

hazards are shaped by that same collective human activity.

Small-scale actions within local communities have consequences, particularly with regard to changing consumption practices and land use, as well as the sharing of knowledge on environmental hazards and the efficacy and reach of diverse responses. Part of this small-scale activity is academic research and practice, combining innovative insights about nature, society and their inseparability. We build up knowledge, combine ideas and work together to produce wisdom. But we can also consider how this academic work is translated into other parts of our life.

Local communities can have global impact when amplified by news and social media, scientific projects and supranational organisations. The key here is inclusion and a commitment to acknowledging and valuing diverse experiences and voices on environmental hazards. How could governance and risk organisations become more inclusive? How could our news media amplify the experiences of those marginalised and excluded from debates of risk reduction, without resorting to sensationalism? How could social media become a useful tool for disaster risk reduction and disaster response, as well as a respected source for verifiable, science-based knowledge?

Small-scale actions and broader social structures are not separate; the former makes up the latter. And so moving forward we must consider the importance of small-scale actions as well as larger networks that push for a continuingly better future. We can and should use the material in this book for our own activities. We can and should use the material in this book to convince others to act. We have ten millennia of modern human history to draw upon, through which people have suffered recurring catastrophes. The time to stop disasters is now. Thank you for being part of this effort.

Index

Note: Numbers in **bold** indicate a table and numbers in *italics* indicate a figure on the corresponding page.